Springer Proceedings in Mathematics & Statistics

Volume 392

This book series features volumes composed of selected contributions from workshops and conferences in all areas of current research in mathematics and statistics, including data science, operations research and optimization. In addition to an overall evaluation of the interest, scientific quality, and timeliness of each proposal at the hands of the publisher, individual contributions are all refereed to the high quality standards of leading journals in the field. Thus, this series provides the research community with well-edited, authoritative reports on developments in the most exciting areas of mathematical and statistical research today.

Mohammad Ashraf · Asma Ali ·
Vincenzo De Filippis
Editors

Algebra and Related Topics with Applications

ICARTA-2019, Aligarh, India, December 17–19

Springer

Editors
Mohammad Ashraf
Department of Mathematics
Aligarh Muslim University
Aligarh, India

Asma Ali
Department of Mathematics
Aligarh Muslim University
Aligarh, India

Vincenzo De Filippis
Department of Engineering
University of Messina
Messina, Italy

ISSN 2194-1009 ISSN 2194-1017 (electronic)
Springer Proceedings in Mathematics & Statistics
ISBN 978-981-19-3897-9 ISBN 978-981-19-3898-6 (eBook)
https://doi.org/10.1007/978-981-19-3898-6

Mathematics Subject Classification: 05C12, 05C25, 05C50, 05C99, 06C05, 06C20, 13A15, 13B02, 13B25, 13B99, 13F10, 13M10, 16D50, 16L60, 16N60, 16P20, 16R50, 16Y30, 16W10, 94B05, 94B15

This Springer imprint is published by the registered company Springer Nature Singapore Pte Ltd.
The registered company address is: 152 Beach Road, #21-01/04 Gateway East, Singapore 189721, Singapore

Organization

ICARTA 2019 was organized by the Department of Mathematics, Aligarh Muslim University, Aligarh, India.

Scientific Advisory Committee

Ahmed Alkenani, King Abdulaziz University, Saudi Arabia
Ayman Badawi, American University of Sharjah, UAE
Ulrich Dempwolff, TU Kaiserslautern, Germany
S. K. Jain, Ohio State University, USA
T. K. Lee, National Taiwan University, Taiwan
Sergio R. Lopez-Permouth, Ohio University, USA
Hidetoshi Marubayashi, Tokushima Bunri University, Japan
Najib Mahdou, S. M. Ben Abdellah University, Morocco
J. K. Park, Busan National University, South Korea
Dushan Pagon, University of Maribor, Slovenia
Murtaza A. Quadri, Aligarh Muslim University, Aligarh
S. Tariq Rizvi, Ohio State University, USA
Jugal Verma, IIT Bombay, Mumbai, India
Joso Vukman, University of Maribor, Slovenia

Academic Program Commitee

Nadeem ur Rehman, Aligarh Muslim University, Aligarh
Shakir Ali, Aligarh Muslim University, Aligarh
Mohammad Aslam Siddeeque, Aligarh Muslim University, Aligarh
Muzibur Rahman Mozumder, Aligarh Muslim University, Aligarh
Ghulam Mohammad, Aligarh Muslim University, Aligarh

Referees

A. Ali
S. Ali
M. Ashraf
T. Asir
A. Badawi
G. K. Bakshi
R. Bandi
M. Bhaintwal
A. Boua
L. Carini
A. K. Chaturvedi
T. T. Chelvam
J. Coykendall
A. Das
B. K. Dass
N. Debshai
V. De Filippis
B. Dhara
F.A. Ebrahim
A. Fošner
M. Fošner
A. J. Gupta
C. Haetinger
S. Huang
A. Jabeen
P. K. Jain
T. Johnsen
A. N. Khan
M. S. Khan
M. Kumar
V. Kumar
G. Lee
T. K. Lee
C. Li
S. R. López-Permouth
N. Mahdou
G. Mohammad
N. Mohammad
M. T. Nair
S. K. Nimbhorkar
L. Oukhtite
A. Pasini

M. K. Patel
S. Pirzada
B. Prajapati
O. Prakash
A. R. Rahimi
R. Raja
A. R. Rajan
M. A. Raza
N. Rehman
G. Scudo
K. Selvakumar
Y. Shang
M. A. Siddique
H. Tanguchi
S. Tiwari
A. Verdur
B. A. Wani
B. N. Waphare
F. Wei
M. K. Yadav
Y. Zhou

Sponsoring Institutions

Department of Science and Technology (DST), India
Council of Scientific and Industrial Research (CSIR), India
Indian National Science Academy, India
Aligarh Muslim University, Aligarh, India

Preface

Algebra is considered a significant milestone in Mathematics. Algebra is not just limited to Mathematics; also, it has a lot of real-world applications be it computer science, chemical science, technology, coding theory, cryptography, graph theory, etc. In fact, the world revolves around the applications of algebra.

The Department of Mathematics, Aligarh Muslim University, Aligarh, India, organized an International Conference on Algebra and Related Topics with Applications (ICARTA-19) with the aim to provide a forum for researchers, eminent academicians, research scholars and students to exchange ideas, and to communicate and discuss research findings and new advances in different branches of algebra, especially Ring theory, Coding theory, Cryptography and Graph theory.

During the conference, world-renowned algebraists gave 8 plenary talks and 20 invited talks which have been potentially affected by the most recent developments in the related areas. This conference covered topics of several new directions and applications. Among the participants of the conference, 90 exuberant younger mathematicians presented their research articles, during proper thematic sessions. More than one dozen participants from various countries like USA, Egypt, Korea, Nigeria, Taiwan, Italy, Germany and Norway together with nearly two hundred delegates from within India participated in this conference.

A special session was devoted in the honour of Prof. M. A. Quadri who is one of the esteemed professors who initiated study and research in the area of modern mathematics in the Department of Mathematics, AMU, Aligarh.

We appreciate the active participation of all young researchers and academicians. Hopefully, the conference also enables participants to explore possible avenues to foster academic and research exchange, as well as scientific activities within and abroad of India. This refereed volume includes papers from renowned algebraists and invited speakers as well as other participants of the conference. All submitted papers are rigorously reviewed, followed by a careful selection process.

In addition to highlighting the latest research being done on the frontiers of algebra, the articles published also provide insights into how ideas have explored and have been connected. The proceeding's overall approach addresses the challenges of abundant topics of algebra particularly semi groups, groups, derivations in rings, rings and modules, group rings, matrix algebra, triangular algebra, polynomial rings and lattice theory. Apart from these topics, we also received research papers which have applications in coding theory and graph theory.

This research volume is distinguished from many others by its variety of topics, methodologies and depth of research. We believe that this volume will thus further expand our understanding and can serve as a reference book in the rapidly expanding field of algebra and related topics with their applications to coding theory, cryptography and graph theory.

We gratefully acknowledge the funding received towards this conference from the Aligarh Muslim University (AMU), Aligarh, Department of Science and Technology (DST), New Delhi, Indian National Science Academy (INSA), New Delhi, and the Council of Scientific and Industrial Research (CSIR), New Delhi. This volume would not have been possible without the support of expert referees who provided their valuable comments through reports diligently and promptly despite their busy schedules. We would like to thank Prof. M. Imdad, Chairman, Department of Mathematics, for his consistent support and guidance during the running of this conference. Furthermore, we would like to thank the rest of the faculty members, research scholars of Mathematics Department, AMU, for their collaborative effort during the conference. Also thanks to committee members, especially Prof. Nadeem ur Rehman, Dr. Shakir Ali, Dr. Mujeebur Rehman, Dr. M. Aslam Siddeeque and Dr. Ghulam Mohammad, who enabled this conference to be possible. We would like to say special thanks to Prof. M. A. Quadri. In spite of his health problems, his support, guidance and overall insights have made this an inspiring experience for us. We would like to express our gratitude to the entire team of Springer for publishing this volume. Thank you Mr. Shamim Ahmad, Senior Editor, Mathematical Sciences, Springer, India for facilitating the publication process, we truly appreciate your hard work and enthusiasm, everything was so intelligible and gave clear guidance. We look forward to continue our relationship.

Aligarh, India — Mohammad Ashraf
Aligarh, India — Asma Ali
Messina, Italy — Vincenzo De Filippis

Contents

Algebra

About the Editors

Mohammad Ashraf is a Professor in the Department of Mathematics at Aligarh Muslim University (AMU). Currently, he is the Dean, Faculty of Science and Chairperson, Department of Mathematics, AMU, Aligarh, India. He received his Ph.D. degree from AMU, and started his teaching carrier at the Department of Mathematics, AMU, Aligarh, in the year 1986. He had made substantial contributions to Ring theory and Algebras, Linear Algebra, Coding theory, Graph theory and Cryptography. He proved several exemplary results which are published in internationally recognized and reputed journals. He has published more than 200 peer-reviewed scientific articles. He has organized several International conferences and edited five Research Volumes and Proceedings. Professor Ashraf has successfully completed many major research projects from the UGC, DST, New Delhi, and NBHM, Mumbai, India. He had visited several foreign countries to deliver plenary and Invited lectures in International conferences and seminars and had been Visiting Scientist for many years at the University of Maribore, Slovenia, under the bilateral exchange program supported by Slovenian-Indian joint working group on Scientific and Technological co-operation under the auspices of the DST, India, and Ministry of Higher Education Science and Technology (MHEST), Republic of Slovenia. Together with all the above achievements, Prof. Ashraf is Editor/Managing Editor of many reputed international Mathematical journals. He is also a life member of several Mathematical Societies.

Asma Ali is a Professor at the Department of Mathematics, Aligarh Muslim University (A.M.U.), Aligarh, India. She joined the Department of Mathematics A.M.U., India, in 1995 as a faculty member and has been serving the department since then. She was awarded Senior Research Fellowship from UGC, New Delhi, and successfully completed her Ph.D. in 1991. Professor Asma has made valuable contributions in her field of research and has published a number of papers in reputed International Journals which are being cited in India and abroad. Very recently, she has been selected for LEAP, Leadership for Academicians Programme, under MHRD Scheme Pandit Madan Mohan Malaviya National Mission on Teachers and Teaching (PMMMNMTT), Domestic training at CALEM, India, December 16–30, 2019, and

Foreign training, February 15–22, 2020, at Monash University, Melbourne, Australia. Professor Asma has received Honour of recognition for her contribution and achievements as a successful professional by the International Society of Women in Science and has been a council member of Indian Mathematical Society. She is also a life member of many organizations including American Mathematical Society and Indian Science Congress.

Vincenzo De Filippis is an Associate Professor of Mathematics at the University of Messina, Italy. He received his Ph.D. degree in Mathematics (1999) from the University of Messina, Italy. He is an author and co-author of more than 100 publications in professional journals and has been on research visits to several Institutions in Europe and Asia. His research interests include Ring Theory, Theory of Associative Algebras, and Linear and Multilinear Algebras. His theoretical work is primarily aimed at studying the structure of algebras satisfying functional identities. He has delivered many invited and plenary lectures at International Conferences on algebraic structures and their applications. He is also a co-editor of various research volumes and a member of the Italian Mathematical Society (UMI) and the National Society of Algebraic and Geometric Structures and their Applications (GNSAGA-INDAM).

Algebra

Characterization of b-generalized Derivations in Rings with Involution

Adnan Abbasi, Muzibur Rahman Mozumder, and Aisha Jabeen

Abstract Let $\mathcal{W}$ be a ring with involution, $\mathcal{Q}$ be the right Martindale quotient ring, and $\mathcal{C}$ be the extended centroid of $\mathcal{W}$. Let $d : \mathcal{W} \to \mathcal{Q}$ be an additive map and $b \in \mathcal{Q}$. An additive map $\mathfrak{F} : \mathcal{W} \to \mathcal{Q}$ is called b-generalized derivation with associative map d if $\mathfrak{F}(xy) = \mathfrak{F}(x)y + bxd(y)$ for all $x, y \in \mathcal{W}$. In this manuscript, we study commuting b-generalized derivations in rings with involution.

Keywords Prime ring · b-generalized derivation · Involution

1 Introduction

Throughout the paper, $\mathcal{W}$ always denotes a prime ring with involution, $\mathcal{Q}$ be the right Martindale quotient ring of $\mathcal{W}$, $\mathcal{C} = \mathcal{Z}(\mathcal{Q})$ be the center of $\mathcal{Q}$ usually known as the extended centroid of $\mathcal{W}$ and is a field. An additive mapping "$* : \mathcal{W} \to \mathcal{W}$ is called an involution if $*$ is an anti-automorphism of order 2; that is, $(x^*)^* = x$ for all $x \in \mathcal{W}$". An element x in a ring with involution is said to be "hermitian if $x^* = x$ and skew-hermitian if $x^* = -x$". The sets of all hermitian and skew-hermitian elements of $\mathcal{W}$ will be denoted by $H(\mathcal{W})$ and $S(\mathcal{W})$, respectively. A ring equipped with an involution is known as ring with involution or $*$-ring. If $\mathcal{W}$ is 2-torsion free, then every $x \in \mathcal{W}$ can be uniquely represented in the form $2x = h + k$, where $h \in H(\mathcal{W})$ and $k \in S(\mathcal{W})$. Note that $S(\mathcal{W}) = H(\mathcal{W})$ if char$(\mathcal{W}) = 2$. The involution is said to be of the first kind if $\mathcal{Z}(\mathcal{W}) \subseteq \mathcal{H}(\mathcal{W})$, otherwise it is said to be of the second kind. In the later case it is worthwhile to see that $\mathcal{S}(\mathcal{W}) \cap \mathcal{Z}(\mathcal{W}) \neq (0)$. We refer the reader

A. Abbasi
Department of Mathematics, Madanpalle Institute of Technology & Science, Madanpalle, India

M. R. Mozumder (✉)
Department of Mathematics, Aligarh Muslim University, Aligarh, India
e-mail: muzibamu81@gmail.com

A. Jabeen
Department of Applied Sciences & Humanities, Jamia Millia Islamia, New Delhi 110025, Delhi, India

M. Ashraf et al. (eds.), *Algebra and Related Topics with Applications*,
Springer Proceedings in Mathematics & Statistics 392,
https://doi.org/10.1007/978-981-19-3898-6_1

to [3, 5] for justification and amplification for the above-mentioned notations and key definitions.

An additive mapping "$d : \mathcal{W} \rightarrow \mathcal{W}$ is said to be a derivation on $\mathcal{W}$ if $d(xy) = d(x)y + xd(y)$ for all $x, y \in \mathcal{W}$". A derivation "d is said to be inner if there exists $a \in \mathcal{W}$ such that $d(x) = ax - xa$ for all $x \in \mathcal{W}$". An additive map "$\mathfrak{F} : \mathcal{W} \rightarrow \mathcal{W}$ is called a generalized derivation of $\mathcal{W}$ if there exists a derivation d of $\mathcal{W}$ such that $\mathfrak{F}(xy) = \mathfrak{F}(x)y + xd(y)$ for all $x, y \in \mathcal{W}$". The derivation d is uniquely determined by $\mathfrak{F}$ and is called the associated derivation of $\mathfrak{F}$. The very recent concept of generalized derivations introduced by Koşan and Lee [6], namely, b-generalized derivation which is defined as follows: An additive mapping "$\mathfrak{F} : \mathcal{W} \rightarrow \mathcal{Q}$ is called a (left) b-generalized derivation of $\mathcal{W}$ associated with d, an additive map from $\mathcal{W}$ to $\mathcal{Q}$, if $\mathfrak{F}(xy) = \mathfrak{F}(x)y + bxd(y)$ for all $x, y \in \mathcal{W}$, where $b \in \mathcal{Q}$". Also Lee proved that if $\mathcal{W}$ is a prime ring and $0 \neq b \in \mathcal{Q}$, then the associated map d is a derivation, i.e., $d(xy) = d(x)y + xd(y)$ for all $x, y \in \mathcal{W}$." It is easy to see that every generalized derivation is a 1-generalized derivation. Also, the mapping $x \in \mathcal{W} \rightarrow ax + bxc \in \mathcal{Q}$ for some fixed $a, b, c \in \mathcal{Q}$ is a b-generalized derivation of $\mathcal{W}$, which is known as inner b-generalized derivation of $\mathcal{W}$. Beside this, they also characterized b-generalized derivation. That is every b-generalized derivation $\mathfrak{F}$ on a semiprime ring $\mathcal{W}$ is of the form $\mathfrak{F}(x) = ax + bd(x)$ for all $x \in \mathcal{W}$, where $a, b \in \mathcal{Q}$.

A map "$f : \mathcal{W} \rightarrow \mathcal{W}$ is said to be centralizing(commuting) on a nonempty subset S of $\mathcal{W}$, if $[f(x), x] \in \mathcal{Z}(\mathcal{W})([f(x), x] = 0)$ for all $x, y \in S$". The study of centralizing(commuting) mappings initially started by Divinsky [4] who proved that a simple artinian ring is commutative if it has a commuting non-trivial automorphism. Two years later, Posner [9] proved that the existence of a nonzero commuting derivation on a prime ring prompts the ring to be commutative. Over the last few decades, several authors have proved commutativity theorems for prime and semiprime rings admitting automorphisms or derivations which are centralizing(commuting) mappings on an appropriate subset of the ring.

Following [1], a mapping "$f{:}\mathcal{W} \rightarrow \mathcal{W}$ is called $*$-centralizing ($*$-commuting) on a nonempty set S of $\mathcal{W}$ if $[f(x), x^*] \in \mathcal{Z}(\mathcal{W})([f(x), x^*] = 0)$ for all $x \in S$". For any central element a the map defined by $x \mapsto ax^*$ is $*$-commuting on $\mathcal{W}$. Very recently Ali and Dar [1] proved the following result as follows: let $\mathcal{W}$ be a prime ring with involution $*$ such that char$(\mathcal{W}) \neq 2$. Let d be a nonzero derivation on $\mathcal{W}$ such that $[d(x), x^*] \in \mathcal{Z}(\mathcal{W})$ for all $x \in \mathcal{W}$, then $\mathcal{W}$ is commutative. Later this result was extended by Najjer et al. [8] for $*$-centralizing derivation. Recently, Alahmadi et al. [2], generalized above result as follows: "Let $\mathcal{W}$ be a prime ring with involution of the second kind such that char$(\mathcal{W}) \neq 2$. If $\mathcal{W}$ admits a nonzero generalized derivation $\mathfrak{F} : \mathcal{W} \rightarrow \mathcal{W}$ such that $[\mathfrak{F}(x), x^*] \in \mathcal{Z}(\mathcal{W})$ for all $x \in \mathcal{W}$, then $\mathcal{W}$ is commutative". Driving motivation from the formal definition of b-generalized derivation and results studied in [1, 2, 8], we proposed investigation in the same vane by studying commuting b-generalized derivation and another $*$-identity on b-generalized derivation have also been studied in our manuscript, we conclude our manuscript with an example in support of our hypothesis of second kind involution, which shows that second kind assumption is essential in our results.

2 Results

Remark 1 Let $\mathcal{W}$ be a prime ring with involution $'*'$ of the second kind such that char$(\mathcal{W}) \neq 2$ and let $\mathfrak{F}$ be a nonzero b-generalized derivation on $\mathcal{W}$ associated with a derivation d on $\mathcal{W}$ such that $d(h_0) = 0$ for all $h_0 \in H(\mathcal{W}) \cap \mathcal{Z}(\mathcal{W})$. Then $d(z) = 0$ for all $z \in \mathcal{Z}(\mathcal{W})$.

Proof Given that $d(h_0) = 0$ for all $h_0 \in H(\mathcal{W}) \cap \mathcal{Z}(\mathcal{W})$. Now replacing h_0 by ${k_0}^2$, where $k_0 \in S(\mathcal{W}) \cap \mathcal{Z}(\mathcal{W})$. This implies that $d({k_0}^2) = 0$ for all $k_0 \in S(\mathcal{W}) \cap \mathcal{Z}(\mathcal{W})$. Thus we have $2k_0d(k_0) = 0$ for all $k_0 \in S(\mathcal{W}) \cap \mathcal{Z}(\mathcal{W})$. Since char$(\mathcal{W}) \neq 2$, then we obtain $k_0d(k_0) = 0$ for all $k_0 \in S(\mathcal{W}) \cap \mathcal{Z}(\mathcal{W})$. Since $S(\mathcal{W}) \cap \mathcal{Z}(\mathcal{W}) \neq (0)$, then by the primeness, we obtain $d(k_0) = 0$ for all $k_0 \in S(\mathcal{W}) \cap \mathcal{Z}(\mathcal{W})$. Now we consider $2d(z) = d(2z) = d(h_0 + k_0) = d(h_0) + d(k_0)$ and we know that $d(h_0) = 0$ for all $h_0 \in H(\mathcal{W}) \cap \mathcal{Z}(\mathcal{W})$ and $d(k_0) = 0$ for all $h_0 \in S(\mathcal{W}) \cap \mathcal{Z}(\mathcal{W})$, this implies that $2d(z) = 0$ for all $z \in \mathcal{Z}(\mathcal{W})$. Since char$(\mathcal{W}) \neq 2$. This implies that $d(z) = 0$ for all $z \in \mathcal{Z}(\mathcal{W})$. □

Theorem 1 *Let $\mathcal{W}$ be a noncommutative prime ring with involution $'*'$ of the second kind such that char$(\mathcal{W}) \neq 2$ and let $\mathfrak{F}$ be a nonzero b-generalized derivation on $\mathcal{W}$ associated with a derivation d on $\mathcal{W}$ such that $[\mathfrak{F}(x), x^*] = 0$ for all $x \in \mathcal{W}$, then $\mathfrak{F}(x) = \lambda x$, where $\lambda \in C$ for all $x \in \mathcal{W}$.*

Proof By the given hypothesis, we have

$$[\mathfrak{F}(x), x^*] = 0 \quad \text{for all } x \in \mathcal{W}. \tag{1}$$

Replacing x by $h + h_1$ in above relation where $h, h_1 \in H(\mathcal{W})$, yields that

$$[\mathfrak{F}(h), h_1] + [\mathfrak{F}(h_1), h] = 0 \quad \text{for all } h, h_1 \in H(\mathcal{W}). \tag{2}$$

Substituting h_1h_0 for h_1 in (2) where $h_0 \in H(\mathcal{W}) \cap \mathcal{Z}(\mathcal{W})$, we obtain

$$[\mathfrak{F}(h), h_1]h_0 + [\mathfrak{F}(h_1), h]h_0 + [bh_1, h]d(h_0) = 0 \quad \text{for all } h, h_1 \in H(\mathcal{W}). \tag{3}$$

and $h_0 \in H(\mathcal{W}) \cap \mathcal{Z}(\mathcal{W})$. By application of (2), we get $[bh_1, h]d(h_0) = 0$, then by the primeness of $\mathcal{W}$, we get either $[bh_1, h] = 0$ for all $h, h_1 \in H(\mathcal{W})$ or $d(h_0) = 0$ for all $h_0 \in H(\mathcal{W}) \cap \mathcal{Z}(\mathcal{W})$. Consider

$$[bh_1, h] = 0 \quad \text{for all } h, h_1 \in H(\mathcal{W}). \tag{4}$$

Taking h_0 for h_1, where $h_0 \in H(\mathcal{W}) \cap \mathcal{Z}(\mathcal{W})$. Since $S(\mathcal{W}) \cap \mathcal{Z}(\mathcal{W}) \neq (0)$, then by the primeness of $\mathcal{W}$, we obtain

$$[b, h] = 0 \quad \text{for all } h \in H(\mathcal{W}). \tag{5}$$

Replacing h by kk_0, where $k \in S(\mathcal{W})$ and $k_0 \in S(\mathcal{W}) \cap \mathcal{Z}(\mathcal{W})$, we get

$$[b, k]k_0 = 0 \text{ for all } k \in S(\mathcal{W}) \text{ and } k_0 \in S(\mathcal{W}) \cap \mathcal{Z}(\mathcal{W}).$$

Since $S(\mathcal{W}) \cap \mathcal{Z}(\mathcal{W}) \neq (0)$, application of primeness of $\mathcal{W}$ implies that

$$[b, k] = 0 \text{ for all } k \in S(\mathcal{W}). \tag{6}$$

Consider $2[b, x] = [b, 2x] = [b, h] + [b, k]$. Using (5) and (6), we obtain $2[b, x] = 0$ for all $x \in \mathcal{W}$. Since char$(\mathcal{W}) \neq 2$, this implies that $b \in \mathcal{Z}(\mathcal{W})$. Using it into (4), we get

$$b[h_1, h] = 0 \text{ for all } h, h_1 \in H(\mathcal{W}). \tag{7}$$

By the primeness of $\mathcal{W}$, we obtain either $b = 0$ or $[h_1, h] = 0$ for all $h, h_1 \in H(\mathcal{W})$. Consider $b = 0$ and on linearizing (1), we obtain

$$[\mathfrak{F}(x), y^*] + [\mathfrak{F}(y), x^*] = 0 \text{ for all } x, y \in \mathcal{W}. \tag{8}$$

Replacing y by yk_0, where $k_0 \in S(\mathcal{W}) \cap \mathcal{Z}(\mathcal{W})$, we get

$$-[\mathfrak{F}(x), y^*]k_0 + [\mathfrak{F}(y), x^*]k_0 = 0 \text{ for all } x, y \in \mathcal{W} \text{ and } k_0 \in S(\mathcal{W}) \cap \mathcal{Z}(\mathcal{W}). \tag{9}$$

Combining (8) and (9), we get $2[\mathfrak{F}(y), x^*]k_0 = 0$ for all $x, y \in \mathcal{W}$ and $k_0 \in S(\mathcal{W}) \cap \mathcal{Z}(\mathcal{W})$. Since char$(\mathcal{W}) \neq 2$ and $S(\mathcal{W}) \cap \mathcal{Z}(\mathcal{W}) \neq (0)$, implies that $[\mathfrak{F}(y), x^*] = 0$ for all $x, y \in \mathcal{W}$. Replacing x by x^* and y by x, we obtain $[\mathfrak{F}(x), x] = 0$ for all $x \in \mathcal{W}$. Hence in view of [7, Theorem 1.1], we get $\mathfrak{F}(x) = \lambda x$ for all $x \in \mathcal{W}$ where $\lambda \in C$. Now consider

$$[h_1, h] = 0 \text{ for all } h, h_1 \in H(\mathcal{W}). \tag{10}$$

Replacing h_1 by kk_0, where $k \in S(\mathcal{W})$ and $k_0 \in S(\mathcal{W}) \cap \mathcal{Z}(\mathcal{W})$, we obtain

$$[k_1, h] = 0 \text{ for all } h \in H(\mathcal{W}) \text{ and } k_1 \in S(\mathcal{W}). \tag{11}$$

Again taking kk_0 for h in (10), we obtain

$$[h_1, k] = 0 \text{ for all } h \in H(\mathcal{W}) \text{ and } k \in S(\mathcal{W}). \tag{12}$$

Replacing h_1 by k_1k_0 in (12), where $k_1 \in S(\mathcal{W})$ and $k_0 \in S(\mathcal{W}) \cap \mathcal{Z}(\mathcal{W})$, we get

$$[k_1, k] = 0 \text{ for all } k, k_1 \in S(\mathcal{W}). \tag{13}$$

Consider $4[x, y] = [2x, 2y] = [h_1 + k_1, h + k] = [h_1, h] + [h_1, k] + [k_1, h] + [k_1, k]$. From the application of Eqs. (10), (11), (12), and (13), we obtain $4[x, y] = 0$ for all $x, y \in \mathcal{W}$. Since char$(\mathcal{W}) \neq 2$, this implies that $[x, y] = 0$ for all $x, y \in \mathcal{W}$. This gives is $\mathcal{W}$ is commutative, which is a contradiction to our assumption. Now suppose $d(h_0) = 0$ for all $h_0 \in H(\mathcal{W}) \cap \mathcal{Z}(\mathcal{W})$. Using Remark 1, $d(z) = 0$ for all

$z \in \mathcal{Z}(\mathcal{W})$. Now follow the same line of proof as we did after (8), we get the required result. □

Theorem 2 *Let $\mathcal{W}$ be a noncommutative prime ring with involution '$*$' of the second kind such that char($\mathcal{W}$) $\neq 2$ and let $\mathfrak{F}$ be a nonzero b-generalized derivation on $\mathcal{W}$ associated with a derivation d on $\mathcal{W}$ such that $\mathfrak{F}(x \circ x^*) = 0$ for all $x \in \mathcal{W}$, then $\mathfrak{F}(x) = \lambda x$ for all $x \in \mathcal{W}$.*

Proof Given that

$$\mathfrak{F}(x \circ x^*) = 0 \text{ for all } x \in \mathcal{W}. \tag{14}$$

Linearization of the above relation yields that

$$[\mathfrak{F}(x \circ y^*), r] + [\mathfrak{F}(y \circ x^*), r] = 0 \text{ for all } x, y, r \in \mathcal{W}. \tag{15}$$

Substituting yh_0 for y in (15), where $h_0 \in H(\mathcal{W}) \cap \mathcal{Z}(\mathcal{W})$, we obtain

$$[\mathfrak{F}(x \circ y^*), r]h_0 + [\mathfrak{F}(y \circ x^*), r]h_0 + [b(x \circ y^*), r]d(h_0) + [b(y \circ x^*), r]d(h_0) = 0 \tag{16}$$

for all $x, y, r \in \mathcal{W}$ and $h_0 \in H(\mathcal{W}) \cap \mathcal{Z}(\mathcal{W})$. By the application of (15), we get

$$[b(x \circ y^*), r]d(h_0) + [b(y \circ x^*), r]d(h_0) = 0 \text{ for all } x, y, r \in \mathcal{W}. \tag{17}$$

Replacing y by yk_0 in (17) and combining the obtain result with (17), we get

$$2[b(y \circ x^*), r]d(h_0)k_0 = 0 \text{ for all } x, y, r \in \mathcal{W} \text{ and } k_0 \in S(\mathcal{W}) \cap \mathcal{Z}(\mathcal{W}).$$

Since char($\mathcal{W}$) $\neq 2$ and $S(\mathcal{W}) \cap \mathcal{Z}(\mathcal{W}) \neq (0)$, thus, we have

$$[b(y \circ x^*), r]d(h_0) = 0 \text{ for all } x, y, r \in \mathcal{W} \text{ and } h_0 \in H(\mathcal{W}) \cap \mathcal{Z}(\mathcal{W}).$$

Again taking x by h_0 where $h_0 \in H(\mathcal{W}) \cap \mathcal{Z}(\mathcal{W}) \neq (0)$ and on solving we have

$$[by, r]d(h_0) = 0 \text{ for all } x, y, r \in \mathcal{W} \text{ and } h_0 \in H(\mathcal{W}) \cap \mathcal{Z}(\mathcal{W}).$$

Applying primeness of $\mathcal{W}$ we get either $[by, r] = 0$ for all $y, r \in \mathcal{W}$ or $d(h_0) = 0$ for all $h_0 \in H(\mathcal{W}) \cap \mathcal{Z}(\mathcal{W})$. Now we consider

$$[by, r] = 0 \text{ for all } y, r \in \mathcal{W}.$$

This can be further written as

$$b[y, r] + [b, r]y = 0 \text{ for all } y, r \in \mathcal{W}. \tag{18}$$

Replacing y by yu where $u \in \mathcal{W}$ in (18), we get

$$by[u,r]+b[y,r]u+[b,r]yu=0 \text{ for all } y,r\in\mathcal{W}. \tag{19}$$

Using (18) in (19), we obtain $by[u,r]=0$ for all $y,r,u\in\mathcal{W}$. By the primeness of $\mathcal{W}$, we get either $b=0$ or $[u,r]=0$ for all $u,r\in\mathcal{W}$. If we consider $[u,r]=0$ for all $u,r\in\mathcal{W}$, which shows that $\mathcal{W}$ is commutative, which is a contradiction to our supposition, this implies that $b=0$. Now replacing x by h in (14), we get

$$[\mathfrak{F}(h^2),r]=0 \text{ for all } r\in\mathcal{W} \text{ and } h\in H(\mathcal{W}). \tag{20}$$

Taking $h+h_0$ for h in (20), where $h_0\in H(\mathcal{W})\cap\mathcal{Z}(\mathcal{W})$ and using (20), we obtain

$$2[\mathfrak{F}(hh_0),r]=0 \text{ for all } r\in\mathcal{W} \text{ and } h\in H(\mathcal{W}).$$

Since $\text{char}(\mathcal{W})\neq 2$, this implies that

$$[\mathfrak{F}(hh_0),r]=0 \text{ for all } r\in\mathcal{W} \text{ and } h\in H(\mathcal{W}).$$

By the definition of $\mathfrak{F}$ and $b=0$, we get

$$[\mathfrak{F}(h)h_0,r]=0 \text{ for all } r\in\mathcal{W} \text{ and } h\in H(\mathcal{W}).$$

Since $S(\mathcal{W})\cap\mathcal{Z}(\mathcal{W})\neq(0)$, this implies that

$$[\mathfrak{F}(h),r]=0 \text{ for all } r\in\mathcal{W} \text{ and } h\in H(\mathcal{W}). \tag{21}$$

Replacing h by kk_0, where $k\in S(\mathcal{W})$, $k_0\in S(\mathcal{W})\cap\mathcal{Z}(\mathcal{W})$ and using $b=0$, we obtain

$$[\mathfrak{F}(k),r]=0 \text{ for all } r\in\mathcal{W} \text{ and } k\in S(\mathcal{W}). \tag{22}$$

Now consider $2[\mathfrak{F}(x),r]=[\mathfrak{F}(2x),r]=[\mathfrak{F}(h),r]+[\mathfrak{F}(k),r]$. Using (21) and (22), we get $2[\mathfrak{F}(x),r]=0$ for all $x,r\in\mathcal{W}$. Since $\text{char}(\mathcal{W})\neq 2$, this implies that $[\mathfrak{F}(x),r]=0$ for all $x,r\in\mathcal{W}$. Hence in view of [7, Theorem 1.1], we have $\mathfrak{F}(x)=\lambda x$ for all $x\in\mathcal{W}$. Now consider $d(h_0)=0$ for all $h_0\in H(\mathcal{W})\cap\mathcal{Z}(\mathcal{W})$. Then by Remark 1, $d(z)=0$ for all $z\in\mathcal{Z}(\mathcal{W})$. Now follow the same steps as we did after (20) and using $d(z)=0$ for all $z\in\mathcal{Z}(\mathcal{W})$, we get the required result. This completes the proof of the theorem. □

The following example shows that the second kind involution condition is essential in Theorem 1.

Example 1 Let $\mathcal{W}=\left\{\begin{pmatrix}\beta_1 & \beta_2\\ \beta_3 & \beta_4\end{pmatrix} \,\middle|\, \beta_1,\beta_2,\beta_3,\beta_4\in\mathbb{Z}\right\}$. Of course $\mathcal{W}$ with matrix addition and matrix multiplication is a prime ring. Define mappings $\mathfrak{F},d,*:\mathcal{W}\longrightarrow\mathcal{W}$ by

$$\mathfrak{F}\begin{pmatrix}\beta_1 & \beta_2\\ \beta_3 & \beta_4\end{pmatrix}=\begin{pmatrix}0 & -\beta_2\\ \beta_3 & 0\end{pmatrix},\quad d\begin{pmatrix}\beta_1 & \beta_2\\ \beta_3 & \beta_4\end{pmatrix}=\begin{pmatrix}0 & -\beta_2\\ \beta_3 & 0\end{pmatrix}$$

and a fixed element $b = \begin{pmatrix} 1 & 0 \\ 0 & 1 \end{pmatrix}$, $\begin{pmatrix} \beta_1 & \beta_2 \\ \beta_3 & \beta_4 \end{pmatrix}^* = \begin{pmatrix} \beta_4 & -\beta_2 \\ -\beta_3 & \beta_1 \end{pmatrix}$.

Obviously, $\mathcal{Z}(\mathcal{W}) = \left\{ \begin{pmatrix} \beta_1 & 0 \\ 0 & \beta_1 \end{pmatrix} \mid \beta_1 \in \mathcal{Z} \right\}$. Then $x^* = x$ for all $x \in \mathcal{Z}(\mathcal{W})$, and hence $\mathcal{Z}(\mathcal{W}) \subseteq H(\mathcal{W})$, which shows that the involution $'*'$ is of the first kind. Moreover, $\mathfrak{F}$, d are nonzero b-generalized derivation and associated derivation with fixed element b defined as above, such that the hypotheses in Theorem 1 is satisfied but $\mathfrak{F}$ is not in the form $\mathfrak{F}(x) = \lambda x$ for all $x \in \mathcal{W}$. Hence, the hypothesis of the second kind involution is crucial in our results.

References

1. Ali, S., Dar, N.A.: On $*$-centralizing mappings in rings with involution. Georgian Math. J. **21**, 25–28 (2014)
2. Alahmadi, A., Alhazmi, H., Ali, S., Dar, N.A., Khan, A.N.: Additive maps on prime and semiprime rings with involution. Hacet. J. Math. Stat. **49**(3), 1126–1133 (2020)
3. Beidar, K.I., Martindale III, W.S., Mikhalev, A.V.: Rings with generalized identities. Monographs and Textbooks in Pure and Applied Mathematics, Monographs and Textbooks in Pure and Applied Mathematics 196, Dekker, New York (1996)
4. Divinsky, N.: On commuting automorphisms of rings. Trans. Roy. Soc. Canada. Sect. III **3**(49), 19–22 (1955)
5. Herstein, I.N.: Rings with involution. Chicago Lectures in Mathematics. The University of Chicago Press, Chicago (1976)
6. Kǫsan, M.T., Lee, T.K.: b-Generalized derivations having nilpotent values. J. Aust. Math. Soc. **96**, 326–337 (2014)
7. Liu, C.K.: An Engel condition with b-generalized derivations. Linear Algebra Appl. **65**, 300–312 (2017)
8. Nejjar, B., Kacha, A., Mamouni, A., Oukhtite, L.: Commutativity theorems in rings with involution. Commun. Algebra **45**(2), 698–708 (2017)
9. Posner, E.C.: Derivations in prime rings. Proc. Am. Math. Soc. **8**, 1093–1100 (1957)

Jordan Generalized n-derivations of Unital Algebras Containing Idempotents

Mohd Shuaib Akhtar, Mohammad Afajal Ansari, and Nazia Parveen

Abstract Suppose that $\mathcal{A}$ is a unital algebra containing a nontrivial idempotent. In this paper, by introducing the notion of Jordan generalized n-derivations, it is shown that under certain conditions every multiplicative Jordan generalized n-derivation on $\mathcal{A}$ is additive. As a consequence, multiplicative Jordan generalized derivations on triangular algebras are characterized.

Keywords Unital algebras · Jordan generalized derivation · Jordan generalized n-derivation

1 Introduction

Let $\mathcal{R}$ be a commutative ring with identity and $\mathcal{A}$ be an algebra over $\mathcal{R}$. Recall that an $\mathcal{R}$-linear mapping $\delta : \mathcal{A} \to \mathcal{A}$ is called a *derivation* if $\delta(ab) = \delta(a)b + a\delta(b)$ holds for all $a, b \in \mathcal{A}$. An $\mathcal{R}$-linear mapping $\delta : \mathcal{A} \to \mathcal{A}$ is said to be a *Jordan derivation* if $\delta(a \circ b) = \delta(a) \circ b + a \circ \delta(b)$ for all $a, b \in \mathcal{A}$, where $a \circ b = ab + ba$ is the usual Jordan product. A *Jordan triple derivation* is an $\mathcal{R}$-linear mapping $\delta : \mathcal{A} \to \mathcal{A}$ which satisfies $\delta((a \circ b) \circ c) = (\delta(a) \circ b) \circ c + (a \circ \delta(b)) \circ c + (a \circ b) \circ \delta(c)$ for all $a, b, c \in \mathcal{A}$. It can be easily seen that every derivation is a Jordan derivation and every Jordan derivation is a Jordan triple derivation. Note that if the mapping $\delta : \mathcal{A} \to \mathcal{A}$ is not necessarily linear in the above definitions, then δ is said to be a multiplicative derivation, multiplicative Jordan derivation and multiplicative Jordan triple derivation, respectively. An $\mathcal{R}$-linear mapping $\Delta : \mathcal{A} \to \mathcal{A}$ is called a *generalized derivation* with associated derivation δ on $\mathcal{A}$ if $\Delta(ab) = \Delta(a)b + a\delta(b)$ holds for all $a, b \in \mathcal{A}$. An $\mathcal{R}$-linear mapping $\Delta : \mathcal{A} \to \mathcal{A}$ is said to be a *Jordan generalized derivation* with associated Jordan derivation δ on $\mathcal{A}$ if $\Delta(a \circ b) = \Delta(a) \circ b + a \circ \delta(b)$ for all $a, b \in \mathcal{A}$. Similarly, an $\mathcal{R}$-linear mapping $\Delta : \mathcal{A} \to \mathcal{A}$ is said to be a *Jordan generalized triple derivation* with associated Jordan triple derivation δ on $\mathcal{A}$

M. S. Akhtar (✉) · M. A. Ansari · N. Parveen
Department of Mathematics, Aligarh Muslim University, Aligarh 202002, India
e-mail: mshuaibakhtar@gmail.com

M. Ashraf et al. (eds.), *Algebra and Related Topics with Applications*,
Springer Proceedings in Mathematics & Statistics 392,
https://doi.org/10.1007/978-981-19-3898-6_2

if $\Delta((a \circ b) \circ c) = (\Delta(a) \circ b) \circ c + (a \circ \delta(b)) \circ c + (a \circ b) \circ \delta(c)$ for all $a, b, c \in \mathcal{A}$.

Now we discuss a more general class of mappings. Let us define the following sequence of polynomials:

$$\begin{aligned} p_1(x_1) &= x_1 \\ p_2(x_1, x_2) &= p_1(x_1) \circ x_2 = x_1 \circ x_2 \\ p_3(x_1, x_2, x_3) &= p_2(x_1, x_2) \circ x_3 = (x_1 \circ x_2) \circ x_3 \\ &\vdots \\ p_n(x_1, x_2, \ldots, x_n) &= p_{n-1}(x_1, x_2, \ldots, x_{n-1}) \circ x_n. \end{aligned}$$

The polynomial $p_n(x_1, x_2, \ldots, x_n)(n \geq 2)$ is called Jordan n-product. Note that

$$p_n(x_1, x_2, \ldots, x_n) = p_{n-1}(x_1 \circ x_2, x_3, \ldots, x_n) \quad (n \geq 2)$$

for all $x_1, x_2, \ldots, x_n \in \mathcal{A}$. An $\mathcal{R}$-linear mapping $\delta : \mathcal{A} \to \mathcal{A}$ is said to be a Jordan n-derivation if

$$\delta(p_n(x_1, x_2, \ldots, x_n)) = \sum_{i=1}^{n} p_n(x_1, x_2, \ldots, x_{i-1}, \delta(x_i), x_{i+1}, \ldots, x_n)$$

for all $x_1, x_2, \ldots, x_n \in \mathcal{A}$. A $\mathcal{R}$-linear mapping $\Delta : \mathcal{A} \to \mathcal{A}$ is said to be a Jordan generalized n-derivation with associated Jordan n-derivation δ on $\mathcal{A}$ if

$$\begin{aligned} \Delta(p_n(x_1, x_2, \ldots, x_n)) = &\; p_n(\Delta(x_1), x_2, \ldots, x_{n-1}, x_n) \\ &+ \sum_{i=2}^{n} p_n(x_1, x_2, \ldots, x_{i-1}, \delta(x_i), x_{i+1}, \ldots, x_n) \end{aligned}$$

for all $x_1, x_2, \ldots, x_n \in \mathcal{A}$. If the condition of linearity is removed in the above definitions, then the corresponding Jordan n-derivation (resp. Jordan generalized n-derivation) is called multiplicative Jordan n-derivation (resp. multiplicative Jordan generalized n-derivation). By the definition, it is clear that a Jordan generalized 2-derivation is the usual Jordan generalized derivation and Jordan generalized 3-derivation is Jordan generalized triple derivation.

Over the past decade, a lot of work has been done on the additivity of mappings on various rings and algebras. In the year 1969, Martindale III [19] proved a remarkable result which states that every multiplicative bijective mapping from a prime ring containing a nontrivial idempotent onto an arbitrary ring is necessarily additive. Inspired by this result, many researchers obtained similar results in various rings and algebras. Daif [7] proved that every multiplicative derivation of a 2-torsion free prime ring containing a nontrivial idempotent is additive. Herstein [11] proved that every Jordan derivation on a prime ring of characteristic different from 2 is a derivation. Brešar [5] extended this result to 2-torsion free semiprime ring. Benkovič [2] proved that every linear Jordan derivation from upper triangular matrix algebras into its bimodule is the sum of a linear derivation and a linear antiderivation. Li and Lu [14] proved that every additive Jordan derivation on reflexive algebras is an additive derivation. Benkovič and Širovnik [4] obtained that under certain conditions every Jordan derivation of a unital algebra is the sum of a derivation and a singular Jordan derivation. Lee and Quynh [20] gave a characterization of additive Jordan triple derivations of arbitrary semiprime rings. Recently, Qi et al. [21] introduced the notion of Jordan n-derivation generalizing the concept of Jordan derivation and characterized Jordan n-derivations of unital rings containing idempotents.

Brešar [6] initiated the study of generalized derivations. Hvala [10] studied generalized derivation on prime rings. Jing and Lu [12] considered generalized Jordan derivations of prime rings and standard operator algebras. Vukman [22] extended this result to semiprime rings and proved that every generalized Jordan derivation of a 2-torsion-free semiprime ring is a generalized derivation. Hou and Qi [9] studied generalized Jordan derivations on nest algebras. Ma and Ji [18] considered generalized Jordan derivations of upper triangular matrix algebras and proved that every generalized Jordan derivation from the algebra of all upper triangular matrices over a commutative ring with identity into its bimodule is the sum of a generalized derivation and an antiderivation. Zhang and Yu [26] proved that every Jordan derivation of a triangular algebra is a derivation. Further, Li and Benkovič [15] generalized this result and proved that every Jordan generalized derivation (that is, Jordan generalized 2-derivation) of a triangular algebra is a generalized derivation. In addition, the characterization of Jordan derivations, Jordan triple derivations and generalized Jordan derivations on various rings and algebras are considered in [1, 8, 13, 16, 17, 23, 24], etc. Motivated by the afore-mentioned work, we study multiplicative Jordan generalized n-derivations of unital algebras containing idempotents and prove that under certain conditions every multiplicative Jordan generalized n-derivation on a unital algebra containing a nontrivial idempotent is additive.

2 Preliminaries

Let $\mathcal{A}$ be a unital algebra with a nontrivial idempotent e, and write $f = 1 - e$. Then $\mathcal{A}$ can be represented in the so-called Peirce decomposition form $\mathcal{A} = e\mathcal{A}e + e\mathcal{A}f + f\mathcal{A}e + f\mathcal{A}f$, where $e\mathcal{A}e$ and $f\mathcal{A}f$ are subalgebras of $\mathcal{A}$ with identity elements e and f, respectively, $e\mathcal{A}f$ is an $(e\mathcal{A}e, f\mathcal{A}f)$-bimodule and $f\mathcal{A}e$ is an $(f\mathcal{A}f, e\mathcal{A}e)$-bimodule. If $\mathcal{A}$ is a unital algebra with a nontrivial idempotent e such that $f\mathcal{A}e = \{0\}$, then $\mathcal{A}$ is a triangular algebra. Throughout the paper, we assume that $\mathcal{A}$ is a 2-torsion free unital algebra with a nontrivial idempotent e satisfying the following conditions:

$$\begin{cases} exe \cdot e\mathcal{A}f = \{0\} = f\mathcal{A}e \cdot exe \quad \text{implies} \quad exe = 0 \\ e\mathcal{A}f \cdot fxf = \{0\} = fxf \cdot f\mathcal{A}e \quad \text{implies} \quad fxf = 0. \end{cases} \qquad (\spadesuit)$$

To simplify the notation, we will use the following convention: $a_{11} = eae \in e\mathcal{A}e = \mathcal{A}_{11}$, $a_{12} = eaf \in e\mathcal{A}f = \mathcal{A}_{12}$, $a_{21} = fae \in f\mathcal{A}e = \mathcal{A}_{21}$ and $a_{22} = faf \in f\mathcal{A}f = \mathcal{A}_{22}$. Then each element $a \in \mathcal{A}$ can be represented in the form $a = a_{11} + a_{12} + a_{21} + a_{22}$. We shall frequently use the following results throughout the paper without further mentioning.

Lemma 1 *Let $\mathcal{A}$ be a unital algebra containing a nontrivial idempotent e, and write $f = 1 - e$. For any $a \in \mathcal{A}$ and for any integer $n \geq 2$, we have*

$$p_n(a, e, \ldots, e) = 2^{n-1}eae + eaf + fae \text{ and } p_n(a, f, \ldots, f) = 2^{n-1}faf + eaf + fae.$$

Proof By a recursive calculation, we have

$$\begin{aligned} p_n(a, e, \ldots, e) &= p_{n-1}(a \circ e, e, \ldots, e) \\ &= p_{n-1}(2eae + eaf + fae, e, \ldots, e) \\ &= p_{n-2}((2eae + eaf + fae) \circ e, e, \ldots, e) \\ &= p_{n-2}(4eae + eaf + fae, e, \ldots, e) \\ &= \cdots \\ &= 2^{n-1}eae + eaf + fae. \end{aligned}$$

Similarly, one can obtain $p_n(a, f, \ldots, f) = 2^{n-1}faf + eaf + fae$.

Lemma 2 *Let Δ be a multiplicative Jordan generalized n-derivation on $\mathcal{A}$. Then there exist an inner derivation g and a multiplicative Jordan generalized n-derivation Δ' on $\mathcal{A}$ such that*

$$\Delta = g + \Delta' \text{ and } e\Delta'(e)f = 0, \ f\Delta'(e)e = 0.$$

Proof Let $x_0 = e\Delta(e)f - f\Delta(e)e$ and define maps $g, \Delta' : \mathcal{A} \to \mathcal{A}$ by

$$g(x) = [x, x_0] \text{ and } \Delta'(x) = \Delta(x) - g(x)$$

for all $x \in \mathcal{A}$. It is easy to see that g is an inner derivation and Δ' is a multiplicative Jordan generalized n-derivation. Since

$$\begin{aligned} \Delta'(e) &= \Delta(e) - [e, e\Delta(e)f - f\Delta(e)e] \\ &= \Delta(e) - e\Delta(e)f - f\Delta(e)e \\ &= e\Delta(e)e + f\Delta(e)f, \end{aligned}$$

we get $e\Delta'(e)f = 0$, $f\Delta'(e)e = 0$. □

Lemma 3 ([21, Theorem 2.1(Claims 1–4)]) *Let δ be a multiplicative Jordan n-derivation on $\mathcal{A}$ such that $e\delta(e)f = 0$, $f\delta(e)e = 0$. Then we have*

$$\delta(0) = 0, \quad \delta(e) \in \mathcal{A}_{11}, \quad \delta(f) \in \mathcal{A}_{22}, \quad \delta(\mathcal{A}_{11}) \subseteq \mathcal{A}_{11}, \quad \delta(\mathcal{A}_{22}) \subseteq \mathcal{A}_{22},$$

$$\delta(\mathcal{A}_{12}) \subseteq \mathcal{A}_{12} + \mathcal{A}_{21}, \quad \textit{and} \quad \delta(\mathcal{A}_{21}) \subseteq \mathcal{A}_{12} + \mathcal{A}_{21}.$$

3 Multiplicative Jordan Generalized n-derivation

In this section, we discuss the additivity of multiplicative Jordan generalized n-derivations on unital algebras. The main result of the paper states as follows:

Theorem 1 *Let $\mathcal{A}$ be a 2-torsion-free unital algebra with a nontrivial idempotent e satisfying (♠). Then every multiplicative Jordan generalized n-derivation $\Delta : \mathcal{A} \to \mathcal{A}$ is additive.*

Proof In view of Lemma 2, it suffices to consider only those multiplicative Jordan generalized n-derivations Δ satisfying $e\Delta(e)f = 0$, $f\Delta(e)e = 0$. We shall establish the theorem by a series of claims.

Claim 1. $\Delta(0) = 0$.

$$\begin{aligned}\Delta(0) &= \Delta(p_n(0, 0, \ldots, 0))\\ &= p_n(\Delta(0), 0, \ldots, 0) + p_n(0, \delta(0), \ldots, 0) + \cdots + p_n(0, 0, \ldots, \delta(0))\\ &= 0.\end{aligned}$$

Claim 2. $e\delta(e)f = 0$, $f\delta(e)e = 0$, $e\delta(f)f = 0$, and $f\delta(f)e = 0$.

Using Claim 1 and the fact that $e \circ f = 0$, we obtain

$$\begin{aligned}0 &= \Delta(p_n(e, f, f, \ldots, f))\\ &= p_n(\Delta(e), f, f, \ldots, f) + p_n(e, \delta(f), f, \ldots, f) + \cdots + p_n(e, f, f, \ldots, \delta(f))\\ &= p_n(\Delta(e), f, f, \ldots, f) + p_{n-1}(e \circ \delta(f), f, \ldots, f)\\ &= 2^{n-1} f\Delta(e)f + e\Delta(e)f + f\Delta(e)e + 2^{n-2} f(e \circ \delta(f))f + e(e \circ \delta(f))f + f(e \circ \delta(f))e\\ &= 2^{n-1} f\Delta(e)f + e\delta(f)f + f\delta(f)e.\end{aligned}$$

Hence, $e\delta(f)f = 0$, $f\delta(f)e = 0$. Since δ is a multiplicative Jordan n-derivation on $\mathcal{A}$, we get

$$\begin{aligned}0 &= \delta(p_n(f, e, e, \ldots, e))\\ &= p_n(\delta(f), e, e, \ldots, e) + p_n(f, \delta(e), e, \ldots, e) + \cdots + p_n(f, e, e, \ldots, \delta(e))\\ &= p_n(\delta(f), e, e, \ldots, e) + p_{n-1}(f \circ \delta(e), e, \ldots, e)\\ &= 2^{n-1} e\delta(f)e + e\delta(f)f + f\delta(f)e + 2^{n-2} e(f \circ \delta(e))e + e(f \circ \delta(e))f + f(f \circ \delta(e))e\\ &= 2^{n-1} e\delta(f)e + e\delta(e)f + f\delta(e)e.\end{aligned}$$

Thus, $e\delta(e)f = 0$, $f\delta(e)e = 0$. Therefore, from now on we can use Lemma 3 in the following claims.

Claim 3. $\Delta(\mathcal{A}_{11}) \subseteq \mathcal{A}_{11}$ and $\Delta(\mathcal{A}_{22}) \subseteq \mathcal{A}_{22}$.

For any $a_{11} \in \mathcal{A}_{11}$, by Claim 1 and Lemma 3, we have

$$\begin{aligned} 0 &= \Delta(p_n(a_{11}, f, f, \ldots, f)) \\ &= p_n(\Delta(a_{11}), f, f, \ldots, f) + p_n(a_{11}, \delta(f), f, \ldots, f) + \cdots + p_n(a_{11}, f, f, \ldots, \delta(f)) \\ &= p_n(\Delta(a_{11}), f, f, \ldots, f) \\ &= 2^{n-1} f\Delta(a_{11})f + e\Delta(a_{11})f + f\Delta(a_{11})e. \end{aligned}$$

On multiplying the above equation from left by e and right by f, we get $e\Delta(a_{11})f = 0$; multiplying the above equation from left by f and right by e, we have $f\Delta(a_{11})e = 0$. Thus, the above equation becomes $2^{n-1} f\Delta(a_{11})f = 0$. Since $\mathcal{A}$ is 2-torsion free, we obtain $f\Delta(a_{11})f = 0$. Therefore, $\Delta(a_{11}) = e\Delta(a_{11})e \in \mathcal{A}_{11}$ for all $a_{11} \in \mathcal{A}_{11}$. Hence, $\Delta(\mathcal{A}_{11}) \subseteq \mathcal{A}_{11}$. In a similar manner, one can prove that $\Delta(\mathcal{A}_{22}) \subseteq \mathcal{A}_{22}$.

Claim 4. $\Delta(\mathcal{A}_{12}) \subseteq \mathcal{A}_{12} + \mathcal{A}_{21}$ and $\Delta(\mathcal{A}_{21}) \subseteq \mathcal{A}_{12} + \mathcal{A}_{21}$.

Since $p_n(a_{12}, f, f, \ldots, f) = a_{12}$ for any $a_{12} \in \mathcal{A}_{12}$, by Lemma 3, we have

$$\begin{aligned} \Delta(a_{12}) &= \Delta(p_n(a_{12}, f, f, \ldots, f)) \\ &= p_n(\Delta(a_{12}), f, f, \ldots, f) + p_n(a_{12}, \delta(f), f, \ldots, f) + \cdots + p_n(a_{12}, f, f, \ldots, \delta(f)) \\ &= 2^{n-1} f\Delta(a_{12})f + e\Delta(a_{12})f + f\Delta(a_{12})e + (n-1)a_{12}\delta(f). \end{aligned}$$

Multiplying by e on both sides of the above equation, we get $e\Delta(a_{12})e = 0$. Similarly, using the relation $p_n(a_{12}, e, e, \ldots, e) = a_{12}$, one can obtain $f\Delta(a_{12})f = 0$. Hence, $\Delta(a_{12}) \in \mathcal{A}_{12} + \mathcal{A}_{21}$ for all $a_{12} \in \mathcal{A}_{12}$. Similarly, we can prove $\Delta(a_{21}) \in \mathcal{A}_{12} + \mathcal{A}_{21}$ for all $a_{21} \in \mathcal{A}_{21}$.

Claim 5. For any $a_{ii} \in \mathcal{A}_{ii}, a_{ij} \in \mathcal{A}_{ij}$ and $a_{ji} \in \mathcal{A}_{ji}$, we have $\Delta(a_{ii} + a_{ij}) = \Delta(a_{ii}) + \Delta(a_{ij})$ and $\Delta(a_{ii} + a_{ji}) = \Delta(a_{ii}) + \Delta(a_{ji})$, $1 \le i \ne j \le 2$.

Let $a_{11} \in \mathcal{A}_{11}$ and $a_{12} \in \mathcal{A}_{12}$. We show that $T = \Delta(a_{11} + a_{12}) - \Delta(a_{11}) - \Delta(a_{12}) = 0$. We compute

$$\begin{aligned} &\Delta(p_n(a_{11} + a_{12}, f, f, \ldots, f)) \\ &= p_n(\Delta(a_{11} + a_{12}), f, f, \ldots, f) + p_n(a_{11} + a_{12}, \delta(f), f, \ldots, f) + \cdots \\ &\quad + p_n(a_{11} + a_{12}, f, f, \ldots, \delta(f)). \end{aligned}$$

Using the relations $p_n(a_{11} + a_{12}, f, f, \ldots, f) = p_n(a_{12}, f, f, \ldots, f)$, $p_n(a_{11}, f, f, \ldots, f) = 0$ and Claim 1, we have

$$\begin{aligned}
&\Delta(p_n(a_{11}+a_{12}, f, f, \ldots, f)) \\
&= \Delta(p_n(a_{11}, f, f, \ldots, f)) + \Delta(p_n(a_{12}, f, f, \ldots, f)) \\
&= p_n(\Delta(a_{11}), f, f, \ldots, f) + p_n(a_{11}, \delta(f), f, \ldots, f) + \cdots + p_n(a_{11}, f, f, \ldots, \delta(f)) \\
&\quad + p_n(\Delta(a_{12}), f, f, \ldots, f) + p_n(a_{12}, \delta(f), f, \ldots, f) + \cdots + p_n(a_{12}, f, f, \ldots, \delta(f)).
\end{aligned}$$

Comparing the above two equations, we get

$$p_n(T, f, f, \ldots, f) = 2^{n-1} fTf + eTf + fTe = 0,$$

which implies that $eTf = 0$, $fTe = 0$ and $fTf = 0$. It remains to show that $eTe = 0$. On the one hand, we have

$$\begin{aligned}
&\Delta(p_n(a_{11}+a_{12}, e-f, e-f, \ldots, e-f)) \\
&= p_n(\Delta(a_{11}+a_{12}), e-f, e-f, \ldots, e-f) + p_n(a_{11}+a_{12}, \delta(e-f), e-f, \ldots, e-f) \\
&\quad + \cdots + p_n(a_{11}+a_{12}, e-f, e-f, \ldots, \delta(e-f)).
\end{aligned}$$

On the other hand, using the facts $p_n(a_{11}+a_{12}, e-f, e-f, \ldots, e-f) = p_n(a_{11}, e-f, e-f, \ldots, e-f)$ and $p_n(a_{12}, e-f, e-f, \ldots, e-f) = 0$, we get

$$\begin{aligned}
&\Delta(p_n(a_{11}+a_{12}, e-f, e-f, \ldots, e-f)) \\
&= \Delta(p_n(a_{11}, e-f, e-f, \ldots, e-f)) + \Delta(p_n(a_{12}, e-f, e-f, \ldots, e-f)) \\
&= p_n(\Delta(a_{11}), e-f, e-f, \ldots, e-f) + p_n(a_{11}, \delta(e-f), e-f, \ldots, e-f) \\
&\quad + \cdots + p_n(a_{11}, e-f, e-f, \ldots, \delta(e-f)) \\
&\quad + p_n(\Delta(a_{12}), e-f, e-f, \ldots, e-f) + p_n(a_{12}, \delta(e-f), e-f, \ldots, e-f) \\
&\quad + \cdots + p_n(a_{12}, e-f, e-f, \ldots, \delta(e-f)).
\end{aligned}$$

Comparing the above two expressions, we obtain

$$p_n(T, e-f, e-f, \ldots, e-f) = p_n(eTe, e-f, e-f, \ldots, e-f) = 2^{n-1} eTe = 0,$$

which implies that $eTe = 0$. Thus, $T = 0$, i.e., $\Delta(a_{11}+a_{12}) = \Delta(a_{11}) + \Delta(a_{12})$. In a similar fashion, we can prove the other cases.

Claim 6. Δ is additive on $\mathcal{A}_{ij}$, $1 \le i \ne j \le 2$.

Observe that, for any $a_{12}, b_{12} \in \mathcal{A}_{12}$,

$$a_{12} + b_{12} = p_n(e + a_{12}, f + b_{12}, f + b_{12}, \ldots, f + b_{12}).$$

Using Claim 5 and the above fact, we have

$$\begin{aligned}
&\Delta(a_{12}+b_{12})\\
&=\Delta(p_n(e+a_{12}, f+b_{12}, f+b_{12}, \ldots, f+b_{12}))\\
&=p_n(\Delta(e+a_{12}), f+b_{12}, f+b_{12}, \ldots, f+b_{12})+p_n(e+a_{12}, \delta(f+b_{12}), f+b_{12}, \ldots, f+b_{12})\\
&\quad+\cdots+p_n(e+a_{12}, f+b_{12}, f+b_{12}, \ldots, \delta(f+b_{12}))\\
&=p_n(\Delta(e), f+b_{12}, f+b_{12}, \ldots, f+b_{12})+p_n(e, \delta(f+b_{12}), f+b_{12}, \ldots, f+b_{12})\\
&\quad+\cdots+p_n(e, f+b_{12}, f+b_{12}, \ldots, \delta(f+b_{12}))\\
&\quad+p_n(\Delta(a_{12}), f+b_{12}, f+b_{12}, \ldots, f+b_{12})+p_n(a_{12}, \delta(f+b_{12}), f+b_{12}, \ldots, f+b_{12})\\
&\quad+\cdots+p_n(a_{12}, f+b_{12}, f+b_{12}, \ldots, \delta(f+b_{12}))\\
&=\Delta(p_n(e, f+b_{12}, f+b_{12}, \ldots, f+b_{12}))+\Delta(p_n(a_{12}, f+b_{12}, f+b_{12}, \ldots, f+b_{12}))\\
&=\Delta(b_{12})+\Delta(a_{12}).
\end{aligned}$$

Similarly, we can prove that Δ is additive on $\mathcal{A}_{21}$.

Claim 7. Δ is additive on $\mathcal{A}_{ii}$, $1 \leq i \leq 2$.

Let $a_{11}, b_{11} \in \mathcal{A}_{11}$. Set $T = \Delta(a_{11}+b_{11}) - \Delta(a_{11}) - \Delta(b_{11})$. In view of Claim 3, $T = eTe \in \mathcal{A}_{11}$. For any $a_{12} \in \mathcal{A}_{12}$, we have

$$\begin{aligned}
&\Delta(p_n(a_{11}+b_{11}, a_{12}+f, \ldots, a_{12}+f))\\
&=p_n(\Delta(a_{11}+b_{11}), a_{12}+f, a_{12}+f, \ldots, a_{12}+f)\\
&\quad+p_n(a_{11}+b_{11}, \delta(a_{12}+f), a_{12}+f, \ldots, a_{12}+f)\\
&\quad+\cdots+p_n(a_{11}+b_{11}, a_{12}+f, a_{12}+f, \ldots, \delta(a_{12}+f)).
\end{aligned}$$

On the other hand, using Claim 6, we get

$$\begin{aligned}
&\Delta(p_n(a_{11}+b_{11}, a_{12}+f, a_{12}+f, \ldots, a_{12}+f))\\
&=\Delta(p_n(a_{11}, a_{12}+f, a_{12}+f, \ldots, a_{12}+f))+\Delta(p_n(b_{11}, a_{12}+f, a_{12}+f, \ldots, a_{12}+f))\\
&=p_n(\Delta(a_{11}), a_{12}+f, a_{12}+f, \ldots, a_{12}+f)+p_n(a_{11}, \delta(a_{12}+f), a_{12}+f, \ldots, a_{12}+f)\\
&\quad+\cdots+p_n(a_{11}, a_{12}+f, a_{12}+f, \ldots, \delta(a_{12}+f))\\
&\quad+p_n(\Delta(b_{11}), a_{12}+f, a_{12}+f, \ldots, a_{12}+f)+p_n(b_{11}, \delta(a_{12}+f), a_{12}+f, \ldots, a_{12}+f)\\
&\quad+\cdots+p_n(b_{11}, a_{12}+f, a_{12}+f, \ldots, \delta(a_{12}+f)).
\end{aligned}$$

Comparing the above two equations, we get

$$p_n(T, a_{12}+f, a_{12}+f, \ldots, a_{12}+f) = p_n(eTe, a_{12}+f, a_{12}+f, \ldots, a_{12}+f) = 0,$$

which implies that $eTea_{12} = 0$, for all $a_{12} \in \mathcal{A}_{12}$. Following a similar calculation as above, one can prove $a_{21}eTe = 0$ for all $a_{21} \in \mathcal{A}_{21}$. Since $\mathcal{A}$ satisfies (♠), we conclude that $T = eTe = 0$, that is, $\Delta(a_{11}+b_{11}) = \Delta(a_{11}) + \Delta(b_{11})$. Similarly, we can prove that Δ is additive on $\mathcal{A}_{22}$.

Claim 8. $\Delta(a_{11}+a_{22}) = \Delta(a_{11}) + \Delta(a_{22})$ for all $a_{11} \in \mathcal{A}_{11}$, $a_{22} \in \mathcal{A}_{22}$.

Let $a_{11} \in \mathcal{A}_{11}$ and $a_{22} \in \mathcal{A}_{22}$. We show that $T = \Delta(a_{11} + a_{22}) - \Delta(a_{11}) - \Delta(a_{22}) = 0$. On the one hand, we have

$$\begin{aligned}&\Delta(p_n(a_{11} + a_{22}, f, f, \dots, f))\\&= p_n(\Delta(a_{11} + a_{22}), f, f, \dots, f) + p_n(a_{11} + a_{22}, \delta(f), f, \dots, f)\\&\quad + \dots + p_n(a_{11} + a_{22}, f, f, \dots, \delta(f)).\end{aligned}$$

On the other hand, using the facts that $p_n(a_{11} + a_{22}, f, f, \dots, f) = p_n(a_{22}, f, f, \dots, f)$ and $p_n(a_{11}, f, f, \dots, f) = 0$, we get

$$\begin{aligned}\Delta(p_n(a_{11} + a_{22}, f, f, \dots, f)) &= \Delta(p_n(a_{11}, f, f, \dots, f)) + \Delta(p_n(a_{22}, f, f, \dots, f))\\&= p_n(\Delta(a_{11}), f, f, \dots, f) + p_n(a_{11}, \delta(f), f, \dots, f)\\&\quad + \dots + p_n(a_{11}, f, f, \dots, \delta(f)) + p_n(\Delta(a_{22}), f, f, \dots, f)\\&\quad + p_n(a_{22}, \delta(f), f, \dots, f) + \dots + p_n(a_{22}, f, f, \dots, \delta(f)).\end{aligned}$$

Comparing the above equations, we get

$$p_n(T, f, f, \dots, f) = 2^{n-1} fTf + eTf + fTe = 0,$$

which implies that $eTf = 0$, $fTe = 0$ and $fTf = 0$. It remains to show that $eTe = 0$. Using a similar technique as used above, one can obtain

$$p_n(T, e, e, \dots, e) = 2^{n-1} eTe + eTf + fTe = 0,$$

which implies that $eTe = 0$. Thus, $T = 0$, i.e., $\Delta(a_{11} + a_{22}) = \Delta(a_{11}) + \Delta(a_{12})$.

Claim 9. $\Delta(a_{ii} + a_{ij} + a_{ji}) = \Delta(a_{ii}) + \Delta(a_{ij} + a_{ji})$ for all $a_{ii} \in \mathcal{A}_{ii}$, $a_{ij} \in \mathcal{A}_{ij}$, $1 \le i \ne j \le 2$.

Let $a_{ij} \in \mathcal{A}_{ij}$, $1 \le i, j \le 2$. Set $T = \Delta(a_{11} + a_{12} + a_{21}) - \Delta(a_{11}) - \Delta(a_{12} + a_{21})$.

$$\begin{aligned}&\Delta(p_n(a_{11} + a_{12} + a_{21}, f, f, \dots, f))\\&= p_n(\Delta(a_{11} + a_{12} + a_{21}), f, f, \dots, f) + p_n(a_{11} + a_{12} + a_{21}, \delta(f), f, \dots, f)\\&\quad + \dots + p_n(a_{11} + a_{12} + a_{21}, f, f, \dots, \delta(f))\end{aligned}$$

and

$$\begin{aligned}&\Delta(p_n(a_{11} + a_{12} + a_{21}, f, f, \dots, f))\\&= \Delta(p_n(a_{11}, f, f, \dots, f)) + \Delta(p_n(a_{12} + a_{21}, f, f, \dots, f))\\&= p_n(\Delta(a_{11}), f, f, \dots, f) + p_n(a_{11}, \delta(f), f, \dots, f)\\&\quad + \dots + p_n(a_{11}, f, f, \dots, \delta(f)) + p_n(\Delta(a_{12} + a_{21}), f, f, \dots, f)\\&\quad + p_n(a_{12} + a_{21}, \delta(f), f, \dots, f) + \dots + p_n(a_{12} + a_{21}, f, f, \dots, \delta(f)).\end{aligned}$$

Comparing the above two equations, we get

$$p_n(T, f, f, \dots, f) = 2^{n-1} fTf + eTf + fTe = 0,$$

which implies that $eTf = 0$, $fTe = 0$ and $fTf = 0$. It remains to show that $eTe = 0$. By a similar technique as used above, one can obtain

$$p_n(T, e-f, e-f, \dots, e-f) = p_n(eTe, e-f, e-f, \dots, e-f) = 2^{n-1} eTe = 0,$$

which implies that $eTe = 0$. Thus, $T = 0$, i.e., $\Delta(a_{11} + a_{12} + a_{21}) = \Delta(a_{11}) + \Delta(a_{12} + a_{21})$. Similarly, computing $\Delta(p_n(a_{22} + a_{12} + a_{21}, f, f, \dots, f))$ and $\Delta(p_n(a_{22} + a_{12} + a_{21}, f-e, f-e, \dots, f-e))$ in two ways, respectively, one can prove that $\Delta(a_{22} + a_{12} + a_{21}) = \Delta(a_{22}) + \Delta(a_{12} + a_{21})$.

Claim 10. $\Delta(a_{11} + a_{12} + a_{21} + a_{22}) = \Delta(a_{11}) + \Delta(a_{12} + a_{21}) + \Delta(a_{22})$ for all $a_{ij} \in \mathcal{A}_{ij}$, $1 \le i, j \le 2$.

Let $T = \Delta(a_{11} + a_{12} + a_{21} + a_{22}) - \Delta(a_{11}) - \Delta(a_{12} + a_{21}) - \Delta(a_{22})$. For any $a_{ij} \in \mathcal{A}_{ij}$, $1 \le i, j \le 2$, we have

$$\begin{aligned}
&\Delta(p_n(a_{11} + a_{12} + a_{21} + a_{22}, e, e, \dots, e)) \\
&= p_n(\Delta(a_{11} + a_{12} + a_{21} + a_{22}), e, e, \dots, e) + p_n(a_{11} + a_{12} + a_{21} + a_{22}, \delta(e), e, \dots, e) \\
&\quad + \dots + p_n(a_{11} + a_{12} + a_{21} + a_{22}, e, e, \dots, \delta(e)).
\end{aligned}$$

On the other hand, using Claim 9 and the fact that $p_n(a_{22}, e, e, \dots, e) = 0$, we get

$$\begin{aligned}
&\Delta(p_n(a_{11} + a_{12} + a_{21} + a_{22}, e, e, \dots, e)) \\
&= \Delta(p_n(a_{11} + a_{12} + a_{21}, e, e, \dots, e)) + \Delta(p_n(a_{22}, e, e, \dots, e)) \\
&= p_n(\Delta(a_{11} + a_{12} + a_{21}), e, e, \dots, e) + p_n(a_{11} + a_{12} + a_{21}, \delta(e), e, \dots, e) \\
&\quad + \dots + p_n(a_{11} + a_{12} + a_{21}, e, e, \dots, \delta(e)) + p_n(\Delta(a_{22}), e, e, \dots, e) \\
&\quad + p_n(a_{22}, \delta(e), e, \dots, e) + \dots + p_n(a_{22}, e, e, \dots, \delta(e)) \\
&= p_n(\Delta(a_{11}), e, e, \dots, e) + p_n(\Delta(a_{12} + a_{21}), e, e, \dots, e) \\
&\quad + p_n(\Delta(a_{22}), e, e, \dots, e) + p_n(a_{11} + a_{12} + a_{21} + a_{22}, \delta(e), e, \dots, e) \\
&\quad + \dots + p_n(a_{11} + a_{12} + a_{21} + a_{22}, e, e, \dots, \delta(e)).
\end{aligned}$$

Comparing the above two equations, we obtain $p_n(T, e, e, \dots, e) = 2^{n-1} eTe + eTf + fTe = 0$ which in turn gives $eTe = eTf = fTe = 0$. In a similar manner, calculating $\Delta(p_n(a_{11} + a_{12} + a_{21} + a_{22}, f, f, \dots, f))$ in two ways, one can obtain

$$p_n(T, f, f, \dots, f) = 2^{n-1} fTf + eTf + fTe = 0$$

which in turn implies that $fTf = 0$. Hence, $T = 0$, that is, $\Delta(a_{11} + a_{12} + a_{21} + a_{22}) = \Delta(a_{11}) + \Delta(a_{12} + a_{21}) + \Delta(a_{22})$.

Claim 11. $\Delta(a_{12} + a_{21}) = \Delta(a_{12}) + \Delta(a_{21})$ for all $a_{12} \in \mathcal{A}_{12}$, $a_{21} \in \mathcal{A}_{21}$.

Let $a_{12} \in \mathcal{A}_{12}, a_{21} \in \mathcal{A}_{21}$. Note that $p_n(e + a_{12}, f + a_{21}, e, \ldots, e) = a_{12} + a_{21} + 2^{n-2}a_{12}a_{21}$, provided $n \geq 3$. Using Claims 5 and 9 and the above fact, we have

$$\begin{aligned}
&\Delta(a_{12} + a_{21}) + \Delta(2^{n-2}a_{12}a_{21}) \\
&= \Delta(a_{12} + a_{21} + 2^{n-2}a_{12}a_{21}) \\
&= \Delta(p_n(e + a_{12}, f + a_{21}, e, \ldots, e)) \\
&= p_n(\Delta(e + a_{12}), f + a_{21}, e, \ldots, e) + p_n(e + a_{12}, \delta(f + a_{21}), e, \ldots, e) \\
&\quad + \cdots + p_n(e + a_{12}, f + a_{21}, e, \ldots, \delta(e)) \\
&= \Delta(p_n(e, f, e, \ldots, e)) + \Delta(p_n(a_{12}, f, e, \ldots, e)) + \Delta(p_n(e, a_{21}, e, \ldots, e)) \\
&\quad + \Delta(p_n(a_{12}, a_{21}, e, \ldots, e)) \\
&= \Delta(a_{12}) + \Delta(a_{21}) + \Delta(2^{n-2}a_{12}a_{21}),
\end{aligned}$$

which implies that $\Delta(a_{12} + a_{21}) = \Delta(a_{12}) + \Delta(a_{21})$. If $n = 2$, then using Claims 8 and 10, we get

$$\begin{aligned}
&\Delta(a_{12} + a_{21}) + \Delta(a_{12}a_{21}) + \Delta(a_{21}a_{12}) \\
&= \Delta(a_{12} + a_{21} + a_{12}a_{21} + a_{21}a_{12}) \\
&= \Delta(p_2(e + a_{12}, f + a_{21})) \\
&= p_2(\Delta(e + a_{12}), f + a_{21}) + p_2(e + a_{12}, \delta(f + a_{21})) \\
&= p_2(\Delta(e), f) + p_2(\Delta(a_{12}), f) + p_2(\Delta(e), a_{21}) + p_2(\Delta(a_{12}), a_{21}) \\
&\quad + p_2(e, \delta(f)) + p_2(a_{12}, \delta(f)) + p_2(e, \delta(a_{21})) + p_2(a_{12}, \delta(a_{21})) \\
&= \Delta(p_2(e, f)) + \Delta(p_2(a_{12}, f)) + \Delta(p_2(e, a_{21})) + \Delta(p_2(a_{12}, a_{21})) \\
&= \Delta(a_{12}) + \Delta(a_{21}) + \Delta(a_{12}a_{21}) + \Delta(a_{21}a_{12}),
\end{aligned}$$

which yields that $\Delta(a_{12} + a_{21}) = \Delta(a_{12}) + \Delta(a_{21})$.

Claim 12. Δ is additive on $\mathcal{A}$.

Let $a = a_{11} + a_{12} + a_{21} + a_{22}, b = b_{11} + b_{12} + b_{21} + b_{22} \in \mathcal{A}$. Using Claims 6, 7, 10 and 11, we obtain

$$
\begin{aligned}
&\Delta(a+b)\\
&=\Delta((a_{11}+a_{12}+a_{21}+a_{22})+(b_{11}+b_{12}+b_{21}+b_{22}))\\
&=\Delta((a_{11}+b_{11})+(a_{12}+b_{12})+(a_{21}+b_{21})+(a_{22}+b_{22}))\\
&=\Delta(a_{11}+b_{11})+\Delta(a_{12}+b_{12})+\Delta(a_{21}+b_{21})+\Delta(a_{22}+b_{22})\\
&=\Delta(a_{11})+\Delta(b_{11})+\Delta(a_{12})+\Delta(b_{12})+\Delta(a_{21})+\Delta(b_{21})+\Delta(a_{22})+\Delta(b_{22})\\
&=\Delta(a_{11}+a_{12}+a_{21}+a_{22})+\Delta(b_{11}+b_{12}+b_{21}+b_{22})\\
&=\Delta(a)+\Delta(b).
\end{aligned}
$$

The proof of the theorem is completed.

4 Applications

In this section, we apply Theorem 1 to certain classes of unital algebras such as triangular algebras, nest algebras and block upper triangular matrix algebras.

Triangular algebras: Let $\mathcal{R}$ be a commutative ring with identity, $\mathcal{A}$, $\mathcal{B}$ unital $\mathcal{R}$-algebras and $\mathcal{M}$ an $(\mathcal{A}, \mathcal{B})$-bimodule. The $\mathcal{R}$-algebra

$$
\mathfrak{A}=Tri(\mathcal{A},\mathcal{M},\mathcal{B})=\left\{\begin{pmatrix} a & m\\ 0 & b\end{pmatrix}\ \middle|\ a\in\mathcal{A}, m\in\mathcal{M}, b\in\mathcal{B}\right\}
$$

under the usual matrix operations is called a triangular algebra. It is easy to see that $\mathfrak{A}$ is a unital algebra containing a nontrivial idempotent $e=\begin{pmatrix} 1 & 0\\ 0 & 0\end{pmatrix}$ satisfying the assumptions (♠). In [15], Li and Benkovič proved that a Jordan generalized derivation on a 2-torsion-free triangular algebra is a generalized derivation. In view of Theorem 1, we have the following result which generalizes the result of Li and Benkovič [15, Theorem 2.5].

Corollary 1 *Let $\mathfrak{A}=Tri(\mathcal{A},\mathcal{M},\mathcal{B})$ be a 2-torsion-free triangular algebra. Then every multiplicative Jordan generalized derivation $\Delta:\mathfrak{A}\to\mathfrak{A}$ is an additive generalized derivation.*

The main examples of triangular algebras are upper triangular matrix algebras, block upper triangular matrix algebras and nest algebras (see [3, 25] for details). Hence, applying Corollary 1, we obtain the following results.

Corollary 2 *Let $T_n(\mathbb{F})(n\geq 2)$ be a upper triangular matrix algebra over the real or complex field $\mathbb{F}$. Then every multiplicative Jordan generalized derivation $\Delta: T_n(\mathbb{F})\to T_n(\mathbb{F})$ is an additive generalized derivation.*

Corollary 3 *Let $B_n^{\bar{d}}(\mathbb{F})(n\geq 2)$ be a block upper triangular matrix algebra over the real or complex field $\mathbb{F}$ with $B_n^{\bar{d}}(\mathbb{F})\neq M_n(\mathbb{F})$. Then every multiplicative Jordan generalized derivation $\Delta: B_n^{\bar{d}}(\mathbb{F})\to B_n^{\bar{d}}(\mathbb{F})$ is an additive generalized derivation.*

Corollary 4 *Let $\mathcal{N}$ be a nest of a Banach space X and $T(\mathcal{N})$ be the associated nest algebra. Let $\Delta : T(\mathcal{N}) \to T(\mathcal{N})$ be a multiplicative Jordan generalized derivation. If there exists a nontrivial element in $\mathcal{N}$ which is complemented in X, then Δ is an additive generalized derivation. Moreover, if $\mathcal{N}$ is a nest of a Hilbert space $\mathcal{H}$, then every multiplicative Jordan generalized derivation of $T(\mathcal{N})$ is an additive generalized derivation.*

Acknowledgements The authors would like to express their sincere thanks to the referees for his/her helpful comments and suggestions which have improved the paper.

References

1. Ashraf, M., Parveen, N.: On Jordan triple higher derivable mappings on rings. Mediterr. J. Math. **13**, 1465–1477 (2016)
2. Benkovič, D.: Jordan derivations and antiderivations on triangular matrices. Linear Algebra Appl. **397**, 235–244 (2005)
3. Benkovič, D., Eremita, D.: Multiplicative Lie n-derivations of triangular rings. Linear Algebra Appl. **436**, 4223–4240 (2012)
4. Benkovič, D., Širovnik, N.: Jordan derivations of unital algebras with idempotents. Linear Algebra Appl. **437**, 2271–2284 (2012)
5. Brešar, M.: Jordan mappings of semiprime rings. J. Algebra **127**, 218–228 (1989)
6. Brešar, M.: On the distance of the composition of two derivations to the generalized derivations. Glasg. Math. J. **33**, 89–93 (1991)
7. Daif, M.N.: When is Multiplicative derivation additive? Int. J. Math. Math. Sci. **14**, 615–618 (1991)
8. Fošner, M., Iliševič, D.: On Jordan triple derivations and related mappings. Mediterr. J. Math. **5**, 415–427 (2008)
9. Hou, J.-C., Qi, X.-F.: Generalized Jordan derivation on nest algebras. Linear Algebra Appl. **430**, 1479–1485 (2009)
10. Hvala, B.: Generalized derivations in rings. Commun. Algebra **26**, 1147–1166 (1998)
11. Herstein, I.N.: Jordan derivations of prime rings. Proc. Am. Math. Soc. **8**, 1104–1110 (1957)
12. Jing, W., Lu, S.: Generalized Jordan derivations on prime rings and standard operator algebras. Taiwanese J. Math. **7**, 605–613 (2003)
13. Jing, W., Lu, F.: Additivity of Jordan (triple) derivations on rings. Commun. Algebra **40**, 2700–2719 (2012)
14. Li, J., Lu, F.: Additive Jordan derivations of reflexive algebras. J. Math. Anal. Appl. **329**, 102–111 (2007)
15. Li, Y., Benkovič, D.: Jordan generalized derivations on triangular algebras. Linear Multilinear Algebra **59**, 841–849 (2011)
16. Lu, F.: Additivity of Jordan maps on standard operator algebras. Linear Algebra Appl. **357**, 123–131 (2002)
17. Lu, F.: Jordan derivable maps of prime rings. Commun. Algebra **38**, 4430–4440 (2010)
18. Ma, F., Ji, G.-X.: Generalized Jordan derivations on triangular matrix algebras. Linear Multilinear Algebra **55**, 355–363 (2007)
19. Martindale, W.S., III.: When are multiplicative mappings additive? Proc. Am. Math. Soc. **21**, 695–698 (1969)
20. Lee, T.-K., Quynh, T.C.: Centralizers and Jordan triple derivations of semiprime rings. Commun. Algebra **47**, 236–251 (2019)
21. Qi, X., Guo, Z., Zhang, T.: Characterizing Jordan n-derivations of unital rings containing idempotents. Bull. Iranian Math. Soc. **46**, 1639–1658 (2020)

22. Vukman, J.: A note on generalized derivations of semiprime rings. Taiwanese J. Math. **11**, 367–370 (2007)
23. Wang, Y.: Additivity of multiplicative maps on triangular rings. Linear Algebra Appl. **434**, 625–635 (2011)
24. Wei, F., Xiao, Z.-K.: Generalized Jordan derivations on semiprime rings and its applications in range inclusion problems. Mediterr. J. Math. **8**, 271–291 (2011)
25. Xiao, Z.-K., Wei, F.: Lie triple derivations of triangular algebras. Linear Algebra Appl. **437**, 1234–1249 (2012)
26. Zhang, J.-H., Yu, W.-Y.: Jordan derivations of triangular algebras. Linear Algebra Appl. **419**, 251–255 (2006)

Variational Analysis of Approximate Defective Eigenvalues

Rafikul Alam

Abstract Consider an $n \times n$ matrix $A \in \mathbb{C}^{n\times n}$ and $\lambda \in \mathbb{C}$. It is easy to show that there is a matrix $\widehat{A} \in \mathbb{C}^{n\times n}$ such that λ is an eigenvalue of $\widehat{A}$ and $\|A - \widehat{A}\|_2 = \sigma_{\min}(\lambda)$, where $\sigma_{\min}(\lambda)$ is the smallest singular value of $A - \lambda I$. The question that we ask is this: Does there exist a matrix D such that λ is a defective eigenvalue of D and $\|A - D\|_2 = \sigma_{\min}(\lambda)$? If such a defective matrix D exists, then we refer to λ as an approximate defective eigenvalue of A. The aim of this paper is to characterize approximate defective eigenvalues. We show that λ is an approximate defective eigenvalue of A if and only if λ is a Clarke stationary point of the function $\phi : z \longmapsto \sigma_{\min}(z)$. As a consequence, when A is simple, we show that

$$\mathrm{d}(A) = \min\{\sigma_{\min}(\lambda) : \lambda \in \mathbb{C} \setminus \Lambda(A) \text{is a Clarke stationary point of } \sigma_{\min}(z)\}$$

is the distance from A to the nearest defective matrix, where $\Lambda(A)$ is the spectrum of A.

Keywords Variational analysis · Approximate defective eigenvalues

1 Introduction

Let $\mathbb{C}^{n\times n}$ denote the set of $n \times n$ matrices with entries in $\mathbb{C}$. Let $A \in \mathbb{C}^{n\times n}$ and $\lambda \in \mathbb{C}$ be an eigenvalue of A. Then there exist nonzero vectors x and y in $\mathbb{C}^n$ such that $Ax = \lambda x$ and $y^*A = \lambda y^*$, where y^* is the conjugate transpose of y. We refer to x (resp., y) as a right (resp., left) eigenvector of A corresponding to λ. We also refer to (λ, x, y) as an eigentriple of A. We denote the spectrum of A by $\Lambda(A)$, that is, $\Lambda(A) := \{\lambda \in \mathbb{C} : \mathrm{rank}(A - \lambda I) < n\}$. Let m (resp., g) be the algebraic (resp., geometric) multiplicity of λ. Then λ is said to be simple if $m = 1$. On the other hand, λ is said to be semisimple (resp., defective) if $m = g$ (resp., $m > g$). Finally, λ is

R. Alam (✉)
Department of Mathematics, Indian Institute of Technology Guwahati, Guwahati 781 039, India
e-mail: rafik@iitg.ac.in; rafikul68@gmail.com

M. Ashraf et al. (eds.), *Algebra and Related Topics with Applications*,
Springer Proceedings in Mathematics & Statistics 392,
https://doi.org/10.1007/978-981-19-3898-6_3

said to be a non-derogatory defective eigenvalue of A if $m > g = 1$. A matrix A is said to be simple if it has n distinct eigenvalues and A is said to be defective if it has fewer than n linearly independent eigenvectors.

We consider the Euclidean norm on $\mathbb{C}^n$ given by $\|x\|_2 := (x^*x)^{1/2}$ and the spectral norm on $\mathbb{C}^{n\times n}$ given by $\|A\|_2 := \max_{\|x\|=1} \|Ax\|_2$. The singular value decomposition (SVD) of A is given by $A = U\mathrm{diag}(\sigma_1, \ldots, \sigma_n)V^*$, where U and V are unitary and $\sigma_1 \geq \cdots \geq \sigma_n \geq 0$ are the singular values of A. A complex number μ can be viewed as an approximate eigenvalue of A, that is, μ as an eigenvalue of $A + E$ for some matrix E. The backward error $\omega(\mu, A)$ of μ as an approximate eigenvalue of A is defined as

$$\omega(\mu, A) := \min\{\|E\|_2 : \mu \in \Lambda(A + E)\}. \tag{1}$$

Consider the SVD $A - \mu I = U\mathrm{diag}(\sigma_1, \ldots, \sigma_n)V^*$. Then setting $E := U\mathrm{diag}(0, \ldots, 0, -\sigma_n)V^*$, it follows that $\mu \in \Lambda(A + E)$ and $\omega(\mu, A) = \sigma_n$. In fact, setting $u := Ue_n$ and $v := Ve_n$, we have $(A + E)v = \mu v$ and $u^*(A + E) = \mu u^*$. Thus, (μ, v, u) is an eigentriple of $A + E$ which we refer to as an *approximate eigentriple* of A.

Definition 1 A complex number μ is said to be an approximate defective (resp., multiple) eigenvalue of A if there exists E such that μ is a defective (resp., multiple) eigenvalue of $A + E$ and $\|E\|_2 = \omega(\mu, A)$.

Given a complex number μ, let

$$\delta_\mu(A) := \min\{\|E\|_2 : \mu \in \Lambda(A + E) \text{ is a multiple eigenvalue}\}.$$

It is shown in [10] that there is a matrix E such that μ is a multiple eigenvalue of $A + E$ and

$$\delta_\mu(A) = \|E\|_2 = \max_{\gamma\geq 0} \sigma_{2n-1}\left(\begin{bmatrix} A - \mu I & \gamma I \\ 0 & A - \mu I \end{bmatrix}\right) \geq \omega(\mu, A), \tag{2}$$

where $\sigma_{2n-1}(\cdot)$ is the $(2n-1)$-th singular value of the $2n$-by-$2n$ matrix. Hence, μ is not an approximate defective eigenvalue of A when $\delta_\mu(A) > \omega(\mu, A)$. So, is it possible to characterize complex numbers which are approximate defective eigenvalues of A?

The main objective of this paper is to analyze approximate defective eigenvalues of A. For $\mu \in \mathbb{C}$, let $\mathscr{D}_\mu \subset \mathbb{C}^{n\times n}$ denote the set of matrices for which μ is a defective eigenvalue, that is,

$$\mathscr{D}_\mu := \{X \in \mathbb{C}^{n\times n} : \mu \in \Lambda(X) \text{ is defective}\}.$$

We characterize μ such that there exists a matrix E such that $A + E \in \mathscr{D}_\mu$ and $\|E\|_2 = \omega(\mu, A)$. Next, we construct E such that $A + E \in \mathscr{D}_\mu$ and $\|E\|_2 = \omega(\mu, A)$. We show that μ is an approximate non-derogatory defective eigenvalue of A $\Longleftrightarrow$

$0 \in \partial\sigma_{\min}(\mu) \iff \mu$ is a Clarke stationary point of the map $\sigma_{\min} : z \longmapsto \omega(z, A)$, where $\partial\sigma_{\min}(\mu)$ is the Clarke subdifferential of $\sigma_{\min}(z)$ at μ.

The problem of characterizing approximate defective eigenvalues is closely related to the Wilkinson problem [17–22]. The Wilkinson problem seeks to determine a defective matrix nearest to a simple matrix and has been studied extensively over the years; see [1–4, 6–11, 17–22].

Wilkinson Problem (1965): *Let* A *be simple and let* $\mathrm{d}(A) := \inf\{\|E\|_2 : A + E \text{ is defective}\}$. *Determine* $\mathrm{d}(A)$ *and construct* E *such that* $A + E$ *is defective and* $\|E\|_2 = \mathrm{d}(A)$.

Observe that if E is such that $A + E$ is defective then $A + E \in \mathscr{D}_\mu$ for some $\mu \in \mathbb{C}$ and, in view of (1), we have $\|E\|_2 \geq \omega(\mu, A)$. We show that

$$\mathrm{d}(A) = \min_{\mu \in \mathbb{C}}\{\omega(\mu, A) : 0 \in \partial\sigma_{\min}(\mu)\}. \tag{3}$$

The importance of the Wilkinson problem stems from the fact that $\mathrm{d}(A)$ provides insight into ill-conditioning of the eigenvalue problem $Au = \lambda u$. Sensitivity analysis of eigenvalues plays an important role in the accuracy assessment of computed eigenvalues. Eigenvalues of matrices are usually computed by employing backward stable algorithms [17]. This means that the computed eigenvalues of A are exact eigenvalues of $A + E$ for some matrix E such that $\|E\|_2$ is small. Let λ be a simple eigenvalue of A with associated left and right eigenvectors y and x, respectively. Then it is well known (see, for example, [17]) that

$$\lambda(A + E) = \lambda + \frac{|y^* E x|}{|y^* x|} + \mathscr{O}(\|E\|_2^2),$$

where $\lambda(A + E)$ is an eigenvalue of $A + E$ closest to λ. This shows that the *sensitivity* of a simple eigenvalue λ to a small perturbation in A is measured by the *condition number* [17]

$$\mathrm{cond}(\lambda, A) := \frac{\|x\|_2\|y\|_2}{|y^* x|}. \tag{4}$$

It follows that λ is *ill-conditioned* when $|y^* x|$ is small, that is, a small perturbation E may cause a large error in the eigenvalue $\lambda(A + E)$. Obviously an extreme case of ill-conditioning occurs when $y^* x = 0$. This is indeed the case when λ is a nonderogatory defective eigenvalue. It turns out that the extreme cases of ill-conditioning are associated with multiple eigenvalues.

Theorem 1 (Wilkinson [20]) *Let* λ *be an eigenvalue of* A. *Then* λ *is a multiple eigenvalue if and only if there exist left and right eigenvectors* y *and* x *of* A *corresponding to* λ *such that* $y^* x = 0$.

It is customary to define $\mathrm{cond}(\lambda, A) := \infty$ when λ is multiple. Thus, numerically an eigenvalue λ is expected to behave like a multiple eigenvalue when $\mathrm{cond}(\lambda, A)$

is large. The fact that $\mathrm{d}(A)$ is closely related to ill-conditioning of eigenvalues of A is confirmed by the following bound due to Wilkinson [18, 20]: If $\mathrm{cond}(\lambda, A) > 1$ then

$$\mathrm{d}(A) \leq \min_{\lambda} \frac{\|A\|_2}{\sqrt{\mathrm{cond}(\lambda, A)^2 - 1}},$$

where the minimum is taken over by all the eigenvalues of A. This shows that $\mathrm{d}(A) \simeq 1/\mathrm{cond}(\lambda, A)$ when $\mathrm{cond}(\lambda, A)$ is large. In other words, matrices with very sensitive eigenvalues are close to a defective matrix and the distance is almost inversely proportional to the condition number of its most sensitive eigenvalue. We mention, however, that the upper bound is not attained by $\mathrm{d}(A)$ and it is easy to construct examples for which the bound is not sharp. The crux of the matter is that having an ill-conditioned eigenvalue is sufficient for a matrix to be close to a defective matrix but it is not necessary. Several upper and lower bounds of $\mathrm{d}(A)$ have been obtained over the years [6–9, 11, 18–22]. See [1] for a comprehensive catalog of upper and lower bounds of $\mathrm{d}(A)$. The Wilkinson problem was open for almost four decades. As a consequence of approximate defective eigenvalues, we provide a brief outline of a solution to the Wilkinson problem obtained in [3, 4].

2 Approximate Multiple Eigenvalues

Let $A \in \mathbb{C}^{n \times n}$. We consider the Frobenius norm defined by $\|A\|_F := \sqrt{\mathrm{Trace}(A^*A)}$. For the rest of the paper, we define $\sigma_{\min} : \mathbb{C} \to \mathbb{R}$ by $\sigma_{\min}(z) := \omega(z, A)$, where $\omega(z, A)$ is the backward error defined in (1). Let $\lambda \in \mathbb{C}$. By considering SVD of $A - \lambda I$, it follows that $\sigma_{\min}(\lambda) = \omega(\lambda, A) = \sigma_n$, where σ_n is the smallest singular value of $A - \lambda I$. Indeed, let u and v, respectively, be left and right singular vectors of $A - \lambda I$ corresponding to $\sigma_{\min}(\lambda)$. Then defining $E := -\sigma_{\min}(\lambda)uv^*$, it is easily seen that $\lambda \in \Lambda(A + E)$ and $\|E\|_2 = \|E\|_F = \sigma_{\min}(\lambda)$. Note that λ need not be a multiple eigenvalue of $A + E$. However, under appropriate assumptions, λ can be induced as a multiple/defective eigenvalue of $A + E$.

We need the following elementary result which will play an important role in the subsequent development. See also [3].

Theorem 2 *Let $\lambda \in \mathbb{C}$ and $A \in \mathbb{C}^{n \times n}$. Suppose that λ is not an eigenvalue of A. Consider the singular value decomposition $A - \lambda I = U \Sigma V^*$.*

(a) Suppose that $\sigma_{\min}(\lambda)$ is a multiple singular value with multiplicity m. Let U_m and V_m, respectively, denote the last m columns of U and V. Define $E := -\sigma_{\min}(\lambda) U_m V_m^*$. Then $\|E\|_2 = \sigma_{\min}(\lambda)$ and λ is a multiple eigenvalue of $A + E$ with geometric multiplicity m. The columns of U_m and V_m, respectively, are orthonormal left and right eigenvectors of $A + E$ corresponding to λ.

(b) Suppose that $A - \lambda I$ has a pair of left and right singular vectors u and v corresponding to $\sigma_{\min}(\lambda)$ such that $u^*v = 0$. Define $E := -\sigma_{\min}(\lambda)uv^*$. Then $\|E\|_2 = \|E\|_F = \sigma_{\min}(\lambda)$ and λ is a non-derogatory defective eigenvalue of

$A + E$. Further, u and v are left and right eigenvectors of $A + E$ corresponding to λ.

Proof (a) By construction $\|E\|_2 = \sigma_{\min}(\lambda)$ and $(A + E)V_m = \lambda V_m$ and $U_m^*(A + E) = \lambda U_m^*$. This shows that the geometric multiplicity of λ as an eigenvalue of $A + E$ is at least m. Since by construction $\text{rank}(A + E - \lambda I) = n - m$, it follows that the geometric multiplicity of λ is exactly m.

(b) By construction $\|E\|_2 = \|E\|_F = \sigma_{\min}(\lambda)$, $(A + E)v = \lambda v$ and $u^*(A + E) = \lambda u^*$. So, we only need to show that λ is a non-derogatory defective eigenvalue. Since $u^*v = 0$, by Theorem 1, λ must be a multiple eigenvalue of $A + E$. Since $\text{rank}(A + E - \lambda I) = n - 1$, it follows that λ is a non-derogatory defective eigenvalue of $A + E$. □

We mention that for any $\lambda \in \mathbb{C}$ there always exists a matrix E, which can be constructed from the SVD of $A - \lambda I$, such that λ is a multiple eigenvalue of $A + E$. However, in such cases we always have $\|E\|_2 > \sigma_{\min}(\lambda)$ unless λ satisfies the assumptions in Theorem 2. Indeed, consider the SVD $A - \lambda I = U\text{diag}(\sigma_1, \cdots, \sigma_n)V^*$ and define

$$E := A - U\text{diag}(\sigma_1, \cdots, \sigma_{n-m}, 0, \cdots, 0)V^* = -U\text{diag}(0, \ldots, 0, \sigma_{n-m+1}, \ldots, \sigma_n)V^*.$$

Then it follows that λ is a multiple eigenvalue of $A + E$ with geometric multiplicity m and $\|E\|_2 = \sigma_{n-m+1} \geq \sigma_n = \sigma_{\min}(\lambda)$. Next, we show that the conditions in Theorem 2 are also necessary for λ to be an approximate multiple eigenvalue of A.

To proceed further, we need the best low rank approximation of a matrix. Given $A \in \mathbb{C}^{n\times n}$ and $\ell < \text{rank}(A)$, consider the rank-ℓ minimization problems

$$A_\ell = \text{argmin}_{\text{rank}(X)=\ell}\|A - X\|_2,$$
$$A_\ell = \text{argmin}_{\text{rank}(X)=\ell}\|A - X\|_F.$$

Theorem 3 (Eckart-Young [12]) *Let $A \in \mathbb{C}^{n\times n}$ and $\ell < r := \text{rank}(A)$. Consider the SVD $A = U\text{diag}(\sigma_1, \cdots, \sigma_r, 0, \cdots, 0)V^*$. Define $A_\ell := U\text{diag}(\sigma_1, \cdots, \sigma_\ell, 0, \cdots, 0)V^*$. Then*

$$A_\ell = \text{argmin}_{\text{rank}(X)=\ell}\|A - X\|_2 \text{ and } \|A - A_\ell\|_2 = \sigma_{\ell+1},$$

$$A_\ell = \text{argmin}_{\text{rank}(X)=\ell}\|A - X\|_F \text{ and } \|A - A_\ell\|_F = \sqrt{\sigma_{\ell+1}^2 + \cdots + \sigma_r^2}.$$

Further, $A_\ell = \text{argmin}_{\text{rank}(X)=\ell}\|A - X\|_F$ is unique.

We say that λ is an approximate multiple eigenvalue of A of geometric multiplicity ℓ if there exists a matrix E such that $\|E\|_2 = \omega(\lambda, A)$ and λ is an eigenvalue of $A + E$ of geometric multiplicity ℓ. The following result characterizes an approximate multiple eigenvalue.

Theorem 4 *Let $\lambda \in \mathbb{C}$ and $A \in \mathbb{C}^{n\times n}$. Suppose that λ is not an eigenvalue of A. Let $\ell \geq 2$. Then λ is an approximate multiple eigenvalue of A of geometric multiplicity ℓ $\iff$ $\sigma_{\min}(\lambda)$ is a multiple singular value of $A - \lambda I$ of multiplicity at least ℓ.*

Proof If the multiplicity of $\sigma_{\min}(\lambda)$ as a singular value of $A - \lambda I$ is ℓ then the result follows from Theorem 2(a). So, suppose that λ is an approximate multiple eigenvalue of A of geometric multiplicity ℓ. Then there exists E such that $\|E\|_2 = \omega(\lambda, A) = \sigma_{\min}(\lambda)$ and λ is an eigenvalue of $A + E$ of geometric multiplicity ℓ. This implies that $\mathrm{rank}(A - \lambda I + E) = n - \ell$. Since $\mathrm{rank}(A - \lambda I) = n$ and $\|(A - \lambda I) - (A - \lambda I + E)\|_2 = \|E\|_2 = \sigma_{\min}(\lambda)$, it follows that $A - \lambda I + E$ is a best rank-$(n - \ell)$ approximation of $A - \lambda I$. Hence by Theorem 3, the smallest singular value of $A - \lambda I$ must have multiplicity at least ℓ. □

The case when λ is an approximate multiple eigenvalue of A of geometric multiplicity $\ell = 1$ requires special treatment. Theorem 2(b) shows that λ is an approximate defective eigenvalue of A whenever $A - \lambda I$ has a pair of orthogonal left and right singular vectors corresponding to $\sigma_{\min}(\lambda)$. We now show that the existence of a pair of orthogonal left and right singular vectors of $A - \lambda I$ corresponding to $\sigma_{\min}(\lambda)$ is also a necessary condition for λ to be an approximate defective eigenvalue of A.

Theorem 5 *Let $\lambda \in \mathbb{C}$ and $A \in \mathbb{C}^{n\times n}$. Suppose that λ is not an eigenvalue of A. Then the following conditions are equivalent.*

(a) *There is a pair of left and right singular vectors u and v of $A - \lambda I$ corresponding to $\sigma_{\min}(\lambda)$ such that $u^*v = 0$.*
(b) *There exists E such that $\|E\|_F = \sigma_{\min}(\lambda)$ and λ is a defective eigenvalue of $A + E$.*
(c) *There exists E such that* $\mathrm{rank}(E) = 1$, $\|E\|_2 = \sigma_{\min}(\lambda)$ *and λ is a defective eigenvalue of $A + E$.*

Proof If (a) holds then (b) also holds by Theorem 2(b). So, suppose that (b) holds. Consider the SVD $A - \lambda I = U\mathrm{diag}(\sigma_1, \ldots, \sigma_n)V^*$. Since $\|(A - \lambda I) - (A - \lambda I + E)\|_F = \sigma_{\min}(\lambda) = \min\{\|(A - \lambda I) - K\|_F : \mathrm{rank}(K) = n - 1\}$, by Theorem 3, $A - \lambda I + E$ is a unique best rank-$(n - 1)$ approximation of $A - \lambda I$ and is given by $A - \lambda I + E = U\mathrm{diag}(\sigma_1, \ldots, \sigma_{n-1}, 0)V^* \Longrightarrow E = U\mathrm{diag}(0, \ldots, 0, -\sigma_n)V^*$. Hence E is a rank-1 matrix. This shows that (c) holds.

Now suppose that (c) holds. Since $\mathrm{rank}(A - \lambda I + E) = n - 1$, it follows that λ is a non-derogatory defective eigenvalue of $A + E$. Also since $\|(A - \lambda I) - (A - \lambda I + E)\|_2 = \sigma_{\min}(\lambda) = \min\{\|(A - \lambda I) - K\|_2 : \mathrm{rank}(K) = n - 1\}$ and $\mathrm{rank}(E) = 1$, by Theorem 3, E must be of the form $E = -\sigma_{\min}(\lambda)uv^*$ for some left and right singular vectors u and v of $A - \lambda I$ corresponding to $\sigma_{\min}(\lambda)$. Then obviously, we have $(A + E)v = \lambda v$ and $u^*(A + E) = \lambda u^*$. Since λ is non-derogatory defective, by Theorem 1, we have $u^*v = 0$. This shows that (a) holds. □

Note that the notion of approximate eigenvalue depends on the choice of a norm. Also note that $\omega(\lambda, A) = \min\{\|E\|_2 : \lambda \in \Lambda(A + E)\} = \min\{\|E\|_F : \lambda \in \Lambda(A + E)\} = \sigma_{\min}(\lambda)$. We say that λ is an approximate multiple (resp., defective) eigenvalue of A with respect to Frobenius norm if there exists E such that $\lambda \in \Lambda(A + E)$ is a multiple (resp., defective) eigenvalue and $\|E\|_F = \sigma_{\min}(\lambda)$. As a consequence of Theorem 5, we have the following result.

Corollary 1 *Let $\lambda \in \mathbb{C}$ and $A \in \mathbb{C}^{n\times n}$. Suppose that λ is not an eigenvalue of A. Then λ is an approximate non-derogatory defective eigenvalue of A with respect to Frobenius norm if and only if $A - \lambda I$ has a pair of left and right singular vectors u and v corresponding to $\sigma_{\min}(\lambda)$ such that $u^*v = 0$. In such a case, $E := -\sigma_{\min}(\lambda)uv^*$ induces λ as a non-derogatory defective eigenvalue of $A + E$.*

Next, we investigate the complex numbers that satisfy the condition in Corollary 1. For this purpose, we identify the function $\sigma_{\min} : \mathbb{C} \longrightarrow \mathbb{R}, x + iy \longmapsto \sigma_{\min}(x + iy)$, with the function $\phi : \mathbb{R}^2 \longrightarrow \mathbb{R}, (x, y) \longmapsto \sigma_{\min}(x + iy)$. We write the gradient $\nabla\phi(a, b) = (\phi_x(a, b), \phi_y(a, b))$ as a complex number and define $\nabla\sigma_{\min}(a + ib) := \phi_x(a, b) + i\phi_y(a, b)$ and refer to $\nabla\sigma_{\min}(a + ib)$ as the gradient of $\sigma_{\min}(z)$ at $a + ib$. We say the $\lambda \in \mathbb{C}$ is a *stationary point* of $\sigma_{\min}(z)$ if $\nabla\sigma_{\min}(\lambda) = 0$.

Theorem 6 (Sun [15]) *Let $A \in \mathbb{C}^{n\times n}$. Suppose that $\lambda \in \mathbb{C}$ is not an eigenvalue of A and that $\sigma_{\min}(\lambda)$ is a simple singular value of $A - \lambda I$. Let u and v be left and right singular vectors of $A - \lambda I$ corresponding to $\sigma_{\min}(\lambda)$. Then $\sigma_{\min}(z)$ is differentiable at λ and the gradient of $\sigma_{\min}(z)$ at λ is given by $\nabla\sigma_{\min}(\lambda) = -v^*u$.*

We now show that the stationary points of $\sigma_{\min}(z)$ are approximate non-derogatory defective eigenvalues for A.

Theorem 7 *Let $A \in \mathbb{C}^{n\times n}$. Suppose that $\lambda \in \mathbb{C}$ is not an eigenvalue of A and that $\sigma_{\min}(\lambda)$ is a simple singular value of $A - \lambda I$. Let u and v be left and right singular vectors of $A - \lambda I$ corresponding to $\sigma_{\min}(\lambda)$. Then λ is an approximate non-derogatory defective eigenvalue of A with respect to Frobenius norm $\Longleftrightarrow$ $\nabla\sigma_{\min}(\lambda) = 0$. Define $E := -\sigma_{\min}(\lambda)uv^*$. Then $\|E\|_F = \|E\|_2 = \sigma_{\min}(\lambda)$ and λ is a non-derogatory defective eigenvalue of $A + E$.*

Proof Suppose that $\nabla\sigma_{\min}(\lambda) = 0$. Then by Theorem 6, $v^*u = -\nabla\sigma_{\min}(\lambda) = 0 \Rightarrow$ u and v are orthogonal. Hence by Corollary 1, λ is a non-derogatory defective eigenvalue of $A + E$. Obviously, $\|E\|_F = \|E\|_2 = \sigma_{\min}(\lambda)$ as E is a rank-1 matrix and u and v are unit vectors.

Conversely, if λ is an approximate non-derogatory defective eigenvalue for A with respect to Frobenius norm then by Corollary 1, $A - \lambda I$ has a pair of left and right singular vectors $\hat{u}$ and $\hat{v}$ corresponding to $\sigma_{\min}(\lambda)$ such that $\hat{u}^*\hat{v} = 0$. Since $\sigma_{\min}(\lambda)$ is simple, we have $u = w\hat{u}$ and $v = w\hat{v}$ for some $w \in \mathbb{C}$ such that $|w| = 1$. Consequently, by Theorem 6, we have $\nabla\sigma_{\min}(\lambda) = -v^*u = -\hat{v}^*\hat{u} = 0$. □

Thus, stationary points of $\sigma_{\min}(z)$ are approximate non-derogatory defective eigenvalues of A. This raises a natural question: What can be said about λ when $\sigma_{\min}(\lambda)$ is a multiple singular value of $A - \lambda I$? By Theorem 4, λ is an approximate multiple eigenvalue of A. But is λ an approximate defective eigenvalue of A? We now investigate this issue.

3 Approximate Defective Eigenvalues

Let $A \in \mathbb{C}^{n\times n}$. Suppose that $\lambda \in \mathbb{C}$ is not an eigenvalue of A. Then by Corollary 1, λ is an approximate defective eigenvalue of A with respect to Frobenius norm if and only if $A - \lambda I$ has a pair of left and right singular vectors u and v corresponding to $\sigma_{\min}(\lambda)$ such that $u^*v = 0$. By Theorem 7, this condition is satisfied when λ is a stationary point of $\sigma_{\min}(z)$. So, does there exist a pair of left and right singular vectors u and v of $A - \lambda I$ corresponding to $\sigma_{\min}(\lambda)$ such that $u^*v = 0$ when $\sigma_{\min}(\lambda)$ is multiple? To answer this question, we need to consider stationary points when $\sigma_{\min}(z)$ is nonsmooth. Note that $\sigma_{\min}(z)$ is not differentiable at λ when $\sigma_{\min}(\lambda)$ is multiple. However, $\sigma_{\min}(z)$ is Lipschitz continuous. Consequently, the notion of the Clarke stationary point of $\sigma_{\min}(z)$ can be utilized to deal with the case when $\sigma_{\min}(\lambda)$ is a multiple singular value of $A - \lambda I$.

The generalized Clarke directional derivative of a locally Lipschitz function $f : \mathbb{C}^n \to \mathbb{R}$ at $x \in \mathbb{C}^n$ in the direction v is defined by [5]

$$\delta f(x; v) := \limsup_{y\to x,\, t\to 0+} \frac{f(y + tv) - f(y)}{t}.$$

Then the Clarke subdifferential of f at x is given by

$$\partial f(x) := \{y \in \mathbb{C}^n : \delta f(x; v) \geq \mathrm{Re}\langle v, y\rangle \text{ for all } v \in \mathbb{C}^n\},$$

where $\langle x, y\rangle := y^*x$ is the usual inner product on $\mathbb{C}^n$. Equivalently, we have [5]

$$\begin{aligned}\partial f(x) &= \text{convex hull}\left\{\lim_{x_k\to x} \nabla f(x_k) : f \text{ is differentiable at } x_k\right\}\\ &= \{y \in \mathbb{C}^n : \delta f(x; v) \geq \mathrm{Re}\langle v, y\rangle \text{ for all } v \in \mathbb{C}^n\}.\end{aligned}$$

We mention that if f is differentiable in a neighborhood of x then $\partial f(x) = \{\nabla f(x)\}$. We equip $\mathbb{C}^{n\times n}$ with the usual inner product $\langle X, Y\rangle := \mathrm{Trace}(Y^*X)$.

Definition 2 (Clarke stationary point [5]) Let $f : \mathbb{C}^n \to \mathbb{R}$ be locally Lipschitz and $\lambda \in \mathbb{C}$. Then λ is said to be a Clarke stationary point of f if $0 \in \partial f(\lambda)$.

Note that if λ is a stationary point of $\sigma_{\min}(z)$ then λ is also a Clarke stationary point of $\sigma_{\min}(z)$. We now determine the Clarke subdifferential of $\sigma_{\min}(z)$. The field of values of a matrix A is given by $\mathscr{F}(A) := \{x^*Ax : x \in \mathbb{C}^n \text{ and } x^*x = 1\}$.

Theorem 8 (Subdifferential) *Let $A \in \mathbb{C}^{n\times n}$ and $\lambda \in \mathbb{C} \setminus \Lambda(A)$. Suppose that the multiplicity of $\sigma_{\min}(\lambda)$ is m. Let $U \in \mathbb{C}^{n\times m}$ and $V \in \mathbb{C}^{n\times m}$ be such that $(A - \lambda I)V = \sigma_{\min}(\lambda)U$ and $(A - \lambda I)^*U = \sigma_{\min}(\lambda)V$ with $V^*V = I_m = U^*U$. Then we have*

$$\partial\sigma_{\min}(\lambda) = -\mathscr{F}(V^*U),$$

where $\mathscr{F}(V^*U)$ is the field of values of V^*U.

Proof Set $G(z) := A - zI$. Then $\sigma_{\min}(z) = \sigma_n(G(z))$ is Lipschitz continuous and $G(z)$ is a smooth function with $\nabla G(\lambda) = -I$. Note that $\nabla G(\lambda) : \mathbb{C} \to \mathbb{C}^{n\times n}$ is a linear map and is given by $\nabla G(\lambda)z = -zI$. Now we determine the adjoint operator $(\nabla G(\lambda))^* : \mathbb{C}^{n\times n} \to \mathbb{C}$. We have $\langle z, (\nabla G(\lambda))^* Y\rangle = \langle \nabla G(\lambda)z, Y\rangle = \langle -zI, Y\rangle = \langle z, -\text{Trace}(Y)\rangle$ for $Y \in \mathbb{C}^{n\times n}$ which shows that $(\nabla G(\lambda))^* Y = -\text{Trace}(Y)$ for $Y \in \mathbb{C}^{n\times n}$.

We now show that $\partial\sigma_{\min}(\lambda) = (\nabla G(\lambda))^*\partial\sigma_n(G(\lambda))$, where $\partial\sigma_n(G(\lambda))$ is the Clarke subdifferential of the map $\mathbb{C}^{n\times n} \to \mathbb{R}$, $X \mapsto \sigma_n(X)$, evaluated at $G(\lambda)$. By the chain rule [5], we have $\partial\sigma_{\min}(\lambda) \subset (\nabla G(\lambda))^*\partial\sigma_n(G(\lambda))$. For the reverse inclusion, we use generalized Clarke directional derivatives of $\sigma_{\min}(z)$ and $\sigma_n(X)$. Since $\nabla G(\lambda)z = -Iz$, it is easily seen that $\delta\sigma_{\min}(\lambda; z) = \delta\sigma_n(G(\lambda); \nabla G(\lambda)z)$ for all $z \in \mathbb{C}$. Hence for $Y \in \partial\sigma_n(G(\lambda))$, we have

$$\text{Re}\langle z, (\nabla G(\lambda))^* Y\rangle = \text{Re}\langle \nabla G(\lambda)z, Y\rangle \le \delta\sigma_n(G(\lambda); \nabla G(\lambda)z) = \delta\sigma_{\min}(\lambda; z)$$

for all $z \in \mathbb{C}$. This shows that $(\nabla G(\lambda))^* Y \in \partial\sigma_{\min}(\lambda)$ and hence the reverse inclusion follows.

Next, we determine the subdifferential $\partial\sigma_{\min}(\lambda)$. Since $G(\lambda)$ is nonsingular, by Corollary 6.4, [14], the subdifferential of $X \mapsto \sigma_n(X)$ evaluated at $G(\lambda)$ is given by

$$\begin{aligned}\partial\sigma_n(G(\lambda)) &= \text{convex hull}\{uv^* : G(\lambda)v = \sigma_{\min}(\lambda)u,\ G(\lambda)^*u = \sigma_{\min}(\lambda)v,\ \|u\|_2 = \|v\|_2 = 1\}\\ &= \text{convex hull}\{Uxx^*V^* : x \in \mathbb{C}^m,\ \|x\|_2 = 1\}.\end{aligned}$$

Hence, we have

$$\begin{aligned}\partial\sigma_{\min}(\lambda) &= (\nabla G(\lambda))^*\partial\sigma_n(G(\lambda))\\ &= \text{convex hull}\{-\text{Trace}(Uxx^*V^*) : x \in \mathbb{C}^m, \|x\|_2 = 1\}\\ &= \{-x^*V^*Ux : x \in \mathbb{C}^m, \|x\|_2 = 1\} = -\mathcal{F}(V^*U).\end{aligned}$$

This completes the proof. □

See [13] for a similar result and a different proof.

Theorem 9 *Let $A \in \mathbb{C}^{n\times n}$ and $\lambda \in \mathbb{C}$. Suppose that λ is not an eigenvalue of A. Then λ is a Clarke stationary point of $\sigma_{\min}(z)$ if and only if there exists a pair of left and right singular vectors u and v of $A - \lambda I$ corresponding to $\sigma_{\min}(\lambda)$ such that $u^*v = 0$. Consequently, λ is an approximate non-derogatory defective eigenvalue of A with respect to Frobenius norm if and only if λ is a Clarke stationary point of $\sigma_{\min}(z)$. Define $E = -\sigma_{\min}(\lambda)uv^*$. Then λ is a non-derogatory defective eigenvalue of $A + E$ and $\|E\|_F = \|E\|_2 = \sigma_{\min}(\lambda)$.*

Proof Suppose that the multiplicity of $\sigma_{\min}(\lambda)$ is m. Let U and V be as in Theorem 8. Then by Theorem 8, $\partial\sigma_{\min}(\lambda) = -\mathcal{F}(V^*U)$. Now suppose that λ is a Clarke stationary point, that is, $0 \in \partial\sigma_{\min}(\lambda)$. Then $0 \in \mathcal{F}(V^*U)$ and hence $0 = x^*V^*Ux$ for some $x \in \mathbb{C}^m$ with $\|x\|_2 = 1$. Setting $v := Vx$ and $u := Ux$, it follows that $v^*u = 0$ and that u and v are left and right singular vectors of $A - \lambda I$ corresponding to $\sigma_{\min}(\lambda)$.

Conversely, let u and v be left and right singular vectors of $A - \lambda I$ corresponding to $\sigma_{\min}(\lambda)$ such that $u^*v = 0$. Then $u = Ux$ and $v = Vx$ for some $x \in \mathbb{C}^m$ such that $\|x\|_2 = 1$. Thus, we have $0 = v^*u \in \mathscr{F}(V^*U) = -\partial\sigma_{\min}(\lambda)$ showing that λ is a Clarke stationary point.

Finally, by Corollary 1, λ is a Clarke stationary point of $\sigma_{\min}(z)$ if and only if λ is an approximate non-derogatory defective eigenvalue of A with respect to Frobenius norm. The fact that λ is a non-derogatory defective eigenvalue of $A + E$ is immediate. Obviously, we have $\|E\|_F = \|E\|_2 = \sigma_{\min}(\lambda)$. □

We conclude that $\lambda \in \mathbb{C} \setminus \Lambda(A)$ is an approximate defective eigenvalue of A whenever λ is a Clarke stationary point of $\sigma_{\min}(z)$. Consequently, when A is simple, we have

$$\mathrm{d}(A) \leq \inf\{\sigma_{\min}(\lambda) : \lambda \in \mathbb{C} \setminus \Lambda(A) \text{ is a Clarke stationary point of } \sigma_{\min}(z)\}. \quad (5)$$

Does the equality hold in (5)? How to determine a Clarke stationary point λ such that $\mathrm{d}(A) = \sigma_{\min}(\lambda)$? To answer these questions, we need to consider coalescence of pseudospectral components of A.

Let $A \in \mathbb{C}^{n \times n}$. For $\varepsilon > 0$, $\Lambda_\varepsilon(A) := \{z \in \mathbb{C} : \omega(z, A) \leq \varepsilon\} = \{z \in \mathbb{C} : \sigma_{\min}(z) \leq \varepsilon\}$ is called the ε-pseudospectrum of A; see [16]. It is easily seen that

$$\Lambda_\varepsilon(A) := \bigcup_{\|E\|_2 \leq \varepsilon} \{\Lambda(A + E) : E \in \mathbb{C}^{n \times n}\} = \bigcup_{\|E\|_F \leq \varepsilon} \{\Lambda(A + E) : E \in \mathbb{C}^{n \times n}\}. \quad (6)$$

Thus, the ε-pseudospectrum of A is the collection of all eigenvalues of all matrices whose distance from A is less than or equal to ε. Some important properties of pseudospectra of A are summarized in the following result.

Theorem 10 (Alam-Bora [3]) *Let $A \in \mathbb{C}^{n \times n}$. Consider the ε-pseudospectrum $\Lambda_\varepsilon(A)$. Then the following results hold.*

(a) For $\varepsilon > 0$, $\Lambda_\varepsilon(A)$ consists of at most n components (i.e., maximal connected subsets) and each component contains at least one eigenvalue of A in its interior.
(b) The boundary $\partial\Lambda_\varepsilon(A)$ of $\Lambda_\varepsilon(A)$ is an algebraic curve and $\partial\Lambda_\varepsilon(A) \subset \{\lambda \in \mathbb{C} : \sigma_{\min}(\lambda) = \varepsilon\}$. Further, $\partial\Lambda_\varepsilon(A)$ consists of finitely many piecewise smooth curves.
(c) Let $\mathrm{int}(\Lambda_\varepsilon(A))$ denote the set of interior points of $\Lambda_\varepsilon(A)$. Then $\mathrm{int}(\Lambda_\varepsilon(A)) = \{\lambda \in \mathbb{C} : \sigma_{\min}(\lambda) < \varepsilon\}$.

Note that $\Lambda_\varepsilon(A)$ consists of at most n components when ε is sufficiently small. As ε grows gradually, the components of $\Lambda_\varepsilon(A)$ enlarge and "coalesce" to form bigger components. For example, if A is normal with k distinct eigenvalues then, for sufficiently small ε, $\Lambda_\varepsilon(A)$ consists of k disjoint disks of radius ε centered at the eigenvalues. The coalescence of pseudospectral components plays an important role in locating (Clarke) stationery points of $\sigma_{\min}(z)$. We, therefore, define what is meant by saying that two components of $\Lambda_\varepsilon(A)$ coalesce.

Observe that if Δ_ε is a component of $\mathrm{int}(\Lambda_\varepsilon(A))$ then $\mathrm{clos}(\Delta_\varepsilon)$, the closure of Δ_ε, is a (possibly bigger) component of $\Lambda_\varepsilon(A)$. Hence for the boundary, we have $\partial\Delta_\varepsilon \subset \partial\mathrm{clos}(\Delta_\varepsilon)$ and the containment may be strict. On the other hand, if Δ_ε is a component of $\Lambda_\varepsilon(A)$ then $\mathrm{int}(\Delta_\varepsilon)$ may be a disjoint union of more than one component of $\mathrm{int}(\Lambda_\varepsilon(A))$.

Definition 3 We say that λ is a point of coalescence of two disjoint components, say, Ω_1 and Ω_2 of $\mathrm{int}(\Lambda_\varepsilon(A))$ if λ is a common boundary point of Ω_1 and Ω_2, that is, if $\lambda \in \partial\Omega_1 \cap \partial\Omega_2$.

We will be lax and say that λ is a point of coalescence of two components of $\Lambda_\varepsilon(A)$.

If λ is a point of coalescence of two or more components of $\Lambda_\varepsilon(A)$ then it is proved in [13] that there is a pair of left and right singular vectors u and v of $A - \lambda I$ corresponding to $\sigma_{\min}(\lambda)$ such that $u^*v = 0$. In other words, λ is a Clarke stationary point, which is referred to as a resolvent critical point in [13]. In fact, the following result holds; see Corollary 8.4 and Theorem 8.8 in [13].

Theorem 11 (Lewis-Pang [13]) *Let $A \in \mathbb{C}^{n\times n}$ and λ be a point of coalescence of two or more components of $\Lambda_\varepsilon(A)$. Then λ is a Clarke stationary point of $\sigma_{\min}(z)$. Moreover, $\{\sigma_{\min}(\lambda) : \lambda$ is a Clarke stationary point of $\sigma_{\min}(z)\}$ is a finite set.*

By Theorem 11, the infimum in (5) is a minimum. We mention that Wilkinson's problem is equivalent to characterizing stability of an eigendecomposition $A = X\mathrm{diag}(\lambda_i)X^{-1}$ when A is a simple matrix. By characterizing stability, we mean determining the radius of the largest open ball centered at A on which the factors X, X^{-1} and $\mathrm{diag}(\lambda_i)$ vary continuously as functions of A. Obviously, $B(A, \varepsilon) := \{Y \in \mathbb{C}^{n\times n} : \|A - Y\|_2 < \varepsilon\}$ is the largest open ball on which the eigendecomposition is stable if and only if $\varepsilon = \mathrm{d}(A)$. Now, let $\#(\Lambda_\varepsilon(A))$ denote the number of components of $\Lambda_\varepsilon(A)$. If $\#(\Lambda_\varepsilon(A)) = n$ then obviously $\mathrm{d}(A) > \varepsilon$. On the other hand, if $\#(\mathrm{int}(\Lambda_\varepsilon(A))) = n$ but $\#(\Lambda_\varepsilon(A)) < n$ then at least two components of $\Lambda_\varepsilon(A)$ must coalesce. Consequently, in such a case we have $\mathrm{d}(A) \geq \varepsilon$. Now, if two components of $\Lambda_\varepsilon(A)$ coalesce say at λ then does there exist E such that $\|E\|_2 = \varepsilon$ and λ is a defective eigenvalue of $A + E$? It is shown in [3] that λ is indeed a multiple eigenvalue $A + E$ for some E such that $\|E\|_2 = \varepsilon = \sigma_{\min}(\lambda)$ thereby providing a solution to the Wilkinson problem.

Theorem 12 (Alam-Bora [3]) *Let $A \in \mathbb{C}^{n\times n}$ be a simple matrix. Let $\#(\Lambda_\varepsilon(A))$ denote the number of components of $\Lambda_\varepsilon(A)$. Let $\varepsilon > 0$ be such that* $(\mathrm{int}(\Lambda_\varepsilon(A))) =$ *n and $\#(\Lambda_\varepsilon(A)) \leq n - 1$. Let λ be a common boundary point of components of* $\mathrm{int}(\Lambda_\varepsilon(A))$. *Then we have* $\mathrm{d}(A) = \varepsilon = \sigma_{\min}(\lambda)$. *Consider the SVD $A - \lambda I = U\Sigma V^*$. Set $u := Ue_n$ and $v := Ve_n$ when $\sigma_{\min}(\lambda)$ is simple, and $u := [Ue_{n-1}, Ue_n]$ and $v := [Ve_{n-1}, Ve_n]$ when $\sigma_{\min}(\lambda)$ is multiple. Define $E = -\sigma_{\min}(\lambda)uv^*$. Then λ is a multiple eigenvalue of $A + E$ and $\|E\|_2 = \varepsilon$. Further, λ is a non-derogatory defective eigenvalue of $A + E$ whenever $\sigma_{\min}(\lambda)$ is a simple singular value of $A - \lambda I$.*

The construction in Theorem 12 shows that the point of coalescence λ is a defective eigenvalue of $A + E$ when $\sigma_{\min}(\lambda)$ is a simple singular value of $A - \lambda I$, otherwise λ is a multiple eigenvalue of $A + E$. By Theorem 11, λ is a Clarke stationary point and hence by Theorem 9, there exists a rank-1 matrix E such that λ is a defective eigenvalue of $A + E$ and $\|E\|_2 = \|E\|_F = \sigma_{\min}(\lambda)$. However, the construction of E in Theorem 9 involves a pair of left and right singular vectors u and v such that $u^*v = 0$, whose existence is guaranteed in Theorem 11, but it is not known how to compute u and v. It is shown [4] that $A - \lambda I$ has a pair of left and right singular vectors u and v such that $u^*v = 0$ when $\sigma_{\min}(\lambda)$ is multiple and that u and v can be computed by an algorithm. We summarize these results in the following theorem.

Theorem 13 *Let $A \in \mathbb{C}^{n\times n}$ be a simple matrix. Let $\#(\Lambda_\varepsilon(A))$ denote the number of components of $\Lambda_\varepsilon(A)$. Let $\varepsilon > 0$ be such that $\#(\mathrm{int}(\Lambda_\varepsilon(A))) = n$ and $\#(\Lambda_\varepsilon(A)) \leq n - 1$. Let λ be a common boundary point of components of $\mathrm{int}(\Lambda_\varepsilon(A))$. Then λ is a Clarke stationary point of $\sigma_{\min}(z)$ and we have $\mathrm{d}(A) = \varepsilon = \sigma_{\min}(\lambda) = \min\{\sigma_{\min}(\mu) : \mu \in \mathbb{C} \setminus \lambda(A)$ is a Clarke stationary point$\}$. Further, $A - \lambda I$ has a pair of normalized left and right singular vectors u and v corresponding to $\sigma_{\min}(\lambda)$ such that $u^*v = 0$. Define $E := -\sigma_{\min}(\lambda)uv^*$. Then λ is a non-derogatory defective eigenvalue of $A + E$ and $\|E\|_2 = \|E\|_F = (A)$.*

We mention that the approach developed in [4], which does not employ variational analysis, leads to an optimization-based quadratically convergent algorithm for the computation of a matrix E such that $A + E$ is defective.

References

1. Alam, R.: Wilkinson's problem revisited. J. Anal. **4**, 176–205 (2006)
2. Alam, R.: On the construction of nearest defective matrices to a normal matrix. Linear Algebra Appl. **395**, 367–370 (2005) & Appl. 395, 367-370 (2005)
3. Alam, R., Bora, S.: On sensitivity of eigenvalues and eigendecompositions of matrices. Linear Algebra Appl. **396**, 273–301 (2005)
4. Alam, R., Bora, S., Byers, R., Overton, M.L.: Characterization and construction of the nearest defective matrix via coalescence of pseudospectral components. Linear Algebra. Appl. **435**, 494–513 (2011)
5. Clarke, F.H.: Optimization and Nonsmooth Analysis. Wiley, New York (1983). Reprinted by SIAM, Philadelphia (1990)
6. Demmel, J.W.: Computing stable eigendecompositions of matrices. Linear Algebra Appl. **79**, 163–193 (1986)
7. Demmel, J.W.: On condition numbers and the distance to the nearest ill-posed problem. Numer. Math. **51**, 251–289 (1987)
8. Demmel, J. W.: Nearest defective matrices and the geometry of ill-conditioning. In: Cox, M.G., Hammarling, S. (eds.), Reliable Numerical Computing, pp. 35–55. Oxford Sci. Pub. (1990)
9. Kahan, W.: On Neighbouring Defective Matrices. Notes for lecture at Gatlinburg VIII, Oxford University (1981)
10. Malyshev, A.N.: A formula for the 2-norm distance from a matrix to the set of matrices with multiple eigenvalues. Numer. Math. **83**, 443–454 (1999)
11. Ruhe, A.: Properties of a matrix with a very ill-conditioned eigenproblem. Numer. Math. **15**, 57–60 (1970)

12. Horn, R.A., Johnson, C.R.: Matirx Analysis. Cambridge University Press (1985)
13. Lewis, A.S., Pang, C.H.J.: Variational analysis of pseudospectra. SIAM J. Optim. **19**, 1048–1072 (2008)
14. Lewis, A.S., Sendov, H.S.: Nonsmooth analysis of singular values Part II: applications. Set-Valued Analysis **13**, 243–264 (2005)
15. Sun, J.G.: A note on simple non-zero singular values. J. Comput. Math. **6**, 166–258 (1988)
16. Trefethen, L.N., Embree, M.: Spectra and Pseudospectra: The Behaviour of Nonnormal Matrices and Operators. Princeton University Press (2005)
17. Wilkinson, J.H.: The Algebraic Eigenvalue Problem. Oxford University Press (1996)
18. Wilkinson, J.H.: Notes on matrices with a very ill-conditioned eigenproblem. Numer. Math. **19**, 175–178 (1972)
19. Wilkinson, J.H.: On neighbouring matrices with quadratic elementary divisors. Numer. Math. **44**, 1–21 (1984)
20. Wilkinson, J.H.: Sensitivity of eigenvalues. Utilitas Mathematica **25**, 5–76 (1984)
21. Wilkinson, J.H.: Sensitivity of eigenvalues II. Utilitas Mathematica **30**, 243–286 (1986)
22. Wilkinson, J.H.: Eigenvlue problems. In: Iserles, A., Powell, M.J.D. (eds.), The State of Art in Numerical Analysis, pp. 1–39. Claredon Press, Oxford (1987)

Structure of Prime Near Rings with Generalized Derivations

Asma Ali and Inzamam ul Huque

Abstract The purpose of this paper is to obtain the structure of a prime near ring N admitting a right generalized derivation F associated with a nonzero derivation d satisfying either of the conditions: (i) $F([x, y]_\sigma) = \pm x^m(x \circ y)_\sigma x^n$, (ii) $F([x, y]_\sigma) = \pm x^m[x, y]_\sigma x^n$, (iii) $F(x \circ y)_\sigma = \pm x^m(x \circ y)_\sigma x^n$, (iv) $F(x \circ y)_\sigma = \pm x^m[x, y]_\sigma x^n$, (v) $F([x, y]_\sigma) = \pm[F(x), y]_\sigma$ and (vi) $F(x \circ y)_\sigma = \pm(F(x) \circ y)_\sigma$ for all $x, y \in U$, where U is a nonzero semigroup ideal of N, $\sigma : N \to N$ is a map such that $\sigma(U) = U$ and m, n are non-negative integers. Moreover, we give a characterization of these mappings.

Keywords Prime near ring · Generalized derivations · Semigroup ideal · Commuting map

1 Introduction

A right near ring N is a triplet $(N, +, \cdot)$, where $+$ and $\cdot$ are two binary operations such that (i) $(N, +)$ is a group (not necessarily abelian), (ii) $(N, \cdot)$ is a semigroup and (iii) $(x + y) \cdot z = x \cdot z + y \cdot z$ for all $x, y, z \in N$. Consonantly, instead of (iii), if N satisfies left distributive law, then N is said to be a left near ring. The most natural example of a right near ring is the set of all identity preserving mappings acting from left of an additive group G (not necessarily abelian) into itself with pointwise addition and composition of mappings as multiplication. If these mappings act from right on G, then we get a left near ring (For more examples, we can refer Pilz [8]). A near ring N is said to be zero-symmetric if $x0 = 0$ for all $x \in N$ (right distributive law yields that $0x = 0$). Throughout the paper, N represents a zero-symmetric right near ring with $Z(N)$ as multiplicative center of N. For any $x, y \in N$, the symbols $[x, y]$ and $(x \circ y)$ denote the Lie product $xy - yx$ and the Jordan product $xy + yx$, respectively. If $\sigma : N \to N$ is any map, then we write $[x, y]_\sigma = \sigma(x)y - yx$ and

A. Ali (✉) · I. Huque
Department of Mathematics, Aligarh Muslim University, Aligarh, India
e-mail: asma_ali2@rediffmail.com

M. Ashraf et al. (eds.), *Algebra and Related Topics with Applications*,
Springer Proceedings in Mathematics & Statistics 392,
https://doi.org/10.1007/978-981-19-3898-6_4

$(x \circ y)_\sigma = \sigma(x)y + yx$ for all $x, y \in N$. A near ring N is said to be prime if $xNy = \{0\}$ for all $x, y \in N$ implies that $x = 0$ or $y = 0$. A nonempty subset U of a near ring N is said to be a semigroup right (resp. semigroup left) ideal of N if $UN \subseteq U$ (resp. $NU \subseteq U$), and if U is both a semigroup right ideal as well as a semigroup left ideal, then it is said to be a semigroup ideal of N. If S is a nonempty subset of N, then a mapping $f : S \to N$ is said to be centralizing (resp. commuting) on S if $[f(x), x] \in Z(N)$ (resp. $[f(x), x] = 0$) for all $x \in S$. The notion of derivation in near rings was initiated by Bell and Mason [1]. An additive mapping $d : N \to N$ is said to be a derivation on N if $d(xy) = d(x)y + xd(y)$ for all $x, y \in N$ or equivalently in [10], $d(xy) = xd(y) + d(x)y$ for all $x, y \in N$. Inspired by the definition of derivation in near rings, Gölbaşi [7] defined generalized derivation in near rings as follows: An additive mapping $F : N \to N$ is said to be a right (resp. left) generalized derivation associated with a derivation d on N if $F(xy) = F(x)y + xd(y)$ (resp. $F(xy) = d(x)y + xF(y)$) for all $x, y \in N$. Moreover, F is said to be a generalized derivation associated with a derivation d on N if it is both a right generalized derivation as well as a left generalized derivation on N. Thus, the notion of generalized derivation covers the notion of multiplier for $d = 0$. There are many results asserting that prime near rings with certain constrained derivations and generalized derivations have ring like behavior.

In [5], Daif and Bell proved that if R is a prime ring, I a nonzero ideal of R and R admits a derivation d such that $d([x, y]) = \pm[x, y]$ for all $x, y \in I$, then R is commutative. Further, Dhara [6] proved that if R is a semiprime ring and F is a generalized derivation associated with a derivation d on R such that $F([x, y]) = \pm[x, y]$ or $F(x \circ y) = \pm(x \circ y)$ for all $x, y \in I$, a nonzero ideal of R, then R contains a nonzero central ideal, provided $d(I) \neq \{0\}$. Moreover, he obtained that if R is a prime ring, R must be commutative, provided $d \neq 0$. Further, Boua and Oukhtite [4] extended these results for prime near rings. More precisely, they proved that if N is a prime near ring with a nonzero derivation d such that $d([x, y]) = \pm[x, y]$ or $d(x \circ y) = \pm(x \circ y)$ for all $x, y \in N$, then N is a commutative ring. In [3], Boua obtained the commutativity of a prime near ring N in case of a semigroup ideal U of N satisfying one of the conditions: (i) $d([x, y]) = [d(x), y]$; (ii) $[d(x), y] = [x, y]$; (iii) $d(x \circ y) = d(x) \circ y$ and (iv) $d(x) \circ y = x \circ y$ for all $x, y \in U$. Recently, Shang [9] considered the more general situations for a generalized derivation F of a prime near ring N satisfying any one of the following: (i) $F([x, y]) = \pm x^k[x, y]x^l$ and (ii) $F(x \circ y) = \pm x^k(x \circ y)x^l$ for all $x, y \in N$; where $k \geq 0, l \geq 0$ are non-negative integers and proved that N is a commutative ring. In this line of investigation, it is natural to look forward for some comparable results for generalized derivation in prime near rings for more general constraints replacing $[x, y]$ and $(x \circ y)$ by $[x, y]_\sigma$ and $(x \circ y))_\sigma$, respectively. In this paper, we obtain the structure of a prime near ring N with generalized derivation $F : N \to N$ associated with a nonzero derivation d on N satisfying certain identities. Moreover, we prove some theorems which give a suitable characterization of these mappings.

2 Main Results

For developing the proofs of our main results, we need the following lemmas.

Lemma 1 ([2], Lemma 1.2(i), (iii) and Lemma 1.3(iii)). *Let N be a prime near ring.*

(i) If $z \in Z(N) \setminus \{0\}$, then z is not a zero divisor.
(ii) If $z \in Z(N) \setminus \{0\}$ and $zx \in Z(N)$, then $x \in Z(N)$.
(iii) If z centralizes a nonzero semigroup left ideal, then $z \in Z(N)$.

Lemma 2 ([2], Lemma 1.4) *Let N be a prime near ring and U be a nonzero semigroup ideal of N. If $x, y \in N$ and $xUy = \{0\}$, then $x = 0$ or $y = 0$.*

Lemma 3 ([2], Lemma 1.5) *If N is a prime near ring and $Z(N)$ contains a nonzero semigroup left ideal or a semigroup right ideal, then N is a commutative ring.*

Lemma 4 ([8], Proposition 1.5) *If N is a near ring, then $-xy = (-x)y$ for all $x, y \in N$.*

Lemma 5 ([2], Lemma 1.3) *Let N be a prime near ring and U be a nonzero semigroup right (resp. semigroup left) ideal of N and x is an element of N such that $Ux = \{0\}$ (resp. $xU = \{0\}$), then $x = 0$.*

Theorem 1 *Let N be a prime near ring, U a nonzero semigroup ideal of N and $\sigma : N \to N$ be a map such that $\sigma(U) = U$. If there exist non-negative integers $m \geq 0$, $n \geq 0$ and N admits a right generalized derivation F associated with a nonzero derivation d such that $F([x, y]_\sigma) = \pm x^m [x, y]_\sigma x^n$ for all $x, y \in U$, then F is a left multiplier on N or N is a commutative ring.*

Proof By hypothesis,

$$F([x, y]_\sigma) = \pm x^m [x, y]_\sigma x^n \quad \text{for all} \quad x, y \in U. \tag{1}$$

Replacing y by yx in (1), we get

$$F([x, y]_\sigma x) = \pm x^m [x, y]_\sigma x^{n+1},$$
$$F([x, y]_\sigma)x + [x, y]_\sigma d(x) = \pm x^m [x, y]_\sigma x^{n+1} \quad \text{for all} \quad x, y \in U.$$

Using hypothesis, we arrive at

$$[x, y]_\sigma d(x) = 0,$$
$$\sigma(x) y d(x) = yxd(x) \quad \text{for all} \quad x, y \in U. \tag{2}$$

Substituting zy for y in (2) and using (2), we obtain

$$\sigma(x) zyd(x) = zyxd(x) = z\sigma(x) yd(x),$$

which gives

$$[\sigma(x), z]yd(x) = 0 \text{ for all } x, y \in U, z \in N.$$

Since $\sigma(U) = U$, we get

$$[t, z]Ud(x) = \{0\} \text{ for all } t, x \in U, z \in N.$$

Applying Lemma 2, we obtain either $d(x) = 0$ or $t \in Z(N)$ for all $t, x \in U$, i.e., $d(x) = 0$ for all $x \in U$ or $U \subseteq Z(N)$. Latter case yields that N is a commutative ring by Lemma 3. Consider the case, $d(x) = 0$ for all $x \in U$. Replacing x by xr for $r \in N$, we get $xd(r) = 0$ for all $x \in U, r \in N$, i.e., $Ud(r) = \{0\}$. Using Lemma 5, we get $d = 0$ on N and hence F is a left multiplier on N. □

Theorem 2 *Let N be a prime near ring, U a nonzero semigroup ideal of N and $\sigma : N \to N$ be a map such that $\sigma(U) = U$. If there exist non-negative integers $m \geq 0$, $n \geq 0$ and N admits a right generalized derivation F associated with a nonzero derivation d such that $F(x \circ y)_\sigma = \pm x^m (x \circ y)_\sigma x^n$ for all $x, y \in U$, then F is a left multiplier on N or N is a commutative ring.*

Proof Suppose that

$$F(x \circ y)_\sigma = \pm x^m (x \circ y)_\sigma x^n \text{ for all } x, y \in U. \tag{3}$$

Substituting yx in place of y in (3) and using $(x \circ yx)_\sigma = (x \circ y)_\sigma x$, we get

$$F((x \circ y)_\sigma x) = \pm x^m (x \circ y)_\sigma x^{n+1} \text{ for all } x, y \in U,$$

which gives

$$F(x \circ y)_\sigma x + (x \circ y)_\sigma d(x) = \pm x^m (x \circ y)_\sigma x^{n+1} \text{ for all } x, y \in U.$$

Now using (3), we find that

$$\begin{aligned} (x \circ y)_\sigma d(x) &= 0, \\ \sigma(x) y d(x) &= -yxd(x) \text{ for all } x, y \in U. \end{aligned} \tag{4}$$

Replacing y by ry for $r \in N$ in (4), using (4) and Lemma 4, we get

$$ryxd(x) = r(-(\sigma(x)yd(x))) = r(-\sigma(x))yd(x) = (-\sigma(x))ryd(x).$$

This implies that

$$[r, -\sigma(x)]yd(x) = 0 \text{ for all } x, y \in U, r \in N.$$

Since $\sigma(U) = U$, we find that

$$[r, -s]Ud(x) = \{0\} \text{ for all } s, x \in U, r \in N.$$

Applying Lemma 2, we get either $d(x) = 0$ or $-s \in Z(N)$ for all $s, x \in U$, i.e., $d(x) = 0$ for all $x \in U$ or $-U \subseteq Z(N)$. Since $-U$ is also a semigroup right ideal of N, for if $x \in U$ and $r \in N$, $(-x)r = -xr \in -U$; therefore, latter case gives that N is a commutative ring by Lemma 3. For the first case, arguing in the similar manner as in Theorem 1, we can obtain F is a left multiplier on N. □

Theorem 3 *Let N be a prime near ring, U a nonzero semigroup ideal of N and $\sigma : N \to N$ be a map such that $\sigma(U) = U$. If there exist non-negative integers $m \geq 0$, $n \geq 0$ and N admits a right generalized derivation F associated with a nonzero derivation d such that $F([x, y]_\sigma) = \pm x^m (x \circ y)_\sigma x^n$ for all $x, y \in U$, then F is a left multiplier on N or N is a commutative ring.*

Proof Suppose that

$$F([x, y]_\sigma) = \pm x^m (x \circ y)_\sigma x^n \text{ for all } x, y \in U. \tag{5}$$

Replacing y by yx in (5), using $[x, yx]_\sigma = [x, y]_\sigma x$ and $(x \circ yx)_\sigma = (x \circ y)_\sigma x$, we obtain

$$F([x, y]_\sigma x) = \pm x^m (x \circ y)_\sigma x^{n+1} \text{ for all } x, y \in U,$$

i.e.,

$$F([x, y]_\sigma)x + [x, y]_\sigma d(x) = \pm x^m (x \circ y)_\sigma x^{n+1} \text{ for all } x, y \in U.$$

By hypothesis, we have

$$[x, y]_\sigma d(x) = 0 \text{ for all } x, y \in U. \tag{6}$$

Since Eq. (6) is same as Eq. (2), arguing in the similar manner as in Theorem 1, we can get the result. □

Theorem 4 *Let N be a prime near ring, U a nonzero semigroup ideal of N and $\sigma : N \to N$ be a map such that $\sigma(U) = U$. If there exist non-negative integers $m \geq 0$, $n \geq 0$ and N admits a right generalized derivation F associated with a nonzero derivation d such that $F(x \circ y)_\sigma = \pm x^m [x, y]_\sigma x^n$ for all $x, y \in U$, then F is a left multiplier on N or N is a commutative ring.*

Proof Let

$$F(x \circ y)_\sigma = \pm x^m [x, y]_\sigma x^n \text{ for all } x, y \in U. \tag{7}$$

Substituting yx for y in (7), we find that

$$F(x \circ yx)_\sigma = F((x \circ y)_\sigma x) = \pm x^m [x, yx]_\sigma x^n = \pm x^m [x, y]_\sigma x^{n+1},$$

which implies that

$$F(x \circ y)_\sigma x + (x \circ y)_\sigma d(x) = \pm x^m [x, y]_\sigma x^{n+1} \text{ for all } x, y \in U.$$

Using the hypothesis, we get

$$(x \circ y)_\sigma d(x) = 0 \text{ for all } x, y \in U$$

which is Eq. (4); therefore, arguing in the similar manner as in Theorem 2, we can get the result. □

Theorem 5 *Let N be a prime near ring and U be a nonzero semigroup ideal of N. If $\sigma : N \to N$ is a map such that $\sigma(U) = U$ and N admits a right generalized derivation F associated with a nonzero derivation d such that F is commuting on U and $F([x, y]_\sigma) = \pm[F(x), y]_\sigma$ for all $x, y \in U$, then F is a left multiplier on N or N is a commutative ring.*

Proof Assume that

$$F([x, y]_\sigma) = [F(x), y]_\sigma \text{ for all } x, y \in U. \tag{8}$$

Replacing y by yx in (8), we get

$$F([x, yx]_\sigma) = F([x, y]_\sigma x) = [F(x), yx]_\sigma \text{ for all } x, y \in U,$$

i.e.,

$$F([x, y]_\sigma)x + [x, y]_\sigma d(x) = \sigma(F(x))yx - yxF(x) \text{ for all } x, y \in U.$$

Since F is commuting on U, therefore the last expression gives that

$$F([x, y]_\sigma)x + [x, y]_\sigma d(x) = \sigma(F(x))yx - yF(x)x = [F(x), y]_\sigma x \text{ for all } x, y \in U,$$

which reduces to

$$[x, y]_\sigma d(x) = 0 \text{ for all } x, y \in U. \tag{9}$$

Since Eq. (9) is same as Eq. (2), arguing in the similar manner as in Theorem 1, we can get the result.

Using the same techniques, we can prove the result for the case $F([x, y]_\sigma) = -[F(x), y]_\sigma$ for all $x, y \in U$. □

Theorem 6 *Let N be a prime near ring and U be a nonzero semigroup ideal of N. If $\sigma : N \to N$ is a map such that $\sigma(U) = U$ and N admits a right generalized derivation F associated with a nonzero derivation d such that F is commuting on U and $F(x \circ y)_\sigma = \pm(F(x) \circ y)_\sigma$ for all $x, y \in U$, then F is a left multiplier on N or N is a commutative ring.*

Proof By hypothesis,

$$F(x \circ y)_\sigma = (F(x) \circ y)_\sigma \text{ for all } x, y \in U. \tag{10}$$

Replacing y by yx in (10), we get

$$F(x \circ yx)_\sigma = F((x \circ y)_\sigma x) = (F(x) \circ yx)_\sigma \text{ for all } x, y \in U,$$

which implies that

$$F(x \circ y)_\sigma x + (x \circ y)_\sigma d(x) = \sigma(F(x))yx + yxF(x) \text{ for all } x, y \in U.$$

Since F is commuting on U, we get

$$F(x \circ y)_\sigma x + (x \circ y)_\sigma d(x) = \sigma(F(x))yx + yF(x)x = (F(x) \circ y)_\sigma x \text{ for all } x, y \in U.$$

Using (10), the last expression reduces to

$$(x \circ y)_\sigma d(x) = 0 \text{ for all } x, y \in U. \tag{11}$$

Since Eq. (11) is same as Eq. (4), arguing in the similar manner as in Theorem 2, we can obtain the result. Using the same techniques, we can prove the result for the case $F(x \circ y)_\sigma = -(F(x) \circ y)_\sigma$ for all $x, y \in U$. □

The following example shows that the primeness hypothesis in Theorems 1–6 is essential.

Example 1 Suppose that S is a zero-symmetric right near ring and let

$$N = \left\{ \begin{pmatrix} 0 & x & y \\ 0 & 0 & 0 \\ 0 & z & 0 \end{pmatrix} \mid 0, x, y, z \in S \right\} \text{ and } U = \left\{ \begin{pmatrix} 0 & x & 0 \\ 0 & 0 & 0 \\ 0 & y & 0 \end{pmatrix} \mid 0, x, y \in S \right\}.$$

It can be seen that N is a non-prime zero-symmetric right near ring with respect to matrix addition and matrix multiplication and U is a nonzero semigroup ideal of N.

Define the mappings $F, d, \sigma : N \to N$ by

$$F \begin{pmatrix} 0 & x & y \\ 0 & 0 & 0 \\ 0 & z & 0 \end{pmatrix} = \begin{pmatrix} 0 & z & y \\ 0 & 0 & 0 \\ 0 & 0 & 0 \end{pmatrix}, \ d \begin{pmatrix} 0 & x & y \\ 0 & 0 & 0 \\ 0 & z & 0 \end{pmatrix} = \begin{pmatrix} 0 & -x & -y \\ 0 & 0 & 0 \\ 0 & -z & 0 \end{pmatrix},$$

and

$$\sigma \begin{pmatrix} 0 & x & y \\ 0 & 0 & 0 \\ 0 & z & 0 \end{pmatrix} = \begin{pmatrix} 0 & 0 & 0 \\ 0 & 0 & 0 \\ 0 & y & 0 \end{pmatrix}.$$

Then F is a right generalized derivation associated with a nonzero derivation d on N satisfying (i) $F([x, y]_\sigma) = \pm x^m (x \circ y)_\sigma x^n$, (ii) $F([x, y]_\sigma) = \pm x^m [x, y]_\sigma x^n$, (iii) $F(x \circ y)_\sigma = \pm x^m (x \circ y)_\sigma x^n$, (iv) $F(x \circ y)_\sigma = \pm x^m [x, y]_\sigma x^n$, (v) $F([x, y]_\sigma) = \pm [F(x), y]_\sigma$ and (vi) $F(x \circ y)_\sigma = \pm (F(x) \circ y)_\sigma$ for all $x, y \in U$. However, neither F is a left multiplier on N nor N is commutative.

Acknowledgements The authors are greatly indebted to the referee for his valuable comments and suggestions.

References

1. Bell, H. E., Mason, G.: On derivations in near rings. Near-rings and Near-fields. N.-Holl. Math. Stud. **137**, 31–35 (1987)
2. Bell, H.E.: On derivations in near rings II. Near-rings, Near-fields and K-Loops, pp. 191–197. Kluwer, Dordrecht (1997)
3. Boua, A.: Some conditions under which prime near rings are commutative rings. Int. J. Open Probl. Comput. Math. **5**, 7–15 (2012)
4. Boua, A., Oukhtite, L.: Derivations on prime near rings. Int. J. Open Probl. Comput. Sci. Math. **4**, 162–167 (2011)
5. Daif, M.N., Bell, H.E.: Remarks on derivations on semiprime rings. Int. J. Mat. Mat. Sci. **15**, 205–206 (1992)
6. Dhara, B.: Remarks on generalized derivations in prime and semiprime rings. Int. J. Mat. Mat. Sci. Article ID 646587, 6 pages (2010). https://doi.org/10.1155/2010/646587
7. Golbasi, O.: On generalized derivations of prime near rings. Hacettepe J. Math. Stat. **35**, 173–180 (2006)
8. Pilz, G.F.: Near-Rings. The Theory and Its Applications, 2nd edn. North-Holland: Amsterdam, The Netherlands, New York, NY, USA (1983)
9. Shang, Y.: A note on the commutativity of prime near rings. Algebra Colloq. **22**(3), 361–366 (2015)
10. Wang, X.K.: Derivations in prime near rings. Proc. Am. Math. Soc. **121**, 361–366 (1994)

w-FP-projective Modules and Dimensions

Refat Abdelmawla Khaled Assaad, El Mehdi Bouba, and Mohammed Tamekkante

Abstract Let R be a ring. An R-module M is said to be an absolutely w-pure module if and only if $\mathrm{Ext}_R^1(F, M)$ is a GV-torsion module for any finitely presented module F. In this paper, we introduce and study the concept of w-FP-projective module which is in some way a generalization of the notion of FP-projective module. An R-module M is said to be w-FP-projective if $\mathrm{Ext}_R^1(M, N) = 0$ for any absolutely w-pure module N. This new class of modules will be used to characterize (Noetherian) DW rings. Hence, we introduce the w-FP-projective dimensions of modules and rings. The relations between the introduced dimensions and other (classical) homological dimensions are discussed. Illustrative examples are given.

Keywords Absolutely pure · Absolutely w-pure · w-flat · w-injective · DW rings and domains · $PvMD$s · Krull domainsm

1 Introduction

Throughout, all rings considered are commutative with unity and all modules are unital. Let R be a ring and M be an R-module. As usual, we use $\mathrm{pd}_R(M)$, $\mathrm{id}_R(M)$, and $\mathrm{fd}_R(M)$ to denote, respectively, the classical projective dimension, injective dimension, and flat dimension of M, and $\mathrm{wdim}(R)$ and $\mathrm{gldim}(R)$ to denote, respectively, the weak and global homological dimensions of R.

Now, we review some definitions and notation. Let J be an ideal of R. Following [9], J is called a *Glaz-Vasconcelos ideal* (a GV-ideal for short) if J is finitely gener-

R. A. Khaled Assaad (✉) · M. Tamekkante
Department of Mathematics, Faculty of Science, University Moulay Ismail Meknes, 11201 Zitoune, Morocco
e-mail: refat90@hotmail.com

E. M. Bouba
Mathematics Department, Pluridisciplinary faculty, Mohammed First University, B.P. 300, Selouane, Nador 62700, Morocco

M. Ashraf et al. (eds.), *Algebra and Related Topics with Applications*, Springer Proceedings in Mathematics & Statistics 392,
https://doi.org/10.1007/978-981-19-3898-6_5

ated and the natural homomorphism $\varphi : R \to J^* = \mathrm{Hom}_R(J, R)$ is an isomorphism. Let M be an R-module and define

$$\mathrm{tor}_{GV}(M) = \{x \in M \mid Jx = 0 \text{ for some } J \in GV(R)\},$$

where $GV(R)$ is the set of GV-ideals of R. It is clear that $\mathrm{tor}_{GV}(M)$ is a submodule of M. Now M is said to be *GV-torsion* (resp., *GV-torsion-free*) if $\mathrm{tor}_{GV}(M) = M$ (resp., $\mathrm{tor}_{GV}(M) = 0$). A GV-torsion-free module M is called a *w-module* if $\mathrm{Ext}^1_R(R/J, M) = 0$ for any $J \in GV(R)$. Projective modules and reflexive modules are w-modules. In the recent paper [17], it was shown that flat modules are w-modules. The notion of w-modules was introduced firstly over a domain [16] in the study of Strong Mori domains and was extended to commutative rings with zero divisors in [9]. Let w-$\mathrm{Max}(R)$ denote the set of maximal w-ideals of R, i.e., w-ideals of R maximal among proper integral w-ideals of R. Following [9, Proposition 3.8], every maximal w-ideal is prime. For any GV-torsion-free module M,

$$M_w := \{x \in E(M) \mid Jx \subseteq M \text{ for some } J \in GV(R)\}$$

is a w-submodule of $E(M)$ containing M and is called the *w-envelope* of M, where $E(M)$ denotes the injective hull of M. It is clear that a GV-torsion-free module M is a w-module if and only if $M_w = M$. Let M and N be R-modules and let $f : M \to N$ be a homomorphism. Following [18], f is called a *w-monomorphism* (resp., *w-epimorphism, w-isomorphism*) if $f_{\mathfrak{p}} : M \to N_{\mathfrak{p}}$ is a monomorphism (resp., an epimorphism, an isomorphism) for all $\mathfrak{p} \in w$-$\mathrm{Max}(R)$. A sequence $0 \to A \to B \to C \to 0$ of R-modules is said to be w-exact if Recall from [12] that an R-module A is called absolutely pure if A is a pure submodule in every R-module which contains A as a submodule. C. Megibben showed in [20], that an R-module A is absolutely pure if and only if $\mathrm{Ext}^1_R(N, A) = 0$ for every finitely presented R-module N. Hence, an absolutely pure module is precisely an FP-injective module in [21]. For more details about absolutely pure (or FP-injective) modules, see [3, 12, 19–21]. In a very recent paper[4], the authors introduced the notion of absolutely w-pure modules as generalization of absolutely pure (FP-injective) modules in the sense of the w-operation theory. As in [5], a w-exact sequence of R-modules $0 \to A \to B \to C \to 0$ is said to be w-pure exact if, for any R-module M, the induced sequence $0 \to A \otimes M \to B \otimes M \to C \otimes M \to 0$ is w-exact. In particular, if A is a submodule of B and $0 \to A \to B \to B/A \to 0$ is a w-pure exact sequence of R-modules, then A is said to be a w-pure submodule of B. If A is a w-pure submodule in every R-module which contains A as a submodule, then A is said to be an absolutely w-pure module. Following [4, Theorem 2.6], an R-module A is absolutely w-pure if and only if $\mathrm{Ext}^1_R(N, A)$ is a GV-torsion R-module for every finitely presented R-module N. In [1], Ding and Mao introduced and studied the notion of FP-projective dimension of modules and rings; the FP-projective dimension of an R-module M, denoted by $\mathrm{fpd}_R(M)$, is the smallest positive integer n for which $\mathrm{Ext}^{n+1}_R(M, A) = 0$ for all absolutely pure (FP-injective) R-modules A, and FP-projective dimension of R, denoted by $\mathrm{fpD}(R)$, is defined as the supremum of the FP-projective dimensions of finitely generated R-modules. These dimensions measure how far away a finitely generated module is from being finitely presented, and how far away a ring is from being Noetherian.

In Sect. 2, we introduce the concept of w-FP-projective modules. Hence, we prove that a ring R is DW ([14]) if and only if every FP-projective R-module is w-FP-projective if and only if every finitely presented R-module is w-FP-projective, and R is a coherent DW-ring if and only if every finitely generated ideal is w-FP-projective.

Section 3 deals with the w-FP projective dimension of modules and rings. After a routine study of these dimensions, we prove that R is a Noetherian DW-ring if and only if every R-module is w-FP-projective and R is FP-hereditary DW-ring if and only if every submodule of projective R-module is w-FP-projective.

2 W-FP-projective Modules

We start with the following definition.

Definition 1 An R-module M is said to be w-FP-projective if $\mathrm{Ext}_R^1(M, A) = 0$ for any absolutely w-pure R-module A.

Since every absolutely pure module is absolutely w-pure ([4, Corollary 2.7]), we have the following inclusions:

$$\{\text{Projective modules}\} \subseteq \{w\text{-FP-projective modules}\} \subseteq \{\text{FP-projective modules}\}$$

Recall that a ring R is called a DW-ring if every ideal of R is a w-ideal, or equivalently every maximal ideal of R is w-ideal [14]. Examples of DW-rings are Prüfer domains, domains with Krull dimension one, and rings with Krull dimension zero. Hence, it is clear that if R is a DW-ring, then w-FP-projective R-modules are just the FP-projective R-modules. Moreover, using [4, Corollary 2.9], it is easy to see that over a von Neumann regular ring, the three classes of modules above coincide.

Remark 1 It is proved in [15] that a finitely generated R-module M is finitely presented if and only if $\mathrm{Ext}_R^1(M, A) = 0$ for any absolutely pure (FP-injective) R-module A. Thus, every finitely generated w-FP-projective R-module is finitely presented.

We need the following lemma.

Lemma 1 *Every GV-torsion R-module is absolutely w-pure.*

Proof Let A be an arbitrary R-module and N be a finitely presented R-module. For any maximal w-ideal $\mathfrak{p}$ of R, the naturel homomorphism

$$\theta : \mathrm{Hom}_R(N, A)_{\mathfrak{p}} \to \mathrm{Hom}_{R_{\mathfrak{p}}}(N_{\mathfrak{p}}, A_{\mathfrak{p}})$$

induces a homomorphism

$$\theta_1 : \mathrm{Ext}^1_R(N, A)_{\mathfrak{p}} \to \mathrm{Ext}^1_{R_{\mathfrak{p}}}(N_{\mathfrak{p}}, A_{\mathfrak{p}})$$

Following [7, Proposition 1.10], θ_1 is a monomorphism. Suppose that A is a GV-torsion R-module. Then, we get $(\mathrm{Ext}^1_R(N, A))_{\mathfrak{p}} = 0$ since $A_{\mathfrak{p}} = 0$ (by [7, Lemma 0.1]). Hence, $\mathrm{Ext}^1_R(N, A)$ is GV-torsion (by [7, Lemma 0.1]). Consequently, A is an absolutely w-pure R-module (by [4, Theorem 2.6]). □

The first main result of this paper characterizes DW-rings in terms of w-FP-projective R-modules.

Proposition 1 *Let R be a ring. Then the following conditions are equivalent:*

(1) Every finitely presented R-module is w-FP-projective.
(2) Every FP-projective R-module is w-FP-projective.
(3) R is a DW-ring.

Proof (3) ⇒ (2) is obvious and (2) ⇒ (1) follows from the fact that finitely presented R-modules are always FP-projective.
(1) ⇒ (3) Suppose that R is not a DW-ring. Then, by [8, Theorem 6.3.12], there exist maximal ideal $\mathfrak{m}$ of R which is not w-ideal, and so by [8, Theorem 6.2.9], $\mathfrak{m}_w = R$. Hence, by [8, Proposition 6.2.5], $R/\mathfrak{m}$ is a GV-torsion R-module (sine $\mathfrak{m}$ is a GV-torsion-free R-module), and so $R/\mathfrak{m}$ is an absolutely w-pure R-module (by Lemma 1). Hence, by hypothesis, for any I finitely generated ideal I of R, we get $\mathrm{Ext}^1_R(R/I, R/\mathfrak{m}) = 0$ since R/I is a finitely presented R-module. Using [10, Lemma 3.1], we obtain that $\mathrm{Tor}^1_R(R/I, R/\mathfrak{m}) = 0$, which means that $R/\mathfrak{m}$ is flat. Accordingly, $\mathfrak{m}$ is a w-ideal, and then $\mathfrak{m}_w = \mathfrak{m}$, a contradiction with $\mathfrak{m}_w = R$. Consequently, R is a DW-ring. □

Next, we will give an example of FP-projective module, which is not w-FP-projective.

Example 1 Let $(R, \mathfrak{m})$ be a regular local ring with $\mathrm{gldim}(R) = n$ $(n \geq 2)$. By [2, Example 2.6], R is not DW ring. Hence, there exists an FP-projective R-module M which is not w-FP-projective.

Next, we give some characterizations of w-FP-projective modules.

Proposition 2 *Let M be an R-module. Then the following are equivalent:*

1. *M is w-FP-projective.*
2. *M is projective with respect to every exact sequence* $0 \to A \to B \to C \to 0$, *where A is absolutely w-pure.*
3. $P \otimes M$ *is w-FP-projective for any projective R-module P.*
4. *Hom(P, M) is w-FP-projective for any finitely generated projective R-module P.*

Proof (1) $\Leftrightarrow$ (2) is straightforward.
(1) $\Rightarrow$ (3) Let A be any absolutely w-pure R-module and P be a projective R-module. Following [8, Theorem 3.3.10], we have the isomorphism:

$$\mathrm{Ext}_R^1(P \otimes M, A) \cong \mathrm{Hom}(P, \mathrm{Ext}_R^1(M, A)).$$

Since M is w-FP-projective, we have $\mathrm{Ext}_R^1(M, A) = 0$. Thus, $\mathrm{Ext}_R^1(P \otimes M, A) = 0$, and so $P \otimes M$ is w-FP-projective.
(1) $\Rightarrow$ (4) Let A be any absolutely w-pure R-module and P be a finitely generated projective R-module. Using [8, Theorem 3.3.12], we have the isomorphism:

$$\mathrm{Ext}_R^1(\mathrm{Hom}(P, M), A) \cong P \otimes \mathrm{Ext}_R^1(M, A) = 0.$$

Hence, $\mathrm{Hom}(P, M)$ is a w-FP-projective R-module.
(3) $\Rightarrow$ (1) and (4) $\Rightarrow$ (1) Follow by letting $P = R$. □

Recall that a fractional ideal I of a domain R is said to be w-invertible if $(II^{-1})_w = R$. A domain R is said to be a Prüfer v-multiplication domain ($PvMD$) when any nonzero finitely generated ideal of R is w-invertible. Equivalently, R is a $PvMD$ if and only if $R_\mathfrak{p}$ is a valuation domain for any maximal w-ideal $\mathfrak{p}$ of R ([23, Theorem 2.1]). The class of $PvMDs$ strictly contains the classes of Prüfer domains, Krull domains, and integrally closed coherent domains.

Proposition 3 *Let R be a $PvMD$. Then* $\mathrm{pd}_R(M) \leq 1$ *for any* w-FP-*projective* R-*module* M.

Proof Let M be a w-FP-projective R-module. Following [4, Theorem 2.10], every h-divisible R-module is absolutely w-pure. Hence, $\mathrm{Ext}_R^1(M, D) = 0$ for any h-divisible R-module D. Hence, by [22, vii, Proposition 2.5], $\mathrm{pd}_R(M) \leq 1$, as desired. □

Proposition 4 *If M is a* w-FP-*projective R-module and* $\mathrm{Ext}_R^1(M, G) = 0$ *for any GV-torsion-free R-module G, then M is projective.*

Proof Let A be an arbitrary R-module. The exact sequence $0 \to \mathrm{tor}_{GV}(A) \to A \to A/\mathrm{tor}_{GV}(A) \to 0$ gives rise to the exact sequence $0=\mathrm{Ext}_R^1(M, \mathrm{tor}_{GV}(A)) \to \mathrm{Ext}_R^1(M, A) \to \mathrm{Ext}_R^1(M, A/\mathrm{tor}_{GV}(A))=0$ Thus $\mathrm{Ext}_R^1(M, A) = 0$, and so M is projective. □

Proposition 5 *Let $(R, \mathfrak{m})$ be a local ring which is not DW-ring (for example, regular local rings R with* $\mathrm{gldim}(R) = n$ $(n \geq 2)$*). Then every finitely generated* w-FP-*projective R-module M is free.*

Proof Let M be a finitely generated w-FP-projective R-module. As in the proof of Proposition 1, there exist a maximal ideal $\mathfrak{m}$ of R which is not w-ideal, and so $R/\mathfrak{m}$ is an absolutely w-pure R-module. we obtain that $\mathrm{Tor}_R^1(M, R/\mathfrak{m}) = 0$. But M is finitely generated, and so finitely presented (by Remark 1). Hence, by [11, Lemma 2.5.8], M is projective. Consequently, M is free since R is local. □

Proposition 6 *The class of all* w-FP*-projective modules is closed under arbitrary direct sums and under direct summands.*

Proof Follows from [8, Theorem 3.3.9(2)]. □

Recall that a ring R is called coherent if every finitely generated ideal of R is finitely presented.

Lemma 2 *Let R be a coherent ring and A be an R-module. Then A is absolutely w-pure if and only if $Ext_R^{n+1}(N, A)$ is a GV-torsion R-module for any finitely presented module N and any integer $n \geq 0$.*

Proof ($\Rightarrow$) suppose that A is absolutely w-pure R-module and let N be a finitely presented R-module. The case $n = 0$ is obvious. Hence, assume that $n > 0$. Consider an exact sequence

$$0 \to N' \to F_{n-1} \to \cdots \to F_0 \to N \to 0$$

where F_0,...,F_{n-1} are finitely generated free R-modules and N' is finitely presented. Such sequence exists since R is coherent. Thus, $(\text{Ext}_R^{n+1}(N, A))_\mathfrak{p} \cong (\text{Ext}_R^1(N', A))_\mathfrak{p}$ $= 0$ for any w-maximal ideal $\mathfrak{p}$ of R. So, $\text{Ext}_R^{n+1}(N, A)$ is a GV-torsion R-module.
($\Leftarrow$) Clear. □

Lemma 3 *Let R be a coherent ring and $0 \to A \to B \to C \to 0$ be an exact sequence of R-modules, where A is absolutely w-pure. Then, B is absolutely w-pure if and only if C is absolutely w-pure.*

Proof Let N be a finitely presented R-module. We have $\text{Ext}_R^1(N, A) \to \text{Ext}_R^1(N, B) \to \text{Ext}_R^1(N, C) \to \text{Ext}_R^2(N, A)$ By Lemma 2, for any maximal w-ideal $\mathfrak{p}$, we get $0 = \text{Ext}_R^1(N, A)_\mathfrak{p} \to \text{Ext}_R^1(N, B)_\mathfrak{p} \to \text{Ext}_R^1(N, C)_\mathfrak{p} \to \text{Ext}_R^2(N, A)_\mathfrak{p} = 0$. Thus, $\text{Ext}_R^1(N, B)_\mathfrak{p} \cong \text{Ext}_R^1(N, C)_\mathfrak{p}$. So, $\text{Ext}_R^1(N, B)$ is a GV-torsion R-module if and only if $\text{Ext}_R^1(N, C)$ is a GV-torsion R-module. Thus, B is absolutely w-pure if and only if C is absolutely w-pure. □

Proposition 7 *Let R be a coherent ring and M be an R-module. Then the following are equivalent:*

1. *M is w-FP-projective.*
2. *$Ext_R^{n+1}(M, A) = 0$ for any absolutely w-pure module A and any integer $n \geq 0$.*

Proof (1) $\Rightarrow$ (2) Let A be an absolutely w-pure R-module. The case $n = 0$ is obvious. So, we may assume $n > 0$. Consider an exact sequence

$$0 \to A \to E^0 \to \cdots \to E^{n-1} \to A' \to 0$$

where E^0,...,E^{n-1} are injective R-modules. By Lemma 3, A' is absolutely w-pure. Hence, $\text{Ext}_R^{n+1}(M, A) \cong \text{Ext}_R^1(M, A') = 0$.
(2) $\Rightarrow$ (1) Obvious. □

Proposition 8 *Let R be a coherent ring and $0 \to M'' \to M' \to M \to 0$ be an exact sequence of R-modules, where M is w-FP-projective. Then, M' is w-FP-projective if and only if M'' is w-FP-projective.*

Proof Follows from Proposition 7. □

We end this section with the following characterizations of a coherent DW-rings.

Proposition 9 *Let R be a ring. Then the following are equivalent:*

1. *R is a coherent DW-ring.*
2. *Every finitely generated submodule of a projective R-module is w-FP-projective.*
3. *Every finitely generated ideal of R is w-FP-projective.*

Proof (1) $\Rightarrow$ (2) Follows immediately from [13, Theorem 3.7] since, over a DW-ring, the classes of w-FP-projective modules and FP-projective modules coincide.
(2) $\Rightarrow$ (3) Obvious.
(3) $\Rightarrow$ (1) R is coherent by Remark 1. Assume that R is not a DW-ring. As in the proof of Proposition 1, there exist a maximal ideal $\mathfrak{m}$ of R such that $R/\mathfrak{m}$ is absolutely w-pure and $\mathfrak{m}_w = R$. So, for any finitely generated ideal I of R, we have

$$0 = \mathrm{Ext}_R^1(I, R/\mathfrak{m}) \to \mathrm{Ext}_R^2(R/I, R/\mathfrak{m}) \to \mathrm{Ext}_R^2(R, R/\mathfrak{m}) = 0,$$

and then $\mathrm{Ext}_R^2(R/I, R/\mathfrak{m}) = 0$. By [10, Lemma 3.1], $\mathrm{Tor}_R^2(R/I, R/\mathfrak{m}) = 0$, which means that $\mathrm{fd}_R(R/\mathfrak{m}) \leq 1$. Then $\mathfrak{m}$ is flat, and so a w-ideal, a contradiction. □

Corollary 1 *Let R be a domain. Then R is a coherent DW-domain if and only if every finitely generated torsion-free R-module is w-FP-projective.*

Proof Following [8, Theorem 1.6.15], every finitely generated torsion-free R-module can be embedded in a finitely generated free module (since R is a domain). Hence, ($\Rightarrow$) follows immediately from Proposition 9. For ($\Leftarrow$), it suffices to see that since R is a domain, every ideal is torsion-free, and then use Proposition 9. □

3 The W-FP-projective Dimension of Modules and Rings

In this section, we introduce and investigate the w-FP-projective dimension for modules and rings.

Definition 2 Let R be a ring. For any R-module M, the w-FP-projective dimension of M, denoted by w-$\mathrm{fpd}_R(M)$, is the smallest integer $n \geq 0$ such that $\mathrm{Ext}_R^{n+1}(M, A) = 0$ for any absolutely w-pure R-module A. If no such integer exists, set w-$\mathrm{fpd}_R(M) = \infty$.
The w-FP-projective dimension of R is defined by

$$w\text{-fpD}(R) = \sup\{w\text{-fpD}_R(M) : \text{M is finitely generated } R\text{-module}\}$$

Clearly, an R-module M is w-FP-projective if and only if $w\text{-fpd}_R(M) = 0$, and $\text{fpd}_R(M) \leq w\text{-fpd}_R(M)$, with equality when R is a DW-ring. However, this inequality may be strict (Remark 1). Also, $\text{fpD}(R) \leq w\text{-fpD}(R)$ with equality when R is a DW-ring, and this inequality may be strict. To see that, consider a regular local ring $(R, \mathfrak{m})$ with $\text{gldim}(R) = n$ ($n \geq 2$). Since R is Noetherian, we get $\text{fpD}(R) = 0$ (by [1, Proposition 2.6]). Moreover, by Remark 1, there exists an (FP-projective) R-module M which is not w-FP-projective. Thus, $w\text{-fpD}(R) > 0$.

First, we give a description of the w-FP-Projective dimension of modules over coherent ring.

Proposition 10 *Let R be a coherent ring. The following statements are equivalent for an R-module M.*

1. $w\text{-fpd}(M) \leqslant n$.
2. $\mathit{Ext}_R^{n+1}(M, A) = 0$ *for any absolutely w-pure R-module A.*
3. $\mathit{Ext}_R^{n+j}(M, A) = 0$ *for any absolutely w-pure R-module A and any $j \geq 1$.*
4. *If the sequence $0 \to P_n \to P_{n-1} \to \cdots \to P_0 \to M \to 0$ is exact with $P_0, \ldots,$ P_{n-1} are w-FP-projective R-modules, then P_n is w-FP-projective.*
5. *If the sequence $0 \to P_n \to P_{n-1} \to \cdots \to P_0 \to M \to 0$ is exact with $P_0, \ldots,$ P_{n-1} are projective R-modules, then P_n is w-FP-projective.*
6. *There exists an exact sequence $0 \to P_n \to P_{n-1} \to \cdots \to P_0 \to M \to 0$ where each P_i is w-FP-projective.*

Proof (3) ⇒ (2) ⇒ (1) and (4) ⇒ (5) ⇒ (6) are trivial.
(1) ⇒ (4) Let $0 \to P_n \to P_{n-1} \to \cdots \to P_0 \to M \to 0$ be an exact sequence of R-modules with $P_0, \ldots, P_{n-1}$ are w-FP-projective, and set $K_0 = \text{Ker}(P_0 \to M)$ and $K_i = \text{Ker}(P_i \to P_{i-1})$, where $i = 1, \ldots, n-1$. Using Proposition 7, we get

$$0 = \text{Ext}_R^{n+1}(M, A) \cong \text{Ext}_R^n(K_0, A) \cong \cdots \cong \text{Ext}_R^1(P_n, A)$$

for all absolutely w-pure R-module A. Thus, P_n is w-FP-projective.
(6) ⇒ (3) We procced by induction on $n \geq 0$. For the $n = 0$, M is w-FP-projective module and so (3) holds by proposition 7. If $n \geq 1$, then there is an exact sequence $0 \to P_n \to P_{n-1} \to \cdots \to P_0 \to M \to 0$ where each P_i is w-FP-projective. Set $K_0 = \text{Ker}(P_0 \to M)$. Then, we have the following exact sequences

$$0 \to P_n \to P_{n-1} \to \cdots \to P_1 \to K_0 \to 0$$

and

$$0 \to K_0 \to P_0 \to M \to 0$$

Hence, by induction $\text{Ext}_R^{n-1+j}(K_0, A) = 0$ for all absolutely w-pure R-module A and all $j \geq 1$. Thus, $\text{Ext}_R^{n+j}(M, A) = 0$, and so we have the desired result. □

The proof of the next proposition is standard homological algebra. Thus we omit its proof.

Proposition 11 *Let R be a coherent ring and $0 \to M'' \to M' \to M \to 0$ be an exact sequence of R-modules. If two of w-fpd$_R(M'')$, w-fpd$_R(M')$ and w-fpd$_R(M)$ are finite, so is the third. Moreover*

1. $w\text{-}fpd_R(M'') \leq \sup\{w\text{-}fpd_R(M'),\ w\text{-}fpd_R(M) - 1\}$.
2. $w\text{-}fpd_R(M') \leq \sup\{w\text{-}fpd_R(M''),\ w\text{-}fpd_R(M)\}$.
3. $w\text{-}fpd_R(M) \leq \sup\{w\text{-}fpd_R(M'),\ w\text{-}fpd_R(M'') + 1\}$.

Corollary 2 *Let R be a coherent ring and $0 \to M'' \to M' \to M \to 0$ be an exact sequence of R-modules. If M' is w-FP-projective and $w\text{-fpd}_R(M) > 0$, then $w\text{-}fpd_R(M) = w\text{-}fpd_R(M'') + 1$.*

Proposition 12 *Let R be a coherent ring and $\{M_i\}$ be a family of R-modules. Then $w\text{-}fpd_R(\oplus_i M_i) = \sup_i\{w\text{-}fpd_R(M_i)\}$.*

Proof The proof is straightforward. □

Proposition 13 *Let R be a ring and $n \geq 0$ be an integer. Then the following statements are equivalent:*

1. $w\text{-}fpD(R) \leq n$.
2. $w\text{-}fpd(M) \leqslant n$ *for all R-modules M.*
3. $w\text{-}fpd(R/I) \leqslant n$ *for all ideals I of R.*
4. $id_R(A) \leqslant n$ *for all absolutely w-pure R-modules A.*

Consequently, we have

$$\begin{aligned} w\text{-}fpD(R) &= \sup\{w\text{-}fpd_R(M) \mid M \text{ is an } R\text{-module}\} \\ &= \sup\{w\text{-}fpd_R(R/I) \mid I \text{ is an ideal of } R\} \\ &= \sup\{id_R(A) \mid A \text{ is an abosolutely } w\text{-pure } R\text{-module}\} \end{aligned}$$

Proof (2) ⇒ (1) ⇒ (3) are trivial.
(3) ⇒ (4) Let A be an absolutely w-pure R-module. For any ideal I of R, we have $\mathrm{Ext}_R^{n+1}(R/I, A) = 0$. Thus, $id_R(A) \leqslant n$.
(4) ⇒ (2) Let M be an R-module. For any absolutely w-pure R-module A, we have $\mathrm{Ext}_R^{n+1}(M, A) = 0$. Hence, $w\text{-fpd}(M) \leqslant n$. □

Note that Noetherian rings need not to be DW (for example, a regular ring with global dimension 2), and DW-rings need not to be Noetherian (for example, a non-Noetherian von Neumann regular ring). Next, we show that rings R with $w\text{-fpD}(R) = 0$ are exactly Noetherian DW-rings.

Proposition 14 *Let R be a ring. Then the following are equivalent:*

1. $w\text{-}fpD(R) = 0$.
2. *Every R-module is w-FP-projective.*
3. *R/I is w-FP-projective for every ideal I of R.*

4. *Every absolutely w-pure R-module is injective.*
5. *R is Noetherian DW-ring.*

Proof The equivalence of (1), (2), (3), and (4) follows from Proposition 13.
(2) $\Leftrightarrow$ (5) Follows from Proposition 1 and [1, Proposition 2.6]. □

Recall from[13], that a ring R is said FP-hereditary if every ideal of R is FP-projective. Note that FP-hereditary rings need not to be DW (for example, a non DW Noetherian ring), and DW-rings need not to be FP-hereditary (for example, a non-Noetherian von Neumann regular ring). Next, we show that rings R with w-fpD$(R) \leq 1$ are exactly FP-hereditary DW-rings.

Proposition 15 *Let R be a ring. Then the following are equivalent:*

1. *w-fpD$(R) \leq 1$.*
2. *Every submodule of w-*FP*-projective R-module is w-*FP*-projective.*
3. *Every submodule of projective R-module is w-*FP*-projective.*
4. *I is w-*FP*-projective for every ideal I of R.*
5. *$id_R(A) \leq 1$ for all absolutely w-pure R-module A.*
6. *R is a (coherent) FP-hereditary DW-ring.*

Proof The implications (2) $\Rightarrow$ (3) $\Rightarrow$ (4) are obvious.
(4) $\Rightarrow$ (5) Let A be an absolutely w-pure R-module and I be an ideal of R. The exact sequence $0 \to I \to R \to R/I \to 0$ gives rise to the exact sequence

$$0 = \mathrm{Ext}^1_R(I, A) \to \mathrm{Ext}^2_R(R/I, A) \to \mathrm{Ext}^2_R(R, A) = 0.$$

Thus, $\mathrm{Ext}^2_R(R/I, A) = 0$, and so $id_R(A) \leq 1$.
(5) $\Rightarrow$ (4) Let I be an ideal of R. For any absolutely w-pure R-module A, we have

$$0 = \mathrm{Ext}^2_R(R/I, A) = \mathrm{Ext}^1_R(I, A).$$

Thus, I is w-FP-projective.
(4) $\Rightarrow$ (6) By hypothesis, R is FP-hereditary. Now, by Proposition 9 R is a coherent DW-ring.
(6) $\Rightarrow$ (2) By [13, Theorem 3.16], since the w-FP-projective R-modules are just the FP-projective R-modules over a DW-ring.
(1) $\Leftrightarrow$ (5) By Proposition 13. □

Remark 2 In the Example 1, the ring R is coherent but not DW. Then, R contains a finitely generated ideal which is not w-FP-projective by Proposition 5.

References

1. Mao, L., Ding, N.: FP-projective dimension. Commun. Algebra **33**, 1153–1170 (2005)
2. Tamekkante, M., Assaad, R.A.K., Bouba, E.: Note On The DW Rings. Inter. Elec. J. Algebra **25** (2019)
3. Pinzon, K.R.: Absolutely Pure Modules. University of Kentucky, Thesis (2005)
4. Xing, S., Wang, F.: Purity over Prfer v-multiplication domains. II, J. Algebra Appl. 1850223 (2018)
5. Xing, S., Wang, F.: Purity over Prfer v-multiplication domains. J. Algebra Appl. **16**(5), 1850100 (2018)
6. Almahdi, F.A.A., Tamekkante, M.: Ding *w*-flat modules and dimensions. Algebra Colloq. **25**(203), 203–216 (2018)
7. Wang, F., Kim, H.: Two generalizations of projective modules and their applications. J. Pure Appl. Algebra **219** (2015)
8. Wang, F., Kim, H.: Foundations of Commutative Rings and Their Modules. Springer Nature Singapore Pte Ltd., Singapore (2016)
9. Yin, H., Wang, F., Zhu, X., Chen, Y.: *w*-Modules over commutative rings. J. Korean Math. Soc. **48**, 207–222 (2011)
10. Ding, N.Q., Chen, J.L.: The homological dimensions of simple modules. Bull. Austral. Math. Soc. **48**(2), 265–274 (1993)
11. Glaz, S.: Commutative Coherent Rings. Lecture Notes Maths, Springer (1989)
12. Maddox, B.H.: Absolutely pure modules. Proc. Am. Math. Soc. **18**, 155–158 (1967)
13. Moradzadeh-Dehkordi Shojaee, S.H.A.: Rings in which every ideal is pure-projective or FP-projective. J. Algebra **478**, 419–436 (2017)
14. Mimouni, A.: Integral domains in which each ideal is a *w*-ideal. Commun. Algebra **33**, 1345–1355 (2005)
15. Enochs, E.E.: A note on absolutely pure modules. Canad. Math. Bull. **19**(3), 361–362 (1976)
16. Wang, F.G., McCasland, R.L.: On *w*-modules over strong Mori domains. Commun. Algebra **25**, 1285–1306 (1997)
17. Zhao, S.Q., Wang, F.G., Chen, H.L.: Flat modules over a commutative ring are *w*-modules. J. Sichuan Normal Univ. **35**, 364–366 (2012). (in Chinese)
18. Wang, F.G.: Finitely presented type modules and *w*-coherent rings. J. Sichuan Normal Univ. **33**, 1–9 (2010)
19. Jain, S.: Flat and FP-injectivity. Proc. Am. Math. Soc. **41**, 437–442 (1973)
20. Megibben, C.: Absolutely pure modules. Proc. Am. Math. Soc. **26**, 561–566 (1970)
21. Stenström, B.: Coherent rings and FP-injective modules. J. Lond. Math. Soc. **2**(2), 323–329 (1970)
22. Fuchs, L., Salce, L.: Modules over Non-Noetherian domains. Math. Surv. Monog. Am. Math. Soc, Providence **84** (2001)
23. Wang, F.G., McCasland, R.L.: On strong Mori domains. J. Pure Appl. Algebra **135**, 155–165 (1999)
24. Tamekkante, M., Chhiti, M., Louartiti, K.: Weak projective modules and dimension. Int. J. Algebra. **5**, 1219–1224 (2011)

Central Values of X-generalized Skew Derivations on Right Ideals in Prime Rings

Luisa Carini and Vincenzo De Filippis

Abstract Let R be a prime ring of characteristic different from 2, Q its right Martindale quotient ring, C its extended centroid, I a right ideal of R, $a \in Q$, G a nonzero X-generalized skew derivation of R, $f(x_1, \ldots, x_n)$ a multilinear polynomial over C with n non-commuting variables, and S the set of the evaluations of $f(x_1, \ldots, x_n)$ on I. If $[f(x_1, \ldots, x_n), x_{n+1}]x_{n+2}$ is not an identity for I and $aG(x)x \in Z(R)$ for all $x \in S$, then we determine all the possible forms of G.

Keywords Generalized skew derivation · Multilinear polynomial · Prime ring

1 Introduction

Let R be a prime ring, $Z(R)$ its center, Q its *right Martindale quotient ring*, C the center of Q, which is called *extended centroid* of R (see [3] for more details about these objects). An additive mapping $d\colon R \longrightarrow R$ is said to be a *derivation* of R if $d(xy) = d(x)y + xd(y)$ for all $x, y \in R$. An additive mapping $F\colon R \longrightarrow R$ is called a *generalized derivation* of R if there exists a derivation d of R such that $F(xy) = F(x)y + xd(y)$ for all $x, y \in R$.

The previous definitions can be extended as follows. Let R be an associative ring and α be an automorphism of an associative ring R. An additive mapping $d\colon R \longrightarrow R$ is said to be a *skew derivation* of R if $d(xy) = d(x)y + \alpha(x)d(y)$ for all $x, y \in R$. An additive mapping $F\colon R \longrightarrow R$ is called a *generalized skew derivation* of R if there exists a skew derivation d of R with associated automorphism α such that $F(xy) = F(x)y + \alpha(x)d(y)$ for all $x, y \in R$. Many papers in literature study generalized derivations and generalized skew derivations of rings that satisfy certain

L. Carini (✉)
University of Messina, 98166 Messina, Italy
e-mail: lcarini@unime.it

V. De Filippis
Department of Engineering, University of Messina, Messina 98166, Italy
e-mail: defilippis@unime.it

M. Ashraf et al. (eds.), *Algebra and Related Topics with Applications*,
Springer Proceedings in Mathematics & Statistics 392,
https://doi.org/10.1007/978-981-19-3898-6_6

identities, in order to describe both the structure of the rings and the form of the maps involved with these identities (see for example [6, 9–14, 32–34, 36–38]).

These results emphasize the strong relationship between the structure of a ring R and the behavior of certain additive maps defined on R.

In a recent paper [31], Koşan and Lee introduced the notion of *left b-generalized derivations*. More precisely an additive map $F\colon R \longrightarrow Q$ is called a *left b-generalized derivation* if there exist $b \in Q$ and a derivation d of R such that $F(xy) = F(x)y + bxd(y)$, for all $x, y \in R$. Clearly, this concept generalizes the ones of derivations and generalized derivations. More recently (see [21, 23–25]), taking a cue from Koşan and Lee's work, the second author and F. Wei define and characterize an additive map in a different point of view known as *X-generalized skew derivation* or *b-generalized skew derivation* which extends the concept of a generalized skew derivation. More precisely, let R be an associative ring, $b \in Q$, $d : R \longrightarrow R$ a linear mapping, and α be an automorphism of R. A linear mapping $F : R \longrightarrow R$ is called an *X-generalized skew derivation* of R, with associated term (b, α, d) if there exist $b \in Q$, a linear mapping $d : R \longrightarrow R$ and an automorphism α of R, such that

$$F(xy) = F(x)y + b\alpha(x)d(y)$$

for all $x, y \in R$. Moreover, in [25, Remark 1.8], it is also proved that if F is a X-generalized skew derivation with associated term (b, α, d), then the linear map d must be a skew derivation of R, with associated automorphism α.

According to the above definition, it is clear that X-generalized skew derivations cover the concepts of derivations, generalized derivations, skew derivations, and generalized skew derivations.

The main goal of this paper is to investigate the set $P(G, f(I)) = \{G(x)x : x \in S\}$, where $G : R \to R$ is an additive map of R, $f(I) = \{f(r_1, \ldots, r_n) : r_1, \ldots, r_n \in I\}$ is the set of all evaluations of a multilinear polynomial $f(x_1, \ldots, x_n)$ over C in n non-commuting indeterminates, and I is a right ideal of R.

In [36, Theorem 2], Lee and Shiue prove that if G is a nonzero derivation of R and $P(G, f(R)) \subseteq C$, then $f(x_1, \ldots, x_n)$ is central valued on R, unless when $char(R) = 2$ and R satisfies s_4. Demir and Argaç [26] extend the above results to the case of generalized derivations. More precisely they prove that, if G is a nonzero generalized derivation of R and $P(G, f(R)) \subseteq C$, then either $f(x_1, \ldots, x_n)$ is central valued on R or there exists $b \in C$ such that $G(x) = bx$ for all $x \in R$ and $f(x_1, \ldots, x_n)^2$ is central valued on R, unless when $char(R) = 2$ and R satisfies s_4. Later on, in [22], the second author and Dhara generalize the previous result to the case of generalized derivations acting on polynomials (not necessarily multilinear) that are evaluated on right ideals.

Following this line of investigation, in [5] the authors prove that, if δ is a nonzero derivation of R, G a nonzero generalized derivation of R, and $\delta(x) = 0$, for all $x \in P(G, f(R))$, then $f(x_1, \ldots, x_n)^2$ is central valued on R and there exist $a, b \in U$ such that $G(x) = ax$ and $\delta(x) = [b, x]$, for any $x \in R$, with $[a, b] = 0$. Later, Argaç and Dhara in [27] extend the result contained in [5] to the case δ is a generalized derivations of R.

Let us fix our attention on the following two results:

Theorem 1 ([1, Theorem 3.7]) *Let R be a prime ring, $f(x_1, \ldots, x_n)$ a multilinear polynomial over C in n non-commuting indeterminates, I a nonzero right ideal of R, and $F : R \to R$ be a nonzero generalized skew derivation of R.*

Suppose that $F(f(r_1, \ldots, r_n))f(r_1, \ldots, r_n) \in C$, for all $r_1, \ldots, r_n \in I$. If the polynomial $f(x_1, \ldots, x_n)$ is not central valued on R, then either $char(R) = 2$ and R satisfies s_4 or one of the following holds:

(i) $f(x_1, \ldots, x_n)x_{n+1}$ *is an identity for I;*
(ii) $F(I)I = (0)$*;*
(iii) $[f(x_1, \ldots, x_n), x_{n+1}]x_{n+2}$ *is an identity for I, there exist $b, c, q \in Q$ with q an invertible element such that $F(x) = bx - qxq^{-1}c$ for all $x \in R$, and $q^{-1}cI \subseteq I$. Moreover, in this case either $(b - c)I = (0)$ or $b - c \in C$ and $f(x_1, \ldots, x_n)^2$ is central valued on R.*

Theorem 2 ([2, Main Theorem]) *Let R be a prime ring of characteristic different from 2, $f(x_1, \ldots, x_n)$ a multilinear polynomial over C in n non-commuting indeterminates, I a nonzero right ideal of R, $0 \neq a \in R$ and $F : R \to R$ be a nonzero generalized derivation of R.*

If $aF(f(r_1, \ldots, r_n))f(r_1, \ldots, r_n) = 0$, for all $r_1, \ldots, r_n \in I$, then one of the following holds:

(i) $aI = aF(I) = (0)$*;*
(ii) $F(x) = bx + [c, x]$*, for all $x \in R$, where $b, c \in Q$. In this case either $[c, I]I = (0) = abI$ or $aI = (0) = a(b + c)I$;*
(iii) $[f(x_1, \ldots, x_n), x_{n+1}]x_{n+2}$ *is an identity for I.*

Motivated by Theorems 1 and 2, here we would like to determine a first approach to the study of some algebraic properties satisfied by the set $P(G, f(I))$, in the case G is a X-generalized skew derivation and I is a right ideal of R.

In this sense, the main result of the present paper is

Theorem 3 *Let R be a prime ring of characteristic different from 2, Q be its right Martindale quotient ring and C be its extended centroid, I a right ideal of R, $a \in Q$, G a nonzero X-generalized skew derivation of R, $f(x_1, \ldots, x_n)$ a multilinear polynomial over C with n non-commuting variables, and S the set of the evaluations of $f(x_1, \ldots, x_n)$ on I. If $f(x_1, \ldots, x_n)$ is not central valued on R and $aG(x)x \in Z(R)$ for all $x \in S$, then $[f(x_1, \ldots, x_n), x_{n+1}]x_{n+2}$ is an identity for I unless when G assumes one of the following forms:*

1. *$G(x) = bx + cqxq^{-1}u$, for all $x \in R$, where $b, c, q, u \in Q$ (q is an invertible element of Q). Moreover, in this case one of the following holds:*

   (a) *there exists $\mu \in C$ such that $q^{-1}uI = \mu I$ and $a(b + cu)I = (0)$;*
   (b) *there exists $\mu \in C$ such that $q^{-1}uI = \mu I$, $f(x_1, \ldots, x_n)^2$ is central valued on R and $a(b + cu) \in C$;*

(c) *there exists* $\mu \in C$ *such that* $q^{-1}uI = \mu I$ *and* $f(x_1, \ldots, x_n)x_{n+1}$ *is an identity for* I;
(d) $acqI = abI = (0)$;
(e) $acqI = (0)$, $f(x_1, \ldots, x_n)^2$ *is central valued on* R *and* $ab \in C$;
(f) $acqI = (0)$ *and* $f(x_1, \ldots, x_n)x_{n+1}$ *is an identity for* I;

2. $G(x) = bx + c\alpha(x)u$, *for all* $x \in R$, *where* $b, c, u \in Q$ *and* α *is an outer automorphismof* R. *In this case, one of the following holds:*

   (a) $ac\alpha(I) = abI = (0)$;
   (b) $ac\alpha(I) = (0)$, $f(x_1, \ldots, x_n)^2$ *is central valued on* R *and* $ab \in C$;
   (c) $ac\alpha(I) = 0$ *and* $f(x_1, \ldots, x_n)x_{n+1}$ *is an identity for* I;
   (d) $uI = abI = (0)$;
   (e) $uI = (0)$, $f(x_1, \ldots, x_n)^2$ *is central valued on* R *and* $ab \in C$;
   (f) $uI = 0$ *and* $f(x_1, \ldots, x_n)x_{n+1}$ *is an identity for* I;

3. $G(x) = bx + cd(x)$, *for all* $x \in R$, *where* $b, c \in Q$ *and* d *is a skew derivation of* R. *In this case one of the following holds:*

   (a) $ac\alpha(I) = acd(I) = abI = (0)$;
   (b) $ac\alpha(I) = acd(I)$, $ab \in C$ *and* $f(x_1, \ldots, x_n)^2$ *is central valued on* R.

To be able to demonstrate our results, we firstly need to list some useful well-known facts:

Fact 4 Let R be a prime ring, then the following statements hold:

1. Every generalized derivation of R can be uniquely extended to Q [33, Theorem 3].
2. Any automorphism of R can be uniquely extended to Q [16, Fact 2].
3. Every generalized skew derivation of R can be uniquely extended to Q [9, Lemma 2].
4. If G is a X-generalized skew derivation of R with associated term (b, α, d), then G can be uniquely extended to Q and assumes the form $G(x) = ax + bd(x)$, where $a \in Q$ [25, Remark 1.9].

Fact 5 Let R be a prime ring and I be a two-sided ideal of R.

1. I, R, and Q satisfy the same generalized polynomial identities with coefficients in Q (see [15]).
2. I, R, and Q satisfy the same generalized polynomial identities with automorphisms (see [17, Theorem 1]).

Fact 6 In [19] Chuang and Lee prove that if $\Phi(x_i, D(x_i))$ is a generalized polynomial identity for R, where R is a prime ring and D is an outer skew derivation of R, then R also satisfies the generalized polynomial identity $\Phi(x_i, y_i)$, where x_i and y_i are distinct indeterminates. Moreover, if $\Phi(x_i, D(x_i), \alpha(x_i))$ is a generalized polynomial identity for a prime ring R, D is an outer skew derivation of R and α is an outer automorphism of R, then R also satisfies the generalized polynomial identity $\Phi(x_i, y_i, z_i)$, where x_i, y_i, and z_i are distinct indeterminates (see [19, Theorem 1]).

We conclude this section with some remarks on matrix algebras:

Fact 7 ([34, Lemma], [39, Lemma 2]) Let T be a K-algebra with 1 and let $R = M_m(T)$, $m \geq 2$. As usual, we denote the matrix unit having 1 in (i, j)-entry and zero elsewhere by e_{ij}.

Suppose that $f(x_1, \ldots, x_n)$ is a multilinear polynomial over K, that is not central valued on R. Then, for any $i \neq j$ there exist $r_1, \ldots, r_n \in R$ and $0 \neq \beta \in K$ such that $f(r_1, \ldots, r_n) = \beta e_{ij} \neq 0$. Moreover, since $f(x_1, \ldots, x_n)$ is a multilinear polynomial and C is a field, we may assume that $\beta = 1$.

Fact 8 ([20, Lemma 1.5]) Let $\mathscr{H}$ be an infinite field and $n \geq 2$. If $A_1, \ldots, A_k$ are not scalar matrices in $M_m(\mathscr{H})$ then there exists some invertible matrix $P \in M_m(\mathscr{H})$ such that each matrix $PA_1P^{-1}, \ldots, PA_kP^{-1}$ has all nonzero entries.

2 An Auxiliary Generalized Polynomial Identity

In this section, we prove the following result:

Proposition 1 *Let R be a prime ring of characteristic different from 2, Q be its right Martindale quotient ring and C be its extended centroid, $a, b, c \in Q$. If R satisfies the following generalized polynomial identity*

$$\Psi(x_1, \ldots, x_n) = \left[af(x_1, \ldots, x_n)^2 + bf(x_1, \ldots, x_n)cf(x_1, \ldots, x_n), x_{n+1}\right] \quad (1)$$

then one of the following holds:

1. *$a = b = 0$;*
2. *$c \in C$ and $a + bc = 0$;*
3. *$c \in C$, $f(x_1, \ldots, x_n)^2$ is central valued on R and $a + bc \in C$.*

We permit the following useful result:

Fact 9 Let R be a prime ring of characteristic different from 2, $f(x_1, \ldots, x_n)$ a non-central multilinear polynomial over C in n non-commuting indeterminates, $G : R \to R$ a nonzero generalized derivation of R and $a \in R$ be a fixed element. If $aG(f(r_1, \ldots, r_n))f(r_1, \ldots, r_n) \in C$ for all $r_1, \ldots, r_n \in R$, then there exists $p \in Q$ such that $G(x) = px$ for all $x \in R$ and either $ap = 0$ or $f(x_1, \ldots, x_n)^2$ is central valued on R and $ap \in C$ (it is a consequence of [28, Theorem 2.6]).

Remark 1 If $b \in C$ or $c \in C$, then the conclusion of Proposition 1 could be obtained as consequence of Fact 9. In fact, both $b \in C$ and $c \in C$ imply that G is an inner generalized derivation of R.

In light of this, in order to prove Proposition 1, our aim will be to prove that either $b \in C$ or $c \in C$.

We begin with

Lemma 1 *Let $R = M_m(C)$, $m \geq 2$. Then either $b \in C$ or $c \in C$.*

Proof We suppose both $b \notin Z(R)$ and $c \notin Z(R)$ and prove that a contradiction follows.

We firstly assume that C is infinite. By Fact 8, there exists some invertible matrix $P \in M_m(C)$ such that $\varphi(x) = PxP^{-1}$ and $\varphi(b)$, $\varphi(c)$ have all nonzero entries. Denote $\varphi(b) = \sum_{hl} b_{hl}e_{hl}$, $\varphi(c) = \sum_{hl} c_{hl}e_{hl}$, where $0 \neq b_{hl}, 0 \neq c_{hl} \in C$. Of course, in the main relation we may replace a, b, and c with $\varphi(a)$, $\varphi(b)$, and $\varphi(c)$, respectively. Hence, for $f(r_1, \ldots, r_n) = \lambda e_{ij} \neq 0$ in (1) and left multiplying by e_{ij}, we obtain $b_{ji}c_{ji} = 0$, which is a contradiction.

Now let E be an infinite field which is an extension of the field C and let $\overline{R} = M_t(E) \cong R \otimes_C E$. The generalized polynomial $\Psi(x_1, \ldots, x_n)$ is multi-homogeneous of multi-degree $(2, \ldots, 2)$ in the indeterminates $x_1, \ldots, x_n$.

Hence the complete linearization of $\Psi(x_1, \ldots, x_n)$ is a multilinear generalized polynomial $\Theta(x_1, \ldots, x_n, y_1, \ldots, y_n)$. Moreover,

$$\Theta(x_1, \ldots, x_n, x_1, \ldots, x_n) = 2^n\Psi(x_1, \ldots, x_n)$$

is a multilinear generalized polynomial identity for R and $\overline{R}$ too. Since $\mathrm{char}(C) \neq 2$, we obtain $\Psi(r_1, \ldots, r_n) = 0$ for all $r_1, \ldots, r_n \in \overline{R}$, and the conclusion follows from the first part of the present Lemma. □

Proof of Proposition 1

Proof Here we assume again $b \notin C$ and $c \notin C$. Clearly, in this case $\Psi(x_1, \ldots, x_n)$ is a non-trivial generalized polynomial identity for R, then, by [15] it follows that $\Psi(x_1, \ldots, x_n)$ is a non-trivial generalized polynomial identity for Q. By the well-known Martindale's theorem of [40], Q is a primitive ring having nonzero socle with the field C as its associated division ring. By [30, Page 35] Q is isomorphic to a dense subring of the ring of linear transformations of a vector space V over C, containing nonzero linear transformations of finite rank. Assume first that $\dim_C V = k \geq 2$ is a finite positive integer, then $Q \cong M_k(C)$ and the conclusion follows from Lemma 1.

On the other hand, if $\dim_C V = \infty$ and by [41, Lemma 2], it follows that Q satisfies the generalized polynomial identity

$$[ax_1^2 + bx_1cx_1, x_2]. \tag{2}$$

Moreover Q is a dense ring of C-linear transformations over a vector space V.

Since $c \notin C$, there exists $v \neq 0$, such that $\{v, cv\}$ are linear C-independent. By the density of Q, there exist $s_1, s_2 \in Q$ such that

$$s_1v = 0; \quad s_1(cv) = v; \quad s_2v = cv$$

hence

$$0 = [as_1^2 + bs_1cs_1, s_2]v = bv.$$

Of course for any $w \in V$ such that $\{w, v\}$ are linearly C-dependent, $bw = 0$. Let now $w \in V$ such that $\{w, v\}$ are linearly C-independent and $bw \neq 0$. By the above argument it follows that w and cw must be linearly C-dependent, as are $\{w + v, c(w + v)\}$ and $\{w - v, c(w - v)\}$. Therefore there exist $\alpha_w, \alpha_{w+v}, \alpha_{w-v} \in C$ such that

$$cw = \alpha_w w, \quad c(w + v) = \alpha_{w+v}(w + v), \quad c(w - v) = \alpha_{w-v}(w - v).$$

Therefore

$$\alpha_w w + cv = \alpha_{w+v} w + \alpha_{w+v} v \tag{3}$$

and

$$\alpha_w w - cv = \alpha_{w-v} w - \alpha_{w-v} v. \tag{4}$$

By comparing (3) with (4) we get both

$$(2\alpha_w - \alpha_{w+v} - \alpha_{w-v})w + (\alpha_{w-v} - \alpha_{w+v})v = 0 \tag{5}$$

and

$$2cv = (\alpha_{w+v} - \alpha_{w-v})w + (\alpha_{w+v} + \alpha_{w-v})v. \tag{6}$$

By (5) and since $\{w, v\}$ are C-independent and $char(C) \neq 2$, we have $\alpha_w = \alpha_{w+v} = \alpha_{w-v}$. Thus by (6) it follows $2cv = 2\alpha_w v$. Since $\{cv, v\}$ are C-independent, the conclusion $\alpha_w = \alpha_{w+v} = 0$ follows, that is $cw = 0$ and $c(w + v) = 0$, which implies the contradiction $cv = 0$.

Hence we may conclude that $bw = 0$, for any $w \in V$. Thus $bV = (0)$, that is $b = 0$ which is again a contradiction. □

3 The Case of Inner X-generalized Skew Derivations

In this section we prove Theorem 3 in the case there exists an automorphism $\alpha \in Aut(R)$ and $b, c, u \in Q$ such that $G(x) = bx + c\alpha(x)u$, for any $x \in R$.

Let us start with the following first result, that is a simple application of Proposition 1:

Proposition 2 *Let R be a prime ring of characteristic different from 2, Q be its right Martindale quotient ring and C be its extended centroid, $a, b, c, u, q \in Q$, such that q is an invertible element of Q. If R satisfies the following generalized polynomial identity*

$$\left[a\left(bf(x_1,\ldots,x_n)+cqf(x_1,\ldots,x_n)q^{-1}u\right)f(x_1,\ldots,x_n),x_{n+1}\right] \quad (7)$$

then one of the following holds:

1. $ab = ac = 0$;
2. $q^{-1}u \in C$ *and* $a(b+cu) = 0$;
3. $q^{-1}u \in C$, $f(x_1,\ldots,x_n)^2$ *is central valued on* R *and* $a(b+cu) \in C$.

Lemma 2 *Let* R *be a noncommutative prime ring,* $a \in R$, $f(x_1,\ldots,x_n)$ *a polynomial over* C. *If* $[af(r_1,\ldots,r_n),r_{n+1}] = 0$, *for all* $r_1,\ldots,r_{n+1} \in R$, *then either* $a = 0$ *or* $f(x_1,\ldots,x_n)$ *is central valued on* R *and* $a \in Z(R)$.

Proof Of course, in case $f(x_1,\ldots,x_n)$ is central valued on R, it follows easily $a \in Z(R)$.

Then we may assume $f(x_1,\ldots,x_n)$ is not central valued on R and, by contradiction, suppose $a \neq 0$.

It is well known that, since $f(x_1,\ldots,x_n)$ is not central and $char(R) \neq 2$, the additive subgroup S of R generated by $\{f(x_1,\ldots,x_n) : x_i \in R\}$ contains a non-central Lie ideal L of R. Therefore $[au,r] = 0$, for any $u \in L$ and $r \in R$. In particular, for any $u \in L$,

$$0 = [au,u] = [a,u]u.$$

Hence, by [36, Theorem 2], $a \in Z(R)$. Thus $a[L,R] = (0)$, which is a contradiction, since $0 \neq a \in Z(R)$ and $L \not\subseteq Z(R)$. □

Lemma 3 *Let* R *be a noncommutative prime ring,* $a,b \in R$, $f(x_1,\ldots,x_n)$ *a polynomial over* C. *If* $af(r_1,\ldots,r_n)b = 0$, *for all* $r_1,\ldots,r_n \in R$, *then either* $a = 0$ *or* $b = 0$, *unless when* $f(x_1,\ldots,x_n)$ *is central valued on* R *and* $ab = 0$.

Proof If $f(x_1,\ldots,x_n)$ is central valued on R, it follows easily $ab = 0$.

Then we may assume $f(x_1,\ldots,x_n)$ is not central valued on R and, by using the same argument as in Lemma 2, we arrive at $aLb = (0)$, where L is a non-central Lie ideal of R. In this case is well known that either $a = 0$ or $b = 0$. □

Proposition 3 *Let* R *be a prime ring of characteristic different from 2,* Q *be its right Martindale quotient ring and* C *be its extended centroid,* $a,b,c,u \in Q$, $\alpha \in Aut(R)$ *and* G *be the inner* X*-generalized skew derivation of* R *defined as follows:*

$$G(x) = bx + c\alpha(x)u, \qquad \forall x \in R.$$

Let $f(x_1,\ldots,x_n)$ *be a non-central multilinear polynomial over* C *with* n *non-commuting variables. If*

$$[aG(f(r_1,\ldots,r_n))f(r_1,\ldots,r_n),r_{n+1}] = 0$$

for all $r_1,\ldots,r_{n+1} \in R$, *then one of the following holds:*

1. $ab = ac = 0$;
2. $ab = u = 0$;
3. $ab \in C$, $ac = 0$ *and* $f(x_1, \ldots, x_n)^2$ *is central valued on* R;
4. $ab \in C$, $u = 0$ *and* $f(x_1, \ldots, x_n)^2$ *is central valued on* R;
5. *there exists an invertible element* $q \in Q$ *such that* $\alpha(x) = qxq^{-1}$, *for any* $x \in R$, *with* $q^{-1}u \in C$ *and* $a(b + cu) = 0$;
6. $f(x_1, \ldots, x_n)^2$ *is central valued on* R *and there exists an invertible element* $q \in Q$ *such that* $\alpha(x) = qxq^{-1}$, *for any* $x \in R$, *with* $q^{-1}u \in C$ *and* $a(b + cu) \in C$.

Proof If there exists an invertible element $q \in Q$, such that $\alpha(x) = qxq^{-1}$ for all $x \in R$, then the conclusion follows from Proposition 1. Thus, we may assume that α is not inner. In what follows we denote $f^{\alpha}(x_1, \ldots, x_n)$ the polynomial obtained from $f(x_1, \ldots, x_n)$ by replacing each coefficient γ_σ with $\alpha(\gamma_\sigma)$. By hypothesis, R satisfies the generalized polynomial identity

$$\left[a\left(bf(x_1, \ldots, x_n) + cf^{\alpha}(\alpha(x_1), \ldots, \alpha(x_n))u\right)f(x_1, \ldots, x_n), x_{n+1}\right]. \tag{8}$$

Since α is not inner, R satisfies the generalized polynomial identity

$$\left[a\left(bf(x_1, \ldots, x_n) + cf^{\alpha}(y_1, \ldots, y_n)u\right)f(x_1, \ldots, x_n), x_{n+1}\right]. \tag{9}$$

Hence, the following are both generalized identities for R:

$$\left[acf^{\alpha}(y_1, \ldots, y_n)uf(x_1, \ldots, x_n), x_{n+1}\right] \tag{10}$$

and

$$\left[abf(x_1, \ldots, x_n)^2, x_{n+1}\right]. \tag{11}$$

By applying Lemmas 2 and 3 to relation (10), and since both $f(x_1, \ldots, x_n)$ and $f^{\alpha}(x_1, \ldots, x_n)$ are not central valued on R, one has that either $ac = 0$ or $u = 0$. Analogously, relation (11) implies that either $ab = 0$ or $f(x_1, \ldots, x_n)^2$ is central valued on R and $ab \in C$.

Crossing all the cases just mentioned, we get each of the required conclusions in the case α is not inner. □

4 The Main Result for Prime Rings

This part of our paper is devoted to the proof of Theorem 3 in the case $I = R$, more precisely we prove the following:

Theorem 10 *Let R be a prime ring of characteristic different from 2, Q be its right Martindale quotient ring and C be its extended centroid, $a \in Q$, G a nonzero X-generalized skew derivation of R, $f(x_1, \ldots, x_n)$ a multilinear polynomial over C with n non-commuting variables, and S the set of the evaluations of $f(x_1, \ldots, x_n)$ on R. If $f(x_1, \ldots, x_n)$ is not central valued on R and $aG(x)x \in Z(R)$ for all $x \in S$, then:*

1. *If $G(x) = bx + c\alpha(x)u$, for all $x \in R$, where $a, b, c, u \in Q$ and $\alpha \in Aut(R)$, one of the following holds:*
   (a) $ab = ac = 0$;
   (b) $ab = u = 0$;
   (c) $ab \in C$, $ac = 0$ *and* $f(x_1, \ldots, x_n)^2$ *is central valued on* R;
   (d) $ab \in C$, $u = 0$ *and* $f(x_1, \ldots, x_n)^2$ *is central valued on* R;
   (e) *there exists an invertible element* $q \in Q$ *such that* $\alpha(x) = qxq^{-1}$, *for any* $x \in R$, *with* $q^{-1}u \in C$ *and* $a(b + cu) = 0$;
   (f) $f(x_1, \ldots, x_n)^2$ *is central valued on* R *and there exists an invertible element* $q \in Q$ *such that* $\alpha(x) = qxq^{-1}$, *for any* $x \in R$, *with* $q^{-1}u \in C$ *and* $a(b + cu) \in C$.
2. *If $G(x) = bx + cd(x)$, for all $x \in R$, where $b, c \in Q$ and d is a skew derivation of R, then one of the following holds:*
   (a) $ab = ac = 0$;
   (b) $ab \in C$, $ac = 0$ *and* $f(x_1, \ldots, x_n)^2$ *is central valued on* R.

Proof We write $G(x) = bx + cd(x)$, for all $x \in R$, where $b, c \in Q$ are suitable fixed elements and d is a skew derivation of R with associated automorphism α. Denote

$$f(x_1, \ldots, x_n) = \sum_{\sigma \in S_n} \gamma_\sigma x_{\sigma(1)} \cdot x_{\sigma(2)} \cdots x_{\sigma(n)}, \quad \gamma_\sigma \in C.$$

Let $f^d(x_1, \ldots, x_n)$ be the polynomial obtained from $f(x_1, \ldots, x_n)$ by replacing each coefficient γ_σ with $d(\gamma_\sigma)$, and $f^\alpha(x_1, \ldots, x_n) = \alpha\big(f(x_1, \ldots, x_n)\big)$. By using this notation, we have

$$d\big(\gamma_\sigma \cdot x_{\sigma(1)} \cdot x_{\sigma(2)} \cdots x_{\sigma(n)}\big) =$$

$$d(\gamma_\sigma)x_{\sigma(1)} \cdot x_{\sigma(2)} \cdots x_{\sigma(n)} + \alpha(\gamma_\sigma)\sum_{j=0}^{n-1} \alpha(x_{\sigma(1)} \cdot x_{\sigma(2)} \cdots x_{\sigma(j)})d(x_{\sigma(j+1)})x_{\sigma(j+2)} \cdots x_{\sigma(n)}$$

and

$$d\big(f(x_1, \ldots, x_n)\big) =$$

$$f^d(x_1, \ldots, x_n) + \sum_{\sigma \in S_n} \alpha(\gamma_\sigma)\sum_{j=0}^{n-1} \alpha(x_{\sigma(1)} \cdot x_{\sigma(2)} \cdots x_{\sigma(j)})d(x_{\sigma(j+1)})x_{\sigma(j+2)} \cdots x_{\sigma(n)}.$$

Firstly we remark that, if either $c = 0$ or $d = 0$ or $0 \neq d$ is an inner skew derivation, then the result is a consequence of Proposition 3.

Therefore, we always assume $c \neq 0$ and $d \neq 0$, moreover the skew derivation d is not inner.

By hypothesis, R satisfies the generalized polynomial identity

$$\left[a\Big(bf(x_1,\ldots,x_n) + cd(f(x_1,\ldots,x_n))\Big)f(x_1,\ldots,x_n), x_{n+1}\right] \tag{12}$$

that is

$$\left[a\Big(bf(x_1,\ldots,x_n) + cf^d(x_1,\ldots,x_n)\Big)f(x_1,\ldots,x_n) + \right.$$
$$\left. ac\Big(\sum_{\sigma\in S_n}\alpha(\gamma_\sigma)\sum_{j=0}^{n-1}\alpha(x_{\sigma(1)}\cdot x_{\sigma(2)}\cdots x_{\sigma(j)})d(x_{\sigma(j+1)})x_{\sigma(j+2)}\cdots x_{\sigma(n)}\Big)f(x_1,\ldots,x_n), x_{n+1}\right] \tag{13}$$

Since d is outer abd by (13), R satisfies

$$\left[a\Big(bf(x_1,\ldots,x_n) + cf^d(x_1,\ldots,x_n)\Big)f(x_1,\ldots,x_n) + \right.$$
$$\left. ac\Big(\sum_{\sigma\in S_n}\alpha(\gamma_\sigma)\sum_{j=0}^{n-1}\alpha(x_{\sigma(1)}\cdot x_{\sigma(2)}\cdots x_{\sigma(j)})y_{\sigma(j+1)}x_{\sigma(j+2)}\cdots x_{\sigma(n)}\Big)f(x_1,\ldots,x_n), x_{n+1}\right] \tag{14}$$

In particular, R satisfies any blended component

$$\left[ac\Big(\sum_{\sigma\in S_n}\alpha(\gamma_\sigma)\sum_{i=1}^{n}\alpha(x_{\sigma(1)}\cdot x_{\sigma(2)}\cdots x_{\sigma(i-1)})y_{\sigma(i)}x_{\sigma(i+1)}\cdots x_{\sigma(n)}\Big)f(x_1,\ldots,x_n), x_{n+1}\right]. \tag{15}$$

Since R and Q satisfy the same generalized polynomial identities, Q satisfies (15).

If $\alpha = \mathrm{id}_R \in \mathrm{Aut}(R)$, then d is an ordinary derivation of R and (15) reduces to

$$\left[ac\Big(\sum_{i=1}^{n} f(x_1,\ldots,y_i,\ldots,x_n)\Big)f(x_1,\ldots,x_n), x_{n+1}\right]. \tag{16}$$

Replacing in (16) any y_i with $[w, x_i]$, for a fixed element $w \in Q \setminus C$, we have that

$$\Big[ac[w, f(x_1,\ldots,x_n)]f(x_1,\ldots,x_n), x_{n+1}\Big]$$

is a generalized identity for Q. Thus, by Proposition 1 and since $w \notin C$, it follows $ac = 0$. Thus, by (12), Q satisfies

$$\left[abf(x_1, \ldots, x_n)^2, x_{n+1}\right] \tag{17}$$

and, once again by Proposition 1, one has that either $ab = 0$ or $f(x_1, \ldots, x_n)^2$ is central valued on Q and $ab \in C$, as required.

Therefore, we may assume that $\alpha \neq \mathrm{id}_R \in \mathrm{Aut}(R)$.

If there exists an invertible element $q \in Q \setminus C$ such that $\alpha(x) = qxq^{-1}$ for all $x \in Q$, by replacing each $y_{\sigma(i)}$ with $qx_{\sigma(i)}$ in (15), it follows that Q satisfies the generalized polynomial identity

$$\left[ac\left(q \sum_{\sigma \in S_n} \gamma_\sigma x_{\sigma(1)} \cdots x_{\sigma(n)}\right) f(x_1, \ldots, x_n), x_{n+1}\right]$$

that is

$$\left[acqf(x_1, \ldots, x_n)^2, x_{n+1}\right].$$

As above, Proposition 1 implies that either $ac = 0$ or $f(x_1, \ldots, x_n)^2$ is central valued on Q and $acq \in C$.

In the first case relation (12) reduces to (17) and we conclude as done previously.

Thus we assume that $ac \neq 0$, $f(x_1, \ldots, x_n)^2$ is central valued on Q and $acq \in C$. By replacing each $y_{\sigma(i)}$ with $q[q^{-1}, x_{\sigma(i)}]$ in (15), it follows that Q satisfies the generalized polynomial identity

$$\left[acq[q^{-1}, f(x_1, \ldots, x_n)]f(x_1, \ldots, x_n), x_{n+1}\right]$$

that is

$$acq\left[[q^{-1}, f(x_1, \ldots, x_n)]f(x_1, \ldots, x_n), x_{n+1}\right].$$

Since $acq \neq 0$ and $q \notin C$, and in light of Proposition 1, this last relation implies a contradiction.

Finally, assume that α is not inner. By (15) it follows that Q satisfies the generalized polynomial identity

$$\left[ac\left(\sum_{\sigma \in S_n} \alpha(\gamma_\sigma) \sum_{i=1}^{n} z_{\sigma(1)} \cdot z_{\sigma(2)} \cdots z_{\sigma(i-1)} y_{\sigma(i)} x_{\sigma(i+1)} \cdots x_{\sigma(n)}\right) f(x_1, \ldots, x_n), x_{n+1}\right].$$

In particular, for any $i = 1, \dots, n$, Q also satisfies the generalized polynomial identity

$$\left[ac\left(\sum_{\sigma \in S_{n-1}} \alpha(\gamma_\sigma) z_{\sigma(1)} \cdot z_{\sigma(2)} \cdots z_{\sigma(i-1)} \cdot z_{\sigma(i+1)} \cdots z_{\sigma(n)} \cdot y_i\right) f(x_1, \dots, x_n), x_{n+1}\right]. \tag{18}$$

Let us write

$$\sum_{\sigma \in S_{n-1}} \alpha(\gamma_\sigma) x_{\sigma(1)} \cdots x_{\sigma(j-i)} x_{\sigma(j+1)} \cdots x_{\sigma(n)} = t_j(x_1, \dots, x_{j-1}, x_{j+1}, \dots, x_n),$$

where any t_j is a multilinear polynomial of degree $n - 1$ and x_j never appears in any monomial of t_j. Thus

$$f^\alpha(x_1, \dots, x_n) = \sum_j t_j(x_1, \dots, x_{j-1}, x_{j+1}, \dots, x_n) x_j$$

moreover $f^\alpha(x_1, \dots, x_n)$ is not an identity for Q. Therefore there exists $j \in \{1, \dots, n\}$ such that t_j is not an identity for Q.

Starting from (18) it follows that, for any $j = 1, \dots, n$, Q satisfies

$$\left[act_j(z_1, \dots, z_{j-1}, z_{j+1}, \dots, z_n) y_j f(x_1, \dots, x_n), x_{n+1}\right].$$

By Lemmas 2 and 3, and since $f(x_1, \dots, x_n)$ is not central valued on Q, it follows that one of the following holds:

- either $ac = 0$
- or $t_j(z_1, \dots, z_{j-1}, z_{j+1}, \dots, z_n) y_j = 0$, for any $z_1, \dots, z_{j-1}, z_{j+1}, \dots, z_n$ and any y_j.

In case $ac = 0$ then Q satisfies (17) and we conclude as above.

On the other hand, if

$$t_j(z_1, \dots, z_{j-1}, z_{j+1}, \dots, z_n) y_j = 0$$

for all $z_1, \dots, z_{j-1}, z_{j+1}, \dots, z_n \in Q$ and all $y_j \in Q$, then $f^\alpha(x_1, \dots, x_n)$ is an identity for Q, which is a contradiction. □

As an easy consequence we have

Corollary 1 *Let R be a prime ring of characteristic different from 2, Q be its right Martindale quotient ring and C be its extended centroid, $a \in Q$, G a nonzero X-generalized skew derivation of R, $f(x_1, \dots, x_n)$ a multilinear polynomial over C with n non-commuting variables, and S the set of the evaluations of $f(x_1, \dots, x_n)$ on R. If $f(x_1, \dots, x_n)$ is not central valued on R and $aG(x)x = 0$ for all $x \in S$, then one of the following holds:*

1. *there exist* $b, c \in Q$ *and a skew derivation* d *of* R *such that* $G(x) = bx + cd(x)$, *for all* $x \in R$, *with* $ab = ac = 0$;
2. *there exists* $b \in Q$ *such that* $G(x) = bx$, *for all* $x \in R$, *with* $ab = 0$.

We would like to conclude this section by providing a shorter and leaner formulation of Theorem 10:

Theorem 11 *Let* R *be a prime ring of characteristic different from 2,* Q *be its right Martindale quotient ring and* C *be its extended centroid,* $a \in Q$, G *a nonzero* X*-generalized skew derivation of* R, $f(x_1, \dots, x_n)$ *a multilinear polynomial over* C *with* n *non-commuting variables, and* S *the set of the evaluations of* $f(x_1, \dots, x_n)$ *on* R. *If* $f(x_1, \dots, x_n)$ *is not central valued on* R *and* $aG(x)x \in Z(R)$ *for all* $x \in S$, *then one of the following holds:*

1. $aG(x) = 0$, *for all* $x \in R$
2. $f(x_1, \dots, x_n)^2$ *is central valued on* R *and there exists* $\lambda \in C$ *such that* $aG(x) = \lambda x$, *for all* $x \in R$.

5 The Main Result

We are finally in the position to prove Theorem 3.

Lemma 4 *Let* R *be a prime ring,* $f(x_1, \dots, x_n)$ *a multilinear polynomial over* C *in* n *non-commuting indeterminates,* I *a nonzero right ideal of* R, *and* $a \in R$ *be a fixed element.*

Suppose that $af(r_1, \dots, r_n)^2 \in C$, *for all* $r_1, \dots, r_n \in I$. *If the polynomial* $f(x_1, \dots, x_n)$ *is not central valued on* R, *then either* $char(R) = 2$ *and* R *satisfies* s_4 *or one of the following holds:*

(i) $f(x_1, \dots, x_n)x_{n+1}$ *is an identity for* I;
(ii) $aI = (0)$;
(iii) $a \in C$ *and* $f(x_1, \dots, x_n)^2$ *is central valued on* R.

Proof It is an easy consequence of Theorem 1. □

Lemma 5 *Let* R *be a prime ring,* $f(x_1, \dots, x_n)$ *a multilinear polynomial over* C *in* n *non-commuting indeterminates,* I *a nonzero right ideal of* R, *and* $a, b \in R$ *be nonzero fixed elements. Suppose that* $af(r_1, \dots, r_n)b = 0$, *for all* $r_1, \dots, r_n \in I$. *Then one of the following holds:*

1. $aI = (0)$;
2. *there exists an idempotent element* $e \in soc(Q)$ *such that* $I = eR$ *and* $[f(x_1, \dots, x_n), x_{n+1}]x_{n+2}$ *is an identity for* I;
3. *there exists an idempotent element* $e \in soc(Q)$ *such that* $I = eR$, $char(R) = 2$ *and* $s_4(x_1, x_2, x_3, x_4)x_5$ *is an identity for* I;
4. $f(x_1, \dots, x_n)$ *is central valued on* R *and* $ab = 0$.

Proof Firstly we remark that, in case $f(x_1, \ldots, x_n)$ is central valued on R and by the primeness of R, we get easily that $ab = 0$. Moreover, if $[f(x_1, \ldots, x_n), x_{n+1}]x_{n+2}$ is an identity for I, then, by [35], there exists an idempotent element in the socle of R such that $I = eR$. Thus, by contradiction, we suppose that the following hold simultaneously:

- $aI \neq (0)$;
- $[f(x_1, \ldots, x_n), x_{n+1}]x_{n+2}$ is not an identity for I;
- if $char(R) = 2$, $s_4(x_1, x_2, x_3, x_4)x_5$ is not an identity for I;
- $f(x_1, \ldots, x_n)$ is not central valued on R.

Our aim is to show that a number of contradictions follows.

Assume firstly that I is not a PI-ring. By [8, Theorem 1] it follows that there exists a non-PI right ideal of R, namely $I_0 \subseteq I$, such that $[I_0, I] \subseteq f(I)$ and $II_0 \subseteq I_0$.

By our hypothesis, $a[I_0, I]b = (0)$, and in particular $a[I_0, IaI]b = (0)$, that is $a[I_0, Ia]Ib = (0)$. A fortiori we get $a[I_0, Ia][I_0, I]b = (0)$. Thus, since $a[I_0, I]b = (0)$, it follows $aIaI_0[I_0, I]b = (0)$. Since $aI \neq (0)$, one has that either $aI_0 = (0)$ or $[I_0, I]b = (0)$.

Notice that, if $aI_0 = (0)$ then also $(0) = aII_0 \subseteq aI_0$, implying the contradiction $aI = (0)$. On the other hand $[I_0, I]b = (0)$ implies $[I_0, I_0]b = (0)$ and, by applying [18] and since $b \neq 0$, we get $[I_0, I_0]I_0 = (0)$ which is again a contradiction, since I_0 is not PI.

Let now I be a PI-ring. Hence, by [35] there exists an idempotent element in the socle of R such that $I = eR$. Since $[f(x_1, \ldots, x_n), x_{n+1}]x_{n+2}$ is not an identity for I and, in case $char(R) = 2$, $s_4(x_1, x_2, x_3, x_4)x_5$ is not an identity for I, then, by [8, Theorem 1], it follows both $eR(1-e)f(I)$ and $[I, I] \subseteq f(I)$, implying $aeR(1-e)b = (0)$ and $a[I, I]b = (0)$.

By the primeness of R and since $aI \neq (0)$, $aeR(1-e)b = (0)$ means $(1-e)b = 0$, that is either $e = 1$ or $b = eb \in I$.

Notice that, if $e = 1$ then $I = R$. Hence, since $f(x_1, \ldots, x_n)$ is not central valued on R, $a[R, R]b = (0)$ and $a = 0$ follows easily. Thus, for the rest of the proof we assume $b = eb \in I$ and $a[I, I]b = (0)$.

In particular

$$(0) = a\Big[[I, I], bR\Big]b = abR[I, I]b$$

and, by the primeness of R, either $[I, I]b = (0)$ or $ab = 0$.

If $[I, I]b = (0)$ and since $b \neq 0$, it follows from [18] the contradiction $[I, I]I = (0)$. Hence $ab = 0$, so that

$$(0) = a\Big[I, bR\Big]b = aIbRb$$

that is $aIb = (0)$, which forcrs $aI = (0)$, again a contradiction. □

Lemma 6 *Let $m \geq 2$ and $R = M_h(K)$ be the ring of all $h \times h$ matrices over a field K of characteristic different from 2, I a nonzero right ideal of R, $f(x_1, \dots, x_n)$ a non-central multilinear polynomial over K. Let $a, b, c, u, q \in R$ be such that q is an invertible element of R and*

$$\Big[a\big(bf(s_1, \dots, s_n) + cqf(s_1, \dots, s_n)q^{-1}u\big)f(s_1, \dots, s_n), r_{n+1}\Big] = 0 \qquad (19)$$

for any $s_1, \dots, s_n \in I$ and $r_{n+1} \in R$. Then one of the following holds:

1. *there exists $\mu \in Z(R)$ such that $q^{-1}uI = \mu I$ and $a(b + cu)I = (0)$;*
2. *there exists $\mu \in Z(R)$ such that $q^{-1}uI = \mu I$, $f(x_1, \dots, x_n)^2$ is central valued on R and $a(b + cu) \in Z(R)$;*
3. *there exists $\mu \in Z(R)$ such that $q^{-1}uI = \mu I$ and $f(x_1, \dots, x_n)x_{n+1}$ is an identity for I;*
4. *$acqI = abI = (0)$;*
5. *$acqI = (0)$, $f(x_1, \dots, x_n)^2$ is central valued on R and $ab \in Z(R)$;*
6. *$acqI = (0)$ and $f(x_1, \dots, x_n)x_{n+1}$ is an identity for I;*
7. *$[f(x_1, \dots, x_n), x_{n+1}]x_{n+2}$ is an identity for I.*

Proof Denote e_{ij} the usual unit matrix with 1 in (i, j)-entry and zero elsewhere, $acq = \sum_{lm} a_{lm}e_{lm}$, $q^{-1}u = \sum_{lm} b_{lm}e_{lm}$, for $a_{lm}, b_{lm} \in K$.

Since there exists a set of matrix units that contains the idempotent generator of a given minimal right ideal, any minimal right ideal is part of a direct sum of minimal right ideals adding to R. Hence we may assume that any minimal right ideal of R is a direct sum of minimal right ideals, each of the form $e_{ii}R$. Moreover I has a number of uniquely determined simple components, that are minimal right ideals of R and I is their direct sum. So we may write $I = eR$ for some $e = \sum_{i=1}^{p} e_{ii}$ and $p \in \{1, 2, \dots, h\}$. Moreover $p \geq 2$, if not $[I, I]I = 0$ and a fortiori $[f(x_1, \dots, x_n), x_{n+1}]x_{n+2}$ is an identity for I.

Notice that, if $h = 2$ then $p = 2$ and $I = R$. In this case, the conclusion follows from Proposition 3. Thus we may assume $h \geq 3$.

By Lemma 3 in [7], if $[f(x_1, \dots, x_n), x_{n+1}]x_{n+2}$ is not an identity for I, then for all $\gamma \in K$, $s \leq p$ and $t \neq s$ there exist $r_1, \dots, r_n \in I$ such that $f(r_1, \dots, r_n) = \gamma e_{st}$. Then, by our hypothesis we have that

$$\gamma^2 acqe_{st}q^{-1}ue_{st} \in Z(R). \qquad (20)$$

Since (20) represent a matrix of rank 1 and $\gamma \neq 0$, one has $acqe_{st}q^{-1}ue_{st} = 0$, that is

$$a_{ks}b_{ts} = 0 \quad \forall s \leq p, t \neq s \quad \text{and} \quad \forall k \geq 1. \qquad (21)$$

Assume firstly there exist $i, j \leq p$ and $i \neq j$ such that $b_{ji} \neq 0$. Then, by (21), $a_{ki} = 0$ for any $k \geq 1$. Consider the following automorphisms of R:

$$\varphi'(x) = (1 + e_{ji})x(1 - e_{ji}) = x + e_{ji}x - xe_{ji} - e_{ji}xe_{ji}$$

$$\varphi''(x) = (1 - e_{ji})x(1 + e_{ji}) = x - e_{ji}x + xe_{ji} - e_{ji}xe_{ji}.$$

Note that $\varphi'(I) \subseteq I$ and $\varphi''(I) \subseteq I$ are right ideals of R satisfying

$$\Big[\varphi'(a)\big(\varphi'(b)f(s_1,\dots,s_n) + \varphi'(cq)f(s_1,\dots,s_n)\varphi'(q^{-1}u)\big)f(s_1,\dots,s_n), r_{n+1}\Big] = 0 \quad (22)$$

for any $s_1, \dots, s_n \in \varphi'(I)$ and $r_{n+1} \in R$, and

$$\Big[\varphi''(a)\big(\varphi''(b)f(s_1,\dots,s_n) + \varphi''(cq)f(s_1,\dots,s_n)\varphi''(q^{-1}u)\big)f(s_1,\dots,s_n), r_{n+1}\Big] = 0 \quad (23)$$

for any $s_1, \dots, s_n \in \varphi''(I)$ and $r_{n+1} \in R$. Denote $\varphi'(acq) = \sum_{lm} a'_{lm}e_{lm}$, $\varphi'(q^{-1}u) = \sum_{lm} b'_{lm}e_{lm}$, $\varphi''(acq) = \sum_{lm} a''_{lm}e_{lm}$, $\varphi''(q^{-1}u) = \sum_{lm} b''_{lm}e_{lm}$, for $a'_{lm}, b'_{lm}, a''_{lm}, b''_{lm} \in K$.

By calculation one has that $b'_{ji} = b_{ji} + b_{ii} - b_{jj} - b_{ij}$ and $b''_{kh} = b_{ji} - b_{ii} + b_{jj} - b_{ij}$. In case both $b'_{ji} = 0$ and $b''_{ji} = 0$, then $b_{ij} = b_{ji} \neq 0$. Hence, by by (21), $a_{kj} = 0$ for any $k \geq 1$.

On the other hand, if $b'_{ji} \neq 0$ (or $b''_{ji} \neq 0$), then, again by (21), one has $a'_{ki} = 0$ (or $a''_{ki} = 0$, respectively) for any $k \geq 1$. In particular, for any $k \neq j$, $0 = a'_{ki} = a_{ki} - a_{kj} = -a_{kj}$ (or $0 = a''_{ki} = a_{ki} + a_{kj} = a_{kj}$, respectively).

Therefore, in any case, $a_{kj} = 0$ for any $k \geq j$ and $a_{ki} = 0$ for any $k \geq 1$.

Let now $r \leq p, r \neq j$ and consider the following automorphisms of R:

$$\chi'(x) = (1 + e_{rj})x(1 - e_{rj}) = x + e_{rj}x - xe_{rj} - e_{rj}xe_{rj}$$

$$\chi''(x) = (1 - e_{rj})x(1 + e_{rj}) = x - e_{rj}x + xe_{rj} - e_{rj}xe_{rj}.$$

As above $\chi'(I) \subseteq I$ and $\chi''(I) \subseteq I$ are right ideals of R satisfying

$$\Big[\chi'(a)\big(\chi'(b)f(s_1,\dots,s_n) + \chi'(cq)f(s_1,\dots,s_n)\chi'(q^{-1}u)\big)f(s_1,\dots,s_n), r_{n+1}\Big] = 0 \quad (24)$$

for any $s_1, \dots, s_n \in \chi'(I)$ and $r_{n+1} \in R$, and

$$\Big[\chi''(a)\big(\chi''(b)f(s_1,\dots,s_n) + \chi''(cq)f(s_1,\dots,s_n)\chi''(q^{-1}u)\big)f(s_1,\dots,s_n), r_{n+1}\Big] = 0 \quad (25)$$

for any $s_1, \dots, s_n \in \chi''I)$ and $r_{n+1} \in R$. Denote $\chi'(acq) = \sum_{lm} a'''_{lm}e_{lm}$, $\chi'(q^{-1}u) = \sum_{lm} b'''_{lm}e_{lm}$, $\chi''(acq) = \sum_{lm} a^{iv}_{lm}e_{lm}$, $\chi''(q^{-1}u) = \sum_{lm} b^{iv}_{lm}e_{lm}$, for $a'''_{lm}, b'''_{lm}, a^{iv}_{lm}, b^{iv}_{lm} \in K$. Since $b'''_{ji} = b^{iv}_{ji} = b_{ji} \neq 0$ and by the above argument, we get both $a'''_{kj} = 0$ and $a^{iv}_{kj} = 0$, for any $k \neq j$.

In particular, for any $k \neq r$, j

$$0 = a'''_{kj} = a_{kj} - a_{kr} = -a_{kr}.$$

Moreover, for $k = r$,

$$0 = a'''_{rj} = a_{rj} + a_{jj} - a_{rr} - a_{jr} = a_{jj} - a_{rr} - a_{jr}$$

and

$$0 = a^{iv}_{rj} = a_{rj} - a_{jj} + a_{rr} - a_{jr} = -a_{jj} + a_{rr} - a_{jr}.$$

These last two relations imply that $a_{jj} = a_{rr}$ and $a_{jr} = 0$.

Summarizing, we obtain the following conclusion:

- *If there exist $i, j \leq p$, $i \neq j$ such that $b_{ji} = 0$, then $a_{kr} = 0$ and $a_{rr} = a_{jj}$, for any $r \leq p$ and $k \neq r$.*

This means that there exists $\lambda \in Z(R)$ such that $acqI = \lambda I$.

If we suppose $\lambda \neq 0$, by (20) it follows

$$\gamma^2 \lambda e_{st} q^{-1} u e_{st} \in Z(R)$$

that is $b_{ts} = 0$, a contradiction. Thus $\lambda = 0$, $acqI = (0)$ and, by our main assumption, we have

$$\left[abf(s_1, \ldots, s_n)^2, r_{n+1}\right] = 0$$

for any $s_1, \ldots, s_n \in I$ and $r_{n+1} \in R$. Hence, by Lemma 4 it follows that one of the following holds:

1. $f(x_1, \ldots, x_n)x_{n+1}$ is an identity for I;
2. $abI = (0)$;
3. $ab \in C$ and $f(x_1, \ldots, x_n)^2$ is central valued on R.

In any case we are done.

Assume finally $b_{ts} = 0$, for any $s, t \leq p$ and $t \neq s$. Then, by (21), $a_{ki} = 0$ for any $k \geq 1$. Let $i, j \leq p$, $i \neq j$ and consider again the above defined automorphism φ' of R:

$$\varphi'(x) = (1 + e_{ji})x(1 - e_{ji}) = x + e_{ji}x - xe_{ji} - e_{ji}xe_{ji}$$

where $\varphi'(acq) = \sum_{lm} a'_{lm} e_{lm}$ and $\varphi'(q^{-1}u) = \sum_{lm} b'_{lm} e_{lm}$.

If $b'_{ji} \neq 0$, by using the above argument, we obtain $\varphi'(acq)\varphi'(I) = (0)$, that is $acqI = (0)$ and, as previously remarked, the concusion follows from Lemma 4.

Thus we may admit that, for any $i, j \leq p$, $i \neq j$, the automorphism φ' of R is defined in such a way that $b'_{ji} = 0$, that is

$$0 = b'_{ji} = b_{ji} + b_{ii} - b_{jj} - b_{ij} = b_{ii} - b_{jj}.$$

Hence, we obtain that

– $b_{ts} = 0$ and $b_{tt} = b_{ss}$, for any $s, t \leq p$ and $t \neq s$.

Therefore there exists $\mu \in Z(R)$ such that $q^{-1}uI = \mu I$ and, by our hypothesis, it follows that

$$\left[a(b+cu)f(s_1,\ldots,s_n)^2, r_{n+1}\right] = 0 \tag{26}$$

for any $s_1, \ldots, s_n \in I$ and $r_{n+1} \in R$. Application of Lemma 4 implies that one of the following conclusions holds:

1. $f(x_1, \ldots, x_n)x_{n+1}$ is an identity for I;
2. $a(b+cu)I = (0)$;
3. $a(b+cu) \in C$ and $f(x_1, \ldots, x_n)^2$ is central valued on R

as required. □

Remark 2 ([4, Lemma]) Let I be a nonzero right ideal of R and $p \in Q$. Then the following conditions are equivalent:

1. $[p, I]I = (0)$;
2. there exists $\beta \in C$ such that $(p - \beta)I = (0)$.

Lemma 7 *Let R be a noncommutative prime ring of characteristic different from 2 with right Martindale quotient ring Q and extended centroid C, I a nonzero right ideal of R. Let $f(x_1, \ldots, x_n)$ be a non-central multilinear polynomial over C, $a, b, c, u, q \in Q$ be such that q is an invertible element of R and I satisfies*

$$a\big(bf(x_1,\ldots,x_n) + cqf(x_1,\ldots,x_n)q^{-1}u\big)f(x_1,\ldots,x_n) \in C. \tag{27}$$

If R does not satisfy any non-trivial generalized polynomial identity, then one of the following holds:

1. *there exists $\mu \in C$ such that $q^{-1}uI = \mu I$ and $a(b+cu)I = (0)$;*
2. *$acqI = abI = (0)$.*

Proof For any $w \in I$, R satisfies the generalized polynomial identity

$$\left[a\big(bf(wx_1,\ldots,wx_n) + cqf(wx_1,\ldots,wx_n)q^{-1}u\big)f(wx_1,\ldots,wx_n), x_{n+1}\right] \tag{28}$$

Since (28) must be trivial generalized polynomial identities for R, by [15] it follows that $abw = \lambda_w acqw$, with $\lambda_w \in C$ is depending on the choice of $w \in I$. Hence (28) reduces to

$$\left[acqf(wx_1,\ldots,wx_n)(\lambda_w + q^{-1}u)f(wx_1,\ldots,wx_n), x_{n+1}\right]. \tag{29}$$

Once again (29) is a trivial identity for R. This implies that either $acqw = abw = 0$ or $(\lambda_w + q^{-1}u)w = 0$.

Therefore each element of I belongs to one of the sets $S_1 = \{x \in I : acqx = abx = 0\}$ and $S_2 = \{x \in I : (\lambda_x + q^{-1}u)x = 0\}$. That is to say, I is the union of its additive subgroups S_1 and S_2. However a group cannot be the union of two proper subgroups, so we have that either $I = S_1$ or $I = S_2$. Hence, either $acqx = abx = 0$ for all $x \in I$, or $[q^{-1}u, x]x = 0$ for all $x \in I$.

In this last case, by (27), it follows that and I satisfies

$$a(b + cu)f(x_1, \ldots, x_n)^2 \in C. \tag{30}$$

Hence, since R does not satisfy any non-trivial generalized polynomial identity, and by Lemma 4, we conclude that $a(b + cu)I = (0)$. □

Proposition 4 *Let R be a noncommutative prime ring of characteristic different from 2 with right Martindale quotient ring Q and extended centroid C, I a nonzero right ideal of R. Let $f(x_1, \ldots, x_n)$ be a non-central multilinear polynomial over C, $a, b, c, u, q \in Q$ be such that q is an invertible element of R and*

$$\left[a\big(bf(s_1, \ldots, s_n) + cqf(s_1, \ldots, s_n)q^{-1}u\big)f(s_1, \ldots, s_n), r_{n+1}\right] = 0 \tag{31}$$

for any $s_1, \ldots, s_n \in I$ and $r_{n+1} \in R$. Then one of the following holds:

1. *there exists $\mu \in C$ such that $q^{-1}uI = \mu I$ and $a(b + cu)I = (0)$;*
2. *there exists $\mu \in C$ such that $q^{-1}uI = \mu I$, $f(x_1, \ldots, x_n)^2$ is central valued on R and $a(b + cu) \in C$;*
3. *there exists $\mu \in C$ such that $q^{-1}uI = \mu I$ and $f(x_1, \ldots, x_n)x_{n+1}$ is an identity for I;*
4. *$acqI = abI = (0)$;*
5. *$acqI = (0)$, $f(x_1, \ldots, x_n)^2$ is central valued on R and $ab \in C$;*
6. *$acqI = (0)$ and $f(x_1, \ldots, x_n)x_{n+1}$ is an identity for I;*
7. *$[f(x_1, \ldots, x_n), x_{n+1}]x_{n+2}$ is an identity for I.*

Proof Since if R does not satisfy any non-trivial generalized polynomial identity the result follows from Lemma 7, in all that follows we may assume that R satisfies some non-trivial generalized polynomial identity, that is R is a GPI-ring. By [40] RC is a primitive ring and so Q has nonzero socle H with nonzero right ideal $J = IH$. Moreover J and I satisfy the same generalized identities with coefficients in Q. Thus replace R by H and I by J, then without loss of generality we may consider that R is a simple ring and equal to its own socle and $I = IR$.

Notice that, if $[q^{-1}u, I]I = (0)$, then, by Remark 2, there exists $\mu \in C$ such that $q^{-1}uI = \mu I$ and

$$\left[a(b + cu)f(s_1, \ldots, s_n)^2, r_{n+1}\right] = 0 \tag{32}$$

for any $s_1, \ldots, s_n \in I$ and $r_{n+1} \in R$. Similarly, if $acqI = (0)$ then

$$\Big[abf(s_1, \ldots, s_n)^2, r_{n+1}\Big] = 0 \tag{33}$$

for any $s_1, \ldots, s_n \in I$ and $r_{n+1} \in R$.

In any case, we obtain the required conclusions as an application of Lemma 4.

Here, by contradiction, we assume that there exist $h_0, h_1, h_2, \ldots, h_{n+4} \in I$ such that

1. $[q^{-1}u, h_0]h_1 \neq 0$;
2. $acqh_2 \neq 0$;
3. $[f(h_3, \ldots, h_{n+2}), h_{n+3}]h_{n+4} \neq 0$.

Moreover choose F to be the algebraic closure of C or C, according to $|C| = \infty$ or $|C| < \infty$. Note that $IH \otimes_C F$ is a completely reducible right $H \otimes_C F$-module such that

$$\Big[a\big(bf(s_1, \ldots, s_n) + cqf(s_1, \ldots, s_n)q^{-1}u\big)f(s_1, \ldots, s_n), r_{n+1}\Big] = 0 \tag{34}$$

for any $s_1, \ldots, s_n \in IH \otimes_C F$ and $r_{n+1} \in H \otimes_C F$. Thus there exists an idempotent $h \in IH \otimes_C F$ such that $h_0, h_1, h_2, \ldots, h_{n+4} \in h(IH \otimes_C F)$. By Litoff's Theorem (for a proof see [29]) there exists $e^2 = e \in H \otimes_C F$ such that

$$h, cqh, hcq, q^{-1}uh, hq^{-1}u, bh, hb, ah, ha, h_i \in e(H \otimes_C F)e \quad \forall i = 0, \ldots, n+4$$

with $e(H \otimes_C F)e \cong M_k(F)$, for $k \geq 2$.

For all $s_1, \ldots, s_n \in he(H \otimes_C F)e \subseteq (IH \otimes_C F) \cap e(H \otimes_C F)e$ and $r_{n+1} \in e(H \otimes_C F)e$, we have

$$\begin{aligned}
0 &= \Big[a\big(bf(s_1, \ldots, s_n) + cqf(s_1, \ldots, s_n)q^{-1}u\big)f(s_1, \ldots, s_n), r_{n+1}\Big] \\
&= \Big[a\big(bhf(s_1, \ldots, s_n) + cqhf(s_1, \ldots, s_n)q^{-1}uh\big)f(s_1, \ldots, s_n), r_{n+1}\Big] \\
&= \Big[(eae)\big((ebe)f(s_1, \ldots, s_n) + (ecqe)f(s_1, \ldots, s_n)(eq^{-1}ue)\big)f(s_1, \ldots, s_n), r_{n+1}\Big] = 0.
\end{aligned}$$

By Lemma 6, we have that one of the following holds:

1. $[e(q^{-1}u)e, he(H \otimes_C F)e]he(H \otimes_C F)e = 0$, which implies the contradiction $0 \neq [q^{-1}u, h_0]h_1 = [eq^{-1}ue, heh_0e]heh_1e = 0$;
2. $(eae)(ecqe)he(H \otimes_C F)e = 0$, which implies the contradiction $0 \neq acqh_2 = (eae)(ecqe)heh_2e = 0$;

3. $he(H \otimes_C F)e$ satisfies $[f(x_1, \ldots, x_n), x_{n+1}]x_{n+2}$, which gives again a contradiction, since

$$[f(heh_3e, \ldots, heh_{n+2}e), heh_{n+3}e]heh_{n+4}e = [f(h_3, \ldots, h_{n+2}), h_{n+3}]h_{n+4} \neq 0 \qquad \square$$

Proposition 5 *Let R be a noncommutative prime ring of characteristic different from 2 with right Martindale quotient ring Q and extended centroid C, I a nonzero right ideal of R. Let $f(x_1, \ldots, x_n)$ be a non-central multilinear polynomial over C, $a, b, c, u \in Q$, $\alpha \in Aut(R)$ be such that*

$$\left[a\left(bf(s_1, \ldots, s_n) + c\alpha\big(f(s_1, \ldots, s_n)\big)u\right)f(s_1, \ldots, s_n), r_{n+1}\right] = 0 \tag{35}$$

for any $s_1, \ldots, s_n \in I$ and $r_{n+1} \in R$. If α is not an inner automorphism of R, then one the following holds:

1. *$ac\alpha(I) = abI = (0)$;*
2. *$ac\alpha(I) = (0)$, $f(x_1, \ldots, x_n)^2$ is central valued on R and $ab \in C$;*
3. *$ac\alpha(I) = 0$ and $f(x_1, \ldots, x_n)x_{n+1}$ is an identity for I;*
4. *$uI = abI = (0)$;*
5. *$uI = (0)$, $f(x_1, \ldots, x_n)^2$ is central valued on R and $ab \in C$;*
6. *$uI = 0$ and $f(x_1, \ldots, x_n)x_{n+1}$ is an identity for I;*
7. *$[f(x_1, \ldots, x_n), x_{n+1}]x_{n+2}$ is an identity for I.*

Proof Clearly, in case either $uI = (0)$ or $ac\alpha(I) = 0$, then $abf(r_1, \ldots, r_n)^2 \in C$, for all $r_1, \ldots, r_n \in I$ and the conclusion follows from Lemma 4. Thus we suppose there are $v_1, v_2 \in I$ such that $ac\alpha(v_1) \neq 0$ and $uv_2 \neq 0$.

We remark that, if $ac\alpha(v_1 - v_2) = 0$, then $ac\alpha(v_2) \neq 0$ follows. Analogously, in case $u(v_1 - v_2) = 0$, one has $uv_1 \neq 0$.

Hence one of the following cases must occur:

- $ac\alpha(v_1) \neq 0$ and $uv_1 \neq 0$;
- $ac\alpha(v_2) \neq 0$ and $uv_2 \neq 0$;
- $ac\alpha(v_1 - v_2) \neq 0$ and $u(v_1 - v_2) \neq 0$.

In any case, there exists a suitable element $w \in I$ such that $ac\alpha(w) \neq 0$ and $uw \neq 0$. For such an element $0 \neq w \in I$, R satisfies

$$\left[a\left(bf(wx_1, \ldots, wx_n) + cf^{\alpha}(\alpha(w)\alpha(x_1), \ldots, \alpha(w)\alpha(x_n))u\right)f(wx_1, \ldots, wx_n), x_{n+1}\right]. \tag{36}$$

Since α is X-outer, by Theorem 3 in [17], R satisfies

$$\left[a\left(bf(wx_1, \ldots, wx_n) + cf^{\alpha}(\alpha(w)y_1, \ldots, \alpha(w)y_n)u\right)f(wx_1, \ldots, wx_n), x_{n+1}\right] \tag{37}$$

and in particular R satisfies the component

$$\left[a\left(cf^{\alpha}(\alpha(w)y_1, \ldots, \alpha(w)y_n)u\right)f(wx_1, \ldots, wx_n), x_{n+1}\right]. \tag{38}$$

Since $ac\alpha(w) \neq 0$ and $uw \neq 0$, we have that (38) is a non-trivial generalized polynomial identity for R. By [40] Q is a primitive ring having nonzero socle H with the field C as its associated division ring. Moreover R and Q satisfy the same generalized polynomial identities with automorphisms [17, Theorem 1]. Therefore Q satisfies (36). Suppose there exist $a_1, \ldots, a_{n+2} \in I$ such that $[f(a_1, \ldots, a_n), a_{n+1}]a_{n+2} \neq 0$. Since Q is a regular GPI-ring, there exists an idempotent element $e \in IQ$ such that $eQ = \sum_{i=1}^{n+1} a_i Q + wQ$ and $w = ew, a_i = ea_i$, for any $i = 1, \ldots, n+1$. Therefore, by (36), Q satisfies

$$\left[a\left(bf(ex_1, \ldots, ex_n) + cf^{\alpha}(\alpha(e)\alpha(x_1), \ldots, \alpha(e)\alpha(x_n))u\right)f(ex_1, \ldots, ex_n), x_{n+1}\right]. \tag{39}$$

We may assume $e \neq 1$, if not $eQ = Q$ and we conclude by Proposition 3. Moreover, as above, relation (39) implies that Q satisfies

$$\left[acf^{\alpha}(\alpha(e)y_1, \ldots, \alpha(e)y_n)uf(ex_1, \ldots, ex_n), x_{n+1}\right].$$

Replacing x_{n+1} with $(1-e)x_{n+1}$ and x_i with $x_i e$, for any $i = 1, \ldots, n$, it follows that Q satisfies

$$(1-e)x_{n+1}acf^{\alpha}(\alpha(e)y_1, \ldots, \alpha(e)y_n)uf(ex_1e, \ldots, ex_ne).$$

Hence, by the primeness of Q and since $e \neq 1$,

$$acf^{\alpha}(\alpha(e)r_1, \ldots, \alpha(e)r_n)uf(es_1e, \ldots, es_ne) = 0$$

for any $r_1, \ldots, r_n, s_1, \ldots, s_n \in Q$.

Since $f(ea_1, \ldots, ea_n)ea_{n+1} \neq 0$ and $uew \neq 0$, by Fact 5 it follows that

$$acf^{\alpha}(\alpha(e)y_1, \ldots, \alpha(e)y_n)$$

is a generalized identity for Q. By using the result in [18] and since $ac\alpha(w) \neq 0$, $f^{\alpha}(\alpha(e)y_1, \ldots, \alpha(e)y_n)$ is also an identity for Q. The last identity clearly leads to the fact that Q satisfies $f(e\alpha^{-1}(y_1), \ldots, e\alpha^{-1}(y_n))$, therefore $f(ex_1, \ldots, ex_n)$ is an identity for Q, a contradiction. $\square$

5.1 The Proof of Theorem 3

Proof Let $b, c \in Q$ and d be a skew derivation of R such that $G(x) = bx + cd(x)$. In the case d is an inner skew derivation of R, the proof of our main result is contained in Propositions 4 and 5. Therefore, in all that follows we assume that d is not inner.

As above, we write $f(x_1, \ldots, x_n) = \sum_{\sigma \in S_n} \gamma_\sigma x_{\sigma(1)} \cdots x_{\sigma(n)}$ with $\gamma_\sigma \in C$ and denote by $f^d(x_1, \ldots, x_n)$ the polynomial obtained from $f(x_1, \ldots, x_n)$ by replacing each coefficient γ_σ with $d(\gamma_\sigma)$, and $f^\alpha(x_1, \ldots, x_n)$ the polynomial obtained from $f(x_1, \ldots, x_n)$ by replacing each coefficient γ_σ with $\alpha(\gamma_\sigma)$.

Since IQ satisfies

$$aG(f(x_1, \ldots, x_n))f(x_1, \ldots, x_n) \in C$$

then, for all $0 \neq u \in I$, Q satisfies

$$\Big[a\Big(bf(ux_1, \ldots, ux_n) + cf^d(ux_1, \ldots, ux_n)\Big)f(ux_1, \ldots, ux_n), x_{n+1}\Big]$$
$$+\Big[ac\Big(\sum_{\sigma \in S_n} \alpha(\gamma_\sigma)\sum_{j=0}^{n-1} \alpha(ux_{\sigma(1)} \ldots ux_{\sigma(j)})d(ux_{\sigma(j+1)})ux_{\sigma(j+2)} \ldots ux_{\sigma(n)}\Big)f(ux_1, \ldots, ux_n), x_{n+1}\Big]. \tag{40}$$

By [19, Theorem 1], Q satisfies

$$\Big[a(bf(ux_1, \ldots, ux_n) + cf^d(ux_1, \ldots, ux_n))f(ux_1, \ldots, ux_n), x_{n+1}\Big]$$
$$+\Big[ac(\sum_{\sigma \in S_n} \alpha(\gamma_\sigma)\sum_{j=0}^{n-1} \alpha(ux_{\sigma(1)} \ldots ux_{\sigma(j)})d(u)x_{\sigma(j+1)} \ldots ux_{\sigma(n)}\Big)f(ux_1, \ldots, ux_n), x_{n+1}\Big]$$
$$+\Big[ac\Big(\sum_{\sigma \in S_n} \alpha(\gamma_\sigma)\sum_{j=0}^{n-1} \alpha(ux_{\sigma(1)} \ldots ux_{\sigma(j)})\alpha(u)y_{\sigma(j+1)}ux_{\sigma(j+2)} \ldots ux_{\sigma(n)}\Big)f(ux_1, \ldots, ux_n), x_{n+1}\Big]. \tag{41}$$

Of course we suppose $[f(x_1, \ldots, x_n), x_{n+1}]x_{n+2}$ is not an identity for I. Thus there exist $a_1, \ldots, a_{n+2} \in I$ such that $[f(a_1, \ldots, a_n), a_{n+1} + a_{n+2} \neq 0$.

Case 1. We firstly assume $ac\alpha(I) = (0)$

In this case, and supposing in addition that $acd(I) = (0)$, one has that I satisfies

$$abf(x_1, \ldots, x_n)^2 \in C.$$

Hence, by Lemma 4, and since $f(x_1, \ldots, x_n)x_{n+1}$ is not an identity for I, it follows that either $abI = (0)$ (that is $aG(x) = 0$, for any $x \in I$) or $ab \in C$ and $f(x_1, \ldots, x_n)^2$ is central valued on R, as required.

Thus, we suppose there exists $w \in I$ such that $acd(w) \neq 0$. By (40), and since $ac\alpha(w) = 0$, it follows that Q satisfies

$$\begin{aligned}&\left[a\Big(bf(wx_1,\ldots,wx_n)+cf^d(wx_1,\ldots,wx_n)\Big)f(wx_1,\ldots,wx_n),x_{n+1}\right]\\&+\left[ac\Big(\sum_{\sigma\in S_n}\alpha(\gamma_\sigma)d(w)x_{\sigma(1)})wx_{\sigma(2)}\ldots wx_{\sigma(n)}\Big)f(wx_1,\ldots,wx_n),x_{n+1}\right].\end{aligned} \tag{42}$$

Since $acd(w) \neq 0$, (42) is a non-trivial generalized polynomial identity for Q, then Q has nonzero socle H which satisfies the same generalized polynomial identities of Q. Since R is prime, $acd(w) \neq 0$ implies $Iacd(w) \neq (0)$, that is there exists $w' \in I$ such that $w'acd(w) \neq 0$. Without loss of generality R is simple and equal to its own socle, $IR = I$ and $a \in I$. In fact, R is GPI and so Q has nonzero socle H with nonzero right ideal $J = IH$ [40]. H is simple and $J = JH$ satisfies the same basic conditions as I. Now just replace R by H and I by J.

Moreover $R = H$ is a regular ring, hence there exists $g = g^2 \in R$ such that $wR + w'R + \sum_{i=1}^{n+1} R = gR$. Then $g \in IR = I$, $w = gw$, $w' = gw'$, and $a_i = ga_i$ for each $i = 1, \ldots, n+1$. Since gR satisfies the same generalized identities with skew derivations and automorphisms of I, we may also assume $g \neq 1$, if not $gR = R$ and the conclusion follows from Theorem 10.

For all $s_1, \ldots, s_n \in gR$ and $r_{n+1} \in R$, we have

$$\begin{aligned}&\left[a\Big(bf(s_1g,\ldots,s_ng)+cf^d(s_1g,\ldots,s_ng)\Big)f(s_1g,\ldots,s_ng),(1-g)r_{n+1}\right]\\&+\left[ac\Big(\sum_{\sigma\in S_n}\alpha(\gamma_\sigma)d(g)s_{\sigma(1)}g)s_{\sigma(2)g}\ldots s_{\sigma(n)g}\Big)f(s_1g,\ldots,s_ng),(1-g)r_{n+1}\right]=0\end{aligned}$$

that is

$$\begin{aligned}(1-g)r_{n+1}\Big\{&a\Big(bf(s_1,\ldots,s_n)+cf^d(s_1,\ldots,s_n)\Big)f(s_1,\ldots,s_n)g\\&+ac\Big(\sum_{\sigma\in S_n}\alpha(\gamma_\sigma)d(g)s_{\sigma(1)})s_{\sigma(2)}\ldots s_{\sigma(n)}\Big)f(s_1,\ldots,s_n)g\Big\}=0\end{aligned} \tag{43}$$

By the primeness of R and since $g \neq 1$,

$$\begin{aligned}&a\Big(bf(s_1,\ldots,s_n)+cf^d(s_1,\ldots,s_n)\Big)f(s_1,\ldots,s_n)g\\&+ac\Big(\sum_{\sigma\in S_n}\alpha(\gamma_\sigma)d(g)s_{\sigma(1)})s_{\sigma(2)}\ldots s_{\sigma(n)}\Big)f(s_1,\ldots,s_n)g=0\end{aligned} \tag{44}$$

for any $s_1, \dots, s_n \in gR$. Since $ac\alpha(g) = 0$ and left multiplying (44) by w', it follows that gR satisfies

$$\left\{(w'ab)f(x_1, \dots, x_n)^2 + (w'ac)d\Big(f(x_1, \dots, x_n)\Big)f(x_1, \dots, x_n)\right\}g \qquad (45)$$

Let $\overline{gR} = \frac{gR}{gR \cap l_R(gR)}$ and notice that $\overline{gR}$ is a prime C-algebra. If we define the following map on R

$$F(x) = (w'ab)x + (w'ac)d(x) \quad \forall x \in R$$

it is easy to see that $F(gR) \subseteq gR$. Thus we may introduce

$$\overline{F} : \overline{gR} \to \overline{gR}$$

such that $\overline{F}(\overline{x}) = \overline{F(x)}$, for all $x \in gR$. Clearly $\overline{F}$ is a X-generalized skew derivation of $\overline{gR}$ and, in light of (45),

$$\left\{\overline{F}\Big(f(x_1, \dots, x_n)\Big)f(x_1, \dots, x_n)\right\}g$$

is an identity for $\overline{gR}$. In particular

$$\overline{F}\Big(\overline{f(x_1, \dots, x_n)}\Big)\overline{f(x_1, \dots, x_n)} = \overline{0}. \qquad (46)$$

Application of Corollary 1 implies that one of the following holds:

- $[f(x_1, \dots, x_n), x_{n+1}]x_{n+2}$ is an identity for gR, which gives a contradiction, since

$$[f(ga_1, \dots, ga_n), ga_{n+1}]ga_{n+2} \neq 0$$

- $w'acg = 0$, which is again a contradiction, since $0 \neq w'acd(w) = w'acd(gw) = w'acgd(w')$.

Case 2. In all that follows we assume there is $u \in I$ such that $ac\alpha(u) \neq 0$

Starting from (41) Q satisfies

$$\left[ac\Big(\sum_{\sigma \in S_n} \alpha(\gamma_\sigma) \sum_{j=0}^{n-1} \alpha(ux_{\sigma(1)} \dots ux_{\sigma(j)})\alpha(u)y_{\sigma(j+1)}ux_{\sigma(j+2)} \dots ux_{\sigma(n)}\Big)f(ux_1, \dots, ux_n), x_{n+1}\right]. \qquad (47)$$

Since $0 \neq ac\alpha(u)$, (47) is a non-trivial generalized polynomial identity for Q, then Q has nonzero socle H which satisfies the same generalized polynomial identities of Q. In order to prove our result, we may replace Q by H and assume that Q is a regular ring. Thus there exists $0 \neq e = e^2 \in IQ$ such that $\sum_{i=1}^{n+1} a_i Q + uQ = eQ$, $u = eu$ and $a_i = ea_i$ for each $i = 1, \ldots, n+1$. As above, we may also assume $e \neq 1$, if not the conclusion follows from Theorem 10.

Assume that α is X-outer. Thus, by (47) it follows that Q satisfies

$$\left[ac\Big(\sum_{\sigma \in S_n} \alpha(\gamma_\sigma) \sum_{j=0}^{n-1} \alpha(e)z_{\sigma(1)} \ldots \alpha(e)z_{\sigma(j)})\alpha(e)y_{\sigma(j+1)}ex_{\sigma(j+2)} \ldots ex_{\sigma(n)}\Big) f(ex_1, \ldots, ex_n), x_{n+1}\right]. \tag{48}$$

In particular, by (48) Q satisfies

$$\left[ac\Big(\sum_{\sigma \in S_n} \alpha(\gamma_\sigma)\alpha(e)y_{\sigma(1)} \cdots \alpha(e)y_{\sigma(n)}\Big) f(ex_1, \ldots, ex_n), x_{n+1}\right]. \tag{49}$$

Thus

$$\left[acf^\alpha(\alpha(e)r_1, \ldots, \alpha(e)r_n) f(es_1e, \ldots, es_ne), (1-e)r_{n+1}\right] = 0 \tag{50}$$

for all $r_1, \ldots, r_n, s_1, \ldots, s_n, r_{n+1} \in Q$. Hence we get

$$(1-e)r_{n+1}acf^\alpha(\alpha(e)r_1, \ldots, \alpha(e)r_n) f(es_1e, \ldots, es_ne) = 0$$

and, by the primeness of Q and since $e \neq 1$, it follows that Q satisfies

$$acf^\alpha(\alpha(e)r_1, \ldots, \alpha(e)r_n) f(es_1e, \ldots, es_ne) = 0$$

Since $f(ea_1, \ldots, ea_n)ea_{n+1} \neq 0$ and by Fact 5, it follows that $acf^\alpha(\alpha(e)y_1, \ldots, \alpha(e)y_n)$ is a generalized identity for Q. Moreover, by [18] and since $ac\alpha(u) \neq 0$, Q satisfies $f^\alpha(\alpha(e)y_1, \ldots, \alpha(e)y_n)$. As in the proof of Proposition 5, this implies that $f(ex_1, \ldots, ex_n)$ is an identity for Q, a contradiction. Finally consider the case when there exists an invertible element $q \in Q$ such that $\alpha(x) = qxq^{-1}$, for all $x \in Q$. Thus from (47) we have that Q satisfies

$$\left[ac\Big(\sum_{\sigma \in S_n} \alpha(\gamma_\sigma) \sum_{j=0}^{n-1} q(ex_{\sigma(1)} \cdots ex_{\sigma(j)})eq^{-1}y_{\sigma(j+1)}ex_{\sigma(j+2)} \cdots ex_{\sigma(n)}\Big) f(ex_1, \ldots, ex_n), x_{n+1}\right]. \tag{51}$$

Since $\alpha(\gamma_\sigma) = \gamma_\sigma$ and by replacing $y_{\sigma(i)}$ with $qx_{\sigma(i)}$, for all $\sigma \in S_n$ and for all $i = 1, \dots, n$, it follows that Q satisfies

$$\left[acq\left(\sum_{\sigma\in S_n} \gamma_\sigma ex_{\sigma(1)} \cdots ex_{\sigma(j)}ex_{\sigma(j+1)}ex_{\sigma(j+2)} \cdots ex_{\sigma(n)}\right) f(ex_1, \dots, ex_n), x_{n+1}\right] \tag{52}$$

that is

$$\left[acqf(ex_1, \dots, ex_n)^2, x_{n+1}\right]. \tag{53}$$

By Lemma 4, it follows that one of the following holds:

(i) $f(x_1, \dots, x_n)x_{n+1}$ is an identity for eQ, which contradicts $[f(a_1, \dots, a_n), a_{n+1}]a_{n+2} \neq 0$;
(ii) $acqe = 0$, which is again a contradiction, since $0 \neq acq(u)q^{-1} = acq(eu)q^{-1}$;
(iii) $acq \in C$ and $f(x_1, \dots, x_n)^2$ is central valued on Q.

In order to complete our proof, we have to discuss only the last case (iii).

Since $0 \neq acq \in C$ and by replacing in (51) $y_{\sigma(i)}$ with $qy_{\sigma(i)}$, for all $\sigma \in S_n$ and for all $i = 1, \dots, n$, it follows that Q satisfies

$$\left[\left(\sum_{\sigma\in S_n} \gamma_\sigma \sum_{j=0}^{n-1} ex_{\sigma(1)} \cdots ex_{\sigma(j)}ey_{\sigma(j+1)}ex_{\sigma(j+2)} \cdots ex_{\sigma(n)}\right) f(ex_1, \dots, ex_n), x_{n+1}\right]$$

and, in particular, for any $i = 1, \dots, n$, Q satisfies

$$\left[\left(\sum_{\sigma\in S_n} \gamma_\sigma ex_{\sigma(1)} \cdots ey_{\sigma(i)} \cdots ex_{\sigma(n)}\right) f(ex_1, \dots, ex_n), x_{n+1}\right]. \tag{54}$$

For $y_{\sigma(i)} = 0$ when $i \neq 1$, (54) leads to the identity

$$\left[ey_1\left(\sum_{\sigma\in S_{n-1}} \gamma_\sigma ex_{\sigma(2)} \cdots ex_{\sigma(n)}\right) f(ex_1, \dots, ex_n), x_{n+1}\right]. \tag{55}$$

We denote $\sum_{\sigma\in S_{n-1}} \gamma_\sigma)x_{\sigma(2)} \cdots x_{\sigma(n)} = t_1(x_2, \dots, x_n)$, then Q satisfies

$$\left[ey_1et_1(ex_2, \dots, ex_n)f(ex_1, \dots, ex_n), x_{n+1}\right]. \tag{56}$$

In particular,

$$\begin{aligned} 0 &= \left[ey_1et_1(ex_2, \dots, ex_n)f(ex_1, \dots, ex_n), x_{n+1}(1-e)\right] \\ &= ey_1et_1(ex_2, \dots, ex_n)f(ex_1, \dots, ex_n)x_{n+1}(1-e) \end{aligned} \tag{57}$$

and, by the primeness of Q,

$$t_1(ex_2, \ldots, ex_n)f(ex_1, \ldots, ex_n) = 0.$$

Repeating the same above process, for any $i = 1, \ldots, n$ we arrive at

$$t_i(ex_1, \ldots, ex_{i-1}, ex_{i+1}, \ldots, ex_n)f(ex_1, \ldots, ex_n) = 0 \quad \forall i \geq 1. \tag{58}$$

Finally notice that

$$f(x_1, \ldots, x_n) = \sum_j x_j t_j(x_1, \ldots, x_{j-1}, x_{j+1}, \ldots, x_n),$$

where any t_j is a multilinear polynomial of degree $n-1$ and x_j never appears in any monomial of t_j. This remark and relation (58) lead to $f(ex_1, \ldots, ex_n)^2 = 0$. By [18], we conclude that $f(ex_1, \ldots, ex_n)e$ must be an identity for Q, which is a contradiction again. □

References

1. Albas, E., Argac, N., De Filippis, V., Demir, C.: Generalized Skew Derivations on Multilinear Polynomials in Right Ideals of Prime Rings. Hacettepe J. Math. Statistic. **43**(1), 69–83 (2014)
2. Argaç, N., Demir, Ç.: Generalized derivations of prime rings on multilinear polynomials with annihilator conditions. Turkish J. Math. **37**, 231–243 (2013)
3. Beidar, K.I., Martindale III, W.S., Mikhalev, A.V.: Rings with generalized identities. Monogr. Textbooks Pure Appl. Math. **196** (1996)
4. Bresar, M.: One-sided ideals and derivations of prime rings. Proc. Am. Math. Soc. **122**, 979–983 (1994)
5. Carini, L., De Filippis, V.: Centralizers of generalized derivations on multilinear polynomials in prime rings. Siberian Math. J. **53**(6), 1051–1060 (2012)
6. Carini, L., De Filippis, V., Scudo, G.: Identities with product of generalized skew derivations on multilinear polynomials. Commun. Algebra **44**(7), 3122–3138 (2016)
7. Chang, C.M.: Power central values of derivations on multilinear polynomials Taiwanese. J. Math. **7**(2), 329–338 (2003)
8. Chang, C.M., Lee, T.-K.: Additive subgroups generated by polynomial values on right ideals. Commun. Algebra **29**(7), 2977–2984 (2001)
9. Chang, J.C.: On the identitity $h(x) = af(x) + g(x)b$. Taiwanese J. Math. **7**, 103–113 (2003)
10. Chang, J.C.: Generalized skew derivations with annihilating Engel conditions. Taiwanese J. Math. **12**, 1641–1650 (2008)
11. Chang, J.C.: Right generalized (α, β)-derivations having power central values. Taiwanese J. Math. **13**, 1111–1120 (2009)
12. Chang, J.C.: Generalized skew derivations with nilpotent values on Lie ideals. Monatsh. Math. **161**, 155–160 (2010)
13. Cheng, H.-W., Wei, F.: Generalized skew derivations of rings. Adv. Math. (China) **35**, 237–243 (2006)

14. Chou, M.-C., Liu, C.-K.: An Engel condition with skew derivations. Monatsh. Math. **158**, 259–270 (2009)
15. Chuang, C.L.: GPIs having coefficients in Utumi quotient rings. Proc. Am. Math. Soc. **103**, 723–728 (1988)
16. Chuang, C.-L.: Differential identities with automorphisms and antiautomorphisms I. J. Algebra **149**, 371–404 (1992)
17. Chuang, C.L.: Differential identities with automorphisms and antiautomorphisms. J. Algebra **160**, 130–171 (1993)
18. Chuang, C.L., Lee, T.K.: Rings with annihilator conditions on multilinear polynomials. Chinese J. Math. **24**, 177–185 (1996)
19. Chuang, C.L., Lee, T.K.: Identities with a single skew derivation. J. Algebra **288**, 59–77 (2005)
20. De Filippis, V.: A product of two generalized derivations on polynomials in prime rings. Collect. Math. **61**, 303–322 (2010)
21. De Filippis, V.: Annihilators and power values of generalized skew derivations on Lie ideals. Canadian Math. Bull. **59**(2), 258–270 (2016)
22. De Filippis, V., Dhara, B.: Generalized derivations on right ideals and polynomials in prime rings. Southeast Asian Bull. Math. **38**, 341–357 (2014)
23. De Filippis, V., Wei, F.: b-generalized (α, β)-derivations and b-generalized (α, β)-biderivations of prime rings. Taiwanese J. Math. **22**(2), 313–323 (2018)
24. De Filippis, V., Wei, F.: An Engel condition with X-Generalized Skew Derivations on Lie ideals. Commun. Algebra **46**(12), 5433–5446 (2018)
25. De Filippis, V., Wei, F.: b-generalized skew derivations on Lie ideals, Mediterr. J. Math. **15/2** (2018). https://doi.org/10.1007/s00009-018-1103-2
26. Demir, Ç., Argaç, N.: Prime rings with generalized derivations on right ideals. Algebra Colloq. **18**, 987–998 (2011)
27. Dhara, B., Argaç, N.: Generalized derivations acting on multilinear polynomials in prime rings and Banach algebras. Commun. Math. Stat. **4**, 39–54 (2016)
28. Dhara, B., Kar, S., Pradhan, K.G.: Co-centralizing generalized derivations acting on multilinear polynomials in prime rings. Bull. Iranian Math. Soc. **42**, 1331–1342 (2016)
29. Faith, C., Utumi, Y.: On a new proof of Litoff's theorem. Acta Math. Acad. Sci. Hung. **14**, 369–371 (1963)
30. Jacobson, N.: Structure of Rings. American Mathematical Society, Providence, RI (1964)
31. Koşan, M.T., Lee, T.K.: b-generalized derivations of semiprime rings having nilpotent values. J. Austral. Math. Soc. **96**, 326–337 (2014)
32. Lanski, C.: An Engel condition with derivation. Proc. Am. Math. Soc. **118**, 731–734 (1993)
33. Lee, T.K.: Generalized derivations of left faithful rings. Commun. Algebra **27**(8), 4057–4073 (1999)
34. Lee, T.K.: Derivations with invertible values on a multilinear polynomial. Proc. Am. Math. Soc. **119**, 1–5 (1993)
35. Lee, T.K.: Power reduction property for generalized identities of onesided ideals. Algebra Colloq. **3**, 19–24 (1996)
36. Lee T.K., Shiue, W.K.: Derivations cocentralizing polynomials .Taiwanese J. Math. **2**, 457–467 (1998)
37. Lee, T.K., Shiue, W.K.: Identities with generalized derivations. Commun. Algebra **29**(10), 4435–4450 (2001)
38. Lee, T.K., Zhou, Y.: An identity with generalized derivations. J. Algebra Appl. **8**(3), 307–317 (2009)
39. Leron, U.: Nil and power central polynomials in rings. Trans. Am. Math. Soc. **202**, 97–103 (1975)

40. Martindale, W.S., III.: Prime rings satisfying a generalized polynomial identity. J. Algebra **12**, 576–584 (1969)
41. Wong, T.L.: Derivations with power central values on multilinear polynomials. Algebra Colloq. **3**, 369–378 (1996)

Commutative Polynomial Rings which are Principal Ideal Rings

Henry Chimal-Dzul

Abstract A well-known result by Zariski and Samuel asserts that a commutative principal ideal ring is a direct sum of finitely many principal ideal domains and Artinian chain rings. Based on this result, it is shown, among other things, that a commutative polynomial ring $R[x]$ is a principal ideal ring if and only if R is a finite direct sum of fields.

Keywords Principal ideal ring · Polynomial ring · Principal ideal domain · Artinian chain ring · Bézout domain

1 Introduction

A first exposure to the theory of rings almost certainly involves a study of various examples of principal ideal rings. The most common examples are the ring of integers $\mathbb{Z}$ and the polynomial ring $K[x]$ with coefficients in a field K. These are also examples of Euclidean domains. In general, it is well known that Euclidean domains are principal ideal rings and that there are principal ideal rings which are not Euclidean domains (see [4] and [3, Example 3.79] for more details). However, even with these results in hand, more than likely $K[x]$ is the only example that we would have come across of a commutative polynomial ring that is a principal ideal ring. This remark brings up the discussion of when a commutative polynomial ring is a principal ideal ring.

The main goal of this paper is to characterize all commutative polynomial rings $R[x]$ which are principal ideal rings. We prove that $R[x]$ is a principal ideal ring if and only if R is a finite direct sum of fields. This result shows that in the commutative case, polynomial rings with coefficients in a field are the building blocks of polynomial

H. Chimal-Dzul (✉)
Institute of Mathematics, University of Zurich, Winterthurerstrasse 190, CH-8057 Zurich, Switzerland
e-mail: henry.chimal-dzul@math.uzh.ch

M. Ashraf et al. (eds.), *Algebra and Related Topics with Applications*,
Springer Proceedings in Mathematics & Statistics 392,
https://doi.org/10.1007/978-981-19-3898-6_7

rings that are principal ideal rings, a fact that has been proved in [1, Theorem 2.3] for finite commutative rings only. Thus, the main result in this paper, Theorem 4, generalizes [1, Theorem 2.3] to arbitrary commutative rings.

In the background of the main contribution of this paper stands a more general result by Zariski and Samuel [5], which establishes that a commutative ring is a principal ideal ring if and only if it is a direct sum of finitely many principal ideal domains and Artinian chain rings. For the reader's convenience, this background material is presented in Sect. 2. Section 3 focuses on the main result of this paper, which will be derived from studying polynomials over integral domains and local rings (Theorems 2 and 3). In view of this study, we obtain that if R is an integral domain or an Artinian local ring that is not a field, then $R[x]$ is not a PIR. In Sect. 4, we show how to construct some families of non-principal ideals in a polynomial ring over an integral domain (resp., over an Artinian chain ring) with irreducible elements (resp., with zero divisors). This leads to simple examples of integral domains and unique factorization domains which are not Bézout domains.

2 Structure Theorem of PIRs

Throughout this paper, unless otherwise stated, all rings are assumed to be commutative with identity. Let R be a ring. As often, $R[x]$ denotes the ring of all polynomials in an indeterminate x with coefficients in R; $a_1R + \cdots + a_nR$ the ideal of R generated by $a_1, \ldots, a_n \in R$; and $U(R)$ the group of units of R. The quotient ring $\mathbb{Z}/m\mathbb{Z}$ will be written as $\mathbb{Z}_m$ and its elements will be identified with the integers $0, 1, \ldots, m-1$.

Recall that a ring in which every ideal is principal is called a *principal ideal ring* (PIR). A *principal ideal domain* (PID) is an integral domain that is a PIR. The following facts about PIRs are well-known and will be frequently used in this paper.

Lemma 1 *1. If R is a PIR then R/I is also a PIR for any ideal I of R.*
2. The direct sum $\bigoplus_{i=1}^{n} R_i$ is a PIR if and only if so are all the rings $R_1, \ldots, R_n$.

A local ring is called an *Artinian chain ring* if its maximal ideal is principal and generated by a nilpotent element. It can be proved that an Artinian chain ring $\mathscr{A}$ with maximal ideal $\theta\mathscr{A}$ is a PIR and that every proper ideal of $\mathscr{A}$ is of the form $\theta^i\mathscr{A}$ for some $i \geq 1$. One may view fields as Artinian chain rings having zero maximal ideal. Artinian chain rings are also referred to as *special PIRs* in [5, Sect. 15, p. 242].

Theorem 1 ([5, Theorem 33, p. 245]) *Every PIR is (isomorphic) to a direct sum of finitely many PIDs and Artinian chain rings.*

3 Characterization of Polynomial Rings that Are PIRs

In this section, we give necessary and sufficient conditions under which $R[x]$ is a PIR. Although, for our purposes, it is sufficient to study polynomials over PIDs and Artinian chain rings, we extend our analysis to polynomials over integral domains and local rings.

We start regarding the case when R is an integral domain. First, recall from [3, Sect. 7.2] that every PID is a unique factorization domain (UFD), every polynomial ring over a UFD is also a UFD, and that prime and maximal ideals coincide in a UFD.

Theorem 2 *For an integral domain $\mathscr{D}$, $\mathscr{D}[x]$ is a PIR if and only if $\mathscr{D}$ is a field.*

Proof By Lemma 1, if $\mathscr{D}[x]$ is a PIR then $\mathscr{D} \cong \mathscr{D}[x]/x\mathscr{D}[x]$ is a PIR. Indeed, note that $\mathscr{D}$ is a PID. Thus, $\mathscr{D}$ is a UFD, where it follows that $\mathscr{D}[x]$ is a UFD. Consider the map $\varphi : \mathscr{D}[x] \to \mathscr{D}$ defined by $\varphi\Big(\sum_{i=0}^{n} f_i x^i\Big) = f_0$. Then φ is a ring epimorphism. Hence $\mathscr{D}[x]/\ker\varphi \cong \mathscr{D}$, and so $\ker\varphi$ is a prime ideal of $\mathscr{D}[x]$. Since $\mathscr{D}[x]$ is a UFD, $\ker\varphi$ is maximal. Therefore $\mathscr{D}$ is a field. The converse is clear.

Let $\mathscr{A}$ be an Artinian chain ring with maximal ideal $\theta\mathscr{A}$ and $K = \mathscr{A}/\theta\mathscr{A}$ (the *residue field* of $\mathscr{A}$). Note that the natural ring epimorphism $\bar{\ } : \mathscr{A} \to K$, defined as $a \mapsto \bar{a} = a + \theta\mathscr{A}$, induces a polynomial ring epimorphism $\Phi : \mathscr{A}[x] \to K[x]$ given by

$$\Phi\Big(\sum_{i=0}^{n} f_i x^i\Big) = \sum_{i=0}^{n} \overline{f}_i x^i. \tag{1}$$

The kernel of Φ is

$$\ker\Phi = \left\{\sum_{i=0}^{n} f_i x^i \in \mathscr{A}[x] \;:\; f_i \in \theta\mathscr{A},\ 0 \le i \le n,\ n \in \mathbb{N}\right\} = \theta\mathscr{A}[x].$$

If e is the nilpotency index of θ, then $(\ker\Phi)^e = (\theta\mathscr{A}[x])^e = \theta^e\mathscr{A}[x] = 0$. Thus $\ker\Phi$ is a nilpotent ideal of $\mathscr{A}[x]$, and so every element of $\ker\Phi$ is nilpotent.

Lemma 2 *Let $\mathscr{A}$ be an Artinian chain ring with residue field K. The group of units of $\mathscr{A}[x]$ is*

$$U(\mathscr{A}[x]) = \{u + f \;:\; u \in U(\mathscr{A}),\ f \in \ker\Phi\}.$$

Moreover, $\Phi(f) \in K[x]$ is a unit if and only if $f \in U(\mathscr{A}[x])$.

Proof First observe that for every $f \in \ker\Phi$, $1 + f \in U(\mathscr{A}[x])$ because $1 = 1 + f^n = (1 + f)(1 - f + f^2 + \cdots + (-1)^{n-1} f^{n-1})$. Thus, for any $u \in U(\mathscr{A})$ and $f \in \ker\Phi$, $u + f \in U(\mathscr{A}[x])$ because $u + f = u(1 + u^{-1} f)$ and $u^{-1} f \in \ker\Phi$. Consequently, $\{u + f \;:\; u \in U(\mathscr{A}),\ f \in \ker\Phi\} \subseteq U(\mathscr{A}[x])$. To prove the reverse inclusion, let $a \in U(\mathscr{A}[x])$. Then $\Phi(a) \in U(K[x])$. An element $g \in K[x]$ is a unit if and only if $0 \neq g \in K$. Thus $0 \neq \Phi(a) \in K$, where it follows that $a = u + f$, where $u \in U(\mathscr{A})$ and $f \in \ker \Phi$.

Theorem 3 *For a local ring R, $R[x]$ is a PIR if and only if R is a field.*

Proof Assume that $R[x]$ is a PIR. Then $R \cong R[x]/xR[x]$ is a PIR by Lemma 1. Hence R is a direct sum of finitely many PIDs and Artinian chain rings. Since R is local, R is not a direct sum of two or more rings. Hence, R is either a PID or an Artinian chain ring. If R is a PID then Theorem 2 implies that R is a field. Thus, assume that R is an Artinian chain ring with residue field K. Let I be the preimage of the ideal $xK[x]$ of $K[x]$ under the ring epimorphism Φ defined in (1). Let $\Psi : R[x] \to K[x]/xK[x]$ be the map given by $a \mapsto \Phi(a) + xK[x]$. Then Ψ is a ring epimorphism, and so Ψ induces a ring isomorphism $\widetilde{\Psi} : R[x]/\ker\Psi \to K[x]/xK[x]$. Since $K[x]/xK[x] \cong K$, it follows that $\ker\Psi$ is a maximal ideal of $R[x]$. Because $R[x]$ is a PIR, there is $f \in R[x]$ such that $\ker\Psi = fR[x]$. Since $x \in \ker\Psi$, then $x = fg$ for some $g \in R[x]$. Hence $x = \Phi(x) = \Phi(f)\Phi(g) \in K[x]$. The polynomial x is irreducible in $K[x]$, so either $\Phi(f)$ or $\Phi(g)$ is unit in $K[x]$. Since $\ker\Psi$ is maximal, $\Phi(f)$ is not a unit in $K[x]$. It follows that $\Phi(g)$ is a unit in $K[x]$. By Lemma 2, $g \in U(R[x])$. Therefore $\ker\Psi = xR[x]$, and so $R \cong R[x]/\ker\Psi \cong K$, where we obtain that R is a field. The converse is simple.

We are now in a position to prove the main result of this paper.

Theorem 4 *The polynomial ring $R[x]$ is a PIR if and only if R is a finite direct sum of fields.*

Proof Assume $R[x]$ is a PIR. By Lemma 1, $R \cong R[x]/xR[x]$ is a PIR. Thus, in light of Theorem 1, R is a direct sum of finitely many PIDs and Artinian chain rings, say $R \cong \mathscr{D}_1 \oplus \cdots \oplus \mathscr{D}_n \oplus \mathscr{A}_1 \oplus \cdots \oplus \mathscr{A}_m$, where each $\mathscr{D}_i$ is a PID and every $\mathscr{A}_j$ is an Artinian chain ring. This decomposition extends to a polynomial ring isomorphism $R[x] \cong \mathscr{D}_1[x] \cdots \oplus \mathscr{D}_n[x] \oplus \mathscr{A}_1[x] \oplus \cdots \oplus \mathscr{A}_m[x]$. Lemma 1 implies that for all i, j, $\mathscr{D}_i[x]$ and $\mathscr{A}_j[x]$ are PIRs. By Theorem 2, $\mathscr{D}_i[x]$ is a PIR if and only if $\mathscr{D}_i$ is a field. Likewise, by Theorem 3, $\mathscr{A}_j[x]$ is a PIR if and only if $\mathscr{A}_j$ is a field. Hence R is a finite direct sum of fields. The converse is evident.

Theorem 4 generalizes [1, Theorem 2.3], where it was shown that for a finite (commutative ring) R, the polynomial ring $R[x]$ is a PIR if and only if R is a direct sum of finitely many finite fields. Some examples of finite rings R are analyzed in [1] to determine whether $R[x]$ is a PIR or not.

The following result is an instance of Theorem 4 and will allow us to provide some examples of polynomial rings that are PIRs (with either finitely many or infinitely many elements).

Proposition 1 *Let $\mathscr{D}$ be a PID, $0 \neq d \in \mathscr{D}$ a non-unit, and $R = \mathscr{D}/d\mathscr{D}$. Then $R[x]$ is a PIR if and only if d is a product of distinct irreducible elements in $\mathscr{D}$.*

Proof Since $d \neq 0$ is a non-unit and $\mathscr{D}$ is a PID (and so a UFD), d can be expressed as a product of irreducible elements in $\mathscr{D}$, say $d = p_1^{\alpha_1} \cdots p_n^{\alpha_n}$. Notice that the ideals $p_i^{\alpha_i}\mathscr{D}$ and $p_j^{\alpha_j}\mathscr{D}$ are comaximal whenever $i \neq j$. Therefore, by the Chinese Remainder Theorem (see [3, Exercise 7.14]), we obtain $R = \mathscr{D}/d\mathscr{D} \cong \mathscr{D}/p_1^{\alpha_1}\mathscr{D} \oplus \cdots \oplus \mathscr{D}/p_n^{\alpha_n}\mathscr{D}$. Let $\mathscr{A}_i = \mathscr{D}/p_i^{\alpha_i}\mathscr{D}$. Then $\mathscr{A}_i$ is an Artinian chain ring with maximal ideal $\overline{p_i}\mathscr{A}_i$. The nilpotency index of $\overline{p_i}$ is α_i. As in the proof of Theorem 4, $R[x]$ is a PIR if and only if $\mathscr{A}_i$ is a field for all $1 \leq i \leq n$. But $\mathscr{A}_i$ is a field if and only if $\alpha_i = 1$. Thus $R[x]$ is a PIR if and only if d is a product of distinct irreducible elements in $\mathscr{D}$.

Remark 1 One may suspect that Propositional 1 can be stated for a UFD instead of a PID. However, the fact that in a UFD irreducible elements could give rise to prime ideals that are not maximal erases this possibility. For example, $\mathbb{Z}[t]$ is a UFD and $f = t$ is irreducible in $\mathbb{Z}[t]$. But $(\mathbb{Z}[t]/t\mathbb{Z}[t])[x] \cong \mathbb{Z}[x]$ is not a PIR. Note that $t\mathbb{Z}[t]$ is a prime ideal in $\mathbb{Z}[t]$ but not maximal.

Corollary 1 *The ring $\mathbb{Z}_m[x]$ is a PIR if and only if m is square-free.*

The previous corollary has been derived in [1] using the characterization of finite rings R for which $R[x]$ is a PIR (see [1, Theorem 2.3 and Sect. 3]). Here we have recovered Corollary 1 from Proposition 1 by taking $\mathscr{D} = \mathbb{Z}$ and $d = m$. Thus Proposition 1 could be considered as a generalization of the remarks in [1].

Corollary 2 *Let K be a field, $f \in K[t]$ and $R = K[t]/fK[t]$. Then $R[x]$ is a PIR if and only if f is a product of distinct irreducible polynomials in $K[t]$.*

For a ring R and a polynomial $f = a_0 + a_1 t + \cdots + a_n t^n \in R[t]$, the *formal derivative* of f is defined to be $f' = a_1 + 2a_2 t + \cdots + n a_n t^{n-1}$. If $R = K$ is a field, then f is a product of distinct irreducible factors provided that f and f' are relatively prime (see [3, Exercise 3.34]).

Example 1 Let $R = \mathbb{Q}[t]/f\mathbb{Q}[t]$, where $f = 2t^4 - 3t^2 + 2t + 4$. We would like to determine whether $R[x]$ is a PIR. The formal derivative of $f = 2t^4 - 3t^2 + 2t + 4$ is $f' = 8t^3 - 6t + 2$, which factorizes as $f' = 2(t+1)(2t-1)^2$. Now f' and f are relatively prime because neither $t + 1$ nor $2t - 1$ divides f. Hence f is a product of distinct irreducible factors in $\mathbb{Q}[t]$. Therefore, by Corollary 2, $R[x]$ is a PIR.

4 Non-principal Ideals in Some Polynomial Rings

In light of Theorem 2, if R is an integral domain that is not a field, then $R[x]$ is not a PIR. Likewise, if R is a local ring with zero divisors then $R[x]$ has at least one non-principal ideal by Theorem 3. In this section, we construct some families of non-principal ideals in $R[x]$ in some of these cases.

Proposition 2 *Let R be an integral domain with at least one irreducible element p, and $f \in R[x]$ of positive degree. If p does not divide f then $I = pR[x] + xfR[x]$ is a non-principal ideal of $R[x]$.*

Proof First, observe that I contains only polynomials in $R[x]$ whose constant term is divisible by p. Since p is irreducible, $1 \notin I$, i.e., I is a proper ideal of $R[x]$. For the sake of contradiction, suppose that $I = gR[x]$ for some $g \in R[x]$. Since $p \in I$ then g divides p. Because p is irreducible in R, then p is also irreducible in $R[x]$. Thus g is a unit or g is an associate of p. Since I is a proper ideal, g is not a unit. Hence g is an associate of p. Consequently, $I = gR[x] = pR[x]$. Because $f \in I = pR[x]$, then p divides f (a contradiction). Therefore I is a non-principal ideal of $R[x]$.

Remark 2 It is important that the assumption on the integral domain R having at least one irreducible element in Proposition 2 is sufficient but not necessary. That is, there are integral domains $\mathscr{D}$ without irreducible elements for which $\mathscr{D}[x]$ is not a PIR. An example of such domains is the *ring of all algebraic integers*:

$$\mathscr{O} = \{z \in \mathbb{C} \mid f(z) = 0 \text{ for some monic polynomial } f(x) \in \mathbb{Z}[x]\}.$$

There are no irreducible elements in $\mathscr{O}$ because any nonzero non-unit $\alpha \in \mathscr{O}$ can be written as $\alpha = \sqrt{\alpha}\sqrt{\alpha}$. Note that $\sqrt{\alpha} \in \mathscr{O}$ since if $f(x) = \sum_{i=0}^{n} a_i x^i$ is such that $f(\alpha) = 0$ then $\sqrt{\alpha}$ is a root of $g(x) = \sum_{i=0}^{n} a_i x^{2i} \in \mathbb{Z}[x]$. Lastly, we observe that $\mathscr{O}[x]$ is not a PIR. Indeed, if $\mathscr{O}[x]$ is a PIR, then from Lemma 1 we obtain that $\mathscr{O}$ is a PIR, and so a Noetherian domain, i.e., it satisfies the Ascending Chain Condition on ideals (see [3, Sect. 7.3] for more details). However, $\mathscr{O}$ is not Noetherian as, for example,

$$3\mathscr{O} \subseteq 3^{1/2}\mathscr{O} \subseteq 3^{1/4}\mathscr{O} \subseteq \cdots \subseteq 3^{1/2^n}\mathscr{O} \subseteq \cdots$$

is an strictly ascending chain of principal ideals of $\mathscr{O}$.

Example 2 Let $\mathbb{Z}[i] = \{a + bi \mid a, b \in \mathbb{Z}\}$ be the ring of Gaussian integers. The norm of a Gaussian integer $z = a + bi$ is defined as $N(a + bi) = a^2 + b^2$. This norm is multiplicative in the sense that $N(zw) = N(z)N(w)$ for all $z, w \in \mathbb{Z}[i]$. Hence, an element $z \in \mathbb{Z}[i]$ is a unit if and only if $N(z) = 1$. Moreover, if $N(z)$ is a prime integer, then z is irreducible in $\mathbb{Z}[i]$. Thus, irreducible elements in $\mathbb{Z}[i]$ are abundant. Consequently, for any irreducible $\omega \in \mathbb{Z}[i]$ and $f \in \mathbb{Z}[i][x]$ such that ω does not divide f, the ideal

$$\omega\mathbb{Z}[i][x] + xf\mathbb{Z}[i][x]$$

of $\mathbb{Z}[i][x]$ is non-principal by Proposition 2. As a particular case, we can take $f = 1 + x + \cdots + x^n$ and $\omega = 1 + i$.

Recall that a *Bézout domain* is an integral domain in which the sum of any two principal ideals is principal. Every PID is a Bézout domain. On the other hand, Bézout domains and UFDs are independent classes of rings in the sense that there are examples of Bézout domains that are not UFDs, and vice-versa. In the same vein, Proposition 2 provides simple examples of integral domains and UFDs which are not Bézout domains. Moreover, note that these rings are not Artinian.

We now turn our attention on non-principal ideals in $\mathscr{A}[x]$, where $\mathscr{A}$ is an Artinian chain ring. The proof of Theorem 3 establishes that the preimage of the ideal $xK[x]$ of $K[x]$, under the ring epimorphism $\Phi : \mathscr{A}[x] \to K[x]$ defined in (1), is non-principal. Note that $\Phi^{-1}(xK[x]) = \theta\mathscr{A}[x] + x\mathscr{A}[x]$, where $\theta\mathscr{A}$ is the maximal ideal of $\mathscr{A}$. Thus, the ideal $\theta\mathscr{A}[x] + x\mathscr{A}[x]$ of $\mathscr{A}[x]$ is non-principal. We now generalize this construction.

Proposition 3 *Let $\mathscr{A}$ be an Artinian chain ring with maximal ideal $\theta\mathscr{A} \neq 0$ and residue field K. Let $\Phi : \mathscr{A}[x] \to K[x]$ be the ring homomorphism defined in* (1)*. For any monic irreducible polynomial $f \in \mathscr{A}[x]$ and $0 \neq m \in \theta\mathscr{A}$, the ideal $m\mathscr{A}[x] + f\mathscr{A}[x]$ of $\mathscr{A}[x]$ is non-principal.*

Proof Let $I = m\mathscr{A}[x] + f\mathscr{A}[x]$. If $1 \in I$ then $1 = ma + fb$ for some $a, b \in \mathscr{A}[x]$. Since $m \in \theta\mathscr{A}$, it follows from Lemma 2 that $f \in U(\mathscr{A}[x])$, which is a contradiction. Thus $1 \notin I$, that is, I is a proper ideal of $\mathscr{A}[x]$. For the sake of contradiction, assume that I is principal. So, $I = g\mathscr{A}[x]$ for some $g \in \mathscr{A}[x]$. Since $f \in I$, we can write $f = gh$ for some $h \in \mathscr{A}[x]$. Because f is irreducible, either g or h is a unit. Since I is a proper ideal, necessarily $h \in U(\mathscr{A}[x])$. Hence $I = g\mathscr{A}[x] = fh^{-1}\mathscr{A}[x] = f\mathscr{A}[x]$. Since f is monic irreducible, for any nonzero polynomial $a \in \mathscr{A}[x]$, the product fa has positive degree. This shows that $m \notin I$, which is a contradiction. Therefore I is non-principal.

The previous result generalizes Theorem 2.2 in [1], which states that $R[x]$ is not a PIR whenever R is a finite local ring which is not a field. On the other hand, the construction of non-principal ideals proposed in Proposition 3 depends on identifying monic irreducible polynomials in $\mathscr{A}[x]$, where $\mathscr{A}$ is an Artinian chain ring. The question of whether a polynomial in $\mathscr{A}[x]$ is irreducible is a bit subtle in many ways because of the presence of zero divisors. For instance, in the ring $\mathbb{Z}_4[x]$, the polynomial $x + 2$ can be written as

$$x + 2 = (2x + 1)(2x^2 + x + 2). \tag{2}$$

Although, we cannot conclude yet that $x + 2$ is reducible in $\mathbb{Z}_4[x]$ because $2x + 1 \in U(\mathbb{Z}_4[x])$ (see Lemma 2). We now present a test for irreducibility for polynomials with coefficients in an Artinian chain ring, which will lead us to the conclusion that, for example, $x + 2$ is irreducible in $\mathbb{Z}_4[x]$.

Lemma 3 *Let $\mathscr{A}$ be an Artinian chain ring with residue field K. Let $f \in \mathscr{A}[x]$ and $\Phi : \mathscr{A}[x] \to K[x]$ be the ring homomorphism defined in* (1)*. If $\Phi(f)$ is irreducible in $K[x]$ then f is irreducible in $\mathscr{A}[x]$.*

Proof Assume that $\Phi(f)$ is irreducible in $K[x]$. If $f = ab$ for some $a, b \in \mathscr{A}[x]$, then $\Phi(f) = \Phi(a)\Phi(b)$. So either $\Phi(a)$ or $\Phi(b)$ is a unit in $K[x]$. Hence, in light of Lemma 2, either a or b is a unit in $\mathscr{A}[x]$. It follows that f is irreducible in $\mathscr{A}[x]$.

Remark 3 The converse of Lemma 3 does not hold in general. For example, $f(x) = x^2 + 2 \in \mathbb{Z}_4[x]$ is irreducible but $\Phi(f) = x^2 \in \mathbb{Z}_2[x]$ is reducible. Note that $\mathbb{Z}_2$ is (isomorphic) to the residue field of $\mathbb{Z}_4$.

Irreducible polynomials in $K[x]$, where K is a field, always exist: degree one polynomials are irreducible. Thus, for an Artinian chain ring $\mathscr{A}$, polynomials of the form $f = a + bx$ for which $b \in U(\mathscr{A})$ are irreducible in $\mathscr{A}[x]$. On the other hand, the existence of irreducible polynomials of degree higher than one in $\mathscr{A}[x]$ depends on the residue field K. If K is a finite field, then there exist irreducible polynomials of any degree [2]. But, if K has characteristic zero then there may be irreducible elements up to certain degree. For example, if $R = \mathbb{R}[a]/a^n\mathbb{R}[a]$ then R is an Artinian chain ring with maximal ideal $\overline{a}R$ and residue field $K \cong \mathbb{R}$. In this case, there are only irreducible polynomials of degree up to 2.

Example 3 The ideal $I = 2\mathbb{Z}_4[x] + (2x^2 + x + 2)\mathbb{Z}_4[x]$ of $\mathbb{Z}_4[x]$ is non-principal. To prove this, recall that $2x + 1 \in U(\mathbb{Z}_4[x])$.Thus, in view of (2), I can be written as

$$I = 2\mathbb{Z}_4[x] + (x + 2)\mathbb{Z}_4[x].$$

Therefore, by Proposition 3, I is non-principal.

Example 4 A polynomial f of degree 2 or 3 with coefficients in a field K is irreducible if and only if $f(k) \neq 0$ for all $k \in K$. It follows that $x^3 + x^2 + 1$ and $x^3 + x + 1$ are the only irreducible polynomials of degree 3 in $\mathbb{Z}_2[x]$. Thus, for a fix integer $k \geq 2$, any polynomial $f = x^3 + ax^2 + bx + c \in \mathbb{Z}_{2^k}[x]$ such that $\Phi(f) \in \{x^3 + x^2 + 1, x^3 + x + 1\}$ is irreducible by Lemma 3. Hence, by Proposition 3, the ideal $2^l\mathbb{Z}_{2^k}[x] + f\mathbb{Z}_{2^k}[x]$ of $\mathbb{Z}_{2^k}[x]$ is non-principal for $1 \leq l \leq k - 1$.

Example 5 Let $f = t - 1 \in \mathbb{Q}[t]$, $k \geq 2$ be an integer, and $\mathscr{A} = \mathbb{Q}[t]/f^k\mathbb{Q}[t]$. Since f is irreducible in $\mathbb{Q}[t]$, the ring $\mathscr{A}$ is an Artinian chain ring with maximal ideal $\overline{f}\mathscr{A}$ and residue field $K = \mathscr{A}/\overline{f}\mathscr{A} \cong \mathbb{Q}$. Let $p \geq 2$ be a prime number and $g_p = 1 + x + x^2 + \cdots + x^{p-1} \in \mathscr{A}[x]$. Then $\Phi(g_p) = 1 + x + x^2 + \cdots + x^{p-1} \in \mathbb{Q}[x]$ is the pth cyclotomic polynomial, which is irreducible in $\mathbb{Q}[x]$ by [3, Corollary 3.104]. Thus, by Lemma 3, g_p is irreducible in $\mathscr{A}[x]$. Therefore, the ideal $I_p = \overline{f}\mathscr{A}[x] + g_p\mathscr{A}[x]$ of $\mathscr{A}[x]$ is non-principal for every prime number $p \geq 2$.

Acknowledgements The author gratefully thanks to the anonymous referees for their valuable comments which substantially improved the presentation of the paper.

References

1. Chimul-Dzul, H., Lopez-Andrade, C.A.: When is $R[x]$ a principal ideal ring?. Rev. Integr. Temas Mat. **35**(2), 143–148 (2017)
2. Lidl, R., Niederreiter, H.: Introduction to Finite Fields and Their Applications, 1st edn. Cambridge University Press, MR1294139 (1994)
3. Rotman, J.J.: A First Course in Abstract Algebra, 3rd ed. Pearson (2005)
4. Wilson, J.C.: A principal ideal ring that is not a euclidean ring. Math. Mag. **46**(1), 34–38 (1973)
5. Zariski, O., Samuel, P.: Commutative Algebra. The University Series in Higher Mathematics, vol. I. D Van Nostrand Company Inc, Princeton, NJ (1958)

Two Remarks on Generalized Skew Derivations in Prime Rings

Vincenzo De Filippis and Francesco Rania

Abstract Let R be a prime ring of characteristic different from 2, Q_r its right Martindale quotient ring, F and G two non-zero generalized skew derivations of R, associated with the same automorphism α and commuting with α. In this work we describe all possible forms of F and G in the following two cases: (a) there exist $a, b \in Q_r$ and a non-central Lie ideal L of R such that $aF(x)b = 0$, for all $x \in L$; (b) there exist $a_1, a_2, b_1, b_2 \in Q_r$ such that $a_1F(x)b_1 + a_2G(x)b_2 = 0$, for all $x \in R$.

Let R be a prime ring with center $Z(R)$, Q_r its *right Martindale quotient ring*, C the center of Q_r, usually called *extended centroid* of R (see [1] for more details).

An additive mapping $d: R \longrightarrow R$ is said to be a *derivation* of R if

$$d(xy) = d(x)y + xd(y)$$

for all $x, y \in R$. An additive mapping $F: R \longrightarrow R$ is called a *generalized derivation* of R if there exists a derivation d of R such that

$$F(xy) = F(x)y + xd(y)$$

for all $x, y \in R$.

Let R be an associative ring and α be an automorphism of R. An additive mapping $d: R \longrightarrow R$ is said to be a *skew derivation* of R if

$$d(xy) = d(x)y + \alpha(x)d(y)$$

V. De Filippis (✉)
De Filippis: Department of Engineering, University of Messina, Messina, Italy
e-mail: defilippis@unime.it

F. Rania
DIGES, Magna Graecia University, Catanzaro, Italy
e-mail: raniaf@unicz.it

M. Ashraf et al. (eds.), *Algebra and Related Topics with Applications*,
Springer Proceedings in Mathematics & Statistics 392,
https://doi.org/10.1007/978-981-19-3898-6_8

for all $x, y \in R$. The automorphisms α is called an *associated automorphism* of d. An additive mapping $F: R \longrightarrow R$ is called a *generalized skew derivation* of R if there exists a skew derivation d of R with associated automorphism α such that

$$F(xy) = F(x)y + \alpha(x)d(y)$$

for all $x, y \in R$. In this case, d is called an *associated skew derivation* of F and α is called an *associated automorphism* of F.

In this paper we investigate some generalized differential identities involving generalized skew derivations of a prime ring of characteristic different from 2.

In [2, Theorem 2.1] Brešar describes the form of three derivations d, g, h of a prime ring R satisfying the condition $d(x) = ag(x) + h(x)b$, for any $x \in R$, where $a, b \in R \setminus Z(R)$. As a consequence he also studies the case when $ag(x) + h(x)b = 0$, for any $x \in R$ [2, Corollary 2.4]. More precisely, in this last case he concludes that there exists $\lambda \in C$ such that $g(x) = [\lambda b, x]$ and $h(x) = [\lambda a, x]$, for any $x \in R$. The results by Brešar extend a theorem of Herstein contained in [12].

Following this line of investigation, J.-C. Chang generalizes the previous results to the case of both skew derivation (see [3]) and generalized skew derivations (see [4]).

Here we would like to continue the study of linear differential identities having the same flavor of the above-cited ones, and involving generalized skew derivations. In this sense, the main goal of the present paper is to prove the following theorems:

Theorem 1 *Let R be a prime ring of characteristic different from 2, F a non-zero generalized skew derivation of R, with associated automorphism α, and a, b non-zero elements of Q_r such that*

$$aF(w)b = 0 \quad \forall w \in L.$$

Then one of the following holds:

(a) the associated automorphism α is not inner and there exist $c, u \in Q_r$ be such that $F(x) = cx + \alpha(x)u$, for any $x \in R$, with $ac = ub = 0$;
(b) there exist $c, u, q \in Q_r$ and $\lambda \in C$ such that $F(x) = cx + \alpha(x)u$, for any $x \in R$, where $\alpha(x) = qxq^{-1}$, for any $x \in R$, with $a(c + \lambda q) = 0$ and $(\lambda + q^{-1}u)b = 0$.

Theorem 2 *Let R be a prime ring of characteristic different from 2, F, G two non-zero generalized skew derivations of R, associated with the same automorphism α and commuting with α. Let a_1, a_2, b_1, b_2 be non-zero elements of Q_r such that*

$$a_1F(x)b_1 + a_2G(x)b_2 = 0 \quad \forall x \in R.$$

Then one of the following cases must occur

(a) There exist $p, u, v, w, q \in Q_r$, where q is an invertible element, such that $F(x) = px + qxq^{-1}u$, $G(x) = vx + qxq^{-1}w$, for any $x \in R$, and one of the following holds:

1. *there exist* $\alpha_1, \alpha_2, \alpha_3, \alpha_4 \in C$ *such that* $b_1 = \alpha_1 b_2 + \alpha_2 q^{-1} w b_2$, $q^{-1} u b_1 = \alpha_3 b_2 + \alpha_4 q^{-1} w b_2$ *and* $\alpha_1 a_1 p + \alpha_3 a_1 q + a_2 v = \alpha_2 a_1 p + \alpha_4 a_1 q + a_2 q = 0$;
2. *there exist* $\lambda, \alpha_1, \alpha_2, \alpha_3, \alpha_4 \in C$ *such that* $q^{-1} w b_2 = \lambda b_2$, $b_1 = (\alpha_1 + \lambda\alpha_2) b_2$, $q^{-1} u b_1 = (\alpha_3 + \lambda\alpha_4) b_2$ *and* $(\alpha_1 + \lambda\alpha_2) a_1 p + (\alpha_3 + \lambda\alpha_4) a_1 q + a_2(v + \lambda q) = 0$;
3. *there exist* $0 \neq \lambda \in C$ *and* $\beta_1, \beta_2 \in C$ *such that* $a_1 p = \lambda a_1 q$, $a_2 v = \beta_1 a_1 q$, $a_2 q = \beta_2 a_1 q$ *and* $\lambda b_1 + q^{-1} u b_1 + \beta_1 b_2 + \beta_2 q^{-1} w b_2 = 0$;
4. *there exist* $0 \neq \lambda \in C$ *and* $\mu, \eta \in C$ *such that* $a_1 p = \lambda a_1 q$, $a_2(v + \mu q) = \eta a_1 q$, $(\lambda + q^{-1} u) b_1 = -\eta b_2$ *and* $q^{-1} w b_2 = \mu b_2$.

(b) *There exist* $p, u, v, w \in Q_r$ *such that* $F(x) = px + \alpha(x) u$, $G(x) = vx + \alpha(x) w$, *for any* $x \in R$, *and one of the following holds:*

5. $a_1 p = a_2 v = u b_1 = w b_2 = 0$;
6. $a_1 p = a_2 v = 0$ *and there exists* $\mu \in C$ *such that* $u b_1 = \mu w b_2$ *and* $a_2 = -\mu a_1$;
7. $u b_1 = w b_2 = 0$ *and there exists* $\lambda \in C$ *such that* $a_1 p = \lambda a_2 v$ *and* $b_2 = -\lambda b_1$;
8. *there exist* $\lambda, \mu \in C$ *such that* $a_1 p = \lambda a_2 v$, $b_2 = -\lambda b_1$, $u b_1 = \mu w b_2$ *and* $a_2 = -\mu a_1$.

(c) *There exist* $p, v \in Q_r$ *and* d, δ *skew derivations of* R *such that* $F(x) = px + d(x)$, $G(x) = vx + \delta(x)$, *for all* $x \in R$, *and one of the following holds:*

9. *there exist* $\vartheta \in C$ *and* $0 \neq \eta \in C$ *such that* $\delta(x) = \eta d(x)$, *for any* $x \in R$, $a_1 p = \vartheta a_2 v$, $b_2 = -\vartheta b_1$, *and* $a_1 = \vartheta\eta a_2$;
10. *there exist* $0 \neq \vartheta \in C$, $0 \neq \eta \in C$ *and* $p_0 \in Q_r$ *such that* $\delta(x) = p_0 x - \alpha(x) p_0 + \eta d(x)$, *for any* $x \in R$, $a_1 = \vartheta\eta a_2$, $b_2 = -\vartheta b_1$, $p_0 b_1 = 0$ *and* $\eta a_2 p - a_2(v + p_0) = 0$;
11. *there exist* $\vartheta \in C$, $0 \neq \eta \in C$ *and* $p_0, q \in Q_r$, *where* q *is an invertible element, such that* $\delta(x) = p_0 x - q x q^{-1} p_0 + \eta d(x)$, *for any* $x \in R$, $a_1 = \vartheta\eta a_2$, $b_2 = -\vartheta b_1$, $q^{-1} p_0 b_1 = \vartheta b_1$ *and* $\eta a_2 p - a_2(v + p_0) + \vartheta a_2 q = 0$.

Let us recall some basic facts which will be useful in the sequel.

Fact 1 Let R be a prime ring, then the following statements hold:

(a) Every generalized derivation of R can be uniquely extended to Q_r [14, Theorem 3].
(b) Any automorphism of R can be uniquely extended to Q_r [7, Fact 2].
(c) Every generalized skew derivation of R can be uniquely extended to Q_r [4, Lemma 2].

Fact 2 A generalized skew derivation having associated automorphism α and skew derivation d assumes the following form:

$$F(x) = ax + d(x) \tag{1}$$

for all $x \in R$ (see [4, Lemma 2], [5, Theorem 3.1 and Corollary 3.2]).

We also need to recall some well-known results on generalized polynomial identities for prime rings involving skew derivations and automorphisms.

Fact 3 ([9]) If $\Phi(x_i, D(x_i))$ is a generalized polynomial identity for R, where R is a prime ring and D is an outer skew derivation of R, then R also satisfies the generalized polynomial identity $\Phi(x_i, y_i)$, where x_i and y_i are distinct indeterminates.

If $\Phi(x_i, D(x_i), \alpha(x_i))$ is a generalized polynomial identity for a prime ring R, D is an outer skew derivation of R and α is an outer automorphism of R, then R also satisfies the generalized polynomial identity $\Phi(x_i, y_i, z_i)$, where x_i, y_i, and z_i are distinct indeterminates.

Fact 4 ([13, Theorem 6.5.9, page 365]) Let a prime ring R obey a polynomial identity of the type $f(x_j^{\alpha_i \Delta_k}) = 0$, where $f(z_j^{i,k})$ is a generalized polynomial with the coefficients from Q_r, $\Delta_1, \ldots, \Delta_n$ are mutually different correct words from a reduced set of skew derivations commuting with all the corresponding automorphisms, and $\alpha_1, \ldots, \alpha_m$ are mutually outer automorphisms. In this case the identity $f(z_j^{i,k}) = 0$ is valid on Q_r.

Fact 5 ([8, Theorem 1]) Let R be a prime ring and I be a two-sided ideal of R. Then I, R, and Q_r satisfy the same generalized polynomial identities with coefficients in Q_r (see [6]). Furthermore, I, R, and Q_r satisfy the same generalized polynomial identities with automorphisms.

Fact 6 ([9, Theorem 2]) Let R be a prime ring and I be a two-sided ideal of R. Then I, R, and Q_r satisfy the same generalized polynomial identities with a single skew derivation.

In the sequel, R will be a non-commutative ring of characteristic different from 2, F and G two non-zero generalized skew derivations of R, associated with the same automorphism α and commuting with α.

1 Annihilating Condition for a Single Generalized Skew Derivation

In this second section our aim will be to prove Theorem 1. More precisely, let F be a generalized skew derivation of R and a, b are non-zero elements of R such that

$$aF(w)b = 0 \quad \forall w \in L \quad \text{a non-central Lie ideal of} \quad R. \tag{2}$$

The study of this result will be useful for the proof of our main Theorem (i.e., Theorem 2).

We permit the following:

Lemma 1 *Let R be a prime and $a_i, b_i \in U$, for $1 \leq i \leq n$. If $\sum_{i=1}^{n} a_i[x, y]b_i = 0$, for all $x, y \in R$. If $a_i \neq 0$ for some i, then $b_1, \ldots, b_n$ are C-dependent. Similarly, if $b_i \neq 0$ for some i, then $a_1, \ldots, a_n$ are C-dependent.*

Proof The result follows easily from [15, Lemma 2.2] and [16, Lemma 1].

Lemma 2 *Let $c, u \in Q_r$ be such that $F(x) = cx + \alpha(x)u$, for any $x \in R$. If*

$$aF([r_1, r_2])b = 0 \quad \forall r_1, r_2 \in R. \tag{3}$$

then one of the following holds:

(a) $ac = ub = 0$;
(b) there exist $q \in Q_r$ and $\lambda \in C$ such that $\alpha(x) = qxq^{-1}$, for any $x \in R$, with $a(c + \lambda q) = 0$ and $(\lambda + q^{-1}u)b = 0$.

Proof By our assumption R satisfies

$$a\Big(c[x_1, x_2] + \alpha([x_1, x_2])u\Big)b. \tag{4}$$

We consider firstly the case $\alpha(x) = qxq^{-1}$, for any $x \in R$, where $q \in Q_r$ is an invertible element. In this case, by (4), R satisfies

$$a\Big(c[x_1, x_2] + q[x_1, x_2]q^{-1}u\Big)b. \tag{5}$$

A direct application of Lemma 1 leads to conclusion (b).

Therefore we may assume that α is not an inner automorphism of Q_r. Thus, by (4) and Fact 3, R satisfies the generalized polynomial identity

$$a\Big(c[x_1, x_2] + [y_1, y_2]u\Big)b. \tag{6}$$

In particular R satisfies both the blended components $ac[x_1, x_2]b$ and $a[y_1, y_2]ub$. Since $a \neq 0$ and $b \neq 0$ and by the primeness of R, we get the required conclusion $ac = ub = 0$.

Proof (Proof of Theorem 1) By Fact 2, $F(x) = cx + d(x)$ for all $x \in R$, where $c \in Q_r$ and d is the skew derivation associated with F.

Since L is not central and $char(R) \neq 2$, it is well known that there exists a non-zero ideal I of R such that $0 \neq [I, R] \subseteq L$ (see [11, pages 4–5]). Therefore, by (2), the ideal I satisfies $aF([x_1, x_2])b$. Since R and I satisfy the same generalized identities with automorphisms and skew derivations, we may assume that R also satisfies $aF([x_1, x_2])b$, that is

$$a\Big(c[x_1, x_2] + d([x_1, x_2])\Big)b. \tag{7}$$

In case d is an inner skew derivation of R, the conclusion follows from Lemma 2. Then we may assume that d is not inner and prove that a contradiction follows. Expansion of (7) says that R satisfies

$$a\Big(c[x_1, x_2] + d(x_1)x_2 + \alpha(x_1)d(x_2) - d(x_2)x_1 - \alpha(x_2)d(x_1)\Big)b. \tag{8}$$

Since d is not inner and by Fact 3, (8) implies that R satisfies

$$a\Big(c[x_1, x_2] + y_1x_2 + \alpha(x_1)y_2 - y_2x_1 - \alpha(x_2)y_1\Big)b \tag{9}$$

and in particular R satisfies

$$a\Big(y_1x_2 - \alpha(x_2)y_1\Big)b. \tag{10}$$

If α is outer, relation (10) implies that R satisfies

$$a\Big(y_1x_2 - z_2y_1\Big)b$$

and, in particular, $a[r_1, r_2]b = 0$, for any $r_1, r_2 \in R$. It follows that either $a = 0$ or $b = 0$, which contradicts the assumption $a, b \neq 0$.

On the other hand, if $\alpha(x) = qxq^{-1}$, where q is an invertible element of Q_r, one may replace in (main-8) y_1 with qx_1. Hence R satisfies $aq[x_1, x_2]b$. Since q is invertible, once again the contradiction that either $a = 0$ or $b = 0$ follows.

2 Annihilating Conditions for Two Generalized Skew Derivations

We conclude our paper giving the description of two generalized skew derivations F and G of a prime ring R satisfying the condition

$$a_1F(x)b_1 + a_2G(x)b_2 = 0 \quad \forall x \in R \tag{11}$$

where $a_1, a_2, b_1, b_2 \in Q_r$.

In light of Theorem 1, we may assume that a_1, a_2, b_1, b_2 are all non-zero elements of Q_r and also that both $F \neq 0$ and $G \neq 0$.

We start with two useful results, that we quote as follows, by applying [6, Theorem 2]:

Lemma 3 *Let R be a prime and $a_i, b_i \in Q_r$, for $1 \le i \le n$. If $\sum_{i=1}^{n} a_i x b_i = 0$, for all $x \in R$, and $b_i \neq 0$ for some i, then $a_1, \dots, a_n$ are C-dependent (see [15, Lemma 2.2]).*

Lemma 4 *Let R be a prime and $a_i, b_i, c_i, d_i \in Q_r$ such that $\sum_{i=1}^{m} a_i x b_i + \sum_{j=1}^{n} c_j x d_j = 0$, for all $x \in R$. If $a_1, \dots, a_m$ are linearly C-independent then each b_i is a linear combination of $d_1, \dots, d_n$ over C. Analogously, if $b_1, \dots, b_m$ are linearly C-independent then each a_i is a linear combination of $c_1, \dots, c_n$ over C. (see [17, Lemma 1.2]).*

Lemma 5 *Let F and G be inner generalized skew derivations of R defined as*

$$F(x) = px + qxq^{-1}u, \quad G(x) = vx + qxq^{-1}w, \quad \forall x \in R$$

where $p, u, v, w, q \in Q_r$ and q is an invertible element. If R satisfies (11), one of the following holds:

(a) there exist $\alpha_1, \alpha_2, \alpha_3, \alpha_4 \in C$ such that $b_1 = \alpha_1 b_2 + \alpha_2 q^{-1} w b_2$, $q^{-1} u b_1 = \alpha_3 b_2 + \alpha_4 q^{-1} w b_2$ and $\alpha_1 a_1 p + \alpha_3 a_1 q + a_2 v = \alpha_2 a_1 p + \alpha_4 a_1 q + a_2 q = 0$;
(b) there exist $\lambda, \alpha_1, \alpha_2, \alpha_3, \alpha_4 \in C$ such that $q^{-1} w b_2 = \lambda b_2$, $b_1 = (\alpha_1 + \lambda\alpha_2) b_2$, $q^{-1} u b_1 = (\alpha_3 + \lambda\alpha_4) b_2$ and $(\alpha_1 + \lambda\alpha_2) a_1 p + (\alpha_3 + \lambda\alpha_4) a_1 q + a_2 (v + \lambda q) = 0$;
(c) there exist $0 \neq \lambda \in C$ and $\beta_1, \beta_2 \in C$ such that $a_1 p = \lambda a_1 q$, $a_2 v = \beta_1 a_1 q$, $a_2 q = \beta_2 a_1 q$ and $\lambda b_1 + q^{-1} u b_1 + \beta_1 b_2 + \beta_2 q^{-1} w b_2 = 0$;
(d) there exist $0 \neq \lambda \in C$ and $\mu, \eta \in C$ such that $a_1 p = \lambda a_1 q$, $a_2(v + \mu q) = \eta a_1 q$, $(\lambda + q^{-1} u) b_1 = -\eta b_2$ and $q^{-1} w b_2 = \mu b_2$.

Proof By our main hypothesis

$$a_1 F(x) b_1 + a_2 G(x) b_2 = 0 \quad \forall x \in R.$$

Under the assumptions of the present Lemma, we have that R satisfies the generalized identity

$$a_1 \big(px + qxq^{-1}u\big) b_1 + a_2 \big(vx + qxq^{-1}w\big) b_2 \tag{12}$$

that is

$$(a_1 p) x b_1 + (a_1 q) x (q^{-1} u b_1) + (a_2 v) x b_2 + (a_2 q) x (q^{-1} w b_2). \tag{13}$$

By Lemma 3 and since a_1, a_2, b_1, b_2 are all non-zero we may divide the proof in two cases.

Case 1. $\{a_1p, a_1q\}$ **is a linearly** C**-independent set**

Application of Lemma 4 implies that there exist $\alpha_1, \alpha_2, \alpha_3, \alpha_4 \in C$ such that

$$\begin{aligned} b_1 &= \alpha_1 b_2 + \alpha_2 q^{-1} w b_2 \\ q^{-1} u b_1 &= \alpha_3 b_2 + \alpha_4 q^{-1} w b_2. \end{aligned} \tag{14}$$

Thus, by (13), R satisfies

$$(a_1p)x(\alpha_1 b_2 + \alpha_2 q^{-1} w b_2) + (a_1q)x(\alpha_3 b_2 + \alpha_4 q^{-1} w b_2) + (a_2v)xb_2 + (a_2q)x(q^{-1}wb_2)$$

that is

$$(\alpha_1 a_1 p + \alpha_3 a_1 q + a_2 v)xb_2 + (\alpha_2 a_1 p + \alpha_4 a_1 q + a_2 q)xq^{-1}wb_2. \tag{15}$$

Firstly we note that, if $\alpha_2 a_1 p + \alpha_4 a_1 q + a_2 q = 0$ then, by the primeness of R and since $b_2 \neq 0$, (15) implies $\alpha_1 a_1 p + \alpha_3 a_1 q + a_2 v = 0$. Hence, in consideration of what is stated in relations (14), we get conclusion (a) of the present Lemma.
On the other hand, if $\alpha_2 a_1 p + \alpha_4 a_1 q + a_2 q \neq 0$ and by Lemma 3, there is $\lambda \in C$ such that $q^{-1}wb_2 = \lambda b_2$. Thus (15) reduces to

$$(\alpha_1 a_1 p + \alpha_3 a_1 q + a_2 v)xb_2 + \lambda(\alpha_2 a_1 p + \alpha_4 a_1 q + a_2 q)xb_2. \tag{16}$$

Again by the primeness of R and since $b_2 \neq 0$, $\alpha_1 a_1 p + \alpha_3 a_1 q + a_2 v + \lambda(\alpha_2 a_1 p + \alpha_4 a_1 q + a_2 q) = 0$ follows.

Case 2. $a_1 p = \lambda a_1 q, 0 \neq \lambda \in C$

In this case, again by (13), R satisfies

$$a_1 q x(\lambda b_1 + q^{-1} u b_1) + (a_2 v)xb_2 + (a_2 q)x(q^{-1} w b_2). \tag{17}$$

Notice that, in case $\{b_2, q^{-1}wb_2\}$ is a linearly C-independent set, by (17) and Lemma 3, it follows

$$a_2 v = \beta_1 a_1 q, \quad a_2 q = \beta_2 a_1 q \quad \beta_1, \beta_2 \in C$$

and (17) reduces to

$$a_1 q x(\lambda b_1 + q^{-1} u b_1 + \beta_1 b_2 + \beta_2 q^{-1} w b_2).$$

Therefore, since $a_1 q \neq 0$, we get $\lambda b_1 + q^{-1} u b_1 + \beta_1 b_2 + \beta_2 q^{-1} w b_2 = 0$.

Assume finally that $\{b_2, q^{-1}wb_2\}$ is a linearly C-dependent set.

Without loss of generality we may write $q^{-1}wb_2 = \mu b_2$, for a suitable $\mu \in C$. Hence, by (17), R satisfies

$$a_1qx(\lambda b_1 + q^{-1}ub_1) + (a_2v + \mu a_2 q)xb_2 \tag{18}$$

implying that there exists $\eta \in C$ such that

$$a_2v + \mu a_2 q = \eta a_1 q$$

$$\lambda b_1 + q^{-1}ub_1 = -\eta b_2.$$

Lemma 6 *Let F and G be inner generalized skew derivations of R defined as*

$$F(x) = px + \alpha(x)u, \quad G(x) = vx + \alpha(x)w, \quad \forall x \in R$$

where $p, u, v, w \in Q_r$ and α is an outer automorphism of R. If R satisfies (11), one of the following holds:

(a) $a_1p = a_2v = ub_1 = wb_2 = 0$;
(b) $a_1p = a_2v = 0$ *and there exists* $\mu \in C$ *such that* $ub_1 = \mu wb_2$ *and* $a_2 = -\mu a_1$;
(c) $ub_1 = wb_2 = 0$ *and there exists* $\lambda \in C$ *such that* $a_1p = \lambda a_2 v$ *and* $b_2 = -\lambda b_1$;
(d) *there exist* $\lambda, \mu \in C$ *such that* $a_1p = \lambda a_2v$, $b_2 = -\lambda b_1$, $ub_1 = \mu wb_2$ *and* $a_2 = -\mu a_1$.

Proof Here R satisfies

$$a_1\big(px + \alpha(x)u\big)b_1 + a_2\big(vx + \alpha(x)w\big)b_2. \tag{19}$$

Since α is outer, by (19), it follows that R satisfies the generalized identity

$$a_1\big(px_1 + x_2u\big)b_1 + a_2\big(vx_1 + x_2w\big)b_2. \tag{20}$$

In particular, both

$$a_1px_1b_1 + a_2vx_1b_2 \tag{21}$$

and

$$a_1x_2ub_1 + a_2x_2wb_2 \tag{22}$$

are satisfied by R. Relation (21) implies that

- either $a_1p = a_2v = 0$
- or there exists $\lambda \in C$ such that $a_1p = \lambda a_2v$ and $b_2 = -\lambda b_1$.

Analogously, (22) implies that

- either $ub_1 = wb_2 = 0$
- or there exists $\mu \in C$ such that $ub_1 = \mu wb_2$ and $a_2 = -\mu a_1$.

Putting together all the previous informations, one of the following cases must occur:

(a) $a_1 p = a_2 v = ub_1 = wb_2 = 0$;
(b) $a_1 p = a_2 v = 0$ and there exists $\mu \in C$ such that $ub_1 = \mu wb_2$ and $a_2 = -\mu a_1$;
(c) $ub_1 = wb_2 = 0$ and there exists $\lambda \in C$ such that $a_1 p = \lambda a_2 v$ and $b_2 = -\lambda b_1$;
(d) there exist $\lambda, \mu \in C$ such that $a_1 p = \lambda a_2 v$, $b_2 = -\lambda b_1$, $ub_1 = \mu wb_2$ and $a_2 = -\mu a_1$.

Before proceeding with the proof of our main result, we need to recall the following:

Lemma 7 *Let R be a prime ring, $\alpha, \beta \in \mathrm{Aut}(Q_r)$ and $d, \delta : R \to R$ be two skew derivations, associated with the same automorphism α. If there exist $0 \neq \eta \in C$, and $u \in Q_r$ such that*

$$\delta(x) = \Big(ux - \beta(x)u\Big) + \eta d(x), \quad \forall x \in R \tag{23}$$

then either $\alpha = \beta$ or $\delta(x) = \eta d(x)$, for all $x \in R$.

Proof By the definition of δ we have

$$\delta(xy) = uxy - \beta(x)\beta(y)u + \eta d(x)y + \eta\alpha(x)d(y). \tag{24}$$

On the other hand, right multiplying relation (23) by $y \in R$, it follows that

$$\delta(x)y = uxy - \beta(x)uy + \eta d(x)y \quad \forall x, y \in R. \tag{25}$$

Therefore, subtracting relation (25) from (24), and using again (23), we get

$$\big\{\alpha(x) - \beta(x)\big\} \cdot \big\{uy - \beta(y)u\big\} = 0 \quad \forall x, y \in R. \tag{26}$$

Replacing y by yt in (26) and then using (26) we have

$$\big\{\alpha(x) - \beta(x)\big\} \cdot \beta(y) \cdot \big\{\beta(t)u - ut\big\} = 0 \quad \forall x, y, t \in R. \tag{27}$$

Then, by the primeness of R, above relation yields either $\alpha(x) - \beta(x) = 0$ for any $x \in R$, or $\beta(t)u - ut = 0$ for any $t \in R$. The last case and (23) imply $\delta(x) = \eta d(x)$, for all $x \in R$, as required.

Lemma 8 ([10, Lemma 3.2]) *Let R be a prime ring, $\alpha, \beta \in \mathrm{Aut}(Q_r)$ and $d : R \to R$ be a skew derivation, associated with the automorphism α. If there exist $0 \neq \theta \in C$, $0 \neq \eta \in C$ and $u, b \in Q_r$ such that*

$$d(x) = \theta\Big(ux - \alpha(x)u\Big) + \eta\Big(bx - \beta(x)b\Big), \quad \forall x \in R$$

then d is an inner skew derivation of R. More precisely, either $b = 0$ or $\alpha = \beta$.

Proof (Proof of Theorem 2) For sake of clearness we recall that we may write $F(x) = px + d(x)$ and $G(x) = vx + \delta(x)$, for all $x \in R$ and suitable $p, v \in Q_r$ and d, δ skew derivations associated with the same automorphism α. Moreover we also recall that both d and δ commute with α.

We also remind that, by our main hypothesis R satisfies

$$a_1\Big(px + d(x)\Big)b_1 + a_2\Big(vx + \delta(x)\Big)b_2. \tag{28}$$

The case $d = 0$ and $\delta \neq 0$

We firstly study the case $F(x) = px$ and $G(x) = vx + \delta(x)$, for all $x \in R$. Since $F \neq 0$, we may assume in what follows $p \neq 0$. Moreover δ is not an inner skew derivation of R, otherwise the conclusion follows by Lemmas 5 and 6. In this situation, by (28) we have that R satisfies

$$a_1 p x_1 b_1 + a_2\Big(vx_1 + x_2\Big)b_2.$$

In particular $a_2 y b_2 = 0$, for any $y \in R$, which is a contradiction, since both $a_2 \neq 0$ and $b_2 \neq 0$.

Analogously, we get a contradiction in the case we assume $\delta = 0$ and $d \neq 0$.

The case $d \neq 0, \delta \neq 0$

Here we study the case when $F(x) = px + d(x)$ and $G(x) = vx + \delta(x)$, for all $x \in R$. We start with the case d, δ are linearly C-independent modulo inner skew derivations. Hence, by (28),

$$a_1\Big(px_1 + x_2\Big)b_1 + a_2\Big(vx_1 + x_3\Big)b_2 \tag{29}$$

is satisfied by R. In particular, $a_1 x_2 b_1$ is a generalized identity for R, which is a contradiction, since both $a_1 \neq 0$ and $b_1 \neq 0$.

Thus we assume that $\{d, \delta\}$ are linearly C-dependent modulo inner skew derivations. Hence there exist $\lambda, \mu \in C$, $u \in Q_r$ and an automorphism β of R such that $\lambda d(x) + \mu\delta(x) = ux - \beta(x)u$, for any $x \in R$.

If $\lambda = 0$ and $\mu \neq 0$, we write

$$\delta(x) = \Big(p_0 x - \beta(x)p_0\Big), \quad \forall x \in R$$

where $p_0 = \mu^{-1}u$. Since the automorphism associated with a skew derivation is unique, in this case $\alpha = \beta$.

If d is also inner, the conclusion follows from Lemmas 5 and 6. Hence we may assume that d is not inner. Thus, by (28), R satisfies

$$a_1\Big(px_1 + x_2\Big)b_1 + a_2\Big(vx_1 + p_0x_1 - \beta(x_1)p_0\Big)b_2 \tag{30}$$

and in particular $a_1x_2b_1$ is an identity for R, which is a contradiction.

Similarly, we get a contradiction in the case $\mu = 0$ and $\lambda \neq 0$.

Hence, in the sequel we assume that both $\lambda \neq 0$ and $\mu \neq 0$. We may write

$$\delta(x) = \Big(p_0x - \beta(x)p_0\Big) + \eta d(x), \quad \forall x \in R \tag{31}$$

where $\eta = -\lambda\mu^{-1} \neq 0$ and, as above, $p_0 = \mu^{-1}u$. By Lemma 7, either $\alpha = \beta$ or $p_0 = 0$ and $\delta(x) = \eta d(x)$, for all $x \in R$.

Moreover, by Lemma 8, if d is an inner skew derivation, then also δ is inner and the conclusion follows again from Lemmas 5 and 6.

Therefore, in what follows we assume that $0 \neq d$ is outer.

In the case $\delta = \eta d$, (28) reduces to

$$a_1\Big(px + d(x)\Big)b_1 + a_2\Big(vx + \eta d(x)\Big)b_2. \tag{32}$$

Thus, since d is not inner, R satisfies

$$a_1\Big(px_1 + x_2\Big)b_1 + a_2\Big(vx_1 + \eta x_2\Big)b_2. \tag{33}$$

In particular, both

$$a_1px_1b_1 + a_2vx_1b_2 \tag{34}$$

and

$$a_1x_2b_1 + \eta a_2x_2b_2 \tag{35}$$

are identities for R. Those relations imply that there exists $\vartheta \in C$ such that

$$a_1p = \vartheta a_2v \quad b_2 = -\vartheta b_1 \quad a_1 = \vartheta\eta a_2.$$

Suppose now $\alpha = \beta$. By relations (31) and (28) R satisfies

$$a_1\Big(px + d(x)\Big)b_1 + a_2\Big(vx + p_0x - \alpha(x)p_0 + \eta d(x)\Big)b_2. \tag{36}$$

Since d is not inner, it follows that

$$a_1\Big(px_1 + x_2\Big)b_1 + a_2\Big(vx_1 + p_0x_1 - \alpha(x_1)p_0 + \eta x_2\Big)b_2 \tag{37}$$

is a generalized identity for R. Hence R satisfies both

$$a_1 p x_1 b_1 + a_2\Big(v x_1 + p_0 x_1 - \alpha(x_1) p_0\Big) b_2 \tag{38}$$

and

$$a_1 x_2 b_1 + \eta a_2 x_2 b_2. \tag{39}$$

By (39) and applying Lemma 3, we have that there exists $0 \neq \vartheta \in C$ such that

$$a_1 = \vartheta \eta a_2 \quad b_2 = -\vartheta b_1.$$

Substituting a_1 and b_2 in relation (38), it follows that

$$\vartheta \eta a_2 p x_1 b_1 - \vartheta a_2\Big(v x_1 + p_0 x_1 - \alpha(x_1) p_0\Big) b_1. \tag{40}$$

If α is not inner, by (40) we have that R satisfies

$$\vartheta \eta a_2 p x_1 b_1 - \vartheta a_2\Big(v x_1 + p_0 x_1 - x_2 p_0\Big) b_1. \tag{41}$$

Thus both $a_2 x_2 p_0 b_1$ and

$$\Big(\vartheta \eta a_2 p - \vartheta a_2(v + p_0)\Big) x_1 b_1$$

are identities for R, implying $p_0 b_1 = 0$ and $\eta a_2 p - a_2(v + p_0) = 0$.

On the other hand, if $\alpha(x) = q x q^{-1}$, for any $x \in R$, by (40) it follows that

$$\Big(\eta a_2 p - a_2(v + p_0)\Big) x_1 b_1 + a_2 q x_1 q^{-1} p_0 b_1$$

is a generalized identity for R. Thus, there exists $\vartheta \in C$ such that

$$q^{-1} p_0 b_1 = \vartheta b_1 \quad \eta a_2 p - a_2(v + p_0) + \vartheta a_2 q = 0.$$

Acknowledgements We would like to express our gratitude to the anonymous referees for their careful reading of the manuscript and for insightful comments and suggestions, which significantly helped us improve the final presentation of this article.

Funding The work of the first author was partially supported by a grant of the italian "National Group for Algebraic and Geometric Structures, and their Applications" (GNSAGA-INDAM).

References

1. Beidar, K.I., Martindale III, W.S., Mikhalev, A.V.: Rings with generalized identities, monographs and textbooks. In: Pure and Applied Mathematics, vol. 196. Marcel Dekker, Inc., New York (1996)
2. Brešar, M.: Centralizing mappings and derivations in prime rings. J. Algebra **156**, 358–394 (1993)
3. Chang, J.-C.: A special identitity of (α, β)-derivations and its consequences. Taiwanese J. Math. **1**(1), 21–30 (1997)
4. Chang, J.-C.: On the identitity $h(x) = af(x) + g(x)b$. Taiwanese J. Math. **7**(1), 103–113 (2003)
5. Cheng, H.-W., Wei, F.: Generalized skew derivations of rings. Adv. Math. **35**(2), 237–243 (2006)
6. Chuang, C.-L.: GPIs having coefficients in Utumi quotient rings. Proc. Am. Math. Soc. **103**, 723–728 (1988)
7. Chuang, C.-L.: Differential identities with automorphisms and antiautomorphisms I. J. Algebra **149**, 371–404 (1992)
8. Chuang, C.-L.: Differential identities with automorphisms and antiautomorphisms II. J. Algebra **160**, 130–171 (1993)
9. Chuang, C.-L., Lee, T.-K.: Identities with a single skew derivation. J. Algebra **288**, 59–77 (2005)
10. De Filippis, V., Di Vincenzo, O.M.: Generalized skew derivations and nilpotent values on Lie ideals. Alg. Coll. **26**(4), 589–614 (2019)
11. Herstein, I.N.: Topics in Ring Theory. University of Chicago Press, Chicago (1969)
12. Herstein, I.N.: A note on derivation II. Canad. Math. Bull. **22**, 509–511 (1979)
13. Kharchenko, V.K.: Automorphisms and derivations of associative rings. Mathematics and Its Applications (Soviet Series), vol. 69, Kluwer Academic Publishers (1991)
14. Lee, T.-K.: Generalized derivations of left faithful rings. Commun. Algebra **27**(8), 4057–4073 (1999)
15. Lee, T.-K.: Posner's theorem for (σ, τ)-derivations and σ-centralizing maps. Houston J. Math. **30**(2), 309–320 (2004)
16. Lee, T.-K.: Identities with generalized derivations. Commun. Algebra **29**(10), 4437–4450 (2001)
17. Lee, P.-H., Wong, T.-L., Lin, J.-S., Wang, J.-R.: Commuting traces of multiadditive mappings. J. Algebra **193**(2), 709–723 (1997)

Dimensional Dual Hyperovals—An Updated Survey

Ulrich Dempwolff

Abstract A survey on dimensional dual hyperovals is the report of Satoshi Yoshiara (Proceedings of a Conference on Pingree Park 2004, Colonel,USA, pp 247–266, 2006). It describes the initial investigations in this field and covers roughly the period from 1995 to 2005. The present report is an update of this survey and tries to explain relevant developments after 2005.

Keywords Dimensional dual hyperovals · Bilinear dual hyperovals · Quotients of hyperovals · Universal covers · Automorphism groups

1 Introduction

Dimensional dual hyperovals are a fairly new topic in finite geometry. There has been a steady output on this subject for the last 25 years. The number of researchers remained to be low. However, the author of this report believes that dimensional dual hyperovals deserve attention. Firstly, the theory is in an early stage meaning that elementary questions can be asked, other elementary questions are still open. Secondly, dimensional dual hyperovals have connections to other topics from finite geometry, to topics from information theory and combinatorics. The very definition of these objects suggests that one also can expect connections to translation planes. We will discuss also connections with APN functions, distance regular graphs and bent functions. As some dimensional dual hyperovals have a large automorphism group, it is not surprising that representation theory of finite groups has sometimes effective applications. In constructions of dimensional dual hyperovals, multilinear algebra and arithmetic in finite fields are useful.

Yoshiara's article *Dimensional dual arcs* [65] is in essence a survey on dimensional dual hyperovals. It provides a nice overview of the research of the first ten years.

U. Dempwolff (✉)
Department of Mathematics, University of Kaiserslautern, Erwin-Schroedinger-Strasse, 67653 Kaiserslautern, Germany
e-mail: dempwolff@mathematik.uni-kl.de

M. Ashraf et al. (eds.), *Algebra and Related Topics with Applications*,
Springer Proceedings in Mathematics & Statistics 392,
https://doi.org/10.1007/978-981-19-3898-6_9

Our aim is to continue this report and explain investigations afterwards. To keep this report self-contained, we will—if necessary—repeat material from [65]. Following Yoshiara I will pose a few problems which in my view deserve closer examination.

1.1 The Definition

A set $\mathcal{S}$ of n-dimensional subspaces, $n \geq 2$, of a finite-dimensional vector space V over a finite field $\mathbb{F}_q$ is called a *dual hyperoval of rank n over* $\mathbb{F}_q$, we use the symbol DHO as an abbreviation, if

(D) Every two members of $\mathcal{S}$ have a nontrivial intersection and for every $X \in \mathcal{S}$ and every 1-space $P \subseteq X$ there exist precisely one $X' \in \mathcal{S}$, such that $P = X \cap X'$.

Axiom (D) is equivalent to

(D1) $\dim X_1 \cap X_2 = 1$ for two distinct $X_1, X_2 \in \mathcal{S}$.
(D2) $\dim X_1 \cap X_2 \cap X_3 = 0$ for three distinct $X_1, X_2, X_3 \in \mathcal{S}$.
(D3) $|\mathcal{S}| = (q^n - 1)/(q - 1) + 1$.

We call a set $\mathcal{A}$ of subspaces of rank n a *dual arc of rank n* and use the symbol DA as an abbreviation, if only axioms (D1) and (D2) hold.

In geometric settings, often the language of projective geometry is preferred and one calls a DHO of rank n a *$(n - 1)$-dimensional dual hyperoval*. However, in this report, we use the term dimension or rank only in the sense of linear algebra. For $n = 2$, a DHO is an ordinary dual hyperoval (see [65, Lemma 2.4]). There exists an extensive literature ([5, 30]) on hyperovals, and thus on ordinary dual hyperovals. In this article, we are **exclusively interested in the generalization**

$$n \geq 3.$$

1.2 Basic Notions

The space $U(\mathcal{S}) = \langle X \mid X \in \mathcal{S} \rangle$ is called the *ambient space of* $\mathcal{S}$. Of course, in studying DHOs, only the ambient spaces are relevant. Sometimes it is convenient to require (as an additional axiom) that the DHO generates the surrounding space.

The DHO $\mathcal{S}$ in U *splits over the subspace Y* if

$$U = X \oplus Y \quad \text{for all} \quad X \in \mathcal{S}.$$

In particular, $\mathcal{S}$ (consider $\mathcal{S}$ as a DHO in $U(\mathcal{S})$) splits over $U(\mathcal{S}) \cap Y$ too. We call a DHO $\mathcal{S}$ *split* if $\mathcal{S}$ splits over some subspace and otherwise *nonsplit*. If $U = U(\mathcal{S})$ and $\mathcal{S}$ splits over Y we call Y a *complement* of $\mathcal{S}$.

Let U and U' be $\mathbb{F}_q$-spaces and $\Phi : U \to U'$ an additive mapping. Let α be an automorphism of $\mathbb{F}_q$. Then Φ is an *α-semilinear map* if $(au)\Phi = a^\alpha(u\Phi)$ for all $a \in \mathbb{F}_q$ and all $u \in U$ (we will write linear (semilinear) mappings on the right-hand side of an argument). Two DHOs $\mathcal{S}$ and $\mathcal{S}'$ are *isomorphic*, if there exists an invertible semilinear operator Φ, that sends the ambient space of $\mathcal{S}$ onto the ambient space of $\mathcal{S}'$, such that $\mathcal{S}' = \mathcal{S}\Phi$.

Again, let $\mathcal{S}$ be a DHO over $\mathbb{F}_q$ with the ambient space $U = U(\mathcal{S})$. A *linear automorphism (α-linear automorphism)* ϕ of $\mathcal{S}$ is a linear (α-linear) isomorphism of U, which maps $\mathcal{S}$ onto $\mathcal{S}$. The linear automorphisms form a group LinAut($\mathcal{S}$), which is normal in the group of all semilinear automorphisms SemAut($\mathcal{S}$). If we denote by $\Gamma\mathrm{L}(U)$ the group of invertible semilinear operators, then SemAut($\mathcal{S}$) is the stabilizer of the *set* $\mathcal{S}$ in $\Gamma\mathrm{L}(U)$. The kernel K of the permutation action of SemAut($\mathcal{S}$) on $\mathcal{S}$ is isomorphic to the multiplicative group of $\mathbb{F}_q$. The group $\mathrm{Aut}(\mathcal{S}) = \mathrm{SemAut}(\mathcal{S})/K$ is the *automorphism group of* $\mathcal{S}$. For $q = 2$, of course, $\mathrm{LinAut}(\mathcal{S}) = \mathrm{SemAut}(\mathcal{S}) \simeq \mathrm{Aut}(\mathcal{S})$.

Let $\mathcal{S}$ and $\mathcal{S}'$ be DHOs of rank n with ambient spaces U and U', respectively. An epimorphism $\pi : U' \to U$ is called a *covering map*, if $\mathcal{S} = \mathcal{S}'\pi$. One says, that $\mathcal{S}$ is a *quotient* of $\mathcal{S}'$ and $\mathcal{S}'$ a *cover* of $\mathcal{S}$. Note, that the restriction of π to any member of $\mathcal{S}'$ is a monomorphism. The cover is *proper*, if π is not an isomorphism. A DHO is *simply connected*, if it has no proper cover.

Let $\mathcal{S}$ be a DHO over $\mathbb{F}_2$ in the ambient space U. Let $\mathcal{B} = \{X + u \mid X \in \mathcal{S},\ u \in U\}$ the set of all cosets of the members of the DHO in U (development of $\mathcal{S}$). The incidence structure $\mathrm{Af}(\mathcal{S}) = (U, \mathcal{B}, \in)$ is the *affine expansion* of the DHO $\mathcal{S}$. We call U the *points* and $\mathcal{B}$ the *blocks* of the affine expansion. The affine expansion is a semibiplane: Any two points are connected by 2 or 0 blocks, any two blocks intersect at 2 or 0 points. The incidence graph of the affine expansion will be denoted by $\Gamma(\mathcal{S})$. For the definition of the affine expansion of a DHO over an arbitrary field; see [65, Sect. 2.7]. This definition is a little bit more complicated than the simplified description of the affine expansion over $\mathbb{F}_2$ given here.

2 DHOs over $\mathbb{F}_q$, $q > 2$

The theory of dimensional dual hyperovals is presently for the most part a theory of DHOs over $\mathbb{F}_2$. We briefly report on what is known about DHOs over $\mathbb{F}_q$, $q > 2$.

2.1 Odd Characteristic

It seems quite likely that there are no DHOs over finite fields of odd characteristic. But the restrictions given in [65, Theorem 2.5] have not been extended afterwards. So

Problem 1 Show that there exists no DHO of odd characteristic or disprove this conjecture by a counterexample.

is completely open.

2.2 *Even Characteristic*

The knowledge about DHOs over $\mathbb{F}_q$, $q > 2$ a 2-power, is also limited. The most intriguing example is the *Mathieu DHO* $\mathcal{M}$ [65, 5.1] of rank 3 whose ambient space is a unitary space rank 6 over $\mathbb{F}_4$. This DHO plays an exceptional role among the DHOs for several reasons. Presently, $\mathcal{M}$ is the only unitary DHO (*unitary* in the sense of Sect. 6.2). To my knowledge, the Mathieu DHO is the only DHO, such that the ambient space is an *irreducible module* for the group of linear automorphisms. It plays an exceptional role among the doubly transitive DHOs (see Sect. 9.3): it is the only doubly transitive DHO whose automorphism group has a simple socle. The existence of $\mathcal{M}$ is an immediate consequence of a six-dimensional, unitary representation over $\mathbb{F}_4$ of the triple cover $3 \cdot \mathrm{M}_{22}$ of the Mathieu group of degree 22. An elementary treatment of $\mathcal{M}$ is Yoshiara [72]. In this article, it is shown that $\mathcal{M}$ is simply connected and split (by similar methods which occur in splitness proofs discussed in Sect. 4.3).

The Veronesean DHOs $\mathcal{V}_n(q)$ are described below. For large rank, one can expect that many nonisomorphic quotients of the Veronesean DHOs exist. Quotients of $\mathcal{V}_n(q)$ will be the topic of Sect. 5.2.

Problem 2 Find DHOs over $\mathbb{F}_q$, $q > 2$ a 2-power, that are not quotients of Veronesean DHOs.

3 Split DHOs and DHO Sets

So far every (concrete) DHO which has been investigated with respect to being split turned out to be split.

Problem 3 Try to find a nonsplit DHO.

Split DHOs have an obvious, but useful property: they can be represented in a simple way within the framework of linear algebra.

Definition 1 Let X, Y be finite-dimensional $\mathbb{F}_q$-spaces and $\Sigma \subseteq \mathrm{Hom}_{\mathbb{F}_q}(X, Y)$, $|\Sigma| = (q^n - 1)/(q - 1) + 1$, with $0 \in \Sigma$. We call Σ a *DHO set* if for every $\phi \in \Sigma$ the mapping

$$\kappa = \kappa_\phi : \Sigma - \{\phi\} \ni \psi \mapsto \kappa(\psi) = \ker(\psi - \phi)$$

is a bijection of $\Sigma - \{\phi\}$ onto the set of 1-spaces of X.

The next two lemmas have a straightforward verification. They show how split DHOs can be coordinatized by DHO sets.

Lemma 1 *Let X, Y be finite-dimensional $\mathbb{F}_q$-spaces and $\Sigma \subseteq \mathrm{Hom}_{\mathbb{F}_q}(X, Y)$ a DHO set. Set $U = X \oplus Y$ and $X(\phi) = \{(x, x\phi) \mid x \in X\}$ for $\phi \in \Sigma$. Set $\mathcal{S} = \{X(\phi) \mid \phi \in \Sigma\}$. Then $\mathcal{S}$ is a DHO over $\mathbb{F}_q$ of rank $\dim_{\mathbb{F}_q} X$ and $\mathcal{S}$ splits over Y.*

Lemma 2 *Let $\mathcal{S}$ a DHO over $\mathbb{F}_q$ with ambient space U that splits over the subspace Y. Let $X \in \mathcal{S}$. Then there exists a DHO set $\Sigma \subseteq \mathrm{Hom}_{\mathbb{F}_q}(X, Y)$ such that $\mathcal{S} = \{X(\phi) \mid \phi \in \Sigma\}$ and $X = X(0)$.*

The DHO sets are called *linear systems* in [72, Sect. 3.2]. We mimic the terminology of translation planes where spreads, the analogues of DHOs, are represented in a similar fashion by spread sets (see Sect. 7.1). The theory of translation planes covers many aspects and has a vast literature (see [2]). So the analogy between DHO sets and spread sets suggests

Problem 4 Look for topics on translation planes (spreads) that can be reformulated for split DHOs.

DHOs, which can be coordinatized by additively closed DHO sets, are called bilinear and have received special attention. Considering the size of a DHO, one observes that a DHO set can be additively closed only if the DHO is defined over $\mathbb{F}_2$.

Definition 2 Let m, n be positive integers, $n \geq 3$ and let $B : \mathbb{F}_2^n \times \mathbb{F}_2^n \to \mathbb{F}_2^m$ be a bilinear mapping. Set $U = \mathbb{F}_2^n \times \mathbb{F}_2^m$ and define for $e \in \mathbb{F}_2^n$

$$X(e) = \{(x, B(x, e)) \mid x \in \mathbb{F}_2^n\} \subseteq U \quad \text{and set} \quad \mathcal{S}_B = \{X(e) \mid e \in \mathbb{F}_2^n\}.$$

If $\mathcal{S}_B$ is a DHO (of rank n over $\mathbb{F}_2$) we call $\mathcal{S}_B$ a *bilinear DHO*.

Let $\mathcal{S}_B$ be a DHO. Note, that if we define $B(e) \in \mathrm{End}(\mathbb{F}_2^n, \mathbb{F}_2^m)$ by $xB(e) = B(x, e)$ then $\Sigma_B = \{B(e) \mid e \in \mathbb{F}_2^n\}$ is an additively closed DHO set. The DHO $\mathcal{S}_B$ splits over $Y = 0 \times \mathbb{F}_2^m$. In [18], some basic theory on bilinear DHOs is developed. Define for $e \in \mathbb{F}_2^n$ the transformation $\tau_e \in \mathrm{GL}(U)$ by

$$(x, y)\tau_e = (x, y + B(x, e)) = (x, y + xB(e)).$$

Then τ_e maps $X(a)$ onto $X(a + e)$ and fixes the ambient space of $\mathcal{S}_B$. So τ_e induces an automorphism of $\mathcal{S}_B$ and the group $T_B = \{\tau_e \mid e \in \mathbb{F}_2^n\} \leq \mathrm{Aut}(\mathcal{S}_B)$ (restricted to the ambient space) is called the *standard translation group*. The general notion *translation group* is taken from [18] and will be introduced in Sect. 9.1. Bilinear DHOs have a purely group theoretic characterization in terms of translation groups.

Define $B^o : \mathbb{F}_2^n \times \mathbb{F}_2^n \to \mathbb{F}_2^m$ by $B^o(x, e) = B(e, x)$. Then B^o defines a bilinear DHO $\mathcal{S}_{B^o}$ too, the DHO *opposite* to $\mathcal{S}_B$. A bilinear DHO $\mathcal{S}_B$ is called *alternating* if $B(e, e) = 0$ for all $e \in \mathbb{F}_2^n$ and it is called *symmetric* if $B = B^o$. Alternating DHOs are in particular symmetric. The Huybrechts DHOs $\mathcal{H}_n$ and the Buratti–Del Fra DHOs $\mathcal{D}_n$ from Sect. 4 are bilinear, $\mathcal{H}_n$ is alternating and $\mathcal{D}_n$ is symmetric. Another important family of bilinear DHOs are the *bilinear DHOs of Yoshiara*

Example 1 (DHOs of Yoshiara [63], [65, Sect. 5.5]) Let h, m, n be positive integers, $1 \leq h, m < n$, $(n, h) = (n, m) = 1$, $F = \mathbb{F}_{2^n}$. Define $B : F \times F \to F$ by

$$B(x, e) = xe^{2^m} + x^{2^h} e.$$

Then $\mathcal{S}_B = \mathcal{S}^n_{m,h}$ is a bilinear DHO of rank n in $U = F \times F$. If $h + m \not\equiv 0 \pmod n$, then U is the ambient space. If however $h + m \equiv 0 \pmod n$, then the ambient space has codimension 1 in U. The full automorphism group (see [58]) has the form $\mathrm{Aut}(\mathcal{S}^n_{m,h}) = T \cdot H$, with $T = T_B$ and $H \simeq \Gamma\mathrm{L}(1, F)$.

These DHOs are contained in the larger class of *Yoshiara DHOs of type* $\mathcal{S}^n_{\sigma,\phi}$ (see [63], [65, Sect. 5.5])). Set again $F = \mathbb{F}_{2^n}$, let σ be a generator of $\mathrm{Gal}(F : \mathbb{F}_2)$, and let ϕ be an o-polynomial on F. For $e \in F$ define $X(e) = \{(x, x^\sigma e + x\phi(e)) \mid x \in F\}$ then $\mathcal{S}^n_{\sigma,\phi} = \{X(e) \mid e \in F\}$ is a DHO in $F \times F$. If ϕ is a *monomial* but not in $\mathrm{Gal}(F : \mathbb{F}_2)$, then $\mathrm{Aut}(\mathcal{S}^n_{\sigma,\phi}) \simeq \Gamma\mathrm{L}(1, F)$ (see [58, Theorem 1.1]).

4 DHOs with Ambient Spaces of Maximal Rank

Clearly, the rank of the ambient space of a DHO of rank n is $\geq 2n - 1$ and there exist many DHOs over $\mathbb{F}_2$ whose rank of the ambient space obtains this lower bound (see Sect. 7.1).

4.1 *Yoshiara's Upper Bound for the Rank of an Ambient Space*

Yoshiara [64], [65, Sect. 2.4] gives an upper bound for the rank of the ambient space of a DHO.

Theorem 1 *Let $\mathcal{S}$ be a DHO of rank n over $\mathbb{F}_q$ with ambient space U. Then*

$$\dim U(\mathcal{S}) \leq \begin{cases} \binom{n+1}{2} + 2, & \text{for } q = 2, \\ \binom{n+1}{2}, & \text{for } q > 2. \end{cases}$$

The family of Veronesean DHOs (see the first example below) shows that the Yoshiara bound is sharp for $q > 2$. Yoshiara conjectures that $\dim U(\mathcal{S}) \leq \binom{n+1}{2}$ holds for $q = 2$ too [65, Problem 2.7].

Problem 5 Prove the conjecture of Yoshiara.

4.2 DHOs Meeting Yoshiara's (Conjectured) Bound for the Rank of an Ambient Space

Presently, four families of simply connected DHOs of rank n in ambient spaces of rank $\binom{n+1}{2}$ are known. Three of them are described already in [65].

Example 2 ([65, Sect. 5.2], [61]) Let q be a 2-power, $n \geq 3$ and V be an n-dimensional $\mathbb{F}_q$-space. Set $U = S^2(V)$ (symmetric square of V). Define the *Veronesean DHO of rank n* as

$$\mathcal{V}_n(q) = \{X(\infty)\} \cup \{X(e) \mid 0 \neq e \in V\}$$

with

$$X(\infty) = \{x \cdot x \mid x \in V\} \quad \text{and} \quad X(e) = \{x \cdot e \mid x \in V\}.$$

Note that $X(e) = X(ke)$ for $0 \neq k \in \mathbb{F}_q$. Denote by $\rho : G \to \mathrm{GL}(U)$ the action of $G = \mathrm{GL}(V)$ induced on the $\mathbb{F}_q G$-module $U = S^2(V)$. Then $X(\infty)\rho(\phi) = X(\infty)$ and $X(e)\rho(\phi) = X(e\phi)$ for $\phi \in G$, i.e. $\rho(G) \leq \mathrm{SemAut}(\mathcal{V}_n(q))$. The action of a field automorphism in $\mathrm{Gal}(\mathbb{F}_q)$ can also be extended to U, such that this automorphism induces an element of $\mathrm{SemAut}(\mathcal{V}_n(q))$. Indeed, $\mathrm{SemAut}(\mathcal{V}_n(q)) \simeq \Gamma\mathrm{L}(n, q)$ and $\mathrm{Aut}(\mathcal{V}_n(q)) \simeq \mathrm{P}\Gamma\mathrm{L}(n, q)$ (see [65, Sect. 5.2]).

Example 3 ([25], [65, Sect. 5.4], [61]) Let $n \geq 3$ and V be an n-dimensional $\mathbb{F}_2$-space. Set $W = \wedge^2(V)$ (alternating square of V), $U = V \oplus W$. Define the *Huybrechts DHO of rank n* as

$$\mathcal{H}_n = \{X(e) \mid e \in V\} \quad \text{with} \quad X(e) = \{(x, x \wedge e) \mid x \in V\}.$$

Since $V \times V \ni (x, e) \mapsto x \wedge e \in W$ is bilinear, and $e \wedge e = 0$ the DHO $\mathcal{H}_n$ is alternating. Denote by T the standard translation group with respect to this bilinear mapping. For $\phi \in G = \mathrm{GL}(V)$ define an action $\overline{\phi}$ on U by $(x, y \wedge z)\overline{\phi} = (x\phi, (y\phi) \wedge (z\phi))$. Then $\overline{\phi} \in \mathrm{Aut}(\mathcal{H}_n)$ and $X(e)\overline{\phi} = X(e\phi)$. The full automorphism group is $\mathrm{Aut}(\mathcal{H}_n) = \mathrm{SemAut}(\mathcal{H}_n) \simeq T \cdot \overline{G}$ with $G \simeq \overline{G} = \{\overline{\phi} \mid \phi \in G\}$ (see [65, Sect. 5.2], [64, Proposition 10], [8, Theorem 2]).

Example 4 ([3], [65, Sect.5.4], [50, 60, 61]) Let $n \geq 4$ and V be an n-dimensional $\mathbb{F}_2$-space. Let $\{e_0, e_1, \ldots, e_{n-1}\}$ be a basis of V and let R be the subspace of $S^2(V)$ which is generated $\{e_0 \cdot e_0, e_0 \cdot e_i + e_i \cdot e_i \mid 1 \leq i < n\}$. Set $W = S^2(V)/R$ and denote by $\overline{v \cdot w}$ the image of $v \cdot w \in S^2(V)$ in W. Set $U = V \oplus W$. Define the *Buratti–Del Fra DHO of rank n* as

$$\mathcal{D}_n = \{X(e) \mid e \in V\} \quad \text{with} \quad X(e) = \{(x, \overline{x \cdot e}) \mid x \in V\}.$$

The mapping $V \times V \ni (x, e) \mapsto \overline{x \cdot e} \in W$ is bilinear and symmetric, i.e. $\mathcal{D}_n$ is a symmetric DHO. Denote by T the standard translation group with respect to this bilinear mapping. Let G_0 be the stabilizer of e_0 in $G = \mathrm{GL}(V)$. For $\phi \in G_0$ define an action $\overline{\phi}$ on U by $(x, \overline{y \cdot z})\overline{\phi} = (x\phi, \overline{(y\phi) \cdot (z\phi)})$. The full automorphism group (see [65, Sect. 5.2], [64, Proposition 10], [8, Theorem 2]) is $\mathrm{Aut}(\mathcal{D}_n) = \mathrm{SemAut}(\mathcal{D}_n) \simeq T \cdot \overline{G_0}$ where $G_0 \simeq \overline{G_0} = \{\overline{\phi} \mid \phi \in G_0\}$.

Example 5 ([44, 50, 61]) Let $n \geq 4$ and V be an n-dimensional $\mathbb{F}_2$-space. Let $\{e_0, e_1, \ldots, e_{n-1}\}$ be a basis of V and set $U = S^2(V)$. For vectors $u = \sum u_i e_i$, $v = \sum_i v_i e_i$ in V define $u \cap v = \sum_i u_i v_i e_i$. Then set $t(u, e_0) = u \cdot e_0$ and for $v \in V - \mathbb{F}_2 e_0$ set

$$t(u, v) = u \cdot u + u \cdot v + (u \cap v) \cdot (u \cap v) + (u + u \cap v) \cdot e_0$$

and further

$$X(\infty) = \{x \cdot x \mid x \in V\}, \quad X(e_0) = \{x \cdot e_0 \mid x \in V\}$$

and for $e \in V - \mathbb{F}_2 e_0$ set

$$X(e) = \{t(x, e) \mid x \in V\}.$$

Then the *Taniguchi DHO of rank n* is defined as

$$\mathcal{T}_n = \{X(\infty)\} \cup \{X(e) \mid e \in V - \{0\}\}.$$

Let G_0 be the stabilizer of e_0 in $G = \mathrm{GL}(V)$. The full automorphism group (see [40, Theorem 2]) is $\mathrm{Aut}(\mathcal{T}_n) = \mathrm{SemAut}(\mathcal{T}_n) \simeq \rho(G_0)$ where the representation $\rho : G \to \mathrm{GL}(U)$ is defined as in Example 2.

The original descriptions of the Buratti–Del Fra DHOs [3] and the Taniguchi DHOs [44] are simplified in [50, 60]. A unified approach to all four families is [61] of Taniguchi and Yoshiara.

4.3 $\mathcal{V}_n(q)$, $\mathcal{H}_n$, $\mathcal{D}_n$ and $\mathcal{T}_n$ Are Split

Since the Huybrechts DHO and the Buratti–Del Fra DHO are bilinear, they are split. The splitness of $\mathcal{V}_n(2)$ and $\mathcal{T}_n$ is proved in [74]. Let $\mathcal{S}$ be $\mathcal{V}_n(2)$ or $\mathcal{T}_n$. Yoshiara shows that each complement Y of $\mathcal{S}$ corresponds to a bilinear map $h = h_Y : V \times V \to V$ satisfying specific conditions. Conversely, any bilinear mapping h, which satisfies the specific conditions, defines a complement $Y = Y(h)$. Then $h_{Y(h)} = h$, $Y(h_Y) = Y$, and in both cases, it is shown that such bilinear mappings exist.

The conditions for the bilinear mappings are somewhat technical if $\mathcal{S} = \mathcal{T}_n$ and they are not repeated here. They are simple in the case $\mathcal{S} = \mathcal{V}_n(2)$: h defines on V

the multiplication of a commutative presemifield (see Sect. 8.2) with the additional property $h(a, a) = a$ for $a \in V$. This sets up a nice correspondence between the set of complements of $\mathcal{V}_n(2)$ with classes of commutative presemifields with regard to strong isotopism (a natural notion of equivalence for presemifields). Yoshiara has shown, that $\mathcal{V}_n(q)$, $q > 2$, splits too, but this result has not been published yet.

4.4 Bilinear DHOs Whose Ambient Space Has Maximal Rank

Yoshiara [73] achieves a remarkable result for bilinear DHOs which can be summarized as

Theorem 2 *Let $\mathcal{S}$ be a bilinear DHO of rank n over $\mathbb{F}_2$. The following hold:*

(a) $\dim U(\mathcal{S}) \leq \binom{n+1}{2}$.
(b) Let $\dim U(\mathcal{S}) = \binom{n+1}{2}$. *Then $\mathcal{S}$ is isomorphic to the Huybrechts DHO or the Buratti–Del Fra DHO of rank n.*

Assertion (a) strengthens the conjecture on the Yoshiara bound. Surprising is part (b): DHOs of rank n whose rank of the ambient space meets the bound $\binom{n+1}{2}$ seem to be rather scarce.

5 Quotients of DHOs

Before we consider examples of DHOs and their quotients, we explain how quotients are determined in a given DHO. The following concrete description is taken from Yoshiara [64, Proposition 13].

Let $\mathcal{S}$ be a DHO over $\mathbb{F}_q$ of rank n with ambient space U. For a subspace $W \subseteq U$ set

$$\mathcal{S}/W = \{(X + W)/W \mid X \in \mathcal{S}\}.$$

Let $\pi : U \to U'$ a covering map sending $\mathcal{S}$ to the DHO $\mathcal{S}'$ of rank n in U'. Let $W = \ker \pi$. Then $\mathcal{S}'$ is isomorphic to the DHO $\mathcal{S}/W$ in U/W. Conversely, if $W \subseteq U$ is a subspace with

$$(X + X') \cap W = 0 \quad \text{for all} \quad X, X' \in \mathcal{S},$$

then $\mathcal{S}/W$ is a DHO of rank n.

5.1 Universal Covers

The role of simply connected DHOs is clarified in [13]:

Theorem 3 *Let $\mathcal{S}$ be a DHO of rank n over $\mathbb{F}_q$ with ambient space U. Then there exists up to a linear isomorphism a uniquely determined, simply connected DHO $\widehat{\mathcal{S}}$ with ambient space $\widehat{U}$ and a linear covering map $\pi : (\widehat{\mathcal{S}}, \widehat{U}) \to (\mathcal{S}, U)$, such that for each α-linear covering map $\phi : (\mathcal{S}', U') \to (\mathcal{S}, U)$, there exists a unique α^{-1}-linear covering map $\psi : (\widehat{\mathcal{S}}, \widehat{U}) \to (\mathcal{S}', U')$, such that*

$$\pi = \psi \circ \phi.$$

One calls $\widehat{\mathcal{S}}$ the *universal cover of* $\mathcal{S}$. The universal cover controls the structure of the automorphism group of its quotients

Corollary 1 *Let $\mathcal{S}$ be a simply connected DHO of rank n over $\mathbb{F}_q$ with ambient space U. Let $W \subseteq U$ be a subspace, such that $\mathcal{S}/W$ is a DHO of rank n. Then the stabilizer* $\mathrm{SemAut}(\mathcal{S})_W$ *of W in* $\mathrm{SemAut}(\mathcal{S})$ *is isomorphic to* $\mathrm{SemAut}(\mathcal{S}/W)$.

One is tempted to restrict the attention mainly to simply connected DHOs. However, the above results are merely existence statements. To determine the universal cover of a DHO or to find the quotients of a DHO requires extra work.

5.2 Quotients of the Veronesean DHOs

Let q be a 2-power, $m \geq 3$ and consider $\mathbb{F}_{q^m}$ as a $\mathbb{F}_q$-space. Let V be a n-dimensional subspace of $\mathbb{F}_{q^m}$ and σ be a generator of $\mathrm{Gal}(\mathbb{F}_{q^n} : \mathbb{F}_q)$. For $0 \neq e \in V$ define in $U = \mathbb{F}_{q^m} \times \mathbb{F}_{q^m}$ the projective subspaces $X(\infty) = \{\mathbb{F}_q(x^2, 0) \mid 0 \neq x \in V\}$ and $X(\mathbb{F}_q e) = \{\mathbb{F}_q(xe, x^\sigma e + xe^\sigma) \mid 0 \neq x \in V\}$. Taniguchi [37], [65, Sect. 5.6] shows that $\mathcal{T}_\sigma(V) = \{X(\infty)\} \cup \{X(\mathbb{F}_q e) \mid 0 \neq e \in V\}$ is a DHO of rank n over $\mathbb{F}_q$. The rank of the ambient space depends on the choice of the space V. In particular, for $n = m$, we get $\dim U(\mathcal{T}_\sigma(\mathbb{F}_{q^n})) = 2n$. It turns out that the DHOs $\mathcal{T}_\sigma(V)$ are quotients of $\mathcal{V}_n(q)$ (see Yoshiara [66, Proposition 1]). Automorphism and isomorphism questions are addressed in [65, Sect. 5.6] and [39].

Let $\sigma \in \mathrm{Gal}(\mathbb{F}_{2^n} : \mathbb{F}_2)$ and H a $\mathbb{F}_2$-hyperplane in $\mathbb{F}_{2^n}$. In [59], Taniguchi and Yoshiara find an other family of DHOs denoted by $\mathcal{S}_{\sigma,H}$, of rank n over $\mathbb{F}_2$ which are quotients of $\mathcal{V}_n(2)$ and not isomorphic to DHOs of type $\mathcal{T}_\sigma(V)$. The isomorphism problem between DHOs of type $\mathcal{S}_{\sigma,H}$ is settled too.

Problem 6 Does $\mathcal{V}_n(q)$ have quotients whose ambient space has rank $2n - 1$?

5.3 Quotients of the Huybrechts DHOs

Let $V = \mathbb{F}_2^n$ and $B : V \times V \to W = \mathbb{F}_2^m$ be a bilinear mapping which defines an alternating DHO $\mathcal{S}_B$. It is easy to see (see [15, Lemma 2.2]) that $\mathcal{S}_B$ is a quotient of $\mathcal{H}_n$. Yoshiara (see [68, 70]) as well as Göloğlu and Pott [24] discovered that quadratic APN functions define alternating DHOs. A function $f : X \to Y$, both X and Y are $\mathbb{F}_2$-spaces, is called *quadratic* if $B_f : X \times X \ni (x, y) \mapsto B_f(x, y) = f(x + y) + f(x) + f(y) + f(0) \in Y$ is biadditive. A function $f : X \to Y$ is an *APN function* if for all $0 \neq a \in X$ and all $b \in Y$ the equation $f(x + a) + f(x) = b$ has at most two solutions. If f is a quadratic APN function then indeed $\mathcal{S}_{B_f}$ is an alternating DHO. On the other hand, Edel [22] shows that vice versa alternating DHOs relate to quadratic APN functions. In its general form, this connection between quadratic APN functions and alternating DHOs can be expressed by

Theorem 4 ([18, Theorem 2.4]) *Let $f : X \to Y$ be a quadratic APN function of $\mathbb{F}_2$-spaces X, Y. Then $B_f : X \times X \to Y$ defines an alternating DHO $\mathcal{S}_{B_f}$ in $U = X \oplus Y$. For any alternating DHO $\mathcal{S}_B$ in $X \oplus Y$, there exists a quadratic APN function $f : X \to Y$ such that $\mathcal{S}_B = \mathcal{S}_{B_f}$. Moreover, $\mathcal{S}_{B_f} = \mathcal{S}_{B_g}$ iff $f + g$ is a linear function.*

This implies that an alternating DHO of rank n has an ambient space of rank $\geq 2n$ (see [18, Lemma 5.12]). Quadratic APN functions $f : \mathbb{F}_{2^n} \to \mathbb{F}_{2^n}$ have received a lot of attention; see [35, Sects. 5 and 6] for references. Quadratic APN functions $f : \mathbb{F}_{2^n} \to \mathbb{F}_{2^m}$, $m \geq n$, are characterized in [18] in group theoretic terms (APN functions with "translation groups"). Some basic theory is developed in this article.

For quotients of $\mathcal{H}_n$ which have a doubly transitive automorphism group; see [14] and Sect. 9.3.

5.4 Quotients of the Buratti–Del Fra DHOs

In [46], Taniguchi shows that a quadratic APN function $f : \mathbb{F}_2^n \to \mathbb{F}_2^n$ gives rise to a bilinear dual hyperoval $\mathcal{D}_f = \mathcal{S}_B$ of rank n in $\mathbb{F}_2^n \oplus \mathbb{F}_2^{2(n-1)}$ where the bilinear mapping $B : \mathbb{F}_2^n \times \mathbb{F}_2^n \to \mathbb{F}_2^{2(n-1)}$ can be seen as a perturbation of the mappimg B_f. The DHO $\mathcal{D}_f$ is a quotient of $\mathcal{D}_n$ and $\mathcal{D}_f \simeq \mathcal{D}_g$ iff f and g are EA equivalent (see [18, p.471] for the definition of EA equivalence).

Let X and Y be $\mathbb{F}_2$-spaces and $B : X \times X \to Y$ a bilinear mapping which defines a DHO $\mathcal{S}_B$ of rank n in $U = X \oplus Y$. For $0 \neq e \in X$ let $0 \neq \kappa(e) \in X$ be the unique vector such that $B(\kappa(e), e) = 0$. In [60, Lemma 4], Taniguchi and Yoshiara give a simple criterion for the kernel function κ which guarantees that $\mathcal{S}_B$ is a quotient of $\mathcal{D}_n$. For instance, if n is odd, $X = \mathbb{F}_{2^n}$ then this criterion shows, that $B(x, y) = x^4y + xy^4 + xy + (xy)^2$ defines a quotient $\mathcal{S}_B$ of $\mathcal{D}_n$, whose ambient space have rank $2n$. In Sect. 7.3, we show that a bilinear DHO $\mathcal{S}_B$ of rank n gives rise to a DHO of rank $n + 1$, which is called the extension of $\mathcal{S}_B$. If $\mathcal{S}_B$ is symmetric, then the extension is bilinear. In [10], extensions of symmetric bilinear DHOs are studied whose twofold extension is bilinear again. It turns out that such DHOs are quotients of the Huybrechts DHOs or the Buratti–Del Fra DHOs. More examples of quotients of $\mathcal{D}_n$ with ambient spaces of rank $2n$ and $2n-1$ are found there.

5.5 *Quotients of the Taniguchi DHOs*

Taniguchi [48] investigates $\mathcal{T}_n$ for possible quotients. He associates $\sigma \in \mathrm{Gal}(\mathbb{F}_{2^{n-1}} : \mathbb{F}_2)$ with some quotient—denoted by T_σ—of $\mathcal{T}_n$ whose ambient space has rank $3n - 2$. He also shows that $T_\sigma \simeq T_\rho$ iff $\rho = \sigma$ or $\rho = \sigma^{-1}$. The rank of the ambient spaces of T_σ is significantly larger than the lower bound $2n - 1$ (for the rank of the ambient space a DHO of rank n). By [1], proper quotients of $\mathcal{T}_4$ have an ambient space of rank 9.

Problem 7 Show or disprove that a quotient of $\mathcal{T}_n$ has an ambient space of rank $\geq 2n + 1$.

6 Duality

For many mathematical theories, the notion of duality is of general interest. Concepts of duality for DHOs is the topic of this Section.

6.1 *Doubly Dual Hyperovals*

Let $\mathcal{S}$ be a DHO of rank n over $\mathbb{F}_q$ with an ambient space U of rank $2n$. Let $U^\star$ be the dual space of U ($\mathbb{F}_q$-space of linear functionals) and define a set $\mathcal{S}^t = \{X^t \mid X \in \mathcal{S}\}$ of n-spaces in $U^\star$ (X^t is the space of linear functionals that vanish on X). One observes, that $\mathcal{S}^t$ is a DHO iff $U = X + Y + Z$ for every three $X, Y, Z \in \mathcal{S}$. In this case, we call $\mathcal{S}$ a DDHO (*doubly dual hyperoval*). In the next subsection, we consider DHOs of polar type. Symplectic DHOs or orthogonal DHOs (in quadratic spaces of maximal Witt index) provide examples of DDHOs. Some DHOs of Yoshiara of odd rank are DDHOs (see Sect. 6.3). In [1], it is shown that there exist precisely 26 DHOs of rank 4 over $\mathbb{F}_2$ whose ambient space have rank 8. None of these DHOs is a DDHO.

Problem 8 Show or disprove that a DDHO has an odd rank.

Let $\mathcal{S}$ be a DDHO in $U \simeq \mathbb{F}_2^{2n}$. In [9], it is observed that the characteristic function of the set $(\bigcup_{X\in\mathcal{S}} X) - \{0\} \subset U$ is a bent function and if $\mathcal{S}$ splits over $Y \subseteq U$, then the characteristic function of the set $Y \cup \bigcup_{X\in\mathcal{S}} X$ is a bent function too.

Taniguchi in [57] discusses connections between distance regular graphs and DDHOs. In [34] the authors consider the incidence graphs of the affine expansions of bilinear Yoshiara DHOs and show that $\Gamma(\mathcal{S}^n_{m,h})$ is a distance regular DHO iff $m+h$ is coprime to n (see [34, Theorem 1.8]). Taniguchi generalizes this result significantly

Theorem 1 ([57, Theorem 2]) *Let $\mathcal{S}_B$ a bilinear DDHO. Then $\Gamma(\mathcal{S}_B)$ is a distance regular graph.*

He conjectures

Problem 9 Let $\mathcal{S}_B$ be a bilinear DHO of rank n with an ambient space of rank $2n$. Assume that $\Gamma(\mathcal{S}_B)$ is distance regular. Then $\mathcal{S}_B$ a DDHO.

has a positive answer.

6.2 DHOs in Polar Spaces

Let U be a finite dimensional, nondegenerate symplectic, orthogonal or unitary space over $\mathbb{F}_q$. If maximal isotropic (symplectic or unitary space) or maximal totally singular spaces (orthogonal space) have rank n, one calls U a *polar space of rank n*. Let $\mathcal{S}$ be a DHO in a polar space U of rank n. We say that $\mathcal{S}$ is a *symplectic or unitary DHO* (U is symplectic or unitary) if all members of $\mathcal{S}$ are isotropic spaces of rank n and we say that $\mathcal{S}$ is a *orthogonal DHO* (U is orthogonal) if all members of $\mathcal{S}$ are totally singular spaces of rank n. Section 4 of Yoshiara's report [65] is devoted to DHOs in polar spaces.

Based on the work of Vanhove [62], Sheekey [36] makes the following important contribution:

Theorem 2 *Let $\mathcal{S}$ be DHO of rank n of isotropic or totally singular subspaces ambient in the polar space U. Then n is odd or $U \simeq V^-(2n+2, q)$.*

Here we denote by $V^-(2n, q)$ (by $V^+(2n, q)$) a nondegenerate, orthogonal space of dimension $2n$ and Witt index $n-1$ (and Witt index n). In particular, Sheekey's Theorem answers question (1) of [65, Problem 4.7].

Problem 10 Let $n > 3$ be even and q be a prime power. Show that $V^-(2(n+1), q)$ does not contain an orthogonal DHO or give a counterexample.

On the other hand, many orthogonal DHOs of rank n, n odd, are constructed as quotients of orthogonal spreads in $V^+(2n+2, 2)$ (see Sect. 7.2). In [32], Nambu and Yoshiara show that a Yoshiara DHO $\mathcal{S}^n_{\sigma,\phi}$ is of polar type iff n is odd, $\sigma, \phi \in \mathrm{Gal}(\mathbb{F}_{2^n} : \mathbb{F}_2)$, and $\sigma \circ \phi^2 = 1$. In this case, the DHO is not only symplectic, but orthogonal too. Reference [9] contains examples of bilinear DHOs (of odd rank) that are symplectic but not orthogonal. The Mathieu DHO $\mathcal{M}$ is a unitary DHO of rank 3 in a unitary $\mathbb{F}_4$-space of rank 6.

6.3 Knuth Operations

Let $B : X \times X \to X$ define a bilinear DHO $\mathcal{S}_B$ with the ambient space $X \oplus X$. It is convenient to identify B with a monomorphism $B : X \to \mathrm{End}(X)$ by defining $xB(y) = B(x, y)$. Let $\cdot : X \times X \to \mathbb{F}_2$ be a nondegenerate, symmetric bilinear form (for instance the ordinary dot product of $X = \mathbb{F}_2^n$). Then X can be identified with the dual space $X^\star$ via the bilinear form and the operator adjoint to $\phi \in \mathrm{End}(X)$ is identified with the operator $\phi^t \in \mathrm{End}(X)$ with $x\phi^t \cdot y = x \cdot y\phi$ for all $x, y \in X$. Then the monomorphism $B^t : X \to \mathrm{End}(X)$ defined by $B^t(e) = B(e)^t$ may or may not define a bilinear DHO. But if B^t defines a DHO (i.e. $\mathcal{S}_B$ is a DDHO) then $\mathcal{S}_{B^t} \simeq \mathcal{S}_B^t$. Recall the definition of the opposite DHO $\mathcal{S}_{B^o}$ from Sect. 3. In analogy to the terminology of semifields, we call the operations $B \mapsto B^t$ and $B \mapsto B^o$ *Knuth operations* and resulting isomorphism types of DHOs a *Knuth class*.

Lemma 1 ([23, Sect. 5], [9, Sect. 3], [47, p. 211]) *Let $B : X \to \mathrm{End}(X)$ be a monomorphism which defines a bilinear DHO over $\mathbb{F}_2$ with ambient space $U \simeq X \oplus X$. Assume that B^t defines a DHO too. Then $\mathcal{S}_{B^{oto}} \simeq \mathcal{S}_{B^{tot}}$ and the Knuth class contains at most six nonisomorphic members coordinatized by B, B^t, B^o, B^{to}, B^{ot} and B^{tot}.*

As an example consider $\mathcal{S}_B = \mathcal{S}^n_{m,h}$, a bilinear DHO of Yoshiara. Then $\mathcal{S}_{B^o} = \mathcal{S}^n_{h,m}$. Assume that $\mathcal{S}_B$ is a DDHO. Then n is odd. In this case, $\mathcal{S}_{B^t} \simeq \mathcal{S}^n_{m+h,n-h}$. Moreover, $\mathcal{S}_{B^{ot}} \simeq \mathcal{S}^n_{m+h,n-m}$, $\mathcal{S}_{B^{to}} \simeq \mathcal{S}^n_{n-h,m+h}$, and $\mathcal{S}_{B^{oto}} \simeq \mathcal{S}^n_{n-m,m+h} \simeq \mathcal{S}^n_{m,n-m-h} \simeq \mathcal{S}_{B^{tot}}$. If, for instance, $h \not\equiv \pm m \pmod n$, then [58, Theorem 1.2] shows that the Knuth class has size 6.

Let n be odd. Taniguchi [47] considers the alternating bilinear DHO $\mathcal{S}_B$ where $B = B_f$ is the bilinear mapping associated with the quadratic APN function $f : \mathbb{F}_{2^n} \ni x \mapsto x^3 + \mathrm{Tr}(x^9)$ (Tr denotes the absolute trace in $\mathbb{F}_{2^n}$) and proves that the DHOs associated with B^t and B^{to} are not isomorphic to bilinear Yoshiara DHOs.

Let $\phi \in \mathrm{End}(X)$. Then ϕ is *selfadjoint* if $(x\phi) \cdot y = x \cdot (y\phi)$ for all $x, y \in X$ and ϕ is *skewsymmetric* if $(x\phi) \cdot x = 0$ for all $x \in X$. Define a nondegenerate quadratic form Q on $U = X \oplus X$ by $Q(x, y) = x \cdot y$. This turns U into an orthogonal space of type $V^+(2n, 2)$. One verifies immediately that a DHO set of skewsymmetric operators define via Lemma 1 in U an orthogonal DHO and a DHO set of selfadjoint operators define in U a symplectic DHO (with respect to the bilinear form associated with Q). So the Knuth operations induce a one-to-one correspondence between symmetric DDHOs (alternating DDHOs) $\mathcal{S}_{B^{ot}}$ and symplectic DHOs (orthogonal DHOs) $\mathcal{S}_B$.

In [43], Taniguchi shows that an alternating DHO of rank n is a DDHO iff n is odd and in [11], it is shown that n is odd for a symmetric DDHO (however, there exist symmetric DHOs of odd rank which are not DDHOs). The above one-to-one correspondence shows that both results follow Sheekey's theorem [36, Corollary 1].

7 Secondary Constructions

In this section, we consider DHOs derived from other structures.

7.1 *Quotients of Spreads*

Let V be a $2n$-dimensional $\mathbb{F}_2$-space, $n \geq 3$ and $\mathcal{T}$ be a *spread* in V, i.e. a collection of n-spaces of V such that any two spaces from $\mathcal{T}$ have a trivial intersection and $V = \bigcup_{S \in \mathcal{T}} S$. Then V together with the cosets $\mathcal{L} = \{S + v \mid v \in V,\ S \in \mathcal{T}\}$ of the elements from $\mathcal{T}$ form a translation plane $\pi(\mathcal{T}) = (V, \mathcal{L}, \in)$.

Let $P \subset V$ be a 1-space and $S_0 \in \mathcal{T}$ the unique space which contains P. Set

$$\mathcal{T}/P = \{(S + P)/P \mid S \in \mathcal{T} - \{S_0\}\}.$$

By [6], [7, Example 1.2], [42, Proposition 6] $\mathcal{T}/P$ is a dimensional dual hyperoval in $U = V/P$. As the examples show, the isomorphism type of such a quotient depends on the choice of P. Concrete investigations of quotients of spreads are Taniguchi [41] and [42]. A set $\Xi \subseteq \mathrm{GL}(n, 2) \cup \{0_{n\times n}\}$, $0_{n\times n} \in \Xi$, is a *spread set* if $|\Xi| = 2^n$ and $\det(\phi - \psi) \neq 0$ for $\phi, \psi \in \Xi$, $\phi \neq \psi$. A spread $\mathcal{T}$ in $V = \mathbb{F}_2^n \times \mathbb{F}_2^n$ can be described by a spread set as $\mathcal{T} = \{V(\infty)\} \cup \{V(\phi) \mid \phi \in \Xi\}$ with $V(\infty) = 0 \times \mathbb{F}_2^n$ and $V(\phi) = \{(x, x\phi) \mid x \in \mathbb{F}_2^n\}$. If $\Xi - \{0_{n\times n}\}$ is a subgroup of $\mathrm{GL}(n, 2)$, one calls the associated plane a *nearfield plane*. Such planes have been classified and Taniguchi studies quotients $\mathcal{T}/P$ for spreads from nearfield planes choosing $P \subseteq V(\infty)$ (this quotient does not depend on the choice of the particular subspace P in $V(\infty)$). He shows [41] that a $\mathcal{T}/P$ is a Yoshiara DHO if $\mathcal{T}$ is the desarguesian spread, but $\mathcal{T}/P$ is not isomorphic to a Yoshiara DHO if $\mathcal{T}$ is the spread of a regular nearfield (Dickson nearfield) plane. In [42], it is shown that the translation planes $\pi(\mathcal{T}_1)$ and $\pi(\mathcal{T}_2)$ are isomorphic if the DHOs $\mathcal{T}_1/P$ and $\mathcal{T}_2/P$ are isomorphic.

The translation planes of order 16 have been classified [21] as well as the DHOs of rank 4 over $\mathbb{F}_2$ with an ambient space of rank 7 [1]. It turns out that 28 of the 37 DHOs are quotients of one of the 8 spreads of order 16. However, there exist nonisomorphic planes that produce isomorphic quotients. The group $G = \{\phi \in \mathrm{GL}(V) \mid \mathcal{T}\phi = \mathcal{T}\}$ is the translation complement. It is clear that $\mathcal{T}/P \simeq \mathcal{T}/P'$ if P and P' lie in the same orbit of the translation complement. However, for the quotients of the translation planes of order 16, one observes $\mathcal{T}/P \not\simeq \mathcal{T}/P'$ if P and P' lie in different orbits of the translation complement.

Problem 11 Let $\mathcal{T}$ be a spread in $V \simeq \mathbb{F}_2^{2n}$ and let P, P' be 1-spaces in V. Suppose $\mathcal{T}/P \simeq \mathcal{T}/P'$. Show or disprove that P and P' lie in the same orbit of the translation complement.

7.2 Quotients of Orthogonal Spreads

Let V be a $(2n+2)$-dimensional $\mathbb{F}_2$-space and Q a nondegenerate quadratic form on V such that (V, Q) is an orthogonal space of type $V^+(2n+2, 2)$. A set $\mathcal{O}$ of totally singular $(n+1)$-spaces is called an *orthogonal spread* if any two members of $\mathcal{O}$ have trivial intersection and $\bigcup_{S\in\mathcal{O}} S$ contains all singular vectors. Orthogonal spreads in $V^+(2n+2, 2)$ only exist if n is odd.

Let $P \subset V$ be a 1-space. If P is a *nonsingular* space then $W = P^{\perp}/P$ is turned in an obvious way into a nondegenerate symplectic space and

$$\mathcal{O}/P = \{(S+P)/P \mid S \in \mathcal{O}\}$$

is a spread of (isotropic) spaces in W (see [27]). Translation planes arising in this way have been studied in particular by Kantor et al. [28, 29].

Suppose now that P is a *singular* space. Then $U = P^{\perp}/P$ is turned in an obvious way into a nondegenerate orthogonal space of type $V^+(2n, 2)$. Let S_0 the unique member of $\mathcal{O}$ which contains P. Then

$$\mathcal{O}/P = \{(S+P)/P \mid S \in \mathcal{O} - \{S_0\}\}$$

is an orthogonal DHO (see [20, Theorem 1.1]). In this way, one can obtain many orthogonal DHOs: Denote by $\rho(n)$ the number of prime factors of the integer n. It is shown [20, Theorem 1.2] that for n odd $V^+(2n, 2)$ contains at least $2^{n(\rho(n)-2)}/n^2$ pairwise nonisomorphic orthogonal DHOs. Moreover, it is shown that this construction leads to orthogonal DHOs in $V^+(2n, 2)$ with a cyclic automorphism group of order $2^n - 1$ fixing one member of the DHO and acting transitively on the remaining members [20, Theorem 1.3]. Finally, one can construct by this method orthogonal DHOs in $V^+(2n, 2)$ having an elementary abelian automorphism group of order 2^n acting regularly on the DHO [20, Example 8.1].

7.3 Extensions of Bilinear DHOs

Let X, Y be finite-dimensional $\mathbb{F}_2$-spaces, $\dim X = n$ and let $B : X \times X \to Y$ be bilinear, such that $\mathcal{S}_B$ is a bilinear DHO in $U = X \oplus Y$. Recall the definition of the opposite DHO $\mathcal{S}_{B^o}$ from Sect. 3. Set $\overline{X} = \mathbb{F}_2 \oplus X$ and $\overline{Y} = X \oplus Y$. For $e \in X$, define two subspaces of $\overline{X} \oplus \overline{Y}$ by

$$X(0, e) = \{(b, be, be + x, (be + x)B(e)) \mid (b, x) \in \overline{X}\},$$

$$X(1, e) = \{(b, be + x, be, (be + x)B^o(e)) \mid (b, x) \in \overline{X}\},$$

and set $\overline{\mathcal{S}} = \overline{\mathcal{S}}_B = \{X(a, e) \mid (a, e) \in \overline{X}\}$.

Theorem 3 *The set* $\overline{\mathcal{S}}$ *is a DHO in* $\overline{X} \oplus \overline{Y}$. *Moreover*

(a) If $\mathcal{S}$ *is symmetric, then* $\overline{\mathcal{S}}$ *is bilinear.*
(b) $X \oplus Y$ *is the ambient space of* $\mathcal{S}$ *iff* $\overline{X} \oplus \overline{Y}$ *is the ambient space of* $\overline{\mathcal{S}}$.

One calls the DHO $\overline{\mathcal{S}}$ the *extension* of $\mathcal{S}$. For symmetric DHOs, this construction is introduced in [18]. The generalization to arbitrary DHOs is [49] where Taniguchi also shows

Theorem 4 *If* $\mathcal{S}$ *is a simply connected bilinear DHO, then* $\overline{\mathcal{S}}$ *is simply connected too.*

The standard translation group of a bilinear DHO $\mathcal{S}_B$ gives rise to the existence of a large elementary abelian group of automorphisms of the extended DHO: Let $e \in X$. Define with respect to the decomposition $U = \mathbb{F}_2 \oplus X \oplus X \oplus Y$ two operators

$$n_{1,e} = \begin{pmatrix} 1 & e & & \\ & 1 & & \\ & & 1 & B(e) \\ & & & 1 \end{pmatrix} \quad \text{and} \quad n_{0,e} = \begin{pmatrix} 1 & & e & \\ & 1 & & B^o(e) \\ & & 1 & \\ & & & 1 \end{pmatrix}.$$

Then $N_a = \{n_{a,e} \mid e \in X\}$, $a \in \mathbb{F}_2$, are elementary abelian 2-subgroups of $\mathrm{Aut}(\overline{\mathcal{S}})$. The group N_a fixes all elements in $\mathcal{S}_a = \{X(a, e) \mid e \in X\}$ and it acts regularly on $\mathcal{S}_{a+1}$. In particular, $X(0, e)n_{1,f} = X(0, e + f)$ and $X(1, e)n_{0,f} = X(1, e + f)$. Moreover, $N = N_0 \times N_1$ is an elementary abelian group of order $|X|^2$. In Sect. 9.2, DHOs (of rank $n + 1$ over $\mathbb{F}_2$) will be considered which admit elementary abelian automorphism groups of order 2^{2n} which have two orbits of length 2^n on the DHO and an action "like the group N". This will lead to characterizations of extensions of DHOs in group theoretic terms.

In [71], Yoshiara develops a more general concept of extensions: Let $\mathcal{A}$ be a DHO of rank $n + 1$, $n \geq 3$, over $\mathbb{F}_q$. Let $\mathcal{B}$ a set of subspaces of rank n in $U(\mathcal{A})$ which forms a dual arc. The author calls $\mathcal{B}$ a *sub-dual arc* or subDA if for each $B \in \mathcal{B}$ there exists a $A \in \mathcal{A}$ with $B \subset A$. Note, that A is uniquely determined by B. Let $\mathcal{A}(\mathcal{B})$ be the set of $A \in \mathcal{A}$ which contain a member from $\mathcal{B}$. Furthermore, $\mathcal{B}$ is a subDHO, if $\mathcal{B}$ is even a DHO of rank n. Yoshiara goes on and writes $\mathcal{A} = \mathcal{B}_1 \sqcup \cdots \sqcup \mathcal{B}_m$ and calls $\mathcal{A}$ the *disjoint union* of the $\mathcal{B}_i$'s if $\mathcal{A} = \mathcal{A}(\mathcal{B}_1) \cup \cdots \cup \mathcal{A}(\mathcal{B}_m)$ and if for all $1 \leq i < j \leq m$ and all $X_i \in \mathcal{B}_i$, $X_j \in \mathcal{B}_j$, we have $X_i \cap X_j = 0$. For instance, we observe that the extension $\overline{\mathcal{S}_B}$ of the bilinear DHO $\mathcal{S}_B$ can be viewed a disjoint union of $\mathcal{S}_B$ and $\mathcal{S}_{B^o}$. Then:

Theorem 5 ([71, Proposition 1.2]) *Let* $\mathcal{A}$ *be a DHO of rank* $n + 1$ *over* $\mathbb{F}_q$ *with* $n \geq 3$. *Assume* $\mathcal{A} = \mathcal{B}_1 \sqcup \cdots \sqcup \mathcal{B}_m$, $m > 1$ *with subDHOs* $\mathcal{B}_i$ *of rank* n. *Then the following hold:*

(1) We have $m = 2$ *and* $q = 2$.
(2) $U(\mathcal{A})$ *is spanned by the members of* $\mathcal{B}_1$ *and one member from* $\mathcal{A}(\mathcal{B}_2)$

(3) If, furthermore, $\mathcal{B}_1$ is simply connected and $\operatorname{rk} U(\mathcal{A}) = \operatorname{rk} U(\mathcal{B}_1) + n + 1$, *then $\mathcal{A}$ is simply connected.*

Based on this result, Yoshiara investigates *split* DHOs $\mathcal{B}_1$ of rank n over $\mathbb{F}_2$ such that there exists a DHO of the form $\mathcal{A} = \mathcal{B}_1 \sqcup \mathcal{B}_2$ of rank $n+1$. Because of a lack of space the concrete, technical conditions (see [71, Theorem 1.3, Corollary 1.4]) for the existence of such DHOs will not be repeated. However, we mention that if a DHO $\mathcal{A}$ of this form exists then its isomorphism type is essentially determined by a pair $(\mathcal{B}_1, Y_1)$, where $\mathcal{B}_1$ splits over Y_1 (see [71, Sect. 1.7]). Yoshiara considers possible extensions of Veronesean, Huybrechts, Bruratti–Del Fra, Taniguchi or Yoshiara type. So far extensions in this generalized form have only been found for bilinear DHOs.

8 Direct Constructions

In a series of papers, Taniguchi gives various constructions of DHOs over $\mathbb{F}_2$. He proves that the number of DHOs (many of them are simply connected) grows exponentially as a function of the rank.

8.1 DHOs of Type $\mathcal{S}_c$

Before the appearance of [51], most of the known DHOs of rank n over $\mathbb{F}_2$ either had an ambient space of rank $\leq 2n$ or were a quotient of one of the "maximal" DHOs of Sect. 4. In [51], Taniguchi produces with one construction, the *DHOs of type* $\mathcal{S}_c$, symmetric DHOs in large numbers over $\mathbb{F}_2$, which have rank n, which have an ambient space of rank $> 2n$ and which are not quotients of "maximal" DHOs. In fact, the DHOs of type $\mathcal{S}_c$ depend on three parameters c, ℓ, r and we denote them here also by $\mathcal{S}_{c,\ell,r}$. We sketch the construction which is a hybrid of the construction of the Buratti–Del Fra DHOs and an example from [10].

Let ℓ, r be positive integers, $\ell r \geq 4$ and $c \in \mathbb{F}_{2^r}$ with absolute trace 1. Consider three spaces: $V_1 = \langle e_1, \ldots, e_\ell \rangle \subset V_2 = \langle e_0, e_1, \ldots, e_\ell \rangle$ are $\mathbb{F}_{2^r}$-spaces of rank ℓ and $\ell + 1$, respectively. We consider the additive group $X = \mathbb{F}_2 e_0 \oplus V_1$ as a $\mathbb{F}_2$-space of rank $\ell r + 1$. Let $S^2(V_2)$ the symmetric square of V_2 and denote by W the $\mathbb{F}_{2^r}$-subspace generated by $e_0 \cdot e_0$ and $c(e_i \cdot e_i) + e_0 \cdot e_i$, $1 \leq i \leq \ell$. Set $Y = S^2(V_2)/W$ and denote by $\overline{x \cdot y}$ the homomorphic image of $x \cdot y$ in Y. Define $B : X \times X \to Y$ by $B(x, y) = \overline{x \cdot y}$. Then $\mathcal{S}_B$ is a symmetric bilinear DHO of rank $1 + \ell r$ over $\mathbb{F}_2$ whose ambient space has rank $1 + r(\ell^2 + 3\ell)/2$. This DHO is denoted by $\mathcal{S}_c = \mathcal{S}_{c,\ell,r}$.

For extreme choices of ℓ and r, known examples are reproduced: If $r = 1, c = 1$, then $\mathcal{S}_c$ is a Buratti–Del Fra DHO, and if $\ell = 1$, then $\mathcal{S}_c$ is the DHO from [10, Example 3.4]. Taniguchi shows

1. Let $c \neq 1$. Then $\mathcal{S}_c$ is not a quotient of a "maximal DHO" from Sect. 4.
2. $\mathcal{S}_{c,\ell,r} \simeq \mathcal{S}_{c',\ell',r'}$ iff $(\ell, r) = (\ell', r')$ and c' is a Galois conjugate of c.

The DHOs of type $\mathcal{S}_c$ are a source of many simply connnected DHOs: Let $\mathcal{S}_c = \mathcal{S}_{c,\ell,r}$ then $\mathcal{S}_c$ is simply connected iff $\mathbb{F}_{2^r} = \mathbb{F}_2[c]$ (see [52]). By [49], the extension $\overline{\mathcal{S}}_c$ of $\mathcal{S}_c$ is simply connected too. If $\mathcal{S}_{c'} = \mathcal{S}_{c',\ell,r}$ is simply connected too, then $\overline{\mathcal{S}}_c \simeq \overline{\mathcal{S}}_{c'}$ iff $\mathcal{S}_c \simeq \mathcal{S}_{c'}$.

Automorphism groups and covering maps of DHOs of type $\mathcal{S}_c$ are studied in [54]: Let T be the translation group of $\mathcal{S}_c = \mathcal{S}_{c,\ell,r}$, then T is normal in $G = \mathrm{Aut}(\mathcal{S}_c)$ and G is the semidirect product of T with a subgroup H, $\mathrm{GL}(\ell, 2^r) \leq H \leq \Gamma\mathrm{L}(\ell, 2^r)$ and $H/\mathrm{GL}(\ell, 2^r) \simeq \mathrm{Gal}(\mathbb{F}_{2^r} : \mathbb{F}_2[c])$. Moreover, $\mathcal{S}_{c',\ell',r'}$ is a cover of $\mathcal{S}_{c,\ell,r}$ iff there exists a positive integer a such that $r = ar'$, $\ell' = a\ell$ and there exists a field automorphism $\sigma : \mathbb{F}_2[c] \to \mathbb{F}_2[c']$ with $c' = c^\sigma$.

8.2 DHOs Constructed with Presemifields

In three articles [53, 55, 56], Taniguchi expands the hybrid construction of the previous subsection by invoking presemifields in the construction.

Let $(S, +)$ be finite additive group. Let $\circ : S \times S \to S$ be a biadditive mapping such that for $0 \neq a \in S$ the mappings $S \ni x \mapsto a \circ x \in S$ and $S \ni x \mapsto x \circ a \in S$ are bijective. Then $(S, +, \circ)$ is called a *presemifield*. This presemifield is *commutative* if $x \circ y = y \circ x$ for all $x, y \in S$. It can be shown that $(S, +)$ is an elementary abelian p-group, p a prime. Therefore, one may identify $(S, +)$ with the additive group of a finite field. The equivalence of presemifields is based on the notion of isotopism. Let S_1 and S_2 be presemifields whose additive groups are identified with $\mathbb{F}_p^n$ and with multiplications $\circ_1$ and $\circ_2$, respectively. The presemifields are *isotopic* if there exist $\lambda, \rho, \mu \in \mathrm{GL}(n, p)$ such that $(x\lambda) \circ_2 (y\rho) = (x \circ_1 y)\mu$. The triple (λ, μ, ρ) is called an *isotopism*.

Let $d > 3$ be a positive integer and $q = 2^d$ and $F = \mathbb{F}_q$. In [53], Taniguchi considers DHOs of rank $n = d + 1$ with a "small" ambient space of rank $2n - 1$. Let $(S, +, \circ)$ be a commutative presemifield whose additive group is $(F, +)$. Let $0 \neq c \in F$ be an element such that

(c1) $c(x \circ x) = (cx) \circ x$ for all $x \in F$.
(c2) $x \circ y + c(x \circ x) + c(y \circ y) \neq 0$ for all $0 \neq x, y \in F$.

One may consider (c2) as a criterion for an irreducible quadratic form over the presemifield S. Let $X = F \oplus \mathbb{F}_2 e_0$ be a $\mathbb{F}_2$-space of rank $n + 1$, $U = X \oplus F$ and define a bilinear map $B : X \times X \to F$ by $B(x + \alpha e_0, y + \beta e_0) = x \circ y + \alpha c(y \circ y) + \beta c(x \circ x)$. Then $\mathcal{S}_c(S) = \mathcal{S}_B$ is a symmetric DHO in U. If $(c, c') \neq (1, 1)$ and $\mathcal{S}_c(S) \simeq \mathcal{S}_{c'}(S')$ then S and S' are isotopic, say by (λ, μ, ρ), and the left multiplications are related by conjugation: $L_{c'} = \lambda^{-1} L_c \lambda$. Surprisingly, one may have $\mathcal{S}_1(S) \simeq \mathcal{S}_1(S')$ for non-isotopic presemifields S and S'. Kantor shows (see [28]) that for d odd and highly composite there exists many nonisotopic commutative presemifields of size 2^d, which, in turn, define many nonisomorphic DHOs of type $\mathcal{S}_c(S)$.

In [55], Taniguchi varies this construction and uses three presemifields: Let $F = \mathbb{F}_{2^d}$. Let $S_1 = (F, +, \circ)$, $S_2 = (F, +, *)$, and $S_3 = (F, +, \star)$ be presemifields, S_1 and S_2 commutative. Let $0 \neq c \in F$ be an element such that:

(c1) $c(x \circ x) = (cx) \circ x$, $c(x * x) = (cx) * x$ for all $x \in F$.
(c2) $(cx) \star y = x \star (cy)$ for all $x, y \in F$.
(c3) $x \circ y + c(x \circ x) + c(y \circ y) \neq 0$ and $x * y + c(x * x) + c(y * y) \neq 0$ for all $0 \neq x, y \in F$.

Set $X = F \oplus F \oplus \mathbb{F}_2$, $Y = F \oplus F \oplus F$ and define a bilinear mapping $B : X \times X \to Y$ by

$$B((x, a, \alpha), (y, b, \beta)) = (x \circ y + \alpha c(y \circ y) + \beta c(x \circ x), a * b + \alpha c(b * b) + \beta c(a * a), x \star b + y \star a).$$

Then $\mathcal{S}_c(S_1, S_2, S_3) = \mathcal{S}_B$ is a symmetric DHO of rank $2d + 1$ over $\mathbb{F}_2$ in $\mathbb{F}_2^{5d+1}$.

In [56], Taniguchi presents a further generalization.

Let $F = \mathbb{F}_{2^d}$. He now considers a family of presemifields $S_{i,j} = (F, +, \star_{ij})$, $1 \leq i < j \leq n$. Let $0 \neq c \in F$ be an element such that for all $1 \leq i < j \leq n$ and all $x, y \in F$

$$(cx) \star_{ij} y = x \star_{ij} (cy)$$

holds. Set $X = \overbrace{(F \oplus \cdots F)}^{n} \oplus \mathbb{F}_2$ and $Y = \overbrace{(F \oplus \cdots F)}^{n} \oplus \overbrace{(F \oplus \cdots F)}^{\binom{n}{2}}$. Define a bilinear mapping $B : X \times X \to Y$ by sending the pair

$$((\ldots x_i, \ldots, x_j, \ldots, \alpha), (\ldots y_i, \ldots, y_j, \ldots, \beta))$$

to

$$(\overbrace{\ldots, x_i y_i + \alpha c y_i^2 + \beta c x_i^2, \ldots,}^{n} \overbrace{\ldots, x_i \star_{ij} y_j + y_i \star_{ij} x_j, \ldots}^{\binom{n}{2}}).$$

Then $\mathcal{S}_c(n, \{S_{ij} \mid 1 \leq i < j \leq n\}) = \mathcal{S}_B$ is a symmetric DHO of rank $nd + 1$ over $\mathbb{F}_2$ which lies in a $\mathbb{F}_2$-space of rank $d(\binom{n}{2} + 2n) + 1$.

For DHOs of type $\mathcal{S}_c(S_1, S_2, S_3)$ and $\mathcal{S}_c(n, \{S_{ij} \mid 1 \leq i < j \leq n\})$, there exist isomorphism theorems similar to the isomorphism theorem for type $\mathcal{S}_c(S)$.

Problem 12 Determine the ambient spaces of DHOs of type $\mathcal{S}_c(S_1, S_2, S_3)$ and $\mathcal{S}_c(n, \{S_{ij} \mid 1 \leq i < j \leq n\})$.

9 DHOs and Groups

A common theme in finite geometry is the classification of geometries that admit specified group actions.

9.1 DHOs with Many Translation Groups

Bilinear DHOs can be characterized by group actions. Let $\mathcal{S}$ be a DHO over $\mathbb{F}_2$ with ambient space U. An elementary abelian 2-subgroup T of the automorphism group of $\mathcal{S}$ is a *translation group*, if T acts regularly on $\mathcal{S}$, such that the DHO splits over $C_U(T) = \{u \in U \mid u\tau = u,\ \tau \in T\}$, the *centralizer of* T *in* U. It is shown in [18, Theorem 3.2], that if the rank of the DHO is ≥ 3, T has *quadratic action on* U, i.e. $[U, T] \subseteq C_U(T)$, where $[U, T] = \langle u(1+\tau) \mid u \in U,\ \tau \in T\rangle$ is the *commutator of* U *and* T. Moreover, $\mathcal{S}$ can be coordinatized as a bilinear DHO $\mathcal{S} = \mathcal{S}_B$ and $T = T_B$ is the standard translation group with respect to B. The translation groups form a conjugate class $\mathcal{C}$ of self-centralizing (TI)-subgroups in the automorphism group of a DHO [18, Theorem 3.11]. In the case of the Huybrechts DHOs, the Buratti–Del Fra DHOs or the bilinear Yoshiara DHOs, the class has a size 1, i.e. the translation groups are normal. Indeed, if there exists more than one translation group, we are in a very tight situation [18, Theorem 4.10] and [18, Theorem 5.10]:

Theorem 6 *Let $\mathcal{S}$ be a bilinear DHO of rank $n \geq 4$ in the ambient space U. Assume, that $\mathcal{S}$ admits at least two translation groups. Then* $\dim U \geq 3(n-1)$ *and $\mathcal{S}$ is the extension of a symmetric, bilinear DHO of rank $n-1$.*

Assume $|\mathcal{C}| = k > 1$, let $\mathcal{S}$ be the extension of $\mathcal{S}_B$ and set $H = \langle \mathcal{C} \rangle$. Then $O_2(H) = N$ where N is defined in Sect. 7.3 and $H/N \simeq \mathrm{D}_{2k}$ is a dihedral group of order $2k$ (see [18, Theorem 5.7]). Let $\mathcal{T}$ be a spread of order 2^{n-1} in $W = \mathbb{F}_2^{2(n-1)}$ associated with a translation plane of a commutative presemifield of order 2^{n-1}. Then one can choose a 1-space P in such a way that $\mathcal{T}/P$ is a symmetric bilinear DHO of rank $n-1$ in W/P. The extension $\overline{\mathcal{T}/P}$ is bilinear and the size of the conjugacy class of translation groups in $\mathrm{Aut}(\overline{\mathcal{T}/P})$ is the same as the size of the middle nucleus of the presemifield associated with $\mathcal{T}$. In [18, Sect. 6], many DHOs are produced that admit more than one translation group.

9.2 DHOs with Many Extension Groups

Dempwolff and Edel [19] provide a group theoretic description of extensions of bilinear DHOs in analogy to the characterization of bilinear DHOs by translation groups. Let $\mathcal{S}$ be a DHO of rank $n+1$ over $\mathbb{F}_2$ with ambient space U. Let $E = \langle E_0, E_1 \rangle$ be a subgroup of $\mathrm{Aut}(\mathcal{S})$. Set $\mathcal{T}_i = \mathrm{Fix}_{\mathcal{S}}(E_i)$, $V_i = \langle S \cap S' \mid S, S' \in \mathcal{T}_i,\ S \neq S' \rangle$ for $i = 0, 1$. One calls E an *extension group* if:

(E1) $\mathcal{S} = \mathcal{T}_0 \cup \mathcal{T}_1$ is a partition and $|\mathcal{T}_i| = 2^n$ for $i = 0, 1$.
(E2) E_i acts regularly on $\mathcal{T}_j$ for $\{i, j\} = \{0, 1\}$.
(E3) Set $Y = C_U(E)$. Then $V_0 + V_1 + Y$ has codimension 1 in U and $\dim(V_0 + V_1 + Y)/Y = 2n$.

The group N defined in Sect. 7.3 is an extension group. If a DHO of rank $n + 1$ admits an extension group, then [19, Theorem 3.2] this DHO is the extension of a bilinear DHO of rank n.

If $\mathcal{S}$ of rank $n + 1$ admits extension groups, then they form a conjugacy class $\mathcal{C}$ in the automorphism group [19, Theorem 3.6]. If $|\mathcal{C}| > 1$, then $\mathcal{S}$ is the result of at least two iterated extensions [19, Theorem 6.1]. The structure of the group $H = \langle \mathcal{C} \rangle$ can be determined [19, Theorem 7.1]: $H/O_2(H) \simeq \mathrm{SL}(k + 1, 2)$ for some $k \in \{1, \ldots, n\}$ and $|O_2(H)| = 2^{(k+1)(n-k+1)}$.

9.3 Doubly Transitive DHOs

A DHO $\mathcal{S}$ is called *doubly transitive* if $\mathrm{Aut}(\mathcal{S})$ contains a subgroup G that acts doubly transitive on the members of $\mathcal{S}$. In that case, the automorphism group of the affine expansion $\mathrm{Af}(\mathcal{S})$ acts flag transitive on this semibiplane (see [33]). A side product of the classification of finite simple groups is the classification of finite doubly transitive groups. See Cameron [4, p. 8] and Liebeck [31, Appendix 1] for a description of these groups. A classification of doubly transitive DHOs requires two steps:

1. Determine doubly transitive groups that can act on a DHO.
2. Determine for these candidates the DHOs whose automorphism groups contain the group in question.

The first step is completed by Yoshiara [69] (we denote by $G^{(\infty)}$ the last member of the derived series of the group G))

Theorem 7 *Assume that $\mathcal{S}$ is a dual hyperoval of rank d with ambient $\mathbb{F}_q$-space V. If a subgroup G of* $\mathrm{Aut}(\mathcal{S})$ *acts on the members of $\mathcal{S}$ doubly transitively, one of the following holds:*

(1) $q = 2$ and $G = N \cdot G_X$, where N is a normal subgroup of G acting regularly on $\mathcal{S}$ and the stabilizer G_X in G of a member $X \in \mathcal{S}$ has one of the following structures:

(a) G is solvable and G_X is isomorphic to a subgroup of $\Gamma\mathrm{L}(1, 2^d)$.
(b) G is non-solvable and $(G_X)^{(\infty)} \simeq \mathrm{PSL}(a, 2^b)$, $a \geq 2$, $n = ab$, $(G_X)^{(\infty)} \simeq \mathrm{PSp}(2a, 2^b)$, $a \geq 2$, $n = 2ab$, $(G_X)^{(\infty)} \simeq \mathrm{G}_2(2^b)$, $n = 6b$, or $(G_X)^{(\infty)} \simeq \mathrm{G}_2(2)'$, $n = 6$, $(G_X)^{(\infty)} \simeq \mathrm{Alt}(6), \mathrm{Alt}(7)$, $n = 4$.

(2) $q = 4$, $d = 3$ and $G^{(\infty)} \simeq \mathrm{M}_{22}$.

Notice that this result is as best as possible. It follows from [26] (see also [7]) that the Mathieu DHO is the unique DHO which satisfies assertion (2). All groups of assertion (1) are doubly transitive automorphism groups of the Huybrechts DHOs.

In [14], it is shown, that the Huybrechts DHO of rank n has doubly transitive quotients $\mathcal{S}$ such that $(\mathrm{Aut}(\mathcal{S})_X)^{(\infty)} \simeq \mathrm{PSL}(a, 2^b), a \geq 2, n = ab$, or $\simeq \mathrm{PSp}(2a, 2^b)$, $n = 2ab$ and the number of such quotients grows exponentially as a function of the rank.

In order to classify doubly transitive DHOs, one has to determine the DHOs of type (1). It turns out that case (1.b) is much easier than case (1.a)

Theorem 8 ([15]) *Let $\mathcal{S}$ be a DHO with a non-solvable automorphism group, that induces a doubly transitive action on its members. Then $\mathcal{S}$ is isomorphic to the Mathieu DHO or $\mathcal{S}$ is a bilinear quotient of a Huybrechts DHO.*

The first contribution to case (1.a) of Theorem 7 is [67]. There Yoshiara classifies all doubly transitive dual hyperovals of rank n with ambient space $\mathbb{F}_2^{2n}$ which admit a doubly transitive group isomorphic to $\mathrm{AGL}(1, 2^n) \simeq \mathbb{F}_2^n \cdot \mathrm{GL}(1, 2^n)$. He shows that these DHOs are DHOs of Yoshiara. In this case, the normal subgroup of order 2^n is a translation group, i.e. the DHOs in question are bilinear. It is natural to ask for the doubly transitive, *bilinear* DHOs of rank n (which admit $\mathrm{AGL}(1, 2^n)$) without the restriction on the ambient space. In [12], it is shown that such a DHO is a DHO of Yoshiara, a quotient of the Huybrechts DHO of rank n, or a member of a new class of DHOs in $\mathbb{F}_2^{2n+n_0}$, $n = 2n_0$, n_0 odd which are denoted by $\mathbf{D}[n, k]$ (the parameter k has four possible values). Although in the generic case of (1.a) a doubly transitive group in $\mathrm{A\Gamma L}(1, 2^n) \simeq \mathbb{F}_2^n \cdot \Gamma\mathrm{L}(1, 2^n)$ contains the group $\mathrm{GL}(1, 2^n)$ there exist positive, composite numbers n and groups $H \leq \Gamma\mathrm{L}(1, 2^n)$ such that $\mathbb{F}_2^n \cdot H$ is doubly transitive and $\mathrm{GL}(1, 2^n)$ is not contained in H. In [17], it is shown that *only for rank $n = 6$ new DHOs* (with this kind of automorphism groups) can *and do* occur. One has

Theorem 9 *Let $n \geq 4$, $n \neq 6$ and $\mathcal{S}$ a doubly transitive, bilinear DHO of rank n over $\mathbb{F}_2$. Then $\mathcal{S}$ is DHO of Yoshiara, a quotient of the Huybrechts DHO of rank n, or a DHO of type* $\mathbf{D}[n, k]$.

The expectation, however, that all doubly transitive DHOs of type (1.a) would be covered by this Theorem turns out to be false. In [16], the existence of non-bilinear DHOs belonging to case (1.a) is established. Let $1 < d < n$ be divisor of n, such that n/d is odd. Let $1 < b < n$, be a multiple of d, such that $(b, n) = d$. Then there exist a DHO $\mathcal{S} = \mathbf{D}[n, d, b]$ non-bilinear of rank n, such that $G = \mathrm{Aut}(\mathcal{S}) \simeq \mathrm{A\Gamma L}(1, 2^n)$ acts doubly transitive on $\mathcal{S}$. For odd n_0, there is a second class of non-bilinear, doubly transitive DHOs of rank $n = 2n_0$ denoted by $\widehat{\mathbf{D}}[2n_0]$. It turns out that $\widehat{\mathbf{D}}[2n_0]$ is the universal cover of $\mathbf{D}[2n_0, -2^{n_0+1} - 1]$ as well as of $\mathbf{D}[2n_0, 2, n_0 \pm 1]$.

Problem 13 Try to find more non-bilinear, doubly transitive DHOs of type (1.a).

The final classification of doubly transitive DHOs appears to be difficult.

10 Miscellaneous Topics

We do not report on the connection between DHOs and diagram geometry[1]: an introduction to diagram geometry would extend this paper far beyond its intended length. But an interested reader may consult [25, 26, 33, 34]. For DHOs with "property (T_i)" and the connection of DHOs with property (T_1) with Steiner Systems, we refer to [65, Sect. 2.6].

10.1 Edel's Characterization of Alternating DHOs

Let $\mathcal{S}$ be a DHO over $\mathbb{F}_2$ and $X, X' \in \mathcal{S}$. Denote by $X \sqcap X'$ the nontrivial vector in $X \cap X'$. For three members $X_1, X_2, X_3 \in \mathcal{S}$, define

$$u(X_1, X_2, X_3) = X_1 \sqcap X_2 + X_1 \sqcap X_3 + X_2 \sqcap X_3$$

and define in the ambient space the subspace

$$P(\mathcal{S}) = \langle u(X_1, X_2, X_3) \mid X_1, X_2, X_3 \in \mathcal{S},\ X_1 \neq X_2 \neq X_3 \neq X_1 \rangle.$$

Edel [22, Theorem 1], [18, Theorem 3.6] gives the following geometric characterization of alternating DHOs:

Theorem 10 *Let $\mathcal{S}$ be a DHO over $\mathbb{F}_2$. Equivalent are:*

(a) $\mathcal{S}$ splits over $P(\mathcal{S})$.
(b) $\mathcal{S}$ is an alternating DHO.

The space $P(\mathcal{S})$ is a supplement of $\mathcal{S}$

Lemma 2 $U(\mathcal{S}) = X + P(\mathcal{S})$ *for every* $X \in \mathcal{S}$.

Proof ix $X_1 \in \mathcal{S}$ and let $X_2, X_3 \in \mathcal{S}$ be two members of $\mathcal{S}$ different from X_1. Then $X_2 \sqcap X_3 = (X_1 \sqcap X_2 + X_1 \sqcap X_3) + u(X_1, X_2, X_3) \in X_1 + P(\mathcal{S})$ and $X_1 \sqcap X_2 \in X_1$ for $X_1 \neq X_2$. Hence, $X_1 + P(\mathcal{S})$ contains a set of generators of $U(\mathcal{S})$.

So the statement "dim $U(\mathcal{S})$ − dim $P(\mathcal{S})$ is the rank of $\mathcal{S}$" is equivalent to Assertion (a) as well as Assertion (b) of Theorem 10.

Problem 14 Let $\mathcal{S}$ be a split DHO over $\mathbb{F}_2$. Show or disprove that $P(\mathcal{S})$ contains a complement.

[1] I quote from the report of one referee: "...DHOs are fall down of diagram geometry; in fact, a fall down of [26]. The expression "d-dimensional dual hyperoval" itself goes back to [26, 63] ...".

10.2 Small DHOs over $\mathbb{F}_2$

Del Fra [7, Theorems 1, and 2] classifies the DHOs of rank 3 over $\mathbb{F}_2$. There exists exactly one DHO with an ambient space of rank 5. It is the bilinear DHO $\mathcal{S}^3_{1,2}$ of Yoshiara. There are exactly two DHOs with an ambient space of rank 6, namely $\mathcal{V}_3(2)$ and $\mathcal{H}_3$.

In [1], a partial classification of DHOs of rank 4 over $\mathbb{F}_2$ is obtained. In contrast to del Fra [7], computer calculations are indispensable. One finds for an ambient space of rank 7 precisely 37 isomorphism types, 7 of which are bilinear. For an ambient space of rank 8, one obtains 26 isomorphism types, 11 of which are bilinear. For ambient spaces of rank ≥ 9, the *information is incomplete.* However, computing covers or quotients is feasible. In this way, one obtains 7 DHOs with an ambient space of rank 9, and for rank 10, one recovers as expected $\mathcal{V}_4(2)$, $\mathcal{H}_4$, $\mathcal{D}_4$ and $\mathcal{T}_4$.

Problem 15 Classify DHOs of rank 4 over $\mathbb{F}_2$ with an ambient space of rank ≥ 9.

For partial classifications of small DHOs over $\mathbb{F}_q$, $q > 2$; see [7] and [65, Sect. 4.2].

References

1. Betten, A., Dempwolff, U., Wassermann, A.: On dual hyperovals of rank 4 over $\mathbb{F}_2$. J. Geom. **108**, 75–98 (2017)
2. Biliotti, M., Jha, V., Johnson, N.: Handbook of Finite Translation Planes. CRC Press (2007)
3. Buratti, M., Del Fra, A.: Semi-Boolean Steiner quadruple systems and dimensional dual hyperovals. Adv. Geom. **3**, 245–253 (2003)
4. Cameron, P.: Primitive permutation groups and finite simple groups. Bull. Lond. Math. Soc. **13**, 1–22 (1981)
5. Cherowitzo, W.: Hyperovals in desarguesian planes: an update. Disc. Math. **155**, 31–38 (1996)
6. Cooperstein, B.N., Thas, J.A.: On generalized k-arcs in PG$(2n, q)$. Ann. Comb. **5**, 141–152 (2001)
7. Del Fra, A.: On d-dimensional dual hyperovals. Geom. Dedicata **79**, 157–178 (2000)
8. Del Fra, A., Yoshiara, S.: Dimensional dual hyperovals associated with Steiner systems. Eur. J. Combin. **26**, 173–194 (2005)
9. Dempwolff, U.: Dimensional doubly dual hyperovals and bent functions. Innov. Incidence Geom. **13**, 146–178 (2013)
10. Dempwolff, U.: Symmetric extensions of bilinear dual hyperovals. Finite Fields Their Appl. **22**, 51–56 (2013)
11. Dempwolff, U.: Symmetric doubly dual hyperovals have an odd rank. Des. Codes Cryptogr. **74**, 153–157 (2015)
12. Dempwolff, U.: Some doubly transitive bilinear dual hyperovals and their ambient spaces. Eur. J. Combin. **44**, 1–22 (2015)
13. Dempwolff, U.: Universal covers of dimensional dual hyperovals. Disc. Math. **338**, 633–636 (2015)
14. Dempwolff, U.: The automorphism groups of doubly transitive, bilinear dual hyperovals. Adv. Geom. **17**, 91–108 (2017)
15. Dempwolff, U.: The non-solvable doubly transitive dimensional dual hyperovals. Adv. Geom. **18**, 1–4 (2018)

16. Dempwolff, U.: Doubly transitive dimensional dual hyperovals: universal covers and non-bilinear examples. To appear in Adv. Geom.
17. Dempwolff, U.: Doubly transitive, bilinear dimensional dual hyperovals: the conclusion. Submitted to European J. Comb.
18. Dempwolff, U., Edel, Y.: Dimensional dual hyperovals and APN functions with translation groups. J. Algebr. Comb. **39**, 457–496 (2014)
19. Dempwolff, U., Edel, Y.: Isomorphisms and automorphisms of extensions of bilinear dimensional dual hyperovals and quadratic APN functions. J. Group Theory **19**, 249–322 (2016)
20. Dempwolff, U., Kantor, W.: Orthogonal dual hyperovals, symplectic spreads and othogonal spreads. J. Algebr. Comb. **41**, 83–108 (2015)
21. Dempwolff, U., Reifart, A.: The classification of the translation planes of order 16. I. Geom. Dedicata **15**, 137–153 (1983)
22. Edel, Y.: On quadratic APN functions and dimensional dual hyperovals. Des. Codes Cryptogr. **57**, 35–44 (2010)
23. Edel, Y.: On some representations of quadratic APN functions and dimensional dual hyperovals. RIMS Kokyuroku **1687**, 118–130 (2010)
24. Göloğlu, F., Pott, A.: Almost perfect nonlinear functions: a possible geometric approach. In: Proceedings of the Contact Forum Coding Theory and Cryptography II, pp. 75–100. The Royal Flemish Academy of Belgium for Science and the Arts (2007)
25. Huybrechts, C.: Dimensional dual hyperovals in projective spaces and $c.AG^*$ geometries. Disc. Math. **255**, 193–223 (2002)
26. Huybrechts, C., Pasini, A.: Flag-transitive extensions of dual affine spaces. Contrib. Algebra Geom. **40**, 503–532 (1999)
27. Kantor, W.M.: Spreads, translation planes and Kerdock sets I. SIAM J. Alg. Disc. Math. **3**, 151–165 (1981)
28. Kantor, W.M.: Commutative semifields and symplectic spreads. J. Algebra **270**, 98–114 (2003)
29. Kantor, W.M., Williams, M.E.: New flag-transitive affine planes of even order. J. Comb. Theory Ser. A **74**, 1–13 (1996)
30. Korchmaros, G.: Old and New Results on Ovals in Finite Projective Planes. Surveys in Combinatorics, 1991 (Guildford, 1991). LMS Lecture Notes Series, vol. 166, pp. 41–72. Cambridge University Press, Cambridge (1991)
31. Liebeck, M.: The affine permutation groups of rank 3. Proc. Lond. Math. Soc. **3**(54), 477–516 (1987)
32. Nambu, N., Yoshiara, S.: Conditions for a class of dimensional dual hyperovals to be of polar type. Finite Fields Their Appl. **13**, 1117–1126 (2007)
33. Pasini, A., Yoshiara, S.: On a new family of flag-transitive semibiplanes. Eur. J. Comb. **22**, 529–545 (2001)
34. Pasini, A., Yoshiara, S.: New distance regular graphs graphs arising from dimensional dual hyperovals. Eur. J. Comb. **22**, 547–560 (2001)
35. Pott, A.: Almost perfect and planar functions. Des. Codes Cryptogr. **78**, 147–195 (2016)
36. Sheekey, J.: Dimensional Dual hyperovals in classical polar spaces. Des. Codes Cryptogr. **80**, 415–420 (2016)
37. Taniguchi, H.: On a family of dual hyperovals over GF(q) with q even. Eur. J. Combin. **26**, 195–199 (2005)
38. Taniguchi, H.: On d-dual hyperovals in $PG(d(d+3)/2, 2)$. Electron. Notes Disc. Math. **26**, 131–138 (2006)
39. Taniguchi, H.: On an Isomorphism Problem of Some Dual Hyperovals in PG (2d + 1, q) with q even. Graphs Comb. **23**, 455–465 (2007)
40. Taniguchi, H.: On automorphisms of some d-dimensional dual hyperovals in $PG(d(d+3)/2, 2)$. Graphs Comb. **24**, 229–236 (2008)
41. Taniguchi, H.: On d-dimensional dual hyperovals in $PG(2d, 2)$. Innov. Incid. Geom. **8**, 137–145 (2008)
42. Taniguchi, H.: On some d-dimensional dual hyperovals in $PG(2d, 2)$. Finite Fields Appl. **14**, 1010–1019 (2008)

43. Taniguchi, H.: On the duals of certain d-dimensional dual hyperovals in $PG(2d+1,2)$. Finite Fields Appl. **15**, 673–681 (2009)
44. Taniguchi, H.: A new family of dual hyperovals in $PG(d(d+3)/2,2)$ with $d \geq 3$. Disc. Math. **309**, 418–429 (2009)
45. Taniguchi, H.: On some d-dimensional dual hyperovals in $PG(d(d+3)/2,2)$. Eur. J. Comb. **31**, 401–410 (2010)
46. Taniguchi, H.: On d-dimensional Buratti-Del Fra type dual hyperovals in $PG(3d,2)$. Disc. Math. **310**, 3633–3645 (2010)
47. Taniguchi, H.: On the dual of the dual hyperoval from APN function $f(x) = x^3 + Tr(x^9)$. Finite Fields Appl. **18**, 210–221 (2012)
48. Taniguchi, H.: Quotients of the deformation of Veronesean dual hyperoval in $PG(3d,2)$. Disc. Math. **312**, 498–508 (2012)
49. Taniguchi, H.: Some examples of simply connected dual hyperovals. Finite Fields Appl. **22**, 45–50 (2013)
50. Taniguchi, H.: Simple expressions of the Buratti-Del Fra dual hyperoval and the deformation of the Veronesean dual hyperoval. Electron. Notes Disc. Math. **40**, 359–364 (2013)
51. Taniguchi, H.: New dimensional dual hyperovals, which are not quotients of the classical dual hyperovals. Disc. Math. **337**, 65–75 (2014)
52. Taniguchi, H.: Some examples of simply connected dual hyperovals II. Finite Fields Appl. **36**, 1–13 (2015)
53. Taniguchi, H.: Bilinear dual hyperovals from binary commutative presemifields. Finite Fields Appl. **42**, 93–101 (2016)
54. Taniguchi, H.: On some bilinear dual hyperovals. Disc. Math. **340**, 3154–3166 (2017)
55. Taniguchi, H.: Bilinear dual hyperovals from binary commutative presemifields II. Finite Fields Appl. **49**, 62–79 (2018)
56. Taniguchi, H.: A variation of the dual hyperoval $\mathcal{S}_c$ using presemifields. Des. Codes Cryptogr. **87**, 895–908 (2019)
57. Taniguchi, H.: Distance regular graphs arising from dimensional dual hyperovals. Finite Fields Appl. **69**, 101776 (2021)
58. Taniguchi, H., Yoshiara, S.: On dimensional dual hyperovals $\mathcal{S}^{d+1}_{\sigma,\phi}$. Innov. Incid. Geom. **1**, 197–219 (2005)
59. Taniguchi, H., Yoshiara, S.: New quotients of the d-dimensional Veronesean dual hyperoval in $PG(2d+1,2)$. Innov. Incid. Geom. **12**, 151–165 (2011)
60. Taniguchi, H., Yoshiara, S.: A new construction of the d-dimensional Buratti-Del Fra dual hyperoval. Eur. J. Comb. **33**, 1030–1042 (2012)
61. Taniguchi, H., Yoshiara, S.: A unified description of four simply connected dimensional dual hyperovals. Eur. J. Combin. **36**, 143–150 (2014)
62. Vanhove, F.: Antidesigns and regularity of partial spreads in dual polar graphs. J. Comb. Des. **19**, 202–216 (2011)
63. Yoshiara, S.: A family of d-dimensional dual hyperovals in $PG(2d+1,2)$. Eur. J. Combin. **20**, 589–603 (1999)
64. Yoshiara, S.: Ambient apaces of dimensional dual arcs. J. Alg. Comb. **19**, 5–23 (2004)
65. Yoshiara, S.: Dimensional dual arcs. In: Finite Geometries, Groups and Computation. In: Proceedings of a Conference on Pingree Park 2004, pp. 247–266. Col. USA (2006)
66. Yoshiara, S.: Notes on Taniguchi's dimensional dual hyperovals. Eur. J. Comb. **28**, 674–684 (2007)
67. Yoshiara, S.: A characterization of a class of dimensional dual hyperovals with doubly transitive automorphism groups and its applications. Eur. J. Combin. **29**, 1521–1534 (2008)
68. Yoshiara, S.: Dimensional dual hyperovals associated with quadratic APN functions. Innov. Incid. Geom. **8**, 147–169 (2008)
69. Yoshiara, S.: Dimensional dual hyperovals with doubly transitive automorphism groups. Eur. J. Comb. **30**, 747–757 (2009)
70. Yoshiara, S.: Notes on APN functions, semibiplanes and dimensional dual hyperovals. Des. Codes Cryptogr. **56**, 197–218 (2010)

71. Yoshiara, S.: Disjoint unions of dimensional dual hyperoval and its splitness. Innov. Incid. Geom. **14**, 43–76 (2015)
72. Yoshiara, S.: An elementary description of the Mathieu dual hyperoval and its splitness. Innov. Incid. Geom. **14**, 81–110 (2015)
73. Yoshiara, S.: Bilinear dual hyperovals in the largest ambient space. J. Algebr. Comb. **46**, 533–548 (2017)
74. Yoshiara, S.: Splitness of the Veronesean and the Taniguchi dual hyperovals. Disc. Math. **342**, 844–854 (2019)

A Note on FMS Modules and FCP Extensions

Atul Gaur and Rahul Kumar

Abstract Let R be a commutative ring with unity and S be a (unital) subring of R such that R is integral over S and $S \subseteq R$ has FCP. Let M be an R-module. For any submodule N of M, it is shown that $R(+)N \subseteq R(+)M$ has FCP if and only if $S(+)N \subseteq S(+)M$ has FCP. We also discuss FMS modules.

Keywords FCP extension · Idealization · Artinian ring

1 Introduction

All rings considered below are commutative with nonzero identity; all ring extensions, ring homomorphisms, and algebra homomorphisms are unital. By a local ring, we mean a ring with a unique maximal ideal. The symbol $\subseteq$ is used for inclusion, while $\subset$ is used for proper inclusion. For a ring extension $S \subseteq R$, the set of all S-subalgebras of R is denoted by $[S, R]$. A chain of S-subalgebras of R is a set of elements of $[S, R]$ that are pairwise comparable with respect to inclusion. The extension $S \subseteq R$ satisfies FCP if each chain of S-subalgebras of R is finite.

Recall [4, cf. Nagata, 1962, p.2] that if R is a ring and M is an R-module, then the idealization $R(+)M$ is the ring defined as follows: Its additive structure is that of the abelian group $R \oplus M$, and its multiplication is defined by $(r_1, m_1)(r_2, m_2) := (r_1r_2, r_1m_2 + r_2m_1)$ for all $r_1, r_2 \in R$ and $m_1, m_2 \in M$. It will be convenient to view R as a subring of $R(+)M$ via the canonical injective ring homomorphism that sends r to $(r, 0)$. If R is a ring, then R is a subring of $R \times R$ via the canonical injective ring homomorphism, $\Delta : R \hookrightarrow R \times R$, given by $\Delta(r) = (r, r)$ for all $r \in R$.

In [5], G. Picavet and M. Picavet-L'Hermitte proved some results on FCP. They investigated that when $R \subseteq R(+)M$ satisfies FCP conditions, for an R-module M. Motivated by their work, we prove that if $S \subseteq R$ is a subring such that R is integral over S and $S \subseteq R$ has FCP and N is a submodule of an R-module M, then $R(+)N \subseteq$

A. Gaur (✉) · R. Kumar
Department of Mathematics, University of Delhi, New Delhi, Delhi, India
e-mail: gaursatul@gmail.com

M. Ashraf et al. (eds.), *Algebra and Related Topics with Applications*,
Springer Proceedings in Mathematics & Statistics 392,
https://doi.org/10.1007/978-981-19-3898-6_10

$R(+)M$ has FCP if and only if $S(+)N \subseteq S(+)M$ has FCP (Proposition 1). We also prove that $R \subseteq R(+)R$ has FCP if and only if $S \subseteq S(+)S$ has FCP, where $S \subseteq R$ is a subring such that R is integral over S and $S \subseteq R$ has FCP (Proposition 2).

For any ring R, let $Spec(R)$ denote the set of all prime ideals of R and $Max(R)$ denote the set of all maximal ideals of R. Recall from [5] that an R-module M is an FMS module if M has finitely many R-submodules. A ring R with finitely many ideals is termed as FMIR [5]. Gabriel Picavet and Martine Picavet-L'Hermitte proved that a faithful R-module M is an FMS module if and only if R is an FMIR and is a direct product of two rings $R' \times R''$, where $|R'| < \infty$ and $|R''/P| = \infty$ for any $P \in Spec(R'')$ and M is the direct product of a finite R'-module and a rank one projective R''-module ([5, Theorem 2.13]). An example is given to show that the above mentioned theorem is not true (Example 4).

Recall that a special principal ideal ring (SPIR) is a principal ideal ring R with a unique nonzero prime ideal M, such that M is nilpotent of index $n > 0$. Note that a SPIR is not a field. As usual, if M is an R-module, then $l_R(M)$ is its length and $|X|$ denotes the cardinality of a set X. If M and N are R-modules, then the set $M \cong_R N$ if M and N are isomorphic as R-modules.

2 Results

We start with the following lemma which is needed for the proof of the main result of the paper.

Lemma 1 *Let M be an R-module and $S \subseteq R$ is a subring such that R is integral over S and $S \subseteq R$ has FCP. Then $l_R(M) < \infty$ if and only if $l_S(M) < \infty$.*

Proof First we claim that $l_S(R/\mathfrak{M}) < \infty$ for every $\mathfrak{M} \in Max(R)$. Let $\mathfrak{M} \in Max(R)$. Set $\mathfrak{m} = \mathfrak{M} \cap S$. Then $\mathfrak{m} \in Max(S)$, since R is integral over S. It follows by [3, Theorem 4.2(a)] that $l_S(R/S) < \infty$. We infer that $l_S(R/\mathfrak{M}) \leq l_S(R/\mathfrak{m}) = l_S(S/\mathfrak{m}) + l_S(R/S) = 1 + l_S(R/S) < \infty$. It is obvious that $l_S(M) < \infty$ implies that $l_R(M) < \infty$.

To prove the converse, let $l_R(M) < \infty$. Set $l = l_R(M)$ and let $0 = M_0 \subseteq \cdots \subseteq M_l = M$ be a composition series of M. Let $j \in [1, l]$. Since M_j/M_{j-1} is a simple R-module, we infer that $M_j/M_{j-1} \cong_R R/\mathfrak{M}$ for some $\mathfrak{M} \in Max(R)$. Clearly, $M_j/M_{j-1} \cong_S R/\mathfrak{M}$, and thus $l_S(M_j/M_{j-1}) = l_S(R/\mathfrak{M}) < \infty$ by the claim. Consequently, $l_S(M) = \sum_{i=1}^{l} l_S(M_i/M_{i-1}) < \infty$.

Now we can state the main result of the paper.

Proposition 1 *Let M be an R-module, N an R-submodule of M and $S \subseteq R$ is a subring such that R is integral over S and $S \subseteq R$ has FCP. Then $R(+)N \subseteq R(+)M$ has FCP if and only if $S(+)N \subseteq S(+)M$ has FCP.*

Proof This is an immediate consequence of [5, Proposition 2.8(2)] and Lemma 1.

The condition R is integral over S cannot be ignored in the Proposition 1, as we have the following example:

Example 1 Let $S = \mathbb{Z}_{2\mathbb{Z}}$, $R = \mathbb{Q}$, $M = \mathbb{Q} \times \mathbb{Q}$, and $N = \mathbb{Q} \times \{0\}$. Then M is an R-module, N is an R-submodule of M, and S is an integrally closed subring of R such that $S \subset R$ has FCP. Clearly $R(+)N \subset R(+)M$ has FCP. However, $S(+)N \subset S(+)M$ does not have FCP.

We now give the immediate corollary to the last result.

Corollary 1 *Let M be an R-module and $S \subseteq R$ is a subring such that R is integral over S and $S \subseteq R$ has FCP. Then $R \subseteq R(+)M$ has FCP if and only if $S \subseteq S(+)M$ has FCP.*

Proof This follows from Proposition 1 with $N = 0$.

The condition R is integral over S in Corollary 1 is necessary as we have the following example:

Example 2 Let $S = \mathbb{Z}_{2\mathbb{Z}}$, $R = \mathbb{Q}$, and $M = \mathbb{Q} \times \mathbb{Q}$. Then M is an R-module and S is an integrally closed subring of R such that $S \subset R$ has FCP. Clearly, $R \subset R(+)M$ has FCP. However, $S \subset S(+)M$ does not have FCP.

Proposition 2 *Let S be a subring of R such that R is integral over S and $S \subseteq R$ has FCP. Then $R \subseteq R(+)R$ has FCP if and only if $S \subseteq S(+)S$ has FCP.*

Proof By [5, Proposition 2.9(1)], it is sufficient to show that $l_R(R) < \infty$ if and only if $l_S(S) < \infty$.

First suppose that $l_R(R) < \infty$. Then $l_S(S) \leq l_S(R) < \infty$ by Lemma 1.

To prove the converse, let $l_S(S) < \infty$. By [3, Theorem 4.2(a)], we obtain that $l_S(R/S) < \infty$. This implies that $l_R(R) \leq l_S(R) = l_S(S) + l_S(R/S) < \infty$.

Note that the condition R is integral over S cannot be ignored in the Proposition 2, as we have the following example:

Example 3 Let $S = \mathbb{Z}_{2\mathbb{Z}}$ and $R = \mathbb{Q}$. Then S is an integrally closed subring of R such that $S \subset R$ has FCP. Clearly, $R \subset R(+)R$ has FCP but $S \subset S(+)S$ has no FCP.

In [5, Theorem 2.13], G. Picavet and M. Picavet-L'Hermitte proved that a faithful R-module M is an FMS module if and only if the following two conditions are satisfied:

(i) R is an FMIR and is a direct product of two rings $R' \times R''$, where $|R'| < \infty$ and $|R''/P| = \infty$ for any $P \in Spec(R'')$.
(ii) M is the direct product of a finite R'-module and a rank one projective R''-module.

If R' or R'' is allowed to be a zero ring, then the above result is true which is discussed in [6]. Otherwise, the above theorem needs some modifications which are evident in the next example.

Example 4 Consider any field F as a module over itself. Then F is an FMS module but $F \neq R' \times R''$ for any rings R' and R'' as $R' \times R''$ is not a domain. Thus, [5, Theorem 2.13] is not true. Also, let $R = M = \mathbb{Z}/6\mathbb{Z}$. Then M is a faithful R-module and an FMS module. Moreover, R is a finite ring and $R \cong \mathbb{Z}/2\mathbb{Z} \times \mathbb{Z}/3\mathbb{Z}$. Suppose that [5, Theorem 2.13] is true. Then there are two rings R' and R'' such that $R \cong R' \times R''$ and $|R''/P| = \infty$ for all $P \in Spec(R'')$. Since $R'' \neq 0$, it is clear that $Spec(R'') \neq \emptyset$. Therefore, R'' is infinite, and hence R is infinite, a contradiction.

Clearly, if a zero ring is not permissible to be taken, then the above result needs a modification. Our next corollary is a modified version of the above result. Though the proof is similar to the proof of [5, Theorem 2.13], for the sake of completeness, we are giving a proof

Corollary 2 *A faithful R-module M is an FMS module if and only if exactly one of the following two conditions hold:*

(i) R is an FMIR which is either a finite ring with M is a finite R-module or $|R/P| = \infty$ for any $P \in Spec(R)$ with M is a rank one projective R-module.
(ii) R is an FMIR which is a direct product of two rings $R' \times R''$, where $|R'| < \infty$ and $|R''/P| = \infty$ for any $P \in Spec(R'')$ and M is the direct product of a finite R-module and a rank one projective R''-module.

Proof If M is an FMS module, then R is an FMIR and M is a finitely generated R-module, by [5, Corollary 2.7]. Now, by [1, Corollary 2.4], we have $R = \prod_{i=1}^{n} R_i$, a product of local rings that are either finite, or an SPIR, or a field. Now, we consider the following three cases:
Case 1: Let R_i be finite for all $i \in [1, n]$. Then R is a finite ring and hence $|M| < \infty$ as M is finitely generated R-module.
Case 2: Let R_i be infinite for all $i \in [1, n]$. Then for any SPIR factor (R_i, P_i) of R, we have $|R_i/P_i| = \infty$ as R_i is local artinian. If R_i is an infinite field, then take $P_i = 0$. Thus, $|R/P| = \infty$ for any $P \in Spec(R)$. Now, by the proof of [5, Theorem 2.13], M is a rank one projective R-module.
Case 3: Let R_i be finite for some $i \in [1, n]$ and R_j be infinite for some $j \in [1, n]$. Let R' be the ring product of the R_i that are finite and R'' be the ring product of the R_i that are infinite. Then the result follows from the proof of [5, Theorem 2.13].
Conversely, if (ii) holds, then result follows from the proof of [5, Theorem 2.13]. We may now assume that (i) holds. If R is finite and M is a finite R-module, then we are done. Now, if $|R/P| = \infty$ for all $P \in Spec(R)$ and M is a rank one projective R-module, then from [2, *Théorèm* 2, *ch. II*, *p.* 141], we have M is finitely generated over R, with M_P is cyclic for each $P \in Max(R)$. Thus, by [5, Corollary 2.7], M is an FMS module.

References

1. Anderson, D.D., Chun, S.: Commutative rings with finitely generated monoids of fractional ideals. J. Algebra **320**(7), 3006–3021 (2008)
2. Bourbaki, N.: Algèbre Commutative, Chs. 1 and 2, Hermann, Paris (1961)
3. Dobbs, D.E., Picavet, G., Picavet-L'Hermitte, M.: Characterizing the ring extensions that satisfy FIP or FCP. J. Algebra **371**, 391–429 (2012)
4. Nagata, M.: Local Rings. Interscience, New York (1962)
5. Picavet, G., Picavet-L'Hermitte, M.: Modules with finitely many submodules. Int. Electron. J. Algebra **19**, 119–131 (2016)
6. Picavet, G., Picavet-L'Hermitte, M.: An addendum to the paper: Modules with finitely many submodules. Int. Electron. J. Algebra **28**, 229 (2020)

A Pair of Derivations of Prime Rings with Involution

Shuliang Huang

Abstract Let $(R,)$ be a prime ring of characteristic different from two with involution of the second kind. Suppose that d and δ are a pair of derivations R such that $[d(x), \delta(x^*)] \pm [x, x^*] = 0$ for all $x \in R$, then R is commutative. Also, some examples are given to show that the restrictions imposed on the hypotheses of the various results are crucial.

Keywords Prime rings · Derivations · Involution · Commutativity

Throughout this present paper, R will represent an associative ring with center $Z(R)$. For any $x, y \in R$, the symbol $[x, y]$ denotes Lie product $xy - yx$ and it is straightforward to check that $[xy, z] = x[y, z] + [x, z]y$ and $[x, yz] = y[x, z] + [x, y]z$ for all $x, y, z \in R$. A ring R is prime if for any $a, b \in R$, $aRb = (0)$ implies $a = 0$ or $b = 0$ and is semiprime if for any $a \in R$, $aRa = \{0\}$ implies $a = 0$. A ring R is called 2-torsion free, if whenever $2x = 0$, with $x \in R$, then $x = 0$. It is easy to check that a prime ring of characteristic different from two is 2-torsion free. By a derivation on R, we mean an additive mapping $d : R \longrightarrow R$ such that $d(xy) = d(x)y + xd(y)$ holds for all $x, y \in R$. The standard identity s_4 in four variables is defined as follows: $s_4 = \sum(-1)^{\tau} X_{\tau(1)} X_{\tau(2)} X_{\tau(3)} X_{\tau(4)}$, where $(-1)^{\tau}$ is the sign of a permutation τ of the symmetric group of degree 4.

Recall that an involution $*$ of a ring R is an anti-automorphism of order 2 (i.e., an additive mapping satisfying $(xy)^* = y^*x^*$ and $(x^*)^* = x$ for all $x, y \in R$). An element r in a ring with involution $(R, *)$ is called to be Hermitian if $r^* = r$ and skew-Hermitian if $r^* = -r$. The sets of all Hermitian and skew-Hermitian elements of R will be denoted by $H(R)$ and $S(R)$, respectively. The involution is said to be the first kind if $Z(R) \subseteq H(R)$, otherwise it is said to be of the second kind. In the latter case, $S(R) \bigcap Z(R) \neq \{0\}$. Following [3], if R is 2-torsion free then every $x \in R$ can be uniquely represented in the form of $2x = h + k$, where $h \in H(R)$ and $k \in S(R)$. Moreover, in this case, x is normal, i.e., $[x, x^*] = 0$, if and only if h and k

S. Huang (✉)
School of Mathematics and Finance, Chuzhou University, Chuzhou Anhui, Chuzhou 239000, People's Republic of China
e-mail: shulianghuang@163.com

M. Ashraf et al. (eds.), *Algebra and Related Topics with Applications*, Springer Proceedings in Mathematics & Statistics 392,
https://doi.org/10.1007/978-981-19-3898-6_11

commute. If all elements in R is normal, then R is said to be a normal ring. A classical example is the ring of Hamilton quaternions. Especially, derivations, biderivations, and superbiderivations of quaternion rings were characterized in [11]. It is worthwhile to note that every prime ring having an involution $*$ is $*$-prime (i.e., $aRb = aRb^* = (0)$ yields that $a = 0$ or $b = 0$) but the converse is, in general, not true. A typical example in [16] is as following: Let R be a prime ring, $S = R \times R^{\circ}$ where R° is the opposite ring of R, define $*_{ex}(x, y) = (y, x)$. From $(0, x)S(x, 0) = 0$, it follows that S is not prime. For the $*$-primeness of S, we suppose that $(a, b)S(x, y) = 0$ and $(a, b)S(x, y)^* = 0$, then we get $aRx \times yRb = 0$ and $aRy \times xRb = 0$, and hence $aRx = yRb = aRy = xRb = 0$, or equivalently $(a, b) = 0$ or $(x, y) = 0$. This example shows that every prime ring can be injected in a $*$-prime ring and from this point of view $*$-prime rings constitute a more general class of prime rings.

A classical problem in ring theory is to investigate and extend conditions under which a ring R becomes commutative. So far the best tools found for this purpose are the derivations on rings and also on their modules. Many results in the literature indicate that the global structure of a ring R is often lightly connected to the behavior of additive mappings defined on R (see [5] for a partial bibliography). There has been an ongoing interest concerning the relationship between the commutativity of a prime ring R and the behavior of some special mappings on R. A number of authors have discussed the commutativity of prime and semiprime rings admitting suitably constrained mappings such as automorphisms, derivations, and multipliers acting on appropriate subsets (for example, one-sided ideals, ideals, Lie ideals, and so on) of the rings (see for example [1, 19]). Moreover, some well-known results on prime rings have been extended to $*$-prime rings (see [6, 15], where further references can be found).

We say a map $f : R \longrightarrow R$ preserves commutativity if $[f(x), f(y)] = 0$ whenever $[x, y] = 0$ for all $x, y \in R$. The study of describing commutativity preserving mappings has been investigated widely in matrix theory, operator algebra theory and ring theory (see [18] for references). As is well known, Bell and Mason in [8] introduced the notion of a certain kind of commutativity preserving maps as follows. For a nonempty subset S of R, a map $f : R \longrightarrow R$ is said to be strong commutativity preserving (SCP) on S if $[f(x), f(y)] = [x, y]$ for all $x, y \in S$. In [7], Bell and Daif showed that if R is a semiprime ring admitting a derivation d such that $[d(x), d(y)] = [x, y]$ for all $x, y \in I$, a nonzero right ideal of R, then I is central. In particular, if $I = R$, then R is commutative. Later in [9], Deng and Ashraf proved if a semiprime ring R admits a derivation d and a map $f : I \longrightarrow R$ defined on a nonzero ideal of R such that $[f(x), d(y)] = [x, y]$ for all $x, y \in I$, then R contains a nonzero central ideal and R is commutative if $I = R$. In the same paper, they studied SCP maps in the case of endomorphisms without involution. In fact, they proved if R is a prime ring of characteristic different from two and φ is a non-trivial endomorphism of R such that $[\varphi(x), \varphi(y)] - [x, y] \in Z(R)$ for all $x, y \in R$, then R is commutative. Taking x^* in place of y in this result, Khan and Ali in [12] proved that if a prime ring R with involution of characteristic different from two admits a non-trivial endomorphism φ satisfying $[\varphi(x), \varphi(x^*)] - [x, x^*] \in Z(R)$ for all $x \in R$, then the involution is of the first kind of R satisfies s_4 and $[\varphi(x), x] = 0$

for all $x \in R$. Recently, in [4], Ali, Dar, and Khan obtained a similar result in $*$-prime rings. To be more precisely, they proved that if R is a prime ring of characteristic different from two with involution of the second kind and d is a nonzero derivation of R such that $[d(x), d(x^*)] = [x, x^*]$ for all $x \in R$, then R is commutative. Motivated by these observations, we continue this line of investigation and we examine what happens in case d and δ are a pair of derivations of a $*$-prime ring R such that $[d(x), \delta(x^*)] \pm [x, x^*] = 0$ for all $x \in R$.

1 Some Preliminaries

In order to prove our result, we need to recall the following known facts:

Lemma 1 ([2, Lemma 2.2]) *Let $(R, *)$ be a prime ring of characteristic different from two with involution of the second kind. If $[x, x^*] = 0$ for all $x \in R$, then R is commutative.*

Lemma 2 ([13, Fact 2]) *Let $(R, *)$ be a prime ring of characteristic different from two with the second involution, then $Z(R) \bigcap H(R) \neq \{0\}$.*

Lemma 3 ([14, Fact 1]) *Let $(R, *)$ be a prime ring of characteristic different from two with involution provided with a derivation d. Then $d(h) = 0$ for all $h \in Z(R) \bigcap H(R)$ implies that $d(z) = 0$ for all $z \in Z(R)$.*

Lemma 4 ([10, Theorem 1]) *Let R be a prime ring of characteristic different from two, I a nonzero ideal of R. If d and δ are nonzero derivations of R such that $[d(x), \delta(y)] = [x, y]$ for all $x, y \in I$, then R is commutative.*

Lemma 5 ([17, Theorem 1]) *Let R be a prime ring with center $Z(R)$. If d is a nonzero derivation of R such that $[d(x), x] \in Z(R)$ for all $x \in R$, then R is commutative.*

2 Main Results

Theorem 1 *Let $(R, *)$ be a prime ring of characteristic different from two with involution of the second kind. Suppose that d and δ are a pair of derivations R such that $[d(x), \delta(x^*)] - [x, x^*] = 0$ for all $x \in R$, then R is commutative.*

Proof. If either $d = 0$ or $\delta = 0$, then $[x, x^*] = 0$ for all $x \in R$. In this case, we are done by Lemma 1. Hence, onward we assume that both $d \neq 0$ and $\delta \neq 0$. We are given that

$$[d(x), \delta(x^*)] - [x, x^*] = 0 \text{ for all } x \in R. \tag{1}$$

The linearization of (1) gives that

$$[d(x), \delta(y^*)] + [d(y), \delta(x^*)] - [x, y^*] - [y, x^*] = 0 \text{ for all } x, y \in R. \quad (2)$$

Writing y^* instead of y in (2), we obtain that

$$[d(x), \delta(y)] + [d(y^*), \delta(x^*)] - [x, y] - [y^*, x^*] = 0 \text{ for all } x, y \in R. \quad (3)$$

Since $*$ is the second kind, by Lemma 2.2, we find that $Z(R) \bigcap H(R) \neq \{0\}$. For all $0 \neq h \in Z(R) \bigcap H(R)$ and $y \in R$, Substituting yh for y in (3) and using (3), we obtain

$$[d(x), y]\delta(h) + [y^*, \delta(x^*)]d(h) = 0 \text{ for all } x, y \in R. \quad (4)$$

For $0 \neq s \in Z(R) \bigcap S(R)$ and $y \in R$, replacing y by ys in (4) yields that

$$\{[d(x), y]\delta(h) - [y^*, \delta(x^*)]d(h)\}s = 0 \text{ for all } x, y \in R. \quad (5)$$

Comparing (4) and (5), we find that $2[d(x), y]\delta(h)s = 0$ for all $x, y \in R$, which means $[d(x), y]\delta(h)s = 0$ since the characteristic of R is different from two. Hence, $[d(x), y]\delta(h)Rs = 0$ for all $x, y \in R$. The primeness of R forces that $[d(x), y]\delta(h) = 0$ for all $x, y \in R$. Use the fact that $\delta(Z(R)) \subseteq Z(R)$ and the primeness of R, we have $\delta(h) = 0$ for all $h \in Z(R) \bigcap H(R)$ or $[d(x), y] = 0$ for all $x, y \in R$. Now we divide the proof into two cases:

Case 1. If $\delta(h) = 0$ for all $h \in Z(R) \bigcap H(R)$, then (4) reduces to $[y^*, \delta(x^*)]d(h) = 0$ for all $x, y \in R$. By the same arguments as above, either $d(h) = 0$ for all $h \in Z(R) \bigcap H(R)$ or $[y^*, \delta(x^*)] = 0$ for all $x, y \in R$.

Subcase 1. Assume that $d(h) = 0$ for all $h \in Z(R) \bigcap H(R)$. In view of Lemma 3, $d(Z(R)) = 0$. Substituting ys for y in (3), we are forced to conclude that

$$\{[d(x), \delta(y)] - [d(y^*), \delta(x^*)] - [x, y] + [y^*, x^*]\}s = 0 \text{ for all } x, y \in R. \quad (6)$$

Combining (3) with (6), we find that $2\{[d(x), \delta(y)] - [x, y]\}s = 0$ which implies that $[d(x), \delta(y)] - [x, y] = 0$ for all $x, y \in R$. By Lemma 4, R is commutative.

Subcase 2. Now suppose that $[y^*, \delta(x^*)] = 0$ for all $x, y \in R$. In particular, $[\delta(x), x] = 0$ for all $x \in R$. By Lemma 5, R is commutative.

Case 2. If $[d(x), y] = 0$ for all $x, y \in R$. Letting $x = y$, we have $[d(x), x] = 0$ for all $x \in R$. Again by virtue of Lemma 5, we get the required result. This completes the proof of the theorem.

Theorem 2 *Let $(R, *)$ be a prime ring of characteristic different from two with involution of the second kind. Suppose that d and δ are a pair of derivations R such that $[d(x), \delta(x^*)] + [x, x^*] = 0$ for all $x \in R$, then R is commutative.*

Proof. As a matter of fact, if $[d(x), \delta(x^*)] + [x, x^*] = 0$ for all $x \in R$, then the derivation $-d$ satisfies the relation $[(-d)(x), \delta(x^*)] - [x, x^*] = 0$ for all $x \in R$. It follows from Theorem 1 that R is commutative.

Remark 1 Our results are still true if we assume that the various conditions are satisfied on a nonzero ideal rather than on the whole ring R.

The following example demonstrates that the condition $*$ is of the second is essential in the hypothesis of Theorems 1 and 2.

Example 1 Let $R = \left\{ \begin{pmatrix} m & n \\ p & q \end{pmatrix} \mid m, n, p, q \in S \right\}$, where S is the ring of integers. It is easy to see that R, under matrix addition and matrix multiplication, is a prime ring of characteristic different from two. Define maps $d\begin{pmatrix} m & n \\ p & q \end{pmatrix} = \begin{pmatrix} 0 & -n \\ p & 0 \end{pmatrix}$, $\delta\begin{pmatrix} m & n \\ p & q \end{pmatrix} = \begin{pmatrix} 0 & -n \\ p & 0 \end{pmatrix}$ and $\begin{pmatrix} m & n \\ p & q \end{pmatrix}^* = \begin{pmatrix} q & -n \\ -p & m \end{pmatrix}$. We can find that $Z(R) = \left\{ \begin{pmatrix} m & 0 \\ 0 & m \end{pmatrix} \mid m \in S \right\}$. Therefore, $x^* = x$ for all $x \in Z(R)$, and hence $Z(R) \subseteq H(R)$. This implies that the involution $*$ is of the first kind not the second kind. Moreover, d and δ are a pair of derivation of R satisfying the property $[d(X), \delta(X^*)] \pm [X, X^*] = 0$ for all $X \in R$. However, R is not a commutative ring.

The following example proves that the primeness hypothesis in Theorem 1 is not superfluous.

Example 2 Take R and d the same as Example 1. Let $S = R \times C$, where C is the ring of complex numbers. Then it is clear that (S, σ) is a semiprime ring with involution of the second kind where $\sigma(r, z) = (r^*, \bar{z})$ for all $(r, z) \in S$. Take $D(r, z) = (d(r), 0) = \Delta(r, z)$ for all $(r, z) \in S$. It is easy to verify that D and Δ is a pair of derivations of S satisfying $[D(X), \Delta(X^*)] - [X, X^*] = 0$ for all $X \in S$, but R is not commutative.

References

1. Ali, A., Kumar, D., Miyan, P.: On generalized derivations and commutativity of prime and semiprime rings. Hacettepe J. Math. Stat. **40**(3), 367–374 (2011)
2. Ali, S., Dar, N.A.: On centralizers of prime rings with involution. Bull. Iranian Math. Soc. **41**(6), 1465–1475 (2015)
3. Ali, S., Dar, N.A., Asci, M.: On derivations and commutativity of rings with involutions. Georgian Math. J. **23**(1), 9–14 (2016)
4. Ali, S., Dar, N.A., Khan, A.N.: On strong commutativity preserving like maps in rings with involution. Miskolc Math. Notes **16**(1), 17–24 (2015)
5. Ashraf, M., De Filippis, V., Pary, S.A., Tiwari, S.K.: Derivations vanishing on commutator identity involving generalized derivation on multilinear polynomials in prime rings. Commun. Algebra **47**(2), 800–813 (2019)
6. Boua, A., Ashraf, M.: Differencial identities and generalized derivations in prime rings with involution. Southeast Asian Bull. Math. **43**, 165–181 (2009)
7. Bell, H.E., Daif, M.N.: On commutativity and strong commutativity preserving maps. Canad. Math. Bull. **37**(4), 443–447 (1994)
8. Bell, H.E., Mason, G.: On derivations in near rings and rings. Math. J. Okayama Univ. **34**, 135–144 (1992)
9. Deng, Q., Ashraf, M.: On strong commutativity preserving maps. Res. Math. **30**, 259–263 (1996)
10. De Filippis, V., Rehman, N., Raza, M.A.: Strong commutativity preserving skew derivations in semiprime rings. Bull. Malays. Math. Sci. Soc. **41**, 1819–1834 (2018)
11. Ghahramani, H., Ghosseiri, M.N., Safari, S.: On derivations, biderivations and superbiderivations of quaternion rings. Rend. Circ. Mat. Palermo Ser. **II**(70), 665–674 (2021)
12. Khan, A.N., Ali, S.: Involution on prime rings with endomorphisms. AIMS Math. **5**(4), 3274–3283 (2020)
13. Mamouni, A., Nejjar, B., Ouhtite, L.: Differential identities on prime rings with involution. J. Algebra Appl. **17**(9), 1850163 (11 pages) (2018)
14. Nejjar, B., Kacha, A., Mamouni, A., Ouhtite, L.: Commuuativity theorems in rings with involution. Commun. Algebra **45**(2), 698–708 (2017)
15. Nejjar, B., Kacha, A., Mamouni, A., Ouhtite, L.: Certain commutativity criteria for rings with involution involving generalized derivations. Georgian Math. J. **27**(1), 133–139 (2020)
16. Oukhtite, L., Salhi, S.: On commutativity of σ-prime rings. Glasnik Mathematicki **41**, 57–64 (2006)
17. Posner, E.C.: Derivations in prime rings. Proc. Am. Math. Soc. **8**, 1093–1100 (1957)
18. Semrl, P.: Commutativity preserving maps. Linear Algebra Appl. **429**, 1052–1070 (2008)
19. Zemazmi, O.A., Ouhtite, L., Ali, S., Muthana, N.: On certain classes of generalized derivations. Mathematica Slovaca **69**(5), 1023–1032 (2019)

Basic One-Sided Ideals of Leavitt Path Algebras over Commutative Rings

Pramod Kanwar, Meenu Khatkar, and R. K. Sharma

Abstract In this article, basic left (right) ideals of Leavitt Path Algebra over a commutative unital ring are studied. We give conditions under which a basic left (right) ideal generated by a vertex is a minimal basic left (right) ideal. It is further shown that if R has no non-zero nilpotent elements, then every minimal basic left ideal $L_R(E)x$ of the Leavitt path algebra $L_R(E)$ contains a vertex. Among other techniques, the proof depends on the fact that a Leavitt Path Algebra over a commutative unital ring R is non-degenerate if and only if R has no non-zero nilpotent elements (equivalently, R is a (commutative) semiprime ring).

Keywords Leavitt path algebra · Basic left ideal · Minimal basic left ideal

1 Introduction

Leavitt path algebras of row-finite graphs, introduced by Abrams and Aranda Pino in [2] and independently by Ara, Moreno, and Pardo in [9], have been of interest to algebraists as well as analysts due to their connections with algebraic structures such as matrix rings, Laurent polynomial rings and also with C^*-algebras (see for example [1, 4–8, 11–14]). Several generalizations of these algebras have also been studied in the last decade. On one hand, Abrams and Aranda Pino generalized the concept to arbitrary graphs (see [3]) and on the other, Tomforde considered these algebras where the coefficients came from a commutative unital ring in place of a field (see [15]). It is not hard to see that several results about these algebras over a field do not remain valid if we replace field with a commutative unital ring. For example, Leavitt path algebra of a finite line graph, being a matrix ring over a field, is simple and that the Leavitt path algebra of a graph over a field is always non-degenerate. In the case

P. Kanwar (✉)
Department of Mathematics, Ohio University-Zanesville, Zanesville, OH, USA
e-mail: kanwar@ohio.edu

M. Khatkar · R. K. Sharma
Department of Mathematics, Indian Institute of Technology Delhi, New Delhi 110016, India

M. Ashraf et al. (eds.), *Algebra and Related Topics with Applications*,
Springer Proceedings in Mathematics & Statistics 392,
https://doi.org/10.1007/978-981-19-3898-6_12

of a commutative unital ring neither of these, however, may not be true. Tomforde showed that some of the well-known results about Leavitt path algebras over a field can be generalized to Leavitt path algebras over commutative unital rings by suitably modifying the statement. To accomplish this, Tomforde introduced the concepts of basic ideals, basically simple Leavitt path algebras, among other things.

In this article, we continue this study of Leavitt path algebras over commutative unital rings. We study basic one-sided ideals in these algebras and show, among other things, that under the condition R has no non-zero nilpotent elements (equivalently, R is a (commutative) semiprime ring), (1) if $v \in E^0$ is a line point, then the left (right) ideal $L_R(E)v$ (resp. $vL_R(E)$) is a minimal basic left (right) ideal and conversely (Theorem 6), (2) the left (right) ideal $L_R(E)v$ (resp. $vL_R(E)$) is a minimal basic left (right) ideal if and only if $vL_R(E)v \cong R$ (Theorem 5), and (3) for any minimal basic left ideal $L_R(E)x$, there exists a line point $v \in E^0$ such that $L_R(E)x \cong L_R(E)v$ (Theorem 7). We also show that the Leavitt path algebra $L_R(E)$ over a commutative unital ring R is non-degenerate if and only if R has no non-zero nilpotent elements, that is, R is a (commutative) semiprime ring (Proposition 2) generalizing the result that the Leavitt path algebra over a field is always non-degenerate. In particular, if R is a (commutative) semiprime ring, then the Leavitt path algebra over R has no non-zero nilpotent left, right, or two-sided ideals.

2 Preliminaries and Notation

Throughout this article, a ring will mean a commutative unital ring and a graph will always mean a directed graph.

A graph with E^0 as the set of vertices, E^1, the set of edges, and the functions $r, s : E^1 \to E^0$ is denoted by $E = (E^0, E^1, r, s)$. For each edge $e \in E^1$, the vertices $s(e)$ and $r(e)$ are called the *source* and *range* of e, respectively. For $v \in E^0$, a *loop* at v is an edge e for which $r(e) = s(e)$. A vertex which does not receive any edge is called a *source*. A vertex which does not emit any edge is called a *sink*. A vertex $v \in E^0$ such that $|s^{-1}(v)| = \infty$ is called an *infinite emitter*. A vertex v which is either a sink or an infinite emitter is called a *singular vertex*. A vertex v which is not a singular vertex is called a *regular vertex*. A vertex which is both a source and a sink is called an *isolated vertex*.

A *path* μ in a graph E is a finite sequence of edges $\mu = e_1e_2 \cdots e_n$ such that $r(e_i) = s(e_{i+1})$ for $i = 1, 2, \ldots, n-1$. In this case, $s(e_1)$ is called the source of μ (denoted by $s(\mu)$), $r(e_n)$ is called the range of μ (denoted by $r(\mu)$), and n is called the *length* of μ. We view the elements of E^0 as paths of length 0. An edge e is an *exit* for a path $\mu = e_1e_2 \cdots e_n$ if there exists i such that $s(e) = s(e_i)$ and $e \neq e_i$. If $\mu = e_1e_2 \ldots e_n$, then we denote the set $\{s(e_i), r(e_i) : i = 1, 2, \ldots, n\}$ by μ^0. If μ is a path such that $v = s(\mu) = r(\mu)$, then μ is called a *closed path* based at v. If $r(\mu) = s(\mu)$ and $s(e_i) \neq s(e_j)$ for every $i \neq j$, then μ is called a *cycle*. A graph without any cycles is called *acyclic*.

We say that a graph E satisfies condition (NE) if no cycle in E has an exit and that it satisfies condition (L) if every cycle in E has an exit.

For $n \geq 2$, we denote the set of paths of length n by E^n and the set of all paths by E^*. We define a relation $\leq$ on E^0 by setting $v \leq w$ if there is a path $\mu \in E^*$ with $s(\mu) = v$ and $r(\mu) = w$. The set $T(v) = \{w \in E^0 \mid v \leq w\}$ is called the *tree* of v.

We say that a vertex $v \in E^0$ is a bifurcation (or that there is a bifurcation at v) if v is the source of at least two edges. A vertex $v \in E^0$ will be called a line point if there are neither bifurcations nor cycles at any vertex $w \in T(v)$, the tree of v. We will denote by $P_l(E)$ the set of all line points in E^0. We say that a path μ contains no bifurcations if the set $\mu^0 \setminus r(\mu)$ contains no bifurcations, that is, if none of the vertices of the path μ, except perhaps $r(\mu)$, is a bifurcation.

A graph is called *row-finite* if every vertex emits only a finite number of edges. Note that, a row-finite graph is finite if E^0 is finite. Throughout this paper, we will be using row-finite graphs only.

Given a graph E, the *extended graph* of E is defined as the graph $\hat{E} = (E^0, E^1 \cup (E^1)^*, r', s')$ where $(E^1)^* = \{e_i^* : e_i \in E^1\}$ and $r'|_{E^1} = r, r'(e_i{}^*) = s(e_i), s'|_{E^1} = s$ and $s'(e_i{}^*) = r(e_i)$. The elements of $(E^1)^*$ are called *ghost edges*.

Let R be a commutative unital ring and E be a graph. Following Tomforde, we define a *Leavitt E-family* to be the set $\{v, e, e^* : v \in E^0,\ e \in E^1\}$ in R such that

1. $vw = \delta_{vw} v$ for all $v, w \in E^0$,
2. $s(e)e = er(e) = e$ for all $e \in E^1$,
3. $r(e)e^* = e^*s(e) = e^*$ for all $e \in E^1$,
4. $e^* f = \delta_{ef} r(e)$ for all $e, f \in E^1$, and
5. $v = \sum_{e \in s^{-1}(v)} ee^*$ for every regular vertex $v \in E^0$.

Note that the Condition (2) and Condition (3) can be combined to read $s'(e)e = er'(e) = e$ for all $e \in E^1 \cup (E^1)^*$. The conditions (4) and (5) are called Cuntz–Kreiger relations and are denoted by (CK1) and (CK2), respectively.

The *Leavitt path algebra of E with coefficients in R*, denoted by $L_R(E)$, is defined as the universal R-algebra generated by a Leavitt E-family.

We remark that if E^0 is finite then $L_R(E)$ is unital R-algebra with unit as sum of all the vertices. If E^0 is infinite, then $L_R(E)$ is an algebra with local units. Also, $L_R(E)$ is a $\mathbb{Z}$-graded algebra with grading induced by degree(v_i) $= 0$, degree(e_i) $= 1$, degree($e_i{}^*$) $= -1$ for all $v_i \in E^0$ and $e_i \in E^1$, that is, $L_R(E) = \bigoplus_{n \in \mathbb{Z}} L_R(E)_n$, where $L_R(E)_0 = RE^0 + A_0$, $L_R(E)_n = A_n$ for $n \neq 0$ and $A_n = \sum\{re_{i_1}e_{i_2} \cdots e_{i_\sigma} e_{j_1}{}^* e_{j_2}{}^* \cdots e_{j_\tau}{}^* : \sigma + \tau \geq 0,\ e_{i_s} \in E^1,\ e_{j_t}{}^* \in (E^1)^*,\ r \in R,\ \sigma - \tau = n\}$ for all n.

3 Basic One-Sided Ideals

An ideal I of $L_R(E)$ is called *basic* if whenever $0 \neq rv \in I$ for $r \in R$, $v \in E^0$, we have $v \in I$ (see [15]). Following this we say that a left ideal I of $L_R(E)$ is a *basic left ideal* if whenever $0 \neq rv \in I$ for $r \in R \setminus \{0\}$ and $v \in E^0$, we have $v \in I$. *Basic right ideals* are defined similarly. A basic left (right) ideal in $L_R(E)$ is called *minimal basic left (resp. right) ideal* if it does not contain any non-zero basic left (resp. right) ideal other than itself.

Example 1 Observe that if $R = \mathbb{Z}_4$ and E is the graph

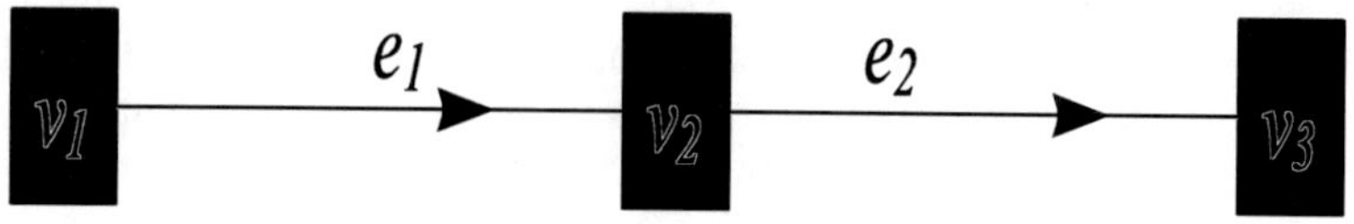

then both $\begin{pmatrix} \mathbb{Z}_4 & 0 & 0 \\ \mathbb{Z}_4 & 0 & 0 \\ \mathbb{Z}_4 & 0 & 0 \end{pmatrix}$ and $\begin{pmatrix} \mathbb{Z}_4 & \mathbb{Z}_4 & 0 \\ \mathbb{Z}_4 & \mathbb{Z}_4 & 0 \\ \mathbb{Z}_4 & \mathbb{Z}_4 & 0 \end{pmatrix}$ are basic left ideals of $L_R(E)$, whereas the left ideals $\begin{pmatrix} 2\mathbb{Z}_4 & 0 & 0 \\ 2\mathbb{Z}_4 & 0 & 0 \\ 2\mathbb{Z}_4 & 0 & 0 \end{pmatrix}$ and $\begin{pmatrix} 2\mathbb{Z}_4 & 2\mathbb{Z}_4 & 0 \\ 2\mathbb{Z}_4 & 2\mathbb{Z}_4 & 0 \\ 2\mathbb{Z}_4 & 2\mathbb{Z}_4 & 0 \end{pmatrix}$ are not basic. Further, the basic left ideal $\begin{pmatrix} \mathbb{Z}_4 & 0 & 0 \\ \mathbb{Z}_4 & 0 & 0 \\ \mathbb{Z}_4 & 0 & 0 \end{pmatrix}$ is minimal and the basic left ideal $\begin{pmatrix} \mathbb{Z}_4 & \mathbb{Z}_4 & 0 \\ \mathbb{Z}_4 & \mathbb{Z}_4 & 0 \\ \mathbb{Z}_4 & \mathbb{Z}_4 & 0 \end{pmatrix}$ is not minimal.

It is clear that every left (right) ideal of a Leavitt path algebra over a field is a basic left (right) ideal. Further, since $v = v \cdot v$ for any $v \in E^0$, the left ideal $L_R(E)v$ contains v. Also, for any $0 \neq r \in R$ and any vertex $w \in E^0$, $w \neq v$, $rw \notin L_R(E)v$. It follows that for any vertex $v \in E^0$, the left ideal $L_R(E)v$ is a basic left ideal. But the basic left ideal $L_R(E)v$ is not necessarily a minimal basic left ideal.

Example 2 If $R = \mathbb{Z}_4$ and E is the graph

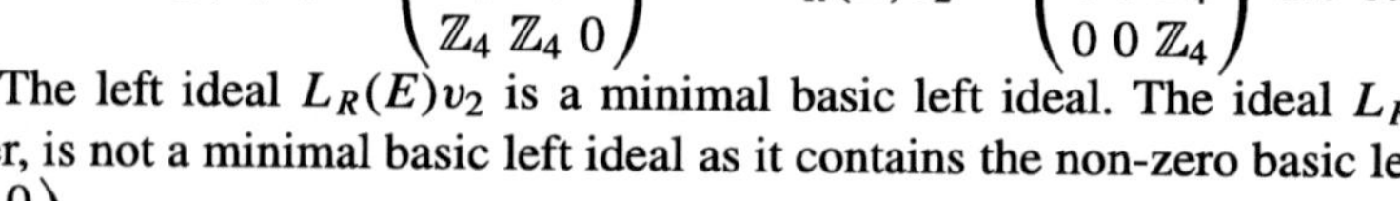

then $L_R(E) = M_3(\mathbb{Z}_4)$.

Here both $L_R(E)v_1 = \begin{pmatrix} \mathbb{Z}_4 & \mathbb{Z}_4 & 0 \\ \mathbb{Z}_4 & \mathbb{Z}_4 & 0 \\ \mathbb{Z}_4 & \mathbb{Z}_4 & 0 \end{pmatrix}$ and $L_R(E)v_2 = \begin{pmatrix} 0 & 0 & \mathbb{Z}_4 \\ 0 & 0 & \mathbb{Z}_4 \\ 0 & 0 & \mathbb{Z}_4 \end{pmatrix}$ are basic left ideals. The left ideal $L_R(E)v_2$ is a minimal basic left ideal. The ideal $L_R(E)v_1$, however, is not a minimal basic left ideal as it contains the non-zero basic left ideal $\begin{pmatrix} \mathbb{Z}_4 & 0 & 0 \\ \mathbb{Z}_4 & 0 & 0 \\ \mathbb{Z}_4 & 0 & 0 \end{pmatrix}$. Further, a minimal basic left ideal need not be minimal as a left ideal, that is, a minimal basic left ideal may contain non-zero non-basic left ideal other than the left ideal itself. For example, the left ideal $L_R(E)v_2$ above is a minimal basic left ideal that contains the non-zero left ideal $\begin{pmatrix} 0 & 0 & 2\mathbb{Z}_4 \\ 0 & 0 & 2\mathbb{Z}_4 \\ 0 & 0 & 2\mathbb{Z}_4 \end{pmatrix}$ which is not basic.

We also observe that for $v_1, v_2, \ldots, v_n \in E^0$, the ideal $\sum_{i=1}^{n} L_R(E)v_i$ is also basic. Further, if $v \in E^0$ is not a sink and $s^{-1}(v) = \{e_1, e_2, \ldots, e_n\}$, then by (CK2), $v = \sum_{i=1}^{n} e_i e_i{}^*$ and for $1 \leq i \leq n$, the idempotents $e_i e_i^*$ are orthogonal. Thus, $L_R(E)v = \bigoplus_{i=1}^{n} L_R(E)e_i e_i{}^*$. We, in fact, have the following theorem.

Theorem 1 *Let R be a commutative unital ring and E be a graph. If v is a vertex in E which is not a sink and $s^{-1}(v) = \{e_1, e_2, \ldots, e_n\}$, then $L_R(E)v = \bigoplus_{i=1}^{n} L_R(E)e_i e_i{}^*$. Moreover, if $r(e_i) \neq r(e_j)$ for $i \neq j$ and $r(e_i) = v_i$, then $L_R(E)v \cong \bigoplus_{i=1}^{n} L_R(E)v_i$.*

Proof We have already shown $L_R(E)v = \bigoplus_{i=1}^{n} L_R(E)e_i e_i{}^*$. Now, let $r(e_i) \neq r(e_j)$ for $i \neq j$ and let $r(e_i) = v_i$. Consider the map $\phi : L_R(E)v \rightarrow \bigoplus_{i=1}^{n} L_R(E)v_i$ defined by $\phi(x) = \sum_{i=1}^{n} xe_i$. The map ϕ is clearly an $L_R(E)$-homomorphism between basic left ideals $L_R(E)v$ and $\bigoplus_{i=1}^{n} L_R(E)v_i$. Moreover, if $\phi(x) = \sum_{i=1}^{n} xe_i = 0$, then $\big(\sum_{i=1}^{n} xe_i\big)r(e_j) = 0$ for each j, that is, $xe_j = 0$ for each j. Thus, $xe_j e_j{}^* = 0$ for each j, and hence $x = xv = \sum_{i=1}^{n} xe_i e_i{}^* = 0$. Thus, ϕ is one-one. The map ϕ is also an epimorphism since for $x_i \in L_R(E)v_i$, $1 \leq i \leq n$, there exists $\sum_{i=1}^{n} x_i e_i{}^* \in \bigoplus_{i=1}^{n} L_R(E)e_i e_i{}^* = L_R(E)v$ and $\phi\big(\sum_{i=1}^{n} x_i e_i{}^*\big) = \sum_{i=1}^{n} x_i$. Thus, $L_R(E)v \cong \bigoplus_{i=1}^{n} L_R(E)v_i$.

Proposition 1 *Let R be a commutative unital ring and E be a graph. If u, v are vertices in E such that $v \in T(u)$ and the path joining u to v contains no bifurcations then $L_R(E)u \cong L_R(E)v$ as basic left $L_R(E)$-modules.*

Proof Let μ be a path from u to v. Clearly, $\mu^*\mu = v$. Since μ does not contain any bifurcations, $\mu\mu^* = u$. Thus, the right multiplication f given by $f(\alpha u) = \alpha u \mu$ gives a $L_R(E)$-homomorphism from $L_R(E)u$ to $L_R(E)v$. Since $\mu\mu^* = u$, $\alpha u \mu = 0$ gives $\alpha u = 0$. Also, for $\alpha v \in L_R(E)v$, $\alpha v \mu^* = \alpha\mu^* u$ and $f(\alpha v \mu^*) = \alpha v \mu^* \mu = \alpha v$. Hence, f is an isomorphism from $L_R(E)u$ to $L_R(E)v$. Thus, $L_R(E)u \cong L_R(E)v$ as basic left $L_R(E)$-modules.

4 Minimal Basic Left Ideals Generated by a Vertex

We observed in the previous section that for any vertex $v \in E^0$, the left ideal $L_R(E)v$ is a basic left ideal and that this is not necessarily a minimal basic left ideal. We now give conditions for the ideal $L_R(E)v$ to be a minimal basic left ideal.

Theorem 2 *Let R be a commutative unital ring and E be a graph. If for $u \in E^0$, $T(u)$ contains some bifurcation then $L_R(E)u$ is not a minimal basic left ideal.*

Proof Let $v \in T(u)$ be a bifurcation and let $\mu = e_1e_2\dots e_n$ be a path from u to v. Let $x \in \mu^0$, the set of vertices of μ, be the first bifurcation in μ. If $x = u$ and $f_1, f_2, \dots f_n$ are the distinct edges with $s(f_i) = u$, then $L_R(E)u = \bigoplus_{i=1}^n L_R(E)f_if_i^*$ giving $L_R(E)u$ is not minimal basic left ideal. If $x \neq u$ then $x = r(e_i)$ for some i. Therefore, the path $e_1e_2\dots e_i$ from u to x contains no bifurcations. Hence, by Proposition 1, $L_R(E)u \cong L_R(E)x$ as $L_R(E)$-modules. Now the path from x to v has a bifurcation at x. Thus, as in the case when $x = u$, $L_R(E)x$ is not a minimal basic left ideal. Since $L_R(E)u \cong L_R(E)x$, $L_R(E)u$ is not a minimal basic left ideal.

We next prove that if there is a closed path based at a vertex u in E^0 then also $L_R(E)u$ is not minimal basic left ideal. We first prove the following result that is also of independent interest.

Theorem 3 *Let R be a commutative unital ring and E be a graph. If for $u \in E^0$, $T(u)$ has no bifurcations and μ is a closed path based at u then the left ideal $L_R(E)(\mu + u)$ is basic.*

Proof Observe that for $0 \neq r \in R$, $ru \neq 0$ ([15], Proposition 3.4) and $r\mu \neq 0$ ([15], Proposition 4.9). Since $T(u)$ has no bifurcations, μ is a cycle and $\mu\mu^* = u = \mu^*\mu$. Clearly, for any vertex $v \in E^0, v \neq u$ and $0 \neq r \in R, rv \notin L_R(E)(\mu + u)$ for if $rv = \alpha(\mu + u)$ then $rv = (rv)v = 0$. We show that for $0 \neq r \in R$, $ru \notin L_R(E)(\mu + u)$. If for $0 \neq r \in R$, $ru \in L_R(E)(\mu + u)$ then $ru = \sum_i r_i\alpha_i(\mu + u)$ where each α_i is a non-zero monomial in $L_R(E)$ and $r_i \in R$. Observe that for each i, $\alpha_i \neq 0$ and $r(\alpha_i) = u = s(\alpha_i)$. Since $T(u)$ contains no bifurcations, each monomial α_i is either a power of μ or a power of μ^* or simply u. Consequently, there exists non-negative integers m, n and $s_i \in R$ for $-n \leq i \leq m$ such that $p(\mu, \mu^*) = s_m\mu^m + s_{m-1}\mu^{m-1} + \cdots + s_1\mu + s_0u + s_{-1}\mu^* + \cdots + s_{-n}(\mu^*)^n$ and $ru = p(\mu, \mu^*)(\mu + u)$. Since $\mu^*\mu = u = \mu\mu^*$, we get $r\mu^n = ru\mu^n = (s_m\mu^{m+n} + \cdots + s_{-n}u)(\mu + u)$. Since the subalgebra of R-algebra $L_R(E)$ generated by μ and u is isomorphic to $R[x]$, we get a polynomial $q(x)$ in $R[x]$ such that $rx^n = q(x)(x + 1)$ which is not possible as for $x = -1$, the right-hand side is 0 but the left-hand side is non-zero. Thus, for $0 \neq r \in R$, $ru \notin L_R(E)(\mu + u)$. Hence, $L_R(E)(\mu + u)$ is a basic left ideal.

Corollary 1 *Let R be a commutative unital ring and E be a graph. If there is some closed path based at $u \in E^0$, then $L_R(E)u$ is not a minimal basic left ideal.*

Proof Let μ be a closed path based at u such that $L_R(E)u$ is a minimal basic left ideal. By Theorem 2, there does not exist any bifurcation at any vertex of μ. Thus, μ is a cycle. Since $\mu + u = (\mu + u)u$, we have $0 \neq L_R(E)(\mu + u) \subseteq L_R(E)u$. Since $L_R(E)u$ is a minimal basic left ideal and $L_R(E)(\mu + u)$ is basic (Theorem 3), we have $L_R(E)(\mu + u) = L_R(E)u$. Therefore, $u \in L_R(E)(\mu + u)$ which is a contradiction as in Theorem 3. Hence, $L_R(E)u$ is not a minimal basic left ideal.

Theorem 4 *Let R be a commutative unital ring and E be a graph. If u is a vertex in E such that $L_R(E)u$ is a minimal basic left ideal then u is a line point.*

Proof Let $v \in T(u)$. If there is a bifurcation at v then, by Theorem 2, $L_R(E)u$ is not a minimal basic left ideal, a contradiction. If there is a cycle based at v then $L_R(E)v$ is not a minimal basic left ideal (Corollary 1). Since $L_R(E)u$ is minimal basic left ideal, by Theorem 2, there does not exist any bifurcations in the unique path joining u to v. But then, by Proposition 1, $L_R(E)u \cong L_R(E)v$, a contradiction as $L_R(E)u$ is not a minimal basic left ideal. Hence, u is a line point.

Before giving necessary and sufficient conditions for the left ideal $L_R(E)v$ to be a minimal basic left ideal, we recall that an algebra A is said to be *non-degenerate* if $aAa = 0$ for $a \in A$ implies $a = 0$. It is known that the Leavitt path algebra over a field is non-degenerate ([14], Proposition 1.1). This is not true when we replace the field with a commutative unital ring. For example, if $R = \mathbb{Z}_4$ and E is the graph

then $2v_1 L_R(E) 2v_1 = 0$ and $2v_1 \neq 0$ so that $L_R(E)$ is not non-degenerate. In fact if R is any commutative unital ring that has non-zero nilpotent elements, equivalently, if R is any commutative semiprime unital ring and E is any graph then the Leavitt path algebra $L_R(E)$ is not non-degenerate, for if r is a non-zero nilpotent element in R having index of nilpotency 2 and v is any vertex in E then $rvL_R(E)rv = 0$ but $rv \neq 0$. If R is a commutative unital ring that has no non-zero nilpotent elements, then $L_R(E)$ is, indeed, non-degenerate as is shown in the following proposition.

Proposition 2 *Let R be a commutative unital ring and E be a graph. Then the Leavitt path algebra $L_R(E)$ is non-degenerate if and only if R has no non-zero nilpotent elements, equivalently, R is a (commutative) semiprime ring. In particular, if R is a commutative ring with no non-zero nilpotent elements, then the Leavitt path algebra over R has no non-zero nilpotent left, right, or two-sided ideals.*

Proof As observed above, if R has non-zero nilpotent elements, then $L_R(E)$ is not non-degenerate. Now let R has no non-zero nilpotent elements. To prove that $L_R(E)$ is non-degenerate, our technique is similar to the one used to prove the result in the case of a field. We first prove that if a is a homogeneous element in the $\mathbb{Z}$ graded R-algebra $L_R(E)$ such that $aL_R(E)a = 0$ then $a = 0$. Let $Z = \{z \in L_R(E) : zL_R(E)z = 0\}$. Then Z is an ideal of $L_R(E)$ and for every $z \in Z$ and $r \in R$, rz and z^* belongs to Z. Further as R has no non-zero nilpotent elements, Z is a basic ideal. In fact, there does not exist any $r \in R$, $v \in E^0$ such that $0 \neq rv \in Z$, for if $0 \neq rv \in Z$ then $rvL_R(E)rv = 0$ and hence $r^2vxv = 0$ for all $x \in L_R(E)$. If E is finite, then $L_R(E)$ is unital, and hence $r^2v = 0$ giving $r^2 = 0$, a contradiction. If E is not finite, then $L_R(E)$ has local units. Therefore, there exists $t \in L_R(E)$ such that $vt = tv = v$. Thus, taking $x = t$ in $r^2vxv = 0$, we get $r^2v = 0$, and hence $r^2 = 0$, a contradiction again. Clearly, Z does not contain any paths. We now show that Z does not contain homogeneous elements of any degree.

Let $x \in L_R(E)_0$ and let $xL_R(E)x = 0$. As observed earlier $x \neq rv$ for any vertex v and $r \in R$. Assume that x is a linear combination of vertices and monomials ab^*, where a and b are paths of same positive degree. Since any vertex u that is not a sink can be replaced with $\sum_{\{e\in E^1|s(e)=u\}} ee^*$ (see CK2), we can rewrite x as linear combination of sinks and monomials ab^*, where a and b are paths of same positive degree. Let $x = x_1 + x_2$, where x_1 is a linear combination of sinks and x_2 is a linear combination of zero degree monomials.

Consider a monomial ab^* appearing in the representation of x_2 such that degree of a is maximum and write $a = ea', b = fb'$, where $e, f \in E^1$ and a', b' are paths of degree 1 less than the degree of a. Combining all monomials pq^* in x_2 where $p = ea_1$ and $q = fb_1$, we can write $x_2 = ex'f^* + y$ where $0 \neq x' \in L_R(E)$ and $e^*yf = 0$. Since x_1 contains only sinks, $e^*x_1f = 0$. Thus

$$e^*xf = e^*x_1f + e^*x_2f = e^*x_1f + e^*ex'f^*f + e^*yf = 0 + x' + 0 = x'.$$

Since Z is an ideal of $L_R(E)$, $x' = e^*xf$ is a non-zero element of Z. Applying this argument recursively to x', we get that Z contains a non-zero linear combination of vertices, which is a contradiction.

Suppose that Z does not contain non-zero homogeneous elements of positive degree $< k$. We prove that Z does not contain non-zero homogeneous elements of degree k. Let $0 \neq x \in L_R(E)_k \cap Z$. For any $e \in E^1$, we have $e^*x \in Z$ and it is a homogeneous element of degree $< k$. Therefore, $e^*x = 0$ for any $e \in E^1$. Consequently, $vx = 0$ for any vertex v such that $s^{-1}(v) \neq \phi$. On the other hand, if v is a vertex such that $s^{-1}(v) = \phi$, then for any $e \in E^1$, we have $ve = vs(e)e = 0$ since $v \neq s(e)$. Thus, $vx = 0$ for any vertex v and this implies $x = 0$ since $L_R(E)$ has local units. Also, since $L_R(E)_{-n} = (L_R(E)_n)^*$, it is clear that $L_R(E)$ does not contain non-zero homogeneous elements of negative degree. Therefore, $L_R(E)$ does not contain any homogeneous elements x such that $xL_R(E)x = 0$.

Now let a be any element of $L_R(E)$ such that $a \in Z$, that is, $aL_R(E)a = 0$. Since $L_R(E)$ is Z-graded, writing $a = a_{\sigma_1} + a_{\sigma_2} + \cdots + a_{\sigma_n}$ where $\sigma_1 < \sigma_2 < \cdots < \sigma_n$ and using $aL_R(E)a = 0$ we get $a_{\sigma_n}L_R(E)a_{\sigma_n} = 0$ (see Proposition II.1.4 in [13]). Since a_{σ_n} is a homogeneous element, we get $a_{\sigma_n} = 0$. Repeating this argument, we get $a = 0$. Hence, $Z = 0$, that is, $L_R(E)$ is non-degenerate.

It is, now, not hard to see that if R is a commutative ring with no non-zero nilpotent elements then the Leavitt path algebra over R has no non-zero nilpotent left, right, or two-sided ideals.

Theorem 5 *Let R be a commutative unital ring having no non-zero nilpotent elements, equivalently, R is a commutative semiprime ring and E be a graph. For any $u \in E^0$, $L_R(E)u$ is a minimal basic left ideal if and only if $uL_R(E)u = Ru \cong R$.*

Proof Let $L_R(E)u$ be a minimal basic left ideal. By Theorem 4, u is a line point. Therefore, for every $v \in T(u)$, there exists only a unique path from u to v. Also, $uL_R(E)u = span_R\{\alpha\beta^* : r(\alpha) = r(\beta)$ and $s(\alpha) = s(\beta) = u\}$. Since there is a unique path from u to v for every $v \in T(u)$ and $r(\alpha) = r(\beta)$, it follows that $\alpha = \beta$. Since $T(u)$ has no bifurcations, and $s(\alpha) = u, \alpha\alpha^* = u$. Thus, $uL_R(E)u = Ru \cong R$. Conversely, let $uL_R(E)u = Ru \cong R$. We first prove that for every $a \in L_R(E)$, $u \in L_R(E)au$. Since R has no non-zero nilpotent elements, $L_R(E)$ is non-degenerate and hence $auL_R(E)au \neq 0$ as $au \neq 0$. Thus, there exists some $x \in L_R(E)$ such that $uxau \neq 0$, for if $uxau = 0$ for all x then $auL_R(E)au = 0$ which is a contradiction. Since $0 \neq uxau \in uL_R(E)u = Ru$, we have $uxau = ru$ for some $r \in R$. Since $L_R(E)au$ is a basic left ideal and $ru = uxau \in L_R(E)au$, we have $u \in L_R(E)au$. Hence, $L_R(E)u$ is a minimal basic left ideal.

Since for any sink u in E^0, $uL_R(E)u = Ru \cong R$, we have the following corollary.

Corollary 2 *Let R be a commutative unital ring with no non-zero nilpotent elements, equivalently, R is a commutative semiprime ring and E be a graph. If $u \in E^0$ is a sink and $v \in E^0$ is a vertex connected to u by a path without bifurcations, then both $L_R(E)u$ and $L_R(E)v$ are minimal basic left ideals.*

Theorem 6 *Let R be a commutative unital ring having no non-zero nilpotent elements, equivalently, R is a commutative semiprime ring and E be a graph. For any $u \in E^0$, $L_R(E)u$ is a minimal basic left ideal if and only if u is a line point in E.*

Proof One direction is given by Theorem 4. For the other direction, let $u \in E^0$ be a line point. Then for every vertex $v \in T(u)$, there is a unique path (say α) from u to v. Thus, $uL_R(E)u = span_R\{\alpha\beta^* : r(\alpha) = r(\beta) \text{ and } s(\alpha) = s(\beta) = u\} = span_R\{\alpha\alpha^*\} = Ru$. Hence, $L_R(E)u$ is a minimal basic left ideal by Theorem 5.

We remark that arguments used to prove the results in this section as well as the previous section for basic left ideals can be suitably modified to prove similar results for basic right ideals. Since the conditions "u is a line point" and "$uL_R(E)u = Ru \cong R$" are symmetric, we have the following corollary.

Corollary 3 *Let R be a commutative unital ring having no non-zero nilpotent elements, equivalently, R is a commutative semiprime ring and E be a graph and let $u \in E^0$. The following conditions are equivalent:*

1. *$L_R(E)u$ is a minimal basic left ideal,*
2. *$uL_R(E)$ is a minimal basic right ideal,*
3. *$uL_R(E)u = Ru \cong R$,*
4. *u is a line point in E.*

Theorem 7 *Let R be a commutative unital ring having no non-zero nilpotent elements, equivalently, R is a commutative semiprime ring and E be a graph. Let $x \in L_R(E)$ be such that $L_R(E)x$ is a minimal basic left ideal. Then there exists a line point v in E such that $L_R(E)x \cong L_R(E)v$ as basic left $L_R(E)$-modules.*

Proof Let $x \in L_R(E)$ be such that $L_R(E)x$ is a minimal basic left ideal. We first claim that there does not exists any vertex u and a cycle c based at u with no exits such that $y = \mu_1\mu_2 \ldots \mu_r x \nu_1\nu_2 \ldots \nu_s$ is a non-zero element in $uL_R(E)u = \{\sum_{i=-m}^{n} r_i c^i;\ m, n \in \mathbb{N} \text{ and } r_i \in R\}$. On the contrary, let there exists a vertex u and a cycle c based at u with no exits such that $y = \mu_1\mu_2 \ldots \mu_r x \nu_1\nu_2 \ldots \nu_s$ is a non-zero element in $uL_R(E)u$. The map $\phi : R[t, t^{-1}] \to L_R(E)$ defined by $\phi(1) = u$, $\phi(t) = c$ and $\phi(t^{-1}) = c^*$ is a monomorphism with image $uL_R(E)u$. Thus, $uL_R(E)u \cong$

$R[t, t^{-1}]$ as R-algebras. Now since $L_R(E)y = L_R(E)\mu_1\mu_2 \ldots \mu_r x \nu_1\nu_2 \ldots \nu_s \cong L_R(E)x$, $L_R(E)y$ is a minimal basic left ideal of $L_R(E)$. Since $uL_R(E)y$ is a minimal basic left ideal of $L_R(E)$ contained in $uL_R(E)u$, the preimage $0 \neq \phi^{-1}(uL_R(E)y)$ is a minimal basic left ideal of $R[t, t^{-1}]$, a contradiction as $R[t, t^{-1}]$ has no minimal basic left ideals.

Since there exists a path $\mu \in L_R(E)$ such that $x\mu$ is in only real edges and is non-zero (see Theorem 2.2.11 and Corollary 2.2.12 in [1] and Theorem 5.1 in [10]), we can assume that x is only in real edges. Then we can write $0 \neq x = \sum_{i=1}^{t} r_i\beta_i$ where β_i's are distinct paths and $r_i \in R$. We will use induction on t to prove that there exists $\mu_1, \mu_2, \ldots, \mu_r, \nu_1, \nu_2, \ldots, \nu_s \in E^0 \cup E^1 \cup E^{1*}$ such that $0 \neq \mu_1\mu_2 \ldots \mu_r x \nu_1\nu_2 \ldots \nu_s = rv$ for some $r \in R$ and $v \in E^0$.

For $t = 1$, we have $0 \neq x = r_1\beta_1$. If β_1 has degree 0 then $0 \neq x = rv$ for some $v \in E^0$. If β_1 has non-zero degree, that is, $x = r_1\beta_1 = r_1e_1e_2 \ldots e_n$(say) where e_i's are edges, then $e_n^*e_{n-1}^* \ldots e_1^*r_1\beta_1 = e_n^*e_{n-1}^* \ldots e_1^*r_1e_1e_2 \ldots e_n = r_1v$ where $v = r(e_n) = r(\beta_1) \in E^0$.

Now let the claim holds for any non-zero element which is a R-linear combination of less than t paths. Let $0 \neq x = \sum_{i=1}^{t} r_i\beta_i$. We can assume that $deg(\beta_i) \leq deg(\beta_{i+1})$ for $1 \leq i \leq t$. Then $0 \neq \beta_1^*x = r_1\beta_1^*\beta_1 + \sum_{i=2}^{t} r_i\beta_1^*\beta_i = r_1v + \sum_{i=2}^{t} r_i\gamma_i$ where $v = r(\beta_1)$ and $\gamma_i = \beta_1^*\beta_i$. If $\gamma_i = 0$ for some i, then applying induction to β_1^*x we are done. Let us now assume that $\gamma_i \neq 0$ for all i. If some γ_i does not start or end in $v = r(\beta_1)$, then applying induction to $0 \neq v\beta_1^*x = r_1v + \sum_{i=2}^{t} r_iv\gamma_i$ or $0 \neq \beta_1^*xv = r_1v + \sum_{i=2}^{t} r_i\gamma_iv$, we get the result. Now, let us assume that $0 \neq \beta_1^*x = r_1v + \sum_{i=2}^{t} r_i\gamma_i = z$ (say), where $0 < deg(\gamma_2) < deg(\gamma_3) < \cdots < deg(\gamma_t)$ and for $2 \leq i \leq t$, $v = r(\gamma_i) = s(\gamma_i) = r(\beta_1)$. If there exists some path μ' such that $\mu'^*\gamma_i = 0$ for some i but not for all, then applying induction to $0 \neq \mu'^*z\mu' = r_1\mu'^*v\mu' + \sum_{i=1}^{t} r_i\mu'^*\gamma_i\mu'$ we get the claim. So now let $\mu'^*\gamma_i = 0$ for all i. Thus, $\gamma_{i+1} = \gamma_i\delta_i$ for some path δ_i. Hence, $z = r_1v + r_2\tau_1 + r_3\tau_1\tau_2 + \cdots + r_t\tau_1 \ldots \tau_{t-1}$, where each τ_i starts and finishes in v. If all the paths τ_i's are not identical, then $\tau_1 \neq \tau_i$ for some i. Hence, $0 \neq \tau_1^*z\tau_i = r_1v$ and we are done. If all the paths τ_i's are identical, then z is a polynomial in a cycle $c = \tau_1$ with an independent term r_1v which is an element of $vL_R(E)v$. But then by what we proved at the beginning of the proof, c must have an exit. Thus, there exists $w \in T(v)$ and two distinct edges e and f having same source w and such that $c = aweb = aeb$ for some paths a and b. Let $\eta = af$. Then $\eta^*c = f^*a^*aeb = f^*eb = 0$. Thus, $\eta^*z\eta = r_1\eta^*v\eta = r_1\eta^*\eta = r_1r(\eta) = r_1r(f) \neq 0$ and is scalar multiple of a vertex.

Hence, there always exists $\mu_1, \mu_2, \ldots, \mu_r, \nu_1, \nu_2, \ldots, \nu_s \in E^0 \cup E^1 \cup (E^1)^*$ and $r \in R$ such that $0 \neq \mu_1\mu_2 \ldots \mu_r x \nu_1\nu_2 \ldots \nu_s = rv$ for some $v \in E^0$. Thus, $rv \in L_R(E)\mu_1\mu_2 \ldots \mu_r x \nu_1\nu_2 \ldots \nu_s$ which is a basic left ideal because it is isomorphic to $L_R(E)x$. Hence, $v \in L_R(E)\mu_1\mu_2 \ldots \mu_r x \nu_1\nu_2 \ldots \nu_s$. Thus, $L_R(E)v = L_R(E)\mu_1\mu_2 \ldots \mu_r x \nu_1\nu_2 \ldots \nu_s \cong L_R(E)x$. Also, $L_R(E)v$, being isomorphic to $L_R(E)x$, is a minimal basic left ideal. Thus, v is a line point.

References

1. Abrams, G., Ara, P., Molina, M.S.: Leavitt path algebras. Lecture Notes in Mathematics. Springer, London (2017)
2. Abrams, G., Pino, G.A.: The Leavitt path algebra of a graph. J. Algebra **293**, 319–334 (2005)
3. Abrams, G., Pino, G.A.: The Leavitt path algebra of arbitrary graphs. Houston J. Math. **34**, 423–442 (2008)
4. Abrams, G., Pino, G.A., Molina, M.S.: Finite dimensional Leavitt path algebras. J. Pure Appl. Algebra **209**(3), 753–762 (2007)
5. Alahmedi, A., Alsulami, H.: Wreath products by a Leavitt path algebra and affinizations. Int. J. Algebra Comput. **24**(5), 707–714 (2014)
6. Alahmedi, A., Alsulami, H.: On the simplicity of the Lie algebra of a Leavitt path algebra. Commun. Algebra **44**, 4114–4120 (2016)
7. Alahmedi, A., Alsulami, H., Jain, S.K., Zelmanov, E.: Leavitt path algebras of finite Gelfand-Kirillov dimension. J. Algebra Appl. **11**(6), 1250225(1)–1250225(6) (2012)
8. Alahmedi, A., Alsulami, H., Jain, S.K., Zelmanov, E.: Structure of Leavitt path algebras of polynomial growth. Proc. Natl. Acad. Sci. USA **110**(38), 15222–15224 (2013)
9. Ara, P., Moreno, M.A., Pardo, E.: Non-stable K-theory for graph algebras. Algebr. Represent. Theory **10**, 157–178 (2007)
10. Gil Canto, C., Nasr-Isfahani, A.: The maximal commutative subalgebra of a Leavitt path algebra. arXiv:1510.03992
11. Larki, H.: Ideal structure of Leavitt path algebras with coefficients in a unital commutative ring. Commun. Algebra **43**, 5031–5058 (2015)
12. Lopatkin, V., Nam, T.G.: On the homological dimensions of Leavitt path algebras with coefficients in commutative rings. J. Algebra **481**, 273–292 (2017)
13. Nastasescu, C., Van Oystaeyen, F.: Graded Ring Theory. North Holland, Amsterdam (1982)
14. Aranda Pino, G., Martı̀n Barquero, D., Martı̀n Gonzı̀lez, C., Siles Molina, M.: The socle of a Leavitt path algebra. J. Pure Appl. Algebra **212**, 500–509 (2008)
15. Tomforde, M.: Leavitt path algebras with coefficients in a commutative ring. J. Pure Appl. Algebra **215**, 471–484 (2011)

On Certain $*$-differential Identities in Prime Rings with Involution

Abdul Nadim Khan, Shakir Ali, Adnan Abbasi, and Mohammed Ayedh

Abstract The purpose of this paper is to study the $*$-differential identities in prime rings with involution $*$ which admits a pair of derivations. In particular, if a prime ring with involution $*$ of the second kind with char$(R) \neq 2$ admits derivations d_1 and d_2 such that

$$d_1([x, x^*]) + [d_2(x), d_2(x^*)] \pm [x, x^*] \in Z(R) \text{ for all } x \in R,$$

then either R is commutative or $dim_C RC = 4$. Apart from proving some other results, we provide some examples to show that the hypotheses imposed on our results are not superfluous.

Keywords Prime ring · Derivation · Involution · $*$-differential identity

1 Introduction

Throughout this article, R will represent an associative ring with center $Z(R)$. An additive mapping $* : R \to R$ is called an involution if $*$ is an anti-automorphism of order 2, that is, $(x^*)^* = x$ for all $x \in R$. An element x in a ring with involution is said to be hermitian if $x^* = x$ and skew-hermitian if $x^* = -x$. The sets of all hermitian and skew-hermitian elements of R will be denoted by $H(R)$ and $S(R)$, respectively. A ring equipped with an involution is known as ring with involution or $*$-ring. The involution is said to be of the first kind if $Z(R) \subseteq H(R)$, otherwise it is said to be

A. N. Khan
Department of Mathematics, Faculty of Science & Arts- Rabigh, King Abdulaziz University, Jeddah, Saudi Arabia

S. Ali (✉) · M. Ayedh
Department of Mathematics, Aligarh Muslim Univerisity, Aligarh, India
e-mail: shakir.ali.mm@amu.ac.in

A. Abbasi
Department of Mathematics, Madanpalle Institute of Technology & Science, Madanpalle, India

M. Ashraf et al. (eds.), *Algebra and Related Topics with Applications*,
Springer Proceedings in Mathematics & Statistics 392,
https://doi.org/10.1007/978-981-19-3898-6_13

of the second kind. In the latter case, $S(R) \cap Z(R) \neq (0)$. If R is 2-torsion free then every $x \in R$ can be uniquely represented in the form $2x = h + k$, where $h \in H(R)$ and $k \in S(R)$. Note that in this case, x is normal, i.e., $xx^* = x^*x$ if and only if h and k commute. If all elements in R are normal, then R is called a normal ring. An example is the ring of quaternions. A description of such rings can be found in [14].

A derivation on R is an additive mapping $d : R \to R$ such that $d(xy) = d(x)y + xd(y)$ for all $x, y \in R$. In last few decades, many algebraists have investigated commutativity of the ring R through some special types of maps on R. A special contribution is due to Posner [26], who established the commutativity of a prime ring R through a nonzero centralizing derivation on R (see also [11]). A growing interest in this field can be seen in [1, 3, 5–7, 9].

In [15], Herstein proved that a prime ring R of characteristic not two with a nonzero derivation d satisfying $d(x)d(y) = d(y)d(x)$ for all $x, y \in R$ must be commutative. Further, Daif [10] showed that if a 2-torsion free semiprime ring R admitting a derivation d such that $d(x)d(y) = d(y)d(x)$ for all $x, y \in I$, where I is a nonzero ideal of R and d is nonzero on I, then R contains a nonzero central ideal (see also [4, 16]).

We say that a map $f : R \to R$ preserves commutativity if $[f(x), f(y)] = 0$ whenever $[x, y] = 0$ for all $x, y \in R$. According to [8], let S be a subset of R, a map $f : R \to R$ is said to be strong commutativity preserving (SCP) on S if $[f(x), f(y)] = [x, y]$ for all $x, y \in S$. In [6], Bell and Daif investigated the commutativity in rings admitting a derivation which is SCP on a nonzero right ideal. Precisely, they proved that if a semiprime ring R admits a derivation d satisfying $[d(x), d(y)] = [x, y]$ for all x, y in a right ideal I of R, then $I \subseteq Z(R)$. In particular, R is commutative if $I = R$. Later, Deng and Ashraf [13] proved that if there exists a derivation d of a semiprime ring R and a map $f : I \to R$ defined on a nonzero ideal I of R such that $[f(x), d(y)] = [x, y]$ for all $x, y \in I$, then R contains a nonzero central ideal. In particular, they showed that R is commutative if $I = R$. Recently, this result was extended to the Lie ideals and symmetric elements of prime rings by Lin and Liu in [20, 21], respectively. Recently, Ali et al. [2] studied strong commutativity-preserving problems in the setting of rings with involution. Many related generalizations of these results can be found in the literature (see for instance [9, 18, 19, 22, 25]).

Our purpose here is to continue this line of investigation by studying commutativity criteria for rings with involution admitting a pair of derivations satisfying certain $*$-differential identities in prime rings with involution. In fact, our results extended and unify several results proved in [23, 24].

2 The Main Results

We begin our discussions with the following known results.

Lemma 1 (Lemma 2.1, [23]) *Let R be a prime ring with involution $*$ of the second kind. Then $[x, x^*] \in Z(R)$ for all $x \in R$ if and only if R is commutative.*

Lemma 2 (Lemma 2.2, [23]) *Let R be a prime ring with involution $*$ of the second kind. Then $x \circ x^* \in Z(R)$ for all $x \in R$ if and only if R is commutative.*

Proposition 1 *Let R be a prime ring and d be a nonzero derivation of R. If $[d(x)^2, x] \in Z(R)$ for all $x \in R$, then $dim_C RC = 4$.*

Proof We have

$$[d(x)^2, x] \in Z(R) \tag{1}$$

for all $x \in R$. This can be rewritten as

$$[[d(x)^2, x], y] = 0 \tag{2}$$

for all $x, y \in R$. Replacing y by $yd(x)$ in above expression, we get

$$[[d(x)^2, x], y]d(x) + y[[d(x)^2, x], d(x)] = 0 \tag{3}$$

for all $x, y \in R$. From (2), we have

$$y[[d(x)^2, x], d(x)] = 0 \tag{4}$$

for all $x, y \in R$. Primeness of R forces that

$$[[x, d(x)^2], d(x)] = 0 \tag{5}$$

for all $x \in R$. In view of [17, Theorem 2.2], we have the required result.

Since $[d(x)^2, x] = [d(x^2), d(x)]$ for all $x \in R$, so we have the following result:

Corollary 1 *Let R be a prime ring and d be a nonzero derivation of R. If $[d(x^2), d(x)] \in Z(R)$ for all $x \in R$, then $dim_C RC = 4$.*

Theorem 1 *Let R be a prime ring with involution $*$ of the second kind such that $char(R) \neq 2$. If R admits derivations d_1 and d_2 such that*

$$d_1([x, x^*]) + [d_2(x), d_2(x^*)] \pm [x, x^*] \in Z(R) \ \textit{for all} \ x \in R,$$

then either R is commutative or $dim_C RC = 4$.

Proof By the assumption, we have

$$d_1([x, x^*]) + [d_2(x), d_2(x^*)] \pm [x, x^*] \in Z(R) \ \text{for all} \ x \in R. \tag{6}$$

We discuss and divide the proof in the following cases.

Case (i) : If $d_1 = 0$, then we have

$$[d_2(x), d_2(x^*)] \pm [x, x^*] \in Z(R) \text{ for all } x \in R.$$

Thus, in view of [23, Theorem 3.1], we obtain R is commutative.
Case (ii) : If $d_2 = 0$, then we have

$$d_1([x, x^*]) \pm [x, x^*] \in Z(R) \text{ for all } x \in R,$$

which is same as [24, Theorem 2.3], and hence result follows.
Case (iii) : If both d_1 and d_2 are zero, then in view of Lemma 1, we get R is commutative.
Case (iv) : If neither d_1 nor d_2 is zero, then the linearization of (6) gives

$$\begin{aligned} &d_1([x, y^*]) + d_1([y, x^*]) + [d_2(x), d_2(y^*)] + [d_2(y), d_2(x^*)] \\ &\pm [x, y^*] \pm [y, x^*] \in Z(R) \text{ for all } x, y \in R. \end{aligned} \tag{7}$$

Replacing y by yh in (7), where $h \in H(R) \cap Z(R)$, we obtain

$$\begin{aligned} &(d_1([x, y^*]) + d_1([y, x^*]) + [d_2(x), d_2(y^*)] + [d_2(y), d_2(x^*)] \\ &\pm[x, y^*] \pm [y, x^*])h + ([x, y^*] + [y, x^*])d_1(h) \\ &+([d_2(x), y^*] + [y, d_2(x^*)])d_2(h) \in Z(R) \text{ for all } x, y \in R. \end{aligned} \tag{8}$$

Therefore, with the help of (7), we get

$$([x, y^*] + [y, x^*])d_1(h) + ([d_2(x), y^*] + [y, d_2(x^*)])d_2(h) \in Z(R) \tag{9}$$

for all $x, y \in R$. Taking y by ky in (9), where $k \in S(R) \cap Z(R)$, we obtain

$$\begin{aligned} &-[x, y^*]kd_1(h) + [y, x^*]kd_1(h) - [d_2(x), y^*]kd_2(h) \\ &+ [y, d_2(x^*)]kd_2(h) \in Z(R) \text{ for all } x, y \in R. \end{aligned} \tag{10}$$

Combining (9) and (10) yields that

$$2([y, x^*]d_1(h) + [y, d_2(x^*)]d_2(h))k \in Z(R) \text{ for all } x, y \in R.$$

Since $\mathrm{char}(R) \neq 2$ and $S(R) \cap Z(R) \neq (0)$, we get

$$[y, x^*]d_1(h) + [y, d_2(x^*)]d_2(h) \in Z(R) \text{ for all } x, y \in R. \tag{11}$$

Taking y by x^* in above, we obtain $[x^*, d_2(x^*)]d_2(h) \in Z(R)$ for all $x \in R$. Applying the primeness of the ring R, we get either $[x^*, d_2(x^*)] \in Z(R)$ for all $x \in R$ or $d_2(h) = 0$ for all $h \in H(R) \cap Z(R)$. If we consider the case $[x^*, d_2(x^*)] \in Z(R)$ for all $x \in R$, thus in view of Posner's result [26], we get R is commutative. Now consider the case in which we have $d_2(h) = 0$ for all $h \in H(R) \cap Z(R)$, using it in (11), we get $[y, x^*]d_1(h) \in Z(R)$ for all $x, y \in R$, and with the help of primeness of the ring R, we obtain either $[y, x^*] \in Z(R)$ for all $x, y \in R$ or $d_1(h) = 0$ for all $h \in H(R) \cap Z(R)$. If $[y, x^*] \in Z(R)$ for all $x, y \in R$, substituting x for y, we get $[x, x^*] \in Z(R)$ for all $x \in R$. Thus, in view of Lemma 1, we obtain R is commutative. If $d_1(h) = 0$ for all $h \in H(R) \cap Z(R)$, this implies that $d_1(z) = 0$ for all $z \in Z(R)$. Now Replacing y by h in (7), we obtain

$$d_1([x, h]) + d_1([h, x^*]) + [d_2(x), d_2(h)] + [d_2(h), d_2(x^*)]$$

$$\pm [x, h] \pm [h, x^*] \in Z(R) \text{ for all } x \in R \text{ and } h \in H(R). \tag{12}$$

Substituting xk_0 for x in (12), where $k_0 \in S(R) \cap Z(R)$, and using $d_1(z) = d_2(z) = 0$ for all $z \in Z(R)$, we get

$$(d_1([x, h]) - d_1([h, x^*]) + [d_2(x), d_2(h)] - [d_2(h), d_2(x^*)]$$

$$\pm [x, h] \mp [h, x^*])k_0 \in Z(R) \text{ for all } x \in R, \tag{13}$$

$h \in H(R)$ and $k_0 \in S(R) \cap Z(R)$. Combining (13) and (12), we get

$$2(d_1([x, h]) + [d_2(x), d_2(h)] \pm [x, h])k_0 \in Z(R) \text{ for all } x \in R,$$

$h \in H(R)$ and $k_0 \in S(R) \cap Z(R)$. Since $\mathrm{char}(R) \neq 2$ and $S(R) \cap Z(R) \neq (0)$, we obtain

$$d_1([x, h]) + [d_2(x), d_2(h)] \pm [x, h] \in Z(R) \text{ for all } x \in R, \tag{14}$$

$h \in H(R)$. Taking $h = kk_0$ in (14), where $k_0 \in S(R) \cap Z(R), k \in S(R)$, using $d_1(z) = d_2(z) = 0$ for all $z \in Z(R)$ and reasoning as above, we get

$$d_1([x, k]) + [d_2(x), d_2(k)] \pm [x, k] \in Z(R) \tag{15}$$

for all $x \in R$ and $k \in S(R)$. Now consider

$$
\begin{aligned}
2d_1([x, y]) + 2[d_2(x), d_2(y)] \pm 2[x, y] &= d_1([x, 2y]) \pm [x, 2y] \\
&\quad + [d_2(x), d_2(2y)] \\
&= d_1[x, h + k] \pm [x, h + k] \\
&\quad + [d_2(x), d_2(h) + d_2(k)] \\
&= d_1([x, h]) + d_1([x, k]) \pm [x, h] \\
&\quad + [d_2(x), d_2(h)] + [d_2(x), d_2(k)] \\
&\quad \pm [x, k].
\end{aligned}
$$

In view of (14) and (15), we have

$$2(d_1([x, y]) + [d_2(x), d_2(y)] \pm [x, y]) \in Z(R) \text{ for all } x, y \in R.$$

Since char$(R) \neq 2$, the last relation implies that

$$d_1([x, y]) + [d_2(x), d_2(y)] \pm [x, y] \in Z(R) \text{ for all } x, y \in R. \tag{16}$$

Replacing y by x^2 in (16), we get $[d_2(x), d_2(x^2)] \in Z(R)$ for all $x \in R$. Hence, in view of Corollary 1, we get the required result.

Corollary 2 (Theorem 3.1, [23]) *Let R be a prime with involution $*$ of the second kind and char$(R) \neq 2$. If R admits a derivation d such that $[d(x), d(x^*)] \pm [x, x^*] \in Z(R)$ for all $x \in R$, then R is commutative.*

Corollary 3 (Theorem 2.3, [24]) *Let R be a prime with involution $*$ of the second kind and char$(R) \neq 2$. If R admits a derivation d such that $d([x, x^*]) \pm [x, x^*] \in Z(R)$ for all $x \in R$, then R is commutative.*

Corollary 4 (Lemma 2.1, [23]) *Let R be a prime with involution $*$ of the second kind and char$(R) \neq 2$. If $[x, x^*] \in Z(R)$ for all $x \in R$, then R is commutative.*

Theorem 2 *Let R be a prime with involution $*$ of the second kind and char$(R) \neq 2$. If R admit derivations d_1 and d_2 such that*

$$d_1([x, x^*]) + d_2(x) \circ d_2(x^*) \pm x \circ x^* \in Z(R)$$

for all $x \in R$, then R is commutative.

Proof We have

$$d_1([x, x^*]) + d_2(x) \circ d_2(x^*) \pm x \circ x^* \in Z(R) \text{ for all } x \in R. \tag{17}$$

Case (i) : If we take $d_1 = 0$, then we have $d_2(x) \circ d_2(x^*) \pm x \circ x^* \in Z(R)$ for all $x \in R$. [Theorem 3.5, [23]] gives us R is commutative.
Case (ii) : If we consider $d_2 = 0$, then (17) becomes $d_1([x, x^*]) \pm x \circ x^* \in Z(R)$ for all $x \in R$. Hence, in view of [Theorem 2.7, [24]], we obtain R is commutative.

Case (iii) : Taking both d_1 and d_2 are zero, then we have $x \circ x^* \in Z(R)$ for all $x \in R$. Lemma 2 implies that R is commutative.

Case (iv) : Consider neither d_1 nor d_2 are zero. Now taking h for x in (17), we obtain

$$(d_2(h))^2 \pm h^2 \in Z(R) \ \text{ for all } \ h \in H(R). \tag{18}$$

Replacing h by $h + h_1$ in (18), where $h \in H(R)$ and $h_1 \in H(R) \cap Z(R)$, we obtain

$$(d_2(h))^2 \pm h^2 + (d_2(h_1))^2 \pm {h_1}^2 + 2d_2(h)d_2(h_1) \pm 2hh_1 \in Z(R)$$

for all $h \in H(R)$ and $h_1 \in H(R) \cap Z(R)$. Application of (18) yields that

$$2d_2(h)d_2(h_1) \pm 2hh_1 \in Z(R) \quad \text{for all } h \in H(R) \text{ and } h_1 \in H(R) \cap Z(R).$$

Since char$(R) \neq 2$, which yields

$$d_2(h)d_2(h_1) \pm hh_1 \in Z(R) \quad \text{for all } h \in H(R) \text{ and } \ h_1 \in H(R) \cap Z(R).$$

This implies that

$$[d_2(h), r]d_2(h_1) \pm [h, r]h_1 = 0 \ \text{ for all } \ r \in R, \ \ h \in H(R) \tag{19}$$

and $h_1 \in H(R) \cap Z(R)$. Replacing r by $d_2(h)$ in (19), we obtain

$$[h, d_2(h)]h_1 = 0 \ \text{ for all } \ h \in H(R).$$

By the hypotheses, we have

$$[d_2(h), h] = 0 \ \text{ for all } \ h \in H(R). \tag{20}$$

Linearization of (20) gives

$$[d_2(h), h'] + [d_2(h'), h] = 0 \ \text{ for all } \ h, h' \in H(R). \tag{21}$$

Substituting kk_0 for h' in (21) where $k \in S(R)$ and $k_0 \in S(R) \cap Z(R)$, we obtain

$$[d_2(h), k]k_0 + [d_2(k), h]k_0 + [k, h]d_2(k_0) = 0 \tag{22}$$

for all $h \in H(R)$ and $k \in S(R)$. Now replacing h by kk_0 in (20), we get

$$[d_2(k), k]{k_0}^2 = 0 \ \text{ for all } \ k \in S(R) \ \text{ and } \ k_0 \in S(R) \cap Z(R). \tag{23}$$

Since $S(R) \cap Z(R) \neq (0)$, the above relation reduces to

$$[d_2(k), k] = 0 \ \text{ for all } \ k \in S(R). \tag{24}$$

Replacing k by $k + k_1$ in (24), we get

$$[d_2(k), k_1] + [d_2(k_1), k] = 0 \text{ for all } k, k_1 \in S(R). \tag{25}$$

Substituting hk_0 for k_1, where $k_0 \in S(R) \cap Z(R)$ in (25), we obtain

$$[d_2(k), h]k_0 + [d_2(h), k]k_0 + [h, k]d_2(k_0) = 0 \tag{26}$$

for all $h \in H(R)$ and $k \in S(R)$ and $k_0 \in S(R) \cap Z(R)$. Combining (22) and (26), we obtain

$$2([d_2(k), h] + [d_2(h), k])k_0 = 0 \text{ for all } h \in H(R), k \in S(R) \text{ and}$$

$k_0 \in S(R) \cap Z(R)$. This implies that

$$[d_2(k), h] + [d_2(h), k] = 0 \text{ for all } h \in H(R) \text{ and } k \in S(R). \tag{27}$$

Consider $4[d_2(x), y] = [d_2(2x), 2y] = [d_2(h + k), h + k] = [d_2(h), h] + [d_2(k), k] + [d_2(k), h] + [d_2(h), k] = 0$. By the application of (20), (24) and (27), we get $4[d_2(x), y] = 0$ for all $x, y \in R$. Since char$(R) \neq 2$ implies that $[d_2(x), y] = 0$ for all $x, y \in R$. Hence, R is commutative by Posner's [26] result.

Corollary 5 (Theorem 3.5, [23]) *Let R be a prime with involution $*$ of the second kind and char$(R) \neq 2$. If R admits a derivation d such that $d(x) \circ d(x^*) \pm x \circ x^* \in Z(R)$ for all $x \in R$, then R is commutative.*

Corollary 6 (Theorem 2.7, [24]) *Let R be a prime with involution $*$ of the second kind and char$(R) \neq 2$. If R admits a derivation d such that $d([x, x^*]) \pm x \circ x^* \in Z(R)$ for all $x \in R$, then R is commutative.*

Corollary 7 (Lemma 2.1, [23]) *Let R be a prime with involution $*$ of the second kind and char$(R) \neq 2$. If $x \circ x^* \in Z(R)$ for all $x \in R$, then R is commutative.*

Theorem 3 *Let R be a prime with involution $*$ of the second kind and char$(R) \neq 2$. If R admits derivations d_1 and d_2 such that*

$$[d_1(x), d_1(x^*)] + d_2(x) \circ d_2(x^*) \pm [x, x^*] \in Z(R)$$

for all $x \in R$, then R is commutative.

Proof In view of our hypotheses, we have

$$[d_1(x), d_1(x^*)] + d_2(x) \circ d_2(x^*) \pm [x, x^*] \in Z(R) \text{ for all } x \in R. \tag{28}$$

Case (i) : If we take $d_1 = 0$, then we have $d_2(x) \circ d_2(x^*) \pm [x, x^*] \in Z(R)$ for all $x \in R$. [Theorem 3.8, [23]] gives us R is commutative.

Case (ii) : If we consider $d_2 = 0$, then (28) becomes $[d_1(x), d_1(x^*)] \pm [x, x^*] \in Z(R)$ for all $x \in R$. This follows from case (i) of Theorem 1; from there, we obtain R is commutative.
Case (iii) : If both d_1 and d_2 are zero, then we have $[x, x^*] \in Z(R)$ for all $x \in R$. Lemma 1 implies that R is commutative.
Case (iv) : Let both d_1 and d_2 be nonzero. Taking h for x in (28), we get

$$(d_2(h))^2 \in Z(R) \text{ for all } h \in H(R). \tag{29}$$

Substituting $h + h_1$ in (29) where $h \in H(R)$ and $h_1 \in H(R) \cap Z(R)$, we have

$$(d_2(h))^2 + (d_2(h_1))^2 + d_2(h)d_2(h_1) \in Z(R) \text{ for all } h \in H(R) \text{ and}$$

$h_1 \in H(R) \cap Z(R)$. Application of (29) yields

$$d_2(h)d_2(h_1) \in Z(R) \text{ for all } h \in H(R) \text{ and } h_1 \in H(R) \cap Z(R). \tag{30}$$

This can be further written as $[d_2(h), r]d_2(h_1) = 0$ for all $r \in R, h \in H(R)$ and $h_1 \in H(R) \cap Z(R)$. Using the primeness of R, we get either $[d_2(h), r] = 0$ for all $r \in R$, $h \in H(R)$ or $d_2(h_1) = 0$ for all $h_1 \in H(R) \cap Z(R)$. If we consider the $[d_2(h), r] = 0$ for all $r \in R$, $h \in H(R)$, which is same (20), follow the same line of proof, we get R is commutative. Consider $d_2(h_1) = 0$ for all $h_1 \in H(R) \cap Z(R)$, which implies that $d_2(z) = 0$ for all $z \in Z(R)$. Linearization of (28) yields that

$$[d_1(x), d_1(y^*)] + d_2(x) \circ d_2(y^*) \pm [x, y^*] + [d_1(y), d_1(x^*)] + d_2(y) \circ d_2(x^*) \pm [y, x^*] \in Z(R) \tag{31}$$

for all $x, y \in R$. Replacing y by yk in (31), where $k \in S(R) \cap Z(R)$, and using $d_2(z) = 0$ for all $z \in Z(R)$, we get

$$-[d_1(x), d_1(y^*)]k - (d_2(x) \circ d_2(y^*))k \mp [x, y^*]k + [d_1(y), d_1(x^*)]k \pm [y, x^*]k + (d_2(y) \circ d_2(x^*)k - [d_1(x), y^*]d_1(k) + [y, d_1(x^*)]d_1(k) \in Z(R) \tag{32}$$

for all $x, y \in R$ and $k \in S(R) \cap Z(R)$. Combining (31) and (32), we obtain

$$-[d_1(x), y^*]d_1(k) + 2[d_1(y), d_1(x^*)]k + [y, d_1(x^*)]d_1(k) + 2(d_2(y) \circ d_2(x^*))k \pm 2[y, x^*]k \in Z(R) \tag{33}$$

for all $x, y \in R$ and $k \in S(R) \cap Z(R)$. Again repeating the same step for (33) and combining it with (33), we get

$$-2[d_1(x), y^*]kd_1(k) - 2[y, d_1(x^*)]kd_1(k) \in Z(R) \text{ for all } x, y \in R \tag{34}$$

and $k \in S(R) \cap Z(R)$. Replacing y by yk in (34) and combining it with (34), we obtain $4[y, d_1(x^*)]k^2 d_1(k) \in Z(R)$ for all $x, y \in R$. This yields $[y, d_1(x^*)]d_1(k) \in Z(R)$ for all $x, y \in R$ and $S(R) \cap Z(R) \neq (0)$. Primeness of the ring R implies that $[y, d_1(x^*)] \in Z(R)$ for all $x, y \in R$ or $d_1(k) = 0$ for all $k \in S(R) \cap Z(R)$. If $[y, d_1(x^*)] \in Z(R)$ for all $x, y \in R$, then R is commutative. Now consider $d_1(k) = 0$ for all $k \in S(R) \cap Z(R)$. This reduces (32) into

$$-[d_1(x), d_1(y^*)]k - (d_2(x) \circ d_2(y^*))k \mp [x, y^*]k + [d_1(y), d_1(x^*)]k$$
$$+ (d_2(y) \circ d_2(x^*)) \pm [y, x^*]k \in Z(R) \text{ for all } x, y \in R \text{ and} \tag{35}$$

$k \in S(R) \cap Z(R)$. In view of (35) and (31), we have

$$2([d_1(y), d_1(x^*)] + d_2(y) \circ d_2(x^*) \pm [y, x^*])k \in Z(R) \text{ for all } x, y \in R.$$

This implies

$$[d_1(y), d_1(x^*)] + d_2(y) \circ d_2(x^*) \pm [y, x^*] \in Z(R) \text{ for all } x, y \in R. \tag{36}$$

Taking x^* for y in (36), we obtain

$$(d_2(x))^2 \in Z(R) \text{ for all } x \in R. \tag{37}$$

Replacing x by $x + y$, we get

$$(d_2(x))^2 + (d_2(y))^2 + d_2(x) \circ d_2(y) \in Z(R)$$

for all $x, y \in R$. Application of (37) yields $d_2(x) \circ d_2(y) \in Z(R)$ for all $x, y \in R$. Taking x^* for y in last relation gives $d_2(x) \circ d_2(x^*) \in Z(R)$ for all $x \in R$. Therefore, in view of [Theorem 3.5, [23]], we get the required conclusion.

Corollary 8 (Theorem 3.8, [23]) *Let R be a prime with involution $*$ of the second kind and char$(R) \neq 2$. If R admits derivations d_1 and d_2 such that $d_2(x) \circ d_2(x^*) \pm [x, x^*] \in Z(R)$ for all $x \in R$, then R is commutative.*

Corollary 9 (Theorem 3.1, [23]) *Let R be a prime with involution $*$ of the second kind and char$(R) \neq 2$. If R admits derivations d_1 and d_2 such that $[d_2(x), d_2(x^*)] \pm [x, x^*] \in Z(R)$ for all $x \in R$, then R is commutative.*

Corollary 10 (Lemma 2.1, [23]) *Let R be a prime with involution $*$ of the second kind and char$(R) \neq 2$. If $\pm[x, x^*] \in Z(R)$ for all $x \in R$, then R is commutative.*

Theorem 4 *Let R be a prime with involution $*$ of the second kind and char$(R) \neq 2$. If R admits derivations d_1 and d_2 such that*

$$d_1(x \circ x^*) + [d_2(x), d_2(x^*)] \pm x \circ x^* \in Z(R)$$

for all $x \in R$, then R is commutative.

Proof We have

$$d_1(x \circ x^*) + [d_2(x), d_2(x^*)] \pm x \circ x^* \in Z(R) \text{ for all } x \in R. \tag{38}$$

Case (i) : Consider $d_1 = 0$, then we have $[d_2(x), d_2(x^*)] \pm x \circ x^* \in Z(R)$ for all $x \in R$. In view of [Theorem 3.8, [23]], we get the required result.
Case (ii) : If $d_2 = 0$, then $d_1(x \circ x^*) \pm x \circ x^* \in Z(R)$ for all $x \in R$. In view of [Theorem 2.5, [24]], we get R is commutative.
Case (iii) : If both d_1 and d_2 are zero, then by Lemma 2, R is commutative.
Case (iv) : Assume that $d_1 \neq 0$ and $d_2 \neq 0$ and substitute h for x in (38), where $h \in H(R)$, we get

$$2(d_1(h^2) \pm h^2) \in H(R) \text{ for all } h \in Z(R).$$

Since char$(R) \neq 2$, this implies that

$$d_1(h^2) \pm h^2 \in H(R) \text{ for all } h \in Z(R).$$

This can be further written as

$$[d_1(h^2), h^2] = 0 \text{ for all } h \in Z(R).$$

$$[d_1(h)h, h^2] + [hd_1(h), h^2] = 0 \text{ for all } h \in R.$$

$$[d_1(h), h^2]h + h[hd_1(h), h^2] = 0 \text{ for all } h \in R. \tag{39}$$

Taking $h + h_0$ in (39), where $h_0 \in H(R) \cap Z(R)$, and using (39), we get

$$2[d_1(h), h^2]h_0 + 4[d_1(h), h]h_0^2 + 2[d_1(h)]hh_0 + 2h[d_1(h), h]h_0 = 0$$

for all $h \in H(R)$ and $h_0 \in H(R) \cap Z(R)$. This implies that

$$2[d_1(h), h^2] + 4[d_1(h), h]h_0 + 2[d_1(h)]h + 2h[d_1(h), h] = 0. \tag{40}$$

Again replacing h by $h + h_0$ in (40) where $h_0 \in H(R) \cap Z(R)$ and by application of (40), we obtain

$$8[d_1(h), h]h_0 = 0 \text{ for all } h \in H(R) \text{ and } h_0 \in H(R) \cap Z(R).$$

This implies that $[d_1(h), h] = 0$ for all $h \in H(R)$, which is same as (20). So the result is done from Theorem 2.

Corollary 11 (Theorem 3.8, [23]) *Let R be a prime with involution $*$ of the second kind and $char(R) \neq 2$. If R admits derivations d_1 and d_2 such that $d_2(x) \circ d_2(x^*) \pm [x, x^*] \in Z(R)$ for all $x \in R$, then R is commutative.*

Corollary 12 (Theorem 2.7, [24]) *Let R be a prime with involution $*$ of the second kind and $char(R) \neq 2$. If R admits derivation d_1 such that $d_1(x \circ x^*) \pm [x, x^*] \in Z(R)$ for all $x \in R$, then R is commutative.*

Corollary 13 (Lemma 2.1, [23]) *Let R be a prime with involution $*$ of the second kind and $char(R) \neq 2$. If $\pm[x, x^*] \in Z(R)$ for all $x \in R$, then R is commutative.*

Theorem 5 *Let R be a prime with involution $*$ of the second kind and $char(R) \neq 2$. If R admits two derivations d_1 and d_2 such that*

$$d_1(x \circ x^*) + d_2(x) \circ d_2(x^*) \pm [x, x^*] \in Z(R)$$

for all $x \in R$, then R is commutative.

Proof Following the same approach as above, we get the required assertion.

In the end, let us write an example which shows that the restriction of the second kind of involution in our main results is not superfluous.

Example 1 Let $R = \left\{ \begin{pmatrix} a_1 & a_2 \\ a_3 & a_4 \end{pmatrix} \middle| \, a_1, a_2, a_3, a_4 \in \mathbb{R} \right\}$ with center $Z(R) = \left\{ \begin{pmatrix} a_1 & 0 \\ 0 & a_1 \end{pmatrix} \middle| \, a_1 \in \mathbb{R} \right\}$. Define mappings $d_1, d_2 : R \longrightarrow R$ and $* : R \longrightarrow R$ such that $d_1 \begin{pmatrix} a_1 & a_2 \\ a_3 & a_4 \end{pmatrix} = \begin{pmatrix} 0 & a_2 \\ -a_3 & 0 \end{pmatrix}, d_2 \begin{pmatrix} a_1 & a_2 \\ a_3 & a_4 \end{pmatrix} = \begin{pmatrix} 0 & -a_2 \\ a_3 & 0 \end{pmatrix}, \begin{pmatrix} a_1 & a_2 \\ a_3 & a_4 \end{pmatrix}^* = \begin{pmatrix} a_4 & -a_2 \\ -a_3 & a_1 \end{pmatrix}$. It can be easily verified that the involution $*$ is of the first kind and R is a prime ring. Moreover, d_1, d_2 are nonzero derivations and the identities of various Theorems 2, 3, 4 and 5 are satisfied. However, R is not commutative. Hence, the hypothesis of second kind of involution is crucial in our results.

We conclude the manuscript with the following example which reveals that our theorems cannot be extended to semiprime rings.

Example 2 Let $(R, *)$ and d_1, d_2 be same as in Example 1 and $R' = \mathbb{C}$, the ring of complex numbers. Define involution on R' by conjugate of complex number. Construct a new ring $\mathbb{A} = R \times R'$ with component-wise addition and multiplication. It is clear that $\mathbb{A}$ is a semiprime ring with char$(R) \neq 2$. Defining a map α on $\mathbb{A}$ as follows $\alpha(x, y) = (x^*, \overline{y})$, it can be easily checked that α is a second kind of involution. Now define derivations D_1 and D_2 on $\mathbb{A}$ by $D_1(x, y) = (d_1(x), 0)$ and $D_2(x, y) = (d_2(x), 0)$, respectively, and are satisfying the requirements of our theorems. Note that R is not commutative. Hence, the primeness hypotheses in our results is not superfluous.

Acknowledgements The authors are greatly indebted to the referee/s for his/her valuable comments which have improved the paper immensely. The research of second named author is supported by SERB-DST MATRICS Project (Grant No. MTR/2019/000603), India.

References

1. Ali, S., Dar, N.A.: On $*$-centralizing mappings in rings with involution. Georgain Math. J. **21**, 25–28 (2014)
2. Ali, S., Dar, N.A., Khan, A.N.: On strong commutativity preserving like maps in rings with involution. Miskolc Math. Notes **16**, 17–24 (2015)
3. Ashraf, M., Ali, M., Ali, S.: Some commutativity theorems for rings with generalized derivations. Southeast Asian Bull. Math. **31**, 415–421 (2007)
4. Ashraf, M., Rehman, N.: On commutativity of rings with derivations. Res. Math. **42**, 3–8 (2002)
5. Bell, H.E., Daif, M.N.: On derivations and commutativity in prime rings. Acta Math. Hungar. **66**, 337–343 (1995)
6. Bell, H.E., Daif, M.N.: On commutativity and strong commutativity preserving maps. Canad. Math. Bull. **37**, 443–447 (1994)
7. Bell, H.E., Martindale, W.S., III.: Centralizing mappings on semiprime rings. Canad. Math. Bull. **30**, 92–101 (1987)
8. Bell, H.E., Mason, G.: On derivations in near rings and rings. Math. J. Okayama Univ. **34**, 135–144 (1992)
9. Brešar, M., Miers, C.R.: Strong commutativity preserving mappings of semiprime rings. Canad. Math. Bull. **37**, 457–460 (1994)
10. Daif, M.N.: Commutativity results for semiprime rings with derivation. Int. J. Math. Math. Sci. **21**, 471–474 (1998)
11. Divinsky, N.: On commuting automorphisms of rings. Trans. Roy. Soc. Canada. Sect. III **49**(3), 19–22 (1955)
12. Dar, N.A., Ali, S.: On $*$-commuting mappings and derivations in rings with involution. Turk. J. Math. **40**(4), 884–894 (2016)
13. Deng, Q., Ashraf, M.: On strong commutativity preserving maps. Res. Math. **30**, 259–263 (1996)
14. Herstein, I.N.: Rings with involution. Chicago Lectures in Mathematics. The university of Chicago Press, Chicago (1976)
15. Herstein, I.N.: A note on derivations. Canad. Math. Bull. **21**, 369–370 (1978)
16. Herstein, I.N.: A Theorem on derivations of prime rings with involution. Canad. Math. Bull. **XXXIV**, 356–369 (1982)
17. Koşan, T.M., Lee, T.K., Zhou, Y.: Identities with Engel conditions on derivations. Monatsh Math. **165**, 543–556 (2012)
18. Lee, T.K., Wong, T.L.: Nonadditive strong commutativity preserving maps. Commun. Algebra **40**, 2213–2218 (2012)
19. Liau, P.K., Liu, C.K.: Strong commutativity preserving generalized derivations on Lie ideals. Linear Multilinear Algebra **59**, 905–915 (2011)
20. Lin, J.S., Liu, C.K.: Strong commutativity preserving maps on Lie ideals. Linear Algebra Appl. **428**, 1601–1609 (2008)
21. Lin, J.S., Liu, C.K.: Strong commutativity preserving maps in prime rings with involution. Linear Algebra Appl. **432**, 14–23 (2010)
22. Liu, C.K.: Strong commutativity preserving generalized derivations on right ideals. Monatsh. Math. **166**, 453–465 (2012)
23. Nejjar, B., Kacha, A., Mamouni, A., Oukhtite, A.L.: Commutativity theorems in rings with involution. Commun. Algebra **45**(2), 698–708 (2017)

24. Nejjar, B., Kacha, A., Mamouni, A.: Some commutativity criteria for rings with involution. Int. Open Probl. Compt. Math. **10**(3) (2017)
25. Qi, X., Hou, J.: Strong commutativity preserving maps on triangular rings. Oper. Matr. **6**, 147–158 (2012)
26. Posner, E.C.: Derivations in prime rings. Proc. Am. Math. Soc. **8**, 1093–1100 (1957)

On b-Generalized Derivations in Prime Rings

Mohammad Salahuddin Khan

Abstract Let R be a noncommutative prime ring with extended centroid C and I a nonzero ideal of R. If R admits a b-generalized derivation G which satisfies $[G(x), y]_k = G(x \circ_k y)$ or $G(x) \circ_k y = x \circ_k y$ for all $x, y \in I$, where k is a fixed positive integer, then there exists $\lambda \in C$ such that $G(x) = \lambda x$ for all $x \in R$.

Keywords b-generalized derivation · Derivation · Ideal · Prime ring

1 Introduction and Notations

In all that follows, unless specially stated, R always denotes an associative ring with center $Z(R)$. A ring R is called prime if $aRb = (0)$ (where $a, b \in R$) implies $a = 0$ or $b = 0$. We denote by Q maximal right ring of quotient of R and C is the center of Q which is called the extended centroid of R; see [6, Chap. 2] for more details. As usual, the symbols $[x, y]$ and $x \circ y$ will denote the commutator $xy - yx$ and anti-commutator $xy + yx$, respectively. Given $x, y \in R$, set $[x, y]_1 = xy - yx$ and inductively $[x, y]_k = [[x, y]_{k-1}, y]$ for $k > 1$. Note that Engel condition is a polynomial $[x, y]_k = \sum_{i=0}^{k}(-1)^i \binom{k}{i} y^i x y^{k-i}$ in noncommutative indeterminates x, y and $[x + y, z]_k = [x, z]_k + [y, z]_k$. For anti-commutator, set $x \circ_0 y = x,\ x \circ_1 y = xy + yx$ and inductively $x \circ_k y = (x \circ_{k-1} y) \circ y$ for $k > 1$.

An additive mapping $d : R \to R$ is said to be a derivation of R if $d(xy) = d(x)y + xd(y)$ for all $x, y \in R$. An additive mapping $G : R \to R$ is called a generalized derivation of R if there exists a derivation d of R such that $G(xy) = G(x)y + xd(y)$ for all $x, y \in R$. Obviously, any derivation is a generalized derivation, but the converse is not true in general. A significative example is a map of the form

M. S. Khan (✉)
Department of Applied Mathematics, Z. H. College of Engineering & Technology, Aligarh Muslim University, Aligarh 202002, India
e-mail: salahuddinkhan50@gmail.com

M. Ashraf et al. (eds.), *Algebra and Related Topics with Applications*, Springer Proceedings in Mathematics & Statistics 392,
https://doi.org/10.1007/978-981-19-3898-6_14

$F(x) = ax + xb$ for some $a, b \in R$; such generalized derivations are called inner. Over the last few decades, several authors have studied prime and semiprime rings with generalized derivations (viz.; [1–3, 7, 9, 13] and references therein).

In a recent paper [11], Koşan and Lee proposed the following definition. Let $d : R \to Q$ be an additive mapping and $b \in Q$. An additive map $G : R \to Q$ is called a left b-generalized derivation, with associated mapping d, if $G(xy) = G(x)y + bxd(y)$ for all $x, y \in R$. In the same paper, it is proved that, if R is prime ring, then d is a derivation of R. For simplicity of notation, this mapping G will be called b-generalized derivation with associated pair (b, d). Clearly, any generalized derivation with associated derivation d is a b-generalized derivation with associated pair $(1, d)$. Similarly, the mapping $x \mapsto ax + b[x, c]$, for $a, b, c \in Q$, is an b-generalized derivation with associated pair $(b, ad(c))$, where $ad(c)(x) = [x, c]$ denotes the inner derivation of R induced by the element c. More generally, the mapping $x \mapsto ax + qxc$, for $a, q, c \in Q$, is a b-generalized derivation with associated pair $(q, ad(c))$. This mapping is called inner b-generalized derivation. In this paper, we characterize b-generalized derivations of prime rings. Precisely, we prove that if a prime ring R admits a b-generalized derivation G such that $[G(x), y]_k = G(x \circ_k y)$ for all $x, y \in I$, where I is a nonzero ideal of R and k is a fixed positive integer, then there exists $\lambda \in C$ such that $G(x) = \lambda x$ for all $x \in R$.

2 The Results

Let R be a prime ring. Then Q is a prime ring and C is a field. Let $Q *_C C\{X\}$ be the free product of Q and the free algebra $C\{X\}$ over C on an infinite set X of indeterminates. Elements of $Q *_C C\{X\}$ are called generalized polynomials and a typical element in $Q *_C C\{X\}$ is a finite sum of monomials of the form $\alpha a_{i_0} x_{j_1} a_{i_1} x_{j_2} \cdots x_{j_n} a_{i_n}$ where $\alpha \in C$, $a_{i_k} \in Q$ and $x_{j_k} \in X$. We say that R satisfies a nonzero generalized polynomial identity (abbreviated as GPI) if there exists a nonzero generalized polynomial $\phi(x_i) \in Q *_C C\{X\}$ such that $\phi(r_i) = 0$ for all $r_i \in R$.

Let V be a right vector space over a field F and let $\mathrm{End}(V_F)$ denote the ring of F-linear transformations on V. A subring R of $\mathrm{End}(V_F)$ is said to be dense if given any finitely many F-independent $v_i \in V$ and arbitrary $w_i \in V$, where $1 \leq i \leq n$, there exists $a \in R$ such that $av_i = w_i$ for all $1 \leq i \leq n$.

We begin with the following lemmas which are essential for developing the proof of our main results.

Lemma 1 *Let R be a noncommutative prime ring, I a nonzero ideal of R and $a \in Q$. If $[ax^m, y^n]_k - a(x^m \circ_k y^n) = 0$ for all $x, y \in I$, where m, n, k are fixed positive integers, then $a \in C$.*

Proof By the assumption, we have

$$[ax^m, y^n]_k - a(x^m \circ_k y^n) = 0 \text{ for all } x, y \in I \tag{1}$$

and hence for all $x, y \in R$ by [6, Theorem 6.4.4]. Suppose on the contrary that $a \notin C$. Then $\psi(x, y) = [ax^m, y^n]_k - a(x^m \circ_k y^n)$ is a nontrivial generalized polynomial identity (GPI) for R and hence for all $x, y \in Q$ by [6, Theorem 6.4.4]. Let F be the algebraic closure of C if C is infinite and set $F = C$ for C finite. Clearly, the map $r \in Q \longmapsto r \otimes 1 \in Q \otimes_C F$ gives a ring embedding. By [14, Proposition], $Q \otimes_C F$ is a prime ring with F as its extended centroid. Thus, $\overline{Q} = Q \otimes_C F$ is a prime ring satisfying a nonzero GPI $\psi(x, y)$ and its extended centroid F is either an algebraically closed field or a finite field. By Martindale's theorem [6, Theorem 6.1.6], $\overline{Q}$ is isomorphic to a dense subring of $\mathrm{End}(V_D)$, where V is a right vector space over a division ring D and D is a finite-dimensional central division algebra over F. Recall that F is either algebraically closed or finite. From the finite-dimensionality of D over F, it follows that $D = F$. Hence, $\overline{Q}$ is isomorphic to a dense subring of $\mathrm{End}(V_F)$. Assume first $\dim V_F = 1$; then $\overline{Q}$ is commutative and hence R is commutative, a contradiction.

Assume next that $\dim V_F \geq 2$. By [5, Lemma 7.1], there exists $v \in V$ such that v and av are linearly independent over F. By the density of R, there exist $x, y \in R$ such that $xv = v,\ yv = 0,\ yav = av$. Therefore, from (1), we have

$$\begin{aligned}
0 &= ([ax^m, y^n]_k - a(x^m \circ_k y^n))v \\
&= \sum_{i=0}^{k}(-1)^i \binom{k}{i} y^{ni} ax^m y^{n(k-i)} v - a(x^m \circ_k y^n)v \\
&= (-1)^k y^{nk} ax^m v = (-1)^k y^{nk} av \\
&= (-1)^k av,
\end{aligned}$$

which gives a contradiction. This completes the proof of lemma.

For $m = n = 1$, we have the following corollary.

Corollary 1 *Let R be a noncommutative prime ring, I a nonzero ideal of R and $a \in Q$. If $[ax, y]_k - a(x \circ_k y) = 0$ for all $x, y \in I$, where k is a fixed positive integer, then $a \in C$.*

Similarly, we can prove the following lemma.

Lemma 2 *Let R be a noncommutative prime ring, I a nonzero ideal of R and $a \in Q$. If $((ax^m) \circ_k y^n) - (x^m \circ_k y^n) = 0$ for all $x, y \in I$, where m, n, k are fixed positive integers, then $a \in C$.*

In particular, for $m = n = 1$, we get the following corollary.

Corollary 2 *Let R be a noncommutative prime ring, I a nonzero ideal of R and $a \in Q$. If $((ax) \circ_k y) - (x \circ_k y) = 0$ for all $x, y \in I$, where k is a fixed positive integer, then $a \in C$.*

Now we are in a position to state our first main result of this paper.

Theorem 1 *Let R be a noncommutative prime ring, I a nonzero ideal of R and $b \in Q$. If R admits a nonzero b-generalized derivation G associated with the map d such that $[G(x), y]_k = G(x \circ_k y)$ for all $x, y \in I$, where k is a fixed positive integer, then there exists $\lambda \in C$ such that $G(x) = \lambda x$ for all $x \in R$.*

Proof Suppose first that $b \neq 0$. By [11, Theorem 2.3], $d : R \to Q$ is a derivation and there exists $a' \in Q$ such that $G(x) = a'x + bd(x)$ for all $x \in R$. It is known that d can be uniquely extended to a derivation of Q [12, Lemma 2]. By the assumption, we have

$$[G(x), y]_k = G(x \circ_k y) \text{ for all } x, y \in I. \tag{2}$$

If $d = 0$, then $[a'x, y]_k = a'(x \circ_k y)$ for all $x, y \in I$ and hence for all $x, y \in Q$ by [6, Theorem 6.4.4]. In view of Corollary 1, we get $a' \in C$, so proof is done.

Assume now that $d \neq 0$. We divide the proof into two cases.

Case 1. d is Q-inner. That is, there exists $c' \in Q$ such that $d(x) = [c', x]$ for all $x \in R$. So $G(x) = a'x + bd(x) = a'x + b[c', x] = ax + bxc$ for all $x \in R$, where $a = a' + bc'$ and $c = -c'$. By (2), we have

$$[ax + bxc, y]_k - a(x \circ_k y) - b(x \circ_k y)c = 0 \text{ for all } x, y \in I. \tag{3}$$

Since I, R and Q satisfy the same polynomial identities [6, Theorem 6.4.4], therefore, $[ax + bxc, y]_k = a(x \circ_k y) + b(x \circ_k y)c$ for all $x, y \in Q$. If $c \in C$, then $d = 0$ as $c = -c' \in C$ and $G(x) = \lambda x$ for all $x \in R$, we get the required form of G. Now we assume that $c \notin C$. Let $\varphi(x, y) = [ax + bxc, y]_k - a(x \circ_k y) - b(x \circ_k y)c$. Clearly, if $c \notin C$, then by (3), $\varphi(x, y)$ is a nonzero GPI of Q. Let F be the algebraic closure of C if C is infinite and set $F = C$ for C finite. Clearly, the map $r \in Q \mapsto r \otimes 1 \in Q \otimes_C F$ gives a ring embedding. So we may assume Q is a subring of $Q \otimes_C F$. By [14, Proposition], $\varphi(x, y)$ is also a nonzero GPI of $Q \otimes_C F$. Moreover, in view of [8, Theorem 3.5], $Q \otimes_C F$ is a prime ring with F as its extended centroid. Thus, $\overline{Q} = Q \otimes_C F$ is a prime ring that satisfies a nonzero GPI $\varphi(x, y)$, and its extended centroid F is either an algebraically closed field or a finite field. By Martindale's Theorem [6, Theorem 6.1.6], $\overline{Q}$ is a primitive ring having nonzero socle with F as its associated division ring. Moreover, $\overline{Q}$ is a dense subring of $\mathrm{End}(V_F)$, where V is a vector space over F. If $\dim V_F = 1$, then $\overline{Q}$ is commutative and hence R is commutative, a contradiction. So $\dim V_F \geq 2$. Assume first that $\dim V_F = 2$, then by [5, Lemma 7.1], there exists $v \in V$ such that v and cv are linearly independent over F. So $bv = v\alpha + cv\beta$ for some $\alpha, \beta \in F$. Since $b \neq 0$, clearly α, β are not all zero. Using the density of R, there exist $x, y \in R$ such that $xv = 0$, $yv = 0$, $xcv = v$, $ycv = v + cv$. Then $(x \circ_k y)v = 0$ and $(x \circ_k y)cv = v$. Therefore, from (3), we have

$$\begin{aligned}
0 &= ([ax+bxc, y]_k - a(x \circ_k y) - b(x \circ_k y)c)v \\
&= \sum_{i=0}^{k} (-1)^i \binom{k}{i} y^i (ax+bxc) y^{k-i} v - a(x \circ_k y)v - b(x \circ_k y)cv \\
&= (-1)^k y^k (ax+bxc)v - bv \\
&= (-1)^k y^k bv - bv = (-1)^k y^k (v\alpha + cv\beta) - (v\alpha + cv\beta) \\
&= (-1)^k v\beta - v\alpha + \left\{(-1)^k - 1\right\} cv\beta,
\end{aligned}$$

which leads a contradiction.

Now if $\dim V_F \geq 3$, then by [5, Lemma 7.1], there exists $v \in V$ such that v and cv are linearly independent over F. Suppose if $bv \in \text{span}_F\{v, cv\}$, then using the same arguments as we have used above, we get a contradiction. Now, if $bv \notin \text{span}_F\{v, cv\}$, then v, cv, bv are linearly independent over F. By the density of R, there exist $x, y \in R$ such that $xv = 0$, $yv = 0$, $xcv = v$, $ycv = 0$ and $ybv = bv$. Then $(x \circ_k y)v = 0$ and $(x \circ_k y)cv = 0$. From (3), we have

$$\begin{aligned}
0 &= ([ax+bxc, y]_k - a(x \circ_k y) - b(x \circ_k y)c)v \\
&= \sum_{i=0}^{k} (-1)^i \binom{k}{i} y^i (ax+bxc) y^{k-i} v - a(x \circ_k y)v - b(x \circ_k y)cv \\
&= (-1)^k y^k (ax+bxc)v \\
&= (-1)^k y^k bv = (-1)^k bv,
\end{aligned}$$

which gives again a contradiction.

Case 2. d is not Q-inner. We have from (2)

$$[a'x + bd(x), y]_k - a'(x \circ_k y) - bd(x \circ_k y) = 0 \tag{4}$$

for all $x, y \in I$ and thus for all $x, y \in R$ [12, Theorem 2]. This can be written as

$$\begin{aligned}
[a'x + bd(x), y]_k &= a'(x \circ_k y) + b \sum_{i=0}^{k} \binom{k}{i} y^i d(x) y^{k-i} \\
&\quad + b \sum_{i=0}^{k-1} \binom{k}{i} y^i x \left(\sum_{r+s=k-i-1} y^r d(y) y^s \right) \\
&\quad + b \sum_{i=1}^{k} \binom{k}{i} \left(\sum_{r+s=i-1} y^r d(y) y^s \right) x y^{k-i}.
\end{aligned}$$

Applying Kharchenko's theorem [10], we obtain

$$
\begin{aligned}
[a'x + bz_1, y]_k &= a'(x \circ_k y) + b\sum_{i=0}^{k}\binom{k}{i} y^i z_1 y^{k-i} \\
&+ b\sum_{i=0}^{k-1}\binom{k}{i} y^i x \left(\sum_{r+s=k-i-1} y^r z_2 y^s\right) \\
&+ b\sum_{i=1}^{k}\binom{k}{i}\left(\sum_{r+s=i-1} y^r z_2 y^s\right) x y^{k-i}
\end{aligned}
$$

for all $x, y, z_1, z_2 \in R$ and hence for all $x, y, z_1, z_2 \in Q$ by [6, Theorem 6.4.4]. In particular, choosing $u \notin C$ and taking $z_1 = [u, x]$, $z_2 = [u, y]$ in the above expression, we get

$$
\begin{aligned}
[a'x + b[u, x], y]_k &= a'(x \circ_k y) + b\sum_{i=0}^{k}\binom{k}{i} y^i [u, x] y^{k-i} \\
&+ b\sum_{i=0}^{k-1}\binom{k}{i} y^i x \left(\sum_{r+s=k-i-1} y^r [u, y] y^s\right) \\
&+ b\sum_{i=1}^{k}\binom{k}{i}\left(\sum_{r+s=i-1} y^r [u, y] y^s\right) x y^{k-i} \\
&= a'(x \circ_k y) + b[u, x \circ_k y].
\end{aligned}
$$

This can be rewritten as

$$
[ax + bxc, y]_k - a(x \circ_k y) - b(x \circ_k y)c = 0 \text{ for all } x, y \in R, \tag{5}
$$

where $a = a' + bu$ and $c = -u$. In view of (3) and (5), since $c = -u \notin C$, the proof is completed by Case 1.

Suppose $b = 0$. Then $G(xy) = G(x)y$ for all $x, y \in R$. By [4, Lemma 2.3], there is $a \in Q$ such that $G(x) = ax$ for all $x \in R$. In this case, from (2), we have $[ax, y]_k = a(x \circ_k y)$ for all $x, y \in R$. Therefore, in view of Corollary 1, $a \in C$. This completes the proof of theorem.

Corollary 3 *Let R be a noncommutative prime ring, I a nonzero ideal of R and $b \in Q$. If R admits a nonzero b-generalized derivation G associated with the map d such that $[G(x), y] = G(x \circ y)$ for all $x, y \in I$, then there exists $\lambda \in C$ such that $G(x) = \lambda x$ for all $x \in R$.*

Theorem 2 *Let R be a noncommutative prime ring, I a nonzero ideal of R and $b \in Q$. If R admits a nonzero b-generalized derivation G associated with the map d such that $G(x) \circ_k y = x \circ_k y$ for all $x, y \in I$, where k is a fixed positive integer, then there exists $\lambda \in C$ such that $G(x) = \lambda x$ for all $x \in R$.*

Proof Suppose first that $b \neq 0$. By [11, Theorem 2.3], $d : R \to Q$ is a derivation d and there exists $a' \in Q$ such that $G(x) = a'x + bd(x)$ for all $x \in R$. It is known that d can be uniquely extended to a derivation of Q [12, Lemma 2]. By the assumption, we have

$$G(x) \circ_k y = x \circ_k y \text{ for all } x, y \in I. \tag{6}$$

If $d = 0$, then $(a'x) \circ_k y = x \circ_k y$ for all $x, y \in I$ and hence for all $x, y \in Q$ by [6, Theorem 6.4.4]. Application of Corollary 2 yields that $a' \in C$ and hence the proof is done.

Assume now that $d \neq 0$. We divide the proof into two cases.

Case 1. d is Q-inner. That is, there exists $c' \in Q$ such that $d(x) = [c', x]$ for all $x \in R$. So $G(x) = a'x + bd(x) = a'x + b[c', x] = ax + bxc$ for all $x \in R$, where $a = a' + bc'$ and $c = -c'$. By (2), we have

$$((ax + bxc) \circ_k y) - (x \circ_k y) = 0 \text{ for all } x, y \in I. \tag{7}$$

In view of [6, Theorem 6.4.4], we have $((ax + bxc) \circ_k y) - (x \circ_k y) = 0$ for all $x, y \in Q$. If $c \in C$, then $d = 0$ as $c = -c' \in C$ and $G(x) = \lambda x$ for all $x \in R$ this proving the theorem. Now we assume that $c \notin C$. Let $\phi(x, y) = ((ax + bxc) \circ_k y) - (x \circ_k y)$. Clearly, if $c \notin C$, then $\phi(x, y)$ is a nonzero GPI of R by (7). So we conclude that $\phi(x, y)$ is a nonzero GPI of Q. Let F be the algebraic closure of C if C is infinite and set $F = C$ for C finite. Clearly, the map $r \in Q \mapsto r \otimes 1 \in Q \otimes_C F$ gives a ring embedding. So we may assume Q is a subring of $Q \otimes_C F$. By [14, Proposition], $\phi(x, y)$ is also a nonzero GPI of $Q \otimes_C F$. Moreover, in view of [8, Theorem 3.5], $Q \otimes_C F$ is a prime ring with F as its extended centroid. Thus, $\overline{Q} = Q \otimes_C F$ is a prime ring satisfies a nonzero GPI $\phi(x, y)$, and its extended centroid F is either an algebraically closed field or a finite field. By Martindale's Theorem [6, Theorem 6.1.6], $\overline{Q}$ is a primitive ring having nonzero socle with F as its associated division ring. Moreover, $\overline{Q}$ is a dense subring of $\mathrm{End}(V_F)$, where V is a vector space over F. If $\dim V_F = 1$, then $\overline{Q}$ is commutative and hence R is commutative, a contradiction. So $\dim V_F \geq 2$. Assume first that $\dim V_F = 2$, then by [5, Lemma 7.1], there exists $v \in V$ such that v and cv are linearly independent over F. So $bv = v\alpha + cv\beta$ for some $\alpha, \beta \in F$. If $\beta = 0$, then we have $bv = v\alpha$ and hence

$$[b, x]v = bxv - xbv = b(xv) - x(bv) = xv\alpha - xv\alpha = 0 \text{ for all } x \in R.$$

This implies that $[b, x]V = (0)$ for all $x \in R$, and since V is faithful, it follows that $[b, x] = 0$ for all $x \in R$. Thus, $b \in Z(R)$. In this case by using the density of R, there exist $x, y \in R$ such that $xv = 0,\ yv = v,\ xcv = v$. Hence, from (7), we have

$$\begin{aligned} 0 &= (((ax + bxc) \circ_k y) - (x \circ_k y))v \\ &= ((ax + bxc) \circ_k y)v - (x \circ_k y)v \\ &= 2^k bv, \end{aligned}$$

a contradiction as $b \neq 0$. So, if $\beta \neq 0$, then again using the density of R, there exist $x, y \in R$ such that $xv = 0,\ yv = 0,\ xcv = v,\ ycv = cv$. Therefore, from (7),

$$\begin{aligned} 0 &= (((ax + bxc) \circ_k y) - (x \circ_k y))v \\ &= ((ax + bxc) \circ_k y)v - (x \circ_k y)v \\ &= cv\beta, \end{aligned}$$

which leads a contradiction.

Now if $\dim V_F \geq 3$, then by [5, Lemma 7.1], there exists $v \in V$ such that v and cv are linearly independent over F. Suppose if $bv \in \mathrm{span}_F\{v, cv\}$, then using the same arguments as we have used above, we get a contradiction. Now, if $bv \notin \mathrm{span}_F\{v, cv\}$, then v, cv, bv are linearly independent over F. By the density of R, there exist $x, y \in R$ such that $xv = 0,\ yv = 0,\ xcv = v,\ ycv = cv$ and $ybv = cv$. So, $((ax + bxc) \circ_k y)v = cv,\ (x \circ_k y)v = 0$. Then, from (7), we have

$$\begin{aligned} 0 &= (((ax + bxc) \circ_k y) - (x \circ_k y))v \\ &= ((ax + bxc) \circ_k y)v - (x \circ_k y)v \\ &= cv, \end{aligned}$$

which gives again a contradiction.

Case 2. d is not Q-inner. We have from (6)

$$((a'x + bd(x)) \circ_k y) - (x \circ_k y) = 0 \tag{8}$$

for all $x, y \in I$ and thus for all $x, y \in R$ [12, Theorem 2]. Applying Kharchenko's theorem [10], we obtain

$$(a'x + bz) \circ_k y = x \circ_k y$$

for all $x, y, z \in R$ and hence for all $x, y, z \in Q$. Choosing $u \notin C$ and taking $z = [u, x]$ in the above expression, we get $(a'x + b[u, x]) \circ_k y = x \circ_k y$ for all $x, y \in Q$. This implies that

$$((ax + bxc) \circ_k y) - (x \circ_k y) = 0 \text{ for all } x, y \in R, \tag{9}$$

where $a = a' + bu$ and $c = -u$. In view of (7) and (9), since $c = -u \notin C$, the proof is completed by Case 1.

Suppose $b = 0$. Then $G(xy) = G(x)y$ for all $x, y \in R$. By [4, Lemma 2.3], there is $a \in Q$ such that $G(x) = ax$ for all $x \in R$. From (6), we have $(ax) \circ_k y = x \circ_k y$ for all $x, y \in R$. Application of Corollary 2 yields that $a \in C$. Thereby the proof is completed.

Corollary 4 *Let R be a noncommutative prime ring, I a nonzero ideal of R and $b \in Q$. If R admits a nonzero b-generalized derivation G associated with the map*

d such that $G(x) \circ y = x \circ y$ for all $x, y \in I$, then there exists $\lambda \in C$ such that $G(x) = \lambda x$ for all $x \in R$.

Acknowledgements The author is very thankful to the referees for their careful reading of the paper and valuable suggestions.

References

1. Alahmadi, A., Ali, S., Khan, A.N., Khan, M.S.: A characterization of generalized derivations on prime rings. Commun. Algebra **44**(8), 32013210 (2016)
2. Albaş, A., Argaç, N., De Filippis, V.: Generalized derivations with Engel conditions on one-sided ideals. Commun. Algebra **36**(6), 2063–2071 (2008)
3. Ashraf, M., Pary, S.A., Raza, M.A.: On generalized derivations in semiprime rings involving anticommutator. Beitr. Algebra Geom. **60**(3), 587–598 (2019)
4. Beidar, K.I.: On functional identities and commuting additive mappings. Commun. Algebra **26**, 1819–1850 (1998)
5. Beidar, K.I., Brešar, M.: Extended Jacobson density theorem for rings with derivations and automorphisms. Israel J. Math. **122**, 317–346 (2001)
6. Beidar, K.I., Martindale, W.S., III., Mikhalev, A.V.: Rings with Generalized Identities. Marcel Dekker Inc, New York (1996)
7. Brešar, M.: On the distance of the composition of two derivations to the generalized derivations. Glasgow Math. J. **33**(1), 89–93 (1991)
8. Erickson, T.S., Martindale, W.S., III., Osborn, J.M.: Prime nonassociative algebras. Pacific J. Math. **60**(1), 49–63 (1975)
9. Dhara, B., De Filippis, V.: Engel conditions of generalized derivations on left ideals and Lie ideals in prime rings. Commun. Algebra **48**(1), 154–167 (2020)
10. Kharchenko, V.K.: Differential identities of semiprime rings. Algebra Logic **18**(1), 86–119 (1979)
11. Koşan, M.T., Lee, T.-K.: b-generalized derivations of semiprime rings having nilpotent values. J. Aust. Math. Soc. **96**(3), 326–337 (2014)
12. Lee, T.-K.: Semiprime rings with differential identities. Bull. Inst. Math. Acad. Sinica **20**(1), 27–38 (1992)
13. Lee, T.-K., Shiue, W.-K.: Identities with generalized derivations. Commun. Algebra **29**(10), 4437–4450 (2001)
14. Lee, P.H., Wong, T.-L.: Derivations centralizing Lie ideals. Bull. Inst. Math. Acad. Sinica **23**(1), 1–5 (1995)

Commutativity of $*$-Prime Rings with Generalized Derivations on $*$-Jordan Ideals

Deepak Kumar and Bharat Bhushan

Abstract The purpose of this paper is to find commutativity conditions in $*$-prime rings with generalized derivations, where $'*'$ is involution of the second kind. More specifically, it is shown that if a 2-torsion free $*$-prime ring $\mathscr{R}$ with involution of the second kind satisfies any of the following assertions: (i) $[\mathscr{F}(\alpha), \alpha^*] \in Z(\mathscr{R})$, (ii) $\mathscr{F}(\alpha) \circ (\alpha^*) \in Z(\mathscr{R})$, (iii) $\mathscr{F}(\alpha) \circ \mathfrak{d}(\alpha^*) \pm \alpha \circ \alpha^* \in Z(\mathscr{R})$ and (iv) $[\mathscr{F}(\alpha), \mathfrak{d}(\alpha^*)] \pm \alpha \circ \alpha^* \in Z(\mathscr{R})$, where $\mathscr{F}$ is a generalized derivation associated with a derivation $\mathfrak{d}$ such that $\mathfrak{d}$ is commuting with $*$ and α varies over a nonzero $*$-Jordan ideal of $\mathscr{R}$, then $\mathscr{R}$ is commutative.

Keywords $*$-prime rings · Generalized derivations · Involution

1 Introduction

In this paper, $\mathscr{R}$ denotes an associative ring with center $Z(\mathscr{R})$. For any $\alpha, \beta \in \mathscr{R}$, commutator (resp. anti-commutator) is defined as $[\alpha, \beta] = \alpha\beta - \beta\alpha$ (resp. $\alpha \circ \beta = \alpha\beta + \beta\alpha$). A ring $\mathscr{R}$ is said to be a n-torsion free, where n is a positive integer, if for any $\alpha \in \mathscr{R}$, $n\alpha = 0$ implies $\alpha = 0$. An anti-automorphism of order two is called an involution $'*'$. A ring $\mathscr{R}$ equipped with an involution $'*'$ is called $*$-ring. An ideal I of $*$-ring is called a $*$-ideal if $I^* = I$. An additive subgroup J of ring $\mathscr{R}$ is called Jordan ideal if $\alpha \circ r \in J \ \forall \ \alpha \in J, r \in \mathscr{R}$. A Jordan ideal J of $*$-ring is called $*$-Jordan ideal if $J^* = J$. An element $\alpha \in \mathscr{R}$ is a symmetric (resp. skew symmetric) element if $\alpha^* = \alpha$ (resp. $\alpha^* = -\alpha$). The sets of symmetric and skew symmetric elements are denoted by $H(\mathscr{R})$ and $S(\mathscr{R})$, respectively. The set $Sa_*(\mathscr{R}) = \{\alpha \in \mathscr{R} \mid \alpha^* = \pm\alpha\}$ is the collection of all symmetric and skew symmetric elements. An involution $'*'$ is said to be of the first kind if $Z(\mathscr{R}) \subseteq H(\mathscr{R})$ and otherwise it is of the second kind.

D. Kumar (✉) · B. Bhushan
Department of Mathematics, Punjabi University, Patiala 147002, India
e-mail: deep_math1@yahoo.com

B. Bhushan
e-mail: bharat_rs18@pbi.ac.in

M. Ashraf et al. (eds.), *Algebra and Related Topics with Applications*,
Springer Proceedings in Mathematics & Statistics 392,
https://doi.org/10.1007/978-981-19-3898-6_15

Thus, in an involution of the second kind, we have $Z(\mathscr{R}) \cap S(\mathscr{R}) \neq \{0\}$. A $*$-ring is called $*$-prime ring if $\alpha\mathscr{R}\beta = \alpha\mathscr{R}\beta^* = \{0\}$ implies either $\alpha = 0$ or $\beta = 0$.

An additive map $\eth : \mathscr{R} \to \mathscr{R}$ is called a derivation if $\eth(\alpha\beta) = \eth(\alpha)\beta + \alpha\eth(\beta)$ $\forall$ $\alpha, \beta \in \mathscr{R}$. A map $I_x : \mathscr{R} \to \mathscr{R}$ defined as $I_x(\alpha) = [x, \alpha]$, where x is any fixed element in $\mathscr{R}$, is an example of derivation known as inner derivation. For fixed elements $x, y \in \mathscr{R}$, a map $I_{x,y} : \mathscr{R} \to \mathscr{R}$ defined as $I_{x,y}(\alpha) = x\alpha + \alpha y$ is a generalized inner derivation. Motivated by it, the notion of a generalized derivation was introduced by Brešar [2] as follows: an additive map $\mathscr{F} : \mathscr{R} \to \mathscr{R}$ is called a generalized derivation associated with a derivation $\eth$ if $\mathscr{F}(\alpha\beta) = \mathscr{F}(\alpha)\beta + \alpha\eth(\beta)$ $\forall$ $\alpha, \beta \in \mathscr{R}$. A derivation $\eth$ is called commuting with $*$ if $\eth(\alpha^*) = (\eth(\alpha))^*$ $\forall$ $\alpha \in \mathscr{R}$.

A map f is called commuting (resp. centralizing) on a set S if $[f(\alpha), \alpha] = 0$ (resp. $[f(\alpha), \alpha] \in Z(\mathscr{R})$) $\forall$ $\alpha \in S$. Motivated by the commuting and centralizing mappings, Ali and Dar [1] defined $*$-centralizing and $*$commuting mappings as follows: a map f is called $*$-centralizing (resp. $*$-commuting) on a set S if $[f(\alpha), \alpha^*] \in Z(\mathscr{R})$ (resp. $[f(\alpha), \alpha^*] = 0$) $\forall$ $\alpha \in S$ and proved that in a 2-torsion free noncommutative prime ring $\mathscr{R}$ with involution of the second kind, there exists no nonzero derivation $\eth : \mathscr{R} \to \mathscr{R}$ such that $\eth$ is $*$-centralizing on $\mathscr{R}$. There has been a lot of work done in $*-$prime rings to find conditions that finally imply commutativity of rings with the help of derivations (see [5–8]). For any prime ring $\mathscr{R}$ with involution of second kind in Nejjar et al., [4] obtained the commutativity of rings if it satisfy any of the following conditions (i) $\eth(\alpha) \circ \eth(\alpha^*) \pm \alpha \circ \alpha^* \in Z(\mathscr{R})$ and (ii) $\eth(\alpha) \circ \eth(\alpha^*) \pm [\alpha, \alpha^*] \in Z(\mathscr{R})$ $\forall$ $\alpha \in \mathscr{R}$. In 2019, Mamouni et al. [3] proved that for any prime ring $\mathscr{R}$ with involution of second kind and J is a Jordan ideal of $\mathscr{R}$, if $\mathscr{F}(\alpha \circ \beta) \in Z(\mathscr{R})$ $\forall$ $\alpha, \beta \in J$, where $\mathscr{F}$ is a generalized derivation, then $\mathscr{R}$ must be a commutative.

In this paper, we continue the line of investigation and study the commutativity of $*$-prime rings satisfying certain differential identities involving generalized derivations acting on $*$-Jordan ideals.

2 Auxiliary Lemmas

Since it is well-known that the center of a prime ring has no proper zero-divisor, this statement is not true in $*$-prime rings. The following example justifies that the center of a $*$-prime ring is not free from the proper zero-divisors.

Example 1 Let $\mathscr{R} = \left\{\begin{pmatrix} \alpha & 0 \\ 0 & \beta \end{pmatrix} | \alpha, \beta \in \mathbb{Z}\right\}$, where $\mathbb{Z}$ is the set of integers. A map $*$: $\mathscr{R} \to \mathscr{R}$ is defined as $\begin{pmatrix} \alpha & 0 \\ 0 & \beta \end{pmatrix}^* = \begin{pmatrix} \beta & 0 \\ 0 & \alpha \end{pmatrix}$. It is easy to verify that $\mathscr{R}$ is a commutative $*$-prime ring. For any nonzero α, $\begin{pmatrix} \alpha & 0 \\ 0 & 0 \end{pmatrix} \in Z(\mathscr{R})$ and $0 \neq \beta$, $\begin{pmatrix} 0 & 0 \\ 0 & \beta \end{pmatrix} \in \mathscr{R}$, we have

$\begin{pmatrix} \alpha & 0 \\ 0 & 0 \end{pmatrix} \begin{pmatrix} 0 & 0 \\ 0 & \beta \end{pmatrix} = \begin{pmatrix} 0 & 0 \\ 0 & 0 \end{pmatrix}$. It shows that the center of $*$-prime ring $\mathscr{R}$ is not free from proper zero-divisors.

As a matter of fact, we now find a certain subset of the center of a $*$-prime ring which is indeed free from proper zero-divisors.

Lemma 1 *Let $\mathscr{R}$ be a $*$-prime ring. Then $Sa_*(\mathscr{R}) \cap Z(\mathscr{R})$ has no proper zero-divisor.*

Proof Let $t \in Sa_*(\mathscr{R}) \cap Z(\mathscr{R})$ be a zero-divisor. Then there exists some $0 \neq r \in \mathscr{R}$ such that $tr = 0$. We have $rts = 0$ for all $s \in \mathscr{R}$. Since $t \in Z(\mathscr{R})$, it gives that

$$t\mathscr{R}r = \{0\}. \tag{1}$$

As $tr = 0$ implies $(rt)^* = 0$, that is, $t^*r^* = 0$ implies $\pm tr^* = 0$. Therefore, we conclude that

$$t\mathscr{R}r^* = \{0\}. \tag{2}$$

From (1) and (2), $*$-primeness of $\mathscr{R}$ yields that $t = 0$. Hence, $Sa_*(\mathscr{R}) \cap Z(\mathscr{R})$ has no proper zero-divisor.

Lemma 2 *Let $\mathscr{R}$ be a 2-torsion free $*$-prime ring and I be a nonzero $*$-ideal. If $I \subseteq Z(\mathscr{R})$, then $\mathscr{R}$ is commutative.*

Proof By the given hypothesis, we find $[\alpha, r] = 0 \ \forall \ \alpha \in I, \ r \in \mathscr{R}$. Replace α by αs, where $s \in \mathscr{R}$, to get $\alpha[s, r] = 0$. Replacing α by $t\alpha$, where $t \in \mathscr{R}$, we have $t\alpha[s, r] = 0$. Replace t by t^* to obtain $t^*\alpha[s, r] = 0 \ \forall \ \alpha \in I, \ r, s, t \in \mathscr{R}$. Application of [7, Lemma 1] yields the desired result.

Lemma 3 *Let $\mathscr{R}$ be a 2-torsion free $*$-prime ring with involution of the second kind and I be a nonzero $*$-ideal. If $[\alpha, \alpha^*] \in Z(\mathscr{R}) \ \forall \ \alpha \in I$, then $\mathscr{R}$ is commutative.*

Proof Polarize given hypothesis to obtain

$$[[\alpha, \beta^*], \alpha] + [[\beta, \alpha^*], \alpha] = 0 \ \forall \ \alpha, \beta \in I. \tag{3}$$

Replace β by β^* in (3) to get

$$[[\alpha, \beta], \alpha] + [[\beta^*, \alpha^*], \alpha] = 0. \tag{4}$$

Replacing β by $\beta\alpha$ in (4) and using it, we conclude

$$[[\alpha, \beta], \alpha]\alpha - \alpha^*[[\alpha, \beta], \alpha] + [\alpha^*, \alpha][\beta^*, \alpha^*] = 0 \ \forall \ \alpha, \beta \in I. \tag{5}$$

Substituting $\beta\alpha$ in place of β in (5) and using it, we conclude

$$-[\alpha^*, \alpha][\beta^*, \alpha^*]\alpha + [\alpha^*, \alpha]\alpha^*[\beta^*, \alpha^*] = 0. \tag{6}$$

Replace α by α^* and β by β^* in (6) to obtain

$$[\alpha, \alpha^*]([\beta, \alpha]\alpha^* - \alpha[\beta, \alpha]) = 0 \ \forall \ \alpha, \beta \in I. \tag{7}$$

Replacing β by $\beta\alpha$ in (7) and using it, we conclude

$$[\alpha, \alpha^*][\beta, \alpha][\alpha, \alpha^*] = 0. \tag{8}$$

Since $[\alpha, \alpha^*]^* = [\alpha, \alpha^*]$ and by the given hypothesis, we have $[\alpha, \alpha^*] \in Z(R)$. Thus, $[\alpha, \alpha^*] \in Sa_*(\mathscr{R}) \cap Z(\mathscr{R})$. Replace β by α^* in (8) to find $[\alpha, \alpha^*]^3 = 0 \ \forall \ \alpha \in \mathscr{R}$. By application of Lemma 1, we conclude $[\alpha, \alpha^*] = 0 \ \forall \ \alpha \in \mathscr{R}$. Replace α by $k + h$, where $k \in S(\mathscr{R}) \cap I, \ h \in H(\mathscr{R}) \cap I$, to obtain

$$[h, k] = 0. \tag{9}$$

Replace k by $\alpha - \alpha^*$, where $\alpha \in I$, in (9) to obtain

$$[h, \alpha - \alpha^*] = 0 \ \forall \ \alpha \in I, \ h \in H(\mathscr{R}) \cap I. \tag{10}$$

For any $0 \neq k_c \in S(\mathscr{R}) \cap Z(\mathscr{R})$, replace k by $k_c(\alpha + \alpha^*)$ in (9) to get

$$[h, \alpha + \alpha^*]k_c = 0 \ \forall \alpha \in I, \ h \in H(\mathscr{R}). \tag{11}$$

As $0 \neq k_c \in Sa_*(\mathscr{R}) \cap Z(\mathscr{R})$, from Lemma 1 in (11), we conclude

$$[h, \alpha + \alpha^*] = 0 \ \forall \ \alpha \in I, \ h \in H(\mathscr{R}) \cap I. \tag{12}$$

Add (10) and (12) to get

$$[h, \alpha] = 0 \ \forall \ \alpha \in I, \ h \in H(\mathscr{R}) \cap I. \tag{13}$$

For any $\beta \in I$, replacing h by $\beta + \beta^*$ in above equation and using similar arguments, we conclude $[\alpha, \beta] = 0 \ \forall \ \alpha, \beta \in I$. Replace β by $r\beta$, where $r \in \mathscr{R}$ to get $[\alpha, r\beta] = 0$. It implies $[\alpha, r]\beta = 0 \ \forall \ \alpha, \beta \in I, \ r \in \mathscr{R}$. Replace β by $s\beta$, where $s \in \mathscr{R}$ to get $[\alpha, r]s\beta = 0$. Replace β by β^* to get $[\alpha, r]s\beta^* = 0 \ \forall \ \alpha, \beta \in I, \ r, s \in \mathscr{R}$. From $*$-primeness of $\mathscr{R}$ and $I \neq \{0\}$, we conclude $[\alpha, r] = 0 \ \forall \ \alpha \in I, \ r \in \mathscr{R}$. Thus, $\mathscr{R}$ is commutative by Lemma 2.

Lemma 4 *Let $\mathscr{R}$ be a 2-torsion free $*$-prime ring with involution of the second kind and I be a nonzero $*$-ideal. If $\alpha \circ \alpha^* \in Z(\mathscr{R}) \ \forall \ \alpha \in I$, then $\mathscr{R}$ is commutative.*

Proof First let

$$\alpha \circ \alpha^* = 0 \ \forall \ \alpha \in I. \tag{14}$$

Polarize (14) to obtain

$$\alpha \circ \beta^* + \beta \circ \alpha^* = 0 \; \forall \; \alpha, \beta \in I. \tag{15}$$

Replace β by βk_c, where $k_c \in S(\mathscr{R}) \cap Z(\mathscr{R})$, to obtain

$$(-\alpha \circ \beta^* + \beta \circ \alpha^*)k_c = 0. \tag{16}$$

As $0 \neq k_c \in Sa_*(\mathscr{R}) \cap Z(\mathscr{R})$, by Lemma 1 in (16), we find

$$-(\alpha \circ \beta^*) + \beta \circ \alpha^* = 0 \; \forall \; \alpha, \beta \in I. \tag{17}$$

Add (15) and (17) to get

$$\beta \circ \alpha^* = 0 \; \forall \; \alpha, \beta \in I. \tag{18}$$

Replace α by α^* and β by βr, where $r \in \mathscr{R}$, in (18) to get $\beta[r, \alpha] = 0$. Replacing β by βs, where $s \in \mathscr{R}$, we get $\beta s[r, \alpha] = 0$. Replace β by β^* to obtain $\beta^* s[r, \alpha] = 0$. From $*$-primeness of $\mathscr{R}$ and $I \neq 0$, we have $I \subseteq Z(\mathscr{R})$. Thus, Lemma 2 gives the desired result.

If $\alpha \circ \alpha^* \neq 0$, then there exists $0 \neq z \in I \cap Z(\mathscr{R})$. By the given hypothesis, we have

$$\alpha \circ \alpha^* \in Z(\mathscr{R}) \; \forall \; \alpha \in I. \tag{19}$$

Polarize (19) to obtain

$$\alpha \circ \beta^* + \beta \circ \alpha^* \in Z(\mathscr{R}) \; \forall \; \alpha, \beta \in I. \tag{20}$$

Replace β by z in (20) to get

$$[\alpha z^* + \alpha^* z, r] = 0 \; \forall \; \alpha \in I, \; r \in \mathscr{R}. \tag{21}$$

Replace α by αk_c, where $0 \neq k_c \in S(\mathscr{R}) \cap Z(\mathscr{R})$, to conclude

$$[-\alpha z^* + \alpha^* z, r]k_c = 0. \tag{22}$$

As $0 \neq k_c \in Sa_*(\mathscr{R}) \cap Z(\mathscr{R})$, using application of Lemma 1 in (22), we find

$$[-\alpha z^* + \alpha^* z, r] = 0 \; \forall \; \alpha \in I, \; r \in \mathscr{R}. \tag{23}$$

Add (21) and (23) to obtain

$$[2\alpha^* z, r] = 0 \; \forall \; \alpha \in I, \; r \in \mathscr{R}. \tag{24}$$

Replace r by rs, where $s \in \mathscr{R}$, in (24) to obtain

$$[\alpha^*, r]sz = 0 \; \forall \; \alpha \in I, r, s \in \mathscr{R}. \tag{25}$$

Subtract (23) from (21) to obtain

$$[\alpha, r]z^* = 0 \;\forall\; \alpha \in I,\; r \in \mathscr{R}. \tag{26}$$

Replace α by α^* and r by rs in (26) to obtain

$$[\alpha^*, r]sz^* = 0 \;\forall\; \alpha \in I, r, s \in \mathscr{R}. \tag{27}$$

From (25) and (27), $*$-primeness of $\mathscr{R}$, we conclude $I \subseteq Z(\mathscr{R})$. Thus, $\mathscr{R}$ is commutative by Lemma 2.

Lemma 5 *Let $\mathscr{R}$ be a 2-torsion free $*$-prime ring with involution of the second kind. If $\mathfrak{d}(h) = 0 \;\forall\; h \in H(\mathscr{R}) \cap Z(\mathscr{R})$, then $\mathfrak{d}(z) = 0 \;\forall\; z \in Z(\mathscr{R})$.*

Proof By the given hypothesis, we have $\mathfrak{d}(h) = 0 \;\forall\; h \in H(\mathscr{R}) \cap Z(\mathscr{R})$. For any $0 \neq t \in Sa_*(\mathscr{R}) \cap Z(\mathscr{R})$ implies $t^2 \in H(\mathscr{R}) \cap Z(\mathscr{R})$. Thus, $\mathfrak{d}(t^2) = 0$ which implies $2\mathfrak{d}(t)t = 0$. Using Lemma 1, we have $\mathfrak{d}(t) = 0 \;\forall\; t \in Sa_*(\mathscr{R}) \cap Z(\mathscr{R})$. For any $z \in Z(\mathscr{R})$, we have $2z = z + z^* + z - z^*$. It implies $\mathfrak{d}(2z) = \mathfrak{d}(z + z^*) + \mathfrak{d}(z - z^*) = 0$. Using 2-torsion free condition, we have $\mathfrak{d}(z) = 0 \;\forall\; z \in Z(\mathscr{R})$.

Lemma 6 *Let $\mathscr{R}$ be a 2-torsion free $*$-prime ring. If $\mathscr{F} : \mathscr{R} \to \mathscr{R}$ is a nonzero generalized derivation associated with a derivation $\mathfrak{d}$ such that $\mathfrak{d}$ is commuting with $*$ and I is a nonzero $*$-ideal. If $[\mathscr{F}(\alpha), \beta] \in Z(\mathscr{R}) \;\forall\; \alpha, \beta \in I$, then $\mathscr{R}$ is a commutative ring.*

Proof Replacing β by β^2 in given hypothesis and using it, we conclude that $2[\mathscr{F}(\alpha), \beta]\beta \in Z(\mathscr{R})$. It implies $\beta[\mathscr{F}(\alpha), \beta] \in Z(\mathscr{R})$ for all $\alpha, \beta \in I$. Now for any $s \in \mathscr{R}$, we have $s\beta[\mathscr{F}(\alpha), \beta] = \beta[\mathscr{F}(\alpha), \beta]s$. By the given hypothesis, we have $s\beta[\mathscr{F}(\alpha), \beta] = \beta s[\mathscr{F}(\alpha), \beta]$. So $[s, \beta][\mathscr{F}(\alpha), \beta] = 0 \;\forall\; \alpha, \beta \in I,\; s \in \mathscr{R}$. Replace s by $\mathscr{F}(\alpha)$ in last equation to obtain $[\mathscr{F}(\alpha), \beta]^2 = 0 \;\forall\; \alpha, \beta \in I$. Since we have $[\mathscr{F}(\alpha), \beta] \in Z(\mathscr{R})$, we conclude $[\mathscr{F}(\alpha), \beta]\mathscr{R}[\mathscr{F}(\alpha), \beta][\mathscr{F}(\alpha), \beta]^* = 0$. We have $([\mathscr{F}(\alpha), \beta][\mathscr{F}(\alpha), \beta]^*)^* = [\mathscr{F}(\alpha), \beta][\mathscr{F}(\alpha), \beta]^*$. From $*$-primeness of $\mathscr{R}$, we conclude either $[\mathscr{F}(\alpha), \beta] = 0$ or $[\mathscr{F}(\alpha), \beta][\mathscr{F}(\alpha), \beta]^* = 0$. Suppose $[\mathscr{F}(\alpha), \beta][\mathscr{F}(\alpha), \beta]^* = 0$. Thereby using $[\mathscr{F}(\alpha), \beta] \in Z(\mathscr{R})$ implies $[\mathscr{F}(\alpha), \beta]\mathscr{R}[\mathscr{F}(\alpha), \beta]^* = 0$. From the equation $[\mathscr{F}(\alpha), \beta]^2 = 0$, we conclude $[\mathscr{F}(\alpha), \beta]\mathscr{R}[\mathscr{F}(\alpha), \beta] = 0$. From $*$-primeness of $\mathscr{R}$, we have $[\mathscr{F}(\alpha), \beta] = 0 \;\forall\; \alpha, \beta \in I$. Replacing α by $\alpha\beta$ in last expression and using it, we conclude $[\alpha, \beta]\mathfrak{d}(\beta) + \alpha[\mathfrak{d}(\beta), \beta] = 0 \;\forall\; \alpha, \beta \in I$. Thereby replacing α by $r\alpha$, where $r \in \mathscr{R}$, we have $[r, \alpha]\beta\mathfrak{d}(\alpha) = 0 \;\forall\; \alpha, \beta \in I,\; r \in \mathscr{R}$. Thence by [8, Lemma 2.2], we conclude $\mathfrak{d}(r) = 0 \;\forall\; r \in \mathscr{R}$. Replace α by αr where $r \in \mathscr{R}$ in $[\mathscr{F}(\alpha), \beta] = 0$ to obtain $\mathscr{F}(\alpha)[r, \beta] = 0 \;\forall\; \alpha, \beta \in I,\; r \in \mathscr{R}$. Replacing r by sr, we obtain $\mathscr{F}(\alpha)s[r, \beta] = 0 \;\forall\; \alpha, \beta \in I,\; r, s \in \mathscr{R}$. Replace r by r^* and β by β^* to obtain $\mathscr{F}(\alpha)s[r, \beta]^* = 0 \;\forall\; \alpha, \beta \in I,\; r, s \in \mathscr{R}$. From $*$-primeness of $\mathscr{R}$, nonzero generalized derivation, we compute $I \subseteq Z(\mathscr{R})$. Thus, $\mathscr{R}$ is commutative by Lemma 2.

3 Main Results

Theorem 1 *Let $\mathscr{R}$ be a 2-torsion free $*$-prime ring with involution of the second kind. If $\mathscr{F} : \mathscr{R} \to \mathscr{R}$ is a nonzero generalized derivation associated with a derivation $\mathfrak{d}$ such that $\mathfrak{d}$ is commuting with $*$ and I is a nonzero $*$-ideal in $\mathscr{R}$, then the following assertions are equivalent:*

(1) $[\mathscr{F}(\alpha), \alpha^*] \in Z(\mathscr{R})\ \forall\, \alpha \in I$;
(2) $\mathscr{F}(\alpha) \circ (\alpha^*) \in Z(\mathscr{R})\ \forall\, \alpha \in I$;
(3) *$\mathscr{R}$ is commutative.*

Proof It is obvious that if $\mathscr{R}$ is commutative, then (1) and (2) hold. We need to prove (1)$\Rightarrow$ (3) and (2)$\Rightarrow$ (3).

First, we prove (1) $\Rightarrow$ (3). By the given hypothesis, we have

$$[\mathscr{F}(\alpha), \alpha^*] \in Z(\mathscr{R})\ \forall\, \alpha \in I. \tag{28}$$

Polarize (28) to obtain

$$[\mathscr{F}(\alpha), \beta^*] + [\mathscr{F}(\beta), \alpha^*] \in Z(\mathscr{R})\ \forall\, \alpha, \beta \in I. \tag{29}$$

Replace β by βh_c, where $h_c \in H(\mathscr{R}) \cap Z(\mathscr{R})$, in (29) to get

$$[\mathscr{F}(\alpha), \beta^*]h_c + [\mathscr{F}(\beta), \alpha^*]h_c + [\beta, \alpha^*]\mathfrak{d}(h_c) \in Z(\mathscr{R}). \tag{30}$$

From (29) and (30), we conclude

$$[[\beta, \alpha], r]\mathfrak{d}(h_c) = 0\ \forall\, \alpha, \beta \in I,\ r \in \mathscr{R},\ h_c \in H(\mathscr{R}) \cap Z(\mathscr{R}). \tag{31}$$

Using $\mathfrak{d}(\alpha^*) = (\mathfrak{d}(\alpha))^*$ and $h_c^* = h_c$, we have $\mathfrak{d}(h_c)^* = \mathfrak{d}(h_c^*) = \mathfrak{d}(h_c)$. Thus, $\mathfrak{d}(h_c) \in Sa_*(\mathscr{R})$ as $h_c \in Z(\mathscr{R})$ implies $\mathfrak{d}(h_c) \in Z(\mathscr{R})$. It implies $\mathfrak{d}(h_c) \in Sa_*(\mathscr{R}) \cap Z(\mathscr{R})$. Let there exists $h_c \in H(\mathscr{R}) \cap Z(\mathscr{R})$ such that $\mathfrak{d}(h_c) \neq 0$. Using Lemma 1 in (31), we conclude $[[\beta, \alpha], r] = 0\ \forall\, \alpha, \beta \in I,\ r \in \mathscr{R}$. Replace β by α^*. By Lemma 3, we conclude $\mathscr{R}$ is commutative. In case $\mathfrak{d}(h_c) = 0\ \forall\, h_c \in H(\mathscr{R}) \cap Z(\mathscr{R})$. From Lemma 5, we have $\mathfrak{d}(t) = 0\ \forall\, t \in Z(\mathscr{R})$. Replace β by βt, where $t \in Z(\mathscr{R})$, in (29) to obtain

$$[\mathscr{F}(\alpha), \beta^*]t^* + [\mathscr{F}(\beta), \alpha^*]t \in Z(\mathscr{R})\ \forall\, \alpha, \beta \in I,\ t \in Z(\mathscr{R}). \tag{32}$$

Using (29) and (32), we conclude

$$[\mathscr{F}(\alpha), \beta](t^* - t) \in Z(\mathscr{R})\ \forall\, \alpha, \beta \in I,\ t \in Z(\mathscr{R}).$$

It implies

$$[[\mathscr{F}(\alpha), \beta], r](t^* - t) = 0\ \forall\, \alpha, \beta \in I,\ r \in \mathscr{R},\ t \in Z(\mathscr{R}). \tag{33}$$

As $\mathscr{R}$ is a ring with involution of the second kind. So, there exists some $t \in Z(\mathscr{R})$ such that $0 \neq t - t^* \in Sa_*(\mathscr{R}) \cap Z(\mathscr{R})$. Using Lemma 1 in (33), we conclude

$$[\mathscr{F}(\alpha), \beta] \in Z(\mathscr{R}) \ \forall\ \alpha, \beta \in I. \tag{34}$$

From (34), Lemma 6, we find $\mathscr{R}$ is a commutative.
(2) $\Rightarrow$ (3)
From given condition, we have

$$\mathscr{F}(\alpha) \circ \alpha^* \in Z(\mathscr{R}) \ \forall\ \alpha \in I. \tag{35}$$

Polarize (35) to obtain

$$\mathscr{F}(\alpha) \circ \beta^* + \mathscr{F}(\beta) \circ \alpha^* \in Z(\mathscr{R}) \ \forall\ \alpha, \beta \in I. \tag{36}$$

Replacing β by βh_c, where $h_c \in H(\mathscr{R}) \cap Z(\mathscr{R})$, in (36) and using it, we conclude

$$(\beta \circ \alpha^*)\mathfrak{d}(h_c) \in Z(\mathscr{R}).$$

It implies

$$[\beta \circ \alpha^*, r]\mathfrak{d}(h_c) = 0 \ \forall\ \alpha, \beta \in I,\ r \in \mathscr{R},\ h_c \in H(R) \cap I. \tag{37}$$

Since, $\mathfrak{d}(h_c)^* = \mathfrak{d}(h_c^*) = \mathfrak{d}(h_c)$. It implies $\mathfrak{d}(h_c) \in Sa_*(\mathscr{R}) \cap Z(\mathscr{R})$. If there exists $h_c \in H(\mathscr{R}) \cap Z(\mathscr{R})$ such that $\mathfrak{d}(h_c) \neq 0$, then application of Lemma 1 in (37) provides $[\beta \circ \alpha^*, r] = 0 \ \forall\ \alpha, \beta \in I,\ r \in \mathscr{R}$. Replacing β by α and thereby from Lemma 4, we conclude $\mathscr{R}$ is commutative. In case $\mathfrak{d}(h_c) = 0 \ \forall\ h_c \in Z(\mathscr{R}) \cap H(\mathscr{R})$. From Lemma 5, we have $\mathfrak{d}(t) = 0 \ \forall\ t \in Z(\mathscr{R})$. Replace β by βk_c, where $0 \neq k_c \in S(\mathscr{R}) \cap Z(\mathscr{R})$, in (36) to get

$$(-\mathscr{F}(\alpha) \circ \beta^* + \mathscr{F}(\beta) \circ \alpha^*)k_c \in Z(\mathscr{R}).$$

It implies

$$[-\mathscr{F}(\alpha) \circ \beta^* + \mathscr{F}(\beta) \circ \alpha^*, r]k_c = 0 \ \ \forall\ \alpha, \beta \in I,\ r \in \mathscr{R},\ k_c \in S(\mathscr{R}) \cap Z(\mathscr{R}). \tag{38}$$

As $0 \neq k_c \in S(\mathscr{R}) \cap Z(\mathscr{R})$, using Lemma 1 in (38), we obtain

$$[-\mathscr{F}(\alpha) \circ \beta^* + \mathscr{F}(\beta) \circ \alpha^*, r] = 0 \ \ \forall\ \alpha, \beta \in I,\ r \in \mathscr{R}. \tag{39}$$

Compare (36) and (39) to obtain

$$[\mathscr{F}(\alpha) \circ \beta^*, r] = 0 \ \ \forall\ \alpha, \beta \in I,\ r \in \mathscr{R}. \tag{40}$$

It implies

$$\mathscr{F}(\alpha) \circ \beta \in Z(\mathscr{R}) \ \ \forall \ \alpha, \beta \in I. \tag{41}$$

As $\beta \in I$, $\mathscr{F}(\alpha) \circ \beta \in I$. We divide the prove into following two parts:
Case 1: If $\mathscr{F}(\alpha) \circ \beta \neq 0$, then there exists $0 \neq z \in I \cap Z(\mathscr{R})$. Replace β by z in (40) to obtain

$$[\mathscr{F}(\alpha), r]z = 0 \ \forall \ \alpha \in I, \ \ r \in \mathscr{R}. \tag{42}$$

Replacing r by rs in (42) and using it, we conclude

$$[\mathscr{F}(\alpha), r]sz = 0 \ \forall \ \alpha \in I, \ \ r, s \in \mathscr{R}. \tag{43}$$

As $0 \neq z \in Z(\mathscr{R}) \cap I$ implies $0 \neq z^* \in Z(\mathscr{R}) \cap I$. Replace β by z^* in (40) and use similar arguments to obtain

$$[\mathscr{F}(\alpha), r]sz^* = 0 \ \forall \ \alpha \in I, \ \ r, s \in \mathscr{R}. \tag{44}$$

From (43) and (44), $*$-primeness of $\mathscr{R}$, we conclude $[\mathscr{F}(\alpha), r] = 0 \ \forall \ \alpha \in I, \ \ r \in \mathscr{R}$. Thus, by Lemma 6, we have the desired result.
Case 2: If $\mathscr{F}(\alpha) \circ \beta = 0$, then replace β by βr, where $r \in \mathscr{R}$ in $\mathscr{F}(\alpha) \circ \beta = 0$ to obtain

$$(\mathscr{F}(\alpha) \circ \beta)r + \beta[r, \mathscr{F}(\alpha)] = 0.$$

It implies

$$\beta[r, \mathscr{F}(\alpha)] = 0 \ \forall \ \alpha, \beta \in I, r \in \mathscr{R}. \tag{45}$$

Replace β by βs, where $s \in \mathscr{R}$ in (45) to get

$$\beta s[r, \mathscr{F}(\alpha)] = 0 \ \forall \ \alpha, \beta \in I, r, s \in \mathscr{R}. \tag{46}$$

Replace β by β^* in (46) to obtain

$$\beta^* s[r, \mathscr{F}(\alpha)] = 0 \ \forall \ \alpha, \beta \in I, r, s \in \mathscr{R}. \tag{47}$$

From (46) and (47), $*$-primeness of $\mathscr{R}$, and $I \neq 0$, we conclude $[\mathscr{F}(\alpha), r] = 0 \ \forall \ \alpha \in I, r \in \mathscr{R}$. Therefore, $\mathscr{R}$ is commutative by Lemma 6.

Theorem 2 *Let $\mathscr{R}$ be a 2-torsion free $*$-prime ring with involution of the second kind. If $\mathscr{F} : \mathscr{R} \to \mathscr{R}$ is a generalized derivation associated with a derivation $\mathfrak{d}$ such that $\mathfrak{d}$ is commuting with $*$ and I is a nonzero $*$-ideal of $\mathscr{R}$, then the following assertions are equivalent:*

(1) $\mathscr{F}(\alpha) \circ \mathfrak{d}(\alpha^*) - \alpha \circ \alpha^* \in Z(\mathscr{R}) \ \forall \ \alpha \in I$;
(2) $\mathscr{F}(\alpha) \circ \mathfrak{d}(\alpha^*) + \alpha \circ \alpha^* \in Z(\mathscr{R}) \ \forall \ \alpha \in I$;
(3) *$\mathscr{R}$ is commutative.*

Proof Clearly, if $\mathscr{R}$ is commutative, then (1) and (2) hold. We need to prove (1)$\Rightarrow$ (3) and (2)$\Rightarrow$ (3).

If $\eth = 0$, then $\pm\alpha \circ \alpha^* \in Z(\mathscr{R})\ \forall\ \alpha \in I$. Therefore, $\mathscr{R}$ is commutative by Lemma 4. So, we take $\eth \neq 0$. First, we prove (1)$\Rightarrow$ (3). By the given hypothesis, we have

$$\mathscr{F}(\alpha) \circ \eth(\alpha^*) - \alpha \circ \alpha^* \in Z(\mathscr{R})\ \forall\ \alpha \in I. \tag{48}$$

Polarize (48) to obtain

$$\mathscr{F}(\alpha) \circ \eth(\beta^*) + \mathscr{F}(\beta) \circ \eth(\alpha^*) - \alpha \circ \beta^* - \beta \circ \alpha^* \in Z(\mathscr{R})\ \forall\ \alpha, \beta \in I. \tag{49}$$

Replacing β by βh_c, where $h_c \in Z(\mathscr{R}) \cap H(\mathscr{R})$, in (49) and using it, we conclude

$$[\mathscr{F}(\alpha) \circ \beta^* + \beta \circ \eth(\alpha^*), r]\eth(h_c) = 0\ \forall\ \alpha, \beta \in I,\ r \in \mathscr{R}. \tag{50}$$

Since, we have $\eth(h_c)^* = \eth(h_c^*) = \eth(h_c)$. Therefore, Lemma 1 implies either $d(h_c) = 0\ \forall\ h_c \in H(\mathscr{R}) \cap Z(\mathscr{R})$ or $[\mathscr{F}(\alpha) \circ \beta^* + \beta \circ \eth(\alpha^*), r] = 0\ \forall\ \alpha, \beta \in I,\ r \in \mathscr{R}$. Let $\eth(h_c) = 0\ \forall\ h_c \in Z(\mathscr{R}) \cap H(\mathscr{R})$. Application of Lemma 5 implies $\eth(t) = 0\ \forall\ t \in Z(\mathscr{R})$. Replace β by βt, where $t \in Z(\mathscr{R})$, in (49) to obtain

$$\mathscr{F}(\alpha) \circ (\eth(\beta^*)t^*) + (\mathscr{F}(\beta)t) \circ \eth(\alpha^*) - \alpha \circ (\beta^* t^*) - (\beta t) \circ \alpha^* \in Z(\mathscr{R}).$$

It implies

$$(\mathscr{F}(\alpha) \circ \eth(\beta^*) - \alpha \circ \beta^*)t^* + (\mathscr{F}(\beta) \circ \eth(\alpha^*) - \beta \circ \alpha^*)t \in Z(\mathscr{R})\ \forall\ \alpha, \beta \in I,\ t \in Z(\mathscr{R}). \tag{51}$$

From (49) and (51), we have

$$(\mathscr{F}(\alpha) \circ \eth(\beta^*) - \alpha \circ \beta^*)(t^* - t) \in Z(\mathscr{R})\ \forall\ \alpha, \beta \in I,\ t \in Z(\mathscr{R}).$$

It implies

$$[\mathscr{F}(\alpha) \circ \eth(\beta^*) - \alpha \circ \beta^*, r](t^* - t) = 0\ \forall\ \alpha, \beta \in I,\ t \in Z(\mathscr{R}),\ r \in \mathscr{R}. \tag{52}$$

As $\mathscr{R}$ is a ring with involution of the second kind. So, there exists some $t \in Z(\mathscr{R})$ such that $0 \neq t - t^* \in Sa_*(\mathscr{R}) \cap Z(\mathscr{R})$. Therefore, using Lemma 1 in (52), we conclude $[\mathscr{F}(\alpha) \circ \eth(\beta^*) - \alpha \circ \beta^*, r] = 0\ \forall\ \alpha, \beta \in I,\ , r \in \mathscr{R}$. Replace β by β^* and α by $\alpha\gamma$ to obtain

$$\mathscr{F}(\alpha\gamma) \circ \eth(\beta) - (\alpha\gamma) \circ \beta \in Z(\mathscr{R})\ \forall\ \alpha, \beta, \gamma \in I. \tag{53}$$

As $\alpha, \beta, \gamma \in I$, we have $(\mathscr{F}(\alpha)\gamma + \alpha\eth(\gamma)) \circ \eth(\beta) - (\alpha\gamma) \circ \beta \in I \cap Z(R)$. We divide the prove into following two parts:

Case 1: If $\mathscr{F}(\alpha\gamma) \circ \eth(\beta) - (\alpha\gamma) \circ \beta \neq 0$, then there exists some $0 \neq t \in I \cap Z(\mathscr{R})$. Replacing β by t in (53) and thereby using $\eth(Z(R)) = \{0\}$, we conclude $[\alpha\gamma, r]t = 0\ \forall\ \alpha, \gamma \in I,\ r \in \mathscr{R}$. Replace r by rs, where $s \in \mathscr{R}$ to obtain

$[\alpha\gamma, r]st = 0 \; \forall \; \alpha, \gamma \in I, r, s \in \mathscr{R}$. Replacing β by t^* in (53) and using similar arguments, we conclude $[\alpha\gamma, r]st^* = 0 \; \forall \; \alpha, \gamma \in I, \; r, s \in \mathscr{R}$. From ∗-primeness of $\mathscr{R}$, we have $[\alpha\gamma, r] = 0 \; \forall \; \alpha, \gamma \in I, \; r \in \mathscr{R}$. It implies $[\alpha, r]\gamma + \alpha[\gamma, r] = 0$. Replacing α by $s\alpha$, where $s \in \mathscr{R}$, we conclude $[s, r]\alpha\gamma = 0$. Since I is ∗-Ideal. It implies $[s, r]\alpha\gamma^* = 0 \; \forall \; \alpha, \gamma \in I, \; r, s \in \mathscr{R}$. From ∗-primeness of $\mathscr{R}$ and [7, Lemma 1], we conclude $\mathscr{R}$ is commutative or $I = 0$. Latter case is not possible. Thus, we have the result.

Case 2: If $\mathscr{F}(\alpha\gamma) \circ \eth(\beta) - (\alpha\gamma) \circ \beta = 0$, then we have

$$(\mathscr{F}(\alpha)\gamma + \alpha\eth(\gamma)) \circ \eth(\beta) - ((\alpha\gamma) \circ \beta) = 0 \; \forall \; \alpha, \beta, \gamma \in I. \tag{54}$$

It implies that

$$(\mathscr{F}(\alpha) \circ \eth(\beta))\gamma + \mathscr{F}(\alpha)[\gamma, \eth(\beta)] + (\alpha \circ \eth(\beta))\eth(\gamma) + \alpha[\eth(\gamma), \eth(\beta)] - (\alpha \circ \beta)\gamma - \alpha[\gamma, \beta] = 0. \tag{55}$$

Replace γ by γs, where $s \in \mathscr{R}$, in above equation to obtain

$$\begin{aligned}(\mathscr{F}(\alpha) \circ \eth(\beta))\gamma s + \mathscr{F}(\alpha)[\gamma, \eth(\beta)]s + \mathscr{F}(\alpha)\gamma[s, \eth(\beta)] + (\alpha \circ \eth(\beta))\eth(\gamma)s \\ +(\alpha \circ \eth(\beta))\gamma\eth(s) + \alpha[\eth(\gamma), \eth(\beta)]s + \alpha\eth(\gamma)[s, \eth(\beta)] + \alpha\gamma[\eth(s), \eth(\beta)] \\ +\alpha[\gamma, \eth(\beta)]\eth(s) - (\alpha \circ \beta)\gamma s - \alpha[\gamma, \beta]s - \alpha\gamma[s, \beta] \\ = 0 \; \forall \; \alpha, \beta, \gamma \in I, \; s \in \mathscr{R}.\end{aligned} \tag{56}$$

Using (55) in (56), we conclude

$$\begin{aligned}\mathscr{F}(\alpha)\gamma[s, \eth(\beta)] + (\alpha \circ \eth(\beta))\gamma\eth(s) + \alpha\eth(\gamma)[s, \eth(\beta)] + \alpha\gamma[\eth(s), \eth(\beta)] \\ +\alpha[\gamma, \eth(\beta)]\eth(s) - \alpha\gamma[s, \beta] = 0 \; \forall \; \alpha, \beta, \gamma \in I, \; s \in \mathscr{R}.\end{aligned}$$

Replacing s by $\eth(\beta)$, we compute

$$(\alpha \circ \eth(\beta))\gamma\eth^2(\beta) + \alpha\gamma[\eth^2(\beta), \eth(\beta)] + \alpha[\gamma, d(\beta)]\eth^2(\beta) - \alpha\gamma[\eth(\beta), \beta] = 0. \tag{57}$$

Replacing α by $s\alpha$, where $s \in \mathscr{R}$, in (57) and comparing with it, we conclude

$$[s, \eth(\beta)]\alpha\gamma\eth^2(\beta) = 0. \tag{58}$$

Replace β by h, where $h \in H(\mathscr{R}) \cap I$, in (58) to obtain

$$[s, \eth(h)]\alpha\gamma\eth^2(h) = 0 \; \forall \; s \in \mathscr{R}, \; \alpha, \gamma \in I, \; h \in H(\mathscr{R}) \cap I. \tag{59}$$

Replacing s by s^* and using $\eth(h)^* = \eth(h)$ in above equation, we conclude

$$[s, \eth(h)]^*\alpha\gamma\eth^2(h) = 0 \; \forall \; s \in \mathscr{R}, \; \alpha, \gamma \in I, \; h \in H(\mathscr{R}) \cap I. \tag{60}$$

For any fixed h, from (59) and (60), $*$-primeness of $\mathscr{R}$, [7, Lemma 1], we conclude either $[s, \mathfrak{d}(h)] = 0\ \forall s \in \mathscr{R}$ or $\mathfrak{d}^2(h) = 0$. Let $\mathscr{U} = \{h \in H(\mathscr{R}) \cap I : [\mathscr{R}, h] = \{0\}\}$, $\mathscr{V} = \{h \in H(\mathscr{R}) \cap I : \mathfrak{d}^2(h) = 0\}$. We note that $H(\mathscr{R}) \cap I$ can be written as the set-theoretic union of the additive subgroups $\mathscr{U}$ and $\mathscr{V}$, which is not possible. Thus, either $H(\mathscr{R}) \cap I = \mathscr{U}$ or $H(\mathscr{R}) \cap I = \mathscr{V}$. We have $\mathfrak{d}(h) \in Z(\mathscr{R})\ \forall\ h \in H(\mathscr{R}) \cap I$ or $\mathfrak{d}^2(h) = 0\ \forall\ h \in I \cap H(\mathscr{R})$. In former case, we have $\mathfrak{d}^2(h) = 0$ as $\mathfrak{d}$ is vanishing on center. Thus, we have $\mathfrak{d}^2(h) = 0\ \forall\ h \in H(\mathscr{R}) \cap I$. Replacing β by k, where $k \in S(\mathscr{R}) \cap I$, in (58) and using similar arguments, we have $\mathfrak{d}^2(k) = 0\ \forall\ k \in S(\mathscr{R}) \cap I$. For any $\alpha \in I$, we have $2\alpha = \alpha + \alpha^* + \alpha - \alpha^*$. It implies $\mathfrak{d}^2(2\alpha) = 0$. That is, $\mathfrak{d}^2(I) = 0$. Using [7, Lemma 4], we have $\mathfrak{d} = 0$, which is a contradiction. If

$$[\mathscr{F}(\alpha) \circ \beta^* + \beta \circ \mathfrak{d}(\alpha^*), r] = 0\ \forall\ \alpha, \beta \in I,\ r \in \mathscr{R}. \tag{61}$$

Replacing β by βt in (61), where $t \in Z(\mathscr{R})$, we find

$$[(\mathscr{F}(\alpha) \circ \beta^*)t^* + (\beta \circ \mathfrak{d}(\alpha^*))t, r] = 0. \tag{62}$$

From (61) and (62), we conclude

$$[\mathscr{F}(\alpha)\beta + \beta\mathscr{F}(\alpha), r](t - t^*) = 0\ \forall\ \alpha, \beta \in I,\ r \in \mathscr{R},\ t \in Z(\mathscr{R}). \tag{63}$$

As $\mathscr{R}$ is a ring with involution of the second kind. So, there exists some $t \in Z(\mathscr{R})$ such that $0 \neq t - t^* \in Sa_*(\mathscr{R}) \cap Z(\mathscr{R})$. Using Lemma 1 in (63), we get

$$[\mathscr{F}(\alpha) \circ \beta, r] = 0\ \forall\ \alpha, \beta \in I,\ r \in \mathscr{R}. \tag{64}$$

Replacing β by α^* and using Theorem 1, we conclude $\mathscr{R}$ is commutative as desired. Now (2)$\Rightarrow$ (3). From given condition, we have

$$\mathscr{F}(\alpha) \circ \mathfrak{d}(\alpha^*) + \alpha \circ \alpha^* \in Z(\mathscr{R})\ \forall\ \alpha \in I. \tag{65}$$

Polarize (65) to obtain

$$\mathscr{F}(\alpha) \circ \mathfrak{d}(\beta^*) + \mathscr{F}(\beta) \circ \mathfrak{d}(\alpha^*) + \alpha \circ \beta^* + \beta \circ \alpha^* \in Z(\mathscr{R})\ \forall\ \alpha, \beta \in I. \tag{66}$$

Replacing β by βh_c, where $h_c \in Z(\mathscr{R}) \cap H(\mathscr{R})$, in (66) and using it, we get

$$(\mathscr{F}(\alpha) \circ \beta^* + \beta \circ \mathfrak{d}(\alpha^*))\mathfrak{d}(h_c) \in Z(\mathscr{R}). \tag{67}$$

Since (67) is same as (50), so using arguments, we obtain $\mathscr{R}$ is commutative.

Theorem 3 *Let $\mathscr{R}$ be a 2-torsion free $*$-prime ring with involution of the second kind. If $\mathscr{F} : \mathscr{R} \to \mathscr{R}$ is a generalized derivation associated with a derivation $\mathfrak{d}$ such that $\mathfrak{d}$ is commuting with $*$ and I is a nonzero $*$-ideal of $\mathscr{R}$, then the following assertions are equivalent:*

(1) $[\mathscr{F}(\alpha), \mathfrak{d}(\alpha^*)] - \alpha \circ \alpha^* \in Z(\mathscr{R})\ \forall\ \alpha \in I$;
(2) $[\mathscr{F}(\alpha), \mathfrak{d}(\alpha^*)] + \alpha \circ \alpha^* \in Z(\mathscr{R})\ \forall\ \alpha \in I$;
(3) $\mathscr{R}$ *is commutative.*

Proof It is obvious that if $\mathscr{R}$ is commutative, then (1) and (2) hold. We need to prove (1)$\Rightarrow$ (3) and (2)$\Rightarrow$ (3). If $\mathscr{F} = 0$, then our theorem is proved by Lemma 4. So, we assume that $\mathscr{F} \neq 0$. First, we prove (1)$\Rightarrow$ (3). We have

$$[\mathscr{F}(\alpha), \mathfrak{d}(\alpha^*)] - \alpha \circ \alpha^* \in Z(\mathscr{R})\ \forall\ \alpha \in I. \tag{68}$$

Polarize (68) to obtain

$$[\mathscr{F}(\alpha), \mathfrak{d}(\beta^*)] + [\mathscr{F}(\beta), \mathfrak{d}(\alpha^*)] - \alpha \circ \beta^* - \beta \circ \alpha^* \in Z(\mathscr{R})\ \forall\ \alpha, \beta \in I. \tag{69}$$

Replacing β by βh_c, where $h_c \in H(\mathscr{R}) \cap Z(\mathscr{R})$, in (69) and using it, we conclude

$$([\mathscr{F}(\alpha), \beta^*] + [\beta, \mathfrak{d}(\alpha^*)])\mathfrak{d}(h_c) \in Z(\mathscr{R}).$$

It implies

$$[[\mathscr{F}(\alpha), \beta^*] + [\beta, \mathfrak{d}(\alpha^*)], r]\mathfrak{d}(h_c) = 0\ \forall\ \alpha, \beta \in I,\ r \in \mathscr{R}\ , h_c \in H(\mathscr{R}) \cap Z(\mathscr{R}). \tag{70}$$

Since $\mathfrak{d}(h_c)^* = \mathfrak{d}(h_c^*) = \mathfrak{d}(h_c)$. Thus, $\mathfrak{d}(h_c) \in Sa_*(\mathscr{R}) \cap Z(\mathscr{R})$. Using Lemma 1, we find either $\mathfrak{d}(h_c) = 0\ \forall\ h_c \in H(\mathscr{R}) \cap Z(\mathscr{R})$ or $[[\mathscr{F}(\alpha), \beta^*] + [\beta, \mathfrak{d}(\alpha^*)], r] = 0\ \forall\ \alpha, \beta \in I,\ r \in \mathscr{R}$. If $\mathfrak{d}(h_c) = 0\ \forall\ h_c \in H(\mathscr{R}) \cap Z(\mathscr{R})$. From Lemma 5, we have $\mathfrak{d}(t) = 0\ \forall\ t \in Z(\mathscr{R})$. Replace β by βt, where $t \in Z(\mathscr{R})$, in (70) to obtain

$$[[\mathscr{F}(\alpha), \beta^*]t^* + [\beta, \mathfrak{d}(\alpha^*)]t, r] = 0\ \forall\ \alpha, \beta \in I,\ r \in \mathscr{R},\ t \in Z(\mathscr{R}). \tag{71}$$

Comparing (70) and (71), we conclude

$$[[\mathscr{F}(\alpha), \beta^*], r](t^* - t) = 0\ \forall\ \alpha, \beta \in I,\ r \in \mathscr{R},\ t \in Z(\mathscr{R}). \tag{72}$$

As $\mathscr{R}$ is a ring with involution of second kind. So, there exists some $t \in Z(\mathscr{R})$ such that $0 \neq t^* - t \in Sa_*(\mathscr{R}) \cap Z(\mathscr{R})$. Using Lemma 1 in (72), we conclude $[[\mathscr{F}(\alpha), \beta^*], r] = 0\ \forall\ \alpha, \beta \in I,\ r \in \mathscr{R}$. Replace β by β^* to obtain

$$[\mathscr{F}(\alpha), \beta] \in Z(\mathscr{R})\ \forall\ \alpha, \beta \in I. \tag{73}$$

From (73) and Lemma 6, we conclude $\mathscr{R}$ is commutative.
If we have

$$[[\mathscr{F}(\alpha), \beta^*] + [\beta, \mathfrak{d}(\alpha^*)], r] = 0\ \forall\ \alpha, \beta \in I,\ r \in \mathscr{R}. \tag{74}$$

Replacing β by $-\beta k_c$, where $k_c \in S(\mathscr{R}) \cap Z(\mathscr{R})$, in (74) and using it, we conclude

$$[[\mathscr{F}(\alpha), \beta^*] - [\beta, \eth(\alpha^*)], r]k_c = 0. \tag{75}$$

As $0 \neq k_c \in Sa_*(\mathscr{R}) \cap Z(\mathscr{R})$, using Lemma 1 in (75), we conclude

$$[[\mathscr{F}(\alpha), \beta^*] - [\beta, \eth(\alpha^*)], r] = 0 \ \forall \ \alpha, \beta \in I, \ r \in \mathscr{R}. \tag{76}$$

Adding (74) and (76), we obtain

$$2[[\mathscr{F}(\alpha), \beta^*], r] = 0 \ \forall \ \alpha, \beta \in I, \ r \in \mathscr{R}. \tag{77}$$

From (77), 2-torsion free condition and Lemma 6, we conclude $\mathscr{R}$ is commutative as desired.
(2) $\Rightarrow$ (3)
From given condition, we have

$$[\mathscr{F}(\alpha), \eth(\alpha^*)] - \alpha \circ \alpha^* \in Z(\mathscr{R}) \ \forall \ \alpha \in I. \tag{78}$$

Polarize (78) to get

$$[\mathscr{F}(\alpha), \eth(\beta^*)] + [\mathscr{F}(\beta), \eth(\alpha^*)] - \alpha \circ \beta^* - \beta \circ \alpha^* \in Z(\mathscr{R}) \ \forall \ \alpha, \beta \in I. \tag{79}$$

Replacing β by βh_c, where $h_c \in H(\mathscr{R}) \cap Z(\mathscr{R})$, in (79) and using it, we get

$$([\mathscr{F}(\alpha), \beta^*] + [\beta, \eth(\alpha^*)])\eth(h_c) \in Z(\mathscr{R}). \tag{80}$$

Since (80) is same as (70), so using same arguments, we obtain $\mathscr{R}$ is commutative.

Theorem 4 *Let $\mathscr{R}$ be a 2-torsion free $*$-prime ring with involution of second kind and $J \not\subset Z(\mathscr{R})$ be a $*$-Jordan ideal of $\mathscr{R}$. Then J contains a nonzero $*$-ideal of $\mathscr{R}$.*

Proof From [10, Lemma 2.1], we have $[[a, b], r] \in J$ for any $a, b \in J$, $r \in \mathscr{R}$. For some $c \in J$, we find $[a, b]cr - cr[a, b] = [a, b]cr + c[a, b]r - c[a, b]r - cr[a, b] = c[[a, b], r] + [[a, b], c]r \in J$. By [11, Lemma 2.4], we have $-2c[ba, r] \in J$ and $2c[ab, r] \in J \ \forall \ a, b, c \in J, \ r \in \mathscr{R}$. Adding these equations, we obtain $2c[a, b], r] \in J$. Using $c[[a, b], r] + [[a, b], c]r \in J$, we conclude $2[[a, b], c]r \in J \ \forall \ a, b, c \in J, \ r \in \mathscr{R}$. Using definition of Jordan ideal, for any $s \in R$, we obtain $2[[a, b], c]rs + 2s[[a, b], c]r \in J$, where $a, b, c \in J$, $r, s \in \mathscr{R}$. It implies $2\mathscr{R}[[J, J], J]\mathscr{R} \subseteq J$. Clearly $(2\mathscr{R}[[J, J], J]\mathscr{R})^* = 2\mathscr{R}[[J, J], J]\mathscr{R}$. Thus, $2\mathscr{R}[[J, J], J]\mathscr{R}$ is a $*$-ideal in J. Now, we will show that $2\mathscr{R}[[J, J], J]\mathscr{R} \neq \{0\}$. If $2\mathscr{R}[[J, J], J]\mathscr{R} = \{0\}$, i.e., $\mathscr{R}[[J, J], J]\mathscr{R}[[J, J], J] = \{0\}$. As $([[J, J], J])^* = [[J, J], J]$, $*$-primeness of $\mathscr{R}$ forces $\mathscr{R}[[J, J], J] = \{0\}$. Again by using $*$-primness of $\mathscr{R}$, we conclude $[[J, J], J] = \{0\}$, i.e., $[[a, b], c] = 0 \ \forall \ a, b, c \in J$. Now, replacing c by $2c[r, s]$, where $r, s \in \mathscr{R}$, we get $c[[a, b], [r, s]] = 0 \ \forall \ a, b, c \in J, \ r, s \in \mathscr{R}$. Now replacing c by $c \circ r$, where $r \in \mathscr{R}$, we get $cr[[a, b], [r, s]] = 0 \ \forall \ a, b, c \in J, \ r, s \in \mathscr{R}$. Replacing c by c^* gives $c^*r[[a, b], [r, s]] = 0$. From the last two equations, $*$-primeness of $\mathscr{R}$, we conclude

$[[a, b], [r, s]] = 0 \ \forall\, a, b \in J,\ r, s \in \mathscr{R}$. Replacing r by rs, where $s \in \mathscr{R}$, we reach at

$$[r, s][[a, b], s] = 0 \ \forall\, a, b \in J,\ r, s \in \mathscr{R}. \tag{81}$$

Replacing r by rt, where $t \in \mathscr{R}$, we have $[r, s]t[[a, b], s] = 0$. Replacing r by h and s by k, where $h \in H(\mathscr{R})$ and $k \in S(\mathscr{R})$, implies $[h, k]t[[a, b], k] = 0$. As $[h, k]^* = [h, k]$, so last equation can also can be written as $[h, k]^* t[[a, b], k] = 0 \ \forall\, a, b \in J,\ t \in \mathscr{R},\ h \in H(\mathscr{R}),\ k \in S(\mathscr{R})$. For any fixed $k \in S(\mathscr{R})$, application of $*$-primeness in last two equations yields to either $[h, k] = 0 \ \forall\, h \in H(\mathscr{R})$ or $[[a, b], k] = 0 \ \forall a, b \in J$. Let $\mathscr{U} = \{k \in S(\mathscr{R}) : \ [H(\mathscr{R}), k] = \{0\}\}$ $\mathscr{V} = \{k \in S(\mathscr{R}) : \ [[J, J], k] = \{0\}\}$. We note that $S(\mathscr{R})$ can be written as the set-theoretic union of the additive subgroups $\mathscr{U}$ and $\mathscr{V}$, which is not possible. Thus, either $S(\mathscr{R}) = \mathscr{U}$ or $S(\mathscr{R}) = \mathscr{V}$. Thus, we get either $[H(\mathscr{R}), k] = \{0\} \ \forall\, k \in S(\mathscr{R})$ or $[[J, J], k] = \{0\} \ \forall\, k \in S(\mathscr{R})$. Now if $[h, k] = 0 \ \forall\, h \in H(\mathscr{R}),\ k \in S(\mathscr{R})$. Replace h by $r + r^*$ and k by $r - r^*$, where $r \in \mathscr{R}$, to get $2[r, r^*] = 0 \ \forall r \in \mathscr{R}$. Using torsion free condition with Lemma 3, we find $\mathscr{R}$ is a commutative ring that implies $J \subseteq Z(\mathscr{R})$, i.e., is a contradiction. So we left with $[[J, J], k] = \{0\} \ \forall\, k \in S(\mathscr{R})$. Replace r by k and s by h, where $k \in S(\mathscr{R})$ and $h \in H(\mathscr{R})$, in (81) and use similar arguments to obtain $[[J, J], h] = \{0\} \ \forall\, h \in H(\mathscr{R})$. Further, the last two equations imply $[[J, J], \mathscr{R}] = \{0\}$. By [9, Lemma 5], we get $\mathscr{R}$ is commutative, which is a contradiction. So $2\mathscr{R}[[J, J], J]\mathscr{R}$ is required nonzero $*$-ideal.

Corollary 1 *Let $\mathscr{R}$ be a $*$-prime ring with involution of second kind. If $\mathscr{F}$ is a generalized derivation associated with a derivation $\mathfrak{d}$ such that $\mathfrak{d}$ is commuting with $*$, and J is nonzero $*$-Jordan ideal of $\mathscr{R}$, then the following assertions are equivalent:*

(1) $[\mathscr{F}(\alpha), \alpha^*] \in Z(\mathscr{R}) \ \forall\, \alpha \in J$.
(2) $\mathscr{F}(\alpha) \circ (\alpha^*) \in Z(\mathscr{R}) \ \forall\, \alpha \in J$.
(3) $\mathscr{F}(\alpha) \circ \mathfrak{d}(\alpha^*) \pm \alpha \circ \alpha^* \in Z(\mathscr{R}) \ \forall\, \alpha \in J$.
(4) $[\mathscr{F}(\alpha), \mathfrak{d}(\alpha^*)] \pm \alpha \circ \alpha^* \in Z(\mathscr{R}) \ \forall\, \alpha \in J$.
(5) *$\mathscr{R}$ is commutative.*

4 Examples

Example 2 Let us consider $\mathscr{R} = \left\{ \begin{pmatrix} a & b \\ c & d \end{pmatrix} | a, b, c, d \in \mathscr{Z} \right\}$, where $\mathscr{Z}$ is the ring of integers. Define mappings $*, \mathfrak{d}, \mathscr{F} : \mathscr{R} \to \mathscr{R}$ as $\begin{pmatrix} a & b \\ c & d \end{pmatrix}^* = \begin{pmatrix} d & -b \\ -c & a \end{pmatrix}$, $\mathscr{F}\begin{pmatrix} a & b \\ c & d \end{pmatrix} = \begin{pmatrix} 0 & -b \\ c & 0 \end{pmatrix}$ and $\mathfrak{d}\begin{pmatrix} a & b \\ c & d \end{pmatrix} = \begin{pmatrix} 0 & -b \\ c & 0 \end{pmatrix}$. Then it is easy to verify that $*$ is an involution of the first kind, $\mathscr{F}$ is a generalized derivation and $\mathfrak{d}$ is a derivation of a noncommutative $*$-prime ring $\mathscr{R}$. Clearly, one can see that $\mathscr{F}(\alpha) \circ \mathfrak{d}(\alpha^*) - \alpha \circ \alpha^* \in Z(\mathscr{R}) \ \forall\, \alpha \in \mathscr{R}$; $\mathscr{F}(\alpha) \circ \mathfrak{d}(\alpha^*) + \alpha \circ \alpha^* \in Z(\mathscr{R}) \ \forall\, \alpha \in \mathscr{R}$; $[\mathscr{F}(\alpha), \mathfrak{d}(\alpha^*)] -$

$\alpha \circ \alpha^* \in Z(\mathscr{R})\ \forall\ \alpha \in \mathscr{R}$; $[\mathscr{F}(\alpha), \mathfrak{d}(\alpha^*)] + \alpha \circ \alpha^* \in Z(\mathscr{R})\ \forall\ \alpha \in \mathscr{R}$. It shows the importance of involution of the second kind in our theorems.

Example 3 Let $\mathscr{R}$ be a ring with involution $*, \mathscr{F}, \mathfrak{d}$ same as in Example 2 and $\mathscr{C}$ be the field of complex numbers. Consider the set $\mathscr{R} = \mathscr{R} \times \mathscr{C}$. We define mappings $\sigma, \mathscr{F}', \mathfrak{d}' : \mathscr{R} \to \mathscr{R}$ by $(r, z)^\sigma = (r^*, \bar{z})$, $\mathscr{F}'(r, z) = (\mathscr{F}(r), 0)$, $\mathfrak{d}'(r, z) = (\mathfrak{d}(r), 0)\ \forall\ (r, z) \in \mathscr{R} \times \mathscr{C}$. It is straight forward to check that σ is involution of second kind, $\mathscr{F}'$ is a generalized derivation, $\mathfrak{d}'$ is a derivation and $\mathscr{R}$ is a semiprime ring. It can be seen that $\mathscr{F}(\alpha) \circ \mathfrak{d}(\alpha^*) - \alpha \circ \alpha^* \in Z(\mathscr{R})\ \forall\ \alpha \in \mathscr{R}$; $\mathscr{F}(\alpha) \circ \mathfrak{d}(\alpha^*) + \alpha \circ \alpha^* \in Z(\mathscr{R})\ \forall\ \alpha \in \mathscr{R}$; $[\mathscr{F}(\alpha), \mathfrak{d}(\alpha^*)] - \alpha \circ \alpha^* \in Z(\mathscr{R})\ \forall\ \alpha \in \mathscr{R}$; $[\mathscr{F}(\alpha), \mathfrak{d}(\alpha^*)] + \alpha \circ \alpha^* \in Z(\mathscr{R})\ \forall\ \alpha \in \mathscr{R}$, which exhibits the importance of $*$-primeness in our theorems.

Acknowledgements The authors are greatly indebted to the referees for their valuable suggestions and careful reading of the article.

References

1. Ali, S., Dar, N.A.: On ∗-centralizing mappings in ring with involution. Georgian Math. J. **21**(1), 25–28 (2014)
2. Brešar, M.: On the distance of composition of two derivation to the generalized derivation. Glasgow Math. J. **33**, 89–93 (1991)
3. Mamouni, A., Oukhtite, L., Nejjar, B., Al Jaraden, J.J.: Some commutativity criteria for prime rings with differential identities on Jordan ideals. Commun. Algebra **47**(1), 355–361 (2019)
4. Nejjar, B., Kacha, A., Mamouni, A., Oukhite, L.: Commutativity theorems in ring with involution. Commun. Algebra **45**(2), 698–708 (2017)
5. Oukhtite, L., Salhi, S.: On commutativity of ∗-prime rings. Glasnik Matematički **41**(1), 57–64 (2006)
6. Oukhtite, L., Salhi, S.: Derivation as homomorphisms and antihomomorphisms in ∗-prime rings. Algebras Groups Geom. **23**(1), 67–72 (2006)
7. Oukhtite, L., Salhi, S.: Derivations and commutativity of ∗-prime rings. Int. J. Contemp. Math. Sci. **1**(9), 439–448 (2006)
8. Oukhtite, L., Salhi, S.: On derivations in ∗-prime rings. Int. J. Algebra **1**(5), 241–246 (2007)
9. Oukhtite, L., Mamouni, A., Mohammad, A.: Commutativity theorems for rings with differential identities on Jordan ideals. Comment. Math. Univ. Carol. **54**(4), 447–457 (2013)
10. Sandhu, G.S., Kumar, D.: A note on derivations and Jordan ideals of prime rings. AIMS Math. **2**(4), 580–585 (2017)
11. Zaidi, S.M.A., Ashraf, M., Ali, S.: On Jordan ideals and left (θ, θ)−derivations in prime rings. Int. J. Math. Sci. **37**, 1957–1964 (2004)

Local Subsemigroups and Variants of Some Classes of Semigroups

Siji Michael and P. G. Romeo

Abstract For an element a in a semigroup S, the local subsemigroup of S with respect to a is the subsemigroup aSa of S. Here we study the structure of local subsemigroups of full transformation semigroups and symmetric inverse monoids. We obtain some results regarding the local subsemigroups and when they are isomorphic to the semigroup itself. Further it is also shown that the set of all local subsemigroups of all finite symmetric inverse monoids and the set of all variants of all finite symmetric inverse monoids is same up to isomorphism.

Keywords Transformation semigroups · Local subsemigroups · Variants

Local subsemigroups and semigroup variants are two well-known constructions in semigroups. In [6], James East studied the link between these two and it is shown that in the case of full transformation semigroup on a set X, the two constructions lead exactly to the same class of semigroups up to isomorphism. In this paper, we discuss the structure of local subsemigroups of finite full transformation semigroups and symmetric inverse monoids. The structure studies are carried out using the egg-box diagrams obtained with the semigroups package (cf. [10]) for GAP (cf. [11]) and (cf. [3, 9]).

1 Preliminaries

In the following we briefly recall some basic notions and results concerning finite transformation semigroups and symmetric inverse monoids. A semigroup S is a nonempty set together with an associative binary operation. An element $x \in S$ is regular if $xyx = x$ and $yxy = y$ for some $y \in S$ and a semigroup S is called regular if all elements of S are regular. An element $x \in S$ is called an idempotent if $x^2 = x$. The collection of all idempotents in S will be denoted by $E(S)$. Two elements of a

S. Michael · P. G. Romeo (✉)
Department of Mathematics, Cochin University of Science and Technology, Kochi, Kerala, India
e-mail: romeopg@cusat.ac.in

M. Ashraf et al. (eds.), *Algebra and Related Topics with Applications*,
Springer Proceedings in Mathematics & Statistics 392,
https://doi.org/10.1007/978-981-19-3898-6_16

semigroup S are said to be $\mathscr{L}$, $\mathscr{R}$, $\mathscr{J}$-equivalent if they generate the same principal left, right, two-sided ideals, respectively.

The join of the equivalence relations $\mathscr{L}$ and $\mathscr{R}$ is denoted by $\mathscr{D}$ and their intersection by $\mathscr{H}$. These equivalence relations are introduced by J.A. Green and are known as Green's relations and are of fundamental importance in the study of the structure of semigroups. The egg-box diagram visualizes $\mathscr{D}$-class structure of semigroup S using rectangular patterns. In each rectangular pattern (which corresponds to each $\mathscr{D}$-class), the rows correspond to the $\mathscr{R}$-classes and the columns to $\mathscr{L}$-classes contained in a $\mathscr{D}$-class.

Definition 1 (*Definition 1.2; cf.* [6]) Let S be a semigroup and a an element of S. The set $aSa = \{axa : x \in S\}$ is a subsemigroup of S called local subsemigroup of S with respect to a.

Definition 2 (*Definition 1.1; cf.* [8]) Let S be a semigroup and a be an element of S. An associative sandwich operation $\star_a$ can be defined on S by $x \star_a y = xay$ for all $x, y \in S$. The semigroup $(S, \star_a)$ is called the variant of S with respect to a and is denoted as S^a.

For a semigroup S and $a \in S$ is invertible, then the variant of S with respect to a, $S^a \cong S$ (cf. [7]). The variants of full transformation semigroups and semigroup of binary relations are widely studied (cf. [2, 5, 12, 13]).

For a finite set X with $|X| = n$, the set of all transformations of X (that is, all functions $X \to X$), under the operation of composition of maps is the full transformation semigroup on X and is denoted as T_X (also denoted as T_n). It is well known that T_X is a regular semigroup. For $f \in T_X$, the image and rank of f will be denoted by

$$im(f) = \{f(x) : x \in X\}$$

$$rank(f) = |im(f)|.$$

Symmetric inverse monoid on a finite set X is the set of all partial bijections on X (that is, all bijections from a subset of X to a subset of X) with composition of maps as the binary operation and is written as IS_X. The domain and range of a partial permutation α is denoted as $dom\alpha$ and $ran\alpha$ respectively. We denote the rank of empty partial permutation as zero. Idempotents of IS_X are the identity mappings on each of the subsets of X, i.e.,

$$E(IS_X) = \{1_A : A \subseteq X\}.$$

2 Local Subsemigroups and Variants of Full Transformation Semigroups

In this section we discuss the local subsemigroups of full transformation semigroup on a finite set X and obtain some results connecting local subsemigroups and variants. It is observed that there is an explicit connection between local subsemigroups and variants of full transformation semigroups as stated in the following theorem.

Theorem 1 (Theorem 1.4; cf. [6]) *Let n be a positive integer and let $a \in T_n$ with $rank(a) = r$. Then*

1. *$aT_na \cong T_r^c$ for some $c \in T_r$ with $rank(c) = rank(a^2)$.*
2. *$T_n^a \cong bT_{2n-r}b$ for some $b \in T_{2n-r}$ with $rank(b) = n$.*

Proposition 1 *Let X be any set with $|X| = n$ and let $\alpha \in T_X$ with $rank(\alpha) = m \leq n$. Then $\alpha T_X \alpha$ is a local subsemigroup of T_X with respect to α and*

$$|\alpha T_X \alpha| = |T_m|.$$

Proof From Theorem 1, $\alpha T_n \alpha \cong T_m^c$ with $rank(c) = rank(\alpha^2)$. Since $\alpha T_n \alpha$ is a variant of T_m, it contains as many elements as in T_m.

Corollary 1 *If $\alpha \in T_X$ with $|X| = n$ and $rank(\alpha) = n$ then $\alpha T_X \alpha$ is same as T_X.*

Proof Since α is a permutation, every element β of T_X is equal to $\alpha(\alpha^{-1}\beta\alpha^{-1})\alpha$. Hence $\alpha T_X \alpha = T_X$.

Comparing egg-box diagrams of local subsemigroups $\alpha T_X \alpha$ with $rank(\alpha) = m$, it can be seen that there are different structures available such as full transformation semigroup of order m and variants of full transformation semigroup of order m.

Definition 3 (*Definition 3.1; cf.* [1]) For $\alpha \in T_X$ we can define the stable image of α denoted as $sim(\alpha)$ by

$$sim(\alpha) = \{x \in X : x \in im(\alpha^n) \quad for \quad every \quad n \geq 0\}.$$

Definition 4 (*Definition 3.3; cf.* [1]) For $\alpha \in T_X$ we define the stabilizer of α as the smallest positive integer $s \geq 0$ such that $\mathrm{im}(\alpha^s) = \mathrm{im}(\alpha^{s+1})$.

From the above two definitions it is clear that if α has the stabilizer s, then $sim(\alpha) = im(\alpha^s)$. Further it is known that the local subsemigroups of full transformation semigroups can be classified using stabilizer and stable image of transformations. By comparing the egg-box diagrams of local subsemigroups of finite full transformation semigroups, we obtain the following results.

Proposition 2 *Let $\alpha \in T_n$ with rank(α)= $m \leq n$ and stabilizer of α is 1. Then $\alpha T_n \alpha$ is isomorphic to T_m.*

Proof From Theorem 1, $\alpha T_n\alpha \cong T_m^c$ with $rank(c) = rank(\alpha^2)$. Since stabilizer of α is 1, $rank(\alpha^2) = rank(\alpha) = m$. c being a permutation in T_m, $T_m^c \cong T_m$. Hence, $\alpha T_n\alpha \cong T_m$.

Example 1 Consider transformation $\alpha = \begin{pmatrix} 1 & 2 & 3 & 4 \\ 2 & 4 & 3 & 2 \end{pmatrix} \in T_4$ which is denoted as (2432). Then $rank(\alpha)$ is 3 and $\alpha^2 = (4234)$. Now $rank(\alpha) = rank(\alpha^2) = 3$ and we get the stabilizer of α is 1. Then by Proposition 2, local subsemigroup of α is isomorphic to T_3 (see Fig. 1).

Fig. 1 Egg-box diagram of local subsemigroup when α = (2432)

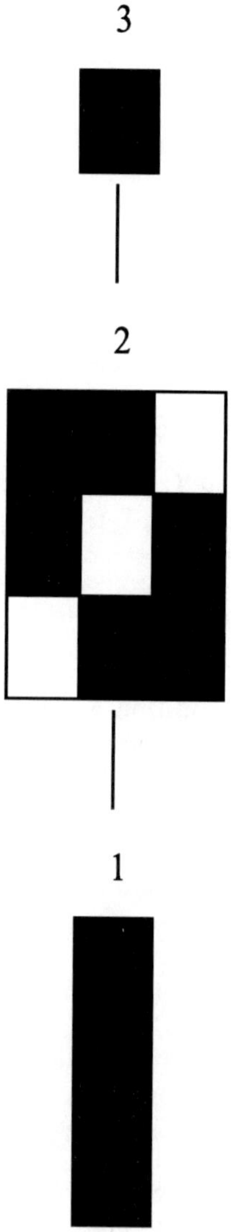

3 Variants of Symmetric Inverse Monoids

In the following, we discuss certain results regarding variants of symmetric inverse monoids and their improvements (cf. [4]) with some advances. Let $X = \{1, 2, \ldots, n\}$ and IS_n denotes symmetric inverse monoid on X. For an element $\alpha \in IS_n$, variant of IS_n with respect to α is denoted as IS_n^α. Now we characterize Green's relations of variants of symmetric inverse monoids on X. Green's relations $\mathscr{R}, \mathscr{L}, \mathscr{J}, \mathscr{H}$ and $\mathscr{D}$ on IS_n^α will be denoted by $\mathscr{R}^\alpha, \mathscr{L}^\alpha, \mathscr{J}^\alpha, \mathscr{H}^\alpha$ and $\mathscr{D}^\alpha$, respectively. Consider the following sets. Let $\alpha \in IS_n$

$$P_1 = \{f \in IS_n : f\alpha \ \ \mathscr{R} \ \ f\}$$

$$P_2 = \{f \in IS_n : \alpha f \ \ \mathscr{L} \ \ f\}$$

and $P = P_1 \cap P_2$.

For an element $\alpha \in IS_n$ there exists an element $\beta \in IS_n$ such that $\alpha\beta$ is an idempotent. As noted earlier, we get $IS_n^\alpha \cong IS_n^{\alpha\beta}$. Since all the idempotents of IS_n are identity mappings on subsets of X, we have α to be $1_A, \quad A \subseteq X$ and $|A| = r$.

Proposition 3 *Let $\alpha = 1_A, \quad A \subseteq X$. Then*

1. $P_1 = \{f \in IS_n : f\alpha \ \ \mathscr{R} \ \ f\} = \{f \in IS_n : im(f) \subseteq A\}$;
2. $P_2 = \{f \in IS_n : \alpha f \ \ \mathscr{L} \ \ f\} = \{f \in IS_n : dom(f) \subseteq A\}$;
3. $P = P_1 \cap P_2 = \{f \in IS_n : dom(f) \subseteq A, im(f) \subseteq A\}$.

Proof Let $f \in IS_n$ and $rank(f) = m$. Since f is a partial bijection, f can be considered as a mapping from $B \mapsto C$ for some $B, C \subseteq X$.

1. Now $f \in P_1 \Leftrightarrow f\alpha \ \ \mathscr{R} \ \ f \Leftrightarrow ker(f\alpha) = ker(f)$ $\Leftrightarrow rank(f\alpha) = rank(f)$. It is clear that $rank(f\alpha) = rank(f)$ only when $m \leq r$, which implies $im(f) \subseteq A$.
2. $f \in P_2 \Leftrightarrow \alpha f \ \ \mathscr{L} \ \ f \Leftrightarrow im(\alpha f) = im(f) \Leftrightarrow rank(\alpha f) = rank(f)$. It can be observed that $rank(\alpha f) = rank(f)$ only when $m \leq r$ which implies $dom(f) \subseteq A$.
3. easily follows from (1) and (2).

By Proposition 3.2 of [5], we obtain Green's relations on IS_n^α as follows.

Theorem 2 *If $f \in IS_n^\alpha$, then*

1. $R_f^\alpha = \begin{cases} R_f \cap P_1 & if \ \ f \in P_1 \\ f & if \ \ f \in IS_n \setminus P_1 \end{cases}$
2. $L_f^\alpha = \begin{cases} L_f \cap P_2 & if \ \ f \in P_2 \\ f & if \ \ f \in IS_n \setminus P_2 \end{cases}$
3. $H_f^\alpha = \begin{cases} H_f & if \ \ f \in P \\ f & if \ \ f \in IS_n \setminus P \end{cases}$

4. $D_f^\alpha = \begin{cases} D_f \cap P & if \quad f \in P \\ L_f^\alpha & if \quad f \in P_2 \setminus P_1 \\ R_f^\alpha & if \quad f \in P_1 \setminus P_2 \\ f & if \quad f \in IS_n \setminus (P_1 \cup P_2) \end{cases}$

Remark 1 Here $P = Reg(IS_X^\alpha) = \{f \in IS_n : dom(f) \subseteq A, im(f) \subseteq A\} = IS_A$. That is, the regular elements of the variant of IS_X with respect to $\alpha = 1_A$, $A \subseteq X$ is IS_A

Next result is about the rank of IS_X^α where rank of a finite semigroup is the minimum cardinality of generating set. That is, $rank(S) = min\{|A| :< A >= S\}$. We know that $rank(S_X) = 2$ and $rank(IS_X) = 3$ for $|X| \geq 3$. The next result states about the minimal generating set and the rank for IS_X^α.

Theorem 3 *Let $M = \{f \in IS_X : rank(f) > r\}$. Then $IS_X^\alpha = \langle M \rangle_\alpha$. Further, any generating set for IS_X^α contains M. Consequently, M is the unique minimal generating set IS_X^α and*

$$rank(IS_X^\alpha) = |M| = \sum_{m=r+1}^{n} \binom{m}{k}^2 k!.$$

Proof We assume that $\alpha = 1_A = \left(\begin{smallmatrix} a_i \\ a_i \end{smallmatrix}\right)$ where $i = 1, \ldots, r$. Since $M = \{f \in IS_X : rank(f) > r\}$. We only need to prove that for $1 \leq rank(f) \leq r \quad f \in \langle M \rangle_\alpha$. For this, we choose any $f \in D_m$ where $1 \leq m \leq r$. Let

$$f = \left(\begin{smallmatrix} x_1 & x_2 & \cdots & x_m \\ y_1 & y_2 & \cdots & y_m \end{smallmatrix}\right)$$

where $m \leq r < n$.

We can choose $g = \left(\begin{smallmatrix} x_1 & x_2 & \cdots & x_m & x_{m+1} \\ a_1 & a_2 & \cdots & a_m & x \end{smallmatrix}\right)$ where $x \notin \{a_1, a_2, \ldots a_r\}$, since $r < n$. Then $g \in D_{m+1} \subseteq M$.

Also let h be the permutation that extends the partial map $\left(\begin{smallmatrix} a_1 & a_2 & \cdots & a_m \\ y_1 & y_2 & \cdots & y_m \end{smallmatrix}\right)$, then $h \in S_X \subseteq M$, so we get $f = g\alpha h = g \star_\alpha h$. Hence $f \in < M >_\alpha$. That is, M generates IS_X^α.

Since any $f \in M$ belongs to singleton maximal D^α-classes, any generating set of IS_X^α must contain M. Hence M is the minimal generating set. Now

$$rank(IS_X^\alpha) = |M| = \sum_{m=r+1}^{n} |D_m| = \sum_{m=r+1}^{n} \binom{m}{k}^2 k!$$

where $|D_m| = \binom{m}{k}^2 k!$. (cf. [7]).

4 Local Subsemigroups and Variants of Symmetric Inverse Monoids

In the following, we compare the structures of local subsemigroups and variants of symmetric inverse monoids.

Proposition 4 *For $\alpha \in IS_n$ with rank(α) = m, $|\alpha IS_n\alpha| = |IS_m| = \sum_{k=0}^{m}\binom{m}{k}^2 k!$.*

Proof Let $\beta \in IS_n$. Then

$$dom(\alpha\beta\alpha) \subseteq dom(\alpha)$$

and $ran(\alpha\beta\alpha) \subseteq ran(\alpha)$. Thus $rank(\alpha\beta\alpha) \leq rank(\alpha)$.

Therefore all the elements of $\alpha IS_n\alpha$ will be of rank $\leq$ rank(α). There will be $\binom{m}{k}$ different choices for $dom(\alpha\beta\alpha)$ with rank k and $\binom{m}{k}$ different choices for $ran(\alpha\beta\alpha)$. For each domain and range, there will be $k!$ different bijections. Hence we have $\binom{m}{k}\binom{m}{k}k!$ bijections of rank k. Since the rank can be varied from 0 to m, we get $\alpha IS_X\alpha$ has $\sum_{k=0}^{m}\binom{m}{k}^2 k!$ elements.

Proposition 5 *For $\alpha \in IS_n$ with rank(α) = n, local subsemigroup $\alpha IS_n\alpha$ is isomorphic to IS_n.*

Proof Clearly, $\alpha IS_n\alpha \subseteq IS_n$.

For the reverse inclusion, let $\beta \in IS_n$. Since $rank(\alpha) = n$, $\alpha \in S_n$. Therefore there exists α^{-1} in IS_n such that $\alpha^{-1}\beta\alpha^{-1} \in IS_n$ which implies $\alpha(\alpha^{-1}\beta\alpha^{-1})\alpha = \beta \in \alpha IS_n\alpha$. So, $IS_n \subseteq \alpha IS_n\alpha$ and hence the proof.

Proposition 6 *If $\alpha \in IS_n$ with $rank(\alpha) < n$ and $rank(\alpha^2) = rank(\alpha)$, then $\alpha IS_X\alpha$ is isomorphic to IS_A, where $A = ran(\alpha)$.*

Proof Let $\alpha \in IS_n$ with $rank(\alpha) < n$ and α be a permutation on a subset A of X.

For $\beta \in IS_n$, $dom(\alpha\beta\alpha) \subseteq A$ and $ran(\alpha\beta\alpha) \subseteq A$ which implies $\alpha\beta\alpha \in IS_A$. Therefore, $\alpha IS_n\alpha \subseteq IS_A$. By result 2, they have the same number of elements. Hence $\alpha IS_n\alpha$ is isomorphic to IS_A.

Now, we describe the relation between local subsemigroups and variants of finite symmetric inverse monoids.

Theorem 4 *Let n be a positive integer and let $\alpha \in IS_n$, with $rank(\alpha) = r$. Then*

1. *$\alpha IS_n\alpha \cong IS_r^c$ for some $c \in IS_r$ with $rank(c) = rank(\alpha^2)$.*
2. *$IS_n^\alpha \cong \beta IS_{2n-r}\beta$ for some $\beta \in IS_{2n-r}$, $rank(\beta) = n$ and $rank(\beta^2) = r$.*

Before proving the theorem, some results of variants of semigroups are recalled below (cf. [6]).

Lemma 1 *Let a and b be regular elements of a semigroup S and define the idempotents $e = ab$ and $f = ba$. Then $aSb = eSe$ and $bSa = fSf$.*

Lemma 2 *If a and b are elements of a semigroup S satisfying $a = aba$ and $b = bab$, then*

$$(aSa, \cdot) \cong (aSb, \star_{aab}) \cong (bSa, \star_{baa}).$$

Lemma 3 *If $\phi : S \to T$ is a semigroup isomorphism and if $c \in S$, then $S^c \cong T^{\phi(c)}$.*

Proof We have for $a, b \in S$, $\phi(ab) = \phi(a)\phi(b)$. Now, $\phi(a \star_c b) = \phi(acb) = \phi(a)\phi(c)\phi(b) = \phi(a) \star_{\phi(c)} \phi(b)$. Hence the result follows.

Proof (of Theorem 4) Let n be a positive integer and fix some $\alpha \in IS_n$ with $rank(\alpha) = r$. Let $X = \{1, 2, \ldots, n\}, Y = \{1, 2, \ldots, r\}$ and $Z = \{1, 2, \ldots, 2n - r\}$, re-labeling if necessary we assume $ran\alpha = Y$ and we can write $\alpha = \binom{x_i}{i}$, where $x_i \in X, i = 1, \ldots, r$.

1. Let β be the unique inverse of α in IS_X.Then $e = \alpha\beta = 1_{dom\alpha} \cdot \alpha = \alpha\beta\alpha$ implies $rank(\alpha^2) = rank(\alpha^2\beta)$. Now, by Lemma 2, $(\alpha IS_n\alpha, \cdot) \cong (\alpha IS_n\beta, \star_{\alpha\alpha\beta})$. Also by Lemma 1, $\alpha IS_n\beta = eIS_ne$. Now by Proposition 6, $(\alpha IS_n\beta, \cdot) \cong (eIS_ne, \cdot) \cong (IS_{dom\alpha}, \cdot)$. Hence, $\alpha IS_n\alpha = (\alpha IS_n\alpha, \cdot) \cong (\alpha IS_n\beta, \star_{\alpha\alpha\beta}) \cong (IS_{dom\alpha}, \star_c) = IS_r^c$ where $c \in IS_r$. Also we get $rank(c) = rank(\alpha^2)$.
2. Let $\beta \in IS_{2n-r}$ with $rank(\beta) = n$ and $rank(\beta^2) = r$. By Part (1), $\beta IS_{2n-r}\beta$ is isomorphic to IS_n^c for some $c \in IS_n$ with $rank(c) = r$. By Theorem 1.1 of [13], $IS^c \cong IS^\alpha$ if $rank(c) = rank(\alpha)$. Hence the proof.

Proposition 7 *If $\alpha, \beta \in IS_X$ with same rank such that $rank(\alpha^2) = rank(\beta^2)$ then the local subsemigroups of α and β are isomorphic.*

Proof Let $rank(\alpha) = rank(\beta) = r$. By Theorem 4, $\alpha IS_n\alpha \cong IS_r^c$ for some $c \in IS_r$ with $rank(c) = rank(\alpha^2)$. Similarly we get $\beta IS_n\beta \cong IS_r^d$ for some $d \in IS_r$ with $rank(d) = rank(\beta^2)$. Theorem 1.1 of [13] states that $IS^c \cong IS^d$ if $rank(c) = rank(d)$. Since $rank(\alpha^2) = rank(\beta^2)$, the local subsemigroups of α and β are isomorphic.

References

1. Araújo, J., Bentz, W., Konieczny, J.: Directed graphs of inner translations of semigroups. Semigroup Forum **94**, 650–673 (2016). https://doi.org/10.1007/s00233-016-9821-x
2. Chase, K.: Sandwich semigroups of binary relations. Discrete Math. **28**(3), 231–236 (1979)
3. Clifford, A.H., Preston, G.B.: The algebraic theory of semigroups, vol. I. Mathematical Surveys. American Mathematical Society, Providence, R.I (1961)
4. Dolinka, I., Durdev, I., East, J., Honyam, P., Sangkhanan, K., Sanwong, J., Sommanee, W.: Sandwich semigroups in locally small categories II: transformations. Algebra Univers. **79** (2018). https://doi.org/10.1007/s00012-018-0539-3
5. Dolinka, I., East, J.: Variants of finite full transformation semigroups. Int. J. Algebra Comput. **25**(8), 1187–1222 (2015). https://doi.org/10.1142/S021819671550037X
6. East, J.: Transformation representations of sandwich semigroups. Exp. Math. **29**(3), 291–295 (2020). https://doi.org/10.1080/10586458.2018.1459963

7. Ganyushkin, O., Mazorchuk, V.: Classical Finite Transformation Semigroups: An Introduction. Algebra and Applications. Springer, London (2009)
8. Hickey, J.B.: Semigroups under a sandwich operation. Proc. Edinburgh Math. Soc.(2) **26**(3), 371–382 (1983)
9. Howie, J.M.: Fundamentals of Semigroup Theory. London Mathematical Society Monographs New Series, Vol. 2. The Clarendon Press, Oxford University Press, New York (1995)
10. Mitchel, J.D. et al.: Semigroups-GAP Package, version 3.0.20 (2018)
11. The GAP Group, GAP-Groups, Algorithms and Programming, version 4.10.0 (2018)
12. Tsyaputa, G.Y.: Transformation semigroups with the deformed multiplication. Bullet. Univ. Kiev. Series: Mech. Math. **3**(1), 82–88 (2003)
13. Tsyaputa, G.Y.: Green's relations on the deformed transformation semigroups. Algebra Discrete Math. **1**, 121–131 (2004)

Closed Weak Supplemented Lattices

Shriram K. Nimbhorkar and Deepali B. Banswal

Abstract We introduce the concept of a Closed Weak Supplemented Lattice (CWS—Lattice), which is a generalization of a supplemented lattice. We show that a finite direct sum of CWS—lattices is a CWS-lattice.

Keywords Supplemented lattice · Weak supplemented lattice · Closed weak supplemented lattice · Refinable lattice

1 Introduction

The concept of a supplemented module and its generalizations are studied by several authors, e.g., Mutlu [6], Tohidi [12], Wang and Ding [13], Wisbauer [14], Zeng et al. [15]. In 2006, Zeng [15] introduced a generalization of the concept of a supplemented module, namely, a closed weak supplemented module.

Călugăreanu [3] translated many concepts from module theory to lattice theory. He introduced the concept of a supplement in terms of elements of a lattice. Alizade and Toksoy [1] introduced the concepts of an ample supplement and an amply supplemented lattice. In [2] they have introduced the concepts of a weak supplement and a weakly supplemented lattice. Nimbhorkar and Shroff [9–11] have introduced some generalizations of extended modules in lattice context. Nimbhorkar and Banswal [7, 8] have studied CESS-lattices and some generalizations of supplemented lattices.

In this paper, we introduce the concept of a closed weak supplemented lattice and obtain some results. This concept generalizes both an extending lattice and a weak supplemented lattice.

S. K. Nimbhorkar (✉)
Department of Mathematics, Dr. Babasaheb Ambedkar Marathwada University, Aurangabad 431004, India
e-mail: sknimbhorkar@gmail.com

D. B. Banswal
Department of Mathematics, B. P. Arts, S. M. A. Science and K. K. C. Commerce College, Chalisgaon 424101, India

M. Ashraf et al. (eds.), *Algebra and Related Topics with Applications*, Springer Proceedings in Mathematics & Statistics 392,
https://doi.org/10.1007/978-981-19-3898-6_18

2 Preliminaries

The undefined terms related to lattices can be found in Grätzer [4]. Throughout this paper L denotes a bounded lattice.

We recall some definitions from Alizade and Toksoy [1, 2] and from Călugăreanu [3].

Definition 1 An element $a \in L$ is said to be small in L if $a \vee b \neq 1$ for every $b \neq 1$. We then write $a \ll L$.

Definition 2 An element $a \in L$ is called a supplement of an element b in L if $a \vee b = 1$ and a is minimal with respect to this property.

Equivalently, a is a supplement of b in L if and only if $a \vee b = 1$ and $a \wedge b$ is small in $[0, a]$.

L is called supplemented if every element $a \in L$ has a supplement in L.

Definition 3 An element a is a weak supplement of b in L if and only if $a \vee b = 1$ and $a \wedge b \ll L$.

L is said to be weakly supplemented if every element $a \in L$ has a weak supplement in L.

Grzeszczuk and Puczylowski [5] developed the concept of an essential element in a lattice with least element 0, see also Călugăreanu [3].

The following definitions are from Nimbhorkar and Shroff [9].

Definition 4 We say that $a \in L$ is essential in L, if there is no nonzero $x \in L$ such that $a \wedge x = 0$.

Let $a, b \in L$, $0 \neq a \leq b$. We say that a is essential in b (or b is an essential extension of a), if there is no nonzero $c \leq b$ with $a \wedge c = 0$. We then write $a \leq_e b$.

Definition 5 If $a \leq_e b$ and for any $c > b$, a is not essential in c, then b is called a maximal essential extension of a.

Definition 6 An element $a \in L$ is called closed (or essentially closed) in L, if a has no proper essential extension in L.

Let $a, b \in L$, $a \leq b$. We say that a is closed in b, if a has no proper essential extension in b. we write $a \leq_{cl} b$.

Definition 7 If $a, b, c \in L$ are such that $a \vee b = c$ and $a \wedge b = 0$ then we say that a, b are direct summand of c and we write $c = a \oplus b$. We say that c is a direct sum of a and b.

The following definition is from Nimbhorkar and Shroff [10].

Definition 8 A bounded lattice L is called CS or extending if every nonzero element is essential in a direct summand of 1.

A nonzero element $a \in L$ is called extending if, every nonzero $b \leq a$ is essential in a direct summand of a.

The proof of the following lemma is the same as that of Proposition 3.1 from Nimbhorkar and Shroff [9].

Lemma 1 *Let* $0 \neq a \in L$. *Then the following statements are equivalent.*
(1) Every closed element $c \leq a$ *in* a *is a direct summand of* a.
(2) For every $d \leq a$, *there exists a direct summand* k *of* a *such that* $d \leq_e k$.

The following proposition is from Nimbhorkar and Shroff [11].

Proposition 1 *Let* L *be a modular lattice with 0. For* $a, b, c \in L$, *if* $a \leq_{cl} b$ *and* $b \leq_{cl} c$ *then* $a \leq_{cl} c$.

3 Closed Weak Supplemented Lattices

Zeng et al. [15] defined the concept of a closed weak supplemented module. We introduce this concept in the context of a lattice.

Definition 9 An element $a \in L$ is called a closed weak supplemented element if for any closed element $c \leq a$, there exists an element $b \leq a$ such that $b \vee c = a$ and $b \wedge c \ll a$.

In short we say that a is a CWS-element.

L is called closed weak supplemented if for any closed element $c \in L$ there exists an element $a \in L$ such that $1 = a \vee c$ and $a \wedge c \ll L$.

In short we say that L is a CWS-lattice.

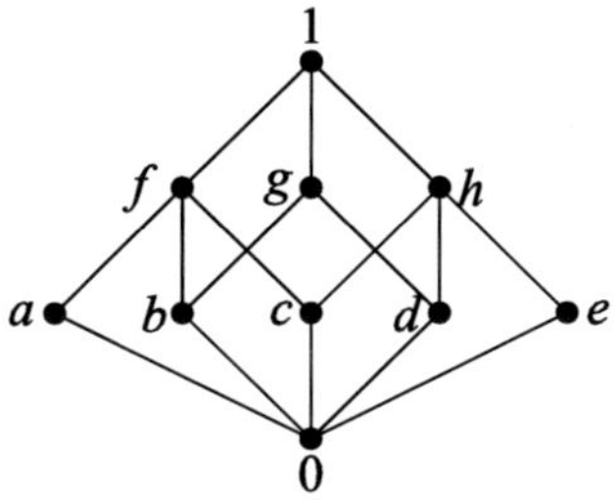

Figure 1

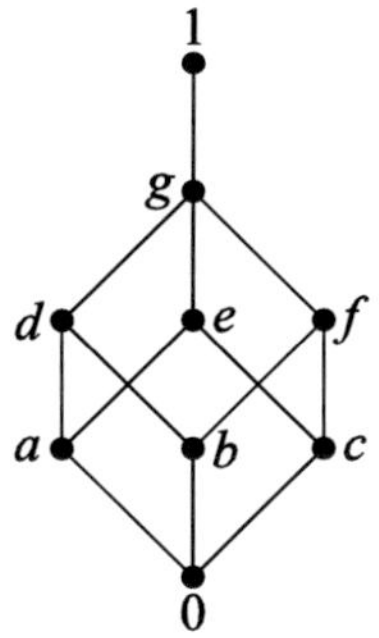

Figure 2

Example 1 Consider the element f in the lattice L shown in Figure 1. We note that $a, b, c \leq f$ are closed elements. We have $a \vee b = a \vee c = b \vee c = f$ and $a \wedge b = a \wedge c = b \wedge c = 0 \ll 1$. Hence f is closed weak supplemented.

Consider the element g in the lattice L shown in Figure 2. We note that d is closed in g and $f \leq g$ is such that $d \wedge f = b$. Since $b \vee e = g$ and $e \neq g$, b is not small in g. Hence g is not closed weak supplemented.

Lemma 2 *Let $a_i \in L$, $1 \leq i \leq n$. Then $a_1 \vee a_2 \vee \ldots \vee a_n \ll 1$ if and only if $a_i \ll 1$ for each $i = 1, \ldots, n$.*

Proof Suppose that $a_1 \vee a_2 \vee \ldots \vee a_n \ll 1$.
Then for any $b \in L$, $b \neq 1$, $(a_1 \vee a_2 \vee \ldots \vee a_n) \vee b \neq 1$.
Hence $a_i \vee b \neq 1$. Thus $a_i \ll 1$ for each i, $1 \leq i \leq n$.
Conversely, suppose that $a_i \ll 1$ for each $i = 1, \ldots, n$.
To show that $a_1 \vee a_2 \vee \ldots \vee a_n \ll 1$.
Without loss of generality, we may assume that $a_1 \vee \ldots \vee a_n \neq 1$.
Suppose that $(a_1 \vee \ldots \vee a_n) \vee b = 1$ for some $b \in L$.
Then $a_1 \vee (a_2 \vee \ldots \vee a_n \vee b) = 1$.
Since $a_1 \ll 1$, we conclude that $a_2 \vee \ldots \vee a_n \vee b = 1$.
By repeating the arguments, we get $a_n \vee b = 1$, a contradiction to $a_n \ll 1$.

Lemma 3 *Let L, L' be lattices. Let $f : L \to L'$ be a one to one and onto homomorphism. If $a \in L$ and $a \ll 1$, then $f(a) \ll 1'$, where $1 \in L$, $1' \in L'$.*

Proof Let $f : L \to L'$ be an onto homomorphism and $a \ll 1$. Let $b \in L'$ be such that $f(a) \vee b = 1'$. Since f is onto $b = f(c)$ for some $c \in L$. Then $f(a) \vee f(c) = 1'$ implies $f(a \vee c) = 1'$. Since f is one to one, we conclude that $a \vee c = 1$. Since $a \ll 1$, we get $c = 1$. Hence $f(c) = b = 1'$.

Lemma 4 *Let L be a modular lattice. Let $a, b \in L$ be such that $b \leq a$. If a is a direct summand of 1, then $b \ll a$ if and only if $b \ll L$.*

Proof Let $b \ll a$ and $b \vee c = 1$ for some $c \in L$. To show that $c = 1$.

We have $a = a \wedge 1 = a \wedge (b \vee c) = b \vee (a \wedge c)$ by using modularity. Since $b \ll a$, we have $a \wedge c = a$. Thus $a \leq c$ and so $b < a \leq c$. Hence $c = b \vee c = 1$.

Conversely, suppose that $b \ll 1$. To show that $b \ll a$.

Let $b \vee d = a$ for some $d \in L$. Since a is a direct summand of 1, there exists $e \in L$ such that $a \oplus e = 1$. Then $b \vee d \vee e = 1$.

Since $b \ll 1$ we get $d \vee e = 1$. Thus $d = d \vee 0 = d \vee (a \wedge e) = a \wedge (d \vee e) = a$. Hence $b \ll a$.

Theorem 1 *Every weak supplemented lattice is a closed weak supplemented lattice.*

Proof Let L be a weak supplemented lattice. Let a be a closed element in L. Since L is a weak supplemented lattice, there exists $b \in L$ such that $a \vee b = 1$ and $a \wedge b \ll L$. Thus L is a closed weak supplemented lattice.

Lemma 5 *Every extending lattice is a closed weak supplemented lattice.*

Proof Let L be an extending lattice. Let c be a closed element in L. Since L is an extending lattice, c is a direct summand of 1 that is $c \oplus b = 1$ for some $b \in L$. Since $c \wedge b = 0$, $c \wedge b \ll L$. Hence L is a closed weak supplemented lattice.

Lemma 6 *Let L be a modular lattice. If L is a supplemented lattice, then L is a weak supplemented lattice.*

Proof Suppose that L is a supplemented lattice. Let $a \in L$. Then there exists a supplement b of a such that $a \vee b = 1$ and $a \wedge b \ll b$. To show that $a \wedge b \ll 1$. Let c be such that $(a \wedge b) \vee c = 1$. Then $[(a \wedge b) \vee c] \wedge b = b$ implies by modularity that $(a \wedge b) \vee (c \wedge b) = b$. Since $a \wedge b \ll b$, we conclude that $b \wedge c = b$ and so $a \wedge b \leq c$. Hence $c = 1$, i.e., $a \wedge b \ll 1$.

Theorem 2 *Let L be a modular CWS-lattice. Then any direct summand of* 1 *is a CWS-element.*

Proof Let $a \in L$ be a direct summand of 1 and b be a closed element in a. Since b is closed in a and a is closed in L, by Proposition 1, b is closed in L. Since L is a closed weak supplemented lattice, there exists $c \in L$ such that $1 = c \vee b$ and $c \wedge b \ll 1$. Thus $a = 1 \wedge a = (b \vee c) \wedge a = (a \wedge c) \vee b$. Since a is a direct summand of 1, by Lemma 4, $a \wedge c \wedge b = c \wedge b \ll a$. Thus a is a CWS-element.

Theorem 3 *Let L be a modular lattice. Let $a, b \in L$. Suppose that c is a weak supplement of $a \vee b$ in L and d is a weak supplement of $a \wedge (b \vee c)$ in a. Then $c \vee d$ is a weak supplement of b in L.*

Proof Since c is a weak supplement of $a \vee b$, we have

$$a \vee b \vee c = 1 \text{ and } (a \vee b) \wedge c \ll 1.$$

As d is a weak supplement of $a \wedge (b \vee c)$ in a, we have

$$[a \wedge (b \vee c)] \vee d = a \text{ and } a \wedge (b \vee c) \wedge d = (b \vee c) \wedge d \ll a.$$

We have

$$\begin{aligned} 1 &= a \vee b \vee c \\ &= [a \wedge (b \vee c)] \vee d \vee b \vee c \\ &= d \vee b \vee c \quad \text{By absorption identity.} \end{aligned}$$

Hence to show that $c \vee d$ is a weak supplement of b in L, it is sufficient to show that $b \wedge (c \vee d) \ll 1$.

From $(a \vee b) \wedge c \ll 1$, $(b \vee c) \wedge d \ll a$ and Lemma 7.5, p. 78 from [3], it follows that

$$[(a \vee b) \wedge c] \vee [(b \vee c) \wedge d] \ll 1 \vee a = 1. \tag{1}$$

We have

$$[(a \vee b) \wedge c] \vee [(b \vee c) \wedge d] \geq b \wedge (c \vee d).$$

From (1) and Lemma 7.3, p. 78 from [3], we conclude that

$$b \wedge (c \vee d) \ll 1.$$

Hence $c \vee d$ is a weak supplement of b in L.

Theorem 4 *Let L be a modular lattice and $b \in L$ be any closed element. Suppose that $1 = a_1 \oplus a_2$ where each a_i $(i = 1, 2)$ is closed weak supplemented. Suppose that c is any weak supplement of $a_i \wedge (a_j \vee b)$ in a_i, $i \neq j$ such that $a_i \wedge (a_j \vee b) \leq_{cl} a_i$ and $a_j \wedge (b \vee c) \leq_{cl} a_j$. Then L is a CWS-lattice.*

Proof Let b be a closed element in L. We note that $1 = a_1 \vee (a_2 \vee b)$ has a weak supplement 0 in L. Since $a_1 \wedge (a_2 \vee b) \leq_{cl} a_1$ and a_1 is closed weak supplemented, there exist an element $c \leq a_1$ such that

$$a_1 = c \vee [a_1 \wedge (a_2 \vee b)] \text{ and } c \wedge (a_1 \wedge (a_2 \vee b)) = c \wedge (a_2 \vee b) \ll a_1.$$

By Theorem 3, c is a weak supplement of $a_2 \vee b$ in L. Hence $1 = c \vee (a_2 \vee b)$. Since $a_2 \wedge (c \vee b) \leq_{cl} a_2$ and a_2 is a closed weak supplemented, $a_2 \wedge (c \vee b)$ has a weak supplement d in a_2. By Theorem 3, $c \vee d$ is a weak supplement of b in L. Hence L is closed weak supplemented.

Theorem 5 *Let L be a modular lattice and $a, b \in L$. Let $1 = a \vee b$. Suppose that a is closed weak supplemented and that for any closed element $c \in L$, $c \wedge a \leq_{cl} a$. Then L is closed weak supplemented if and only if every closed element $c \in L$ with $b \leq c$ has a weak supplement.*

Proof Suppose that $c \in L$ is a closed element such that $b \leq c$. We have $1 = a \vee b = a \vee c$ and $a \vee c$ has a weak supplement 0. Since $c \wedge a \leq_{cl} a$ and a is closed weak supplemented, $c \wedge a$ has a weak supplement e in a. By Theorem 3, e is a weak supplement of c in L.

The converse is obvious.

The following definitions are from Nimbhorkar and Banswal [7].

Definition 10 Let L be a lattice with 0 and $a, b \in L$. If b is a maximal element in the set $\{x \mid x \in L \text{ and } a \leq_e x\}$, then we say that b is an essential closure of a in L.

Definition 11 A lattice L is called a UC-lattice if each of its nonzero elements has a unique essential closure in L.

The following theorem is from Nimbhorkar and Banswal [7].

Theorem 6 *A lattice L is a UC-lattice if and only if for any closed element a in L and for any $b \in L$, $a \wedge b$ is closed in b.*

Theorem 7 *Let L be a distributive UC-lattice. Let $1 = a_1 \vee a_2$ such that $a_1, a_2 \in L$. Then L is closed weak supplemented if and only if a_i is closed weak supplemented.*

Proof Let $c \in L$ be any closed element. Then for each i, $c \wedge a_i$ is closed in a_i. Suppose that $c \wedge a_1 \leq_e d \leq a_1$. Clearly, $a_2 \wedge c \leq_e a_2 \wedge c$. We have $c = (c \wedge a_1) \oplus (c \wedge a_2) \leq_e d \oplus (a_2 \wedge c)$. Since c is closed in L, $c = (a_1 \wedge c) \oplus (a_2 \wedge c) = d \oplus (a_2 \wedge c)$.

So $d = a_1 \wedge c$ and $a_1 \wedge c$ is closed in a_1.

Therefore, there exist elements $d_i \leq a_i$ such that $a_i = d_i \vee (c \wedge a_i)$ and $c \wedge a_i \wedge d_i = c \wedge d_i \ll a_i$, $i = 1, 2$.

Hence $1 = a_1 \vee a_2 = d_1 \vee d_2 \vee (c \wedge a_1) \vee (c \wedge a_2)$ and $a_1 \wedge a_2 = d_1 \wedge d_2 \wedge (c \wedge a_1) \wedge (c \wedge a_2) = 0$. Thus $1 = a_1 \oplus a_2 = d_1 \oplus d_2 \oplus (c \wedge a_1) \oplus (c \wedge a_2) = d_1 \oplus d_2 \oplus c$ and $c \wedge (d_1 \oplus d_2) = (c \wedge a_1) \oplus (c \wedge a_2) \ll a_1 \oplus a_2 = 1$.

Thus L is a closed weak supplemented lattice.

The converse follows from Theorem 2.

Theorem 8 *Let L be a modular lattice. Suppose that for any element $a \in L$ there is a closed element $b \in L$ such that $b = a \vee c$ for some $c \ll L$. Then L is weak supplemented if and only if L is closed weak supplemented.*

Proof Suppose that there is a closed element $b \in L$ such that $b = a \vee c$ for some $c \ll 1$. Since L is closed weak supplemented lattice, there exists an element $d \in L$ such that $1 = d \vee b$ and $d \wedge b \ll L$. So $1 = d \vee a \vee c$. Since $c \ll L$, $1 = d \vee a$. Now, $d \wedge a \leq d \wedge b \ll 1$. Thus L is weak supplemented.

The converse follows by Theorem 1.

Definition 12 A lattice L is called $\oplus$-supplemented if for every element $a \in L$ there is a direct summand $b \in L$ of 1 which is a supplement of a in L.

Example 2 Consider the lattice L shown in Figure 1. We note that for $a \in L, d \in L$ is a direct summand of 1 such that $a \vee d = 1$ and d is minimal with this property. Thus d is a supplement of a. Similarly, we can check for all elements of L. Hence L is a $\oplus$-supplemented lattice.

Zeng et al. [15] have defined the concept of a refinable module. We introduce this concept in the context of a lattice.

Definition 13 A lattice L is called refinable if for elements $a, b \in L$ with $a \vee b = 1$, there is a direct summand c of 1 with $c \leq a$ and $c \vee b = 1$.

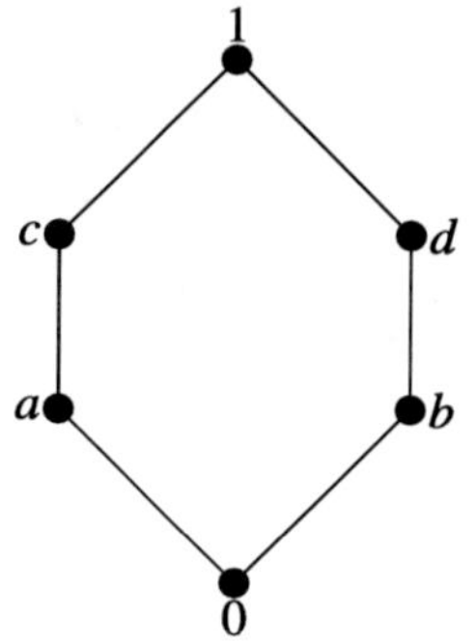

Figure 3

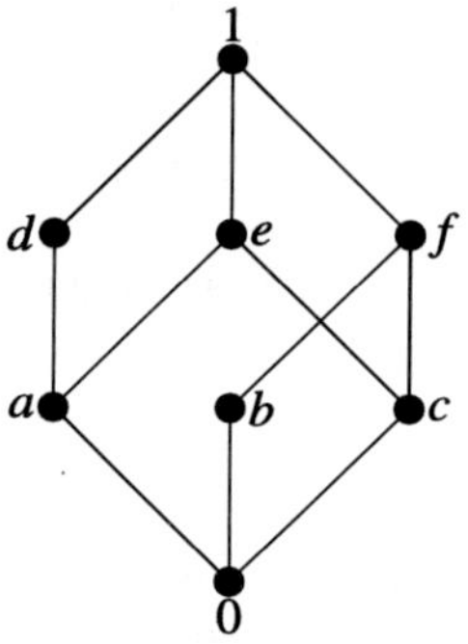

Figure 4

Example 3 Consider the elements c, b in the lattice L shown in Figure 3. We note that $c \vee b = 1$ and there exists a direct summand $a \leq c$ such that $a \vee b = 1$. Similarly, we can verify for all elements of L. Hence L is refinable.

Example 4 Consider the elements d, e in the lattice L shown in Figure 4. We note that $d \vee e = 1$. But there does not exist any direct summand x of 1 such that $x \leq d$ and $x \vee c = 1$. Hence L is not refinable.

Theorem 9 *Let L be a refinable lattice. Suppose that for any element $a \in L$, there is a closed element $b \in L$ such that either $a = b \vee c$ or $b = a \vee d$ for some, $c, d \ll L$. Then the following statements are equivalent,*

1. *L is $\oplus$-supplemented.*
2. *L is supplemented.*
3. *L is weak supplemented.*
4. *L is closed weak supplemented.*

Proof (1) $\Rightarrow$ (2): Follows from the definition of $\oplus$-supplemented.
(2) $\Rightarrow$ (3) : Follows by Lemma 6
(3) $\Rightarrow$ (4) : Follows by Theorem 1.
(4) $\Rightarrow$ (1) : Let L be closed weak supplemented. To show L is $\oplus$-supplemented.

Case (I): Suppose that there is a closed element $b \in L$ such that $a = b \vee c$ for some $c \ll 1$.

Since L is closed weak supplemented, there is an element $e \in L$ such that $1 = b \vee e$ and $e \wedge b \ll 1$.

Hence $1 = b \vee e = a \vee e$ and $a \wedge e \ll 1$.

Since L is refinable, there exist a direct summand $f \in L$ such that $f \leq a$ and $1 = f \vee e$. So $f \wedge e \leq a \wedge e \ll 1$.

As f is a direct summand of 1, we have $f \wedge e \ll f$, which shows that L is $\oplus$-supplemented.

Case (II): Suppose that there is a closed element $a \in L$ such that $b = a \vee d$ for some $d \ll 1$.

Since L is a closed weak supplemented, there is an element $e \in L$ such that $1 = b \vee e$ and $b \wedge e \ll 1$.

Thus $1 = b \vee e = a \vee d \vee e = a \vee e$ and $a \wedge e \ll 1$.

As L is refinable, then there is a direct summand $f \in L$ such that $f \leq a$ and $1 = f \vee e$.

Therefore $f \wedge e \leq a \wedge e \ll f$, since f is a direct summand of 1. Thus L is $\oplus$-supplemented lattice.

References

1. Alizade, R., Toksoy, S.E.: Cofinitely supplemented modular lattices. Arab. J. Sci. Eng. **36**(6), 919 (2011)
2. Alizade, R., Toksoy, S.E.: Cofinitely weak supplemented lattices. Indian J. Pure Appl. Math. **40**(5), 337–346 (2009)
3. Călugăreanu, G.: Lattice Concepts of Module Theory. Kluwer, Dordrecht (2000)
4. Grätzer, G.: Lattice Theory: First Concepts and Distributive Lattices. W. H. Freeman and Company, San Francisco (1971)
5. Grzeszczuk, P., Puczylowski, E.R.: On Goldie and dual Goldie dimension. J. Pure Appl. Algebra **31**, 47–54 (1984)
6. Figen Takil Mutlu: Amply weak semisimple-supplemented-modules. Int. J. Pure Appl. Math. **83**(4), 613–621 (2013)
7. Nimbhorkar, S.K., Banswal, D.B.: A note on CESS-lattices. In: Proceeding of International Conference on Algebra, De Gryuter. https://doi.org/10.1515/9783110542400-014
8. Nimbhorkar, S.K., Banswal, D.B.: Generalizations of supplemented lattices. AKCE Inter. J. Graphs Combin. **16**, 8–17 (2019)
9. Nimbhorkar, S.K., Shroff, R.: Ojective ideals in modular lattices. Czech. Math. J. **65**(140), 161–178 (2015)
10. Nimbhorkar, S.K., Shroff, R.: Generalized extending ideals in modular lattices. J. Indian Math. Soc. **82**(3–4), 127–146 (2015)
11. Nimbhorkar, S.K., Shroff, R.: Goldie extending elements in modular lattices. Math. Bohemica **142**(2), 163–180 (2017)
12. Tohidi, M.: Soc-⊕-supplemented modules. Int. J. Math. Sci. Appl. **2**(2), 803–812 (2012)
13. Wang, Y., Ding, N.: Generalized supplemented modules. Taiwan J. Math. **10**(6), 1589–1601 (2006)
14. Wisbauer, R.: Foundations of Module and Ring Theory. Gordon and Breach Science Publishers, Reading (1991)
15. Zeng, Q.-Y., Shi, M.-H.: On closed weak supplemented modules. J. Zhejiang Univ. - Sci. A **7**(2), 210–215 (2006)

On Compatible Ring Structures of the Injective Hull of a Ring

Jae Keol Park and S. Tariq Rizvi

Abstract For a ring R, let $E(R_R)$ and $Q(R)$ be the injective hull of R_R and the maximal right ring of quotients of R, respectively. If R is right nonsingular, then $E(R_R) = Q(R)$ and $E(R_R)$ has a unique overring structure of R. If R is not right nonsingular, $E(R_R)$ does not necessarily have a compatible overring structure with the ring R, in general. We discuss the disparity between the right rings of quotients and the right essential overrings of a ring R. Several examples and counterexamples for the disparity are given. An example of a right Kasch ring R (hence, $Q(R) = R$) for which $E(R_R)$ has (even infinitely many) distinct compatible overring structures of R is constructed. We discuss such compatible overring structures on $E(R_R)$ in detail. When rings U and T are isomorphic, we show that U is right compatible if and only if T is right compatible.

Keywords Right essential overrings · Right rings of quotients · Compatible ring structures · Right compatible rings

1 Introduction

In this paper, we study the compatibility of overring structures of the injective hull $E(R_R)$ of a ring R. It is well known that a ring R is right nonsingular if and only if $E(R_R) = Q(R)$ and $Q(R)$ is right self-injective (von Neumann) regular. Moreover, in this case the overring structure of $E(R_R)$ is unique (see Corollary 2.4). However,

---

Dedicated to Professor Murtaza A. Quadri for his contributions and dedication to the development of Algebra at Aligarh Muslim University continuing the legacy of his supervisor Professor M. A. Kazim Rizvi

---

J. K. Park (✉)
Department of Mathematics, Busan National University, Busan 46241, South Korea
e-mail: jkp1128@yahoo.com

S. T. Rizvi
Department of Mathematics, The Ohio State University, Lima, OH 45804-3576, USA
e-mail: rizvi.1@osu.edu

M. Ashraf et al. (eds.), *Algebra and Related Topics with Applications*,
Springer Proceedings in Mathematics & Statistics 392,
https://doi.org/10.1007/978-981-19-3898-6_19

if a ring R is not right nonsingular, disparities between overring structures on $E(R_R)$ and $Q(R)$ occur.

First, we study the disparity of right rings of quotients and right essential overrings of a ring R. We present several examples and counterexamples. Motivated by such examples and counterexamples, right essential overring structures of the injective hull $E(R_R)$ of the module R_R are discussed.

If R is a right Kasch ring which is not right self-injective, then $Q(R) = R$ but $E(R_R) \neq R$. Beyond the ring $Q(R)$, we study overring structures on $E(R_R)$, where R is a right Kasch ring which is not right self-injective.

There exists a right Kasch ring R which is not right self-injective such that $E(R_R)$ has an overring structure of the ring R. From this overring structure of $E(R_R)$, construction and explicit description of other compatible ring structures on $E(R_R)$ are exhibited in details by using an R-isomorphism extension to $E(R_R)$ of the identity of R_R (Theorem 2.14 and Example 2.16). As a consequence of our method, we show that there exist even infinitely many distinct compatible ring structures on $E(R_R)$.

Motivated by Remark 2.15, we investigate relationships between compatible ring structures and noncompatible ring structures of the injective hull of a ring in Sect. 3. Moreover, when rings U and T are isomorphic, we show that U is right compatible if and only if T is right compatible.

Ideals of a ring mean two-sided ideals. For a ring R, $I \trianglelefteq R$ denotes that I is an ideal of R. For a nonempty subset X of a ring R, we use $\ell(X)$ to denote the left annihilator of X in R. For a given module M, $N \leq M, N \leq^{\text{ess}} M$, and $N \leq^{\text{den}} M$ denote N is a submodule of M, N is essential in M, and N is dense in M, respectively. When n is an integer n such that $n > 1$, $\mathbb{Z}_n$ denotes the ring of integers modulo n. For a ring R and a positive integer n, $T_n(R)$ stands for the $n \times n$ upper triangular matrix ring over the ring R. Let X be a set. Then $|X|$ denotes the cardinal number of X.

2 Essential Overrings Versus Rings of Quotients

The disparity between the right ring of quotients of R and the right essential overrings of R is important and we will discuss it first. For a ring R, we use $E(R_R)$ to denote the injective hull of R_R. In this section, we study compatible ring structures of $E(R_R)$ (when they exist) with the ring structure of R. Examples of rings R for which $E(R_R)$ has distinct compatible ring structures are exhibited. Furthermore, an example of a ring R for which $E(R_R)$ has infinitely many distinct compatible ring structures is provided. For a given ring R, right ring of quotients of R and right essential overrings of R are compared. Various examples and counterexamples are provided for illustration.

Let M_R be a module and $N \leq M$. Then we say that N_R is *dense* in M_R if for $x, y \in M$ with $y \neq 0$, there exists $r \in R$ such that $xr \in N$ and $yr \neq 0$. We use

$N \leq^{\text{den}} M$ to denote that N is dense in M. We note that if $N \leq^{\text{den}} M$, then $N \leq^{\text{ess}} M$. The converse does not hold in general.

Definition 2.1 An overring S of a ring R is called a *right essential overring* of R if $R_R \leq^{\text{ess}} S_R$. For a ring R, $Q(R)$ denotes the maximal right ring of quotients of R. An intermediate ring between a ring R and $Q(R)$ is called a *right ring of quotients* of R. Thereby, an overring T of a ring R is said to be a *right ring of quotients* of R if and only if $R_R \leq^{\text{den}} T_R$.

Since every dense overmodule is an essential overmodule, any right ring of quotients of a ring R is a right essential overring of R. In Example 2.5, we provide a right essential overring of a ring which is not a right ring of quotients.

Definition 2.2 Assume that R is a ring and $R_R \leq T_R$. Let $(T, +, \circ)$ be a ring structure on T, where $+$ is the given addition on T_R and $\circ$ is the multiplication on T. We say that the ring structure $(T, +, \circ)$ is *compatible* with R if $\circ$ extends the scalar multiplication of T over R. In other words,

$$t \circ r = tr \quad \text{for} \quad t \in T \quad \text{and} \quad r \in R.$$

Thereby, T is an overring of R.

We note that, for a ring R, if S is either a right essential overring of R or S is a right ring of quotients of R, then the ring structure on S is compatible with R.

The next result shows that if R_R is dense in T_R and T has a compatible ring structure (i.e., T is a right ring of quotients of R), then it is unique (see [2, Proposition 7.1.6]).

Proposition 2.3 *Let R be a ring and $R_R \leq^{den} T_R$. If T has a compatible ring structure, then all of the compatible ring structures on T_R coincide with each other. Thereby, T becomes an intermediate ring between R and $Q(R)$ under this unique compatible ring structure on T.*

Proof Let $(T, +, \circ_1)$ and $(T, +, \circ_2)$ be two compatible ring structures on T_R. Assume on the contrary that there are $x, y \in T$ with $x \circ_1 y - x \circ_2 y \neq 0$. Then there exists $r \in R$ such that $yr \in R$ and $(x \circ_1 y - x \circ_2 y)r \neq 0$ because $R_R \leq^{\text{den}} T_R$. Thus

$$(x \circ_1 y - x \circ_2 y)r = x \circ_1 (yr) - x \circ_2 (yr) = x(yr) - x(yr) = 0,$$

which is a contradiction. Therefore, $\circ_1 = \circ_2$. □

From Proposition 2.3, we have the following corollary which shows $Q(R)$ has a unique overring structure of a ring R.

Corollary 2.4 *Let R be a ring. Then $Q(R)$ has a unique overring structure of R.*

Proof The proof follows immediately from Proposition 2.3 because $R_R \leq^{\text{den}} Q(R)_R$ and $Q(R)$ has an overring structure of R. □

A ring R is called *right Kasch* if every simple right R-module is embedded in R_R. It is well known that a ring R is right Kasch if and only if the only dense right ideal of R is R itself. Thus, if R is a right Kasch ring, then $R = Q(R)$. (See [4, Corollary 8.28, p. 281 and Corollary 13.24, p. 371].)

In the following example (see [2, Example 7.1.8]), there exists a ring R for which there is an essential overmodule T_R of R_R such that T has two distinct compatible overring structures of R.

Example 2.5 Let $R = \begin{bmatrix} \mathbb{Z}_4 & 2\mathbb{Z}_4 \\ 0 & \mathbb{Z}_4 \end{bmatrix}$. The addition on R is the usual componentwise addition. For $\begin{bmatrix} u_1 & 2v_1 \\ 0 & w_1 \end{bmatrix}$, $\begin{bmatrix} u_1 & 2v_1 \\ 0 & w_1 \end{bmatrix} \in R$ with $u_i, v_i, w_i \in \mathbb{Z}_4$, $1 \le i \le 2$, the multiplication is defined by

$$\begin{bmatrix} u_1 & 2v_1 \\ 0 & w_1 \end{bmatrix} \begin{bmatrix} u_2 & 2v_2 \\ 0 & w_2 \end{bmatrix} = \begin{bmatrix} u_1u_2 & 2u_1v_2 + 2v_1w_2 \\ 0 & w_1w_2 \end{bmatrix}.$$

Then R is a ring. Let

$$T = \begin{bmatrix} \mathbb{Z}_4 & \mathbb{Z}_4 \\ 0 & \mathbb{Z}_4 \end{bmatrix}.$$

The addition on T is also the usual componentwise addition. For $\begin{bmatrix} a & b \\ 0 & c \end{bmatrix} \in T$ (where $a, b, c \in \mathbb{Z}_4$) and $\begin{bmatrix} u & 2v \\ 0 & w \end{bmatrix} \in R$ (where $u, v, w \in \mathbb{Z}_4$), the R-module scalar multiplication on T is defined by

$$\begin{bmatrix} a & b \\ 0 & c \end{bmatrix} \begin{bmatrix} u & 2v \\ 0 & w \end{bmatrix} = \begin{bmatrix} au & 2av + bw \\ 0 & cw \end{bmatrix}.$$

Then T_R is a right R-module, and $R_R \le^{\text{ess}} T_R$.

From [2, Example 7.1.8], there exist exactly two distinct compatible ring multiplications on T, which are:

$$\begin{bmatrix} a_1 & b_1 \\ 0 & c_1 \end{bmatrix} \diamond_1 \begin{bmatrix} a_2 & b_2 \\ 0 & c_2 \end{bmatrix} = \begin{bmatrix} a_1a_2 & a_1b_2 + b_1c_2 \\ 0 & c_1c_2 \end{bmatrix}$$

and

$$\begin{bmatrix} a_1 & b_1 \\ 0 & c_1 \end{bmatrix} \diamond_2 \begin{bmatrix} a_2 & b_2 \\ 0 & c_2 \end{bmatrix} = \begin{bmatrix} a_1a_2 & a_1b_2 + 2b_1b_2 + b_1c_2 \\ 0 & c_1c_2 + 2a_1b_2 + 2c_1b_2 \end{bmatrix}.$$

Note that $(T, +, \diamond_1) = T_2(\mathbb{Z}_4)$, the 2×2 upper triangular matrix ring over the ring $\mathbb{Z}_4$. Further, the ring structure given on R is $(R, +, \diamond_1)$.

We can check that the ring R is right Kasch (see Example 2.9 and Remark 2.10). Therefore, we have that $R = (R, +, \diamond_1) = Q(R, +, \diamond_1)$. Note that $\diamond_1$ on T extends the R-module scalar multiplication of T over R. As before, we see that

$$R_R = (R, +, \diamond_1)_{(R,+,\diamond_1)} \leq^{\text{ess}} (T, +, \diamond_1)_{(R,+,\diamond_1)} = T_R.$$

Hence, the ring $(T, +, \diamond_1)$ is a right essential overring of $R = (R, +, \diamond_1)$.

Now, $R = (R, +, \diamond_1) = Q(R, +, \diamond_1) \subsetneq (T, +, \diamond_1)$, hence, we have that $(R, +, \diamond_1)_{(R,+,\diamond_1)}$ is not dense in $T_R = (T, +, \diamond_1)_{(R,+,\diamond_1)}$. So, the ring $(T, +, \diamond_1)$ is not a right ring of quotients of $R = (R, +, \diamond_1)$.

Next, $(T, +, \diamond_2)$ is a right essential overring of the ring $R = (R, +, \diamond_1) = (R, +, \diamond_2)$ and $\diamond_1 \neq \diamond_2$ on T. The ring $(T, +, \diamond_2)$ is a right essential overring of the ring $(R, +, \diamond_1)$, which is not a right ring of quotients of the ring $R = (R, +, \diamond_1)$.

Because $\diamond_1 \neq \diamond_2$, T has no unique compatible ring structure with the ring R. By Proposition 2.3, R_R is not dense in T_R.

In a different way, we show that R_R is not dense in T_R. Indeed, take

$$x = \begin{bmatrix} 0 & 1 \\ 0 & 0 \end{bmatrix} \text{ and } y = \begin{bmatrix} 0 & 0 \\ 0 & 2 \end{bmatrix}$$

in T. Assume on the contrary that $R_R \leq^{\text{den}} T_R$. Then there exists $r \in R$ such that $xr \in R$ and $yr \neq 0$. Say $r = \begin{bmatrix} u & 2v \\ 0 & w \end{bmatrix} \in R$, where $u, v, w \in \mathbb{Z}_4$. Because $xr \in R$, $w \in 2\mathbb{Z}_4$. Thus, $yr = 0$, which is a contradiction.

Remark 2.6 Let $(T, +, \diamond_1)$ and $(T, +, \diamond_2)$ be as in Example 2.5. Then $(T, +, \diamond_1) \cong (T, +, \diamond_2)$ (as rings) with an isomorphism $f : (T, +, \diamond_1) \to (T, +, \diamond_2)$ defined by

$$f \begin{bmatrix} a & b \\ 0 & c \end{bmatrix} = \begin{bmatrix} a & b \\ 0 & 2b + c \end{bmatrix},$$

for $\begin{bmatrix} a & b \\ 0 & c \end{bmatrix} \in (T, +, \diamond_1)$.

Definition 2.7 A ring R is called *right compatible* (also called *right Osofsky compatible*) if some injective hull $E(R_R)$ of R_R has a ring structure, where the ring multiplication of $E(R_R)$ extends the R-module scalar multiplication of $E(R_R)$ over R. Similarly, a left compatible ring is defined.

Every ring R satisfying $Q(R) = E(R_R)$ is right compatible. Hence, a ring R is right nonsingular, then R is right compatible. The following are well known:

(1) A ring R is right nonsingular if and only if $Q(R) = E(R_R)$ and $Q(R)$ is a right self-injective (von Neumann) regular ring (see [2, Theorem 2.1.31]).

(2) Let T be a right ring of quotients of a ring R. Then T is right self-injective if and only if $T = E(R_R)$ (see [2, Theorem 7.1.3]).

The next result (see [2, Proposition 7.1.10]) shows that if one injective hull of R_R has a compatible ring structure, then every injective hull of R_R has a compatible ring structure.

Proposition 2.8 *Let R be a ring. If R is right compatible, then every injective hull of R_R has a compatible ring structure.*

Proof Assume that there is an injective hull $E(R_R)$ of R_R such that $(E(R_R), +, \star)$ is a compatible ring structure. Let E_R be an arbitrary injective hull of R_R. Then there is an isomorphism $\phi : E_R \to E(R_R)$ such that $\phi(r) = r$ for each $r \in R$. For $x, y \in E_R$,

Define $x \circ y = \phi^{-1}(\phi(x) \star \phi(y))$ for $x, y \in E_R$. Then $(E_R, +, \circ)$ is a ring. Further,

$$\begin{aligned} x \circ r &= \phi^{-1}(\phi(x) \star \phi(r)) = \phi^{-1}(\phi(x) \star r) \\ &= \phi^{-1}(\phi(x)r) = \phi^{-1}(\phi(xr)) \\ &= xr, \end{aligned}$$

for $x \in E_R$ and $r \in R$. Thus $(E_R, +, \circ)$ is a compatible ring structure. In this case, $(E_R, +, \circ) \cong (E(R_R), +, \star)$ via ϕ.

□

Recall that a ring R is said to be *QF* (quasi-Frobenius) if R is right (or left) Artinian and right (or left self-injective). It is well known that if R is a Dedekind domain, then $A = R/I$ is a QF-ring for any nonzero proper ideal I of R (see [8, Theorem 6.14, p. 174]). Also note that if A is right self-injective and G is a finite group, then the group ring $A[G]$ is right self-injective (see [3]). Hence, the group algebra $F[G]$ of a finite group G over a field F is QF.

In the following, there exists a ring R such that $Q(R)$ is not equal to $E(R_R)$, and $E(R_R)$ has no compatible ring structure with the R-module scalar multiplication on $E(R_R)$ over R.

We use $J(-)$ to denote the Jacobson radical of a ring and $\mathrm{Soc}(M_R)$ to denote the socle of a module M_R. When A is a commutative ring, $\mathrm{Soc}(A)$ denotes $\mathrm{Soc}(A_A)$.

Example 2.9 Let A be a commutative local QF-ring with $J(A) \neq 0$ (e.g., $A = \mathbb{Z}_4$), and let

$$R = \begin{bmatrix} A & \mathrm{Soc}(A) \\ 0 & A \end{bmatrix}.$$

Then we have the following.

(i) $Q(R) = R$. Since A is Artinian, $\mathrm{Soc}(A)_A \leq^{\text{ess}} A_A$. Note that $E(\mathrm{Soc}(A)_A) = A$ because A_A is injective. Since A is local and $\mathrm{End}(A_A) \cong A$, A_A is uniform. Thus, $\mathrm{Soc}(A)_A$ is a simple A-module.

Let $0 \neq V \trianglelefteq A$. Now $0 \neq \mathrm{Soc}(V_A) = V \cap \mathrm{Soc}(A)$, so $\mathrm{Soc}(A) \subseteq V$ because $\mathrm{Soc}(A)_A$ is simple. Therefore, $\mathrm{Soc}(A)$ is the nonzero smallest ideal of A. Let $0 \neq s \in \mathrm{Soc}(A)$. Then $\mathrm{Soc}(A) = sA$. We note that

$$J(R) = \begin{bmatrix} J(A) & \mathrm{Soc}(A) \\ 0 & J(A) \end{bmatrix}.$$

Let M be a maximal right ideal of R. Then $J(R) \subseteq M$, and hence, M is either

$$M_1 = \begin{bmatrix} J(A) & \mathrm{Soc}(A) \\ 0 & A \end{bmatrix} \text{ or } M_2 = \begin{bmatrix} A & \mathrm{Soc}(A) \\ 0 & J(A) \end{bmatrix}.$$

We show that M_1 is not a dense right ideal of R. For this, consider

$$\begin{bmatrix} 1 & 0 \\ 0 & 0 \end{bmatrix} \in R \text{ and } \begin{bmatrix} s & 0 \\ 0 & 0 \end{bmatrix} \in R.$$

If M_1 is a dense right ideal of R, then there exists

$$\begin{bmatrix} a & sb \\ 0 & c \end{bmatrix} \in R \text{ with } a,\ b,\ c \in A$$

such that

$$\begin{bmatrix} 1 & 0 \\ 0 & 0 \end{bmatrix}\begin{bmatrix} a & sb \\ 0 & c \end{bmatrix} \in M_1 \text{ and } \begin{bmatrix} s & 0 \\ 0 & 0 \end{bmatrix}\begin{bmatrix} a & sb \\ 0 & c \end{bmatrix} \neq 0.$$

Therefore, $a \in J(A)$. Note that $s^2 b = 0$ because $s^2 \in \mathrm{Soc}(A)^2 \subseteq \mathrm{Soc}(A)J(A) = 0$. Thus, $sa \neq 0$. But $sa \in \mathrm{Soc}(A)J(A) = 0$, a contradiction. Hence, M_1 is not a dense right ideal of R.

Next to prove that M_2 is not a dense right ideal, consider

$$x = \begin{bmatrix} 0 & 0 \\ 0 & 1 \end{bmatrix} \in R \text{ and } y = \begin{bmatrix} 0 & 0 \\ 0 & s \end{bmatrix} \in R.$$

Then as in the preceding argument, there does not exist $r \in R$ such that $xr \in M_2$ and $yr \neq 0$. Hence, M_2 is not a dense right ideal of R.

Now, let I be a proper right ideal of R such that $I_R \leq^{\mathrm{den}} R_R$. Since M_1 and M_2 are the only maximal right ideals of R, $I \subseteq M_1$ or $I \subseteq M_2$. Neither M_1 nor M_2 is a dense right ideal of R, so I is not a dense right ideal of R. Therefore, the only dense right ideal of R is R itself. Hence, $Q(R) = R$.

(ii) For $f \in \mathrm{Hom}(\mathrm{Soc}(A)_A, A_A)$ and $x \in A$, we let $f \cdot x$ defined by $(f \cdot x)(a) = f(xa)$ for $a \in \mathrm{Soc}(A)$. Then $f \cdot x \in \mathrm{Hom}(\mathrm{Soc}(A)_A, A_A)$. We now let

$$E = \begin{bmatrix} A \oplus \mathrm{Hom}(\mathrm{Soc}(A)_A, A_A) & A \\ \mathrm{Hom}(\mathrm{Soc}(A)_A, A_A) & A \end{bmatrix},$$

where the addition is componentwise and the R-module scalar multiplication is given by

$$\begin{bmatrix} a + f & b \\ g & c \end{bmatrix}\begin{bmatrix} x & y \\ 0 & z \end{bmatrix} = \begin{bmatrix} ax + f \cdot x & ay + f(y) + bz \\ g \cdot x & g(y) + cz \end{bmatrix}$$

for $\begin{bmatrix} a+f & b \\ g & c \end{bmatrix} \in E$ and $\begin{bmatrix} x & y \\ 0 & z \end{bmatrix} \in R$. Then E_R is an injective hull of R_R (see [2, Theorem 7.1.14]). Furthermore, since $Q(R) = R$ by part (i), we have that $Q(R) \neq E_R$.

(iii) Since $\mathrm{Soc}(A)$ is the smallest nonzero ideal of A by the argument in part (i), we have that $\mathrm{Soc}(A) = sA$, where $0 \neq s \in \mathrm{Soc}(A)$. Let $f_0 \in \mathrm{Hom}(\mathrm{Soc}(A)_A, A_A)$ such that $f_0(b) = b$ for every $b \in \mathrm{Soc}(A)$.

Assume that E has a compatible ring structure. Note that $(f_0 \cdot s)(sa) = f_0(s^2a) = 0$ for each $sa \in sA = \mathrm{Soc}(A)$, where $a \in A$ because $s^2 \in \mathrm{Soc}(A)^2 \subseteq \mathrm{Soc}(A)J(A) = 0$. Thus, $f_0 \cdot s = 0$. Therefore,

$$\begin{aligned} \begin{bmatrix} 0 & s \\ 0 & 0 \end{bmatrix} &= \begin{bmatrix} f_0 & 0 \\ 0 & 0 \end{bmatrix} \begin{bmatrix} 0 & s \\ 0 & 0 \end{bmatrix} = \begin{bmatrix} f_0 & 0 \\ 0 & 0 \end{bmatrix} \left(\begin{bmatrix} s & 0 \\ 0 & 0 \end{bmatrix} \begin{bmatrix} 0 & 1 \\ 0 & 0 \end{bmatrix} \right) \\ &= \left(\begin{bmatrix} f_0 & 0 \\ 0 & 0 \end{bmatrix} \begin{bmatrix} s & 0 \\ 0 & 0 \end{bmatrix} \right) \begin{bmatrix} 0 & 1 \\ 0 & 0 \end{bmatrix} = \begin{bmatrix} f_0 \cdot s & 0 \\ 0 & 0 \end{bmatrix} \begin{bmatrix} 0 & 1 \\ 0 & 0 \end{bmatrix} \\ &= \begin{bmatrix} 0 & 0 \\ 0 & 0 \end{bmatrix} \begin{bmatrix} 0 & 1 \\ 0 & 0 \end{bmatrix} \\ &= 0, \end{aligned}$$

a contradiction. Hence, R is not right compatible. By Theorem 2.8, there is no injective hull of R_R with a compatible ring structure.

Remark 2.10 Let R be the ring of Example 2.9. Then R is right Kasch because the only dense right ideal of R is R itself as is shown in Example 2.9. So, every simple right R-module is embedded in R_R. In fact, from the argument used in Example 2.9, M_1 and M_2 are the only maximal right ideals of R. Now let N be a simple right R-module. Then $N \cong R/M_1$ or $N \cong R/M_2$ as R-modules.

As was shown in Example 2.9(i), $\mathrm{Soc}(A)$ is the smallest nonzero ideal of A and $\mathrm{Soc}(A)_A$ is simple. Now take $0 \neq s \in \mathrm{Soc}(A)$. Then $\mathrm{Soc}(A) = sA$. Consider

$$\varphi : A \to \mathrm{Soc}(A) = sA$$

defined by $\varphi(a) = sa$ for $a \in A$. Assume $sa = 0$ with $a \in A$. Then $\mathrm{Soc}(A)a = 0$. If $a \notin J(A)$, then a is invertible because A is local, and hence, $\mathrm{Soc}(A) = 0$, a contradiction. Thus, $a \in J(A)$. Conversely, if $a \in J(A)$, then $\mathrm{Soc}(A)a = 0$. So $sa = 0$. Therefore, $\mathrm{Ker}(\varphi) = J(A)$.

We define

$$\gamma : R/M_1 \to R \text{ by } \gamma\left(\begin{bmatrix} a & 0 \\ 0 & 0 \end{bmatrix} + M_1 \right) = \begin{bmatrix} sa & 0 \\ 0 & 0 \end{bmatrix},$$

where $a \in A$. Then γ is an R-monomorphism from R/M_1 to R_R because $\mathrm{Ker}(\varphi) = J(A)$. Next define

$$\lambda : R/M_2 \to R \text{ by } \lambda\left(\begin{bmatrix} 0 & 0 \\ 0 & c \end{bmatrix} + M_2\right) = \begin{bmatrix} 0 & 0 \\ 0 & sc \end{bmatrix},$$

where $c \in A$. Then λ is also an R-monomorphism from R/M_2 to R_R. Consequently, every simple right R-module is embedded in R_R.

Remark 2.11 In Example 2.9, when $A = \mathbb{Z}_4$, it is shown that the ring R is not right compatible without explicitly constructing an injective hull of R_R (see [6]).

The following, due to Lang [5], shows that a commutative Artinian ring is (right) compatible precisely when R is self-injective.

Theorem 2.12 *Let R be a commutative Artinian ring. Then R is right compatible if and only if $R = E(R_R)$.*

Proof See [2, Theorem 7.3.12] for the proof. □

For an illustration of Theorem 2.12, we provide the next example.

Example 2.13 Assume that A is a commutative QF-ring with $J(A) \neq 0$ (e.g., $A = \mathbb{Z}_4$), where $J(A)$ is the Jacobson radical of A. Let R be the trivial extension of A by $J(A)$, that is,

$$R = \left\{\begin{bmatrix} x & y \\ 0 & x \end{bmatrix} \mid x \in A \text{ and } y \in J(A)\right\}.$$

Then R is a commutative Artinian ring. We claim that R is not right compatible. For this, take $0 \neq y_0 \in \mathrm{Soc}(A)$. We put

$$I = \begin{bmatrix} 0 & y_0 \\ 0 & 0 \end{bmatrix} R = \left\{\begin{bmatrix} 0 & y_0 x \\ 0 & 0 \end{bmatrix} \mid x \in A\right\}.$$

Consider $f : I \to R$ defined by

$$f\begin{bmatrix} 0 & y_0 x \\ 0 & 0 \end{bmatrix} = \begin{bmatrix} y_0 x & y_0 x \\ 0 & y_0 x \end{bmatrix}.$$

Then $f \in \mathrm{Hom}(I_R, R_R)$ because $\mathrm{Soc}(A)J(A) = 0$. Assume on the contrary that R_R is injective. Then by Baer's Criterion, there exists

$$\begin{bmatrix} a & b \\ 0 & a \end{bmatrix} \in R \text{ such that } \begin{bmatrix} a & b \\ 0 & a \end{bmatrix}\begin{bmatrix} 0 & y_0 \\ 0 & 0 \end{bmatrix} = \begin{bmatrix} y_0 & y_0 \\ 0 & y_0 \end{bmatrix}.$$

So, we get a contradiction, and hence, R_R is not injective. By Theorem 2.12, $E(R_R)$ has no compatible ring structure, i.e., R is not right compatible.

The following is useful to construct and describe explicitly other compatible overring structures on the injective hull E_R of R_R if $(E_R, +, \cdot)$ has a compatible overring structure.

Theorem 2.14 *Assume that R is a ring and E_R is an injective hull of R_R with a given compatible ring structure $(E_R, +, \cdot)$ on E_R. Let $f : E_R \to (E_R, +, \cdot)$ be a right R-module isomorphism such that $f(r) = r$ for all $r \in R$. Define*

$$v_1 \diamond v_2 = f^{-1}[f(v_1) \cdot f(v_2)]$$

for $v_1, v_2 \in E_R$. Then $(E_R, +, \diamond)$ is also a compatible ring structure on E_R and $(E_R, +, \diamond) \cong (E_R, +, \cdot)$ as rings.

Proof Say $v_1, v_2 \in E_R$. Then since $v_1 \diamond v_2 = f^{-1}[f(v_1) \cdot f(v_2)]$, we have that

$$f(v_1 \diamond v_2) = f(v_1) \cdot f(v_2).$$

Also, $f(v_1 + v_2) = f(v_1) + f(v_2)$ as f is additive. Because $(E_R, +, \cdot)$ is a ring and f is one-to-one and onto, $(E_R, +, \diamond)$ is a ring. Moreover, $f : (E_R, +, \diamond) \to (E_R, +, \cdot)$ is a ring isomorphism.

Next, for $v \in E_R$ and $r \in R$,

$$\begin{aligned} v \diamond r &= f^{-1}(f(v) \cdot f(r)) = f^{-1}(f(v) \cdot r) \\ &= f^{-1}(f(v)r) = f^{-1}(f(vr)) \\ &= vr \end{aligned}$$

as $f(r) = r$ and $f(v) \cdot r = f(v)r = f(vr)$. Hence, $(E_R, +, \diamond)$ is a compatible ring structure on E_R. □

In the following, we remark that the assumption
"$f(r) = r$ for all $r \in R$"
of Theorem 2.14 is crucial for the compatibility of the ring structure $(E_R, +, \diamond)$ in Theorem 2.14.

Remark 2.15 Assume that R is a ring and E_R is an injective hull of R_R with a compatible ring structure $(E_R, +, \cdot)$ on E_R. Let $u \in R$ be an invertible element such that $u \neq 1_R$, where 1_R is the identity of the ring R. Define

$$h : E_R \to (E_R, +, \cdot) \text{ by } h(y) = u^{-1} \cdot y \text{ for } y \in E_R.$$

Since u is invertible in $(E_R, +, \cdot)$, h is an additive abelian group isomorphism. Furthermore, h is an R-isomorphism of E_R. For $y_1, y_2 \in E_R$, let

$$y_1 \star y_2 = h^{-1}[h(y_1) \cdot h(y_2)].$$

By the proof of Theorem 2.14, $(E_R, +, \star)$ is a ring and $(E_R, +, \star) \cong (E_R, +, \cdot)$ (as rings) via h.

But $(E_R, +, \star)$ is not compatible. To show this, let 1_E be the identity of the ring $(E_R, +, \cdot)$. Then $1_E = 1_R$. In fact, if $1_E \neq 1_R$, then $1_E - 1_R \neq 0$. Since R_R is essential in E_R, there exists $r \in R$ such that

$$0 \neq (1_E - 1_R)r = (1_E - 1_R) \cdot r = 1_E \cdot r - 1_T r = r - r = 0,$$

which is a contradiction. Hence, $1_E = 1_R$. Now

$$\begin{aligned} 1_E \star 1_R &= h^{-1}[h(1_E) \cdot h(1_R)] = h^{-1}[h(1_E) \cdot h(1_E)] \\ &= h^{-1}(u^{-1} \cdot 1_E \cdot u^{-1} \cdot 1_E) = h^{-1}(u^{-1} \cdot u^{-1}) = u^{-1} \\ &\neq 1_E \end{aligned}$$

because $u \neq 1_R = 1_E$. Therefore, the ring structure $(E_R, +, \star)$ is not compatible.

We note that $h(1_R) = u^{-1} \neq 1_E = 1_R$, so $h(1_R) \neq 1_R$. Therefore, the assumption "$f(r) = r$ for all $r \in R$" is crucial for the compatibility of the ring structure $(E_R, +, \diamond)$ in Theorem 2.14.

When R is a right nonsingular ring, $E(R_R) = Q(R)$ and R is right compatible. Furthermore, in this case, $E(R_R)$ has a unique compatible structure by Proposition 2.3 and Corollary 2.4.

In contrast to Theorem 2.12 and Example 2.13, there exists a *noncommutative Artinian ring* R which is right compatible in the following Example 2.16 for which, there exists a right compatible ring R such that R is right Kasch (hence, $Q(R) = R$) but $E(R_R)$ has (even infinitely many) distinct compatible ring structures.

Example 2.16 is based on [1] and [2, Proposition 7.3.16 and Theorem 7.3.17], and here we describe compatible ring structures on the injective hull $E(R_R)$ of a ring R in details as much as possible by using a useful tool established in Theorem 2.14.

Example 2.16 Assume that A is a local commutative QF-ring with $J(A) \neq 0$, where $J(A)$ is the Jacobson radical of A. Let

$$R = \begin{bmatrix} A & A/J(A) \\ 0 & A/J(A) \end{bmatrix} \text{ and } E = \begin{bmatrix} A \oplus (A/J(A)) & A/J(A) \\ A/J(A) & A/J(A) \end{bmatrix}.$$

(i) $Q(R) = R$. There are exactly two maximal right ideals of R, which are

$$M_1 = \begin{bmatrix} J(A) & A/J(A) \\ 0 & A/J(A) \end{bmatrix} \text{ and } M_2 = \begin{bmatrix} A & A/J(A) \\ 0 & 0 \end{bmatrix}.$$

Let I be a dense right ideal of R such that $I \neq R$. Then either $I \subseteq M_1$ or $I \subseteq M_2$. First, assume that $I \subseteq M_1$. Hence, M_1 is a dense right ideal of R. As A is a commutative local QF-ring, $\mathrm{Soc}(A)$ is the smallest nonzero ideal and $\mathrm{Soc}(A)$ is a simple A-module (from the argument used in Example 2.9(i)). Let $0 \neq v \in \mathrm{Soc}(A)$. Then $\mathrm{Soc}(A) = vA$. Take

$$x=\begin{bmatrix}1&0\\0&0\end{bmatrix}\in R \text{ and } 0\neq y=\begin{bmatrix}v&0\\0&0\end{bmatrix}\in R.$$

Since M_1 is dense in R_R, there exists $r \in R$ such that $xr \in M_1$ and $yr \neq 0$. Say

$$r=\begin{bmatrix}a&\overline{b}\\0&\overline{c}\end{bmatrix},$$

where $\overline{a}$ and $\overline{b}$ are images of a and b in $A/J(A)$, respectively. As $xr \in M_1$, $a \in J(A)$. Note that $\mathrm{Soc}(A)J(A)=0$ and $\mathrm{Soc}(A)\subseteq J(A)$. Hence, $yr=0$, a contradiction. Therefore, M_1 is not dense in R_R.

Next, we see that $M_2 \cap \begin{bmatrix}0&0\\0&\overline{1}\end{bmatrix} R=0$. Thus, M_2 is not essential in R_R, so M_2 is not dense in R_R. Thus, I_R is not dense in R_R. Therefore, R itself is the only dense right ideal of R, and hence, R is a right Kasch ring. So $Q(R)=R$.

(ii) The addition $+$ of E is componentwise. The right R-module scalar multiplication of E over R is given as follows:

For $v=\begin{bmatrix}s+\overline{x}&\overline{y}\\\overline{z}&\overline{w}\end{bmatrix}\in E$ and $r=\begin{bmatrix}a&\overline{b}\\0&\overline{c}\end{bmatrix}\in R$, where $s, x, y, z, w, a, b, c \in A$ and $\overline{x}, \overline{b} \in A/J(A)$, etc., denote the images of $x,\ b \in A$, etc., respectively, we define

$$vr=\begin{bmatrix}s+\overline{x}&\overline{y}\\\overline{z}&\overline{w}\end{bmatrix}\begin{bmatrix}a&\overline{b}\\0&\overline{c}\end{bmatrix}=\begin{bmatrix}sa+\overline{xa}&\overline{sb}+\overline{xb}+\overline{yc}\\\overline{za}&\overline{zb}+\overline{wc}\end{bmatrix}.$$

Then E is a right R-module.

We show that the right R-module E is the injective hull of R_R. It was shown by G. F. Birkenmeier, J. K. Park, and S. T. Rizvi in their unpublished paper "An injective hull with distinct ring structures". In the next parts (iii) and (iv), we give their proof in detail as follows (cf. [2, Proposition 7.3.16]).

(iii) First, we show that $R_R \leq^{\mathrm{ess}} E_R$. For this, take $0\neq v=\begin{bmatrix}s+\overline{a}&\overline{b}\\\overline{c}&\overline{d}\end{bmatrix}\in E$. We consider the following four cases.

- $\overline{d}\neq 0$. Since $A/J(A)$ is a field, there exists $\overline{d}_1 \in A/J(A)$ such that $\overline{d}\,\overline{d}_1=\overline{1}$. Then we have that $\begin{bmatrix}0&0\\0&\overline{d}_1\end{bmatrix}\in R$ and $0\neq v\begin{bmatrix}0&0\\0&\overline{d}_1\end{bmatrix}=\begin{bmatrix}0&\overline{b}\,\overline{d}_1\\0&\overline{1}\end{bmatrix}\in R$.
- $\overline{d}=0$ and $\overline{b}\neq 0$. Then there exists $\overline{b}_1\in A/J(A)$ such that $\overline{b}\,\overline{b}_1=\overline{1}$. Hence, $\begin{bmatrix}0&0\\0&\overline{b}_1\end{bmatrix}\in R$ and $0\neq v\begin{bmatrix}0&0\\0&\overline{b}_1\end{bmatrix}=\begin{bmatrix}0&\overline{1}\\0&0\end{bmatrix}\in R$.
- $\overline{d}=0$, $\overline{b}=0$, and $\overline{c}\neq 0$. Take $\overline{c}_1\in A/J(A)$ such that $\overline{c}\,\overline{c}_1=\overline{1}$. Then $\begin{bmatrix}0&\overline{c}_1\\0&0\end{bmatrix}\in R$ and $0\neq v\begin{bmatrix}0&\overline{c}_1\\0&0\end{bmatrix}=\begin{bmatrix}0&(\overline{s}+\overline{a})\overline{c}_1\\0&\overline{1}\end{bmatrix}\in R$.

• $\overline{d} = 0$, $\overline{b} = 0$, $\overline{c} = 0$, and $s + \overline{a} \neq 0$. In this case, if $\overline{a} = 0$, then $s \neq 0$. Hence, $0 \neq v \in R$. Suppose that $\overline{a} \neq 0$. If $\overline{s} + \overline{a} \neq 0$, then $0 \neq v \begin{bmatrix} 0 & \overline{1} \\ 0 & 0 \end{bmatrix} = \begin{bmatrix} 0 & \overline{s} + \overline{a} \\ 0 & 0 \end{bmatrix} \in R$ with $\begin{bmatrix} 0 & \overline{1} \\ 0 & 0 \end{bmatrix} \in R$. If $\overline{s} + \overline{a} = 0$, then $\overline{s} \neq 0$ because $\overline{a} \neq 0$. Hence, $s \notin J(A) = \ell_A(\mathrm{Soc}(A))$. Thus, there exists $w \in \mathrm{Soc}(A)$ such that $sw \neq 0$. Thus, $0 \neq v \begin{bmatrix} w & 0 \\ 0 & 0 \end{bmatrix} = \begin{bmatrix} sw & 0 \\ 0 & 0 \end{bmatrix} \in R$ with $\begin{bmatrix} w & 0 \\ 0 & 0 \end{bmatrix} \in R$.

From the previous cases, for each $0 \neq v \in E$, there exists $r \in R$ such that $0 \neq vr \in R$. Therefore, $R_R \leq^{\mathrm{ess}} E_R$.

(iv) Next, we show that $E = E(R_R)$, the injective hull of R_R. For this, let

$$V = \begin{bmatrix} A \oplus ((A/J(A)) & A/J(A) \\ 0 & 0 \end{bmatrix} \text{ and } W = \begin{bmatrix} 0 & 0 \\ A/J(A) & A/J(A) \end{bmatrix}.$$

Then V and W are R-submodules of E_R and $E_R = V_R \oplus W_R$.

Step 1. V_R is an injective R-module. For this, let I be a proper essential right ideal of R. Then $\begin{bmatrix} 0 & 0 \\ 0 & \overline{1} \end{bmatrix} R \cap I \neq 0$. Since $A/J(A)$ is a field, we have that $\begin{bmatrix} 0 & 0 \\ 0 & A/J(A) \end{bmatrix} \subseteq I$. Also as $\begin{bmatrix} 0 & \overline{1} \\ 0 & 0 \end{bmatrix} R \cap I \neq 0$, and so $\begin{bmatrix} 0 & A/J(A) \\ 0 & 0 \end{bmatrix} \subseteq I$. Further, for any $0 \neq a \in J(A)$, $\begin{bmatrix} a & 0 \\ 0 & 0 \end{bmatrix} R \cap I \neq 0$. Thus, there exists a nonzero proper ideal K of A such that

$$I = \begin{bmatrix} K & A/J(A) \\ 0 & A/J(A) \end{bmatrix}.$$

Let $\phi : I_R \to V_R$ be an R-homomorphism. For each $k \in K$, we have that

$$\phi \begin{bmatrix} k & 0 \\ 0 & 0 \end{bmatrix} = \begin{bmatrix} s_k + \overline{a}_k & \overline{b}_k \\ 0 & 0 \end{bmatrix}$$

with $s_k \in A$ and $\overline{a}_k, \overline{b}_k \in A/J(A)$. Then

$$0 = \phi\left(\begin{bmatrix} k & 0 \\ 0 & 0 \end{bmatrix}\begin{bmatrix} 0 & 0 \\ 0 & \overline{1} \end{bmatrix}\right) = \left(\phi\begin{bmatrix} k & 0 \\ 0 & 0 \end{bmatrix}\right)\begin{bmatrix} 0 & 0 \\ 0 & \overline{1} \end{bmatrix} = \begin{bmatrix} 0 & \overline{b}_k \\ 0 & 0 \end{bmatrix},$$

and so $\overline{b}_k = 0$. Note that since K is a proper ideal of A and thus $K \subseteq J(A)$. As $k \in K$, it follows that $\overline{k} = 0$. Hence, we have that

$$0 = \phi\begin{bmatrix} 0 & \overline{k} \\ 0 & 0 \end{bmatrix} = \phi\left(\begin{bmatrix} k & 0 \\ 0 & 0 \end{bmatrix}\begin{bmatrix} 0 & \overline{1} \\ 0 & 0 \end{bmatrix}\right) = \left(\phi\begin{bmatrix} k & 0 \\ 0 & 0 \end{bmatrix}\right)\begin{bmatrix} 0 & \overline{1} \\ 0 & 0 \end{bmatrix}$$
$$= \begin{bmatrix} 0 & \overline{s}_k + \overline{a}_k \\ 0 & 0 \end{bmatrix}.$$

Thus, $\overline{s}_k + \overline{a}_k = 0$. Now let

$$f : K_A \to A_A \text{ defined by } f(k) = s_k.$$

Then f is an A-homomorphism. For this, say $k,\ \ell \in K$. Then

$$\phi\begin{bmatrix} k+\ell & 0 \\ 0 & 0 \end{bmatrix} = \begin{bmatrix} s_{k+\ell} + \overline{a}_{k+\ell} & 0 \\ 0 & 0 \end{bmatrix}$$

and

$$\phi\begin{bmatrix} k+\ell & 0 \\ 0 & 0 \end{bmatrix} = \phi\begin{bmatrix} k & 0 \\ 0 & 0 \end{bmatrix} + \phi\begin{bmatrix} \ell & 0 \\ 0 & 0 \end{bmatrix} = \begin{bmatrix} s_k + \overline{a}_k & 0 \\ 0 & 0 \end{bmatrix} + \begin{bmatrix} s_\ell + \overline{a}_\ell & 0 \\ 0 & 0 \end{bmatrix}$$
$$= \begin{bmatrix} s_k + s_\ell + \overline{a}_k + \overline{a}_\ell & 0 \\ 0 & 0 \end{bmatrix}.$$

Therefore $s_{k+\ell} = s_k + s_\ell$. Thus $f(k+\ell) = f(k) + f(\ell)$. Next take $k \in K$ and $w \in A$. Then

$$\phi\begin{bmatrix} kw & 0 \\ 0 & 0 \end{bmatrix} = \begin{bmatrix} s_{kw} + \overline{a}_{kw} & 0 \\ 0 & 0 \end{bmatrix}$$

and

$$\phi\begin{bmatrix} kw & 0 \\ 0 & 0 \end{bmatrix} = \phi\left(\begin{bmatrix} k & 0 \\ 0 & 0 \end{bmatrix}\begin{bmatrix} w & 0 \\ 0 & 0 \end{bmatrix}\right) = \left(\phi\begin{bmatrix} k & 0 \\ 0 & 0 \end{bmatrix}\right)\begin{bmatrix} w & 0 \\ 0 & 0 \end{bmatrix}$$
$$= \begin{bmatrix} s_k + \overline{a}_k & 0 \\ 0 & 0 \end{bmatrix}\begin{bmatrix} w & 0 \\ 0 & 0 \end{bmatrix} = \begin{bmatrix} s_k w + \overline{a}_k\overline{w} & 0 \\ 0 & 0 \end{bmatrix}.$$

Therefore $s_{kw} = s_k w$. Hence $f(kw) = f(k)w$. So f is an A-homomorphism.

Since A_A is injective, there exists $t \in A$ such that $tk = f(k) = s_k$ for all $k \in K$. So $\overline{s}_k = \overline{t}\,\overline{k} = 0$, and hence, $\overline{a}_k = 0$ because $\overline{s}_k + \overline{a}_k = 0$. Therefore,

$$\phi\begin{bmatrix} k & 0 \\ 0 & 0 \end{bmatrix} = \begin{bmatrix} s_k & 0 \\ 0 & 0 \end{bmatrix} = \begin{bmatrix} tk & 0 \\ 0 & 0 \end{bmatrix}$$

for all $k \in K$. Next, we let $\phi\begin{bmatrix} 0 & \overline{1} \\ 0 & 0 \end{bmatrix} = \begin{bmatrix} s + \overline{a} & \overline{b} \\ 0 & 0 \end{bmatrix}$ with $s \in A$ and $\overline{a},\ \overline{b} \in A/J(A)$. Then

$$0 = \phi(0) = \phi\left(\begin{bmatrix} 0 & \overline{1} \\ 0 & 0 \end{bmatrix}\begin{bmatrix} 1 & 0 \\ 0 & 0 \end{bmatrix}\right) = \left(\phi\begin{bmatrix} 0 & \overline{1} \\ 0 & 0 \end{bmatrix}\right)\begin{bmatrix} 1 & 0 \\ 0 & 0 \end{bmatrix} = \begin{bmatrix} s + \overline{a} & 0 \\ 0 & 0 \end{bmatrix}.$$

So $\phi \begin{bmatrix} 0 & \overline{1} \\ 0 & 0 \end{bmatrix} = \begin{bmatrix} 0 & \overline{b} \\ 0 & 0 \end{bmatrix}$. Further, we let $\phi \begin{bmatrix} 0 & 0 \\ 0 & \overline{1} \end{bmatrix} = \begin{bmatrix} w + \overline{c} & \overline{d} \\ 0 & 0 \end{bmatrix}$ with $w \in A$ and $\overline{c}$, $\overline{d} \in A/J(A)$. Then, it follows that

$$0 = \phi(0) = \phi\left(\begin{bmatrix} 0 & 0 \\ 0 & \overline{1} \end{bmatrix}\begin{bmatrix} 1 & 0 \\ 0 & 0 \end{bmatrix}\right) = \left(\phi\begin{bmatrix} 0 & 0 \\ 0 & \overline{1} \end{bmatrix}\right)\begin{bmatrix} 1 & 0 \\ 0 & 0 \end{bmatrix} = \begin{bmatrix} w + \overline{c} & 0 \\ 0 & 0 \end{bmatrix}.$$

Hence, $w + \overline{c} = 0$, so we have that $\phi \begin{bmatrix} 0 & 0 \\ 0 & \overline{1} \end{bmatrix} = \begin{bmatrix} 0 & \overline{d} \\ 0 & 0 \end{bmatrix}$. Now we take

$$v_0 = \begin{bmatrix} t + \overline{(-t)} + \overline{b} & \overline{d} \\ 0 & 0 \end{bmatrix} \in V.$$

Then $\phi(x) = v_0 x$ for each $x \in I$. Let $\varphi : R_R \to V$ such that $\varphi(r) = v_0 r$ for all $r \in R$. Then $\varphi \in \mathrm{Hom}(R_R, V_R)$ and φ is an extension of ϕ.

In general, let B be a right ideal of R and left $f \in \mathrm{Hom}(B_R, V_R)$. Let C_R be a complement of B_R in R_R. Then $(B \oplus C)_R \leq^{\text{ess}} R_R$. Define $g : B \oplus C \to V$ by $g(b + c) = f(b)$ for $b \in B$ and $c \in C$. Then $g \in \mathrm{Hom}((B \oplus C)_R, V_R)$.

From the preceding arguments, there exists $h \in \mathrm{Hom}(R_R, V_R)$, which is an extension of g. Then for $b \in B$, $h(b) = g(b) = f(b)$ and therefore h is an extension of f. Hence, V_R is an injective R-module by Baer's Criterion.

Step 2. W_R is an injective R-module. For this, let I be a proper essential right ideal of R. Then as in Step 1,

$$I = \begin{bmatrix} K & A/J(A) \\ 0 & A/J(A) \end{bmatrix}$$

for some proper ideal K of A. So $K \subseteq J(A)$. Let $\psi : I_R \to W_R$ be an R-homomorphism. Then

$$\psi \begin{bmatrix} 0 & \overline{1} \\ 0 & 0 \end{bmatrix} = \begin{bmatrix} 0 & 0 \\ \overline{c}_1 & \overline{d}_1 \end{bmatrix}$$

with $\overline{c}_1$, $\overline{d}_1 \in A/J(A)$. Then

$$0 = \psi(0) = \psi\left(\begin{bmatrix} 0 & \overline{1} \\ 0 & 0 \end{bmatrix}\begin{bmatrix} 1 & 0 \\ 0 & 0 \end{bmatrix}\right) = \left(\psi\begin{bmatrix} 0 & \overline{1} \\ 0 & 0 \end{bmatrix}\right)\begin{bmatrix} 1 & 0 \\ 0 & 0 \end{bmatrix} = \begin{bmatrix} 0 & 0 \\ \overline{c}_1 & 0 \end{bmatrix}.$$

Thus, $\overline{c}_1 = 0$, so it follows that

$$\psi \begin{bmatrix} 0 & \overline{1} \\ 0 & 0 \end{bmatrix} = \begin{bmatrix} 0 & 0 \\ 0 & \overline{d}_1 \end{bmatrix}.$$

Next, say

$$\psi \begin{bmatrix} 0 & 0 \\ 0 & \overline{1} \end{bmatrix} = \begin{bmatrix} 0 & 0 \\ \overline{c}_2 & \overline{d}_2 \end{bmatrix},$$

where $\overline{c}_2, \overline{d}_2 \in A/J(A)$. Then it can be checked that $\overline{c}_2 = 0$, and hence $\psi \begin{bmatrix} 0 & 0 \\ 0 & \overline{1} \end{bmatrix} = \begin{bmatrix} 0 & 0 \\ 0 & \overline{d}_2 \end{bmatrix}$. Finally, by routine computation, we have that $\psi \begin{bmatrix} k & 0 \\ 0 & 0 \end{bmatrix} = 0$ for all $k \in K$ since $\overline{k} = 0$. Now we take

$$w_0 = \begin{bmatrix} 0 & 0 \\ \overline{d}_1 & \overline{d}_2 \end{bmatrix} \in W.$$

Then $\psi(x) = w_0 x$ for each $x \in I$. Let $\mu : R_R \to W$ defined by $\mu(r) = w_0 r$ for all $r \in R$. Then $\mu \in \mathrm{Hom}(R_R, W_R)$ and μ is an extension of ψ.

In general, let B be a right ideal of R and let $\alpha \in \mathrm{Hom}(B_R, W_R)$. Then by the argument used in the proof of Step 1, there exists $\beta \in \mathrm{Hom}(R_R, W_R)$, which is an extension of α. Therefore, W_R is an injective R-module by Baer's Criterion.

From Step 1 and Step 2, V_R and W_R are injective R-modules. Hence, $E_R = V_R \oplus W_R$ is an injective R-module. Because $R_R \leq^{\text{ess}} E_R$ by (iii), E_R is the injective hull of R_R.

(v) We establish a compatible ring structure on $E = E(R_R)$. For this, we need to establish multiplication so that the associativity under the induced multiplication as well as distributive laws under the given addition on E and the induced multiplication hold.

It is useful to adopt the idea in the proof of Theorem 2.14. Hence, it is necessary to consider first a ring whose additive abelian group is isomorphic to the additive abelian group $(E, +)$.

Consider the following $\mathbb{E}$ adopted from private communications with B. L. Osofsky, which is

$$\mathbb{E} = \left\{ \begin{bmatrix} s & 0 & 0 \\ 0 & \overline{x} & \overline{y} \\ 0 & \overline{z} & \overline{w} \end{bmatrix} \mid s, x, y, z, w \in A \right\},$$

where $\overline{x}, \overline{y}, \overline{z}$, and $\overline{w}$ are images of x, y, z, and w in $A/J(A)$, respectively. The addition of $\mathbb{E}$ is componentwise and the multiplication $\diamond$ is defined as follows:

For $v_1 = \begin{bmatrix} s_1 & 0 & 0 \\ 0 & \overline{x}_1 & \overline{y}_1 \\ 0 & \overline{z}_1 & \overline{w}_1 \end{bmatrix} \in \mathbb{E}$ and $v_2 = \begin{bmatrix} s_2 & 0 & 0 \\ 0 & \overline{x}_2 & \overline{y}_2 \\ 0 & \overline{z}_2 & \overline{w}_2 \end{bmatrix} \in \mathbb{E}$, we define

$$v_1 \diamond v_2 = \begin{bmatrix} s_1 s_2 & 0 & 0 \\ 0 & \overline{x}_1\overline{x}_2 + \overline{y}_1\overline{z}_2 & \overline{x}_1\overline{y}_2 + \overline{y}_1\overline{w}_2 \\ 0 & \overline{z}_1\overline{x}_2 + \overline{w}_1\overline{z}_2 & \overline{z}_1\overline{y}_2 + \overline{w}_1\overline{w}_2 \end{bmatrix}.$$

Then $(\mathbb{E}, +, \diamond)$ is ring. Further, $(\mathbb{E}, +, \diamond) \cong A \oplus \mathrm{Mat}_2(A/J(A))$ as rings, and so $(\mathbb{E}, +, \diamond)$ is a QF-ring because $A/J(A)$ is a field.

To induce a compatible ring structure on E from the ring $(\mathbb{E}, +, \diamond)$, first, consider $\eta : E \to \mathbb{E}$ defined by

$$\eta\begin{bmatrix} s+\overline{x} & \overline{y} \\ \overline{z} & \overline{w} \end{bmatrix} = \begin{bmatrix} s & 0 & 0 \\ 0 & \overline{x} & \overline{y} \\ 0 & \overline{z} & \overline{w} \end{bmatrix},$$

where $s, x, y, z, w \in A$, and $\overline{x}, \overline{y}$, etc., are images of x, y, etc., in $A/J(A)$. Then η is an additive abelian group isomorphism. By adopting the idea in the proof of Theorem 2.14, we define a ring multiplication $\bullet$ on E. For this, let $v_1 = \begin{bmatrix} s_1+\overline{x}_1 & \overline{y}_1 \\ \overline{z}_1 & \overline{w}_1 \end{bmatrix} \in E$ and $v_2 = \begin{bmatrix} s_2+\overline{x}_2 & \overline{y}_2 \\ \overline{z}_2 & \overline{w}_2 \end{bmatrix} \in E$. Define

$$v_1 \bullet v_2 = \eta^{-1}(\eta(v_1) \diamond \eta(v_2)).$$

Then we have that

$$v_1 \bullet v_2 = \eta^{-1}(\eta(v_1) \diamond \eta(v_2)) = \eta^{-1}\left(\begin{bmatrix} s_1 & 0 & 0 \\ 0 & \overline{x}_1 & \overline{y}_1 \\ 0 & \overline{z}_1 & \overline{w}_1 \end{bmatrix} \diamond \begin{bmatrix} s_2 & 0 & 0 \\ 0 & \overline{x}_2 & \overline{y}_2 \\ 0 & \overline{z}_2 & \overline{w}_2 \end{bmatrix}\right)$$

$$= \eta^{-1}\begin{bmatrix} s_1s_2 & 0 & 0 \\ 0 & \overline{x}_1\overline{x}_2+\overline{y}_1\overline{z}_2 & \overline{x}_1\overline{y}_2+\overline{y}_1\overline{w}_2 \\ 0 & \overline{z}_1\overline{x}_2+\overline{w}_1\overline{z}_2 & \overline{z}_1\overline{y}_2+\overline{w}_1\overline{w}_2 \end{bmatrix} = \begin{bmatrix} s_1s_2+\overline{x}_1\overline{x}_2+\overline{y}_1\overline{z}_2 & \overline{x}_1\overline{y}_2+\overline{y}_1\overline{w}_2 \\ \overline{z}_1\overline{x}_2+\overline{w}_1\overline{z}_2 & \overline{z}_1\overline{y}_2+\overline{w}_1\overline{w}_2 \end{bmatrix}.$$

Thus, $(E, +, \bullet) \cong (\mathbb{E}, +, \diamond)$ (as rings) via η since $\eta$is an additive abelian group isomorphism and $\eta(v_1 \bullet v_2) = \eta(v_1) \diamond \eta(v_2)$ for $v_1, v_2 \in E$.

Now for $v = \begin{bmatrix} s+\overline{x} & \overline{y} \\ \overline{z} & \overline{w} \end{bmatrix} \in E$ with $s, x, y, z, w \in A$ and $r = \begin{bmatrix} a & b \\ 0 & c \end{bmatrix} \in R$ with $a, b, c \in A$, we note that $v \bullet r = \eta^{-1}(\eta(v) \diamond \eta(r))$. Hence we have that

$$v \bullet r = \begin{bmatrix} sa & \overline{xb}+\overline{yc} \\ 0 & \overline{zb}+\overline{wc} \end{bmatrix}, \quad \text{while } vr = \begin{bmatrix} sa+\overline{xa} & \overline{sb}+\overline{xb}+\overline{yc} \\ \overline{za} & \overline{zb}+\overline{wc} \end{bmatrix}.$$

So $v \bullet r \neq vr$ in general. Thus, $(E, +, \bullet)$ is not compatible with the given R-module scalar multiplication of E over R.

Because the ring multiplication $\bullet$ on E is disqualified to be compatible with the given R-module scalar multiplication of $E = E(R_R)$ over R, first, we let $\sigma : R \to \mathbb{E}$ defined by

$$\sigma\begin{bmatrix} a & b \\ 0 & c \end{bmatrix} = \begin{bmatrix} a & 0 & 0 \\ 0 & \overline{a} & \overline{b} \\ 0 & \overline{0} & \overline{c} \end{bmatrix} \quad \text{where } \begin{bmatrix} a & b \\ 0 & c \end{bmatrix} \in R.$$

Then σ is a one-to-one ring homomorphism. Consider an extension $\theta : E(R_R) \to \mathbb{E}$ of σ defined by

$$\theta\begin{bmatrix} s+\overline{x} & \overline{y} \\ \overline{z} & \overline{w} \end{bmatrix} = \begin{bmatrix} s & 0 & 0 \\ 0 & \overline{s}+\overline{x} & \overline{y} \\ 0 & \overline{z} & \overline{w} \end{bmatrix} \quad \text{for} \quad \begin{bmatrix} s+\overline{x} & \overline{y} \\ \overline{z} & \overline{w} \end{bmatrix} \in E(R_R).$$

Then θ is an additive abelian group isomorphism. For $v_1,\ v_2 \in E$, define

$$v_1 \circ v_2 = \theta^{-1}(\theta(v_1) \diamond \theta(v_2))$$

by adopting the idea in the proof of Theorem 2.14.

Say $v_1 := \begin{bmatrix} s_1+\overline{x_1} & \overline{y_1} \\ \overline{z_1} & \overline{w_1} \end{bmatrix}$, $v_2 := \begin{bmatrix} s_2+\overline{x_2} & \overline{y_2} \\ \overline{z_2} & \overline{w_2} \end{bmatrix} \in E$, where $s_1,\ s_2, x_1,\ x_2$, etc., are in A. Then

$$v_1 \circ v_2 = \theta^{-1}(\theta(v_1) \diamond \theta(v_2)) = \theta^{-1}\left(\begin{bmatrix} s_1 & 0 & 0 \\ 0 & \overline{s}_1+\overline{x}_1 & \overline{y}_1 \\ 0 & \overline{z}_1 & \overline{w}_1 \end{bmatrix} \diamond \begin{bmatrix} s_2 & 0 & 0 \\ 0 & \overline{s}_2+\overline{x}_2 & \overline{y}_2 \\ 0 & \overline{z}_2 & \overline{w}_2 \end{bmatrix}\right)$$

$$= \theta^{-1}\begin{bmatrix} s_1s_2 & 0 & 0 \\ 0 & \overline{s}_1\overline{s}_2+\overline{s}_1\overline{x}_2+\overline{x}_1\overline{s}_2+\overline{x}_1\overline{x}_2+\overline{y_1z_2} & \overline{s}_1\overline{y}_2+\overline{x}_1\overline{y}_2+\overline{y}_1\overline{w}_2 \\ 0 & \overline{z}_1\overline{s}_2+\overline{z}_2\overline{x}_2+\overline{w}_1\overline{z}_2 & \overline{z}_1\overline{y}_2+\overline{w}_1\overline{w}_2 \end{bmatrix}$$

$$= \begin{bmatrix} s_1s_2+\overline{s}_1\overline{x}_2+\overline{x}_1\overline{s}_2+\overline{x}_1\overline{x}_2+\overline{y}_1\overline{z}_2 & \overline{s}_1\overline{y}_2+\overline{x}_1\overline{y}_2+\overline{y}_1\overline{w}_2 \\ \overline{z}_1\overline{s}_2+\overline{z}_1\overline{x}_2+\overline{w}_1\overline{z}_2 & \overline{z}_1\overline{y}_2+\overline{w}_1\overline{w}_2 \end{bmatrix}.$$

Consequently, we have that

$$v_1 \circ v_2 = \begin{bmatrix} s_1s_2+\overline{s}_1\overline{x}_2+\overline{x}_1\overline{s}_2+\overline{x}_1\overline{x}_2+\overline{y}_1\overline{z}_2 & \overline{s}_1\overline{y}_2+\overline{x}_1\overline{y}_2+\overline{y}_1\overline{w}_2 \\ \overline{z}_1\overline{s}_2+\overline{z}_1\overline{x}_2+\overline{w}_1\overline{z}_2 & \overline{z}_1\overline{y}_2+\overline{w}_1\overline{w}_2 \end{bmatrix}.$$

Furthermore, since θ is an additive abelian group isomorphism and $\theta(v_1 \circ v_2) = \theta(v_1) \diamond \theta(v_2)$ for $v_1,\ v_2 \in E$, it follows that $(E,\ +,\ \circ) \cong (\mathbb{E},\ +,\ \diamond)$ (as rings) via θ. Therefore, $(E, +, \circ)$ is a QF-ring because $(\mathbb{E}, +, \diamond)$ is a QF-ring.

Finally, for the compatibility of the multiplication $\circ$ of E with the R-module scalar multiplication of E over R, take $v = \begin{bmatrix} s+\overline{x} & \overline{y} \\ \overline{z} & \overline{w} \end{bmatrix} \in E$, where $s,\ x,\ y,\ z,\ w \in A$, and $r = \begin{bmatrix} a & \overline{b} \\ 0 & \overline{c} \end{bmatrix} \in R$, where $a,\ b,\ c \in A$. Then

$$v \circ r = \begin{bmatrix} s+\overline{x} & \overline{y} \\ \overline{z} & \overline{w} \end{bmatrix} \circ \begin{bmatrix} a & \overline{b} \\ 0 & \overline{c} \end{bmatrix} = \begin{bmatrix} sa+\overline{xa} & \overline{sb}+\overline{xb}+\overline{yc} \\ \overline{za} & \overline{zb}+\overline{wc} \end{bmatrix} = vr.$$

Therefore, the ring structure $(E,\ +,\ \circ)$ is compatible with the R-module scalar multiplication of E over R.

(vi) It is interesting to remark from part (v):

(1) $(E,\ +,\ \circ) \cong (E,\ +,\ \bullet)$ (as rings), which are QF-rings. Indeed, note that $\theta : (E, +, \circ) \to (\mathbb{E}, +, \diamond)$ and $\eta : (E, +, \bullet) \to (\mathbb{E}, +, \diamond)$ are ring isomorphisms.

Thus, $\eta^{-1}\theta : (E, +, \circ) \to (E, +, \bullet)$ is a ring isomorphism. Further, as $(\mathbb{E}, +, \diamond)$ is a QF-ring, so are $(E, +, \circ)$ and $(E, +, \bullet)$.

(2) $(E, +, \circ)$ is compatible with the right R-module scalar multiplication of E, but $(E, +, \bullet)$ is not compatible with the R-module scalar multiplication of E over R by part (v).

(vii) We construct other compatible ring structures on E which are distinct from $(E, +, \circ)$. For this, first, note that the identity map of R_R can be extended to an R-isomorphism of E since E is an injective hull of R_R. Now let $f : E \to E$ be an R-isomorphism such that $f(r) = r$ for all $r \in R$, that is, f is an R-isomorphism of E which is an extension of the identity map of R_R. Then there exist $\alpha, \beta \in \mathrm{Soc}(A)$ such that

$$f\begin{bmatrix}\overline{1} & 0\\ 0 & 0\end{bmatrix} = \begin{bmatrix}\alpha+\overline{1} & 0\\ 0 & 0\end{bmatrix} \quad \text{and} \quad f\begin{bmatrix}0 & 0\\ \overline{1} & 0\end{bmatrix} = \begin{bmatrix}\beta & 0\\ \overline{1} & 0\end{bmatrix}.$$

For this, first, we put $f\begin{bmatrix}\overline{1} & 0\\ 0 & 0\end{bmatrix} = \begin{bmatrix}s_1+\overline{a} & \overline{b}\\ \overline{c} & \overline{d}\end{bmatrix} \in E$, where $s_1, a, b, c, d \in A$. Then

$$\begin{aligned} f\begin{bmatrix}\overline{1} & 0\\ 0 & 0\end{bmatrix} &= f\left(\begin{bmatrix}\overline{1} & 0\\ 0 & 0\end{bmatrix}\begin{bmatrix}1 & 0\\ 0 & 0\end{bmatrix}\right) = \left(f\begin{bmatrix}\overline{1} & 0\\ 0 & 0\end{bmatrix}\right)\begin{bmatrix}1 & 0\\ 0 & 0\end{bmatrix} \\ &= \begin{bmatrix}s_1+\overline{a} & \overline{b}\\ \overline{c} & \overline{d}\end{bmatrix}\begin{bmatrix}1 & 0\\ 0 & 0\end{bmatrix} = \begin{bmatrix}s_1+\overline{a} & 0\\ \overline{c} & 0\end{bmatrix}\end{aligned}$$

because $\begin{bmatrix}1 & 0\\ 0 & 0\end{bmatrix} \in R$. Note that $\begin{bmatrix}0 & \overline{1}\\ 0 & 0\end{bmatrix} \in R$. As $f(r) = r$ for all $r \in R$, we have that

$$\begin{aligned} \begin{bmatrix}0 & \overline{1}\\ 0 & 0\end{bmatrix} &= f\begin{bmatrix}0 & \overline{1}\\ 0 & 0\end{bmatrix} = f\left(\begin{bmatrix}\overline{1} & 0\\ 0 & 0\end{bmatrix}\begin{bmatrix}0 & \overline{1}\\ 0 & 0\end{bmatrix}\right) = \left(f\begin{bmatrix}\overline{1} & 0\\ 0 & 0\end{bmatrix}\right)\begin{bmatrix}0 & \overline{1}\\ 0 & 0\end{bmatrix} \\ &= \begin{bmatrix}s_1+\overline{a} & 0\\ \overline{c} & 0\end{bmatrix}\begin{bmatrix}0 & \overline{1}\\ 0 & 0\end{bmatrix} = \begin{bmatrix}0 & \overline{s_1}+\overline{a}\\ 0 & \overline{c}\end{bmatrix}.\end{aligned}$$

Therefore, $\overline{s_1}+\overline{a} = \overline{1}$ and $\overline{c} = 0$. So $f\begin{bmatrix}\overline{1} & 0\\ 0 & 0\end{bmatrix} = \begin{bmatrix}s_1+\overline{a} & 0\\ 0 & 0\end{bmatrix}$ with $\overline{s_1}+\overline{a} = \overline{1}$. Take $x \in J(A)$. Then $f\left(\begin{bmatrix}\overline{1} & 0\\ 0 & 0\end{bmatrix}\begin{bmatrix}x & 0\\ 0 & 0\end{bmatrix}\right) = f\begin{bmatrix}\overline{x} & 0\\ 0 & 0\end{bmatrix} = f(0) = 0$. On the other hand, as $\begin{bmatrix}x & 0\\ 0 & 0\end{bmatrix} \in R$, we have that

$$\begin{aligned} f\left(\begin{bmatrix}\overline{1} & 0\\ 0 & 0\end{bmatrix}\begin{bmatrix}x & 0\\ 0 & 0\end{bmatrix}\right) &= \left(f\begin{bmatrix}\overline{1} & 0\\ 0 & 0\end{bmatrix}\right)\begin{bmatrix}x & 0\\ 0 & 0\end{bmatrix} = \begin{bmatrix}s_1+\overline{a} & 0\\ 0 & 0\end{bmatrix}\begin{bmatrix}x & 0\\ 0 & 0\end{bmatrix} \\ &= \begin{bmatrix}s_1x+\overline{ax} & 0\\ 0 & 0\end{bmatrix} = \begin{bmatrix}s_1x & 0\\ 0 & 0\end{bmatrix}.\end{aligned}$$

Thus, $0 = \begin{bmatrix} s_1x & 0 \\ 0 & 0 \end{bmatrix}$ because $\overline{ax}$=0, so $s_1x = 0$ for any $x \in J(A)$. Since Soc(A)= $\ell(J(A))$, it follows that $s_1 \in \mathrm{Soc}(A)$. Now we put α=s_1. Then $f\begin{bmatrix} \overline{1} & 0 \\ 0 & 0 \end{bmatrix} = \begin{bmatrix} \alpha + \overline{a} & 0 \\ 0 & 0 \end{bmatrix}$ with $\overline{\alpha} + \overline{a} = \overline{1}$. Since $\alpha \in \mathrm{Soc}(A) \subseteq J(A)$, $\overline{a} = \overline{1}$ from $\overline{\alpha} + \overline{a} = \overline{1}$. Therefore, we have that

$$f\begin{bmatrix} \overline{1} & 0 \\ 0 & 0 \end{bmatrix} = \begin{bmatrix} \alpha + \overline{1} & 0 \\ 0 & 0 \end{bmatrix} \quad \text{with} \quad \alpha \in \mathrm{Soc}(A).$$

Next, we put $f\begin{bmatrix} 0 & 0 \\ \overline{1} & 0 \end{bmatrix} = \begin{bmatrix} s_2 + \overline{a} & \overline{b} \\ \overline{c} & \overline{d} \end{bmatrix} \in E$ with s_2, a, b, c, $d \in A$. Then since $\begin{bmatrix} 1 & 0 \\ 0 & 0 \end{bmatrix} \in R$,

$$f\begin{bmatrix} 0 & 0 \\ \overline{1} & 0 \end{bmatrix} = f\left(\begin{bmatrix} 0 & 0 \\ \overline{1} & 0 \end{bmatrix}\begin{bmatrix} 1 & 0 \\ 0 & 0 \end{bmatrix}\right) = \left(f\begin{bmatrix} 0 & 0 \\ \overline{1} & 0 \end{bmatrix}\right)\begin{bmatrix} 1 & 0 \\ 0 & 0 \end{bmatrix}$$
$$= \begin{bmatrix} s_2 + \overline{a} & \overline{b} \\ \overline{c} & \overline{d} \end{bmatrix}\begin{bmatrix} 1 & 0 \\ 0 & 0 \end{bmatrix} = \begin{bmatrix} s_2 + \overline{a} & 0 \\ \overline{c} & 0 \end{bmatrix}.$$

On the other hand, since $\begin{bmatrix} 0 & 0 \\ 0 & \overline{1} \end{bmatrix} \in R$ and $f(r) = r$ for all $r \in R$, it follows that

$$\begin{bmatrix} 0 & 0 \\ 0 & \overline{1} \end{bmatrix} = f\begin{bmatrix} 0 & 0 \\ 0 & \overline{1} \end{bmatrix} = f\left(\begin{bmatrix} 0 & 0 \\ \overline{1} & 0 \end{bmatrix}\begin{bmatrix} 0 & \overline{1} \\ 0 & 0 \end{bmatrix}\right) = \left(f\begin{bmatrix} 0 & 0 \\ \overline{1} & 0 \end{bmatrix}\right)\begin{bmatrix} 0 & \overline{1} \\ 0 & 0 \end{bmatrix}$$
$$= \begin{bmatrix} s_2 + \overline{a} & 0 \\ \overline{c} & 0 \end{bmatrix}\begin{bmatrix} 0 & \overline{1} \\ 0 & 0 \end{bmatrix} = \begin{bmatrix} 0 & \overline{s_2} + \overline{a} \\ 0 & \overline{c} \end{bmatrix}.$$

So $\overline{s_2} + \overline{a} = 0$ and $\overline{c} = \overline{1}$. Therefore $f\begin{bmatrix} 0 & 0 \\ \overline{1} & 0 \end{bmatrix} = \begin{bmatrix} s_2 + \overline{a} & 0 \\ \overline{1} & 0 \end{bmatrix}$ with $\overline{s_2} + \overline{a} = 0$.

Take $x \in J(A)$. Then $\begin{bmatrix} x & 0 \\ 0 & 0 \end{bmatrix} \in R$ and $\begin{bmatrix} 0 & 0 \\ \overline{1} & 0 \end{bmatrix}\begin{bmatrix} x & 0 \\ 0 & 0 \end{bmatrix} = 0$. Hence

$$0 = f\left(\begin{bmatrix} 0 & 0 \\ \overline{1} & 0 \end{bmatrix}\begin{bmatrix} x & 0 \\ 0 & 0 \end{bmatrix}\right) = \left(f\begin{bmatrix} 0 & 0 \\ \overline{1} & 0 \end{bmatrix}\right)\begin{bmatrix} x & 0 \\ 0 & 0 \end{bmatrix} = \begin{bmatrix} s_2 + \overline{a} & 0 \\ \overline{1} & 0 \end{bmatrix}\begin{bmatrix} x & 0 \\ 0 & 0 \end{bmatrix}$$
$$= \begin{bmatrix} s_2x + \overline{a}\,\overline{x} & 0 \\ \overline{x} & 0 \end{bmatrix} = \begin{bmatrix} s_2x & 0 \\ 0 & 0 \end{bmatrix}$$

as $x \in J(A)$. Thus, $s_2x = 0$ for any $x \in J(A)$ and $s_2 \in \mathrm{Soc}(A)$ since $\mathrm{Soc}(A) = \ell(J(A))$. Further, $0 = \overline{s_2} + \overline{a}$ and $s_2 \in \mathrm{Soc}(A) \subseteq J(A)$. So $\overline{s_2} = 0$ and thus $\overline{a} = 0$. $f\begin{bmatrix} 0 & 0 \\ \overline{1} & 0 \end{bmatrix} = \begin{bmatrix} s_2 + \overline{a} & 0 \\ \overline{1} & 0 \end{bmatrix} = \begin{bmatrix} s_2 & 0 \\ \overline{1} & 0 \end{bmatrix}$.

Now put $\beta = s_2$. Then

$$f\begin{bmatrix}0 & 0\\ \overline{1} & 0\end{bmatrix} = \begin{bmatrix}\beta & 0\\ \overline{1} & 0\end{bmatrix} \quad \text{with} \quad \beta \in \mathrm{Soc}(A).$$

For $\begin{bmatrix}s+\overline{a} & \overline{b}\\ \overline{c} & \overline{d}\end{bmatrix} \in E$, we have that

$$\begin{aligned}
f\begin{bmatrix}s+\overline{a} & \overline{b}\\ \overline{c} & \overline{d}\end{bmatrix} &= f\begin{bmatrix}s & \overline{b}\\ 0 & \overline{d}\end{bmatrix} + f\begin{bmatrix}\overline{a} & 0\\ 0 & 0\end{bmatrix} + f\begin{bmatrix}0 & 0\\ \overline{c} & 0\end{bmatrix}\\
&= \begin{bmatrix}s & \overline{b}\\ 0 & \overline{d}\end{bmatrix} + f\left(\begin{bmatrix}\overline{1} & 0\\ 0 & 0\end{bmatrix}\begin{bmatrix}a & 0\\ 0 & 0\end{bmatrix}\right) + f\left(\begin{bmatrix}0 & 0\\ \overline{1} & 0\end{bmatrix}\begin{bmatrix}c & 0\\ 0 & 0\end{bmatrix}\right)\\
&= \begin{bmatrix}s & \overline{b}\\ 0 & \overline{d}\end{bmatrix} + \left(f\begin{bmatrix}\overline{1} & 0\\ 0 & 0\end{bmatrix}\right)\begin{bmatrix}a & 0\\ 0 & 0\end{bmatrix} + \left(f\begin{bmatrix}0 & 0\\ \overline{1} & 0\end{bmatrix}\right)\begin{bmatrix}c & 0\\ 0 & 0\end{bmatrix}\\
&= \begin{bmatrix}s & \overline{b}\\ 0 & \overline{d}\end{bmatrix} + \begin{bmatrix}\alpha+\overline{1} & 0\\ 0 & 0\end{bmatrix}\begin{bmatrix}a & 0\\ 0 & 0\end{bmatrix} + \begin{bmatrix}\beta & 0\\ \overline{1} & 0\end{bmatrix}\begin{bmatrix}c & 0\\ 0 & 0\end{bmatrix}\\
&= \begin{bmatrix}s & \overline{b}\\ 0 & \overline{d}\end{bmatrix} + \begin{bmatrix}\alpha a+\overline{a} & 0\\ 0 & 0\end{bmatrix} + \begin{bmatrix}\beta c & 0\\ \overline{c} & 0\end{bmatrix}\\
&= \begin{bmatrix}s+\alpha a+\beta c+\overline{a} & \overline{b}\\ \overline{c} & \overline{d}\end{bmatrix}.
\end{aligned}$$

Conversely, for $(\alpha, \beta) \in \mathrm{Soc}(A) \times \mathrm{Soc}(A)$, let

$$f_{(\alpha,\beta)}\begin{bmatrix}s+\overline{a} & \overline{b}\\ \overline{c} & \overline{d}\end{bmatrix} = \begin{bmatrix}s+\alpha a+\beta c+\overline{a} & \overline{b}\\ \overline{c} & \overline{d}\end{bmatrix}.$$

Then $f_{(\alpha,\beta)}$ is an R-homomorphism of E such that $f_{(\alpha,\beta)}(r) = r$ for all $r \in R$ because $\alpha, \beta \in \mathrm{Soc}(A)$. So $f_{(\alpha,\beta)}$ is an R-isomorphism of E. Therefore

$$\{f_{(\alpha,\beta)} \mid (\alpha, \beta) \in \mathrm{Soc}(A) \times \mathrm{Soc}(A)\}$$

is the set of all R-isomorphisms of E for which each R-isomorphism of E is an extension of the identity map of R_R.

Take $(\alpha, \beta) \in \mathrm{Soc}(A) \times \mathrm{Soc}(A)$. Define $\circ_{(\alpha,\beta)}$ on E by adopting the idea of Theorem 2.14:

For $v_1 = \begin{bmatrix}s_1+\overline{a_1} & \overline{b_1}\\ \overline{c_1} & \overline{d_1}\end{bmatrix}$, $v_2 = \begin{bmatrix}s_2+\overline{a_2} & \overline{b_2}\\ \overline{c_2} & \overline{d_2}\end{bmatrix} \in E$, let

$$v_1 \circ_{(\alpha,\beta)} v_2 = f^{-1}_{(\alpha,\beta)}[f_{(\alpha,\beta)}(v_1) \circ f_{(\alpha,\beta)}(v_2)],$$

where $\circ$ is the ring multiplication of E defined in part (v). Then

$$v_1 \circ_{(\alpha,\beta)} v_2 = f_{(-\alpha,-\beta)}[f_{(\alpha,\beta)}(v_1) \circ f_{(\alpha,\beta)}(v_2)]$$

because $f_{(\alpha,\beta)}^{-1} = f_{(-\alpha,-\beta)}$. We can check that $\circ_{(\alpha,\beta)}$ extends the R-module scalar multiplication of E over R. Therefore, $(E, +, \circ_{(\alpha,\beta)})$ is a compatible ring structure on E.

By routine calculation,

$$\begin{bmatrix} s_1 + \overline{a_1} & \overline{b_1} \\ \overline{c_1} & \overline{d_1} \end{bmatrix} \circ_{(\alpha,\beta)} \begin{bmatrix} s_2 + \overline{a_2} & \overline{b_2} \\ \overline{c_2} & \overline{d_2} \end{bmatrix} = \begin{bmatrix} x & y \\ z & w \end{bmatrix},$$

where

$$\begin{aligned} x &= s_1 s_2 + (-\alpha) a_1 a_2 + (-\beta) c_1 a_2 + \beta s_1 c_2 + (-\alpha) b_1 c_2 + (-\beta) d_1 c_2 \\ &\quad + \overline{a_1 a_2} + \overline{a_1 s_2} + \overline{s_1 a_2} + \overline{b_1 c_2}, \\ y &= \overline{a_1 b_2} + \overline{s_1 b_2} + \overline{b_1 d_2}, \\ z &= \overline{c_1 s_2} + \overline{c_1 a_2} + \overline{d_1 c_2}, \text{ and } w = \overline{c_1 b_2} + \overline{d_1 d_2}. \end{aligned}$$

We see that $\circ = \circ_{(0,0)}$. Further, as $v_1 \circ_{(\alpha,\beta)} v_2 = f_{(\alpha,\beta)}^{-1}[f_{(\alpha,\beta)}(v_1) \circ f_{(\alpha,\beta)}(v_2)]$, it follows that

$$f_{(\alpha,\beta)}(v_1 \circ_{(\alpha,\beta)} v_2) = f_{(\alpha,\beta)}(v_1) \circ f_{(\alpha,\beta)}(v_2).$$

Therefore, $f_{(\alpha,\beta)} : (E, +, \circ_{(\alpha,\beta)}) \to (E, +, \circ)$ is a ring isomorphism. So the ring $(E, +, \circ_{(\alpha,\beta)})$ is a QF-ring for any $(\alpha, \beta) \in \text{Soc}(A) \times \text{Soc}(A)$ because the ring $(E, +, \circ)$ is a QF-ring by part (v).

(viii) For (α, β), $(\gamma, \delta) \in \text{Soc}(A) \times \text{Soc}(A)$, we show that

$$\circ_{(\alpha,\beta)} = \circ_{(\gamma,\delta)} \text{ if and only if } (\alpha, \beta) = (\gamma, \delta).$$

First, suppose that $\circ_{(\alpha,\beta)} = \circ_{(\gamma,\delta)}$. As $\text{Soc}(A) \subseteq J(A)$, $\text{Soc}(A) = \ell_A(J(A))$, and $f_{(\alpha,\beta)}^{-1} = f_{(-\alpha,-\beta)}$, it follows that

$$\begin{aligned} \begin{bmatrix} \overline{1} & 0 \\ 0 & 0 \end{bmatrix} \circ_{(\alpha,\beta)} \begin{bmatrix} \overline{1} & 0 \\ 0 & 0 \end{bmatrix} &= f_{(-\alpha,-\beta)} \left(f_{(\alpha,\beta)} \begin{bmatrix} \overline{1} & 0 \\ 0 & 0 \end{bmatrix} \circ f_{(\alpha,\beta)} \begin{bmatrix} \overline{1} & 0 \\ 0 & 0 \end{bmatrix} \right) \\ &= f_{(-\alpha,-\beta)} \left(\begin{bmatrix} \alpha + \overline{1} & 0 \\ 0 & 0 \end{bmatrix} \circ \begin{bmatrix} \alpha + \overline{1} & 0 \\ 0 & 0 \end{bmatrix} \right) \\ &= f_{(-\alpha,-\beta)} \begin{bmatrix} \overline{1} & 0 \\ 0 & 0 \end{bmatrix} = \begin{bmatrix} (-\alpha) + \overline{1} & 0 \\ 0 & 0 \end{bmatrix}. \end{aligned}$$

Similarly,

$$\begin{bmatrix} \overline{1} & 0 \\ 0 & 0 \end{bmatrix} \circ_{(\gamma,\delta)} \begin{bmatrix} \overline{1} & 0 \\ 0 & 0 \end{bmatrix} = \begin{bmatrix} (-\gamma) + \overline{1} & 0 \\ 0 & 0 \end{bmatrix}.$$

Since $\circ_{(\alpha,\beta)} = \circ_{(\gamma,\delta)}$, it follows that $(-\alpha) + \overline{1} = (-\gamma) + \overline{1}$, and hence, $-\alpha = -\gamma$. Therefore, $\alpha = \gamma$.

On the other hand, we have that

$$\begin{aligned}\begin{bmatrix} 0 & 0 \\ \bar{1} & 0 \end{bmatrix} \circ_{(\alpha,\beta)} \begin{bmatrix} \bar{1} & 0 \\ 0 & 0 \end{bmatrix} &= f_{(-\alpha,-\beta)} \left(f_{(\alpha,\beta)} \begin{bmatrix} 0 & 0 \\ \bar{1} & 0 \end{bmatrix} \circ f_{(\alpha,\beta)} \begin{bmatrix} \bar{1} & 0 \\ 0 & 0 \end{bmatrix} \right) \\ &= f_{(-\alpha,-\beta)} \left(\begin{bmatrix} \beta & 0 \\ \bar{1} & 0 \end{bmatrix} \circ \begin{bmatrix} \alpha + \bar{1} & 0 \\ 0 & 0 \end{bmatrix} \right) \\ &= f_{(-\alpha,-\beta)} \begin{bmatrix} 0 & 0 \\ \bar{1} & 0 \end{bmatrix} = \begin{bmatrix} -\beta & 0 \\ \bar{1} & 0 \end{bmatrix}.\end{aligned}$$

Similarly,
$$\begin{bmatrix} 0 & 0 \\ \bar{1} & 0 \end{bmatrix} \circ_{(\gamma,\delta)} \begin{bmatrix} \bar{1} & 0 \\ 0 & 0 \end{bmatrix} = \begin{bmatrix} -\delta & 0 \\ \bar{1} & 0 \end{bmatrix}.$$

Since $\circ_{(\alpha,\beta)} = \circ_{(\gamma,\delta)}$, it follows that $-\beta = -\delta$, and hence, $\beta = \delta$. Consequently, $(\alpha, \beta) = (\gamma, \delta)$. Conversely, if $(\alpha, \beta) = (\gamma, \delta)$, then $\alpha = \gamma$ and $\beta = \delta$, so obviously $\circ_{(\alpha,\beta)} = \circ_{(\gamma,\delta)}$.

(ix) Let $\mathscr{F}$ be the set of all R-isomorphisms f of E such that $f(r) = r$ for any $r \in R$. Define

$$\theta : \mathrm{Soc}(A) \times \mathrm{Soc}(A) \to \mathscr{F} \text{ by } \theta(\alpha, \beta) = f_{(\alpha,\beta)}$$

for $(\alpha, \beta) \in \mathrm{Soc}(A) \times \mathrm{Soc}(A)$. Then by parts (vii) and (viii), θ is a one-to-one onto map. Therefore, it follows that $\mathscr{F} = \{f_{(\alpha,\beta)} \mid (\alpha, \beta) \in \mathrm{Soc}(A) \times \mathrm{Soc}(A)\}$ and $|\mathrm{Soc}(A) \times \mathrm{Soc}(A)| = |\mathscr{F}|$. Also note that the map

$$\nu : \mathscr{F} \to \{\circ_{\alpha,\beta)} \mid (\alpha, \beta) \in \mathrm{Soc}(A) \times \mathrm{Soc}(A)\} \text{ defined by } \nu(f_{(\alpha,\beta)}) = \circ_{(\alpha,\beta)}$$

is a one-to-one onto map. Hence

$$|\{(E, +, \circ_{(\alpha,\beta)}) \mid (\alpha, \beta) \in \mathrm{Soc}(A) \times \mathrm{Soc}(A)\}| = |\mathscr{F}| = |\mathrm{Soc(A)}|^2.$$

Therefore, E has $|\mathrm{Soc}(A)|^2$ compatible ring structures which are

$$\{(E, +, \circ_{(\alpha,\beta)}) \mid (\alpha, \beta) \in \mathrm{Soc}(A) \times \mathrm{Soc}(A)\}.$$

(x) From the preceding arguments, as a byproduct we can construct a ring R for which $E(R_R)$ has infinitely many distinct compatible ring structures. For this, let F be an infinite field, $p(x) \in F[x]$ be an irreducible polynomial, and let n be an integer such that $n > 1$. We put
$$A = F[x]/p(x)^n F[x].$$

Since $F[x]$ is a Dedekind domain, the ring A is a commutative QF-ring (see [8, Theorem 6.14, p. 174]). Because $J(A) = p(x)F[x]/p(x)^n F[x]$, $A/J(A) \cong F[x]/p(x)F[x]$ (so $A/J(A)$ is a field), and thus the ring A is local.

Note that $|\mathrm{Soc}(A)| \geq |F|$ because $\mathrm{Soc}(A) = p(x)^{n-1}F[x]/p(x)^n F[x]$ is a vector space over the field F. Since F is infinite, so is $|\mathrm{Soc}(A)|$. Let

$$R = \begin{bmatrix} A & A/J(A) \\ 0 & A/J(A) \end{bmatrix}$$

as in part (i). Then $E(R_R)$ has $|\mathrm{Soc}(A)|^2$ compatible ring structures by the preceding arguments.

(xi) Let p be a prime integer, G be the cyclic group of order p, and let K be an infinite field of characteristic p. Consider $A = K[G]$, the group ring of the group G over the field K. Then from [7, Lemma 1.17, p. 314],

$$J(A) = \left\{ \sum a_g g \mid \sum a_g = 0, \quad \text{where } a_g \in K \text{ and } g \in G \right\},$$

which is the augmentation ideal $\omega(K[G])$. Hence, $A/J(A) \cong K$, and so A is local and commutative. Further, A is QF by [3]. Hence, $\mathrm{Soc}(A) \neq 0$, so $\mathrm{Soc}(A)$ is a nonzero vector space over the infinite field K. Therefore, $|\mathrm{Soc}(A)|$ is infinite. As in part (i), put

$$R = \begin{bmatrix} A & A/J(A) \\ 0 & A/J(A) \end{bmatrix}.$$

In this case, also from the preceding arguments, $E(R_R)$ has $|\mathrm{Soc}(A)|^2$ compatible ring structures.

(xii) When $A = \mathbb{Z}_{p^m}$, where p is a prime integer and m is an integer such that $m \geq 2$. Then A is a local commutative QF-ring. Also let

$$R = \begin{bmatrix} A & A/J(A) \\ 0 & A/J(A) \end{bmatrix}.$$

In this case, if $(E, +, \cdot)$ is a compatible ring structure, then there exists $(\alpha, \beta) \in \mathrm{Soc}(A) \times \mathrm{Soc}(A)$ such that $\cdot = \circ_{(\alpha,\beta)}$. Thus,

$$\{(E, +, \circ_{(\alpha,\beta)}) \mid (\alpha, \beta) \in \mathrm{Soc}(A) \times \mathrm{Soc}(A)\}$$

is the set of all compatible ring structures on E. Note that $|\mathrm{Soc}(A)| = p$ since $\mathrm{Soc}(A) = p^{m-1}\mathbb{Z}_{p^m}$, Therefore, E has exactly $|\mathrm{Soc}(A)|^2 = p^2$ compatible ring structures (see [2, Theorem 7.3.17]).

3 More on Compatible Ring Structures of $E(R_R)$

In this section, the right compatibility of isomorphic rings is investigated further. It is shown that if U and T are isomorphic rings, then U is right compatible if and only if T

is right compatible (Theorem 3.2). Motivated by Remark 2.15, relationships between compatible ring structures and noncompatible ring structures on the injective hull of a ring are studied (Proposition 3.7).

We begin with the following which exhibits relationships between $E(U_U)$ and $E(T_T)$ when rings U and T are isomorphic.

Lemma 3.1 *Let U and T be isomorphic rings with $\lambda : U \to T$ a ring isomorphism. Then there exists an additive abelian group isomorphism $\sigma : E(U_U) \to E(T_T)$ such that:*

(i) σ is an extension λ;

(ii) $\sigma(yu) = \sigma(y)\lambda(u)$ for $y \in E(U_U)$ and $u \in U$.

Proof For $x \in E(T_T)$ and $u \in U$, we define

$$x \circledcirc u = x\lambda(u).$$

Then $E(T_T)$ is a right U-module for which $\circledcirc$ is the scalar multiplication of $E(T_T)$ over U. Thus, for $t \in T$ and $u \in U$, we have $t \circledcirc u = t\lambda(u)$. Hence, T_U is a submodule of the U-module $E(T_T)_U$.

We show that $E(T_T)_U$ is an injective hull of T_U. For this, assume that I is a right ideal of U and $f \in \mathrm{Hom}_U(I_U, E(T_T)_U)$. Then $\lambda(I)$ is a right ideal of T. Let $g : \lambda(I)_T \to E(T_T)$ be defined by $g(\lambda(a)) = f(a)$ for $\lambda(a) \in \lambda(I)$ with $a \in I$. Let $a \in I$ and $t = \lambda(u) \in T$ with $u \in U$. Then

$$\begin{aligned} g(\lambda(a)t) &= g(\lambda(a)\lambda(u)) = g(\lambda(au)) = f(au) = f(a) \circledcirc u \\ &= f(a)\lambda(u) = f(a)t \\ &= g(\lambda(a))t. \end{aligned}$$

Therefore, $g \in \mathrm{Hom}_T(\lambda(I)_T, E(T_T))$ because g is additive. Hence there exists $g_0 \in \mathrm{Hom}_T(T_T, E(T_T))$, which is an extension of g.

Now let $f_0 : U_U \to E(T_T)_U$ defined by $f_0(u) = g_0(\lambda(u))$ for $u \in U$. Then obviously f_0 is additive. For $u_1, u_2 \in U$, we have that

$$\begin{aligned} f_0(u_1u_2) &= g_0(\lambda(u_1u_2)) = g_0(\lambda(u_1)\lambda(u_2)) = g_0(\lambda(u_1))\lambda(u_2) \\ &= f_0(u_1) \circledcirc u_2. \end{aligned}$$

So $f_0 \in \mathrm{Hom}_U(U_U, E(T_T)_U)$. Further, for $a \in I$, we see that $f_0(a) = g_0(\lambda(a)) = g(\lambda(a)) = f(a)$. Hence f_0 is an extension of f. Consequently, $E(T_T)_U$ is an injective U-module.

Recall that T_U is a submodule of the U-module $E(T_T)_U$. To show that T_U is essential in $E(T_T)_U$, let $0 \neq y \in E(T_T)_U$. Since T_T is essential in $E(T_T)$, there exists $t \in T$ such that $0 \neq yt \in T$. Say $t = \lambda(u)$ with $u \in U$. Then

$$0 \neq yt = y\lambda(u) = y \circledcirc u \in T.$$

So T_U is essential in $E(T_T)_U$. As $E(T_T)_U$ is an injective U-module, $E(T_T)_U$ is an injective hull of T_U.

Next for $u_1, u_2 \in U$, we see that $\lambda(u_1u_2) = \lambda(u_1)\lambda(u_2) = \lambda(u_1) \circledcirc u_2$. Thus, $\lambda : U_U \to T_U$ is a U-module isomorphism because λ is one-to-one, onto, and additive. Since $E(T_T)_U$ is an injective hull of T_U, there exists $\sigma \in \mathrm{Hom}_U(E(U_U), E(T_T)_U)$, which is an extension of λ, and σ is an isomorphism.

Now, for $y \in E(U_U)$ and $u \in U$,

$$\sigma(yu) = \sigma(y) \circledcirc u = \sigma(y)\lambda(u)$$

because $\sigma \in \mathrm{Hom}_U(E(U_U), E(T_T)_U)$. □

The following result is naturally expected to be true. We here provide the proof explicitly.

Theorem 3.2 *Let U and T be isomorphic rings. Then $E(U_U)$ has $\aleph$ distinct compatible ring structures with U if and only if $E(T_T)$ has $\aleph$ distinct compatible ring structures with T, where $\aleph$ is a cardinal number. Thereby, U is right compatible if and only if T is right compatible.*

Proof Assume that $E(T_T)$ has $\aleph$ distinct compatible ring structures. Say Ω is a set with the cardinal number $\aleph$, and let

$$\mathbb{M} = \{\star_\omega \mid \omega \in \Omega\}$$

be the set of $\aleph$ distinct compatible ring multiplications on $E(T_T)$.

Let $\lambda : U \to T$ be a ring isomorphism. By Lemma 3.1, there exists an additive abelian group isomorphism $\sigma : E(U_U) \to E(T_T)$ such that σ is an extension of λ and $\sigma(yu) = \sigma(y)\lambda(u)$ for $y \in E(U_U)$ and $u \in U$.

For a given compatible ring structure $(E(T_T), +, \star_\omega)$ on $E(T_T)$, define a multiplication $\bullet_\omega$ on $E(U_U)$, by adopting the idea in Theorem 2.14, as follows: For $y_1, y_2 \in E(U_U)$, let

$$y_1 \bullet_\omega y_2 = \sigma^{-1}(\sigma(y_1) \star_\omega \sigma(y_2)).$$

Then $\sigma(y_1 \bullet_\omega y_2) = \sigma(v_1) \star_\omega \sigma(y_2)$ for $y_1,\ y_2 \in E(U_U)$.

Since σ is additive, $\sigma(y_1 + y_2) = \sigma(y_1) + \sigma(y_2)$ for $y_1,\ y_2 \in E(U_U)$. Further, as σ is one-to-one and onto, and $(E(T_T), +, \star_\omega)$ is a ring, it follows that $(E(U_U), +, \bullet_\omega)$ is also a ring. Therefore,

$$\sigma : (E(U_U), +, \bullet_\omega) \to (E(T_T), +, \star_\omega)$$

is a ring isomorphism. To show that $(E(U_U), +, \bullet_\omega)$ is a compatible ring structure, let $y \in E(U_U)$ and $u \in U$. Then we see that $\sigma(u) = \lambda(u) \in T$, so

$$\sigma(y) \star_\omega \sigma(u) = \sigma(y) \star_\omega \lambda(u) = \sigma(y)\lambda(u) = \sigma(yu)$$

because $\star_\omega$ is a ring multiplication of $E(T_T)$ which is compatible with the T-module scalar multiplication of $E(T_T)$ over T and $\sigma(yu) = \sigma(y)\lambda(u)$ from the preceding argument (by using Lemma 3.1). Now as $\sigma(y) \star_\omega \sigma(u) = \sigma(yu)$, we have that

$$y \bullet_\omega u = \sigma^{-1}(\sigma(y) \star_\omega \sigma(u)) = \sigma^{-1}(\sigma(yu)) = yu.$$

Thus, $(E(U_U), +, \bullet_\omega)$ is a compatible ring structure.

Next suppose $\star_\omega \neq \star_\nu$ in $\mathbb{M}$. In other words, $(E(T_T), +, \star_\omega)$ and $(E(T_T), +, \star_\nu)$ are distinct compatible ring structures on $E(T_T)$. Then there exist x and y in $E(T_T)$ such that $x \star_\omega y \neq x \star_\nu y$. Say $x = \sigma(v)$ and $y = \sigma(w)$ with $v, w \in E(U_U)$. Then

$$\sigma(v) \star_\omega \sigma(w) \neq \sigma(v) \star_\nu \sigma(w),$$

and hence, $\sigma^{-1}(\sigma(v) \star_\omega \sigma(w)) \neq \sigma^{-1}(\sigma(v) \star_\nu \sigma(w))$. So $v \bullet_\omega w \neq v \bullet_\nu w$. Therefore, $(E(U_U), +, \bullet_\omega)$ and $(E(U_U), +, \bullet_\nu)$ are distinct compatible ring structures on $E(U_U)$. Consequently, $E(U_U)$ has $\aleph$ distinct compatible ring structures.

Conversely, assume that $E(U_U)$ has $\aleph$ distinct compatible ring structures. Using the preceding arguments, $E(T_T)$ has $\aleph$ distinct compatible ring structures. □

Proposition 3.3 *Let U and T be two rings, and let W_T be an overmodule of T_T. Assume that there exists an additive abelian group isomorphism $\sigma : E(U_U) \to W_T$ such that:*
(i) $\sigma_{|U} : U \to T$ is a ring isomorphism;
(ii) $\sigma(yu) = \sigma(y)\sigma(u)$ for $y \in E(U_U)$ and $u \in U$.
Then W_T is an injective hull of T_T.

Proof We observe first that $\sigma^{-1}(wt) = \sigma^{-1}(w)\sigma^{-1}(t)$ for $w \in W$ and $t \in T$. Indeed, we note that $\sigma(\sigma^{-1}(wt)) = wt$ and $\sigma(\sigma^{-1}(w)\sigma^{-1}(t)) = \sigma(\sigma^{-1}(w))\sigma(\sigma^{-1}(t)) = wt$ by the assumption (ii). Therefore, $\sigma^{-1}(wt) = \sigma^{-1}(w)\sigma^{-1}(t)$.

We show that T_T is an essential submodule of W_T. For this, let $0 \neq y \in W$. Then $0 \neq \sigma^{-1}(y) \in E(U_U)$, so there exists $u \in U$ such that $0 \neq \sigma^{-1}(y)u \in U$. Hence, $0 \neq \sigma(\sigma^{-1}(y)u) \in T$. From the condition (ii), $0 \neq \sigma(\sigma^{-1}(y)u) = \sigma(\sigma^{-1}(y))\sigma(u) = y\sigma(u) \in T$ with $\sigma(u) \in T$. Therefore $T_T \leq^{\text{ess}} W_T$.

Next, we prove that W_T is an injective module. Say I is a right ideal of T and $f \in \text{Hom}_T(I_T, W_T)$. Note that $\sigma^{-1}(I)$ is a right ideal of U. Define

$$g : \sigma^{-1}(I) \to E(U_U) \text{ by } g(r) = (\sigma^{-1} f \sigma)(r) \text{ for } r \in \sigma^{-1}(I).$$

Then $g \in \text{Hom}_U(\sigma^{-1}(I)_U, E(U_U))$. Indeed, clearly g is additive. Now, for $r \in \sigma^{-1}(I)$ and $a \in U$, note that $\sigma(ra) = \sigma(r)\sigma(a)$ by the assumption (ii) and $\sigma^{-1}[f(\sigma(r))\sigma(a)] = \sigma^{-1}(f(\sigma(r))\sigma^{-1}(\sigma(a)) = \sigma^{-1}(f\sigma(r))a$ from the preceding argument. Hence, we have that

$$\begin{aligned} g(ra) &= (\sigma^{-1} f \sigma)(ra) = \sigma^{-1} f(\sigma(ra)) \\ &= \sigma^{-1}[f(\sigma(r)\sigma(a))] = (\sigma^{-1} f \sigma(r))a \\ &= g(r)a. \end{aligned}$$

As $E(U_U)$ is an injective hull of U_U, there exists $g_0 \in \mathrm{Hom}_U(U_U, E(U_U))$, an extension of g. Define

$$f_0 : T_T \to W_T \text{ by } f_0(t) = (\sigma g_0 \sigma^{-1})(t) \text{ for } t \in T.$$

Now, we show that $f_0 \in \mathrm{Hom}_T(T_T, W_T)$. Obviously, f_0 is additive. Let $t_1, t_2 \in T$. Then there exist $u_1, u_2 \in U$ such that $t_1 = \sigma(u_1)$ and $t_2 = \sigma(u_2)$. Thus

$$\begin{aligned} f_0(t_1 t_2) &= f_0(\sigma(u_1)\sigma(u_2)) = f_0(\sigma(u_1 u_2)) = (\sigma g_0 \sigma^{-1})(\sigma(u_1 u_2)) \\ &= \sigma(g_0(u_1 u_2)) = \sigma(g_0(u_1)u_2) = \sigma(g_0(u_1))\sigma(u_2) \\ &= (\sigma g_0 \sigma^{-1})(t_1)\sigma(u_2) = f_0(t_1)\sigma(u_2) \\ &= f_0(t_1)t_2 \end{aligned}$$

because $\sigma(g_0(u_1)u_2) = \sigma(g_0(u_1))\sigma(u_2)$ from the assumption (ii). Therefore, we have that $f_0 \in \mathrm{Hom}_T(T_T, W_T)$.

Finally, to show that f_0 is an extension of f, let $s \in I$, and put $s = \sigma(r)$. Then $r = \sigma^{-1}(s) \in \sigma^{-1}(I) \subseteq U$. Hence

$$\begin{aligned} f_0(s) &= (\sigma g_0 \sigma^{-1})(s) = \sigma g_0(r) = \sigma g(r) \\ &= \sigma(\sigma^{-1} f \sigma(r)) = f\sigma(r) = f(\sigma(r)) \\ &= f(s). \end{aligned}$$

Hence, f_0 is an extension of f. Consequently, W_T is injective by Baer's Criterion. So W_T is an injective hull of T_T. □

The following example illustrates Theorem 3.2 and Proposition 3.3.

Example 3.4 Assume that A is a commutative local QF-ring with $J(A) \neq 0$. As in Example 2.16, let

$$R = \begin{bmatrix} A & A/J(A) \\ 0 & A/J(A) \end{bmatrix} \text{ and } E = \begin{bmatrix} A \oplus (A/J(A)) & A/J(A) \\ A/J(A) & A/J(A) \end{bmatrix}.$$

The addition $+$ of E is componentwise. The right R-module scalar multiplication of E over R is given as in Example 2.16:

For $\begin{bmatrix} s + \overline{x} & \overline{y} \\ \overline{z} & \overline{w} \end{bmatrix} \in E$ and $\begin{bmatrix} a & \overline{b} \\ 0 & \overline{c} \end{bmatrix} \in R$, where $s, x, y, z, w, a, b, c \in A$ and $\overline{x}, \overline{y} \in A/J(A)$, etc., denote the images of $x, y \in A$, etc., respectively, we define

$$\begin{bmatrix} s+\overline{x} & \overline{y} \\ \overline{z} & \overline{w} \end{bmatrix}\begin{bmatrix} a & b \\ 0 & c \end{bmatrix} = \begin{bmatrix} sa+\overline{xa} & sb+\overline{xb}+\overline{yc} \\ \overline{za} & \overline{zb}+\overline{wc} \end{bmatrix}.$$

Then E is a right R-module, and $E = E(R_R)$ (see Example 2.16).

(i) Consider the following E_1 adopted from private communications with B. L. Osofsky, which is

$$E_1 = \left\{ s + \begin{bmatrix} \overline{s}+\overline{x} & \overline{y} \\ \overline{z} & \overline{w} \end{bmatrix} \mid s,\ x,\ y,\ z,\ w \in A \right\},$$

where $\overline{x}$, $\overline{y}$, etc., are images of x, y in $A/J(A)$, respectively. Addition on E_1 is defined as follows:

For $v_1 = s_1 + \begin{bmatrix} \overline{s_1}+\overline{x_1} & \overline{y_1} \\ \overline{z_1} & \overline{w_1} \end{bmatrix}$ and $v_2 = s_2 + \begin{bmatrix} \overline{s_2}+\overline{x_2} & \overline{y_2} \\ \overline{z_2} & \overline{w_2} \end{bmatrix}$ in E_1,

$$v_1 + v_2 = (s_1+s_2) + \begin{bmatrix} \overline{s_1}+\overline{s_2}+\overline{x_1}+\overline{x_2} & \overline{y_1}+\overline{y_2} \\ \overline{z_1}+\overline{z_2} & \overline{w_1}+\overline{w_2} \end{bmatrix}.$$

Let

$$B = \left\{ a + \begin{bmatrix} \overline{a} & \overline{b} \\ 0 & \overline{c} \end{bmatrix} \mid a,\ b,\ c \in A \right\} \subseteq E_1.$$

For $a_1 + \begin{bmatrix} \overline{a}_1 & \overline{b}_1 \\ 0 & \overline{c}_1 \end{bmatrix}$ and $a_2 + \begin{bmatrix} \overline{a}_2 & \overline{b}_2 \\ 0 & \overline{c}_2 \end{bmatrix}$ in B, the addition is defined as follows:

$$\left(a_1 + \begin{bmatrix} \overline{a}_1 & \overline{b}_1 \\ 0 & \overline{c}_1 \end{bmatrix}\right) + \left(a_2 + \begin{bmatrix} \overline{a}_2 & \overline{b}_2 \\ 0 & \overline{c}_2 \end{bmatrix}\right) = (a_1+a_2) + \begin{bmatrix} \overline{a_1}+\overline{a_2} & \overline{b_1}+\overline{b_2} \\ 0 & \overline{c_1}+\overline{c_2} \end{bmatrix}.$$

Then B is an additive subgroup of E. Next, the multiplication is defined by

$$\left(a_1 + \begin{bmatrix} \overline{a}_1 & \overline{b}_1 \\ 0 & \overline{c}_1 \end{bmatrix}\right)\left(a_2 + \begin{bmatrix} \overline{a}_2 & \overline{b}_2 \\ 0 & \overline{c}_2 \end{bmatrix}\right) = a_1a_2 + \begin{bmatrix} \overline{a_1a_2} & \overline{a_1b_2}+\overline{b_1c_2} \\ 0 & \overline{c_1c_2} \end{bmatrix}.$$

Then B forms a ring. Furthermore, $R \cong B$ as rings by a ring isomorphism $\tau : R \to B$ defined by

$$\tau \begin{bmatrix} a & b \\ 0 & c \end{bmatrix} = a + \begin{bmatrix} \overline{a} & \overline{b} \\ 0 & \overline{c} \end{bmatrix}.$$

Now for $s + \begin{bmatrix} \overline{s}+\overline{x} & \overline{y} \\ \overline{z} & \overline{w} \end{bmatrix} \in E_1$ and $a + \begin{bmatrix} \overline{a} & \overline{b} \\ 0 & \overline{c} \end{bmatrix} \in B$, define the B-module scalar multiplication

$$\left(s + \begin{bmatrix} \overline{s}+\overline{x} & \overline{y} \\ \overline{z} & \overline{w} \end{bmatrix}\right)\left(a + \begin{bmatrix} \overline{a} & \overline{b} \\ 0 & \overline{c} \end{bmatrix}\right) = sa + \begin{bmatrix} \overline{sa}+\overline{xa} & \overline{sb}+\overline{xb}+\overline{yc} \\ \overline{za} & \overline{zb}+\overline{wc} \end{bmatrix}.$$

Then E_1 is a right B-module. Further, E_1 is an overmodule of B_B.

We let $\sigma : E(R_R) \to E_1$ defined by

$$\sigma \begin{bmatrix} s + \overline{x} & \overline{y} \\ \overline{z} & \overline{w} \end{bmatrix} = s + \begin{bmatrix} \overline{s} + \overline{x} & \overline{y} \\ \overline{z} & \overline{w} \end{bmatrix}$$

for $\begin{bmatrix} s + \overline{x} & \overline{y} \\ \overline{z} & \overline{w} \end{bmatrix} \in E(R_R)$. Then σ is an additive abelian group isomorphism and $\sigma|_R = \tau$. Furthermore, for $v \in E(R_R)$ and $r \in R$, $\sigma(vr) = \sigma(v)\tau(r) = \sigma(v)\sigma(r)$. Hence, Proposition 3.3 yields that E_1 is the injective hull of B_B. Further, as $E = E(R_R)$ has $|\mathrm{Soc}(A)|^2$ compatible ring structures by Example 2.16, E_1 has also $|\mathrm{Soc}(A)|^2$ distinct compatible ring structures from Theorem 3.2.

(ii) As in [2, Theorem 7.3.14], we put

$$\mathfrak{A} = \{(a, -\overline{a}) \mid a \in A\} \subseteq A \times (A/J(A)).$$

The addition on $\mathfrak{A}$ is componentwise. Define

$$(a, -\overline{a})b = (ab, -\overline{ab}) \text{ for } (a, -\overline{a}) \in \mathfrak{A} \text{ and } b \in A.$$

Then $\mathfrak{A}$ is a right A-module and $\mathfrak{A}_A \cong A_A$ via corresponding $(a, -\overline{a})$ to a. Let

$$\mathbb{A} = \begin{bmatrix} \mathfrak{A} & 0 \\ 0 & 0 \end{bmatrix} = \left\{ \begin{bmatrix} (a, -\overline{a}) & 0 \\ 0 & 0 \end{bmatrix} \mid a \in A \right\}.$$

Then $\mathbb{A}$ is a right R-module under the componentwise addition and the R-module scalar multiplication is defined by

$$\begin{bmatrix} (a, -\overline{a}) & 0 \\ 0 & 0 \end{bmatrix} \begin{bmatrix} x & \overline{y} \\ 0 & \overline{z} \end{bmatrix} = \begin{bmatrix} (ax, -\overline{ax}) & 0 \\ 0 & 0 \end{bmatrix}$$

for $\begin{bmatrix} (a, -\overline{a}) & 0 \\ 0 & 0 \end{bmatrix} \in \mathbb{A}$ and $\begin{bmatrix} x & \overline{y} \\ 0 & \overline{z} \end{bmatrix} \in R$. We put

$$E_2 = \mathbb{A} \oplus \begin{bmatrix} A/J(A) & A/J(A) \\ A/J(A) & A/J(A) \end{bmatrix}.$$

For $u_1 := \begin{bmatrix} (s_1, -\overline{s}_1) & 0 \\ 0 & 0 \end{bmatrix} + \begin{bmatrix} \overline{x}_1 & \overline{y}_1 \\ \overline{z}_1 & \overline{w}_1 \end{bmatrix}$ and $u_2 := \begin{bmatrix} (s_2, -\overline{s}_2) & 0 \\ 0 & 0 \end{bmatrix} + \begin{bmatrix} \overline{x}_2 & \overline{y}_2 \\ \overline{z}_2 & \overline{w}_2 \end{bmatrix}$ in E_2, define

$$u_1 + u_2 = \begin{bmatrix} (s_1 + s_2, -(\overline{s}_1 + \overline{s}_2)) & 0 \\ 0 & 0 \end{bmatrix} + \begin{bmatrix} \overline{x}_1 + \overline{x}_2 & \overline{y}_1 + \overline{y}_2 \\ \overline{z}_1 + \overline{z}_2 & \overline{w}_1 + \overline{w}_2 \end{bmatrix}.$$

Then E_2 is an additive abelian group. Next, let

$$C = \left\{ \begin{bmatrix} (a, -\overline{a}) & 0 \\ 0 & 0 \end{bmatrix} + \begin{bmatrix} \overline{a} & \overline{b} \\ 0 & \overline{c} \end{bmatrix} \mid a, b, c \in A \right\},$$

which is an additive abelian subgroup of $(E_2, +)$.

For $k_1 := \begin{bmatrix} (a_1, -\overline{a_1}) & 0 \\ 0 & 0 \end{bmatrix} + \begin{bmatrix} \overline{a_1} & \overline{b_1} \\ 0 & \overline{c_1} \end{bmatrix}$ and $k_2 := \begin{bmatrix} (a_2, -\overline{a_2}) & 0 \\ 0 & 0 \end{bmatrix} + \begin{bmatrix} \overline{a_2} & \overline{b_2} \\ 0 & \overline{c_2} \end{bmatrix}$ in C, define the multiplication by

$$k_1 k_2 = \begin{bmatrix} (a_1 a_2, -\overline{a_1 a_2}) & 0 \\ 0 & 0 \end{bmatrix} + \begin{bmatrix} \overline{a_1 a_2} & \overline{a_1 b_2} + \overline{b_1 c_2} \\ 0 & \overline{c_1 c_2} \end{bmatrix}.$$

Then C is a ring. Moreover, $R \cong C$ as rings by a ring isomorphism $\kappa : R \to C$ defined by

$$\kappa \begin{bmatrix} a & b \\ 0 & c \end{bmatrix} = \begin{bmatrix} (a, -\overline{a}) & 0 \\ 0 & 0 \end{bmatrix} + \begin{bmatrix} \overline{a} & \overline{b} \\ 0 & \overline{c} \end{bmatrix}.$$

Let $\alpha := \begin{bmatrix} (s, -\overline{s}) & 0 \\ 0 & 0 \end{bmatrix} + \begin{bmatrix} \overline{s} + \overline{x} & \overline{y} \\ \overline{z} & \overline{w} \end{bmatrix} \in E_2$ and $\gamma := \begin{bmatrix} (a, -\overline{a}) & 0 \\ 0 & 0 \end{bmatrix} + \begin{bmatrix} \overline{a} & \overline{b} \\ 0 & \overline{c} \end{bmatrix} \in C$. We define

$$\alpha\gamma = \begin{bmatrix} (sa, -\overline{sa}) & 0 \\ 0 & 0 \end{bmatrix} + \begin{bmatrix} \overline{sa} + \overline{xa} & \overline{sb} + \overline{xb} + \overline{yc} \\ \overline{za} & \overline{zb} + \overline{wc} \end{bmatrix}.$$

Then E_2 is a right C-module and E_2 is an overmodule of C_C.

Consider $\mu : E \to E_2$ defined by

$$\mu\left(\begin{bmatrix} s + \overline{x} & \overline{y} \\ \overline{z} & \overline{w} \end{bmatrix} \right) = \begin{bmatrix} (s, -\overline{s}) & 0 \\ 0 & 0 \end{bmatrix} + \begin{bmatrix} \overline{s} + \overline{x} & \overline{y} \\ \overline{z} & \overline{w} \end{bmatrix}$$

for $\begin{bmatrix} s + \overline{x} & \overline{y} \\ \overline{z} & \overline{w} \end{bmatrix} \in E$. Then μ is an additive abelian group isomorphism. Also we see that $\kappa = \mu|_R : R \to C$ is a ring isomorphism.

Now, for $v \in E$ and $r \in R$, we see that $\mu(vr) = \mu(v)\kappa(r) = \mu(v)\mu(r)$. Since E is an injective hull of R_R, by Proposition 3.3, E_2 is an injective hull of C_C. As $E = E(R_R)$ has $|\mathrm{Soc}(A)|^2$ compatible ring structures from Example 2.16, Theorem 3.2 yields that $E_2 = E(C_C)$ has $|\mathrm{Soc}(A)|^2$ distinct compatible ring structures.

(iii) (see Example 2.16(v)) We let

$$E_3 = \left\{ \begin{bmatrix} s & 0 & 0 \\ 0 & \overline{x} & \overline{y} \\ 0 & \overline{z} & \overline{w} \end{bmatrix} \mid s, x, y, z, w \in A \right\}.$$

Then we see that

$$E_3 = \left\{ \begin{bmatrix} s & 0 & 0 \\ 0 & \overline{s} + \overline{x} & \overline{y} \\ 0 & \overline{z} & \overline{w} \end{bmatrix} \mid s, x, y, z, w \in A \right\}.$$

Note that $E_3 = \mathbb{E}$, where $\mathbb{E}$ is in Example 2.16(v). The addition in E_3 is defined componentwise.

Next, let

$$D = \left\{ \begin{bmatrix} a & 0 & 0 \\ 0 & \overline{a} & \overline{b} \\ 0 & 0 & \overline{c} \end{bmatrix} \mid a, b, c \in A \right\} \subseteq E_3.$$

In D, the addition is componentwise and the multiplication is defined by

$$\begin{bmatrix} a_1 & 0 & 0 \\ 0 & \overline{a_1} & \overline{b_1} \\ 0 & 0 & \overline{c_1} \end{bmatrix} \cdot \begin{bmatrix} a_2 & 0 & 0 \\ 0 & \overline{a_2} & \overline{b_2} \\ 0 & 0 & \overline{c_2} \end{bmatrix} = \begin{bmatrix} a_1a_2 & 0 & 0 \\ 0 & \overline{a_1a_2} & \overline{a_1b_2 + b_1c_2} \\ 0 & 0 & \overline{c_1c_2} \end{bmatrix}$$

for $\begin{bmatrix} a_1 & 0 & 0 \\ 0 & \overline{a_1} & \overline{b_1} \\ 0 & 0 & \overline{c_1} \end{bmatrix}$ and $\begin{bmatrix} a_2 & 0 & 0 \\ 0 & \overline{a_2} & \overline{b_2} \\ 0 & 0 & \overline{c_2} \end{bmatrix}$ in D. Then $(D, +)$ is a subgroup of $(E_3, +)$ and $(D, +, \cdot)$ is a ring.

Define $\lambda : R \to D$ by

$$\lambda \begin{bmatrix} a & b \\ 0 & c \end{bmatrix} = \begin{bmatrix} a & 0 & 0 \\ 0 & \overline{a} & \overline{b} \\ 0 & 0 & \overline{c} \end{bmatrix}.$$

Then λ is a ring isomorphism. The scalar multiplication of E_3 over D is defined by

$$\begin{bmatrix} s & 0 & 0 \\ 0 & \overline{s} + \overline{x} & \overline{y} \\ 0 & \overline{z} & \overline{w} \end{bmatrix} \begin{bmatrix} a & 0 & 0 \\ 0 & \overline{a} & \overline{b} \\ 0 & 0 & \overline{c} \end{bmatrix} = \begin{bmatrix} sa & 0 & 0 \\ 0 & \overline{sa} + \overline{xa} & \overline{sb} + \overline{xb} + \overline{yc} \\ 0 & \overline{za} & \overline{zb} + \overline{wc} \end{bmatrix}$$

for $\begin{bmatrix} s & 0 & 0 \\ 0 & \overline{s} + \overline{x} & \overline{y} \\ 0 & \overline{z} & \overline{w} \end{bmatrix} \in E_3$ and $\begin{bmatrix} a & 0 & 0 \\ 0 & \overline{a} & \overline{b} \\ 0 & 0 & \overline{c} \end{bmatrix} \in D$.

We can check that E_3 is a right D-module and D_D is a submodule of E_3. Consider $\theta : E = E(R_R) \to E_3$ defined by

$$\theta\left(\begin{bmatrix} s + \overline{x} & \overline{y} \\ \overline{z} & \overline{w} \end{bmatrix}\right) = \begin{bmatrix} s & 0 & 0 \\ 0 & \overline{s} + \overline{x} & \overline{y} \\ 0 & \overline{z} & \overline{w} \end{bmatrix}$$

for $\begin{bmatrix} s + \overline{x} & \overline{y} \\ \overline{z} & \overline{w} \end{bmatrix} \in E(R_R)$. Then θ is an additive abelian group isomorphism from $E(R_R)$ to E_3. Note that $\theta|_R = \lambda$, which is a ring isomorphism from R to D. Furthermore, for $v \in E(R_R)$ and $r \in R$, we have $\theta(vr) = \theta(v)\lambda(r) = \theta(v)\theta(r)$. As E is the injective hull of R_R from Example 2.16, E_3 is the injective hull of D_D by Proposition 3.3. Recall that E has $|\mathrm{Soc}(A)|^2$ distinct compatible ring structures from Example 2.16. Therefore, Theorem 3.2 yields that $E(D_D)$ has $|\mathrm{Soc}(A)|^2$ distinct compatible ring structures.

(iv) As a consequence, by parts (i), (ii), and (iii), the rings R, B, C, and D are mutually ring isomorphic and right compatible. Furthermore, each of $E(B_B)$, $E(C_C)$, and $E(D_D)$ has $|\mathrm{Soc}(A)|^2$ distinct compatible ring structures.

Remark 3.5 It is shown in [1, Theorem 1] that E_1 is the injective hull of B_B, while it is shown in [2, Theorem 7.3.14] that E_2 is the injective hull of C_C.

Remark 3.6 Let A be a commutative local QF-ring with $J(A) \neq 0$ and let R be the ring as in Example 2.16. Then there exists a compatible ring structure, say $(E(R_R), +, \circ)$ on $E(R_R)$ from Example 2.16. We take $0 \neq \overline{a} \in A/J(A)$ with $a \in A$, and put

$$u = \begin{bmatrix} 1 & \overline{a} \\ 0 & \overline{1} \end{bmatrix} \in R.$$

Then u is invertible in R such that $u \neq 1_R$ and the inverse of u is

$$u^{-1} = \begin{bmatrix} 1 & -\overline{a} \\ 0 & \overline{1} \end{bmatrix}.$$

In Remark 2.15, there exists a ring structure $E(R_R), +, \star)$ on $E(R_R)$ which is not compatible with the R-module scalar multiplication of $E(R_R)$ over R. However, $(E(R_R), +, \star) \cong E(R_R), +, \cdot)$.

Motivated by Remarks 2.15 and 3.6, we obtain the following from our previous results of this section, which exhibits the relationships between compatible ring structures and noncompatible ring structures of the injective hull of a given ring.

Proposition 3.7 *Let $(R, +, \cdot)$ be a ring and E the injective hull of $(R, +, \cdot)_{(R,+,\cdot)}$ such that the ring $(E, +, \cdot)$ is compatible with the $(R, +, \cdot)$-module scalar multiplication of E over $(R, +, \cdot)$.*

Let $E_+ = (E, +)$ be the underlying additive abelian group of E. Let $u \in (R, +, \cdot)$ be an invertible element. Then we have the following.

(i) There exists a ring structure $(R, +, \diamond)$ on R for which u is the identity of $(R, +, \diamond)$.

(ii) There exists a ring isomorphism $\lambda : (R, +, \cdot) \to (R, +, \diamond)$ such that $\lambda(1_R) = u$, where 1_R is the identity of the ring $(R, +, \cdot)$.

(iii) There exists a scalar multiplication $\circledcirc$ of E_+ over the ring $(R, +, \diamond)$ so that E_+ is the injective hull of $(R, +, \diamond)_{(R,+,\diamond)}$.

(iv) E has $\aleph$ distinct compatible ring structures with the $(R, +, \cdot)$-module scalar multiplication over the ring $(R, +, \cdot)$ if and only if E has $\aleph$ distinct compatible ring

structures with the $(R, +, \diamond)$*-module scalar multiplication over the ring* $(R, +, \diamond)$*, where* $\aleph$ *is a cardinal number.*

(v) If $u \neq 1_R$*, then the ring structure* $(E, +, \circledcirc)$ *is not compatible with the* $(R, +, \cdot)$*-module scalar multiplication of* E *over* $(R, +, \cdot)$*.*

Proof For (i) and (ii), define

$$\lambda : (R, +, \cdot) \to (R, +) \quad \text{by } \lambda(a \cdot u^{-1}) = a \quad \text{for } a \in R.$$

Then λ is an additive abelian group isomorphism. We provide a new ring multiplication $\diamond$ on $(R, +)$ as follows: for $a_1,\ a_2 \in R$, let

$$a_1 \diamond a_2 = \lambda(\lambda^{-1}(a_1) \cdot \lambda^{-1}(a_2)).$$

Then $a_1 \diamond a_2 = \lambda(a_1 u^{-1} a_2 u^{-1}) = a_1 u^{-1} a_2$. Since $\lambda^{-1}(a_1 \diamond a_2) = \lambda^{-1}(a_1)\lambda^{-1}(a_2)$ and further λ^{-1} is also an additive abelian group isomorphism. Thus, $(R, +, \diamond)$ is a ring and $\lambda^{-1} : (R, +, \diamond) \to (R, +, \cdot)$ is a ring isomorphism. Hence,

$$\lambda : (R, +, \cdot) \to (R, +, \diamond)$$

is also a ring isomorphism. Also note that u is the identity of the ring $(R, +, \diamond)$. Moreover, say 1_R is the identity of the ring $(R, +, \cdot)$. Then $\lambda(1_R) = \lambda(uu^{-1}) = u$.

(iii) We provide a right $(R, +, \diamond)$-module scalar multiplication on E_+ so that E_+ becomes an injective hull of $(R, +, \diamond)_{(R,+,\diamond)}$. For this, first, we observe that $E = \{vu^{-1} \mid v \in E\}$. Define

$$\sigma : E \to E_+ \quad \text{by} \quad \sigma(vu^{-1}) = v \quad \text{for} \quad vu^{-1} \in E.$$

Then σ is an additive abelian group isomorphism. Next, for $v \in E_+$ and $a \in R$, let

$$v \circledcirc a = \sigma(\sigma^{-1}(v)\sigma^{-1}(a)) = \sigma(vu^{-1}au^{-1}) = vu^{-1}a.$$

It can be checked that E_+ becomes a right $(R, +, \diamond)$-module under the scalar multiplication $\circledcirc$. For example, say $v \in E_+$ and $a_1,\ a_2 \in R$. Then we have that

$$v \circledcirc (a_1 \diamond a_2) = v \circledcirc (a_1 u^{-1} a_2) = vu^{-1}(a_1 u^{-1} a_2),$$

and

$$(v \circledcirc a_1) \circledcirc a_2 = (vu^{-1}a_1) \circledcirc a_2 = (vu^{-1}a_1)u^{-1}a_2 = vu^{-1}(a_1 u^{-1} a_2).$$

Thus $v \circledcirc (a_1 \diamond a_2) = (v \circledcirc a_1) \circledcirc a_2$.

Furthermore, σ is an extension of λ and $\lambda : (R, +, \cdot) \to (R, +, \diamond)$ is a ring isomorphism from part (ii). Also for $vu^{-1} \in E$ and $au^{-1} \in R$, we have

$$\sigma((vu^{-1})(au^{-1})) = \sigma(vu^{-1}au^{-1}) = vu^{-1}a = v \circledcirc a.$$

On the other hand, $\sigma(vu^{-1}) \circledcirc \sigma(au^{-1}) = \sigma(vu^{-1}) \circledcirc \lambda(au^{-1}) = v \circledcirc a$. Therefore,

$$\sigma((vu^{-1})(au^{-1})) = \sigma(vu^{-1}) \circledcirc \sigma(au^{-1}).$$

Since E is an injective hull of $(R, +, \cdot)_{(R,+,\cdot)}$, Proposition 3.3 yields E_+ is an injective hull of $(R, +, \diamond)_{(R,+,\diamond)}$ under the scalar multiplication $\circledcirc$ over the ring $(R, +, \diamond)$.

(iv) Since $(R, +, \cdot) \cong (R, +, \diamond)$, Theorem 3.2 yields that E has $\aleph$ distinct compatible ring structures if and only if E_+ has $\aleph$ distinct compatible ring structures, where $\aleph$ is a cardinal number.

(v) The proof follows from Remark 2.15. □

Let R be the ring in Example 2.16. Then R is also left compatible (see [1, Theorem 1] or [2, Theorem 7.3.14]). Let B, C, and D be rings in Example 3.4(i), (ii), and (iii), respectively. Then $R \cong B \cong C \cong D$ and further B, C, and D are right compatible from Example 3.4. Since C is left comptible (see [1, Theorem 1] or [2, Theorem 7.3.14]), the left-hand side version of Theorem 3.2 yields that R, B, and D are also left compatible. In view of this, the following question may be raised.

Question 3.8 Let V be a ring. Then is it true that V is right compatible if and only if V is left compatible?

Acknowledgements The authors appreciate a thorough report from the referee and are thankful to him/her for helpful comments and suggestions for improvement in the presentation of this work. We are grateful for the partial grant support received from The Ohio State University at Lima, Mathematics Research Institute, Office of International Affairs ,and ASC College, The Ohio State University. The authors also thank each other's institutions for the hospitality received during the research work on this paper.

References

1. Birkenmeier, G.F., Osofsky, B.L., Park, J.K., Rizvi, S.T.: Injective hulls with distinct ring structures. J. Pure Appl. Algebra **213**, 732–735 (2009)
2. Birkenmeier, G.F., Park, J.K., Rizvi, S.T.: Extensions of Rings and Modules. Monograph. Birkhäuser/Springer, New York (2013)
3. Connell, I.G.: On the group rings. Canad. J. Math. **15**, 650–685 (1963)
4. Lam, T.Y.: Lectures on Modules and Rings. Springer, Berlin (1999)
5. Lang, N.C.: On ring properties of injective hulls. Canad. Math. Bull. **18**, 233–239 (1975)
6. Osofsky, B.L.: A non-trivial ring with non-rational injective hull. Canad. Math. Bull. **10**, 275–282 (1967)
7. Passman, D.S.: The Algebraic Structure of Group Rings. Wiley, New York (1977)
8. Sharpe, D.W., Vámos, P.: Injective Modules. Cambridge University Press, Cambridge (1972)

On Commutators Involving Derivations and Automorphisms in Prime Rings

Mohd Arif Raza, Mohammad Shadab Khan, and Nadeem ur Rehman

Abstract Let $\mathscr{R}$ be a prime ring of characteristics different from two, $\mathscr{L}$ a noncentral Lie ideal of $\mathscr{R}$, d a derivation of $\mathscr{R}$, and ξ an automorphism of $\mathscr{R}$. The goal of this manuscript is to discuss the behavior of derivation and structure of $\mathscr{R}$ satisfying $[[u^d, u], u^\xi] = 0$ on noncentral Lie ideal $\mathscr{L}$ of $\mathscr{R}$. This results in the spirit of Posner's theorem.

Keywords Prime ring · Automorphisms · Maximal right ring of quotient · Generalized polynomial identity (GPI)

1 Motivation

This work is inspired by the work of several algebraist in which they have evaluated certain identities having commutators with derivations or automorphisms. In the last few decades, there has been a continuing interest pertaining to the relationship between structure of rings and the existence of certain specific types of mappings, viz., derivations, automorphisms, etc. In [20], Posner discussed the commutativity of prime rings. Absolutely, he showed that if $\mathscr{R}$ is a prime ring and d a derivation of $\mathscr{R}$ such that $[x^d, x] = 0$ for all $x \in \mathscr{R}$, then either $\mathscr{R}$ is commutative or $d = 0$. Many researchers have studied and made an effort to generalize the results obtained on derivations to automorphisms. In [19], Mayne studied Posner's second theorem on derivations [20] for automorphisms of prime rings. Precisely, he proved that let $\mathscr{R}$ be a prime ring with center $\mathscr{Z}(\mathscr{R})$ and ξ be a nontrivial automorphism of $\mathscr{R}$.

M. A. Raza
Department of Mathematics, Faculty of Sciences & Arts-Rabigh, King Abdulaziz University, Jeddah, Kingdom of Saudi Arabia

M. S. Khan
Department of Commerce, Aligarh Muslim University, Aligarh 202002, India

N. Rehman (✉)
Department of Mathematics, Aligarh Muslim University, Aligarh 202002, India
e-mail: rehman100@gmail.com

M. Ashraf et al. (eds.), *Algebra and Related Topics with Applications*,
Springer Proceedings in Mathematics & Statistics 392,
https://doi.org/10.1007/978-981-19-3898-6_20

If $[x^{\xi}, x] \in \mathscr{Z}(\mathscr{R})$ for every $x \in \mathscr{R}$, then $\mathscr{R}$ is a commutative integral domain. In [15], Lee and Lee established that if characteristic not equal to 2 and $[x^d, x] \in \mathscr{Z}$ for all x in a noncentral Lie ideal $\mathscr{L}$ of $\mathscr{R}$, then $\mathscr{R}$ is commutative. An analogous extension for Lie ideals in the automorphism case was obtained by Mayne [17]. He was able to accurately draw a conclusion that let $\mathscr{R}$ be a prime ring of characteristic not equal to 2 and ξ be an automorphism of $\mathscr{R}$. If $\mathscr{L}$ is a Lie ideal of $\mathscr{R}$ such that ξ is nontrivial on $\mathscr{L}$ and $[x^{\xi}, x]$ is in the center of $\mathscr{R}$ for every x in $\mathscr{L}$, then $\mathscr{L}$ is contained in the center of $\mathscr{R}$. In 1990, Vukman [23] showed that $\mathscr{R}$ is commutative if $[[x^d, x], x] = 0$ for all $x \in \mathscr{R}$, where d is a nonzero derivation of $\mathscr{R}$ and characteristic of $\mathscr{R}$ is 2. In 2005, Cheng [9] discussed Vukman problem [23] in case of derivations on prime rings. More precisely, he proved that if $\mathscr{R}$ is a 2-torsion free noncommutative prime ring and d a derivation of $\mathscr{R}$ such that $[[x^d, x], x^d] = 0$, for all $x \in \mathscr{R}$, then $d = 0$. Recently, Ashraf and Pary [1] obtained the analogous result of Cheng [9] for nontrivial automorphisms of prime rings. In fact, they proved that if $\mathscr{R}$ is a prime ring of characteristic different from two which admits a nontrivial automorphism ξ such that $[[x^{\xi}, x], x^{\xi}] \in \mathscr{Z}(\mathscr{R})$ for all $x \in \mathscr{L}$, a noncentral Lie ideal of $\mathscr{R}$, then $\mathscr{R}$ satisfies s_4, the standard identity in four variables. Recently, a lot of work has been done considering derivations/automorphisms on rings, which is fascinating the courtesy of many algebrists, see [2–4, 10, 18, 21–24] and other references in their bibliographic content.

Inspired by the above-mentioned works, the goal of this manuscript is to discuss the behavior of derivation and structure of $\mathscr{R}$ satisfying $[[u^d, u], u^{\xi}] = 0$ on noncentral Lie ideal $\mathscr{L}$ of $\mathscr{R}$. This result generalized several theorems in the literature.

2 Preliminaries and Results

For a given $x, y \in \mathscr{R}$, the commutator of x, y is denoted by $[x, y]$ and defined by $[x, y] = xy - yx$. Recall that a ring $\mathscr{R}$ is prime, if for any $a, b \in \mathscr{R}$, $a\mathscr{R}b = (0)$ implies either $a = 0$ or $b = 0$. Throughout, $\mathscr{R}$ is a prime ring with center $\mathscr{Z}$ and $\mathscr{Q} = \mathscr{Q}_{mr}(\mathscr{R})$ is the maximal right ring of quotient of $\mathscr{R}$. To be noted that $\mathscr{Q}$ is also a prime ring and the center $\mathscr{C}$ of $\mathscr{Q}$, which is called the extended centroid of $\mathscr{R}$, is a field. Moreover, $\mathscr{Z} \subseteq \mathscr{C}$ (further explanation refer to [6]). It is well known that any automorphism of $\mathscr{R}$ can be uniquely extended to an automorphism of $\mathscr{Q}$. An automorphism ξ of $\mathscr{R}$ is called $\mathscr{Q}$-inner if there exists an invertible element $g \in \mathscr{Q}$ such that $x^{\xi} = gxg^{-1}$ for all $x \in \mathscr{R}$. Moreover, ξ is called $\mathscr{Q}$-outer if it is not inner. An additive mapping $d : \mathscr{R} \longrightarrow \mathscr{R}$ is said to be a derivation if $(xy)^d = x^d y + xy^d$ holds for every $x, y \in \mathscr{R}$. A derivation $d : \mathscr{R} \longrightarrow \mathscr{R}$ is inner in case d is of the form $x^d = [q, x]$ for every $x \in \mathscr{R}$ and some fixed element $q \in \mathscr{R}$. A derivation of $\mathscr{R}$ is called $\mathscr{Q}$-inner if its extension to $\mathscr{Q}$ is inner. Also, d is called $\mathscr{Q}$-outer if it is not inner. Herein, we present some well-known facts and results that will be used in the follow-up which are indispensable to establish our principle theorem.

Fact 1 ([8, Theorem 3]) *Suppose that $\mathscr{R}$ is a prime ring and $\mathfrak{A}$ an independent subset of G modulo A_i. Let $\phi = \chi(x_i^{a_j}) = 0$ be a generalized identity with automorphisms of $\mathscr{R}$ reduced with respect to $\mathfrak{A}$. If for all $x_i \in X, a_j \in \mathfrak{A}$, the $x_i^{a_j}$-degree of $\phi = \chi(x_i^{a_j})$ is strictly less than char$(\mathscr{R})$ when char$(\mathscr{R}) \neq 0$, then $\chi(z_{ij}) = 0$ is also a generalized polynomial identity of $\mathscr{R}$.*

Fact 2 *Let $\mathscr{R}$ be a prime ring and $\mathscr{L}$ a noncentral Lie ideal of $\mathscr{R}$. If char$(\mathscr{R}) \neq 2$, then there exists a nonzero ideal I of $\mathscr{R}$ such that $0 \neq [I, \mathscr{R}] \subseteq \mathscr{L}$. If char$(\mathscr{R}) = 2$ and $dim_{\mathscr{C}}\mathscr{R}\mathscr{C} > 4$, then there exists a nonzero ideal I of $\mathscr{R}$ such that $0 \neq [I, \mathscr{R}] \subseteq \mathscr{L}$. Thus, if either char$(\mathscr{R}) \neq 2$ or $dim_{\mathscr{C}}\mathscr{R}\mathscr{C} > 4$, then we may conclude that there exists a nonzero ideal I of $\mathscr{R}$ such that $[I, I] \subseteq \mathscr{L}$.*

Fact 3 ([5, Lemma 7.1]) *Let $\mathscr{V}_{\mathscr{D}}$ be a vector space over a division ring $\mathscr{D}$ with $dim\mathscr{V}_{\mathscr{D}} \geq 2$ and $\mathscr{S} \in End(\mathscr{V})$. If s and $\mathscr{S}s$ are $\mathscr{D}$-dependent for every $s \in \mathscr{V}$, then there exists $\chi \in \mathscr{D}$ such that $\mathscr{S}s = \chi s$ for every $s \in \mathscr{V}$.*

Fact 4 ([11, Lemma 1.5]) *Let $\mathscr{C}$ be an infinite field and $n \geq 2$. Let $\mathscr{A}_1, \ldots, \mathscr{A}_n$ are not scalar matrices in $\mathscr{M}_m(\mathscr{H})$, then there exists some invertible matrix $\mathscr{P} \in \mathscr{M}_m(\mathscr{H})$ such that each matrix $\mathscr{P}\mathscr{A}_1\mathscr{P}^{-1}, \ldots, \mathscr{P}\mathscr{A}_k\mathscr{P}^{-1}$ has all nonzero entries.*

Fact 5 ([12, Proposition 1]) *Let $\mathscr{C}$ be a field of characteristic different from 2, $\mathscr{R} = \mathscr{M}_t(\mathscr{C})$ be the matrix ring over $\mathscr{C}$ and $t \geq 3$. Assume that a and b are elements of $\mathscr{R}$ with $a = \sum_{r,s=1}^{t} a_{rs}e_{rs}$ and $b = \sum_{r,s=1}^{t} b_{rs}e_{rs}$ for $a_{rs}, b_{rs} \in \mathscr{C}$ and further suppose that a and b satisfy the following condition. If $i \neq j$ are fixed integers such that $a_{ij}b_{ij} = 0$, then for any inner automorphism ϕ of $\mathscr{R}$ follows $a'_{ij}b'_{ij} = 0$, where $\phi(a) = \sum_{r,s=1}^{t} a'_{rs}e_{rs}$ and $\phi(b) = \sum_{r,s=1}^{t} b'_{rs}e_{rs}$. Therefore, if $a_{ij}b_{ij} = 0$ for all $i \neq j$, then either $a \in \mathscr{C}$ or $b \in \mathscr{C}$.*

Remark 1 Let $\mathscr{R}$ be a prime ring of characteristic different from 2 and ξ be an automorphism of $\mathscr{R}$ such that for every $x_1, y_1, x_2, y_2 \in \mathscr{R}$, $[[[x_2, y_2], [x_1, y_1]], [x_1, y_1]^{\xi}] = 0$. Then $\mathscr{R}$ is commutative.

Proof If one choose any two fixed elements $x_1, y_1 \in \mathscr{R}$ and denote $a = [x_1, y_1]$, $b = [x_1, y_1]^{\xi}$, then our identity can be rewritten as

$$\Big[\big[[x_2, y_2], a\big], b\Big] = 0$$

for all $x_2, y_2 \in \mathscr{R}$, which gives

$$[x_2, y_2]^{\delta_a \delta_b} = 0$$

for all $x_2, y_2 \in \mathscr{R}$, where δ_a and δ_b are the inner derivations induced, respectively, by elements a and b. In light of [7, Theorem 4], it follows that either $a \in \mathscr{Z}(\mathscr{R})$ or $b \in \mathscr{Z}(\mathscr{R})$. In any case, $[x_1, y_1] \in \mathscr{Z}(\mathscr{R})$. By the arbitrariness of x_1, y_1, the commutativity of $\mathscr{R}$ follows.

From now on, we are going to investigate the generalized polynomial identity

$$[[[x_1, y_1]^d, [x_1, y_1]], [x_1, y_1]^\xi] = 0$$

in prime ring $\mathscr{R}$ involving derivation and automorphism. In case of inner derivation and automorphism, one can rewrite as

$$[[[q, [x_1, y_1]]_2, \mathscr{T}[x_1, y_1]\mathscr{T}^{-1}] = 0 \text{ for all } x_1, y_1 \in \mathscr{R}$$

or

$$\begin{aligned}\Psi(x_1, y_1) = {} & a_1[x_1, y_1]^2a_2[x_1, y_1]a_3 - 2[x_1, y_1]a_1[x_1, y_1]a_2[x_1, y_1]a_3 \\ & +[x_1, y_1]^2a_4[x_1, y_1]a_3 - a_2[x_1, y_1]a_5[x_1, y_1]^2 \\ & +2a_2[x_1, y_1]a_3[x_1, y_1]a_1[x_1, y_1] - a_2[x_1, y_1]a_3[x_1, y_1]^2a_1, \quad (1)\end{aligned}$$

where $a_1 = q, a_2 = \mathscr{T}, a_3 = \mathscr{T}^{-1}, a_4 = \mathscr{T}q\mathscr{T}$, and $a_5 = \mathscr{T}^{-1}q$.

Lemma 1 *Let $\mathscr{R} = \mathscr{M}_2(\mathscr{C})$, the 2×2 matrix ring over $\mathscr{C}$ such that $\mathscr{R}$ satisfies (1). Then either $q \in \mathscr{C}$ or $\mathscr{R} \subseteq \mathscr{M}_2(\mathscr{C})$, where $\mathscr{C}$ is a finite field of characteristic* 3.

Proof Assume first that $\mathscr{C}$, then, by Fact 4, there exists some invertible matrix $\mathscr{B} \in \mathscr{M}_k(\mathscr{C})$ such that each matrix $\mathscr{B}a_1\mathscr{B}^{-1}$, $\mathscr{B}a_2\mathscr{B}^{-1}$, $\mathscr{B}a_3\mathscr{B}^{-1}$, $\mathscr{B}a_4\mathscr{B}^{-1}$, $\mathscr{B}a_5\mathscr{B}^{-1}$ has all nonzero entries. Denote by $\Theta(x) = \mathscr{B}x\mathscr{B}^{-1}$ the inner automorphism induced by $\mathscr{B}$. Say $\Theta(a_1) = \sum_{hl} a_{1hl}e_{hl}$, $\Theta(a_2) = \sum_{hl} a_{2hl}e_{hl}$, $\Theta(a_3) = \sum_{hl} a_{3hl}e_{hl}$, $\Theta(a_4) = \sum_{hl} a_{4hl}e_{hl}$, and $\Theta(a_5) = \sum_{hl} a_{5hl}e_{hl}$ for $0 \neq a_{1hl}, 0 \neq a_{2hl}, 0 \neq a_{3hl}, 0 \neq a_{4hl}, 0 \neq a_{5hl} \in \mathscr{C}$. Without loss of generality, we may replace a_1, a_2, a_3, a_4, and a_5 with $\Theta(a_1)$, $\Theta(a_2)$, $\Theta(a_3)$, $\Theta(a_4)$, and $\Theta(a_5)$, respectively. As above, in relation (1), let $i \neq j$, $[x_1, y_1] = e_{ij}$ and multiply on the right by e_{ij}. Thus, it follows $e_{ij}a_2e_{ij}a_3e_{ij}a_1e_{ij} = 0$ which gives a contradiction.

Now, let $\mathscr{E}$ be an infinite field which is an extension of the field $\mathscr{C}$ and let $\overline{\mathscr{R}} = \mathscr{M}_t(\mathscr{E}) \cong \mathscr{R} \otimes \mathscr{C}\mathscr{E}$. The generalized identity $\Psi(x_1, y_1)$ is homogeneous in both x_1 and y_1 of degree 3. Hence, the complete linearization of $\Psi(x_1, y_1)$ is a multilinear generalized polynomial $\Omega(x_1, y_1, x_2, y_2)$, and

$$\Omega(x_1, y_1, x_1, y_1) = 3^2\Psi(x_1, y_1)$$

Clearly, the multilinear polynomial $\Omega(x_1, y_1, x_2, y_2)$ is a generalized polynomial identity for $\mathscr{R}$ and $\overline{\mathscr{R}}$ too. If $char(\mathscr{C}) \neq 3$, we obtain $\Psi(x_1, y_1)$ for all $x_1, y_1 \in \overline{\mathscr{R}}$ and the conclusion $a_1 = q \in \mathscr{C}$ follows from the first part of the proof.

Proposition 1 *Let $\mathscr{R}$ be a prime ring of characteristic different from* 2 *and ξ be an automorphism of $\mathscr{R}$ such that $[[q, [x_1, y_1]]_2, [x_1, y_1]^\xi] = 0$ for all $x_1, y_1 \in \mathscr{R}$ and $q \in \mathscr{Q}$. Then either $q \in \mathscr{C}$ or $\mathscr{R} \subseteq \mathscr{M}_2(\mathscr{C})$, where $\mathscr{C}$ is a finite field of characteristic* 3.

Proof We are given that $[[q, [x_1, y_1]]_2, [x_1, y_1]^{\xi}] = 0$ for all $x_1, y_1 \in \mathscr{R}$. If ξ is an identity automorphism, then $\mathscr{R}$ satisfies $[[q, [x_1, y_1]]_3 = 0$ and hence application of [14, Theorem 3] yields the required conclusion. Next, we suppose that ξ is non-identity automorphism. Further, if ξ is an outer automorphism and as x_i, y_i-degree is less than $char(\mathscr{R})$, therefore by Fact 1, $\mathscr{R}$ satisfies $[[q, [x_1, y_1]]_2, [x_2, y_2]] = 0$ for all $x_1, y_1, x_2, y_2, \in \mathscr{R}$. In particular, $[[q, [x_1, y_1]]_3 = 0$ for all $x_1, y_1 \in \mathscr{R}$ and hence again we get the required conclusion, by considering the above presentation. Finally, if ξ is an inner automorphism, then there exists an invertible element $\mathscr{T} \in \mathscr{Q}$, such that $x^{\xi} = \mathscr{T} x \mathscr{T}^{-1}$ for all $x \in \mathscr{R}$. Therefore (1) a nontrivial generalized polynomial identity as $q \notin \mathscr{C}$ and ξ is non-identity automorphism. Hence, by [16], $\mathscr{Q}$ is a primitive ring, which is isomorphic to a dense subring of the ring of linear transformations of a vector space $\mathscr{V}$ over $\mathscr{C}$, containing nonzero linear transformations of finite rank. Consider the case $dim_{\mathscr{C}} \mathscr{V} = k$ with k a finite positive integer greater than or equal to 3. Notice that, for $k \geq 2$, we get our conclusion by Lemma 1. Now, we assume that $k \geq 3$, in this condition, $\mathscr{Q}$ is a simple ring which satisfies a nontrivial generalized polynomial identity, moreover $\mathscr{M}_k(\mathscr{C})$ satisfies the same generalized identity of $\mathscr{Q}$. Now, in (1), we choose $[x_1, y_1] = e_{ij}$ for i, j different indices. Then, by computations, it follows that $2e_{ij}a_2e_{ij}a_3e_{ij}a_1e_{ij} = 0$. Since $a_3 = \mathscr{T}^{-1} \neq 0$ and the (j, i)-entry of the above matrix is zero with the application of Fact 5, we get $a_1 = q \in \mathscr{C}$ as $a_2 = \mathscr{T} \notin \mathscr{C}$.

Let us now consider $dim_{\mathscr{C}} \mathscr{V} = \infty$. As we know the fact that (1) is a generalized polynomial identity of $\mathscr{Q}$. Since $a_1 = q$ does not centralize the nonzero ideal $Soc(\mathscr{R})$ of $\mathscr{R}$. Thus, there exists $h_0 \in Soc(\mathscr{R})$ such that $[q, h_0] \neq 0$. In view of Litoff's theorem (see Theorem 4.3.11 in [6]), there exists an idempotent element $e \in Soc(\mathscr{R})$, such that $h_0, a_1, a_2, a_3, a_4 \in e\mathscr{Q}e \cong \mathscr{M}_k(\mathscr{C})$ for some integer k. Also, it is easy to see that (1) is a generalized polynomial identity for $e\mathscr{Q}e$. Then by the finite-dimensional case, we have that $eqe \in Z(e\mathscr{Q}e)$. Thus, the following is a contradiction:

$$qh_0 = eqh_0 = eqeh_0 = h_0eqe = h_0qe = h_0q.$$

Theorem 1 *Let $\mathscr{R}$ be a prime ring of characteristic different from* 2*, $\mathscr{L}$ a noncentral Lie ideal of $\mathscr{R}$, d a derivation of $\mathscr{R}$, and ξ be an automorphism of $\mathscr{R}$ such that $[[x^d, x], x^{\xi}] = 0$ for all $x \in \mathscr{L}$. Then either $d = 0$ or $\mathscr{R} \subseteq \mathscr{M}_2(\mathscr{C})$, where $\mathscr{C}$ is a finite field of characteristic* 3.

Proof In view of Fact 2, there exist a nonzero two-sided ideal I of $\mathscr{R}$ such that $0 \neq [I, \mathscr{R}] \subseteq \mathscr{L}$. Therefore, I satisfies $[[[x_1, y_1]^d, [x_1, y_1]], [x_1, y_1]^{\zeta}] = 0$. We will divide proof into two cases: If d is an inner derivation induced by an element $q \in \mathscr{Q}$, i.e., $x^d = [q, x]$ for all $x \in \mathscr{R}$ and hence it follows that $[[q, [x_1, y_1]]_2, [x_1, y_1]^{\xi}] = 0$. In view of Proposition 1, we get the required conclusion. Next, we assume that d is an outer derivation. In light of Kharchenko's theory [13] and as $I, \mathscr{R}$ satisfy the same differential identities [14, Theorem 3], so we have $[[[x_2, y_2], [x_1, y_1]], [x_1, y_1]^{\xi}] = 0$ for all $x, y \in \mathscr{R}$. In view of Remark 1, we have nothing to prove.

Acknowledgements The authors wish to thank the referee for his/her constructive comments and suggestions, which improve the quality of the paper a lot. For the third author, this research is supported by the National Board of Higher Mathematics (NBHM), India, Grant No. 02011/16/2020 NBHM (R. P.) R & D II/ 7786.

References

1. Ashraf, M., Pary, S.A.: An identity involving automorphisms of prime rings inspired by Posner's theorem. J. Taibah Univ. Sci. **12**(3), 344–347 (2018)
2. Ali, S., Ashraf, M., Raza, M.A., Khan, A.N.: n-commuting mappings on (semi)-prime rings with applications. Commun. Algebra **5**(47), 2262–2270 (2019)
3. Ashraf, M., Raza, M.A., Parry, S.A.: Commutators having idempotent values with automorphisms in semi-prime rings. Math. Rep. (Bucur.) **20**(70), 51–57 (2018)
4. Ashraf, M., Rehman, N.: On commutativity of rings with derivations. Results Math. **42**(1–2), 3–8 (2002)
5. Beidar, K.I., Brešar, M.: Extended Jacobson density theorem for rings with automorphisms and derivations. Israel J. Math. **122**, 317–346 (2001)
6. Beidar, K.I., Martindale III, W.S., Mikhalev, A.V.: Rings with generalized identities. Pure and Applied Mathematics, vol. 196. Marcel Dekker, New York (1996)
7. Bergen, J., Herstein, I.N., Kerr, J.W.: Lie ideals and derivations of prime rings. J. Algebra **71**, 259–267 (1981)
8. Chuang, C.L.: Differential identities with automorphism and anti-automorphism-II. J. Algebra **160**, 291–335 (1993)
9. Cheng, H.: Some results about derivations of prime rings. J. Math. Res. Expos. **25**(4), 625–633 (2005)
10. Davvaz, B., Raza, M.A.: A note on automorphisms of Lie ideals in prime rings. Math. Slovaca **68**(5), 1223–1229 (2018)
11. De Filippis, V.: A product of two generalized derivations on polynomials in prime rings. Collect. Math. **61**, 303–322 (2010)
12. De Filippis, V., Scudo, G.: Strong commutativity and Engel condition preserving maps in prime and semiprime rings. Linear Multilinear Algebra **61**, 917–938 (2013)
13. Kharchenko, V.K.: Differential identities of prime rings. Algebra Logic **17**, 155–168 (1979)
14. Lanski, C.: An Engel condition with derivation. Proc. Amer. Math. Soc. **118**, 731–734 (1993)
15. Lee, P.H., Lee, T.K.: Lie ideals of prime rings with derivations. Bull. Inst. Math. Acad. Sin. **11**, 75–80 (1983)
16. Martindale, W.S., III.: Prime rings satisfying a generalized polynomial identity. J. Algebra **12**, 576–584 (1969)
17. Mayne, J.H.: Centralizing automorphisms of Lie ideals in prime rings. Canad. Math. Bull. **35**, 510–514 (1992)
18. Mayne, J.H.: Centralizing mappings of prime rings. Canad. Math. Bull. **27**(1), 122–126 (1984)
19. Mayne, J.H.: Centralizing automorphisms of prime rings. Canad. Math. Bull. **19**, 113–115 (1976)
20. Posner, E.C.: Derivations in prime rings. Proc. Amer. Math. Soc. **8**, 1093–1100 (1957)
21. Raza, M.A., Rehman, N.: An identity on automorphisms of Lie ideals in prime rings. Ann. Univ. Ferrara **62**(1), 143–150 (2016)
22. Rehman, N., Raza, M.A.: On m-commuting mappings with skew derivations in prime rings. St. Petersburg Math. J. **27**(4), 641–650 (2016)
23. Vukman, J.: Commuting and centralizing mappings in prime rings. Proc. Amer. Math. Soc. **109**, 47–52 (1990)
24. Wang, Y.: Power-centralizing automorphisms of Lie ideals in prime rings. Commun. Algebra **34**, 609–615 (2006)

Modules Invariant Under Clean Endomorphisms of Their Injective Hulls

Check for updates

Jane Roseline and Manoj Kumar Patel

Abstract A module is quasi-injective if and only if it is invariant under endomorphisms of its injective hull. In this paper, we study the class of modules which are invariant under all clean endomorphisms of their injective hulls and show that this class of modules coincide with the class of quasi-injective modules. Some facts and results of this class of modules are obtained. We also establish some relations of clean-invariant modules with automorphism-invariant modules, idempotent-invariant modules, pseudo-continuous modules, and Utumi modules. Apart from these, we have given several sufficient conditions under which automorphism-invariant modules can be clean-invariant.

Keywords Automorphism-invariant modules · Idempotent-invariant modules · Clean-invariant modules

1 Introduction

Let R be an associative ring with unity. Johnson and Wong [15] proved that a module is quasi-injective if and only if it is invariant under endomorphisms of its injective hull; a module M is called quasi-injective [10], if for any submodule A of M, every homomorphism $f : A \to M$ can be extended to an endomorphism of M. Jeremy [14] characterized quasi-continuous modules as those modules that are invariant under idempotent-endomorphisms of their injective hulls. Goel and Jain [11] call a module M, π-injective if for every pair of submodules M_1 and M_2 with $M_1 \cap M_2 = 0$, each projection $\pi_i : M_1 \oplus M_2 \to M_i, i = 1, 2$, can be lifted to an endomorphism of M. π-injective modules are precisely the quasi-continuous modules defined by Jeremy [14]. Thus, the class of modules which are invariant under projections of their injective hulls coincide with the class of modules which are invariant under

J. Roseline · M. K. Patel (✉)
Department of Mathematics, National Institute of Technology Nagaland, Dimapur 797103, Nagaland, India
e-mail: mkpitb@gmail.com

M. Ashraf et al. (eds.), *Algebra and Related Topics with Applications*,
Springer Proceedings in Mathematics & Statistics 392,
https://doi.org/10.1007/978-981-19-3898-6_21

idempotent-endomorphisms of their injective hulls. Noyan et al. ([9], Theorem 16) proved that automorphism-invariant modules [12] are precisely the pseudo-injective modules, where a module M is called pseudo-injective [7], if for any submodule A of M, every monomorphism $f : A \to M$ can be extended to an endomorphism of M.

Clean rings were introduced by Nicholson [18]. An element $a \in R$ is said to be *clean* if $a = e + u$ where e is an idempotent and u is a unit in R. If every element of R is clean, then R is called a clean ring. Nicholson proved that every clean ring is an exchange ring, and a ring with central idempotents is clean if and only if it is an exchange ring. A ring is said to be clean (almost clean) [1] if each of its elements is the sum of a unit (regular element) and an idempotent. A module is clean (almost clean) if its endomorphism ring is clean (almost clean). A module which is invariant under automorphisms of its injective hull is called an *automorphism-invariant module*, i.e., M is called an automorphism-invariant module if $f(M) \subseteq M$ for all $f \in Aut(E(M))$ [12]. A module M is called an idempotent-invariant module if $f(M) \subseteq M$ for all idempotent $f \in End(E(M))$.

The objective of this paper is to characterize the class of modules which are invariant under clean endomorphisms of their injective hulls and show that this class coincides with the class of quasi-injective modules [10]. By a clean endomorphism, we mean an endomorphism which is the sum of an automorphism and an idempotent-endomorphism, i.e., for an R-module M, $f \in End(M)$ is said to be a clean endomorphism if $f = g + h$, where $g^2 = g \in End(M)$ and $h \in Aut(M)$. We shall denote the class of clean endomorphisms of M by $ClEnd(M)$. A module which is invariant under clean endomorphisms of its injective hull will be called a *clean-invariant module*, i.e., M will be called a clean-invariant module if $f(M) \subseteq M$ for all $f \in ClEnd(E(M))$. A submodule $N \subseteq M$ is said to be clean-invariant if $f(N) \subseteq N$ for all $f \in ClEnd(M)$. A submodule N of a module M is said to be essential in M if $N \cap A \neq 0$ for all $A \subseteq M$.

Consider the following conditions for an R-module M [3, 13]:
(C_1) Every submodule of M is essential in a direct summand of M.
(C_2) Every submodule of M isomorphic to a direct summand of M is itself a direct summand of M.
(C_3) If A and B are summands of M with $A \cap B = 0$ then $A \oplus B$ is also a direct summand of M.
M is called a CS (or an *extending*) module if it satisfies (C_1); M is called *continuous* if it satisfies (C_1) and (C_2); M is called *quasi-continuous* if it satisfies (C_1) and (C_3). Modules satisfying (C_1), (C_2), and (C_3) are called $C1$-, $C2$-, and $C3$-modules, respectively. It is well known that the following implications hold:
Injective $\Longrightarrow$ quasi-injective $\Longrightarrow$ continuous $\Longrightarrow$ quasi-continuous $\Longrightarrow$ CS.

But none of the converses hold in general. We refer to [8, 17] for background on (quasi-)injective, (quasi-)continuous, and CS modules.

A right R-module M is said to satisfy the *exchange property* [19] if for every right R-module A and any two direct sum decompositions $A = M_1 \oplus N = \oplus_{i \in I} A_i$ with $M_1 \simeq M$ there exist submodules B_i of A_i such that $A = M_1 \oplus (\oplus_{i \in I} B_i)$. If this holds only for $|I| < \infty$, then M is said to satisfy the finite exchange property.

Clean-invariant modules satisfy the exchange property. Similar results do not hold for idempotent-invariant modules. We provide examples to show that a direct sum of clean-invariant modules need not be clean-invariant although summands of clean-invariant modules inherit the property. Finally, we also establish some of the relations of clean-invariant modules with automorphism-invariant modules, idempotent-invariant modules, and Utumi modules.

Throughout, all rings R are associative with unity and all modules are unitary R-modules, unless otherwise stated. For a module M, we use $E(M)$, $End(M)$, and $Aut(M)$ to denote the injective hull, the endomorphism ring, and the set of automorphisms of M, respectively. $kerf$ and Imf denote the kernel of f and the image of f, respectively. We write $N \subseteq M$ if N is a submodule of M, $N \subseteq^{ess} M$ if N is an essential submodule of M, and $N \subseteq^{\oplus} M$ if N is a direct summand of M.

2 Clean-Invariant Modules

A module which is invariant under clean endomorphisms of its injective hull is called a clean-invariant module, i.e., M is called a clean-invariant module if $f(M) \subseteq M$ for all $f \in ClEnd(E(M))$.

By the following theorem we will prove that the class of clean-invariant modules coincide with the class of quasi-injective modules.

Theorem 1 *A module M is quasi-injective if and only if it is clean-invariant.*

Proof By [15], we know that a module M is quasi-injective if and only if M is invariant under any endomorphism of its injective hull. Since $E(M)$ is an injective module, it is clean [5] and so $End(E(M))$ is a clean ring. Thus, every $f \in End(E(M))$ is the sum of an idempotent-endomorphism and an automorphism. Thus, module M is quasi-injective if and only if M is invariant under any clean endomorphism of its injective hull.

Clean-invariant modules are automorphism-invariant but the converse is not true, in general.

Example 1 If R is the ring of all eventually constant sequences $(x_n)_{n\in\mathbb{N}}$ of elements in $\mathbb{Z}_2$, then $E(R_R) = \prod_{n\in\mathbb{N}} \mathbb{Z}_2$ has only one automorphism, namely, the identity automorphism. Thus R_R is an automorphism-invariant module but R_R is not clean-invariant.

Clean-invariant modules are idempotent-invariant but the converse is not true, in general.

Example 2 If $\mathbb{Z}$ denotes the ring of integers, $\mathbb{Z}_{\mathbb{Z}}$ is an idempotent-invariant module but not clean-invariant.

However, it is worth mentioning that the class of automorphism-invariant modules and the class of idempotent-invariant modules are not contained in one another as shown by the following examples:

(i) If $\mathbb{Z}$, $\mathbb{Q}$ denote the ring of integers and rational numbers, respectively, $\mathbb{Z}_{\mathbb{Z}}$ is an idempotent-invariant module which is not automorphism-invariant because the injective hull $\mathbb{Q}_{\mathbb{Z}}$ of $\mathbb{Z}_{\mathbb{Z}}$ has the automorphism $\varphi : q \to \frac{q}{2}$ but $\varphi(\mathbb{Z}) \nsubseteq \mathbb{Z}$.

(ii) If R is the ring of all eventually constant sequences $(x_n)_{n\in\mathbb{N}}$ of elements in $\mathbb{Z}_2$, then $E(R_R) = \prod_{n\in\mathbb{N}} \mathbb{Z}_2$ has only one automorphism, namely, the identity automorphism. Thus, R_R is an automorphism-invariant module but R_R cannot be CS because by Theorem 1, a module M is quasi-injective if and only if it is automorphism-invariant CS. So if R_R is CS, then R_R would be quasi-injective which is not so. Hence, R_R is not an idempotent-invariant module.

It is to be noted that a summand of a clean-invariant module is also clean-invariant. However, a direct sum of clean-invariant modules need not be clean-invariant as shown by the following examples:

(i) The $\mathbb{Z}$-module $\mathbb{Z}_p \oplus \mathbb{Q}$ is not clean-invariant although $\mathbb{Z}_p$ and $\mathbb{Q}$ are clean-invariant as $\mathbb{Z}$-modules.

(ii) Let $R = \begin{pmatrix} F & F \\ 0 & F \end{pmatrix}$ where F is a field. Then $A = \begin{pmatrix} F & F \\ 0 & 0 \end{pmatrix}$ is an injective R-module and $B = \begin{pmatrix} 0 & 0 \\ 0 & F \end{pmatrix}$ is a clean-invariant R-module. But $A \oplus B = R$ is not a clean-invariant R-module.

(iii) The $\mathbb{Z}$-modules $\mathbb{Z}_p$ and $\mathbb{Z}_{p^2}$, where p is a prime number, are clean-invariant modules. However $\mathbb{Z}_p \oplus \mathbb{Z}_{p^2}$ is not a clean-invariant $\mathbb{Z}$-module.

Theorem 2 *Every submodule N of a clean-invariant module M is also clean-invariant.*

Proof We need to show that N is invariant under all clean endomorphisms of its injective hull $E(N)$. Let $f \in ClEnd(E(N))$. Since $E(N)$ is injective, it is a direct summand of $E(M)$ and so there exists a $g \in ClEnd(E(M))$ which extends f. Since M is clean-invariant, M is invariant under all clean endomorphisms of $E(M)$ and so $g(M) \subseteq M$. Thus, we have $g|_M \in ClEnd(M)$. Since N is a clean-invariant submodule of M, we have $g(N) \subseteq N$. From $g|_{E(N)} = f$, we get $f(N) \subseteq N$.

An R-module M is called *co-Hopfian* if any injective endomorphism of M is an automorphism. A module M is said to be *Dedekind finite* (or directly finite) if $M \cong M \oplus X$ implies $X = 0$, i.e., M is not isomorphic to a proper summand of itself [2, 20].

Theorem 3 *If M is a clean-invariant module, then M is non-co-Hopfian if and only if there exists a decomposition $M = N_r \oplus (\oplus_{i=1}^{r} M_i)$ for any positive integer r, where $N_r \cong M$ and $M_i \neq 0$ for $i = 1, 2, 3, \ldots, r$.*

Proof Let M be non-co-Hopfian, then we have an injective endomorphism $f : M \to M$ which is not an automorphism. Let $N_1 = f(M)$, $N_1 \neq M$ and $g : N_1 \to M$ be an isomorphism; then there exists an endomorphism $h : M \to M$ such that $h|_{N_1} = g$, because M is clean-invariant. Therefore $M = N_1 \oplus ker(h) = N_1 \oplus M_1$, where $M_1 = ker(h)$. It is clear that $M_1 \neq 0$. Since N_1 is non-co-Hopfian, by a similar argument we get $N_1 = N_2 \oplus M_2$, with $N_2 \cong N_1$ and $M_2 \neq 0$. Thus we have $M =$

$N_2 \oplus (M_1 \oplus M_2)$. Now applying the principle of mathematical induction and the definition of co-Hopfian module we attain the desired result $M = N_r \oplus (\oplus_{i=1}^{r} M_i)$ for $r \in Z^+$, where $N_r \cong M$ and $M_i \neq 0$ for $i = 1, 2, 3, \ldots, r$. The converse is clear.

A module M is said to satisfy the cancellation property if whenever $M \oplus N \cong M \oplus K$, then $N \cong K$ [2, 20].

Theorem 4 *Let M be a clean-invariant module, then M is co-Hopfian if and only if it satisfies the cancellation property.*

Proof Assume that M is co-Hopfian. As every co-Hopfian module is directly finite, so M is directly finite and it is clear that clean-invariant directly finite modules satisfy the cancellation property. Conversely, assume that M satisfies the cancellation property. Suppose that M is non-co-Hopfian. Then by Theorem 3, we have a decomposition $M = M \oplus 0 = N_1 \oplus M_1$ such that $N_1 \cong M$ and $M_1 \neq 0$. So by the cancellation property, we get $M_1 = 0$, which is a contradiction to our assumption that $M_1 \neq 0$. Hence, our supposition that M is non-co-Hopfian is wrong.

A nonzero module M is called uniform if any two nonzero submodules of M have nonzero intersection (intersect crossrefer). An R-module M is said to have uniform dimension n (written u.dim$M = n$) if there is an essential submodule $V \subseteq^{ess} M$ that is a direct sum of n uniform submodules [3, 13].

Theorem 5 *Every automorphism-invariant module M with finite uniform dimension is Dedekind finite.*

Proof Let M be an automorphism-invariant module with finite uniform dimension and let $f : M \to M$ be an injective endomorphism of M. Then f extends to an endomorphism $g : E(M) \to E(M)$, which is injective. Since M has finite uniform dimension, $g \in Aut(E(M))$ and because M is automorphism-invariant, $g(M) = M$ ([21], Corollary 2.3). Thus $f(M) = M$ which implies that f is also surjective and so f is an automorphism. Since every injective $f \in End(M)$ is an automorphism, M is co-Hopfian. As every co-Hopfian module is Dedekind finite, so M is Dedekind finite.

Corollary 1 *Every clean-invariant module with finite uniform dimension is Dedekind finite.*

A module M is called square-free [2, 20] if it contains no nonzero submodules isomorphic to a square $A \oplus A$. Equivalently, a module M is square-free if whenever $N \subseteq M$ and $N = Y_1 \oplus Y_2$ with $Y_1 \cong Y_2$, then $Y_1 = Y_2 = (0)$.

Theorem 6 *Every square-free automorphism-invariant module is Dedekind finite.*

Proof Let M be an automorphism-invariant module which is square-free and let $f : M \to M$ be an injective endomorphism of M. Then f extends to an endomorphism $g : E(M) \to E(M)$, which is injective and so $g(E(M))$ is a direct summand of $E(M)$. Thus there exists a submodule A of M such that $E(M) = g(E(M)) \oplus A$, so that $E(M) = g^2(E(M)) \oplus g(A) \oplus A$, where $g(A) \cong A$. $E(M)$ being the injective hull of the square-free module M, is also square-free and so $A = 0$. Thus $E(M) = g(E(M))$ shows that g is surjective and so g is an automorphism of $E(M)$. But M being automorphism-invariant, $g(M) = M$ [21]. Thus $f(M) = M$ which implies that f is also surjective and so f is an automorphism. Hence M is co-Hopfian and so Dedekind finite.

Corollary 2 *Every clean-invariant square-free module is Dedekind finite.*

An R-module M satisfies (C_4) [6] if, whenever A and B are submodules of M with $M = A \oplus B$ and $f : A \to B$ is an R-homomorphism with $kerf \subseteq^{\oplus} A$, then $Imf \subseteq^{\oplus} B$. An R-module M satisfying (C_4) is called a *C_4-module* [6]. A C_4-module which is also CS is called a *pseudo-continuous* module. Some examples of C_4-modules are automorphism-invariant modules and square-free modules. We know that $(C_2) \Rightarrow (C_3) \Rightarrow (C_4)$. Thus, clean-invariant module satisfies C_4 as it satisfies C_3, which further yields the proper implications:
clean-invariant $\Rightarrow$ continuous $\Rightarrow$ idempotent-invariant $\Rightarrow$ pseudo-continuous.

A right R-module M is called a Utumi module (U-module in short) if, whenever A and B are submodules of M with $A \cong B$ and $A \cap B = 0$, there exist two summands P and Q of M such that $A \subseteq^{ess} P$, $B \subseteq^{ess} Q$, and $P \oplus Q \subseteq^{\oplus} M$.

Theorem 7 *Every clean-invariant module is a U-module.*

Proof Let M be a clean-invariant module, then M is an automorphism-invariant module. Since every automorphism-invariant module is a U-module. Thus M is a U-module.

Theorem 8 *If $M = A \oplus B$ is an automorphism-invariant module, where both A and B are clean-invariant modules, then M is a clean-invariant module.*

Proof Since $M = A \oplus B$ is automorphism-invariant, A and B are relatively injective [16]. Since both A and B are clean-invariant modules by Theorem 1, A and B are quasi-injective and so M is quasi-injective ([17], Proposition 1.17). Thus $M = A \oplus B$ is clean-invariant by Theorem 1.

Corollary 3 *If $M = M_1 \oplus M_2 \oplus \cdots \oplus M_n$ is an automorphism-invariant module, where each M_i is a clean-invariant module, then M is clean-invariant.*

Theorem 9 *If $M = A \oplus B$ is an automorphism-invariant module, where both A and B are uniform modules, then M is a clean-invariant module.*

Proof Since $M = A \oplus B$ is an automorphism-invariant module, both A and B being summands of M are automorphism-invariant. Also A and B being uniform modules are clean-invariant ([16], Corollary 13) as every uniform module is CS. Thus M is clean-invariant by Theorem 8.

Corollary 4 *If $M = M_1 \oplus M_2 \oplus \cdots \oplus M_n$ is an automorphism-invariant module, where each M_i is a uniform module, then M is clean-invariant.*

Theorem 10 *If 2 is a unit of R, then every automorphism-invariant R-module is clean-invariant.*

Proof Let M be an automorphism-invariant R-module and let $E(M)$ be its injective hull. Since 2 is a unit of R, 2 is a unit of $End(E(M))$. $E(M)$ being an injective module, it is clean by [5] and so $End(E(M))$ is a clean ring. Thus, every $f \in ClEnd(E(M))$ will be the sum of two automorphisms [4] and since M is automorphism-invariant, $f(M) \subseteq M$ for all $f \in Aut(E(M))$. As $f(M) \subseteq M$ for all $f \in ClEnd(E(M))$, M is clean-invariant.

Acknowledgements The authors would like to thank the referees for careful reading of paper and their valuable suggestions. Also M. K. Patel wishes to thank National Board for Higher Mathematics, with file No.: 02211/3/2019 NBHM (R.P.) RD-II/1439, for financial assistantship.

References

1. Akalan, E., Vaš, L.: Classes of almost clean rings. Algebra represent. Theory **16**, 843–857 (2013)
2. Alahmadi, A., Jain, S.K., Leroy, A.: ADS modules. J. Algebra **352**, 215–222 (2012)
3. Alkan, M., Harmanci, A.: On summand sum and summand intersection property of modules. Turk. J. Math. **26**, 131–147 (2002)
4. Camillo, V.P., Yu, H.P.: Exchange rings, units and idempotents. Commun. Algebra **22**(12), 4737–4749 (1994)
5. Camillo, V.P., Khurana, D., Lam, T.Y., Nicholson, W.K., Zhou, Y.: Continuous modules are clean. J. Algebra **304**, 94–111 (2006)
6. Ding, N.V., Ibrahim, Y., Yousif, M., Zhou, Y.: C4-modules. Commun. Algebra **45**(4), 1727–1740 (2017)
7. Dinh, H.Q.: A note on pseudo-injective modules. Commun. Algebra **33**, 361–369 (2005)
8. Dung, N.V., Huynh, D.V., Smith, P.F., Wisbauer, R.: Extending Modules. Pitman Research Notes in Mathematics Series, vol. 313. Longman Scientific and Technical, Harlow (1994)
9. Er, N., Singh, S., Srivastava, A.K.: Rings and modules which are stable under automorphisms of their injective hulls. J. Algebra **379**, 223–229 (2013)
10. Fuchs, L.: On quasi-injective modules. Ann. Sc. Norm. Super. Pisa **23**, 541–546 (1969)
11. Goel, V.K., Jain, S.K.: π-injective modules and rings whose cyclic are π-injective. Commun. Algebra **6**(1), 59–73 (1978)
12. Guil Asensio, P.A., Srivastava, A.K.: Automorphism-invariant modules satisfy the exchange property. J. Algebra **388**, 101–106 (2013)
13. Hamdouni, A., Özcan, A.Ç., Harmanci, A.: Characterization of modules and rings by the summand intersection property and the summand sum property. J. Algebra Number Theory Appl. **5**, 469–490 (2005)
14. Jeremy, L.: Modules et anneaux quasi-continus. Canad. Math. Bull. **17**, 217–228 (1974)
15. Johnson, R.E., Wong, E.T.: Quasi-injective modules and irreducible rings. J. London Math. Soc. **36**, 260–268 (1961)
16. Lee, T.K., Zhou, Y.: Modules which are invariant under automorphisms of their injective hulls. J. Algebra Appl. **12**(2), 1250159 9 pp. (2013)
17. Mohamed, S.H., Müller, B.J.: Continuous and Discrete Modules. London Mathematical Society Lecture Note Series, vol. 147. Cambridge University Press, Cambridge (1990)
18. Nicholson, W.K.: Lifting idempotents and exchange rings. Trans. Amer. Math. Soc. **229**, 269–278 (1977)

19. Oshiro, K., Rizvi, S.T.: The exchange property of quasi-continuous modules with the finite exchange property. Osaka J. Math. **33**, 217–234 (1996)
20. Quynh, T.C., Koşan, M.T.: On ADS modules and rings. Commun. Algebra **42**, 3541–3551 (2014)
21. Quynh, T.C., Koşan, M.T.: On automorphism-invariant modules. J. Algebra Appl. **14**(5), 1550074, 11 pp. (2015)

A Note on Central Idempotents in Finite Group Rings of Symmetric Groups

Anuradha Sabharwal, Pooja Yadav, and R. K. Sharma

Abstract The number of central idempotents in the group ring $\mathbb{Z}_n[S_4]$ has been computed, for $n = p_1 p_2 \cdots p_l$ where $p_i's$ are distinct primes. It is proved that the number of central idempotents is 2^{5l-2} when $2 \nmid n$; 2^{5l-4} when $3 \nmid n$; 2^{5l} when $6 \nmid n$, and 2^{5l-6} when $6|n$. Further, it has been proved that the number of central idempotents is 2^{kl} in $\mathbb{Z}_n[S_m]$ if $(n, m!) = 1$. Here k is the number of disjoint conjugacy classes in S_m and l is the number of distinct primes involved in the prime factorization of n.

Keywords Group ring · Symmetric group · Central idempotent

1 Introduction

Idempotents in rings and group rings play a very vital role in determining their structure. Therefore, numerous efforts have been made to compute different types of idempotents. Important contributions have been made for computing primitive central idempotents (for example, see [1, 2, 4]). Meyer [2] computed primitive central idempotents of $F_q[G]$ for arbitrary prime powers q, and arbitrary finite groups G. Also, a well-known result of Osima [4, p. 178] gives the explicit form for the primitive central idempotents in $K[G]$, when K is a field. Martinez [1] computed central irreducible idempotents of dihedral group algebra $\mathbb{F}_q[D_{2n}]$.

In this paper, we prove the following two main theorems.

A. Sabharwal (✉)
Department of Mathematics, University of Delhi, Delhi 110 007, India
e-mail: anuradha.sabharwal@gmail.com

P. Yadav
Department of Mathematics, Kamala Nehru College, University of Delhi, Delhi 110 007, India

R. K. Sharma
Department of Mathematics, Indian Institute of Technology Delhi, Delhi 110 016, India

M. Ashraf et al. (eds.), *Algebra and Related Topics with Applications*,
Springer Proceedings in Mathematics & Statistics 392,
https://doi.org/10.1007/978-981-19-3898-6_22

Theorem 1 *Let $n = p_1 p_2 \cdots p_l$ where p_i's are distinct primes. Then the number of central idempotents in $\mathbb{Z}_n[S_4]$ is*

(i) 2^{5l}, *if* $p_i > 3\ \forall 1 \leq i \leq l$;
(ii) 2^{5l-4}, *if* $p_1 = 2$ *and* $p_i > 3\ \forall\ 2 \leq i \leq l$;
(iii) 2^{5l-2}, *if* $p_1 = 3$ *and* $p_i > 3\ \forall\ 2 \leq i \leq l$;
(iv) 2^{5l-6}, *if* $p_1 = 2$, $p_2 = 3$ *and* $p_i > 3\ \forall\ 3 \leq i \leq l$.

Theorem 1 can be partly generalized as follows:

Theorem 2 *Let $n = p_1 p_2 \cdots p_l$ where p_i's are distinct primes. If $(n, m!) = 1$, then the number of central idempotents in $\mathbb{Z}_n[S_m]$ is 2^{kl} where k is the number of conjugacy classes in S_m.*

In this article, we study the number of central idempotents of the symmetric group S_m over $\mathbb{Z}_n$, the ring of integers modulo n, in the case when every prime divisor of n does not divide $|S_m|$. Also, we compute the number of central idempotents in $\mathbb{Z}_n[S_4]$ for the case when n is the product of distinct primes. We have already calculated the number of central idempotents in $\mathbb{Z}_n[S_3]$ for every positive integer n, and further we have provided an explicit form of these central idempotents [*A Note on Central Idempotents in Group Ring of Symmetric Group over $\mathbb{Z}_n$*. Anuradha Sabharwal, R. K. Sharma, Pooja Yadav (Preprint)]. Let G be a group and R be a ring. We denote by $\mathscr{Z}(R[G])$ the set of central elements of the group ring of G over R. For any ring R, $E(R)$ will denote the set of all idempotents in R.

Let $n = p_1^{n_1} p_2^{n_2} \cdots p_l^{n_l}$ be the prime factorization of n. Then by Chinese Remainder Theorem, we see that

$$\phi : \mathbb{Z}_n \longrightarrow \mathbb{Z}_{p_1^{n_1}} \oplus \mathbb{Z}_{p_2^{n_2}} \oplus \cdots \oplus \mathbb{Z}_{p_l^{n_l}}$$

is an isomorphism. This isomorphism maps the residue class of an integer $a(\text{mod } n)$ to a vector with all the components equal to the residue class of $a(\text{mod } p_i^{n_i})$:

$$\bar{a} \mapsto (\bar{a}, \bar{a}, \ldots, \bar{a}).$$

So the residue class of a in $\mathbb{Z}_n$ is an idempotent, if and only if, for each i, $a(\text{mod } p_i^{n_i})$ is an idempotent in $\mathbb{Z}_{p_i^{n_i}}$. The congruence $a^2 \equiv a(\text{mod } p_i^{n_i})$ holds, if and only if $p_i^{n_i} | (a^2 - a)$. Here only one of the factors of a or $(a - 1)$ can be divisible by p_i, and hence either a or $(a - 1)$ has to be divisible by $p_i^{n_i}$. Thus $a \equiv 0, 1(\text{mod } p_i^{n_i})$ are the only idempotents in $\mathbb{Z}_{p_i^{n_i}}$. By Chinese Remainder Theorem these congruences are independent for different i and hence we have the following lemma:

Lemma 1 *The number of pairwise non-congruent idempotents in $\mathbb{Z}_n$ is equal to 2^l, where l is the number of distinct prime factors of n.*

Lemma 2 ([3], p. 134) *Let $\{R_i\}_{i \in I}$ be a family of rings, and let $R = \oplus_{i \in I} R_i$. Then for any group G, we have $R[G] \simeq (\oplus_{i \in I} R_i)[G] \simeq \oplus_{i \in I}(R_i[G])$.*

Recall that a ring R is called a **simple ring** if it has no non-trivial ideals.

Note that $M_n(D)$, the ring of $n \times n$ matrices over any division ring D is a simple ring. A ring R is said to be **semisimple** if it can be decomposed as a direct sum of finitely many minimal left ideals.

We now state some well-known results that will be needed later.

Theorem 3 ([3], Maschke's Theorem) *Let G be a group and R a ring. Then $R[G]$ is semisimple if the following conditions hold:*

(i) R is semisimple.
(ii) G is finite.
(iii) $|G|$ is invertible in R.

Corollary 1 ([3], Corollary 3.4.8) *Let G be a group and K be a field. Then $K[G]$ is semisimple if and only if G is finite and the $char(K) \nmid |G|$.*

The following Wedderburn–Artin theorem applied to group rings helps to determine the structure of a group algebra.

Theorem 4 ([3], Theorem 3.4.9) *Let G be a finite group and K be a field such that $char(K) \nmid |G|$. Then:*

(i) $K[G]$ is a direct sum of a finite number of two-sided ideals $\{B_i\}_{1 \leq i \leq s}$, the simple components of $K[G]$. Each B_i is a simple ring.
(ii) Any two-sided ideal of $K[G]$ is a direct sum of some of the members of the family $\{B_i\}_{1 \leq i \leq s}$.
(iii) Each simple component B_i is isomorphic to a full matrix ring of the form $M_{n_i}(D_i)$, where D_i is a division ring containing an isomorphism copy of K in its center, and the isomorphism $K[G] \simeq \oplus_{i=1}^{s} M_{n_i}(D_i)$ is an isomorphism of K-algebras.
(iv) In each matrix ring $M_{n_i}(D_i)$, the set

$$I = \left\{ \begin{bmatrix} x_1 & 0 & \dots & 0 \\ x_1 & 0 & \dots & 0 \\ & \dots & & \\ x_1 & 0 & \dots & 0 \end{bmatrix} : x_1, x_2, \dots, x_{n_i} \in D_i \right\} \simeq D_i^{n_i}$$

is a minimal left ideal.
(v) $I_i \not\simeq I_j$ if $i \neq j$.
(vi) Any simple $K[G]$-module is isomorphic to some I_i, $1 \leq i \leq s$.

Note that this decomposition is unique.

Corollary 2 ([3], Corollary 3.4.8) *Let G be a finite group and K an algebraically closed field, where $char(K) \nmid |G|$. Then*

$$K[G] \simeq \oplus_{i=1}^{s} M_{n_i}(K) \text{ and } |G| = \sum_{i=1}^{s} n_i^2.$$

Corollary 3 ([3], Proposition 3.6.3) *Let G be a finite group and K an algebraically closed field such that char(K) $\nmid |G|$. Then, the number of simple components of KG is equal to the number of conjugacy classes of G.*

Lemma 3 *Let G be a finite group and p be a prime number such that $p \nmid |G|$. Then*

$$\mathbb{Z}_p[G] \simeq \oplus_{i=1}^{s} M_{n_i}(\mathbb{Z}_p) \text{ and } |G| = \sum_{i=1}^{s} n_i^2.$$

Proof Since $char(\mathbb{Z}_p) = p$ and $p \nmid |G|$, we have that $\mathbb{Z}_p[G] \simeq \oplus_{i=1}^{s} M_{n_i}(D_i)$, where D_i is a division ring containing a copy of Z_p in its center. As $\mathbb{Z}_p$ is a field, $\mathbb{Z}_p[G] \simeq \oplus_{i=1}^{s} M_{n_i}(\mathbb{Z}_p)$. Computing dimensions over $\mathbb{Z}_p$ on both sides of the equation above, we have that $|G| = \sum_{i=1}^{s} n_i^2$.

Lemma 4 *The center* $\mathscr{Z}(M_n(K)) = \{I_{n\times n}\}K$.

Lemma 5 *Let G be a finite group and p be a prime number such that $p \nmid |G|$. Then, the number of simple components of $\mathbb{Z}_p[G]$ is equal to the number of conjugacy classes of G.*

Proof As we know that the set of all class sums of G over $\mathbb{Z}_p$ forms a basis of $\mathscr{Z}(\mathbb{Z}_p[G])$ over $\mathbb{Z}_p$ ([3],Theorem 3.6.2, p. 151), it will suffice to show that the dimension of $\mathscr{Z}(\mathbb{Z}_p[G])$ over $\mathbb{Z}_p$ is equal to the number of simple components of $\mathbb{Z}_p[G]$. From Lemma 3 we know that $\mathbb{Z}_p[G] \simeq \oplus_{i=1}^{s} M_{n_i}(\mathbb{Z}_p)$ and thus $\mathscr{Z}(\mathbb{Z}_p[G]) \simeq \oplus_{i=1}^{s} \mathscr{Z}(M_{n_i}(\mathbb{Z}_p))$. From Lemma 4, $\mathscr{Z}(M_n(\mathbb{Z}_p)) \simeq \mathbb{Z}_p$. Therefore, $\mathscr{Z}(\mathbb{Z}_p[G]) \simeq \underbrace{\mathbb{Z}_p \oplus \cdots \oplus \mathbb{Z}_p}_{s\ times}$; consequently, $[\mathscr{Z}(\mathbb{Z}_p[G]) : \mathbb{Z}_p] = s$.

Example 1 It is easy to see that $\mathbb{Z}_{35}[S_4]$ decomposes as follows:

$$\mathbb{Z}_{35}[S_4] \simeq \mathbb{Z}_5[S_4] \oplus \mathbb{Z}_7[S_4]$$

Since $5 \nmid |S_4|$, we get $\mathbb{Z}_5[S_4] \simeq \mathbb{Z}_5 \oplus \mathbb{Z}_5 \oplus M_2(\mathbb{Z}_5) \oplus M_3(\mathbb{Z}_5) \oplus M_3(\mathbb{Z}_5)$ Again $7 \nmid |S_4|$, gives $\mathbb{Z}_7[S_4] \simeq \mathbb{Z}_7 \oplus \mathbb{Z}_7 \oplus M_2(\mathbb{Z}_7) \oplus M_3(\mathbb{Z}_7) \oplus M_3(\mathbb{Z}_7)$. This finally gives that $\mathbb{Z}_{35}[S_4] \simeq (\mathbb{Z}_5 \oplus \mathbb{Z}_5 \oplus M_2(\mathbb{Z}_5) \oplus M_3(\mathbb{Z}_5) \oplus M_3(\mathbb{Z}_5)) \oplus (\mathbb{Z}_7 \oplus \mathbb{Z}_7 \oplus M_2(\mathbb{Z}_7) \oplus M_3(\mathbb{Z}_7) \oplus M_3(\mathbb{Z}_7))$.

2 Proof (Theorem 1)

We first begin with the following proposition

Proposition 1 *The number of central idempotents in* $\mathbb{Z}_n[S_4]$ *is*

(i) 2, *if* $n = 2$;
(ii) 8, *if* $n = 3$.

Proof The conjugacy classes of $S_4 = \langle \sigma, \tau | \tau^2 = \sigma^4 = 1, \sigma\tau = \tau^{-1}\sigma \rangle$ are

$$\begin{aligned}
C_1 &= \{1\}\\
C_2 &= \{\tau, \sigma^2\tau\sigma^2, \sigma^3\tau\sigma^2\tau, \sigma\tau\sigma^3, \sigma^3\tau\sigma, \tau\sigma^2\tau\sigma^3\}\\
C_3 &= \{\sigma^2, \tau\sigma^2\tau, \sigma^2\tau\sigma^2\tau\}\\
C_4 &= \{\sigma\tau, \tau\sigma, \tau\sigma^3, \sigma^2\tau\sigma^3, \sigma\tau\sigma^2, \sigma^3\tau\sigma^2, \sigma^2\tau\sigma, \sigma^3\tau\}\\
C_5 &= \{\sigma, \sigma^3, \sigma^3\tau\sigma^3, \sigma^2\tau, \sigma\tau\sigma, \tau\sigma^2\}.
\end{aligned}$$

The corresponding class sums are

$$\begin{aligned}
\gamma_1 &= 1\\
\gamma_2 &= \tau + \sigma^2\tau\sigma^2 + \sigma^3\tau\sigma^2\tau + \sigma\tau\sigma^3 + \sigma^3\tau\sigma + \tau\sigma^2\tau\sigma^3\\
\gamma_3 &= \sigma^2 + \tau\sigma^2\tau + \sigma^2\tau\sigma^2\tau\\
\gamma_4 &= \sigma\tau + \tau\sigma + \tau\sigma^3 + \sigma^2\tau\sigma^3 + \sigma\tau\sigma^2 + \sigma^3\tau\sigma^2 + \sigma^2\tau\sigma + \sigma^3\tau\\
\gamma_5 &= \sigma + \sigma^3 + \sigma^3\tau\sigma^3 + \sigma^2\tau + \sigma\tau\sigma + \tau\sigma^2.
\end{aligned}$$

Recall ([3], Theorem 3.6.2, p. 151) that the set of all class sums forms a basis of the center $\mathscr{Z}(R[G])$ of $R[G]$, over a commutative ring R. Hence γ_i's form a basis of center of $\mathbb{Z}_n[S_4]$ over $\mathbb{Z}_n$.

Let e be a central idempotent in $\mathbb{Z}_n[S_4]$. Then, e can be expressed as

$$e = \alpha \cdot \gamma_1 + \beta \cdot \gamma_2 + \gamma \cdot \gamma_3 + \delta \cdot \gamma_4 + \omega \cdot \gamma_5 \text{ for some } \alpha, \beta, \gamma, \delta, \omega \in \mathbb{Z}_n.$$

Since e is an idempotent, comparing the coefficients of class sums in the equation $e^2 = e$, we get the following relations:

$$\alpha = \alpha^2 + 6\beta^2 + 3\gamma^2 + 8\delta^2 + 6\omega^2 \tag{1}$$
$$\beta = 2\alpha\beta + 2\beta\gamma + 8\beta\delta + 4\gamma\omega + 8\delta\omega \tag{2}$$
$$\gamma = 2\beta^2 + 2\gamma^2 + 8\delta^2 + 2\omega^2 + 2\alpha\gamma + 8\beta\omega \tag{3}$$
$$\delta = 3\beta^2 + 4\delta^2 + 3\omega^2 + 2\alpha\delta + 6\beta\omega + 6\gamma\delta \tag{4}$$
$$\omega = 2\alpha\omega + 4\beta\gamma + 8\beta\delta + 2\gamma\omega + 8\delta\omega. \tag{5}$$

The values of $\alpha, \beta, \gamma, \delta, \omega$ give all possible central idempotents in $\mathbb{Z}_n[S_4]$.
Case (i) : $n = 2$
The above equations are reduced to

$$\begin{aligned}
\alpha &= \alpha^2 + \gamma^2\\
\beta &= 0\\
\gamma &= 0\\
\delta &= \beta^2 + \omega^2\\
\omega &= 0.
\end{aligned}$$

Solving these equations in $\mathbb{Z}_2$, we get 2 solutions:

$$\alpha = 0, \beta = 0, \gamma = 0, \delta = 0, \omega = 0$$

$$\alpha = 1, \beta = 0, \gamma = 0, \delta = 0, \omega = 0$$

Hence, there are 2 central idempotents in $\mathbb{Z}_2[S_4]$.

Case (ii): n=3

Equations 1–5 are reduced to

$$\alpha = \alpha^2 + 2\delta^2 \tag{6}$$
$$\beta = 2\alpha\beta + 2\beta\gamma + 2\beta\delta + \gamma\omega + 2\delta\omega \tag{7}$$
$$\gamma = 2\beta^2 + 2\gamma^2 + 2\delta^2 + 2\omega^2 + 2\alpha\gamma + 2\beta\omega \tag{8}$$
$$\delta = \delta^2 + 2\alpha\delta \tag{9}$$
$$\omega = 2\alpha\omega + \beta\gamma + 2\beta\delta + 2\gamma\omega + 2\delta\omega. \tag{10}$$

Subtracting (9) from (6), we get $\alpha - \delta = (\alpha - \delta)^2$ which gives $\alpha - \delta = 0$ or 1. That is, $\alpha = \delta$ or $\alpha = \delta + 1$.

When $\alpha = \delta$, Eqs. (6) and (9) imply

$$\alpha = 0 \text{ and } \delta = 0$$

and Eqs. (7), (8), (10) become

$$\beta = 2\beta\gamma + \gamma\omega$$
$$\gamma = 2\beta^2 + 2\gamma^2 + 2\omega^2 + 2\beta\omega$$
$$\omega = \beta\gamma + 2\gamma\omega$$

$\implies \beta + \omega = 0$ and $\gamma = 2\gamma^2 + 2\beta\omega$.

Solving these equations we get the following values of β, γ, ω:

$$\beta = 0, \ \omega = 0, \ \gamma = 0 \text{ or } 2$$

$$\beta = 1, \ \omega = 2, \ \gamma = 1$$

$$\beta = 2, \ \omega = 1, \ \gamma = 1.$$

Similarly, when $\alpha = \delta + 1$, we get

$$\beta = 0, \ \omega = 0, \ \gamma = 0 \text{ or } 1$$

$$\beta = 1,\ \omega = 2,\ \gamma = 2$$

$$\beta = 2,\ \omega = 1,\ \gamma = 2.$$

Combining both the cases, we get 8 solutions, and hence there are 8 central idempotents in $\mathbb{Z}_3[S_4]$.

We observe that an integer a is invertible in $\mathbb{Z}_p$ if and only if a is invertible in $\mathbb{Z}_{p^n}$. Thus the number of solutions of Eqs. 1–5 in $\mathbb{Z}_p$ is same as the number of solutions of Eqs. 1–5 in $\mathbb{Z}_{p^n}$. Hence the number of central idempotents in $\mathbb{Z}_p[S_4]$ and $\mathbb{Z}_{p^n}[S_4]$ is same. Therefore, $|E(\mathscr{Z}(\mathbb{Z}_{2^n}[S_4]))| = 2$ and $|E(\mathscr{Z}(\mathbb{Z}_{3^n}[S_4]))| = 8$.

We are now ready to prove our main theorem case by case as follows:

Case (i): $p_i > 3\ \forall\ 1 \leq i \leq l$

If $n = p_1 p_2 \cdots p_l$,

$$\mathbb{Z}_n \simeq \mathbb{Z}_{p_1} \oplus \mathbb{Z}_{p_2} \oplus \cdots \oplus \mathbb{Z}_{p_l}.$$

Using Lemma 2,

$$\begin{aligned}\mathbb{Z}_n[S_4] &\simeq (\mathbb{Z}_{p_1} \oplus \mathbb{Z}_{p_2} \oplus \cdots \oplus \mathbb{Z}_{p_l})[S_4]\\ &\simeq \mathbb{Z}_{p_1}[S_4] \oplus \mathbb{Z}_{p_2}[S_4] \oplus \cdots \oplus \mathbb{Z}_{p_l}[S_4].\end{aligned}$$

Also since each $p_i \nmid |S_4|$, the decomposition of $\mathbb{Z}_{p_i}[S_4]$ can be written in the following form:

$$\mathbb{Z}_{p_i}[S_4] \simeq \mathbb{Z}_{p_i} \oplus \mathbb{Z}_{p_i} \oplus M_2(\mathbb{Z}_{p_i}) \oplus M_3(\mathbb{Z}_{p_i}) \oplus M_3(\mathbb{Z}_{p_i}).$$

Now, for matrix ring $M_n(K)$,

$$\mathscr{Z}(M_n(K)) = \{\alpha I : \alpha \in K\}$$

and thus $\mathscr{Z}(M_n(K)) \simeq K$. Therefore,

$$\mathscr{Z}(\mathbb{Z}_{p_i}[S_4]) \simeq \mathbb{Z}_{p_i} \oplus \mathbb{Z}_{p_i} \oplus \mathbb{Z}_{p_i} \oplus \mathbb{Z}_{p_i} \oplus \mathbb{Z}_{p_i}.$$

Hence,

$$\begin{aligned}\mathscr{Z}(\mathbb{Z}_n[S_4]) \simeq\ & (\mathbb{Z}_{p_1} \oplus \mathbb{Z}_{p_1} \oplus \mathbb{Z}_{p_1} \oplus \mathbb{Z}_{p_1} \oplus \mathbb{Z}_{p_1})\\ &\oplus (\mathbb{Z}_{p_2} \oplus \mathbb{Z}_{p_2} \oplus \mathbb{Z}_{p_2} \oplus \mathbb{Z}_{p_2} \oplus \mathbb{Z}_{p_2})\\ &\oplus \cdots \oplus (\mathbb{Z}_{p_l} \oplus \mathbb{Z}_{p_l} \oplus \mathbb{Z}_{p_l} \oplus \mathbb{Z}_{p_l} \oplus \mathbb{Z}_{p_l}).\end{aligned}$$

Since there are only 2 idempotents in each $\mathbb{Z}_{p_i}$, $|E(\mathscr{Z}(\mathbb{Z}_n[S_4]))| = \underbrace{2^5 \times 2^5 \times \cdots \times 2^5}_{l-times}$.

That is, the number of central idempotents in $\mathbb{Z}_n[S_4] = 2^{5l}$.

Case (ii): $p_1 = 2$ and $p_i > 3\ \forall\ 2 \leq i \leq l$

Proceeding as in case (i), we have

$$\begin{aligned}
\mathbb{Z}_n[S_4] &\simeq \mathbb{Z}_{p_1}[S_4] \oplus \mathbb{Z}_{p_2}[S_4] \oplus \cdots \oplus \mathbb{Z}_{p_l}[S_4] \\
&\simeq \mathbb{Z}_{p_1}[S_4] \oplus (\mathbb{Z}_{p_2} \oplus \mathbb{Z}_{p_2} \oplus M_2(\mathbb{Z}_{p_2}) \oplus M_3(\mathbb{Z}_{p_2}) \oplus M_3(\mathbb{Z}_{p_2})) \\
&\quad \oplus \cdots \oplus (\mathbb{Z}_{p_l} \oplus \mathbb{Z}_{p_l} \oplus M_2(\mathbb{Z}_{p_l}) \oplus M_3(\mathbb{Z}_{p_l}) \oplus M_3(\mathbb{Z}_{p_l})) \\
\mathscr{Z}(\mathbb{Z}_n[S_4]) &\simeq \mathscr{Z}(\mathbb{Z}_{p_1}[S_4]) \oplus (\mathbb{Z}_{p_2} \oplus \mathbb{Z}_{p_2} \oplus \mathbb{Z}_{p_2} \oplus \mathbb{Z}_{p_2} \oplus \mathbb{Z}_{p_2}) \\
&\quad \oplus \cdots \oplus (\mathbb{Z}_{p_l} \oplus \mathbb{Z}_{p_l} \oplus \mathbb{Z}_{p_l} \oplus \mathbb{Z}_{p_l} \oplus \mathbb{Z}_{p_l}).
\end{aligned} \tag{11}$$

Since there are only 2 idempotents in $\mathbb{Z}_{p_i}$ for all $2 \leq i \leq l$ and 2 central idempotents in $\mathbb{Z}_2[S_4]$, $|E(\mathscr{Z}(\mathbb{Z}_n[S_4]))| = 2 \times \underbrace{2^5 \times \cdots \times 2^5}_{(l-1)-times}$.

That is, the number of central idempotents in $\mathbb{Z}_n[S_4] = 2^{5l-4}$.

Case (iii): $p_1 = 3$ and $p_i > 3\ \forall\ 2 \leq i \leq l$

Since there are only 2 idempotents in $\mathbb{Z}_{p_i}$ for all $2 \leq i \leq l$ and 8 central idempotents in $\mathbb{Z}_3[S_4]$, then from Eq. (11) we get $|E(\mathscr{Z}(\mathbb{Z}_n[S_4]))| = 2^3 \times \underbrace{2^5 \times \cdots \times 2^5}_{(l-1)-times}$.

That is, in this case, the number of central idempotents in $\mathbb{Z}_n[S_4] = 2^{5l-2}$.

Case (iv): $p_1 = 2$, $p_2 = 3$ and $p_i > 3\ \forall\ 3 \leq i \leq l$

$$\begin{aligned}
\mathscr{Z}(\mathbb{Z}_n[S_4]) \simeq\ & \mathscr{Z}(\mathbb{Z}_{p_1}[S_4]) \oplus \mathscr{Z}(\mathbb{Z}_{p_2}[S_4]) \\
& \oplus (\mathbb{Z}_{p_3} \oplus \mathbb{Z}_{p_3} \oplus \mathbb{Z}_{p_3} \oplus \mathbb{Z}_{p_3} \oplus \mathbb{Z}_{p_3}) \\
& \oplus \cdots \oplus (\mathbb{Z}_{p_l} \oplus \mathbb{Z}_{p_l} \oplus \mathbb{Z}_{p_l} \oplus \mathbb{Z}_{p_l} \oplus \mathbb{Z}_{p_l})
\end{aligned}$$

Since $|E(\mathscr{Z}(\mathbb{Z}_2[S_4]))| = 2$, $|E(\mathscr{Z}(\mathbb{Z}_3[S_4]))| = 8$, and $|E(\mathbb{Z}_{p_i})| = 2$ for all $3 \leq i \leq l$. Therefore, from Eq. (11) we get

$|E(\mathscr{Z}(\mathbb{Z}_n[S_4]))| = 2 \times 2^3 \times \underbrace{2^5 \times \cdots \times 2^5}_{(l-2)-times}$.

That is, in this case, the number of central idempotents in $\mathbb{Z}_n[S_4] = 2^{5l-6}$.

This proves Theorem 1.

Proof (Proof of Theorem 2) As $(n, m!) = 1$ and $n = p_1 p_2 \cdots p_l$, we get $p_i \nmid |S_m|$ for each i. By using Lemmas 3 and 5, the decomposition of $\mathbb{Z}_{p_i}[S_m]$ can be written in the following form:

$$\mathbb{Z}_{p_i}[S_m] \simeq \sum_{j=1}^{k} M_{m_j}(\mathbb{Z}_{p_i}), \tag{12}$$

where $m_1^2 + m_2^2 + \cdots + m_k^2 = |S_m|$ and $k =$ number of distinct conjugacy classes in S_m.

By fundamental theorem of finite Abelian groups, we have

$$\mathbb{Z}_n \simeq \mathbb{Z}_{p_1} \oplus \mathbb{Z}_{p_2} \oplus \cdots \oplus \mathbb{Z}_{p_l}.$$

Therefore,

$$\begin{aligned}\mathbb{Z}_n[S_m] &\simeq (\mathbb{Z}_{p_1} \oplus \mathbb{Z}_{p_2} \oplus \cdots \oplus \mathbb{Z}_{p_l})[S_m] \\ &\simeq \mathbb{Z}_{p_1}[S_m] \oplus \mathbb{Z}_{p_2}[S_m] \oplus \cdots \oplus \mathbb{Z}_{p_l}[S_m].\end{aligned}$$

Using Eq. 12, we get

$$\mathbb{Z}_n[S_m] \simeq \sum_{j=1}^{k} M_{m_j}(\mathbb{Z}_{p_1}) \oplus \sum_{j=1}^{k} M_{m_j}(\mathbb{Z}_{p_2}) \oplus \cdots \oplus \sum_{j=1}^{k} M_{m_j}(\mathbb{Z}_{p_l}).$$

Since $\mathscr{Z}(M_n(K)) \simeq K$ (by using Lemma 4),

$$\mathscr{Z}(\mathbb{Z}_n[S_m]) \simeq \sum_{j=1}^{k} \mathbb{Z}_{p_1} \oplus \sum_{j=1}^{k} \mathbb{Z}_{p_2} \oplus \cdots \oplus \sum_{j=1}^{k} \mathbb{Z}_{p_l}.$$

As each $\mathbb{Z}_{p_i}$ contains only 2 idempotents, $|E(\mathscr{Z}(\mathbb{Z}_n[S_m]))| = \underbrace{2^k \times 2^k \times \cdots \times 2^k}_{l-times}$.

That is, the number of central idempotents in $\mathbb{Z}_n[S_m] = 2^{kl}$.

Theorem 5 *Let K be an algebraically closed field such that $char(K) \nmid |S_n|$. Then the number of central idempotents in $K[S_n]$ is 2^k where k is the number of conjugacy classes in S_n.*

Proof From Corollaries 2 and 3, we have that $K[S_n] \simeq \oplus_{i=1}^{k} M_{n_i}(K)$, where k is the number of conjugacy classes of S_n and $n_1^2 + n_2^2 + \cdots + m_k^2 = |S_n|$. Also by Lemma 4, $\mathscr{Z}(M_n(K)) \simeq K$. Therefore,

$$\mathscr{Z}(K[S_n]) \simeq \oplus_{i=1}^{k} K.$$

Since there are only 2 idempotents in K, the number of central idempotents in $K([S_n])$ is 2^k.

Corollary 4 *Let K be an algebraically closed field such that $char(K) \nmid |S_4|$. Then the number of central idempotents in $K[S_4]$ is 2^5.*

Acknowledgements The authors thank the reviewer for his/her valuable comments.

References

1. Brochero Martínez, F.E.: Structure of finite dihedral group algebra. Finite Fields Appl. **35**, 204–214 (2015)
2. Meyer, H.: Primitive central idempotents of finite group rings of symmetric groups. Math. Comp. **77**(263), 1801–1821 (2008)
3. Polcino Milies, C., Sehgal, S.K.: An introduction to group rings. Algebra and Applications 1. Kluwer Academic Publishers, Dordrecht (2002)
4. Osima, M.: On blocks of characters of the symmetric group. Proc. Jpn. Acad. **31**, 131–134 (1955)

Jordan Product Preserving Generalized Skew Derivations on Lie Ideals

Giovanni Scudo

Abstract Let R be a non-commutative prime ring of characteristic different from 2, Q_r be its right Martindale quotient ring and C be its extended centroid, L a non-central Lie ideal of R and F and G non-zero generalized skew derivations of R such that $F(x)G(y) + F(y)G(x) = x \circ y$, for all $x, y \in L$. Then one of the following holds:

1. there exists $a, b \in Q_r$ such that $F(x) = xa$ and $G(x) = bx$, for any $x \in R$, with $ab = 1_C$;
2. R satisfies $s_4(x_1, \ldots, x_4)$ the standard polynomial identity on 4 non-commuting variables and there exist $a, b \in Q_r$ such that $F(x) = ax$ and $G(x) = xb$, for any $x \in R$, with $ab = 1_C$;
3. R satisfies $s_4(x_1, \ldots, x_4)$ and there exist an invertible element $q \in Q_r$ and $0 \neq \beta \in C$ such that $F(x) = \beta qxq^{-1}$ and $G(x) = \beta^{-1}qxq^{-1}$, for any $x \in R$.

Keywords Generalized skew derivation · Prime ring

1 Introduction

Let R be a prime ring with center $Z(R)$, Q_r the *right Martindale quotient ring* of R, C the center of Q_r, which is called *extended centroid* of R. We recall that, since R is prime, then Q_r is a prime ring and C is a field (for more details about these objects, we refer the reader to [2, Chap. 2]). An additive mapping $d\colon R \longrightarrow R$ is said to be a *derivation* of R if $d(xy) = d(x)y + xd(y)$, for all $x, y \in R$. An additive mapping $F\colon R \longrightarrow R$ is called a *generalized derivation* of R if there exists a derivation d of R such that $F(xy) = F(x)y + xd(y)$ for all $x, y \in R$. The derivation d is uniquely determined by F, which is called an *associated derivation* of F.

G. Scudo (✉)
Department of Engineering, University of Messina, Messina, Italy
e-mail: gscudo@unime.it

M. Ashraf et al. (eds.), *Algebra and Related Topics with Applications*,
Springer Proceedings in Mathematics & Statistics 392,
https://doi.org/10.1007/978-981-19-3898-6_23

The definition of generalized skew derivation is a unified notion of skew derivation and generalized derivation, which are considered as classical additive mappings of non-commutative algebras. Let R be an associative ring and α be an automorphism of R, an additive mapping $d\colon R \longrightarrow R$ is said to be a *skew derivation* of R if $d(xy) = d(x)y + \alpha(x)d(y)$, for all $x, y \in R$; the automorphisms α is called an *associated automorphism* of d. An additive mapping $F\colon R \longrightarrow R$ is called a *generalized skew derivation* of R if there exists a skew derivation d of R, with associated automorphism α, such that $F(xy) = F(x)y + \alpha(x)d(y)$, for all $x, y \in R$. In this case, d is called an *associated skew derivation* of F and α is called an *associated automorphism* of F. It was Chang who first introduced this notion and initiated the study of generalized skew derivations of (semi-)prime rings in [7]. Therein, he described the identity of the form $h(x) = af(x) + g(x)b$, where f, g and h are the so-called generalized (α, β)-derivations of a prime ring R, a and b are some fixed non-central elements of R. In this paper, we study the structure of the prime ring R and the form of generalized (skew) derivations satisfying particular conditions. Many authors follow this line of investigation studying some commutativity preserving conditions on a subset of the ring; specifically, if $S \subseteq R$, the map $F : R \to R$ is called commutativity preserving on S if $[x, y] = 0$ implies $[F(x), F(y)] = 0$; it is called strong commutativity preserving on S if $[F(x), F(y)] = [x, y]$, for all $x, y \in S$.

Many results have been obtained for prime and semiprime algebras by applying the technique of Functional Identities developed by Beidar, Bresar, Chebotar and Martindale (see [4] for details).

Following these ideas, in [3], Bell and Daif proved that if R is a semiprime ring admitting a derivation d strong commutativity preserving on a right ideal I of R, then $I \subseteq Z(R)$, the center of R. Bresar and Miers studied the case in which any additive map is strong commutativity preserving on a semiprime ring R; more precisely, in [5], they showed that the map assumes the form $F(x) = \lambda x + \mu(x)$, where $\lambda \in C$, $\lambda^2 = 1$ and $\mu : R \to C$ is an additive map of R into C.

Later some authors began to consider generalized derivations that are strong commutativity preserving on some subsets of prime and semiprime ring. Moreover, in some papers, the authors studied new conditions generalizing the strong commutativity preserving conditions; for example, in [25], Liu studied the case when I is a right ideal of R, $F : I \to R$ is a map and G is a generalized derivation of R, such that $[F(x), G(y)] = [x, y]$ for all $x, y \in I$; he described the form F and G also in the case when both F and G are generalized derivations of R.

In [24], Liu and Liau studied the same condition satisfied by L, a non-central Lie ideal, where F and G are both generalized derivations of R. They proved that either $R \subseteq M_2(K)$, the ring of all 2×2 matrices over a field K, or there exists $0 \neq \lambda \in C$ such that $G(x) = \lambda x$ and $F(x) = \lambda^{-1}x$, for all $x \in R$. Recently, the authors, in [1], considered F and G two non-zero generalized derivations on R such that $F(x)G(y) - F(y)G(x) = [x, y]$, for all $x, y \in T$, where T is the set of all evaluations of a multilinear polynomial over C; they proved that either $R \subseteq M_2(K)$, for a field K, or there exist $0 \neq a, b \in R$ such that $F(x) = xa$ and $G(x) = bx$, for all $x \in R$, where $ab = 1_K$.

Rania, in [27], analyzed a similar condition involving the Jordan product of two non-zero generalized derivations of R; in particular, he described the form of the maps F and G when the following condition is satisfied: $F(x) \circ G(y) = x \circ y$, for all $x, y \in S$, where S is the set of all evaluations of a non-central polynomial over C. In light of this line of investigation, in this paper, we will study a condition, that generalizes the Jordan product preserving condition when the additive maps are generalized skew derivations on R.

The main goal of the present paper is to prove the following theorem:

Theorem 1 *Let R be a non-commutative prime ring of characteristic different from 2, Q_r be its right Martindale quotient ring and C be its extended centroid, L a non-central Lie ideal of R and F and G non-zero generalized skew derivations of R such that $F(x)G(y) + F(y)G(x) = x \circ y$, for all $x, y \in L$. Then one of the following holds:*

1. *there exist $a, b \in Q_r$ such that $F(x) = xa$ and $G(x) = bx$, for any $x \in R$, with $ab = 1_C$;*
2. *R satisfies $s_4(x_1, \ldots, x_4)$ the standard polynomial identity on 4 non-commuting variables and there exist $a, b \in Q_r$ such that $F(x) = ax$ and $G(x) = xb$, for any $x \in R$, with $ab = 1_C$;*
3. *R satisfies $s_4(x_1, \ldots, x_4)$ and there exist an invertible element $q \in Q_r$ and $0 \neq \beta \in C$ such that $F(x) = \beta q x q^{-1}$ and $G(x) = \beta^{-1} q x q^{-1}$, for any $x \in R$.*

In order to proceed with our proofs, we need to recall some well-known results on skew derivations and automorphisms involved in generalized polynomial identities for prime rings.

Let us denote by $\mathrm{SDer}(Q_r)$ the set of all skew-derivations of Q_r. By a skew-derivation word we mean an additive mapping Δ of the form $\Delta = d_1 d_1 \ldots d_m$, where $d_i \in \mathrm{SDer}(Q_r)$. A skew-differential polynomial is a generalized polynomial with coefficients in Q_r of the form $\Phi(\Delta_j(x_i))$ involving noncommutative indeterminates x_i on which the skew derivation words Δ_j act as unary operations. The skew-differential polynomial $\Phi(\Delta_j(x_i))$ is said to be a *skew-differential identity* on a subset T of Q_r if it vanishes on any assignment of values from T to its indeterminates x_i.

Let R be a prime ring, $\mathrm{SD}_{\mathrm{int}}$ be the C-subspace of $\mathrm{SDer}(Q_r)$ consisting of all inner skew-derivations of Q_r, and let d and δ be two non-zero skew-derivations of Q_r. The following results follow as special cases from results in [8–11].

Fact 2 *Assume that d and δ are skew derivations on R, associated with the same automorphism α of R. If d and δ are $C-$linearly independent modulo $\mathrm{SD}_{\mathrm{int}}$ and $\Phi(\Delta_j(x_i))$ is a skew-differential identity on R, where Δ_j are skew-derivations words of the following form δ, d, then $\Phi(y_{ji})$ is a generalized polynomial identity of R, where y_{ji} are distinct indeterminates.*

In particular, we have

Fact 3 *In [13], Chuang and Lee investigate polynomial identities with a single skew derivation. They prove that if $\Phi(x_i, D(x_i))$ is a generalized polynomial identity* *for R, where R is a prime ring and D is an outer skew derivation of R, then R also satisfies the generalized polynomial identity $\Phi(x_i, y_i)$, where x_i and y_i are distinct indeterminates. Furthermore, they observe [13, Theorem 1] that in the case $\Phi(x_i, D(x_i), \alpha(x_i))$ is a generalized polynomial identity for a prime ring R, D is an outer skew derivation of R and α is an outer automorphism of R, then R also satisfies the generalized polynomial identity $\Phi(x_i, y_i, z_i)$, where x_i, y_i, and z_i are distinct indeterminates.*

Fact 4 *If d and δ are C−linearly dependent modulo $\mathrm{SD_{int}}$, then there exist $\lambda, \mu \in C$, $a \in Q_r$ and $\alpha \in Aut(Q)$ such that $\lambda d(x) + \mu\delta(x) = ax - \alpha(x)a$, for all $x \in R$.*

Fact 5 *By [13] we can state the following result. If d is a non-zero skew-derivation of R and*

$$\Phi\left(x_1, \ldots, x_n, d(x_1), \ldots, d(x_n)\right)$$

is a skew-differential polynomial identity of R, then one of the following statements holds:

1. *either $d \in \mathrm{SD_{int}}$;*
2. *or R satisfies the generalized polynomial identity $\Phi(x_1, \ldots, x_n, y_1, \ldots, y_n)$.*

Fact 6 *Let R be a prime ring and I be a two-sided ideal of R. Then I, R, and Q_r satisfy the same generalized polynomial identities with coefficients in Q_r (see [8]). Furthermore, I, R, and Q_r satisfy the same generalized polynomial identities with automorphisms (see [10, Theorem 1]).*

Fact 7 *Let R be a prime ring, then the following statements hold:*

1. *Every generalized derivation of R can be uniquely extended to Q_r ([23, Theorem 3]).*
2. *Any automorphism of R can be uniquely extended to Q_r ([9, Fact 2]).*
3. *Every generalized skew derivation of R can be uniquely extended to Q_r ([7, Lemma 2]) as follows: by a (right) generalized skew derivation we mean an additive mapping $G : Q_r \longrightarrow Q_r$ such that $G(xy) = G(x)y + \alpha(x)d(y)$ for all $x, y \in Q_r$, where d is a skew derivation of R and α is an automorphism of R. Moreover, there exists $G(1) = a \in Q_r$ such that $G(x) = ax + d(x)$ for all $x \in R$.*

Fact 8 *Let L be a non-central Lie ideal of R and $char(R) \neq 2$, then there exists a non-zero ideal I of R such that $0 \neq [I, R] \subseteq L$ ([19, pp. 4–5], [17, Lemma 2 and Proposition 1], [21, Theorem 4]).*

Fact 9 ([14, Lemma 1.5]) *Let H be an infinite field and $m \geq 2$. If $A_1, \ldots, A_k$ are not scalar matrices in $M_m(H)$ then there exists some invertible matrix $P \in M_m(H)$ such that each matrix $PA_1P^{-1}, \ldots, PA_kP^{-1}$ has all non-zero entries.*

Fact 10 ([16, Proposition 1]) *Let $a, b \in R = M_t(H)$, the ring of all $t \times t$ matrices over a field H of characteristic different from 2, where $t \geq 3$. Denote $a = \sum a_{lk}e_{lk}$*

and $b = \sum b_{lk}e_{lk}$, *where* $a_{lk}, b_{lk} \in K$, *and* e_{lk} *is the unit matrix with* 1 *in* (l, k)*-entry and* 0 *elsewhere.*

Assume that for any fixed integers $i \neq j$ *such that* $a_{ij}b_{ij} = 0$, *it implies that, for any inner automorphism* φ *of* R, $a'_{ij}b'_{ij} = 0$, *where* $\varphi(a) = \sum_{r,s=1}^{t} a'_{rs}e_{rs}$, $\varphi(b) = \sum_{r,s=1}^{t} b'_{rs}e_{rs}$.

Then if $a_{ij}b_{ij} = 0$ *for all* $i \neq j$, *it follows either* $a \in H$ *or* $b \in H$.

Fact 11 ([12, Lemma 3]) *Let* R *be a prime ring,* C *its extended centroid,* $f(x_1, \ldots, x_n)$ *a polynomial over* C *(not necessarily multilinear),* $a \in R$ *such that* $af(r_1, \ldots, r_n) = 0$, *for any* $r_1, \ldots, r_n \in R$. *Then either* $a = 0$ *or* $f(r_1, \ldots, r_n) = 0$, *for all* $r_1, \ldots, r_n \in R$. *(Analogously for* $f(r_1, \ldots, r_n)a = 0$*).*

2 Inner Generalized Skew Derivations

In this section, we prove a reduced version of the Theorem 1; in particular, we consider the case in which F and G are both inner generalized skew derivations. The first step is the case when α is an inner automorphism of R, therefore we prove the following Proposition:

Proposition 1 *Let* R *be a non-commutative prime ring of characteristic different from 2,* Q_r *be its right Martindale quotient ring and* C *be its extended centroid. Suppose that* F *and* G *are non-zero generalized skew derivations of* R*, respectively, defined as*

$$F(x) = ax + qxq^{-1}b, \qquad G(x) = cx + qxq^{-1}u$$

for all $x \in R$ *and fixed* $a, b, c, u \in Q_r$, *with an invertible element* q *of* Q_r. *If* $F(x)G(y) + F(y)G(x) = x \circ y$, *for all* $x, y \in [R, R]$, *then one of the following holds:*

1. *there exist* $a', b' \in Q_r$ *such that* $F(x) = xa'$ *and* $G(x) = b'x$, *for any* $x \in R$, *with* $a'b' = 1_C$;
2. R *satisfies* $s_4(x_1, \ldots, x_4)$ *and there exist* $a', b' \in Q_r$ *such that* $F(x) = a'x$ *and* $G(x) = xb'$, *for any* $x \in R$, *with* $a'b' = 1_C$;
3. R *satisfies* $s_4(x_1, \ldots, x_4)$ *and there exists* $0 \neq \beta \in C$ *such that* $F(x) = \beta qxq^{-1}$ *and* $G(x) = \beta^{-1}qxq^{-1}$, *for any* $x \in R$.

Then we extend the previous result to any automorphism α, proving the following Proposition:

Proposition 2 *Let* R *be a non-commutative prime ring of characteristic different from 2,* Q_r *be its right Martindale quotient ring and* C *be its extended centroid. Suppose that* $\alpha \in Aut(R)$ *and* F *and* G *are non-zero generalized skew derivations of* R *respectively defined as*

$$F(x) = ax + \alpha(x)b, \qquad G(x) = cx + \alpha(x)u$$

for all $x \in R$ *and fixed* $a, b, c, u \in Q_r$. *If* $F(x)G(y) + F(y)G(x) = x \circ y$, *for all* $x, y \in [R, R]$, *then one of the following holds:*

1. *there exist* $a', b' \in Q_r$ *such that* $F(x) = xa'$ *and* $G(x) = b'x$, *for any* $x \in R$, *with* $a'b' = 1_C$;
2. R *satisfies* $s_4(x_1, \dots, x_4)$ *and there exist* $a', b' \in Q_r$ *such that* $F(x) = a'x$ *and* $G(x) = xb'$, *for any* $x \in R$, *with* $a'b' = 1_C$;
3. R *satisfies* $s_4(x_1, \dots, x_4)$ *and there exist an invertible element* $q \in Q_r$ *and* $0 \neq \beta \in C$ *such that* $F(x) = \beta q x q^{-1}$ *and* $G(x) = \beta^{-1} q x q^{-1}$, *for any* $x \in R$.

In order to prove the Proposition 1, assume that R satisfies the following generalized polynomial identity

$$\Psi(x_1, x_2, x_3, x_4) = \left(a[x_1, x_2] + q[x_1, x_2]q^{-1}b\right)\left(c[x_3, x_4] + q[x_3, x_4]q^{-1}u\right) + \left(a[x_3, x_4] + q[x_3, x_4]q^{-1}b\right)\left(c[x_1, x_2] + q[x_1, x_2]q^{-1}u\right) - [x_1, x_2][x_3, x_4] - [x_3, x_4][x_1, x_2]. \tag{1}$$

Lemma 1 *Let* $R = M_t(C)$ *be the ring of all* $t \times t$ *matrices over* C, *with* $char(C) \neq 2$ *and* $t \geq 2$. *If* C *is infinite, then one of the following hold:*

1. $q \in Z(R)$
2. $q^{-1}u \in Z(R)$
3. $a + q^{-1}bq \in Z(R)$

Proof We firstly assume that $q \notin Z(R)$, $q^{-1}u \notin Z(R)$ and $a + q^{-1}bq \notin Z(R)$ and prove that a contradiction follows. Since q, $q^{-1}u$ and $a + q^{-1}bq$ are not scalar matrices, by Fact 9, there exists some invertible matrix $P \in M_t(C)$ such that each matrix PqP^{-1}, $Pq^{-1}uP^{-1}$ and $P(a + q^{-1}bq)P^{-1}$ has all non-zero entries. Denote by $\phi(x) = PxP^{-1}$ the inner automorphism induced by P. Put $\phi(q) = \sum_{hl} q_{hl}e_{hl}$, $\phi(q^{-1}u) = \sum_{hl} v_{hl}e_{hl}$ and $\phi(a + q^{-1}bq) = \sum_{hl} s_{hl}e_{hl}$, for $0 \neq q_{hl}$, $0 \neq v_{hl}$ and $0 \neq s_{hl} \in C$. Without to loss of generality, we may replace q, $q^{-1}u$ and $a + q^{-1}bq$ with $\phi(q)$, $\phi(q^{-1}u)$ and $\phi(a + q^{-1}bq)$, respectively. Let e_{ij} be the usual matrix unit with 1 in the $(i, j)-$entry and zero elsewhere. Hence, for any $i \neq j$, $[x_1, x_2] = [x_3, x_4] = e_{ij}$ in (2) and right multiplying by e_{ij} we get

$$2(ae_{ij} + qe_{ij}q^{-1}b)qe_{ij}q^{-1}ue_{ij} = 0$$

It implies that $q_{ji}v_{ji}s_{ji} = 0$, a contradiction.

Lemma 2 *Let* $R = M_t(C)$ *be the ring of all* $t \times t$ *matrices over* C, *with* $char(C) \neq 2$ *and* $t \geq 2$. *If* C *is infinite and* $q \in Z(R)$ *then one of the following holds:*

1. $a, u \in Z(R)$ *and* $(a+b)(c+u) = 1_C$*;*
2. $t = 2$ *and* $b, c \in Z(R)$ *and* $(a+b)(c+u) = 1_C$*;*

Proof Since $q \in Z(R)$, by (2), we get that R satisfies

$$\Big(a[x_1,x_2]+[x_1,x_2]b\Big)\Big(c[x_3,x_4]+[x_3,x_4]u\Big)+$$

$$\Big(a[x_3,x_4]+[x_3,x_4]b\Big)\Big(c[x_1,x_2]+[x_1,x_2]u\Big)-[x_1,x_2][x_3,x_4]-[x_3,x_4][x_1,x_2]. \tag{2}$$

Fix, for any $i \neq j$, $[x_1,x_2]=[x_3,x_4]=e_{ij}$, in (2), then we have $X = ae_{ij}ce_{ij} + e_{ij}bce_{ij} + e_{ij}be_{ij}u = 0$ Assume that $a, c \notin Z(R)$ and prove that a contradiction follows; they are not scalar matrices and, as above, we apply Fact 9, therefore there exists some invertible matrix $P \in M_t(C)$ such that each matrix PaP^{-1} and PcP^{-1} has all non-zero entries. Denote by $\phi(x) = PxP^{-1}$ the inner automorphism induced by P. Put $\phi(a) = \sum_{hl} a_{hl}e_{hl}$ and $\phi(c) = \sum_{hl} c_{hl}e_{hl}$, for $0 \neq a_{hl}, 0 \neq c_{hl} \in C$. Without to loss of generality, we may replace a and c with $\phi(a)$ and $\phi(c)$, respectively. By $(j, j)-$entry of X, we get $a_{ji}c_{ji} = 0$, for all $i \neq j$, that is a contradiction. It means that either $a \in Z(R)$ or $c \in Z(R)$.

Analogously, it is possible to prove that either $b \in Z(R)$ or $u \in Z(R)$.

Case 1.:$b, c \in Z(R)$. In this case, by (2), R satisfies

$$(a+b)[x_1,x_2][x_3,x_4](c+u)+(a+b)[x_3,x_4][x_1,x_2](c+u)-[x_1,x_2][x_3,x_4]$$

$$-[x_3,x_4][x_1,x_2].$$

If $t = 2$, the polynomial $[x_1,x_2][x_3,x_4]+[x_3,x_4][x_1,x_2]$ is central valued on R, so that $(a+b)(c+u) = 1_C$, we are done.

Assume $t \geq 3$, for any $i \neq j \neq k$, $[x_1,x_2] = e_{ij}$ and $[x_3,x_4] = e_{jk}$, then

$$Y = (a+b)e_{ik}(c+u) - e_{ik} = 0$$

By $(k, i)-$entry of Y, we get $(a+b)_{ki}(c+u)_{ki} = 0$, for all $k \neq i$; by Fact 10, either $(a+b) \in Z(R)$ or $(c+u) \in Z(R)$. In both case, since $[x_1,x_2][x_3,x_4] + [x_3,x_4][x_1,x_2]$ is a non-zero polynomial, by Fact 11, $(a+b)(c+u) = 1_C$.

Case 2.:$b, a \in Z(R)$. In this case, by (2), R satisfies

$$\beta[x_1,x_2]\Big(c[x_3,x_4]+[x_3,x_4]u\Big)+\beta[x_3,x_4]\Big(c[x_1,x_2]+[x_1,x_2]u\Big)-$$

$$[x_1,x_2][x_3,x_4]-[x_3,x_4][x_1,x_2]$$

where $0 \neq \beta = a+b \in Z(R)$. For any $i \neq j$, $[x_1,x_2]=[x_3,x_4]=e_{ij}$, then $2\beta e_{ij}ce_{ij} = 0$. Since $char(R) \neq 2$, $c_{ji} = 0$, for all $i \neq j$, by standard calculations, we get $c \in Z(R)$ and, by Fact 11 $c + u = \beta^{-1}$.

Case 3.:$a, u \in Z(R)$. In this case, by (2), R satisfies

$$[x_1, x_2](a+b)(c+u)[x_3, x_4] + [x_3, x_4](a+b)(c+u)[x_1, x_2] - [x_1, x_2][x_3, x_4]$$
$$-[x_3, x_4][x_1, x_2].$$

For any $i \neq j$, $[x_1, x_2] = [x_3, x_4] = e_{ij}$, then $2p_{ji} = 0$, where $(a+b)(c+u) = \sum_{hl} p_{hl}e_{hl}$, for $p_{hl} \in C$. By standard arguments, we have $(a+b)(c+u) \in Z(R)$, therefore R satisfies

$$(a+b)(c+u)\Big([x_1, x_2][x_3, x_4] + [x_3, x_4][x_1, x_2]\Big) - [x_1, x_2][x_3, x_4] - [x_3, x_4][x_1, x_2]$$

By Fact 11, it means $(a+b)(c+u) = 1_C$.
Case 4.:$c, u \in Z(R)$. In this case, by (2), R satisfies

$$\beta\Big(a[x_1, x_2] + [x_1, x_2]b\Big)[x_3, x_4] + \beta\Big(a[x_3, x_4] + [x_3, x_4]b\Big)[x_3, x_4] -$$
$$[x_1, x_2][x_3, x_4] - [x_3, x_4][x_1, x_2]$$

where $0 \neq \beta = c + u \in Z(R)$. For any $i \neq j$, $[x_1, x_2] = [x_3, x_4] = e_{ij}$, then $2\beta e_{ij} b e_{ij} = 0$. Since $char(R) \neq 2$, $b_{ji} = 0$, for all $i \neq j$; as above, $b \in Z(R)$ and, by Fact 11, $a + b = \beta^{-1}$.

Lemma 3 *Let $R = M_t(C)$ be the ring of all $t \times t$ matrices over C, with $char(C) \neq 2$ and $t \geq 2$. If C is infinite and $q^{-1}u \in Z(R)$ then one of the following holds:*

1. *$a, q \in Z(R)$ and $(a+b)(c+u) = 1_C$;*
2. *$t = 2$ and $q^{-1}b, c+u \in Z(R)$ and $(a+b)(c+u) = 1_C$;*

Proof Since $q^{-1}u \in Z(R)$, by (2), we get that R satisfies

$$\Big(a[x_1, x_2] + q[x_1, x_2]q^{-1}b\Big)(c+u)[x_3, x_4] + \Big(a[x_3, x_4] + q[x_3, x_4]q^{-1}b\Big)(c+u)[x_1, x_2]$$
$$- [x_1, x_2][x_3, x_4] - [x_3, x_4][x_1, x_2]. \tag{3}$$

Assume that $q^{-1}a, c+u \notin Z(R)$ and prove that a contradiction follows. They are not scalar matrices and we apply Fact 9, therefore there exists some invertible matrix $P \in M_t(C)$ such that each matrix $Pq^{-1}aP^{-1}$ and $P(c+u)P^{-1}$ has all non-zero entries. Denote by $\phi(x) = PxP^{-1}$ the inner automorphism induced by P. Put $q^{-1}a = \sum_{hl} p'_{hl}e_{hl}$ and $c + u = \sum_{hl} q'_{hl}e_{hl}$, for $0 \neq p'_{hl}$, $0 \neq q'_{hl} \in C$. Without to loss of generality, we may replace $q^{-1}a$ and $c+u$ with $\phi(q^{-1}a)$ and $\phi(c+u)$ respectively. Fix, for any $i \neq j$, $[x_1, x_2] = [x_3, x_4] = e_{ij}$, in (3); left multiplying by $e_{ij}q^{-1}$, then we have $p'_{ji}q'_{ji} = 0$, that is a contradiction. Then we have two cases:
Case 1.:$c + u \in Z(R)$. In this case, by (3), R satisfies

$$\beta\Big(a[x_1,x_2]+q[x_1,x_2]q^{-1}b\Big)[x_3,x_4]+\beta\Big(a[x_3,x_4]+q[x_3,x_4]q^{-1}b\Big)[x_1,x_2]-$$
$$[x_1,x_2][x_3,x_4]-[x_3,x_4][x_1,x_2]$$

where $0\neq\beta=c+u\in Z(R)$. Again, fix, for any $i\neq j$, $[x_1,x_2]=[x_3,x_4]=e_{ij}$; left multiplying by q^{-1}, then we have $t'_{ij}=0$, where $q^{-1}b=\sum_{hl}t'_{hl}e_{hl}$, for $t'_{hl}\in C$. As above, we get $q^{-1}b\in Z(R)$ and R satisfies, by (3),

$$\beta(a+b)\Big([x_1,x_2][x_3,x_4]+[x_3,x_4][x_1,x_2]\Big)-[x_1,x_2][x_3,x_4]-[x_3,x_4][x_1,x_2]$$

By Fact 11, $(a+b)=\beta^{-1}$.
Case 2.:$q^{-1}a\in Z(R)$. In this case, there exists $\lambda\in C$ such that $a=\lambda q$. By (3), R satisfies

$$\Big(\lambda aq[x_1,x_2]+q[x_1,x_2]q^{-1}b\Big)(c+u)[x_3,x_4]+$$

$$\Big(\lambda q[x_3,x_4]+q[x_3,x_4]q^{-1}b\Big)(c+u)[x_1,x_2]-[x_1,x_2][x_3,x_4]-[x_3,x_4][x_1,x_2]. \quad (4)$$

Consider, for any $i\neq j$, $[x_1,x_2]=[x_3,x_4]=e_{ij}$ and left multiply by q^{-1} in (4), then $2r'_{ji}=0$ for all $i\neq j$, where $(\lambda+q^{-1}b)(c+u)=\sum_{hl}r'_{hl}e_{hl}$, for $r'_{hl}\in C$. We get that $(\lambda+q^{-1}b)(c+u)$ is a diagonal matrix; by standard calculations $(\lambda+q^{-1}b)(c+u)\in Z(R)$, therefore R satisfies

$$\beta q\Big([x_1,x_2][x_3,x_4]+[x_3,x_4][x_1,x_2]\Big)-[x_1,x_2][x_3,x_4]-[x_3,x_4][x_1,x_2]$$

where $\beta=(\lambda+q^{-1}b)(c+u)$. By Fact 11, $q=\beta^{-1}\in Z(R)$, $(a+b)(c+u)=1_C$ and we are done.

Lemma 4 *Let $R=M_t(C)$ be the ring of all $t\times t$ matrices over C, with $char(C)\neq 2$ and $t\geq 2$. If C is infinite and $a+q^{-1}bq\in Z(R)$ then one of the following holds:*

1. *$q^{-1}b, q^{-1}u\in Z(R)$ and $(a+b)(c+u)=1_C$*
2. *$q,c,u\in Z(R)$ and $(a+b)(c+u)=1_C$*
3. *$t=2$ and there exists $\beta\in Z(R)$ such that $F(x)=\beta qxq^{-1}$ and $G(x)=\beta^{-1}qxq^{-1}$, for all $x\in Z(R)$.*

Proof Since $a+q^{-1}bq\in Z(R)$, there exists $\lambda\in C$ such that $q^{-1}bq=\lambda-a$; by (2), we get that R satisfies

$$a[x_1,x_2]c[x_3,x_4]+a[x_1,x_2]q[x_3,x_4]q^{-1}u+q[x_1,x_2]q^{-1}bc[x_3,x_4]+$$

$$\lambda q[x_1,x_2][x_3,x_4]q^{-1}u-q[x_1,x_2]a[x_3,x_4]q^{-1}u+a[x_3,x_4]c[x_1,x_2]+$$

$$a[x_3,x_4]q[x_1,x_2]q^{-1}u+q[x_3,x_4]q^{-1}bc[x_1,x_2]+\lambda q[x_3,x_4][x_1,x_2]q^{-1}u-$$

$$q[x_3,x_4]a[x_1,x_2]q^{-1}u-[x_1,x_2][x_3,x_4]-[x_3,x_4][x_1,x_2]. \tag{5}$$

Assume that $q, q^{-1}u \notin Z(R)$, otherwise we conclude thanks to Lemmas 2 and 3. As above, by using Fact 10, assume that q and $q^{-1}u$ have all non-zero entries. Fix, for any $i \neq j$, $[x_1, x_2] = [x_3, x_4] = e_{ij}$, in (5); left multiplying by $e_{jj}q^{-1}$ and right multiplying by e_{ii}, we get $2p'_{ji}q_{ji}q'_{ji} = 0$, where $q^{-1}a = \sum_{hl} p'_{hl}e_{hl}$, $q = \sum_{hl} q_{hl}e_{hl}$ and $q^{-1}u = \sum_{hl} q'_{hl}e_{hl}$, for $0 \neq q'_{hl}$, $0 \neq q_{hl}$, $p'_{hl} \in C$. Since q and $q^{-1}u$ have all non-zero entries, we get $p'_{ji} = 0$, for all $i \neq j$; it implies that $q^{-1}a$ is a diagonal matrix and, by standard calculations, we get $q^{-1}a \in Z(R)$. Then there exists $\beta \in C$ such that $a = \beta q$. In this case, $F(x) = \lambda qxq^{-1}$, then, by (5), R satisfies

$$\lambda q[x_1,x_2]q^{-1}c[x_3,x_4]+\lambda q[x_1,x_2][x_3,x_4]q^{-1}u+\lambda q[x_3,x_4]q^{-1}c[x_1,x_2]+$$

$$\lambda q[x_3,x_4][x_1,x_1]q^{-1}u-[x_1,x_2][x_3,x_4]-[x_3,x_4][x_1,x_2]$$

Assume $\lambda \neq 0$ and fix, for any $i \neq j$, $[x_1, x_2] = [x_3, x_4] = e_{ij}$. Left multiplying by q^{-1}, as above, by similar calculations, we get $q^{-1}c \in Z(R)$. Then R satisfies

$$\lambda c\Big([x_1,x_2][x_3,x_4]+[x_3,x_4][x_1,x_1]\Big)+\lambda q\Big([x_1,x_2][x_3,x_4]+[x_3,x_4][x_1,x_1]\Big)q^{-1}u-$$

$$[x_1,x_2][x_3,x_4]-[x_3,x_4][x_1,x_2].$$

If $t = 2$, we conclude $c + u = \lambda^{-1} \in Z(R)$. Assume $t \geq 3$, fix, for any $i \neq j, \neq k$, $[x_1, x_2] = e_{ij}$ and $[x_3, x_4] = e_{ji}$ and left multiplying by $e_{kk}q^{-1}$, then R satisfies

$$e_{kk}q^{-1}(e_{ii}+e_{jj}).$$

It implies $q^{-1} \in Z(R)$, a contradiction.

Lemma 5 *Let $R = M_t(C)$ be the ring of all $t \times t$ matrices over C, with char$(C) \neq 2$ and $t \geq 2$. If C is infinite, then one of the following holds:*

1. *$q, a, u \in Z(R)$ snd $(a + b)(c + u) = 1_C$*
2. *$t = 2$, $q, b, c \in Z(R)$ snd $(a + b)(c + u) = 1_C$*
3. *$t = 2$ and there exists $\beta \in Z(R)$ such that $F(x) = \beta qxq^{-1}$ and $G(x) = \beta^{-1}qxq^{-1}$, for all $x \in Z(R)$.*

Proof In light of Lemma 1, if R satisfies the polynomial identity (2), then one of the following holds:

1. $q \in Z(R)$
2. $q^{-1}u \in Z(R)$
3. $a + q^{-1}bq \in Z(R)$

In all cases, we conclude thanks, respectively, to Lemmas 2, 3, 4.

Lemma 6 *Let $R = M_t(C)$ be the ring of all $t \times t$ matrices over C, with $char(C) \neq 2$ and $t \geq 2$. Then one of the following holds:*

1. *$q, a, u \in Z(R)$ snd $(a + b)(c + u) = 1_C$*
2. *$t = 2$, $q, b, c \in Z(R)$ snd $(a + b)(c + u) = 1_C$*
3. *$t = 2$ and there exists $\beta \in Z(R)$ such that $F(x) = \beta q x q^{-1}$ and $G(x) = \beta^{-1} q x q^{-1}$, for all $x \in Z(R)$.*

Proof If C is infinite, the conclusion follows from Lemma 5. Now let E be an infinite field which is an extension of the field C and let $\overline{R} = M_t(E) \cong R \otimes_C E$. Consider the generalized polynomial $\Psi(x_1, x_2, x_3, x_4)$, which is a generalized identity for R, and its complete linearization $\theta(x_1, x_2, \ldots, x_8)$; it is a multilinear generalized polynomial and $\theta(x_1, x_2, x_3, x_4, x_1, x_2, x_3, x_4) = 2^4 \Psi(x_1, x_2, x_3, x_4)$. Obviously, the multilinear polynomial $\theta(x_1, x_2, \ldots, x_8)$ is a generalized polynomial identity for R and $\overline{R}$ too. Since $char(R) \neq 2$, we obtain $\Psi(r_1, r_2, r_3, r_4) = 0$, for all $r_1, r_2, r_3, r_4 \in \overline{R}$ and the conclusion follows from Lemma 5.

Lemma 7 *Either $\Psi(x_1, x_2, x_3, x_4)$ is a non-trivial generalized polynomial identity for R or Proposition 1 holds.*

Proof Assume that $\Psi(x_1, x_2, x_3, x_4)$ is a trivial generalized polynomial identity for R. Let $T = Q_{r*C}C\{X\}$ be the free product over C of the C−algebra Q_r and the free C−algebra $C\{X\}$ with X the set consisting of non-commuting indeterminates x_1, x_2, x_3, x_4. For brevity, we write $X, Y, \Psi(X, Y)$ instead of $[x_1, x_2], [x_3, x_4]$, $\Psi(x_1, x_2, x_3, x_4)$.

Consider the generalized polynomial $\Psi(X, Y) \in T$. By hypothesis,

$$\Psi(X, Y) = \left(aX + qXq^{-1}b\right)\left(cY + qYq^{-1}u\right) + \left(aY + qYq^{-1}b\right)\left(cX + qXq^{-1}u\right)$$

$$-XY - YX = 0 \in T.$$

Assume first that $\{1, a, q\}$ are linearly C−independent. Since $\Psi(X, Y) = 0 \in T$, we have $XY + YX = 0 \in T$, a contradiction. Therefore there exist $\lambda, \beta \in C$, such that $a = \lambda q + \beta$, that is

$$\Psi(X, Y) = \left(\lambda qX + \beta X + qXq^{-1}b\right)\left(cY + qYq^{-1}u\right) +$$

$$\left(\lambda qY + \beta Y + qYq^{-1}b\right)\left(cX + qXq^{-1}u\right) - XY - YX = 0 \in T.$$

If $q \in C$, we have

$$\Psi(X, Y) = X\left(\lambda q + \beta + b\right)\left(cY + Yu\right) + Y\left(\lambda q + \beta + b\right)\left(cX + Xu\right) - XY - YX = 0$$

$\in T$. Since it is a trivial generalized polynomial identity, we have $u \in C$ and $(\lambda q + \beta + b)(c + u) = 1_C$, that is $F(x) = a'x$ and $G(x) = c'x$, for all $x \in R$, where $a' = a + b$, $c' = c + u$ and $a'c' = 1_C$; we are done.

Assume that $\{1, q\}$ are linearly $C-$independent, then, since $\Psi(X, Y) = 0 \in T$, we have

$$\left(\lambda qX + qXq^{-1}b\right)\left(cY + qYq^{-1}u\right) + \left(\lambda qY + qYq^{-1}b\right)\left(cX + qXq^{-1}u\right) = 0 \in T. \quad (6)$$

and

$$\beta X\left(cY + qYq^{-1}u\right) + \beta Y\left(cX + qXq^{-1}u\right) - XY - YX = 0 \in T. \quad (7)$$

Consider the case when $\{1, q^{-1}u\}$ are linearly $C-$independent, then by (7)

$$\beta XcY + \beta YcX - XY - YX = 0 \in T. \quad (8)$$

and

$$\beta XqYq^{-1}u + \beta YqXq^{-1}u = 0 \in T. \quad (9)$$

By (8), we get $c = \beta^{-1}$ and, by (9), we get $\beta = 0$, a contradiction. Then $q^{-1}u \in C$ and (6) and (7) become

$$\left(\lambda qX + qXq^{-1}b\right)\left(c'Y\right) + \left(\lambda qY + qYq^{-1}b\right)\left(c'X\right) = 0 \in T.$$

and

$$\beta X\left(c'Y\right) + \beta Y\left(c'X\right) - XY - YX = 0 \in T$$

where $c' = c + u$. Comparing both trivial generalized polynomial identities, we obtain $c' = \beta^{-1} \in C$, $q^{-1}b \in C$ and $\lambda q + b = 0$. It means that $F(x) = \beta x$ and $G(x) = \beta^{-1}x$, for all $x \in R$.

Proof of Proposition 1

In light of Lemma 7, we may assume that the generalized polynomial $\Psi(x_1, x_2, x_3, x_4)$ is a non-trivial generalized polynomial identity for R. By [8], it follows that $\Psi(x_1, x_2, x_3, x_4)$ is a non-trivial generalized polynomial identity for Q_r. In view of [18, Theorems 2.5 and 3.5], we know that both Q_r and $Q_r \otimes_C \overline{C}$ are centrally closed, where $\overline{C}$ is the algebraic closure of C. We may replace Q_r by itself or $Q_r \otimes_C \overline{C}$ according as C is finite or infinite. Therefore, we may assume that Q_r is centrally closed over C, which is either finite or algebraically closed. By Martindale's theorem [26], Q_r is a primitive ring having a non-zero socle H, with C as the associated division ring. In light of Jacobson's theorem [20, p. 75], Q_r is isomorphic to a dense ring of linear transformations on some vector space V over C.

Assume first that $dim_C V = t \geq 2$ is a finite positive integer. Then $Q \cong M_t(C)$ and the conclusion follows from Lemma 6.

Let now $dim_C V = \infty$. As in [28, Lemma 2], the set $[Q_r, Q_r]$ is dense on R. Since $\Psi(x_1, x_2, x_3, x_4) = 0$ is a generalized polynomial identity of R, we obtain that Q_r satisfies

$$(ax + qxq^{-1}b)(cy + qyq^{-1}u) + (ay + qyq^{-1}b)(cx + qxq^{-1}u) - xy - yx \quad (10)$$

Suppose there exists $v \in V$ such that $\{v, q^{-1}uv\}$ are linearly C—independent. By the density of Q_r, there exist $r_1, r_2 \in Q_r$ such that

$$r_1 v = r_2 v = v \qquad r_1 q^{-1}uv = r_2 q^{-1}uv = -q^{-1}cv.$$

Right multiplying by v in (10), we get $2v = 0$; since $char(R) \neq 2$, we get $v = 0$, a contradiction. Then, for any $v \in V$, $\{v, q^{-1}uv\}$ are linearly C—dependent, that is $q^{-1}u \in C$. Now, by (10), Q_r satisfies

$$(ax + qxq^{-1}b)(c + u)y + (ay + qyq^{-1}b)(c + u)x - xy - yx. \quad (11)$$

Fix $x = y = 1_C$ in (11), then we get $(a + b)(c + u) = 1_C$. Again, fix $x = 1_C$ in (11), then, for all $y \in Q_r$

$$(ay + qyq^{-1}b)(c + u) - y = 0$$

Right multiplying by $(a + b)$, we get

$$ay + qyq^{-1}b = y(a + b).$$

It means that $F(x)$ is a generalized derivation of R, therefore either $q, a \in C$ or $q^{-1}b, a + b \in C$. In both cases we are done.

As a reduced case, we have the following easy consequence of Proposition 1:

Corollary 1 *Let R be a non-commutative prime ring of characteristic different from 2, Q_r be its right Martindale quotient ring and C be its extended centroid. Suppose that F and G are non-zero generalized derivations of R respectively defined as*

$$F(x) = ax + xb, \qquad G(x) = cx + xu$$

for all $x \in R$ and fixed $a, b, c, u \in Q_r$. If $F(x)G(y) + F(y)G(x) = x \circ y$, for all $x, y \in [R, R]$, then one of the following holds:

1. *$a, u \in C$ and $(a + b)(c + u) = 1_C$;*
2. *R satisfies $s_4(x_1, \ldots, x_4)$, $b, c \in C$ and $(a + b)(c + u) = 1_C$.*

Lemma 8 ([6], Reduced version of the Main Theorem) *Let R be a non-commutative prime ring of characteristic different from 2. If there exist $a, b \in R$ such that $[x_1, x_2]a[x_1, x_2]b = 0$ for all $x_1, x_2 \in R$, then either $a = 0$ or $b = 0$.*

2.1 *Proof of Proposition 2*

By our assumption, R satisfies

$$\Big(a[x_1, x_2] + \alpha([x_1, x_2])b\Big)\Big(c[x_3, x_4] + \alpha([x_3, x_4])u\Big)+$$

$$\Big(a[x_3, x_4] + \alpha([x_3, x_4])b\Big)\Big(c[x_1, x_2] + \alpha([x_1, x_2])u\Big) - [x_1, x_2][x_3, x_4] - [x_3, x_4][x_1, x_2]. \quad (12)$$

If α is the identity map on R, then we conclude by Corollary 1. Moreover, in the case that there exists an invertible element $q \in Q_r$, with $q \notin C$, such that $\alpha(x) = qxq^{-1}$, for all $x \in R$, the conclusion follows from Proposition 1. Thus, in all that follows we assume that α in not inner, then, by Fact 3 and relation (12), R satisfies

$$\Big(a[x_1, x_2] + [y_1, y_2]b\Big)\Big(c[x_3, x_4] + [y_3, y_4]u\Big)+$$

$$\Big(a[x_3, x_4] + [y_3, y_4]b\Big)\Big(c[x_1, x_2] + [y_1, y_2]u\Big) - [x_1, x_2][x_3, x_4] - [x_3, x_4][x_1, x_2]. \quad (13)$$

For $y_i = 0$, for all $i = 1, \dots, 4$, by (13), R satisfies

$$a[x_1, x_2]c[x_3, x_4] + a[x_3, x_4]c[x_1, x_2] - [x_1, x_2][x_3, x_4] - [x_3, x_4][x_1, x_2] \quad (14)$$

Thanks to Corollary 1, we get $a, c \in C$ and $ac = 1_C$. Then there exists $0 \neq \beta \in C$ such that $a = \beta$, $c = \beta^{-1}$ and R satisfies

$$\Big(\beta[x_1, x_2] + [y_1, y_2]b\Big)\Big(\beta^{-1}[x_3, x_4] + [y_3, y_4]u\Big)+$$

$$\Big(\beta[x_3, x_4] + [y_3, y_4]b\Big)\Big(\beta^{-1}[x_1, x_2] + [y_1, y_2]u\Big) - [x_1, x_2][x_3, x_4] - [x_3, x_4][x_1, x_2]. \quad (15)$$

Consider $x_1 = x_3 = 0$, $y_1 = y_3$ and $y_2 = y_4$, then

$$2[y_1, y_2]b[y_1, y_2]u = 0$$

for all $y_1, y_2 \in R$. By Lemma 8 either $b = 0$ or $u = 0$.

Assume $b = 0$, then by (15), R satisfies

$$\Big([x_1, x_2][y_3, y_4] + [x_3, x_4][y_1, y_2]\Big)u$$

By Fact 11, $u = 0$. (Analogously if one assume $u = 0$, it implies $b = 0$).

3 The Proof of Theorem 1

We are now ready to prove the main Theorem of the paper. As remarked in Fact 7, we assume that there exist $a, c \in Q_r$ such that $F(x) = ax + d(x)$ and $G(x) = cx + \delta(x)$, for all $x \in R$, where d, δ are skew derivations on R and α the automorphism associated with d and δ; it means that $d(xy) = d(x)y + \alpha(x)d(y)$ and $\delta(xy) = \delta(x)y + \alpha(x)\delta(y)$, for all $x, y \in R$.

By Fact 8, since L is a non-central Lie ideal and $Char(R) \neq 2$, there exists a non-zero ideal I of R such that $F(x)G(y) + F(y)G(x) = x \circ y$, for all $x, y \in [I, I]$. Since R and I satisfy the same generalized differential indentities with automorphisms, we assume that $F(x)G(y) + F(y)G(x) = x \circ y$, for all $x, y \in [R, R]$.

3.1 Proof of Theorem 1

In light of Propositions 1 and 2, we get the required conclusion if one of the following occurs:

1. $d = \delta = 0$;
2. α is an identity map on R;
3. d and δ are inner skew derivations and α is an inner automorphism of R.

Then, in all that follows we assume that

1. either $d = 0$ or $\delta = 0$;
2. α is not the identity map;
3. if d and δ are inner skew derivations, then α is not an inner automorphism of R.

We prove that a number of contradictions occur.

Since R, as well as Q_r, satisfies

$$\Big(a[x_1, x_2] + d([x_1, x_2])\Big)\Big(c[x_3, x_4] + \delta([x_3, x_4])\Big) +$$
$$\Big(a[x_3, x_4] + d([x_3, x_4])\Big)\Big(c[x_1, x_2] + \delta([x_1, x_2])\Big) - [x_1, x_2][x_3, x_4] - [x_3, x_4][x_1, x_2]. \quad (16)$$

then Q_r satisfies

$$\Big(a[x_1,x_2]+d(x_1)x_2+\alpha(x_1)d(x_2)-d(x_2)x_1-\alpha(x_2)d(x_1)\Big)\Big(c[x_3,x_4]+$$

$$\delta(x_3)x_4+\alpha(x_3)\delta(x_4)-\delta(x_4)x_3-\alpha(x_4)\delta(x_3)\Big)+$$

$$\Big(a[x_3,x_4]+d(x_3)x_4+\alpha(x_3)d(x_4)-d(x_4)x_3-\alpha(x_4)d(x_3)\Big)\Big(c[x_1,x_2]+$$

$$\delta(x_1)x_2+\alpha(x_1)\delta(x_2)-\delta(x_2)x_1-\alpha(x_2)\delta(x_1)\Big)-[x_1,x_2][x_3,x_4]-[x_3,x_4][x_1,x_2]. \tag{17}$$

Now we divide the proof into five parts.

• **Let $d=0$ and $\delta\neq 0$.**

By (17) Q_r satisfies

$$a[x_1,x_2]\Big(c[x_3,x_4]+\delta(x_3)x_4+\alpha(x_3)\delta(x_4)-\delta(x_4)x_3-\alpha(x_4)\delta(x_3)\Big)+$$

$$a[x_3,x_4]\Big(c[x_1,x_2]+\delta(x_1)x_2+\alpha(x_1)\delta(x_2)-\delta(x_2)x_1-\alpha(x_2)\delta(x_1)\Big)-[x_1,x_2][x_3,x_4]-$$

$$[x_3,x_4][x_1,x_2]. \tag{18}$$

Of course, we may assume that δ is not inner, otherwise we are done by Proposition 1. In this case Q_r satisfies

$$a[x_1,x_2]\Big(c[x_3,x_4]+z_3x_4+\alpha(x_3)z_4-z_4x_3-\alpha(x_4)z_3\Big)+$$

$$a[x_3,x_4]\Big(c[x_1,x_2]+z_1x_2+\alpha(x_1)z_2-z_2x_1-\alpha(x_2)z_1\Big)-[x_1,x_2][x_3,x_4]-$$

$$[x_3,x_4][x_1,x_2]. \tag{19}$$

Fix $z_i=0$, for all $i=1,\dots,4$, then Q_r satisfies,

$$a[x_1,x_2]c[x_3,x_4]+a[x_3,x_4]c[x_1,x_2]-[x_1,x_2][x_3,x_4]-[x_3,x_4][x_1,x_2].$$

By Proposition 1, we get that there exists $0\neq\lambda\in C$, such that $a=\lambda$ and $c=\lambda^{-1}$. Now, for $z_1=0$, the blended component of (19) is the following

$$[x_3, x_4]\Big(z_1x_2 - \alpha(x_2)z_1\Big) = 0 \tag{20}$$

for all $x_2, x_3, x_4, z_1 \in Q_r$. Replacing x_4 with x_4t, we know that Q_r satisfies

$$[x_3, x_4]t\Big(z_1x_2 - \alpha(x_2)z_1\Big)$$

Since R is a non-commutative prime ring, we get

$$z_1x_2 - \alpha(x_2)z_1 = 0 \tag{21}$$

for all $z_1, x_2 \in Q_r$. If α, is outer, by Fact 3, we get $[Q_r, Q_r] = 0$, a contradiction; if α is an inner automorphism, then there exists an invertible element $q \in Q_r$, where $q \notin C$, such that $\alpha(x) = qxq^{-1}$, for all $x \in R$ and, by (21), $z_1x_2 - qx_2q^{-1}z_1 = 0$, for all $x_2, z_1 \in Q_r$. Left multiplying by q^{-1} and replacing z_1 with qz_1, we get $[Q_r, Q_r] = 0$, a contradiction.

- **Let $d \neq 0$ and $\delta = 0$.**

By using a similar argument, as in the previous case, we have a number of contradictions. We omit the proof for brevity.

- **Let $d \neq 0$ and $\delta \neq 0$ be $C-$ linearly independent modulo SD_{int}.**

By (17) and Fact 2, Q_r satisfies

$$\Big(a[x_1, x_2] + z_1x_2 + \alpha(x_1)z_2 - z_2x_1 - \alpha(x_2)z_1\Big)\Big(c[x_3, x_4]+$$

$$z_7x_4 + \alpha(x_3)z_8 - z_8x_3 - \alpha(x_4)z_7\Big)+$$

$$\Big(a[x_3, x_4] + z_3x_4 + \alpha(x_3)z_4 - z_4x_3 - \alpha(x_4)z_3\Big)\Big(c[x_1, x_2]+$$

$$z_5x_2 + \alpha(x_1)z_6 - z_6x_1 - \alpha(x_2)z_5\Big) - [x_1, x_2][x_3, x_4] - [x_3, x_4][x_1, x_2]. \tag{22}$$

Fix $z_i = 0$, for all $i = 1, \ldots, 8$, then Q_r satisfies,

$$a[x_1, x_2]c[x_3, x_4] + a[x_3, x_4]c[x_1, x_2] - [x_1, x_2][x_3, x_4] - [x_3, x_4][x_1, x_2].$$

By Proposition 1, we get that there exists $0 \neq \beta \in C$, such that $a = \beta$ and $c = \beta^{-1}$. Now, for $z_5 = 0$, the blended component of (22), is the following

$$\Big(\beta[x_3, x_4] + z_3x_4 + \alpha(x_3)z_4 - z_4x_3 - \alpha(x_4)z_3\Big)\Big(z_5x_2 - \alpha(x_2)z_5\Big) = 0$$

for all $z_3, z_4, z_5, x_2, x_3, x_4 \in Q_r$. For $z_3 = z_4 = 0$, R satisfies

$$[x_3, x_4]\Big(z_5x_2 - \alpha(x_2)z_5\Big).$$

We conclude as in the case in which R satisfies (20).

• **Let d and δ be $C-$ linearly dependent modulo SD_{int}.**

We firstly assume that there exist $0 \neq \lambda \in C, 0 \neq \mu \in C, b \in Q_r$ and $\gamma \in Aut(R)$ such that $\lambda d(x) + \mu\delta(x) = bx - \gamma(x)b$, for all $x \in R$. If we define $\eta = -\mu^{-1}\lambda$ and $p = \mu^{-1}b$, then $\delta(x) = \eta d(x) + px - \gamma(x)p$, for all $x \in R$. Therefore, by (16), Q_r satisfies

$$\Big(a[x_1, x_2] + d([x_1, x_2])\Big)\Big(c[x_3, x_4] + \eta d([x_3, x_4]) + p[x_3, x_4] - \gamma([x_3, x_4])p\Big)+$$
$$\Big(a[x_3, x_4] + d([x_3, x_4])\Big)\Big(c[x_1, x_2] + \eta d([x_1, x_2]) + p[x_1, x_2] - \gamma([x_1, x_2])p\Big)-$$
$$[x_1, x_2][x_3, x_4] - [x_3, x_4][x_1, x_2]. \tag{23}$$

Notice that, if d is an inner skew derivation of R , then there exists $a' \in Q_r$ such that $d(x) = a'x - \alpha(x)a'$ and $\delta(x) = (\eta a')x - \alpha(x)(\eta a') + px - \gamma(x)p$, for all $x \in R$. Hence, by [15, Lemma 3.2], δ is an inner skew derivation and we are done by Proposition 1. Therefore, in all that follows, we assume that d is outer. By (23), Q_r satisfies

$$\Big(a[x_1, x_2] + z_1x_2 + \alpha(x_1)z_2 - z_2x_1 - \alpha(x_2)z_1\Big)\Big(c[x_3, x_4]+$$
$$(\eta z_3 + px_3 - \gamma(x_3)p)x_4 + \alpha(x_3)(\eta z_4 + px_4 - \gamma(x_4)p) - (\eta z_4 + px_4 - \gamma(x_4)p)x_3-$$
$$\alpha(x_4)(\eta z_3 + px_3 - \gamma(x_3)p)\Big) + \Big(a[x_3, x_4] + z_3x_4 + \alpha(x_3)z_4 - z_4x_3 - \alpha(x_4)z_3\Big)\Big(c[x_1, x_2]+$$
$$(\eta z_1 + px_1 - \gamma(x_1)p)x_2 + \alpha(x_1)(\eta z_2 + px_2 - \gamma(x_2)p) - (\eta z_2 + px_2 - \gamma(x_2)p)x_1-$$
$$\alpha(x_2)(\eta z_1 + px_1 - \gamma(x_1)p)\Big) - [x_1, x_2][x_3, x_4] - [x_3, x_4][x_1, x_2]. \tag{24}$$

Fix $z_i = 0$, for all $i = 1, \ldots, 4$, then Q_r satisfies

$$a[x_1,x_2]\Big((c+p)[x_3,x_4]-\gamma([x_3,x_4])p\Big)+$$

$$a[x_3,x_4]\Big((c+p)[x_1,x_2]-\gamma([x_1,x_2])p\Big)-[x_1,x_2][x_3,x_4]-[x_3,x_4][x_1,x_2]. \tag{25}$$

If γ is outer, by Proposition 2, $p=0$, $a,c\in C$ and there exists $0\neq\beta\in C$ such that $a=\beta$ and $c=\beta^{-1}$. Now, by (24), Q_r satisfies

$$\Big(\beta[x_1,x_2]+z_1x_2+\alpha(x_1)z_2-z_2x_1-\alpha(x_2)z_1\Big)\Big(\beta^{-1}[x_3,x_4]+$$

$$\eta z_3x_4+\eta\alpha(x_3)z_4-\eta z_4x_3-\eta\alpha(x_4)z_3\Big)+$$

$$\Big(\beta[x_3,x_4]+z_3x_4+\alpha(x_3)z_4-z_4x_3-\alpha(x_4)z_3\Big)\Big(\beta^{-1}[x_1,x_2]+\eta z_1x_2+$$

$$\eta\alpha(x_1)z_2-\eta z_2x_1-\eta\alpha(x_2)z_1\Big)-[x_1,x_2][x_3,x_4]-[x_3,x_4][x_1,x_2]. \tag{26}$$

If α is outer, for $x_1=x_3=0$, Q_r satisfies the following polynomial identity

$$\Big(z_1x_2-y_1z_1\Big)\Big(z_3x_4-y_2z_3\Big)+\Big(z_3x_4-y_2z_3\Big)\Big(z_1x_2-y_1z_1\Big). \tag{27}$$

It is a polynomial identity for Q_r, then we may assume that there exists $k\geq 2$, such that $Q_r\subseteq M_k(C)$.

For $y_1=y_2=0$, $x_2=z_3=e_{21}$, $z_1=e_{12}$, $x_4=e_{11}$ in (27), we get the contradiction $e_{21}=0$. If α is an inner automorphism of R, there exists an invertible element $q\in Q_r$, with $q\notin C$, such that $\alpha(x)=qxq^{-1}$, for all $x\in R$. In this case, by (26), again for $x_1=x_3=0$, Q_r satisfies

$$\Big(z_1x_2-qx_2q^{-1}z_1\Big)\Big(z_3x_4-qx_4q^{-1}z_3\Big)+\Big(z_3x_4-qx_4q^{-1}z_3\Big)\Big(z_1x_2-qx_2q^{-1}z_1\Big). \tag{28}$$

Since $q\notin C$, (28) is a non-trivial generalized polynomial identity for R as well as Q_r. Moreover R is a primitive ring with C as the associated division ring and then, by Jacobson's Theorem, R is isomorphic to a dense subring of the ring of linear transformations of a vector space V over a C. Since R is not commutative, assume $dim_C V\geq 2$ and let $v\in V$ be such that $\{q^{-1}v,v\}$ are linearly C-independent. By the density of R, there exists $r_1,r_2,r_3,r_4\in R$ such that

$$r_3q^{-1}v=r_4q^{-1}v=0 \qquad r_1v=r_2v=r_3v=r_4v=v.$$

Right multiplying by v in (28), we get the contradiction $2v = 0$. It means that, for all $v \in V$, $\{q^{-1}v, v\}$ are linearly C-dependent; by a standard argument, $q^{-1} \in C$, a contradiction.

If γ is an inner automorphism, there exists an invertible element $q' \in Q_r$, with $q' \notin C$, such that $\gamma(x) = q'xq'^{-1}$, for all $x \in R$ and, by (25), Q_r satisfies

$$a[x_1, x_2]\Big((c+p)[x_3, x_4] - q'[x_3, x_4]q'^{-1}p\Big) + a[x_3, x_4]\Big((c+p)[x_1, x_2]- q'[x_1, x_2]q'^{-1}p\Big) - [x_1, x_2][x_3, x_4] - [x_3, x_4][x_1, x_2]. \tag{29}$$

By Proposition 1, since $q' \notin C$, $a, c, q^{-1}p, \in C$ and $ac = 1_C$; therefore there exists $0 \neq \beta \in C$ such that $a = \beta,\ c = \beta^{-1}$ and, by (24), Q_r satisfies

$$\Big(\beta[x_1, x_2] + z_1x_2 + \alpha(x_1)z_2 - z_2x_1 - \alpha(x_2)z_1\Big)\Big(\beta^{-1}[x_3, x_4]+ \eta z_3x_4 + \eta\alpha(x_3)z_4 - \eta z_4x_3 - \eta\alpha(x_4)z_3\Big)+ \Big(\beta[x_3, x_4] + z_3x_4 + \alpha(x_3)z_4 - z_4x_3 - \alpha(x_4)z_3\Big)\Big(\beta^{-1}[x_1, x_2] + \eta z_1x_2+ \eta\alpha(x_1)z_2 - \eta z_2x_1 - \eta\alpha(x_2)z_1\Big) - [x_1, x_2][x_3, x_4] - [x_3, x_4][x_1, x_2]. \tag{30}$$

We conclude with the same argument used for relation (26).

• **The final case.**

Assume again d and δ are $C-$linearly dependent modulo SD_{int}. That is $\lambda d(x) + \mu\delta(x) = bx - \gamma(x)b$, for all $x \in R$. In light of the previous case, suppose either $\lambda = 0$ or $\mu = 0$. We analyze the case in which $\lambda = 0$ (we omit the symmetrical case $\mu = 0$). Under this assumption we have $\delta(x) = px - \gamma(x)p$, for all $x \in R$, where $p = \mu^{-1}b$. Moreover, assume that d is outer, otherwise we conclude thanks to Proposition 2. Therefore Q_r satisfies

$$\Big(a[x_1, x_2] + d([x_1, x_2])\Big)\Big(c[x_3, x_4] + p[x_3, x_4] - \gamma([x_3, x_4])p\Big)+ \Big(a[x_3, x_4] + d([x_3, x_4])\Big)\Big(c[x_1, x_2] + p[x_1, x_2] - \gamma([x_1, x_2])p\Big) - [x_1, x_2][x_3, x_4] -[x_3, x_4][x_1, x_2]. \tag{31}$$

Then

$$\Big(a[x_1,x_2]+z_1x_2+\alpha(x_1)z_2-z_2x_1-\alpha(x_2)z_1\Big)\Big(c[x_3,x_4]+$$

$$(px_3-\gamma(x_3)p)x_4+\alpha(x_3)(px_4-\gamma(x_4)p)-(px_4-\gamma(x_4)p)x_3-$$

$$\alpha(x_4)(px_3-\gamma(x_3)p)\Big)+\Big(a[x_3,x_4]+z_3x_4+\alpha(x_3)z_4-z_4x_3-\alpha(x_4)z_3\Big)\Big(c[x_1,x_2]+$$

$$(px_1-\gamma(x_1)p)x_2+\alpha(x_1)(px_2-\gamma(x_2)p)-(px_2-\gamma(x_2)p)x_1-$$

$$\alpha(x_2)(px_1-\gamma(x_1)p)\Big)-[x_1,x_2][x_3,x_4]-[x_3,x_4][x_1,x_2]. \tag{32}$$

Fix $z_i=0$, for all $i=1,\ldots,4$, then Q_r satisfies

$$a[x_1,x_2]\Big((c+p)[x_3,x_4]-\gamma([x_3,x_4])p\Big)+$$

$$a[x_3,x_4]\Big((c+p)[x_1,x_2]-\gamma([x_1,x_2])p\Big)-[x_1,x_2][x_3,x_4]-[x_3,x_4][x_1,x_2]. \tag{33}$$

If γ is outer, by Proposition 2, $p=0$, $a,c\in C$ and there exists $0\neq\beta\in C$ such that $a=\beta$ and $c=\beta^{-1}$. Now, by (32), Q_r satisfies

$$\Big(\beta[x_1,x_2]+z_1x_2+\alpha(x_1)z_2-z_2x_1-\alpha(x_2)z_1\Big)\beta^{-1}[x_3,x_4]+$$

$$\Big(\beta[x_3,x_4]+z_3x_4+\alpha(x_3)z_4-z_4x_3-\alpha(x_4)z_3\Big)\beta^{-1}[x_1,x_2]-[x_1,x_2][x_3,x_4]$$

$$-[x_3,x_4][x_1,x_2]. \tag{34}$$

For $x_1=0$, $(z_1x_2-\alpha(x_2)z_1)[x_3,x_4]=0$, for all $x_1,x_2,x_3,x_4\in Q_r$. By using similar calculations, as in the relation (20), we get a contradiction.

If γ is an inner automorphism, there exists an invertible element $q'\in Q_r$, with $q'\notin C$, such that $\gamma(x)=q'xq'^{-1}$, for all $x\in R$ and, by (33), Q_r satisfies

$$a[x_1,x_2]\Big((c+p)[x_3,x_4]-q'[x_3,x_4]q'^{-1}p\Big)+a[x_3,x_4]\Big((c+p)[x_1,x_2]-$$

$$q'[x_1,x_2]q'^{-1}p\Big)-[x_1,x_2][x_3,x_4]-[x_3,x_4][x_1,x_2]. \tag{35}$$

By Proposition 1, since $q' \notin C, a, c, q^{-1}p, \in C$ and $ac = 1_C$; therefore there exists $0 \neq \beta \in C$ such that $a = \beta,\ c = \beta^{-1}$ and, by (32), Q_r satisfies

$$\Big(\beta[x_1, x_2] + z_1x_2 + \alpha(x_1)z_2 - z_2x_1 - \alpha(x_2)z_1\Big)\beta^{-1}[x_3, x_4]+$$

$$\Big(\beta[x_3, x_4] + z_3x_4 + \alpha(x_3)z_4 - z_4x_3 - \alpha(x_4)z_3\Big)\beta^{-1}[x_1, x_2] - [x_1, x_2][x_3, x_4]$$

$$-[x_3, x_4][x_1, x_2].$$

We get a contradiction with the same argument used in (34).

Corollary 2 *Let R be a non-commutative prime ring of characteristic different from 2, Q_r be its right Martindale quotient ring and C be its extended centroid, L a non-central Lie ideal of R and F a non-zero generalized skew derivation of R. If F preserves the Jordan Product on L, then one of the following holds:*

1. *there exists $0 \neq \beta \in C$ such that $F(x) = \beta x$, for any $x \in R$, with $\beta^2 = 1_C$;*
2. *R satisfies $s_4(x_1, \ldots, x_4)$ and there exist an invertible element $q \in Q_r$ and $0 \neq \beta \in C$ such that $F(x) = \beta q x q^{-1}$, for any $x \in R$, with $\beta^2 = 1_C$;*

Proof The result follows directly from the Theorem 1, for $F = G$.

References

1. Argaç, N., Scudo, G.: Strong commutativity preserving generalized derivations on multilinear polynomials. Quaest. Math. **40**(8), 1075–1094 (2017)
2. Beidar, K.I., Martindale III, W.S., Mikhalev, A.V.: Rings with Generalized Identities. Monographs and Textbooks in Pure and Applied Math. 196, Marcel Dekker, Inc., New York (1996)
3. Bell, H.E., Daif, M.N.: On commutativity and strong commutativity-preserving maps. Can. Math. Bull. **37**(4), 443–447 (1994)
4. Bresar, M., Chebotar, M.A., Martindale III, W.S.: Functional Identities. Birkhauser, Frontiers in Mathematics (2007)
5. Bresar, M., Miers, C.R.: Strong commutativity preserving maps of semiprime rings. Can. Math. Bull. **37**(4), 457–460 (1994)
6. Carini, L., De Filippis, V., Scudo, G.: Identities with product of generalized derivations of prime rings. Algebra Colloq. **20**(4), 711–720 (2013)
7. Chang, J.C.: On the identity $h(x) = af(x) + g(x)b$. Taiwan. J. Math. **7**, 103–113 (2003)
8. Chuang, C.L.: GPIs having coefficients in Utumi quotient rings. Proc. Am. Math. Soc. **103**, 723–728 (1988)
9. Chuang, C.L.: Differential identities with automorphisms and antiautomorphisms I. J. Algebra **149**, 371–404 (1992)
10. Chuang, C.L.: Differential identities with automorphisms and antiautomorphisms II. J. Algebra **160**, 130–171 (1993)
11. Chuang, C.L.: Identities with skew derivations. J. Algebra **224**, 292–335 (2000)

12. Chuang, C.L., Lee, T.K.: Rings with annihilator conditions on multilinear polynomials. Chin. J. Math. **24**(2), 177–185 (1996)
13. Chuang, C.L., Lee, T.K.: Identities with a single skew derivation. J. Algebra **288**, 59–77 (2005)
14. De Filippis, V.: A product of two generalized derivations on polynomials in prime rings. Collect. Math. **61**, 303–322 (2010)
15. De Filippis, V., Di Vincenzo, O.M.: Generalized skew derivations and nilpotent values on Lie ideals. Algebra Colloq. **26**(4), 589–614 (2019)
16. De Filippis, V., Scudo, G.: Strong commutativity and Engel condition preserving maps in prime and semiprime rings. Linear Multilinear Algebra **61**(7), 917–938 (2013)
17. Di Vincenzo, O.M.: On the n-th centralizer of a Lie ideal. Boll. UMI A(7-3), pp. 77–85 (1989)
18. Erickson, T.S., Martindale, W.S., III., Osborne, J.M.: Prime nonassociative algebras. Pac. J. Math. **60**, 49–63 (1975)
19. Herstein, I.N.: Topics in Ring Theory. The University of Chicago Press (1969)
20. Jacobson, N.: Structure of Rings. American Mathematical Society, Providence, RI (1964)
21. Lanski, C., Montgomery, S.: Lie structure of prime rings of characteristic 2. Pac. J. Math. **42**(1), 117–135 (1972)
22. Lee, T.K.: Derivations with invertible values on a multilinear polynomial. Proc. Am. Math. Soc. **119**, 1–5 (1993)
23. Lee, T.K.: Generalized derivations of left faithful rings. Commun. Algebra **27**(8), 4057–4073 (1999)
24. Liau, P.H., Liu, C.K.: Strong commutativity preserving generalized derivations on Lie ideals. Linear Multilinear Algebra **59**(8), 905–915 (2011)
25. Liu, C.K.: Strong commutativity preserving generalized derivations on right ideals. Monatsh. Math. **166**(3–4), 453–465 (2012)
26. Martindale, W.S., III.: Prime rings satisfying a generalized polynomial identity. J. Algebra **12**, 576–584 (1969)
27. Rania, F.: Generalized derivations with skew-commutativity conditions on polynomials in prime rings. Beitrage zur Algebra und Geometrie **60**(4) (2018)
28. Wong, T.L.: Derivations with power central values on multilinear polynomials. Algebra Colloq. **3**, 369–378 (1996)

Prime Rings with Generalized Derivations and Power Values on Lie Ideals

Mohammad Aslam Siddeeque, Ali Ahmed Abdullah, and Nazim Khan

Abstract Suppose $\mathscr{R}$ is a prime ring that is non-commutative in structure and characteristic of $\mathscr{R}$ is a positive integer apart from 2 and $M = (-2)^{k-1} - 1$, where k is any odd positive integers greater than one. Let the Utumi ring of quotients be denoted by $\mathscr{Q}$, the extended centroid of $\mathscr{R}$ by $\mathscr{C}$. Consider $\mathscr{L}$ to be Lie ideal of $\mathscr{R}$ non-central in nature and $\mathscr{T}$ be a non-zero generalized derivation of $\mathscr{R}$. If $[\mathscr{T}(u^s), u^t]^m = [\mathscr{T}(u), u]$, for every $u \in \mathscr{L}$, where m, s and t be the fixed positive integers such that $m > 1$, $s \geq 1$ and $t \geq 1$, then one of the following situations prevails:

(*i*) The standard identity $s_4(x_1, \ldots, x_4)$ is satisfied by $\mathscr{R}$ and there exists $a \in \mathscr{Q}$ and $\beta \in \mathscr{C}$ such that $\mathscr{T}(x) = \beta x + ax + xa$, for every $x \in \mathscr{R}$.

(*ii*) there exists certain $\theta \in \mathscr{C}$ such that $\mathscr{T}(x) = \theta x$, for every $x \in \mathscr{R}$.

Keywords Prime rings · Generalized derivation · Maximal right ring of quotients · Generalized polynomial identity (GPI) · Polynomial identity (PI)

1 Introduction

In the entire article put forth, $\mathscr{R}$ always depicts prime ring that is non-commutative and associative in nature and its center is given by $\mathscr{Z}(\mathscr{R})$. Further, Q is the Martindale ring of quotients and $\mathscr{Q}$ denotes the Utumi ring of quotients with $\mathscr{C} = \mathscr{Z}(\mathscr{Q})$ as the center of $\mathscr{Q}$ called as the extended centroid of $\mathscr{R}$ and ρ the dense ideal of $\mathscr{R}$. Also, the well-known theory from [1] establishes that Q and $\mathscr{Q}$ share the same center.

A right ideal ρ is a right dense ideal if whenever $x_1, x_2 \in \rho$ with $x_1 \neq 0$, there exists $r \in \mathscr{R}$ with the condition that $x_1 r \neq 0$ and $x_2 r \in \rho$. A left dense ideal is defined likewise. An ideal ρ is called a dense ideal if it is both a left as well as a right dense ideal. Throughout the paper, by a left faithful ring we mean a ring whose

M. A. Siddeeque (✉) · A. A. Abdullah · N. Khan
Department of Mathematics, Aligarh Muslim University, Aligarh, India
e-mail: aslamsiddeeque@gmail.com

M. Ashraf et al. (eds.), *Algebra and Related Topics with Applications*,
Springer Proceedings in Mathematics & Statistics 392,
https://doi.org/10.1007/978-981-19-3898-6_24

left annihilator is zero. Similarly, right faithful ring is also defined and thus, a ring is faithful if it is both left and a right faithful. We observe that since $\mathscr{R}$ is prime, the Utumi ring of quotients $\mathscr{Q}$ is also a prime ring. With respect to a dense right ideal, the right Utumi ring of quotients of $\mathscr{R}$ can be characterized as the ring $\mathscr{Q}(\mathscr{R})$. We state some properties of $\mathscr{Q}(\mathscr{R})$ as follows:

(i) $\mathscr{R} \subseteq \mathscr{Q}(\mathscr{R})$;
(ii) For every $q \in \mathscr{Q}(\mathscr{R})$, there exists a right dense ideal $\mathscr{H}$ of $\mathscr{R}$ such that $q\mathscr{H} \subseteq \mathscr{R}$;
(iii) If $q \in \mathscr{Q}(\mathscr{R})$ and for certain non-zero right dense ideal $\mathscr{H}$ of $\mathscr{R}$ with $q\mathscr{H} = 0$, then $q = 0$;
(iv) If $\mathscr{H}$ is right dense ideal of $\mathscr{R}$ and $\mathscr{F} : \mathscr{H} \longrightarrow \mathscr{R}$ is a right $\mathscr{R}$-module map, then there exists certain $q \in \mathscr{Q}(\mathscr{R})$ such that $\mathscr{F}(x) = qx$, for every $x \in \mathscr{H}$.

These are the characterizing properties of $\mathscr{Q}(\mathscr{R})$. See [2] and [1] for the salient features of special rings like $\mathscr{Q}$, Q and $\mathscr{C}$.

For a prime ring $\mathscr{R}$, the extended centroid $\mathscr{C}$ of $\mathscr{R}$ is notably a field also called as the field of quotients of $\mathscr{Z}(\mathscr{R})$. Let $Y = \{y_1, y_2, \ldots\}$, be the set consisting of the non-commuting indeterminates say $y_1, y_2, \ldots$ which are countable. Let $\mathscr{C}\{Y\}$ be the free $\mathscr{C}$ algebra of the set Y. Consider $\mathscr{Q}\{Y\} = \mathscr{Q} *_{\mathscr{C}} \mathscr{C}\{Y\}$, the free $\mathscr{C}$-product of $\mathscr{Q}$ and $\mathscr{C}\{Y\}$. The elements of $\mathscr{Q}\{Y\}$ are called the GP (the generalized polynomials). By a non-trivial GP, we mean a non-zero element of $\mathscr{Q}\{Y\}$. Every element $w \in \mathscr{Q}\{Y\}$ is of the peculiar form $w = j_o t_1 j_1 t_2 j_2 \ldots t_n j_n$, where $\{j_o, \ldots, j_n\} \subseteq \mathscr{Q}$ and $\{t_1, \ldots, t_n\} \subseteq Y$, is called a monomial where $j_o, \ldots, j_n$ are called the coefficients of w. Each $g \in \mathscr{Q}\{Y\}$ constitutes of such monomials as a finite sum. Such representation is easily seen to be not unique. For a detailed study see [4].

For a lucid explanation of the notion of a non-triviality of a GPI, let us look at the following simple example.

Exam: Let W be the ring of real quaternions and $\mathscr{Z}(W) = \mathbb{R}$ be its center, that is the ring of real numbers. Then, for every $w \in W$, where $w = w_o + w_1 i + w_2 j + w_3 k$ and $w_i \in \mathbb{R}$, for every $i \in \{0, 1, 2, 3\}$, the following relation holds $w^2 = 2w_o w - w\overline{w}$. Follow-up of the above relation, results in the identity below,

$$w^2 iwi - wiw^2 i + iw^2 iw - iwiw^2 = 0, \text{ for every } w \in W,$$

called the non-trivial GPI satisfied by W where we can see $i \neq 0$.

The following definitions concerning commutators and prime rings shall be utilized in the present paper without emphasizing specifically each time. The commutator for every $x, y \in \mathscr{R}$ is given as $[x, y] =: xy - yx$ and anticommutators is given by $xoy =: xy + yx$ and the definition of a prime ring $\mathscr{R}$ viz. if $a\mathscr{R}b = (0)$, where $a, b \in \mathscr{R}$ then $a = 0$ or $b = 0$. Similarly, a ring $\mathscr{R}$ in which $a\mathscr{R}a = (0)$, then it implies that a equates to zero, is termed as semiprime ring.

An additive map $\mathscr{F} : \mathscr{R} \to \mathscr{R}$ is called as derivation if $\mathscr{F}(xy) = x\mathscr{F}(y) + \mathscr{F}(x)y$ stands true for every $x, y \in \mathscr{R}$. By saying derivation is inner derivation induced by an element $q \in \mathscr{R}$, we mean an additive map $\mathscr{G} : \mathscr{R} \to \mathscr{R}$ such that $\mathscr{G}(x) = [q, x]$ for every $x \in \mathscr{R}$. Furthermore, many researchers authored papers in the scenario of generalized derivations satisfying special identities in the presence of prime rings. An additive map $\mathscr{S}$ on $\mathscr{R}$ defined by $\mathscr{S}(w) = aw + wb$, for every $w \in \mathscr{R}$ and for some fixed $a, b \in \mathscr{R}$ is called the generalized inner derivation on the ring $\mathscr{R}$. Such maps prompt the definition of generalized derivation say $\mathscr{T} : \mathscr{R} \to \mathscr{R}$ which is expressed as the following,

$$\mathscr{T}(xy) = x[y, b] + \mathscr{T}(x)y = xI_b(y) + \mathscr{T}(x)y, \text{ for every } x,\ y \in \mathscr{R}$$

where I_b is an inner derivation induced by b. Further, we pen down the definition of a generalized derivation say $\mathscr{T}$ related to a derivation μ of $\mathscr{R}$ as following $\mathscr{T}(xy) = x\mu(y) + \mathscr{T}(x)y$, for every $x, y \in \mathscr{R}$. It is obvious that every inner generalized derivation is a generalized derivation and if $\mu = \mathscr{T}$ in the above definition of a generalized derivation then, $\mathscr{T}$ is an ordinary derivation.

We remark that Lee in [15] discussed the extension of a generalized derivation on any right dense ideal ρ to $\mathscr{Q}$, the Utumi ring of quotients. That is, if we have $\mathscr{T} : \rho \to \mathscr{Q}$, then due to Lee $\mathscr{T}$ is uniquely extended as $\mathscr{T} : \mathscr{Q} \to \mathscr{Q}$ and has the form $\mathscr{T}(x) = \mu(x) + ax$, for every $x \in \mathscr{Q}$, where $a \in \mathscr{Q}$. This definition of a generalized derivation shall be used in the entire paper without any special reference.

Throughout the paper, we will study the situation when the generalized derivations are acted upon the Lie ideals. By a Lie ideal $\mathscr{L}$, we mean an additive group $\mathscr{L}$ where the commutator $[\mathscr{L}, \mathscr{R}]$ is contained in $\mathscr{L}$. Obviously $[\mathscr{L}, \mathscr{L}]$ is a Lie ideal. We will consider only a non-central Lie ideal $\mathscr{L}$. A Lie ideal is called non-central if the commutator $[\mathscr{L}, \mathscr{L}]$ is not zero.

The prolific work in this direction clearly dictates that the global structure of the ring $\mathscr{R}$ is intimately related to the action of additive maps defined on the ring $\mathscr{R}$. For instance, derivation equipped with some properties intrigued many authors to investigate the structure of ring-like commutativity and even characterization of such additive maps. Some of the important works in this direction include [6] and [7].

In [13], the work of Lanski gives that, if for a derivation d of $\mathscr{R}$ such that $d(x)$ is n-nilpotent for some n a positive integer that is $d(x)^n = 0$ for every x from $\mathscr{L}$ a non-central Lie ideal, then d is vanishing on $\mathscr{R}$. An analogue fair result was developed by Lee in [15] for the more complex case of generalized derivations. More precisely, he contributed that if G is a generalized derivation on the prime ring $\mathscr{R}$ and $\mathscr{L}$ a Lie ideal which is non-central. If $G(x)^n = 0$ for every $x \in \mathscr{L}$, for n a certain positive integer, then $\mathscr{R}$ is bound to be commutative.

De Filippis and Carini in [3] studied a non-zero derivation d of $\mathscr{R}$ such that n-power of commutator $[d(x), x]$ is vanishing viz., $[d(x), x]^n = 0$, for every x in a Lie ideal $\mathscr{L}$ of $\mathscr{R}$ which is non-central, for certain positive integer n. They made the conclusion that if $\text{char}(R) \neq 2$, then $\mathscr{R}$ is forced to be commutative.

De Filippis later on in [5] improved this result by taking a generalized derivation G in the attempt to give a more general result instead of d. He concluded that either $\mathscr{R}$ satisfies the standard identity in four non-commuting variables s_4, and there exist

$a \in \mathcal{Q}$, $\alpha \in \mathcal{C}$ such that $G(x) = ax + xa + \alpha x$, for every $x \in \mathcal{R}$ or for particular $\gamma \in \mathcal{C}$, $G(x) = \gamma x$, for every x from $\mathcal{R}$.

Recently, Scudo and Ansari [18] took the task of investigating the set $A = \{[G(u), u] : u \in L\}$. They proved that if $A \neq \{0\}$, then it is void of any non-trivial idempotent element of $\mathcal{R}$. In this investigation, they focused the study of a generalized derivation on prime rings and provided the following result:

Theorem 1 ([18, Theorem]) *Let $\mathcal{R}$ be a prime ring that is non-commutative and char($\mathcal{R}$) $\neq 2$. Suppose associated with $\mathcal{R}$, the Utumi ring of quotients and the extended centroid of $\mathcal{R}$ is denoted by $\mathcal{Q}$ and $\mathcal{C}$, respectively. Then for $\mathcal{L}$ the non-central Lie ideal of $\mathcal{R}$ and $\mathcal{T}$ the non-zero generalized derivation of $\mathcal{R}$,*

if $[\mathcal{T}(u), u]^m = [\mathcal{T}(u), u]$, for every $u \in \mathcal{L}$, where m is a fixed positive integer such that $m > 1$, then one of the following conditions stands true:

(*i*) *The standard identity $s_4(x_1, \ldots, x_4)$ is satisfied by $\mathcal{R}$ and there exists certain $a \in \mathcal{Q}$ and $\beta \in \mathcal{C}$ such that $\mathcal{T}(x) = \beta x + ax + xa$, for every $x \in \mathcal{R}$.*
(*ii*) *There exists $\theta \in \mathcal{C}$ such that $\mathcal{T}(x) = \theta x$, for every $x \in \mathcal{R}$.*

In view of the results above, it is reasonable to raise the following question.

Question: What can we say about the ring $\mathcal{R}$ admitting the generalized derivation $\mathcal{T}$ of $\mathcal{R}$ and $[\mathcal{T}(u^s), u^t]^m = [\mathcal{T}(u), u]$, for every $u \in \mathcal{L}$, where m, s and t are the fixed positive integers such that $m > 1$, $s \geq 1$ and $t \geq 1$, where $\mathcal{L}$ is a Lie ideal which is not central?

Here, in non-commutative prime ring, characteristic of $\mathcal{R}$ different from two, if we choose $s = t = 1$, then we have the same as case of [18, Theorem]. Our goal is to answer the above question in two cases.

Firstly, the case of inner generalized derivation, which we deal in Sect. 3 and then the study of general case, that is, the case of any generalized derivation, which we will discuss in the last section of this paper (Sect. 4).

2 Preliminary Results

Recall that Der($\mathcal{Q}$) is the set of all derivations on Utumi quotient ring $\mathcal{Q}$. The derivation word is an additive map δ given by $\delta = d^1 d^2 \ldots d^m$, with each $d^i \in$ Der($\mathcal{Q}$). Every differential polynomial is a GP taking coefficients out of $\mathcal{Q}$, of peculiar type $\psi(\delta^j x_i)$ utilizing non-commuting indeterminates x_i on which the δ^j words acts as the unary operator. The polynomial $\psi(\delta^j x_i)$ is termed as a DI (differential identity) over a subset T of $\mathcal{Q}$ if whenever a value is assigned from T to x_i, $\psi(\delta^j x_i)$ vanishes to zero. We refer the reader to [1, Chaps. 6–7] for the in depth and presentable description of GPI-Theory involving derivations. The $\mathcal{C}$-subspace of Der($\mathcal{Q}$) that is D_{int} comprises of all inner derivations on $\mathcal{Q}$. For d a nonzero derivation d on $\mathcal{R}$. By [12, Theorem 2], we pen down the non-trivial Kharchenkos effect. See also [14, Theorem 1].

If $\psi(x_1, \ldots x_n, d(x_1), \ldots, d(x_n))$ is a DI on $\mathscr{R}$, then one of the result listed below stands true:

1. The derivation d is inner i.e. $d \in D_{int}$;
2. The following GPI is satisfied by $\mathscr{R}$, $\psi(x_1, \ldots, x_n, y_1, \ldots, y_n)$.

Before we commence to establish our results, we pen down some well-known facts. Precisely we shall use the following facts incessantly wherever applicable.

Fact 2 *Every generalized derivation of $\mathscr{R}$ can be uniquely extended to a generalized derivation of $\mathscr{Q}$ and assumes the form that $\mathscr{T}(x) = \mu(x) + ax$, for some $a \in \mathscr{Q}$ and a derivation μ of $\mathscr{Q}$. That is, every generalized derivation of $\mathscr{R}$ can be defined on $\mathscr{Q}$ explicitly. See [15].*

Fact 3 *Let A be an ideal of $\mathscr{R}$. Then A, $\mathscr{R}$ and $\mathscr{Q}$ satisfy the same GPI with coefficients in $\mathscr{Q}$. See [4].*

Fact 4 *Let A be an ideal of $\mathscr{R}$. Then A, $\mathscr{R}$ and $\mathscr{Q}$ satisfy the same DI with coefficients in $\mathscr{Q}$. See [14].*

Fact 5 *Let $\mathscr{L}$ be a non-central Lie ideal of $\mathscr{R}$. If char($\mathscr{R}$) $\neq 2$ or $\mathscr{R}$ does not satisfy s_4, then there exists a non-zero ideal $\mathscr{H}$ of $\mathscr{R}$ such that $0 \neq [\mathscr{H}, \mathscr{R}] \subseteq \mathscr{L}$. For a simple ring $\mathscr{R}$, $[\mathscr{R}, \mathscr{R}] \subseteq \mathscr{L}$ drawing arguments from [10, pp. 4–5] and also see [8, Lemma 2, Proposition 1].*

Fact 6 *For a prime ring $\mathscr{R}$ having $\mathscr{C}$ as the extended centroid, the following equivalent conditions shall hold:*

(*a*) *The linear space $\mathscr{R}\mathscr{C}$ over $\mathscr{C}$ has atmost four dimension;*
(*b*) *The standard identity $s_4(x_1, \ldots, x_4)$ is satisfied by $\mathscr{R}$;*
(*c*) *For a certain field $\mathscr{F}$, $\mathscr{R}$ embeds in $M_2(\mathscr{F})$ or $\mathscr{R}$ is commutative;*
(*d*) *Degree of algebraicness of $\mathscr{R}$ over $\mathscr{C}$ is two;*
(*e*) *$\mathscr{R}$ satisfies the PI $[[x^2, y], [x, y]] = 0$.*

3 When $\mathscr{T}$ Is Generalized Inner Derivations

We consider in this segment of the proof that, every generalized inner derivation induced by the elements $a, b \in \mathscr{Q}$ takes the form as $\mathscr{T}(x) = ax + xb$, for every $x \in \mathscr{R}$. Henceforth, it is supposed that the following GPI is satisfied by $\mathscr{R}$

$$\Pi(w_1, w_2) = \{[a[w_1, w_2]^s + [w_1, w_2]^s b, [w_1, w_2]^t]\}^m - [a[w_1, w_2] + [w_1, w_2]b, [w_1, w_2]].$$

In an attempt to establish the main result of the article, we shall need the support of the following crucial fact.

Fact 7 *By the pivotal assumption of the article, the following relation*

$$\{[a[w_1, w_2]^s + [w_1, w_2]^s b, [w_1, w_2]^t]\}^m - [a[w_1, w_2] + [w_1, w_2]b, [w_1, w_2]] = 0,$$

holds for every $w_1, w_2 \in \mathscr{R}$. *Further, for every automorphism* ψ *of* $\mathscr{R}$ *which is inner, we have*

$$[\psi(a)[w_1, w_2]^s + [w_1, w_2]^s \psi(b), [w_1, w_2]^t]^m - [\psi(a)[w_1, w_2] + [w_1, w_2]\psi(b), [w_1, w_2]] = 0,$$

holds for every $w_1, w_2 \in \mathscr{R}$. *Clearly,* $a + b, a - b, a, b$ *are the central elements of the ring* $\mathscr{R}$ *if and only if* $\psi(a + b)$, $\psi(a - b)$, $\psi(a)$, $\psi(b)$ *are the central elements of the ring* $\mathscr{R}$. *Thus, whenever it is demanded, we can use* $\psi(a)$ *and* $\psi(b)$ *instead of* a *and* b, *respectively.*

Proposition 1 *Suppose* $\mathscr{R}$ *is a prime ring that is non-commutative in structure. If char(*$\mathscr{R}$*) is a positive integer apart from 2 and* $M = (-2)^{k-1} - 1$ *where* $k(> 1)$ *is any odd positive integers. Let the Utumi ring of quotients be denoted by* $\mathscr{Q}$, *the extended centroid of* $\mathscr{R}$ *by* $\mathscr{C}$. *Consider* $\mathscr{L}$ *to be Lie ideal of* $\mathscr{R}$ *non-central in nature and* $\mathscr{T}$ *be a non-zero inner generalized derivation of* $\mathscr{R}$ *induced by elements* $a, b \in \mathscr{Q}$. *If* $[\mathscr{T}(u^s), u^t]^m = [\mathscr{T}(u), u]$, *for every* $u \in \mathscr{L}$, *where* m, s *and* t *be the fixed positive integers such that* $m > 1$, $s \geq 1$ *and* $t \geq 1$, *then one of the following situations prevails:*

(*i*) *The standard identity* $s_4(x_1, \ldots, x_4)$ *is satisfied by* $\mathscr{R}$ *and there exists* $a \in \mathscr{Q}$ *and* $\beta \in \mathscr{C}$ *such that* $\mathscr{T}(x) = \beta x + ax + xa$, *for every* $x \in \mathscr{R}$.
(*ii*) *there exists certain* $\theta \in \mathscr{C}$ *such that* $\mathscr{T}(x) = \theta x$, *for every* $x \in \mathscr{R}$.

Proof The following GPI is satisfied by $\mathscr{R}$

$$\Pi(w_1, w_2) = \{[a[w_1, w_2]^s + [w_1, w_2]^s b, [w_1, w_2]^t]\}^m - [a[w_1, w_2] + [w_1, w_2]b, [w_1, w_2]],$$

for every $w_1, w_2 \in \mathscr{R}$. Owing to Beidar [2, Theorem 2] and also by Fact 3, this GPI is satisfied by $\mathscr{Q}$ also. When $\mathscr{C}$ is infinite, then $\Pi(r_1, r_2) = 0$, for every $r_1, r_2 \in \mathscr{Q} \otimes \overline{\mathscr{C}}$, where $\overline{\mathscr{C}}$ is the algebraic closure of $\mathscr{C}$. We note that since both $\mathscr{Q} \otimes \overline{\mathscr{C}}$ and $\mathscr{Q}$ are centrally closed (see [9, Theorems 2.5 and 3.5]), we may replace $\mathscr{R}$ by $\mathscr{Q} \otimes \overline{\mathscr{C}}$ or $\mathscr{Q}$, in accordance with the situation whether $\mathscr{C}$ is infinite or finite. Thus we may assume that $\mathscr{R}$ is centrally closed over $\mathscr{C}$ which is either algebraically closed or finite.
Case I: If $a, b \in \mathscr{C}$, then for certain $\theta \in \mathscr{C}$, $\mathscr{T}(x) = \theta x$ for every $x \in [\mathscr{R}, \mathscr{R}]$.
Case II: If either $a \notin \mathscr{C}$ or $b \notin \mathscr{C}$, in this situation by [4], $\Pi(w_1, w_2)$ is a non-trivial GPI satisfied by $\mathscr{R}$. Hence, by Martindale's strong result known as Martindale's Theorem [16], $\mathscr{R}$ is bound to be primitive ring with non-zero socle $\mathscr{S}$ with $\mathscr{C}$ as the division ring. Under the awe of Jacobson's Theorem [11, p. 75], $\mathscr{R}$ is isomorphic to a dense ring of linear transformation on certain linear space $\mathscr{V}$ over $\mathscr{C}$.

If $dim_{\mathscr{C}}(\mathscr{V}) = 2$, then $\mathscr{R} \cong M_2(C)$, the ring of all 2×2 matrices over $\mathscr{C}$. Thus, we may observe from the Fact 6 that, $s_4(x_1, \ldots, x_4)$ is satisfied by $\mathscr{R}$. Now take $b - a = \sum_{ij} \gamma_{ij} e_{ij}$, where $\gamma_{ij} \in \mathscr{C}$ and e_{ij} are the unit standard matrices.

Let $[w_1, w_2] = [e_{ii}, e_{ij}] = e_{ij}$, for any i different from j, from our assumption and suppose χ denote the following

$$\chi = [a(e_{ij})^s + (e_{ij})^s b, (e_{ij})^t]^m - [ae_{ij} + e_{ij}b, e_{ij}].$$

When $s = t = 1$. Then $\chi = \{e_{ij}(b-a)e_{ij}\}^m - \{e_{ij}(b-a)e_{ij}\} = 0$. Consequently, we have that

$$e_{ij}(b-a)e_{ij} = 0, \text{ since } m > 1. \tag{1}$$

This gives that $\gamma_{ji} = 0$, for every i different from j. That is, off-diagonal entries of $b - a$ vanishes to zero. Hence, $b - a$ is diagonal matrix. Also, when $s > 1,\ t = 1$ we get the same conclusion that $b - a$ is a diagonal matrix. Indeed, we can easily see that χ gives the following relation

$$[ae_{ij} + e_{ij}b, e_{ij}] = 0.$$

This implies that

$$e_{ij}(b-a)e_{ij} = 0, \quad \text{which is Eq. (1) thus it gives the same conclusion.}$$

Lastly, for every s and $t > 1$, there still holds the same conclusion by similar tactic as above.

Thus in all, let $\psi(x) = (1 + e_{ij})x(1 - e_{ij})$, for every $x \in \mathscr{R}$, be an inner automorphism induced by matrix $(1 + e_{ij})$ where i different from j. By Fact 7, $\psi(b - a)$ is also a diagonal matrix. Therefore, the (i, j) entry of $\psi(b - a)$ is indeed zero.

$$0 = [\psi(b-a)]_{ij} = \gamma_{jj} - \gamma_{ii}, \text{ that is } \gamma_{jj} = \gamma_{ii}.$$

The above calculation establishes that $b - a \in \mathscr{C}$. Thus, we take opportunity to write the generalized derivation $\mathscr{T}(x) = ax + xb$ as, $\mathscr{T}(x) = ax + \beta x + xa$, where $\beta = b - a \in \mathscr{C}$.

Now we assume that $\dim_{\mathscr{C}}(\mathscr{V}) \geq 3$. The following relation holds for every $u \in [\mathscr{R}, \mathscr{R}]$

$$\{au^{s+t} + u^s bu^t - u^t au^s - u^{s+t}b\}^m - \{au^2 + ubu - uau - u^2 b\} = 0. \tag{2}$$

Further, we claim that for every $v \in \mathscr{V}$, the set $\{v, bv\}$ is linearly $\mathscr{C}$-dependent. For that we suppose on contrary that there exits non-zero $v_o \in \mathscr{V}$ such that $\{v_o, bv_o\}$ are linearly $\mathscr{C}$-independent. Since $\dim_{\mathscr{C}}(\mathscr{V}) \geq 3$, there exists non-zero $w_o \in \mathscr{V}$ such that $\{v_0, bv_o, w_o\}$ are linearly $\mathscr{C}$-independent. With the gratitude towards Jacobson's Theorem, we see that there exists $u_1, u_2 \in \mathscr{R}$ so that the following relations hold

$u_1 v_o = 0, \quad u_2 v_o = v_o, \quad u_1 bv_o = w_o;$

$u_2 bv_o = bv_o, \quad u_1 w_o = -2v_o$ and $u_2 w_o = 0$.

Hence, for some $u \in [\mathscr{R}, \mathscr{R}]$ say $u = [u_1, u_2]$, we have that

$uv_o = 0, \;\; ubv_o = w_o, \quad uw_o = 2v_o.$
Now right multiplying by v_o in relation (2), we obtain that

$$\{au^{s+t} + u^s bu^t - u^t au^s - u^{s+t}b\}^m v_o - \{au^2 + ubu - uau - u^2 b\}v_o = 0.$$

If either $s > 1$ and $t \geq 1$ or $s = 1$ and $t > 1$, we face the same contradiction that $2v_o = 0$. Now, for $s = t = 1$, we have

$$\{au^2 + ubu - uau - u^2 b\}^m v_o - \{au^2 + ubu - uau - u^2 b\}v_o = 0.$$

Making proper use of Density theorem, one can see there exists certain u_1 and $u_2 \in \mathscr{R}$ due to which
$u_1 v_o = 0, \;\; u_2 v_o = v_o, \;\; u_1 bv_o = w_o;$
$u_2 bv_o = bv_o, \;\; u_1 w_o = -\alpha v_o,$ where $0 \neq \alpha \in \mathscr{C}$ and $u_2 w_o = 0.$
Hence, for some $u \in [\mathscr{R}, \mathscr{R}]$ say $u = [u_1, u_2]$, we have that
$uv_o = 0, \;\; ubv_o = w_o, \quad uw_o = \alpha v_o.$
Now right multiplying by v_o in above relation, we obtain

$$\{au^2 + ubu - uau - u^2 b\}^{m-1}\{au^2 + ubu - uau - u^2 b\}v_o - \{au^2 + ubu - uau - u^2 b\}v_o = 0.$$

This implies that

$$\{au^2 + ubu - uau - u^2 b\}^{m-1}(-\alpha v_o) + \alpha v_o = 0.$$

Applying this linear transformation $m - 1$ times on v_o, we have the following simple consequence

$$((-\alpha)^m + \alpha)v_o = 0.$$

Since $\alpha \in \mathscr{C}$, for $\alpha = 1$ we get $((-1)^m + 1)v_o = 0$, which gives contradiction for every even positive integers as char$(\mathscr{R}) \neq 2$. For $\alpha = 1 + 1$ (say 2), $((-2)^{m-1} - 1)v_o = 0$, wherein a contradiction is prompted as char$(\mathscr{R}) \neq M$. Therefore, for every $v \in \mathscr{V}$, the set of vectors $\{bv, v\}$ is linearly $\mathscr{C}$-dependent and for every $v \in \mathscr{V}, \; bv = \beta_v v$, for certain $\beta_v \in \mathscr{C}$. It is easy consequence that β_v does not depends on the vector $v \in \mathscr{V}$ and thus we consider $bv = \beta v$, for every $v \in \mathscr{V}$ and for some fixed $\beta \in \mathscr{C}$. Further, assume that for every $u \in \mathscr{R}$ and for any $v \in \mathscr{V}$, we have

$$[u, b]v = u(bv) - b(uv) = u(\beta v) - \beta uv = 0.$$

Hence, we have $[u, b]\mathscr{V} = 0$, as $[u, b]$ is a linear transformation that acts faithfully on the linear space $\mathscr{V}$. Therefore, $[u, b] = 0$, for every $u \in \mathscr{R}$. Thus, $b \in \mathscr{Z}(\mathscr{R}) \subseteq \mathscr{C}$. Therefore, relation (2) reduces to the following relation

$$\{au^{s+t} - u^t au^s\}^m - \{au^2 - uau\} = 0, \;\; \text{for every } u \in [\mathscr{R}, \mathscr{R}]. \qquad (3)$$

Let us now try to prove that $\{av, v\}$ are linearly $\mathscr{C}$-dependent. In that attempt, we suppose on contrary that for some non-zero $v' \in \mathscr{V}$, $\{av', v'\}$ are linearly $\mathscr{C}$-independent. Since $\dim_{\mathscr{C}}(\mathscr{V}) \geq 3$, there exists $w' \in \mathscr{V}$ such that $\{av', v', w'\}$ are linearly $\mathscr{C}$-independent. Owing to Jacobson's Theorem, there exists $u_1', u_2' \in \mathscr{R}$ so that following relations hold

$u_1'v' = v', \ u_2'v' = v', \ u_1'av' = -2v'$;
$u_2'av' = 2v', \ u_1'w' = -v'$ and $u_2'w' = v'$.

Hence, for some $u \in [\mathscr{R}, \mathscr{R}]$, say $u = [u_1', u_2']$, we have $uv' = [u_1', u_2']v' = 0$, $uav' = [u_1', u_2']av' = 4v'$, $uw' = [u_1', u_2']w' = 2v'$. Now right multiplying by w' in relation (3), we obtain that

$$\{au^{s+t} - u^t au^s\}^m w' - \{au^2 - uau\}w' = 0, \text{ for every } u \in [\mathscr{R}, \mathscr{R}]. \quad (4)$$

When $s = t = 1$. Then we obtain from the above relation that, $8v' = 0$, which is a contradiction. Now if either $s > 1, \ t \geq 1$ or $s \geq 1, \ t > 1$, we arrive at the same contradiction. Thus, for every vector $v \in \mathscr{V}$, the set of vectors $\{av, v\}$ is linearly $\mathscr{C}$-dependent and by same technique utilized to show $b \in \mathscr{C}$, we get $a \in \mathscr{C}$. Thus in all, for $\dim_{\mathscr{C}}(\mathscr{V}) \geq 3$, we get a contradiction that both a and b are in $\mathscr{C}$.

Finally, we conclude that if $\dim_{\mathscr{C}}(\mathscr{V}) = 2$, then $s_4(x_1, \ldots, x_4)$ is satisfied by $\mathscr{R}$ and $\mathscr{T}(x) = \beta x + ax + xa$, for every $x \in \mathscr{R}$, where $\beta = b - a \in \mathscr{C}$.

4 The Study of General Case

In this segment of the proof, we begin by considering that $\mathscr{T}$ is a generalized derivation. In an attempt to prove the main result, we consider for certain $a \in \mathscr{Q}$ and μ a derivation of $\mathscr{R}$, we have $\mathscr{T}(x) = \mu(x) + ax$ by using Lee [15].

Theorem 8 *Suppose $\mathscr{R}$ is a prime ring that is non-commutative in structure and characteristic of $\mathscr{R}$ is a positive integer apart from 2 and $M = (-2)^{k-1} - 1$ where k is any odd positive integers greater than one. Let the Utumi ring of quotients be denoted by $\mathscr{Q}$, the extended centroid of $\mathscr{R}$ by $\mathscr{C}$. Consider $\mathscr{L}$ to be Lie ideal of $\mathscr{R}$ non-central in nature and $\mathscr{T}$ be a non-zero generalized derivation of $\mathscr{R}$. If $[\mathscr{T}(u^s), u^t]^m = [\mathscr{T}(u), u]$, for every $u \in \mathscr{L}$, where $m, \ s$ and t be the fixed positive integers such that $m > 1, \ \ s \geq 1$ and $t \geq 1$, then one of the following situations prevails:*

- (*i*) *The standard identity $s_4(x_1, \ldots, x_4)$ is satisfied by $\mathscr{R}$ and there exists $a \in \mathscr{Q}$ and $\beta \in \mathscr{C}$ such that $\mathscr{T}(x) = \beta x + ax + xa$, for every $x \in \mathscr{R}$.*
- (*ii*) *there exists certain $\theta \in \mathscr{C}$ such that $\mathscr{T}(x) = \theta x$, for every $x \in \mathscr{R}$.*

Proof We find that for certain $a \in \mathscr{Q}$ and μ a derivation of $\mathscr{R}$ related to $\mathscr{T}$ a generalized derivation where $\mathscr{T}(x) = \mu(x) + ax$, for every $x \in \mathscr{R}$. Owing to the Fact 2, we may extend the definition of a generalized derivation on $\mathscr{R}$ to that on the Utumi ring of quotients $\mathscr{Q}$. Further, by the Fact 5, there exists a non-zero ideal $\mathscr{H}$ of $\mathscr{R}$

such that $0 \neq [\mathscr{H}, \mathscr{R}] \subseteq \mathscr{L}$. Also by Fact 4, $\mathscr{H}$ and $\mathscr{Q}$ satisfy the same differential identity, hence for every $w_1, w_2 \in \mathscr{Q}$,

$$[a[w_1, w_2]^s + \mu([w_1, w_2]^s), [w_1, w_2]^t]^m = [a[w_1, w_2] + \mu([w_1, w_2]), [w_1, w_2]]. \tag{5}$$

Under the effect of Kharchenko theory (See [12]), we bifurcate our situation as follows.

(1) **When μ is an inner derivation.**
Then there exists c from $\mathscr{Q}$ such that μ can be expressed as $\mu(w) = [c, w]$ for every $w \in \mathscr{R}$. Hence $\mathscr{T}(w) = (a + c)w - wc$, for every $w \in \mathscr{R}$. Therefore, from differential identity (5), we have
$[(a + c)[w_1, w_2]^s - [w_1, w_2]^s c, [w_1, w_2]^t]^m = [(a + c)[w_1, w_2] - [w_1, w_2]c, [w_1, w_2]]$. Hence, on using Proposition 1 for $(a + c)$ and c, we are done.

(2) **When μ is an outer derivation.**
We observe that

$$\mu([w_1, w_2]^s) = \sum_{i=0}^{s-1}[w_1, w_2]^i\{[\mu(w_1), w_2] + [w_1, \mu(w_2)]\}[w_1, w_2]^{s-i-1}. \tag{6}$$

Using relation (6) in differential identity (5), we have

$$[(a[w_1, w_2]^s + \sum_{i=0}^{s-1}[w_1, w_2]^i\{[\mu(w_1), w_2] + [w_1, \mu(w_2)]\}[w_1, w_2]^{s-i-1}, [w_1, w_2]^t]^m$$
$$= [a[w_1, w_2] + [\mu(w_1), w_2] + [w_1, \mu(w_2)], [w_1, w_2]].$$

The following GPI is satisfied by $\mathscr{Q}$

$$[(a[w_1, w_2]^s + \sum_{i=0}^{s-1}[w_1, w_2]^i\{[y_1, w_2] + [w_1, y_2]\}[w_1, w_2]^{s-i-1}, [w_1, w_2]^t]^m \tag{7}$$

$$= [a[w_1, w_2] + [y_1, w_2] + [w_1, y_2], [w_1, w_2]].$$

Assume $y_1 = y_2 = 0$, thus, using Proposition 1, we recollect that a is central. Under this privilege, we rewrite relation (7) as the following

$$\left[\sum_{i=0}^{s-1}[w_1, w_2]^i\{[y_1, w_2] + [w_1, y_2]\}[w_1, w_2]^{s-i-1}, [w_1, w_2]^t\right]^m \tag{8}$$

$$= [[y_1, w_2] + [w_1, y_2], [w_1, w_2]].$$

The above relation (8), is a PI for $\mathscr{R}$, then by a well-known Posner's result [17], we observe that there exists certain field $\mathscr{F}$ and an integer $l \geq 1$ such that $\mathscr{Q}$

and $M_l(\mathscr{F})$ satisfy the same PI (polynomial identities). It is evident that $l \geq 2$ as $\mathscr{R}$ is non-commutative. From now onwards, we will consider the following matrices

$$w_1 = e_{pp}, \quad w_2 = e_{qp}, \quad y_1 = 0, \quad y_2 = \gamma e_{pq}, \text{ where } p \neq q \text{ and } 0 \neq \gamma \in \mathscr{F}.$$

Therefore,

$$[w_1, w_2] = [e_{pp}, e_{qp}] = -e_{qp} \text{ and } [w_1, y_2] = [e_{pp}, \gamma e_{pq}] = \gamma e_{pq}.$$

Thus relation (8) reduces to the following

$$[\sum_{i=0}^{s-1}(-e_{qp})^i\{\gamma e_{pq}\}(-e_{qp})^{s-i-1}, (-e_{qp})^t]^m = [\gamma e_{pq}, -e_{qp}].$$

In above equation, put $s = t = 1$, we have

$$[\gamma e_{pq}, -e_{qp}]^m = [\gamma e_{pq}, -e_{qp}].$$

Now, after a simple calculation, we obtain that $\{-\gamma(e_{pp} - e_{qq})\}^m = -\gamma(e_{pp} - e_{qq})$, right multiplying above equation by e_{pp} we are in the receipt of the following relation $((-\gamma)^m + \gamma)e_{pp} = 0$. When γ is chosen as 1, we get a contradiction for every even positive integers as char$(\mathscr{R}) \neq 2$. When $\gamma = 1 + 1$ (say 2) we again face a contradiction $((-2)^m + 2)e_{pp} = 0$ as char$(\mathscr{R}) \neq M$. In the same way, put $s = 2$ and $t = 1$ in above, we get

$$[\{\gamma e_{pq}\}(-e_{qp}) + (-e_{qp})\{\gamma e_{pq}\}, (-e_{qp})]^m = [\gamma e_{pq}, -e_{qp}].$$

This implies that

$$0 = -\gamma(e_{pp} - e_{qq}).$$

That is, $e_{pp} = e_{qq}$ if and only if $p = q$, which is a contradiction to our assumption. Similarly, for $s \geq 3$ and $t = 1$ gives contradiction that can be easily verified. At last, it can be easily seen that when $t > 1$, $0 = [\gamma e_{pq}, -e_{qp}]$ which still gives a contradiction.

References

1. Beidar, K.I., Martindale, W.S., III., Mikhalev, A.V.: Rings with generalized identities. Pure Appl. Math, Dekker, New York (1996)
2. Beidar, K.I.: Rings with generalized identities. Moscow Univ. Math. Bull. **33**, 53–58 (1978)
3. Carini, L., De Filippis, V.: Commutators with power central values on a Lie ideals. Pac. J. Math. **193**, 269–278 (2000)
4. Chuang, C.L.: GPIs' having coefficients in Utumi quotient rings. Proc. Am. Math. Soc. **103**, 723–728 (1988)
5. De Filippis, V.: Generalized derivations and commutators with nilpotent values on Lie ideals. Tamsui Oxf. J. Math. Sci. **22**, 167–175 (2006)
6. De Filippis, V.: Scudo, G., Strong commutativity and Engel condition preserving maps in prime and semiprime rings. Linear Multilinear Algebra **61**(7), 917–938 (2013)
7. Dhara, B., De Filippis, V.: Engel conditions of generalized derivations on left ideals and Lie ideals in prime rings. Commun. Algebra **48**(1), 154–167 (2019). https://doi.org/10.1080/00927872.2019.1635608
8. Di Vincenzo, O.M.: On the n-th centralizer of a Lie ideal. Boll. Unione Mat. Ital. **7**, 77–85 (1989)
9. Ericson, T.S., Martindale III, W.S., Osborn, J.M.: Prime non associative algebras. Pac. J. Math. **60**, 49–63 (1975)
10. Herstein, I.N.: Topics in Ring Theory. University of Chicago Press, Chicago IL (1969)
11. Jacobson, N.: Structure of Rings. American Mathematical Society, Providence, RI (1964)
12. Kharchenko, V.K.: Differential identities of prime rings. Algebra Log. **17**, 155–168 (1978)
13. Lanski, C.: Derivations with nilpotent values on Lie ideals. Proc. Am. Math. Soc. **108**, 31–37 (1990)
14. Lee, T.K.: Semiprime rings with differential identities. Bull. Inst. Math. Acad. Sin. **20**, 27–38 (1992)
15. Lee, T.K.: Generalized derivations of left faithful rings. Commun. Algebra **27**, 4057–4073 (1999)
16. Martindale, W.S., III.: Prime rings satisfying a generalized polynomial identity. J. Algebra **12**, 576–584 (1969)
17. Posner, E.C.: Prime rings satisfying a polynomial identity. Arch. Math. **11**, 180–183 (1960)
18. Scudo, G., Ansari, A.Z.: Generalized derivations on Lie Ideals and power values on prime rings. Math. Slovaca **65**(5), 975–980 (2015)

Coding Theory

On the Purity of Resolutions of Stanley-Reisner Rings Associated to Reed-Muller Codes

Sudhir R. Ghorpade and Rati Ludhani

Abstract Following Johnsen and Verdure (2013), we can associate to any linear code C an abstract simplicial complex and in turn, a Stanley-Reisner ring R_C. The ring R_C is a standard graded algebra over a field and its projective dimension is precisely the dimension of C. Thus R_C admits a graded minimal free resolution and the resulting graded Betti numbers are known to determine the generalized Hamming weights of C. The question of purity of the minimal free resolution of R_C was considered by Ghorpade and Singh (2020) when C is the generalized Reed-Muller code. They showed that the resolution is pure in some cases and it is not pure in many other cases. Here we give a complete characterization of the purity of graded minimal free resolutions of Stanley-Reisner rings associated with generalized Reed-Muller codes of an arbitrary order.

Keywords Ring theory · Coding theory

1 Introduction

This article concerns a topic that is at the interface of homological aspects of commutative algebra and the theory of linear error-correcting codes. Our motivation comes from the work of Johnsen and Verdure [11] and the more recent work [8]. In [11], the notion of *Betti numbers* of a linear code is introduced. The Betti numbers of a linear code C of length n are, in fact, the graded Betti numbers of the Stanley-

S. R. Ghorpade (✉) · R. Ludhani
Department of Mathematics, Indian Institute of Technology Bombay, Powai, Mumbai 400076, India
e-mail: srg@math.iitb.ac.in

R. Ludhani
e-mail: ratiludhani@gmail.com

M. Ashraf et al. (eds.), *Algebra and Related Topics with Applications*,
Springer Proceedings in Mathematics & Statistics 392,
https://doi.org/10.1007/978-981-19-3898-6_26

Reisner ring R_C of the simplicial complex Δ_C on $[n] := \{1, \ldots, n\}$ whose faces are precisely the subsets $\{i_1, \ldots, i_t\}$ of $[n]$ for which the columns $H_{i_1}, \ldots, H_{i_t}$ of a parity check matrix H of C are linearly independent. In [11], it was shown that the Betti numbers of a linear code determine its generalized Hamming weights. Further, Johnsen, Roksvold and Verdure [13] showed that the Betti numbers of a linear code (and its elongations) determine its generalized weight polynomials and hence the extended weight enumerators. On the other hand, the work of Jurrius and Pellikaan [14] shows that the extended weight enumerators of a linear code determine its generalized weight enumerator. So it is clear that the Betti numbers of a linear code (and its elongations) are also closely related to several classical parameters of that code. Thus it is useful to know them explicitly. Computation of these Betti numbers is in general, a difficult problem, but it becomes easy, by a formula of Herzog and Kühl [10], when the corresponding minimal free resolutions are pure. An intrinsic characterization of purity of the graded minimal free resolutions of Stanley-Reisner rings associated with arbitrary linear codes was obtained in [8]. As a consequence, known results about the Betti numbers of MDS codes (cf. [11]) and constant weight codes (cf. [12]) were easily deduced.

One of the most important and widely studied classes of linear codes is that of Reed-Muller codes. These codes were introduced by Reed [18] in the binary case and several of their properties were established by Muller [17]; see also [4, pp. 20–38]. We shall consider Reed-Muller codes in the most general sense, as given by Kasami, Lin and Peterson [15] and by Delsarte, Goethals and MacWilliams [6]. Generalized Hamming weights of (generalized) Reed-Muller codes are explicitly known, thanks to the work of Heijnen and Pellikaan [9] (see also [2] and [3]). It is, therefore, natural to ask for an explicit determination of the Betti numbers of Reed-Muller codes. The problem would be tractable if we know when the graded minimal free resolutions of Stanley-Reisner rings of simplicial complexes corresponding to Reed-Muller codes are pure. This question about purity was considered in [8] and an answer was provided in many, but not all, cases. In this article we build upon the work in [8] and complete it to give a characterization of purity of graded minimal free resolutions of Stanley-Reisner rings associated with arbitrary Reed-Muller codes.

This paper is organized as follows. In Sect. 2, we review (generalized) Reed-Muller codes and discuss their properties that are relevant to us. Next, in Sect. 3, the notion of purity of a minimal free resolution is recalled and some key results in [8], such as the intrinsic characterization mentioned above and results about the purity or nonpurity of resolutions corresponding to Reed-Muller codes, are stated. Our main result on a characterization of purity of free resolutions of Stanley-Reisner rings associated with Reed-Muller codes is also proved here. As a corollary, we give a characterization of Reed-Muller codes that are MDS codes.

2 Reed-Muller Codes

Standard references for (generalized) Reed-Muller codes are the book of Assmus and Key [1] (especially Chap. 5) and the seminal paper of Delsarte, Goethals and MacWilliams [6]. Let us begin by setting some basic notation and terminology.

Fix throughout this paper a prime power q and a finite field $\mathbb{F}_q$ with q elements. Let n, k be integers with $1 \leq k \leq n$. We write $[n, k]_q$*-code* to mean a q-ary linear code of length n and dimension k, i.e., a k-dimensional $\mathbb{F}_q$-linear subspace of $\mathbb{F}_q^n$. Recall that the *Hamming weight* of an element $c = (c_1, \dots, c_n) \in \mathbb{F}_q^n$ is defined by

$$\mathrm{wt}(c) := |\{i \in \{1, \dots, n\} : c_i \neq 0\}|.$$

The *minimum distance* of an $[n, k]_q$-code C can be defined by

$$d(C) := \min\{\mathrm{wt}(c) : c \in C\}$$

and if $d(C) = d$, then C may be referred to as an $[n, k, d]_q$-code. In this case, the elements of C of Hamming weight d will be referred to as the *minimum weight codewords* of C. An $[n, k]_q$-code is said to be *nondegenerate* if it is not contained in a coordinate hyperplane of $\mathbb{F}_q^n$. We denote by $\mathbb{N}$ the set of nonnegative integers.

Let m, r be integers such that $m \geq 1$ and $0 \leq r \leq m(q-1)$. Define

$$V_q(r, m) := \{f \in \mathbb{F}_q[X_1, \dots, X_m] : \deg(f) \leq r \text{ and } \deg_{X_i}(f) < q \text{ for } i = 1, \dots, m\}.$$

Note that $V_q(r, m)$ is a $\mathbb{F}_q$-linear subspace of the polynomial ring $\mathbb{F}_q[X_1, \dots, X_m]$. Fix an ordering $\mathsf{P}_1, \dots, \mathsf{P}_{q^m}$ of the elements of $\mathbb{F}_q^m$ and consider the evaluation map

$$\mathrm{Ev} : V_q(r, m) \to \mathbb{F}_q^{q^m} \quad \text{defined by} \quad f \mapsto c_f := (f(\mathsf{P}_1), \dots, f(\mathsf{P}_{q^m})). \quad (1)$$

Clearly, Ev is a linear map and its image is a nondegenerate linear code of length q^m; this code is called the *(generalized) Reed-Muller code of order* r, and it is denoted by $\mathsf{RM}_q(r, m)$. The dimension of $\mathsf{RM}_q(r, m)$ is given by the following formula that can be found in Assmus and Key [1, Theorem 5.4.1]:

$$\dim \mathsf{RM}_q(r, m) = \sum_{s=0}^{r} \sum_{i=0}^{m} (-1)^i \binom{m}{i} \binom{s - iq + m - 1}{s - iq}. \quad (2)$$

In [8, Eq. (13)], a somewhat simpler formula for the dimension is stated (without proof). It is not difficult to derive it from (2). However, we give an independent and direct proof of the simpler formula below.

Lemma 1 *Let m, r be integers such that $m \geq 1$ and $0 \leq r \leq m(q-1)$. Then*

$$\dim \mathsf{RM}_q(r,m) = \sum_{i=0}^{m} (-1)^i \binom{m}{i}\binom{m+r-iq}{m}. \tag{3}$$

Proof It is well known that the map Ev given by (1) is injective. This follows, for instance, from [7, Lemma 2.1]. Also, if $E := \{(v_1, \dots, v_m) \in \mathbb{N}^m : v_1 + \cdots + v_m \le r\}$, then it is easily seen that a basis of $V_q(r,m)$ is given by

$$B := \{X_1^{v_1} \cdots X_m^{v_m} : (v_1, \dots, v_m) \in E \text{ and } 0 \le v_j < q \text{ for } 1 \le j \le m\}.$$

Let $E_j := \{(v_1, \dots, v_m) \in E : v_j \ge q\}$ for $1 \le j \le m$. The set B is clearly in bijection with $E \setminus (E_1 \cup \cdots \cup E_m)$. It is elementary and well known that $|E| = \binom{m+r}{m}$. By changing v_j to $v_j' = v_j - q$, we also see that $|E_j| = \binom{m+r-q}{m}$ for $1 \le j \le m$, and more generally, $|E_{j_1} \cap \cdots \cap E_{j_i}| = \binom{m+r-iq}{m}$ for $1 \le j_1 < \cdots < j_i \le m$. It follows that dim $\mathsf{RM}_q(r,m) =$ dim $V_q(r,m) = |B|$, and this is equal to

$$\begin{aligned} |E| - |E_1 \cup \cdots \cup E_m| &= \binom{m+r}{m} - \sum_{i=1}^{m} (-1)^{i-1} \sum_{1 \le j_1 < \cdots < j_i \le m} |E_{j_1} \cap \cdots \cap E_{j_i}| \\ &= \binom{m+r}{m} - \sum_{i=1}^{m} (-1)^{i-1} \binom{m}{i}\binom{m+r-iq}{m}. \end{aligned}$$

The last expression is clearly equal to the desired formula in (3).

Remark 1 In case $0 \le r < q$, formula (3) simplifies to $\dim \mathsf{RM}_q(r,m) = \binom{m+r}{m}$. This can also be seen by noting that the set E_j in the proof above is empty for each $j = 1, \dots, m$ when $r < q$. On the other hand, if $r = m(q-1)$, then the map Ev given by (1) is also surjective. To see this, write $\mathsf{P}_\nu = (a_{\nu 1}, \dots, a_{\nu m})$ and consider

$$F_\nu(X_1, \dots, X_m) := \prod_{j=1}^{m} \left(1 - (X_j - a_{\nu j})^{q-1}\right) \quad \text{for } \nu = 1, \dots, q^m. \tag{4}$$

Note that for any $\nu \in \{1, \dots, q^m\}$, the polynomial F_ν is in $V_q(m(q-1), m)$ and it has the property that $F_\nu(\mathsf{P}_\nu) = 1$ and $F_\nu(\mathsf{P}_\mu) = 0$ for any $\mu \in \{1, \dots, q^m\}$ with $\mu \ne \nu$. Hence any $\lambda = (\lambda_1, \dots, \lambda_{q^m}) \in \mathbb{F}_q^{q^m}$ can be written as $\lambda = \mathrm{Ev}(F)$, where $F = \lambda_1 F_1 + \cdots + \lambda_{q^m} F_{q^m}$. It follows that $\mathsf{RM}_q(m(q-1), m) = \mathbb{F}_q^{q^m}$. In particular, Lemma 1 yields the following curious identity:

$$\sum_{i=0}^{m} (-1)^i \binom{m}{i}\binom{(m-i)q}{m} = q^m \quad \text{or equivalently,} \quad \sum_{i=0}^{m} (-1)^i \binom{m}{i}\binom{iq}{m} = (-q)^m.$$

It may be interesting to obtain a direct proof of the above identity.

We now recall the following important result about the minimum distance and the minimum weight codewords of Reed-Muller codes.

Proposition 1 *Let m, r be integers such that $m \geq 1$ and $0 \leq r \leq m(q-1)$. Then there are unique $t, s \in \mathbb{N}$ such that*

$$r = t(q-1) + s \quad \text{and} \quad 0 \leq s \leq q-2. \tag{5}$$

With t, s as above, the minimum distance of $\mathsf{RM}_q(r, m)$ is given by

$$d = (q-s)q^{m-t-1}. \tag{6}$$

Further, if $f \in V_q(r, m)$ is given by

$$f(X_1, \ldots, X_m) = \omega_0 \prod_{i=1}^{t} \left(1 - (X_i - \omega_i)^{q-1}\right) \prod_{j=1}^{s} (X_{t+1} - \omega_j') \tag{7}$$

where $\omega_0, \omega_1, \ldots, \omega_t \in \mathbb{F}_q$ with $\omega_0 \neq 0$ and $\omega_1', \ldots, \omega_s'$ are any distinct elements of $\mathbb{F}_q$, then $\mathrm{Ev}(f)$ is a minimum weight codeword of $\mathsf{RM}_q(r, m)$. Moreover, every minimum weight codeword of $\mathsf{RM}_q(r, m)$ is of the form $\mathrm{Ev}(g)$, where g is obtained from a polynomial of the form (7) by substituting for $X_1, \ldots, X_{t+1}$ any $(t+1)$ linearly independent linear forms in $\mathbb{F}_q[X_1, \ldots, X_m]$.

Proof The formula in (6) follows from [6, Theorem 2.6.1] and [15, Theorem 5]. The assertion about the minimum weight codewords is proved in [6, Theorem 2.6.3] (see also [16, Theorem 1]).

We end this section by observing that the Reed-Muller code $\mathsf{RM}_q(r, m)$ is a particularly nice code when m is small or when r is either very small or very large.

Lemma 2 *Let m, r be integers such that $m \geq 1$ and $0 \leq r \leq m(q-1)$. Then $\mathsf{RM}_q(r, m)$ is an MDS code in each of the following cases: (i) $m = 1$, (ii) $r = 0$, (iii) $r = m(q-1)$, and (iv) $r = m(q-1) - 1$.*

Proof (i) If $0 \leq r < q$, then in view of Remark 1 and Proposition 1, we see that $\mathsf{RM}_q(r, 1)$ is a $[q,\ r+1,\ q-r]_q$-code, and hence it is an MDS code.

(ii) Clearly, $\mathsf{RM}_q(0, m)$ is the one-dimensional code of length q^m spanned by the all-1 vector, and this is evidently an MDS code.

(iii) From Remark 1, $\mathsf{RM}_q(m(q-1), m) = \mathbb{F}_q^{q^m}$, which is obviously an MDS code.

(iv) Suppose $r = m(q-1) - 1$. We will show that

$$\mathsf{RM}_q(r, m) = \Lambda, \ \text{ where } \Lambda := \left\{(\lambda_1, \ldots, \lambda_{q^m}) \in \mathbb{F}_q^{q^m} : \lambda_1 + \cdots + \lambda_{q^m} = 0\right\}. \tag{8}$$

This would imply that $\mathsf{RM}_q(r, m)$ is a $[q^m, q^m - 1, 2]_q$-code, and hence an MDS code. To prove (8), first note that the monomial $X_1^{q-1} \cdots X_m^{q-1}$ is in $V_q(m(q-1), m)$, but not in the subspace $V_q(r, m)$. Since we have seen in Remark 1 that Ev gives an isomorphism of $V_q(m(q-1), m)$ onto $\mathbb{F}_q^{q^m}$, it follows that $\dim_{\mathbb{F}_q} V_q(r, m) \leq q^m - 1$. Hence it suffices to show that $\Lambda \subseteq \mathsf{RM}_q(r, m)$. To this end, we assume without loss of generality that the ordering $\mathsf{P}_1, \ldots, \mathsf{P}_{q^m}$ of points of $\mathbb{F}_q^m$ is such that P_1 is the origin. For $1 \le \nu \le q^m$, consider the polynomial F_ν given by (4), and write

$$F_\nu = F_1 + G_\nu, \quad \text{where} \quad F_1 = \prod_{j=1}^{m} \left(1 - X_j^{q-1}\right) \quad \text{and} \quad G_\nu := F_\nu - F_1.$$

Note that $G_\nu \in V_q(r, m)$ for each $\nu = 1, \ldots, q^m$. Also, $F_1(\mathsf{P}_1) = 1$ and $F_1(\mathsf{P}_\mu) = 0$ for $2 \le \mu \le q^m$. So in view of the properties of F_ν noted in Remark 1, we see that $G_1(\mathsf{P}_1) = 0$ while $G_\nu(\mathsf{P}_1) = -1$ and $G_\nu(\mathsf{P}_\nu) = 1$ for $2 \le \nu \le q^m$, and moreover, $G_\nu(\mathsf{P}_\mu) = 0$ for $2 \le \nu, \mu \le q^m$ with $\nu \ne \mu$. Thus given any $\lambda = (\lambda_1, \ldots, \lambda_{q^m}) \in \Lambda$, the polynomial $G := \sum_{\nu=1}^{q^m} \lambda_\nu G_\nu \in V_q(r, m)$ and $\mathrm{Ev}(G) = \lambda$. This proves (8).

Remark 2 In [8, pp. 8–9], the results in Lemma 2, especially (iv), were deduced by appealing to the structure of duals of Reed-Muller codes. Here we have chosen to give a more direct and elementary proof. We remark also that the converse of the result in Lemma 2 is true. An indirect proof of this is given later; see Corollary 1.

3 Characterizations of Purity

Let $n, k \in \mathbb{N}$ with $1 \le k \le n$ and let C be an $[n, k]_q$-code. We have explained in the introduction how one can associate an abstract simplicial complex Δ_C to C. Note that this complex is independent of the choice of a parity check matrix of C. Let $R := \mathbb{F}_q[x_1, \ldots, x_n]$ denote the polynomial ring in n variables over $\mathbb{F}_q$ and let I_C denote the ideal of R generated by the monomials $x_{i_1} \cdots x_{i_t}$ where $\{i_1, \ldots, i_t\}$ vary over nonfaces, i.e., over subsets of $[n] := \{1, \ldots, n\}$ that are not in Δ_C. The Stanley-Reisner ring R_C corresponding to Δ_C (with the base field[1] $\mathbb{F}_q$) is, by definition, the quotient R/I_C. We call R_C the *Stanley-Reisner ring* associated to C. Clearly, R_C is a finitely generated standard graded $\mathbb{F}_q$-algebra and as noted in [8, Sect. 1], R_C is Cohen-Macaulay and it admits an $\mathbb{N}$-graded minimal free resolution of the form

$$F_k \longrightarrow F_{k-1} \longrightarrow \cdots \longrightarrow F_1 \longrightarrow F_0 \longrightarrow R_\Delta \longrightarrow 0 \tag{9}$$

where $F_0 = R$ and each F_i is a graded free R-module of the form

$$F_i = \bigoplus_{j \in \mathbb{Z}} R(-j)^{\beta_{i,j}} \quad \text{for } i = 0, 1, \ldots, k. \tag{10}$$

---

[1] It is only for the sake of definitiveness that we take the base field to be $\mathbb{F}_q$. We could in fact replace $\mathbb{F}_q$ by an arbitrary field. Indeed, it is known that for Stanely-Reisner rings associated with linear codes, and more generally, matroids, the Betti numbers are independent of the choice of a base field; see, e.g., [11, Remark 1]. On the other hand, there are examples of simplicial complexes for which the Betti numbers of their Stanely-Reisner rings do depend on the choice of the base field even when the complex is shellable (see, e.g., [19, Examples 3.3, 3.4]) or stronger still, vertex decomposable (see, e.g., [5, p. 567]).

The nonnegative integers $\beta_{i,j}$ thus obtained are called the *Betti numbers* of C. The resolution (9) is said to be *pure* of type $(d_0, d_1, \ldots, d_k)$ if for each $i = 0, 1, \ldots, k$, the Betti number $\beta_{i,j}$ is nonzero if and only if $j = d_i$. If, in addition, $d_1, \ldots, d_k$ are consecutive, then the resolution is said to be *linear*. We remark that the Betti numbers $\beta_{i,j}$ as well as the properties of purity and linearity depend only on C and they are independent of the choice of a minimal free resolution of R_C.

The result below is due to Johnsen and Verdure [11]; see also [8, Corollary 3.9].

Proposition 2 *Let C be an $[n, k]_q$-code. Then C is an MDS code if and only if C is nondegenerate and every $\mathbb{N}$-graded minimal free resolution of R_C is linear.*

We will now recall the intrinsic characterization of purity given in [8] and alluded to in the Introduction. But first, we review some relevant terminology about codes.

Let n, k and C be as above. By a *subcode* of C we mean a $\mathbb{F}_q$-linear subspace of C. Given a subcode D of C, the *support* of D and the *weight* of D are defined by

$$\mathrm{Supp}(D) := \{i \in [n] : \exists (c_1, \ldots, c_n) \in D \text{ with } c_i \neq 0\} \quad \text{and} \quad \mathrm{wt}(D) := |\mathrm{Supp}(D)|.$$

Given any $c \in C$, we often denote by $\mathrm{Supp}(c)$ and $\mathrm{wt}(c)$ the support of $\langle c \rangle$ and the weight of $\langle c \rangle$, respectively, where $\langle c \rangle$ denotes the subcode of C spanned by c. For $1 \leq i \leq k$, the ith *generalized Hamming weight* of C is defined by

$$d_i(C) := \min\{\mathrm{wt}(D) : D \text{ a subcode of } C \text{ with } \dim D = i\}.$$

It is obvious that $d_1(C) = d(C)$ and it is well known that $d_i(C) < d_{i+1}(C)$ for $1 \leq i \leq k - 1$; see, e.g., [20, Theorem 1]. Note that C is nondegenerate if and only if $d_k(C) = n$. An i-dimensional subcode D of C is said to be *i-minimal* if its support is minimal among the supports of all i-dimensional subcodes of C, i.e., $\mathrm{Supp}(D') \not\subseteq \mathrm{Supp}(D)$ for any i-dimensional subcode D' of C, with $D' \neq D$.

We are now ready to state (an equivalent version of) the intrinsic characterization of purity given in [8, Theorem 3.6].

Proposition 3 *Let C be an $[n, k]_q$-code and let $d_1 < \cdots < d_k$ be its generalized Hamming weights. Also, let R_C be the Stanley-Reisner ring associated to C. Then every $\mathbb{N}$-graded minimal free resolution of R_C is not pure if and only if for some $i \in \{1, \ldots, k\}$, there exists an i-minimal subcode D_i of C such that $\mathrm{wt}(D_i) > d_i$.*

We summarize below the results in [8] about the purity and nonpurity of graded minimal free resolutions of Stanley-Reisner rings associated to Reed-Muller codes.

Proposition 4 *Let m, r be integers such that $m \geq 1$ and $0 \leq r \leq m(q-1)$. Also, let t, s be unique nonnegative integers satisfying* (5). *Then every $\mathbb{N}$-graded minimal free resolution of the Stanley-Reisner ring associated to $\mathsf{RM}_q(r, m)$ is*

(i) pure if $r = 1$,
(ii) not pure if $q = 2$, $m \geq 4$, and $1 < r \leq m - 2$, and
(iii) not pure if $m \geq 2$, $1 < r < m(q-1) - 1$, and $s \neq 1$.

Proof The assertion in (i) is proved in [8, Theorem 4.1], while the assertions in (ii) and (iii) are proved in [8, Proposition 4.4] and [8, Theorem 4.11], respectively.

The values of q, m, r not covered by (i)–(iv) in Lemma 2 and (i)–(iii) in Proposition 4 are precisely $q \geq 3$, $m \geq 2$, and $r = q,\ 2q-1, \ldots, (m-1)q-(m-2)$, except that $(m-1)q-(m-2)$ is excluded if $q = 3$. This is taken care of by the following.

Lemma 3 *Let m, r be integers such that $m \geq 2$ and $1 < r < m(q-1)-1$. Also let t, s be unique integers satisfying* (5)*. Assume that $q \geq 3$ and also that $s = 1$. Then every $\mathbb{N}$-graded minimal free resolution of the Stanley-Reisner ring associated to the Reed-Muller code $\mathsf{RM}_q(r, m)$ is not pure.*

Proof The conditions on m, r and our assumptions imply that $1 \leq t \leq m-1$ and moreover if $q = 3$, then $1 \leq t \leq m-2$. Also note that by Proposition 1, the minimum distance of $\mathsf{RM}_q(r, m)$ is given by $d = (q-1)q^{m-t-1}$. We will divide the proof into two cases according to $q > 3$ and $q = 3$.

Case 1. $q > 3$.

Write $\mathbb{F}_q = \{\omega_1, \ldots, \omega_q\}$, and let ω_1', ω_2' be two distinct elements of $\mathbb{F}_q$. Define

$$Q(X_1, \ldots, X_m) := \left(\prod_{i=1}^{t-1}(X_i^{q-1}-1)\right)\left(\prod_{j=3}^{q}(X_t-\omega_j)\right)\left(\prod_{k=1}^{2}(X_{t+1}-\omega_k')\right).$$

Then $\deg(Q) = (t-1)(q-1)+(q-2)+2 = (t-1)(q-1)+q = t(q-1)+1 = r$, and thus $Q \in V_q(r, m)$. For $i = 1, 2$, let

$$A_i := \left\{\mathbf{a} = (a_1, \ldots, a_m) \in \mathbb{F}_q^m : a_1 = \cdots = a_{t-1} = 0,\ a_t = \omega_i \text{ and } a_{t+1} \notin \{\omega_1', \omega_2'\}\right\}.$$

Then $\mathrm{Supp}(c_Q) = A_1 \cup A_2$. Observe that A_1 and A_2 are disjoint. Consequently,

$$\mathrm{wt}(c_Q) = 2(q-2)q^{m-t-1} \quad \text{and therefore} \quad \mathrm{wt}(c_Q) > d = (q-1)q^{m-t-1},$$

where the last inequality follows since $q > 3$. Thus c_Q is not a minimum weight codeword. If the one-dimensional subcode $\langle c_Q \rangle$ is 1-minimal, then Proposition 3 would imply the desired result. Suppose $\langle c_Q \rangle$ is not 1-minimal. Then there is $F \in V_q(r, m)$ such that $\mathrm{Supp}(c_F) \subsetneq \mathrm{Supp}(c_Q)$ and $\langle c_F \rangle$ is 1-minimal. If c_F is not a minimum weight codeword of $\mathsf{RM}_q(d, m)$, then again Proposition 3 implies the desired result. Thus, suppose c_F is a minimum weight codeword of $\mathsf{RM}_q(d, m)$. By Proposition 1, F must be of the form

$$F(X_1, \ldots, X_m) = \omega_0 \left(\prod_{i=1}^{t}(1-L_i^{q-1})\right)(L_{t+1}-\omega) \tag{11}$$

for some $\omega_0, \omega \in \mathbb{F}_q$ with $\omega_0 \neq 0$ and some linearly independent linear polynomials $L_1, \ldots, L_{t+1}$ in $\mathbb{F}_q[X_1, \ldots, X_m]$, with L_{t+1} homogeneous (while $L_1, \ldots, L_t$ are not necessarily homogeneous). Note that $\mathrm{Supp}(c_F) = A'$, where

$$A' := \left\{\mathbf{a} = (a_1, \ldots, a_m) \in \mathbb{F}_q^m : L_i(\mathbf{a}) = 0 \text{ for } 1 \le i \le t \text{ and } L_{t+1}(\mathbf{a}) \ne \omega\right\}. \tag{12}$$

Since $\mathrm{Supp}(c_F) \subset \mathrm{Supp}(c_Q)$, we obtain $A' \subset A_1 \cup A_2$. We now assert that A' is disjoint from one of the A_i. Indeed, if the assertion is not true, then we can choose $P_i \in A' \cap A_i$ for $i = 1, 2$. Write $b_i := L_{t+1}(P_i)$ for $i = 1, 2$. Since $P_i \in A'$, we see that $b_i \ne \omega$ for $i = 1, 2$. Now pick $\lambda \in \mathbb{F}_q$ such that $\lambda \ne 0, 1$ and $(1-\lambda)b_1 + \lambda b_2 \ne \omega$, which is possible because $q \ge 4$.[2] Define $P_\lambda := (1-\lambda)P_1 + \lambda P_2$. Then $P_\lambda \in A'$, and this contradicts the inclusion $A' \subset A_1 \cup A_2$ because the tth coordinate of P_λ is neither ω_1 nor ω_2. This proves the above assertion. Thus $\mathrm{Supp}(c_F) = A' \subseteq A_i$ for some i. But then $(q-1)q^{m-t-1} \le (q-2)q^{m-t-1}$, which is a contradiction. This proves the claim and hence the desired result when $q > 3$.

Case 2. $q = 3$.

In this case $1 \le t \le m-2$, as noted earlier. Write $\mathbb{F}_q = \{\omega_1, \omega_2, \omega_3\}$. Define

$$Q(X_1, \ldots, X_m) := \left(\prod_{i=1}^{t-1}(X_i^{q-1} - 1)\right)(X_t - \omega_3)(X_{t+1} - \omega_3)(X_{t+2} - \omega_3).$$

Then $\deg(Q) = (t-1)(q-1) + 3 = t(q-1) + 1 = r$, since $q = 3$, and so $Q \in V_q(r, m)$. Let $E := \left\{\mathbf{a} = (a_1, \ldots, a_m) \in \mathbb{F}_q^m : a_1 = \cdots = a_{t-1} = 0\right\}$, and for $i = 1, 2$, let

$$\begin{aligned}
A_i &:= \{\mathbf{a} = (a_1, \ldots, a_m) \in E : a_t = \omega_i \text{ and } a_{t+1}, a_{t+2} \in \{\omega_1, \omega_2\}\},\\
A_i' &:= \{\mathbf{a} = (a_1, \ldots, a_m) \in E : a_{t+1} = \omega_i \text{ and } a_t, a_{t+2} \in \{\omega_1, \omega_2\}\}, \text{ and}\\
A_i'' &:= \{\mathbf{a} = (a_1, \ldots, a_m) \in E : a_{t+2} = \omega_i \text{ and } a_t, a_{t+1} \in \{\omega_1, \omega_2\}\}.
\end{aligned}$$

Then $\mathrm{Supp}(c_Q) = A_1 \cup A_2 = A_1' \cup A_2' = A_1'' \cup A_2''$ and $\mathrm{wt}(c_Q) = 2^3 q^{m-t-2}$. Note that $\mathrm{wt}(c_Q) > (q-1)q^{m-t-1}$, since $q = 3$. Thus, as in Case 1, it suffices to show that there does not exist any $F \in V_q(r, m)$ such that c_F is a minimum weight codeword and $\mathrm{Supp}(c_F) \subsetneq \mathrm{Supp}(c_Q)$. Suppose, if possible, there is such F. Then it must be of the form (11), and its support is given by the set A' in (12). Now write $\mathbb{F}_q \setminus \{\omega\} = \{u_1, u_2\}$, and for $i = 1, 2$, let

$$B_i := \left\{\mathbf{a} = (a_1, \ldots, a_m) \in \mathbb{F}_q^m : L_i(\mathbf{a}) = 0 \text{ for } 1 \le i \le t \text{ and } L_{t+1}(\mathbf{a}) = u_i\right\}.$$

Note that each B_i is an affine space (i.e., a translate of a linear subspace) in $\mathbb{F}_q^m$ and $\mathrm{Supp}(c_F) = B_1 \cup B_2$. Thus $B_1 \cup B_2 \subset A_1 \cup A_2$. We claim that $B_1 \subseteq A_i$ for some $i \in \{1, 2\}$. Indeed, if this is not true, then we can find $P_i \in B_1 \cap A_i$ for each $i = 1, 2$. Since $q = 3$, we can choose $\lambda \in \mathbb{F}_q$ such that $\lambda \ne 0, 1$. Consider $P_\lambda := (1-\lambda)P_1 + \lambda P_2$. Since B_1 is an affine space, $P_\lambda \in B_1$. On the other hand, the tth

---

[2] If $b_1 = b_2$, then the only condition on λ is that $\lambda \ne 0, 1$, whereas if $b_1 \ne b_2$, then it suffices to choose $\lambda \in \mathbb{F}_q$ such that $\lambda \ne 0, 1$ and $\lambda \ne (\omega - b_1)/(b_2 - b_1)$.

coordinate of P_λ is neither ω_1 nor ω_2, and hence $P_\lambda \notin A_1 \cup A_2$. This contradicts the inclusion $B_1 \subset A_1 \cup A_2$, and so the claim is proved. In a similar manner, we see that $B_1 \subseteq A'_j$ and $B_1 \subseteq A''_k$ for some $j, k \in \{1, 2\}$. It follows that $B_1 \subseteq A_i \cap A'_j \cap A''_k$. But clearly, $|B_1| = q^{m-t-1}$ and $|A_i \cap A'_j \cap A''_k| = q^{m-t-2}$. So we obtain $q^{m-t-1} \leq q^{m-t-2}$, which is a contradiction. This completes the proof.

We are now ready to prove the main result of this article.

Theorem 1 *Let $m, r \in \mathbb{N}$ be such that $m \geq 1$ and $0 \leq r \leq m(q-1)$. Then every $\mathbb{N}$-graded minimal free resolution of the Stanley-Reisner ring associated to the Reed-Muller code $\mathsf{RM}_q(r, m)$ is pure if and only if $m = 1$ or $r \leq 1$ or $r \geq m(q-1) - 1$.*

Proof Follows from Lemma 2, Propositions 2, 4, and Lemma 3.

As an application, we show that the converse of the result in Lemma 2 is true.

Corollary 1 *Let $m, r \in \mathbb{N}$ be such that $m \geq 1$ and $0 \leq r \leq m(q-1)$. Then the Reed-Muller code $\mathsf{RM}_q(r, m)$ is an MDS code if and only if $m = 1$ or $r = 0$ or $r \geq m(q-1) - 1$.*

Proof If $m = 1$ or $r = 0$ or $r \geq m(q-1) - 1$, then by Lemma 2, $\mathsf{RM}_q(r, m)$ is an MDS code. Conversely, suppose $\mathsf{RM}_q(r, m)$ is an MDS code. Then by Proposition 2, every $\mathbb{N}$-graded minimal free resolution of its Stanley-Reisner ring is pure. So by Theorem 1, we must have $m = 1$ or $r \leq 1$ or $r \geq m(q-1) - 1$. If $m \geq 2$, then the case $r = 1$ is ruled out because by [8, Theorem 4.1], the generalized Hamming weights (which coincide with the "shifts" in the resolution) of $\mathsf{RM}_q(1, m)$ are given by $d_i = q^m - \lfloor q^{m-i} \rfloor$ for $1 \leq i \leq m+1$, and these are clearly nonconsecutive if $m \geq 2$, and so by Proposition 2, $\mathsf{RM}_q(1, m)$ cannot be an MDS code if $m \geq 2$. Thus we must have $m = 1$ or $r = 0$ or $r \geq m(q-1) - 1$.

Acknowledgements During the course of this work, Sudhir Ghorpade was partially supported by DST-RCN grant INT/NOR/RCN/ICT/P-03/2018 from the Dept. of Science & Technology, Govt. of India, MATRICS grant MTR/2018/000369 from the Science & Engineering Research Board, and IRCC award grant 12IRAWD009 from IIT Bombay.
Rati Ludhani is supported by Prime Minister's Research Fellowship PMRF-192002-256 at IIT Bombay.

References

1. Assmus Jr., E.F., Key, J.D.: Designs and Their Codes. Cambridge Tracts in Mathematics, vol. 103. Cambridge University Press (1992)
2. Beelen, P., Datta, M.: Generalized Hamming weights of affine cartesian codes. Finite Fields Appl. **51**, 130–145 (2018)
3. Beelen, P.: A note on the generalized Hamming weights of Reed-Muller codes. Appl. Algebra Eng. Commun. Comput. **30**, 233–242 (2019)
4. Berlekamp (Ed.), E.R.: Key Papers in the Development of Coding Theory. IEEE Press, New York (1974)

5. Dalili, K., Kummini, M.: Dependence of Betti numbers on characteristic. Commun. Algebra **42**, 563–570 (2014)
6. Delsarte, P., Goethals, J.M., MacWilliams, F.J.: On generalized Reed-Muller codes and their relatives. Inf. Control **16**, 403–442 (1970)
7. Ghorpade, S.R.: A note on Nullstellensatz over finite fields in Contributions in Algebra and Algebraic Geometry. Contemporary Mathematics, vol. 738, pp. 23–32. American Mathematical Society, Providence (2019)
8. Ghorpade, S.R., Singh, P.: Pure Resolutions, linear codes, and Betti numbers. J. Pure Appl. Algebra **224**(10), Art. 106385, 22 pp. (2020)
9. Heijnen, P., Pellikaan, R.: Generalized Hamming weights of q-ary Reed-Muller codes. IEEE Trans. Inf. Theory **44**, 181–196 (1998)
10. Herzog, J., Kühl, M.: On the Betti numbers of finite pure and linear resolutions. Commun. Algebra **12**, 1627–1646 (1984)
11. Johnsen, T., Verdure, H.: Hamming weights and Betti numbers of Stanley-Reisner rings associated to matroids. Appl. Algebra Eng. Commun. Comput. **24**, 73–93 (2013)
12. Johnsen, T., Verdure, H.: Stanley-Reisner resolution of constant weight codes. Des. Codes Cryptogr. **72**, 471–481 (2014)
13. Johnsen, T., Roksvold, J., Verdure, H.: A generalization of weight polynomials to matroids. Discret. Math. **339**, 632–645 (2016)
14. Jurrius, R., Pellikaan, R.: Extended and generalized weight enumerators. In: Proceedings of International Workshop on Coding and Cryptography WCC 2009, pp. 76–91. Selmer Center, Univ. Bergen, Norway (2009)
15. Kasami, T., Lin, S., Peterson, W.W.: New generalization of the Reed-Muller codes-part I: primitive codes. IEEE Trans. Inf. Theory **14**, 189–199 (1968)
16. Leducq, E.: A new proof of Delsarte, Goethals and Mac Williams theorem on minimal weight codewords of generalized Reed-Muller codes. Finite Fields Appl. **18**, 581–586 (2012)
17. Muller, D.E.: Application of Boolean algebra to switching circuit design and to error detection. IRE Trans. Electron. Comput. **EC-3**(3), 6–12 (1954)
18. Reed, I.S.: A class of multiple-error-correcting codes and the decoding scheme. IRE Trans. Inf. Theory PGIT-**4**, 38–49 (1954)
19. Terai, N., Hibi, T.: Some results on Betti numbers of Stanley-Reisner rings. Discret. Math. **157**, 311–320 (1996)
20. Wei, V.K.: Generalized Hamming weights for linear codes. IEEE Trans. Inf. Theory **37**, 1412–1418 (1991)

Skew Constacyclic Codes over $F_q + vF_q$

Ghulam Mohammad, Naim Khan, and Almas Khan

Abstract In this paper, we study v-skew constacyclic codes over the ring $F_q + vF_q$, where $v^2 = 1$, $q = p^m$ and p is an odd prime. We obtain the structural properties of v-skew constacyclic codes over $F_q + vF_q$ using decomposition method. The generator polynomials of v-skew constacyclic codes and their dual codes over R are obtained. Moreover, some examples of v-skew constacyclic codes over $F_q + vF_q$ have also been constructed.

Keywords Gray map · Skew polynomial rings · Skew constacyclic codes

1 Introduction

It has been given the attention to the study of linear and cyclic codes over finite rings because of their new role in algebraic coding theory and their successful applications in information theory, communication, electrical engineering and computer science. The class of cyclic codes is a very important class of linear codes from both theoretical and practical point of view which are easier to implement due to their rich algebraic structure. Cyclic codes have been studied for the past six decades. Based on these facts, cyclic codes have become one of the most important class in coding theory. Hammons et al. in a landmark paper [12] showed that some good nonlinear codes over $\mathbb{Z}_2$ can be obtained as binary images under the Gray map of linear cyclic codes over $\mathbb{Z}_4$. But all this work is restricted to codes that are defined in a commutative ring.

Boucher et al. [7] gave the study of skew cyclic codes and their structural properties over a noncommutative ring $F[x, \theta]$, called skew polynomial ring, where F is a finite field and θ is a field automorphism of F. They gave the generalization of the class of linear and cyclic codes to the class of skew cyclic codes by using the ring $F[x, \theta]$, where the generator polynomials of skew cyclic codes come from the ring

---

G. Mohammad (✉) · N. Khan · A. Khan
Department of Mathematics, Aligarh Muslim University, Aligarh, Uttar Pradesh 202002, India
e-mail: mohdghulam202@gmail.com

M. Ashraf et al. (eds.), *Algebra and Related Topics with Applications*,
Springer Proceedings in Mathematics & Statistics 392,
https://doi.org/10.1007/978-981-19-3898-6_27

$F[x, \theta]$. They also provided some examples of skew cyclic codes with Hamming distances larger than the best-known linear codes with the same parameters. After that, Bhaintwal [6] studied skew quasi-cyclic codes over Galois rings. Since polynomials in skew polynomial rings have many factorizations and hence there are many ideals in skew polynomial ring than in the commutative ring, this was the main aim of studying codes in this setting. But all this work has a restriction to the condition that the order of the automorphism must be a factor of the length n of the code. Siap et al. in [14] removed this restriction and studied the structural properties of these classes of codes of arbitrary length over finite fields. A lot of work has been done in this direction (see Refs. [1, 3, 10]).

Jitman et al. [13] studied skew constacyclic codes by defining the skew polynomial ring with coefficients from finite chain rings, especially the ring $F_{p^m} + uF_{p^m}$, where $u^2 = 0$. Further, Gursoy et al. [11] derived the constructional properties of skew cyclic codes by using the decomposition method over $F_q + vF_q$, where $v^2 = v$ and $q = p^m$. Later on, the authors in [3] obtain the constructional properties of skew cyclic codes over the ring $F_3 + vF_3$ with $v^2 = 1$ by taking the automorphism as $\theta : v \mapsto -v$. They showed that skew cyclic codes over $F_3 + vF_3$ are equivalent to either cyclic codes or quasi-cyclic codes. Further, they studied skew cyclic codes over the ring $F_q + vF_q$ with $v^2 = 1$ in [5] by using decomposition method. Very recently, Al-Ashker and Abu-Jafar [2] investigated the structural properties of skew constacyclic codes over the ring $F_p + vF_p$ with $v^2 = v$. Motivated by the study of Al-Ashker and Abu-Jafar [2], in the present paper, for the first time, we study v-skew constacyclic codes over the ring $F_q + vF_q$, where $v^2 = 1$, $q = p^m$ and p is an odd prime.

2 Preliminaries

Let $R = F_q + vF_q$ where $q = p^m$ and p is an odd prime. Then R is a commutative and nonchain ring with characteristic p which contains q^2 elements. The ring is endowed with the natural addition and multiplication with the property $v^2 = 1$ and it can be viewed as the quotient ring $F_q[v]/\langle v^2 - 1\rangle$. The elements of R can be uniquely written as $a + vb$, where $a, b \in F_q$. It is a semi-local ring having two maximal ideals $\langle 1 - v\rangle$ and $\langle 1 + v\rangle$.

Let $\theta : F_q \longrightarrow F_q$ be the Frobenius automorphism defined as $\theta(a) = a^p$. Define a mapping $\theta_t : F_q \longrightarrow F_q$ such that $\theta_t(a) = a^{p^t}$ for all $a \in F_q$. One can verify that θ_t is an automorphism on F_q and $\theta_t = \theta^t$. It can be observed that the order of θ_t is $|\langle\theta_t\rangle| = m/t$ and the subring F_{p^t} of F_q is invariant under θ_t.

Definition 1 Let α be the given automorphism of R defined by $\alpha(a + vb) = a^{p^t} + vb^{p^t}$ for all $a, b \in F_q$. The set $R[x, \alpha] = \{a_0 + a_1x + a_2x^2 + \cdots + a_nx^n |\ a_i \in R, n \geq 0\}$ of formal polynomials under usual addition of polynomials and multiplication defined by the rule $(ax^i)(bx^j) = a\alpha^i(b)x^{i+j}$ forms a ring. The ring $R[x, \alpha]$ is called skew polynomial ring over R.

It may be observed that $R[x, \alpha]$ is noncommutative ring unless α is the identity automorphism on R. Therefore, when an ideal of $R[x, \alpha]$ is taken, one should clarify whether it is a left ideal or a right ideal. The skew polynomial ring $R[x, \alpha]$ is not left or right Euclidean. However, the division algorithm holds for some polynomials whose leading coefficients are invertible (for detail see Refs. [8] and [13]).

3 Gray Map and Linear Codes over R

Cengellenmis [9] initiated the study of cyclic codes over the ring $F_3 + vF_3$ where $v^2 = 1$. Later on, the authors [4] generalized this study to constacyclic codes over the ring $F_p + vF_p$, where $v^2 = 1$ and p is an odd prime. Let R^n be the set of all n-tuples over the ring R. Then any nonempty subset C of R^n is called a code of length n over R. C is called linear code of length n over R if it is an R-submodule of R^n. Elements of C are called codewords and therefore each codeword c in such a code C is just an n-tuple of the form $c = (c_0, c_1, \dots, c_{n-1}) \in R^n$.

Let C be a linear code of length n over R. Then C is said to be cyclic if for every $(c_0, c_1, \dots, c_{n-1}) \in C$ implies that $(c_{n-1}, c_0, \dots, c_{n-2}) \in C$, negacyclic if $(c_0, c_1, \dots, c_{n-1}) \in C$ implies that $(-c_{n-1}, c_0, \dots, c_{n-2}) \in C$ and v-constacyclic if $(c_0, c_1, \dots, c_{n-1}) \in C$ implies that $(vc_{n-1}, c_0, \dots, c_{n-2}) \in C$.

The Hamming weight $w_H(r)$ of a codeword $r = (r_0, r_1, \dots, r_{n-1})$ is the number of nonzero components. The minimum weight $w_H(C)$ of a code C is the smallest weight among all its nonzero codewords. For $r = (r_0, r_1, \dots, r_{n-1})$, $s = (s_0, s_1, \dots, s_{n-1})$, $d_H(r, s) = |\{i \mid r_i \neq s_i\}|$ is called the Hamming distance between r and s and is denoted by $d_H(r, s) = w_H(r - s)$.

The minimum Hamming distance between distinct pairs of codewords of a code C is called the minimum distance of C and is denoted by $d_H(C)$ or shortly d_H.

Now, we define the Lee weight of an element $r = a + vb \in R$ as follows:

$$w_L(r) = w_H(a, b),$$

where w_H denotes the usual Hamming weight on F_q. Let $r = (r_0, r_1, \dots, r_{n-1})$ be a vector in R^n. Then the Lee weight of r is the rational sum of Lee weights of its components, that is, $w_L(r) = \sum_{i=0}^{n-1} w_L(r_i)$. For any two elements $r, s \in R^n$, the Lee distance is given by $d_L(r, s) = w_L(r - s)$. The minimum Lee distance of a code C is the smallest nonzero Lee distance between all pairs of distinct codewords. The minimum Lee weight of C is the smallest nonzero Lee weight among all codewords. If C is linear, then the minimum Lee distance is the same as the minimum Lee weight.

The Gray map ϕ from R to F_q^2 is defined as $\phi(a + vb) = (a, b)$. It can be easily seen that ϕ is linear. The Gray map ϕ can be extended to R^n in a natural way, that is, $\phi : R^n \longrightarrow F_q^{2n}$ such that $\phi(r_0, r_1, \dots, r_{n-1}) = (a_0, a_1, \dots, a_{n-1}, b_0, b_1, \dots, b_{n-1})$, where $r_i = a_i + vb_i$ for $i = 0, 1, \ldots, n-1$. The map ϕ is a distance-preserving map from (R^n, Lee distance) to (F_q^{2n}, Hamming distance) and it is also F_q-linear.

For a code C over R, let

$$C_{1-v} = \{a \in F_q^n \mid (1+v)a + (1-v)b \in C, \text{ for some } b \in F_q^n\},$$

$$C_{1+v} = \{b \in F_q^n \mid (1+v)a + (1-v)b \in C, \text{ for some } a \in F_q^n\}$$

be two q-ary codes such that $(1+v)C_{1-v}$ is equal to $C\ mod\ (1-v)$ and $(1-v)C_{1+v}$ is equal to $C\ mod\ (1+v)$, respectively. Therefore, any code C over R can be written as $C = (1+v)C_{1-v} \oplus (1-v)C_{1+v}$. According to the generator matrix G, the code C_{1-v} is permutation equivalent to a code with generator matrix of the form

$$\begin{pmatrix} I_{k_1} & 0 & 2A_1 & 2A_2 & 2A_3 \\ 0 & 2I_{k_2} & 0 & 2A_4 & 0 \end{pmatrix}$$

and the code C_{1+v} is permutation equivalent to a code with generator matrix of the form

$$\begin{pmatrix} I_{k_1} & 2B_1 & 0 & 2B_2 & 2B_3 \\ 0 & 0 & 2I_{k_3} & 0 & 2B_4 \end{pmatrix},$$

where A_i, B_j are q-ary matrices with $1 \le i, \ j \le 4$. It is easy to see that $|C_{1-v}||C_{1+v}| = q^{k_1}q^{k_2}q^{k_1}q^{k_3} = q^{2k_1+k_2+k_3} = |C|$ (for details see[15]).

Let $r = (r_0, r_1, \ldots, r_{n-1})$ and $s = (s_0, s_1, \ldots, s_{n-1})$ be two elements of R^n. Then the Euclidean inner product of r and s in R^n is defined as

$$r \cdot s = r_0s_0 + r_1s_1 + \cdots + r_{n-1}s_{n-1}.$$

The dual code $C^{\perp}$ of C is defined as

$$C^{\perp} = \{r \in R^n \mid r \cdot c = 0, \text{ for all } c \in C\}.$$

A code C is called self-orthogonal if $C \subseteq C^{\perp}$ and self dual if $C = C^{\perp}$.

Theorem 1 ([4, Theorem 8]) *Let $C = (1+v)C_{1-v} \oplus (1-v)C_{1+v}$ be a code of length n over R. Then C is a v-constacyclic code if and only if $C^{\perp}$ is also a v-constacyclic code and $C^{\perp} = (1+v)C_{1-v}^{\perp} \oplus (1-v)C_{1+v}^{\perp}$.*

4 v-Skew Constacyclic Codes over R

Skew cyclic codes over the ring R were studied by the authors [5]. In the present section, we generalize this study to v-skew constacyclic codes over R. Let α be an automorphism on R given by $\alpha(a + vb) = a^{p^t} + vb^{p^t}$. Then a linear code C of length n over R is called a skew cyclic code or α-cyclic code if for each $c = (c_0, c_1, \dots, c_{n-1}) \in C$ implies that $\sigma(c) = (\alpha(c_{n-1}), \alpha(c_0), \dots, \alpha(c_{n-2})) \in C$, where $\sigma(c)$ denotes the skew cyclic shift of c, C is called skew negacyclic code if for each $c = (c_0, c_1, \dots, c_{n-1}) \in C$ implies that $\nu(c) = (-\alpha(c_{n-1}), \alpha(c_0), \dots, \alpha(c_{n-2})) \in C$, where $\nu(c)$ denotes the skew negacyclic shift of c and C is called v-skew constacyclic code if for each $c = (c_0, c_1, \dots, c_{n-1}) \in C$ implies that $\tau(c) = (v\alpha(c_{n-1}), \alpha(c_0), \dots, \alpha(c_{n-2})) \in C$, where $\tau(c)$ denotes the skew constacyclic shift of c.

Now, consider $R[x, \alpha]/\langle x^n - v\rangle$. It can be easily seen that $R[x, \alpha]/\langle x^n - v\rangle$ is a left $R[x, \alpha]$ module under the following operations:

$$(f(x) + \langle x^n - v\rangle) + (g(x) + \langle x^n - v\rangle) = (f(x) + g(x)) + \langle x^n - v\rangle,$$

$$r(x)(f(x) + \langle x^n - v\rangle) = r(x)f(x) + \langle x^n - v\rangle$$

for any $r(x) \in R[x, \alpha]$. By the definition of $R[x, \alpha]/\langle x^n - v\rangle$, we can identify each codeword $c = (c_0, c_1, \dots, c_{n-1})$ of the v-skew constacyclic code C by a polynomial $c(x) = c_0 + c_1x + c_2x^2 + \cdots + c_{n-1}x^{n-1}$.

In the following lemma, we give a module structure of v-skew constacyclic codes of arbitrary length.

Lemma 1 *A code C of length n over R is a v-skew constacyclic code if and only if C is a left $R[x, \alpha]$-submodule of $R[x, \alpha]/\langle x^n - v\rangle$.*

Now, we give the characterization of v-skew constacyclic codes over R as follows:

Theorem 2 *If $C = (1 + v)C_{1-v} \oplus (1 - v)C_{1+v}$ is a linear code of length n over R, then C is a v-skew constacyclic code over R with respect to automorphism α if and only if C_{1-v} is a skew cyclic code and C_{1+v} is a skew negacyclic codes of length n over F_q, respectively, with respect to the automorphism θ_t.*

Proof Take $r = (r_0, r_1, \dots, r_{n-1}) \in C$, and we can write its coordinates as $r_i = (1 + v)a_i + (1 - v)b_i$, where $a_i,\ b_i \in F_q$, $0 \le i \le n - 1$. Let $a = (a_0, a_1, \dots, a_{n-1})$ and $b = (b_0, b_1, \dots, b_{n-1})$. Then $a \in C_{1-v}$ and $b \in C_{1+v}$. Now, suppose C_{1-v} is a skew cyclic code and C_{1+v} is a skew negacyclic codes over F_q, respectively, with respect to the automorphism θ_t. This means that

$$\begin{aligned}\sigma(a) &= (\theta_t(a_{n-1}), \theta_t(a_0), \dots, \theta_t(a_{n-2}))\\ &= (a_{n-1}^{p^t}, a_0^{p^t}, \dots, a_{n-2}^{p^t}) \in C_{1-v},\end{aligned}$$

$$\begin{aligned}\nu(b) &= (-\theta_t(b_{n-1}), \theta_t(b_0), \ldots, \theta_t(b_{n-2}))\\ &= (-b_{n-1}^{p^t}, b_0^{p^t}, \ldots, b_{n-2}^{p^t}) \in C_{1+v}.\end{aligned}$$

Thus $(1+v)\sigma(a) + (1-v)\nu(b) \in C$. It is easy to be seen that

$$(1+v)\sigma(a) + (1-v)\nu(b) = \tau(r).$$

Hence $\tau(r) \in C$, which means that C is a v-skew constacyclic code over R with respect to the automorphism α.

Conversely, suppose that C is a v-skew constacyclic code over R with respect to the automorphism α. Let $r_i = (1+v)a_i + (1-v)b_i$, for any $a = (a_0, a_1, \ldots, a_{n-1}) \in C_{1-v}$, $b = (b_0, b_1, \ldots, b_{n-1}) \in C_{1+v}$. Then $r = (r_0, r_1, ..., r_{n-1}) \in C$. By hypothesis $\tau(r) \in C$. Since $(1+v)\sigma(a) + (1-v)\nu(b) = \tau(r)$, we get $(1+v)\sigma(a) + (1-v)\nu(b) \in C$. Thus $\sigma(a) \in C_{1-v}$, $\nu(b) \in C_{1+v}$, which implies that C_{1-v} is a skew cyclic code and C_{1+v} is a skew negacyclic codes of length n over F_q with respect to the automorphism θ_t.

Corollary 1 *If C is a v-skew constacyclic code of length n over R, then the dual code $C^{\perp}$ is also a v-skew constacyclic code of length n over R.*

Proof Using Theorem 1, we have $C^{\perp} = (1+v)C_{1-v}^{\perp} \oplus (1-v)C_{1+v}^{\perp}$. Since the dual code of every skew cyclic and skew negacyclic code over F_q is also a skew cyclic and skew negacyclic code ([8]), by Theorem 2, $C^{\perp}$ is a v-skew constacyclic code.

Lemma 2 *Let τ be the v-skew constacyclic shift of R^n and σ be the skew cyclic shift of F_q^{2n}. If ϕ is the Gray map of R^n into F_q^{2n}, then $\phi\tau = \sigma\phi$.*

Proof Let $r = (r_0, r_1, \ldots, r_{n-1}) \in R^n$, where $r_i = a_i + vb_i$ with $a_i,\ b_i \in F_q$ for $0 \le i \le n-1$. Taking v-skew constacyclic shift of r, we have

$$\begin{aligned}\tau(r) &= (v\theta_t(r_{n-1}), \theta_t(r_0), \ldots, \theta_t(r_{n-2}))\\ &= (v(a_{n-1}^{p^t} + vb_{n-1}^{p^t}), a_0^{p^t} + vb_0^{p^t}, \ldots, a_{n-2}^{p^t} + vb_{n-2}^{p^t})\\ &= (b_{n-1}^{p^t} + va_{n-1}^{p^t}, a_0^{p^t} + vb_0^{p^t}, \ldots, a_{n-2}^{p^t} + vb_{n-2}^{p^t}).\end{aligned}$$

Now, using the definition of Gray map ϕ, we can deduce that

$$\phi(\tau(r)) = (b_{n-1}^{p^t}, a_0^{p^t}, \ldots, a_{n-2}^{p^t}, a_{n-1}^{p^t}, b_0^{p^t}, \ldots, b_{n-2}^{p^t}).$$

On the other hand, $\phi(r) = (a_0, a_1, \ldots, a_{n-1}, b_0, b_1, \ldots, b_{n-1})$. Hence, $\sigma(\phi(r)) = (b_{n-1}^{p^t}, a_0^{p^t}, \ldots, a_{n-2}^{p^t}, a_{n-1}^{p^t}, b_0^{p^t}, \ldots, b_{n-2}^{p^t})$. Therefore, $\phi\tau = \sigma\phi$.

As a consequence of Lemma 2, we get the following main result:

Theorem 3 *Let C be a code of length n over R. Then C is v-skew constacyclic code if and only if $\phi(C)$ is a skew cyclic code of length $2n$ over F_q.*

The following theorem gives the generators polynomials of v-skew constacyclic codes over R:

Theorem 4 *If $C = (1+v)C_{1-v} \oplus (1-v)C_{1+v}$ is a v-skew constacyclic code of length n over R, then $C = \langle (1+v)g_1(x), (1-v)g_2(x) \rangle$ and $|C| = q^{2n-deg(g_1(x))-deg(g_2(x))}$, where $g_1(x)$, $g_2(x)$ are the generator polynomials of C_{1-v}, C_{1+v}, respectively.*

Proof Since $C_{1-v} = \langle g_1(x) \rangle \subseteq F_q[x, \theta_t]/\langle x^n - 1 \rangle$, $C_{1+v} = \langle g_2(x) \rangle \subseteq F_q[x, \theta_t]/\langle x^n + 1 \rangle$ and $C = (1+v)C_{1-v} \oplus (1-v)C_{1+v}$, we find that $C = \{c(x) \mid c(x) = (1+v)f_1(x) + (1-v)f_2(x),\ f_1(x) \in C_{1-v},\ f_2(x) \in C_{1+v}\}$. Therefore,

$$C \subseteq \langle (1+v)g_1(x), (1-v)g_2(x) \rangle \subseteq R[x, \alpha]/\langle x^n - v \rangle.$$

For any

$$(1+v)k_1(x)g_1(x) + (1-v)k_2(x)g_2(x) \in \langle (1+v)g_1(x), (1-v)g_2(x) \rangle \subseteq R[x, \alpha]/\langle x^n - v \rangle,$$

where $k_1(x) \in R[x, \alpha]/\langle x^n - 1 \rangle, k_2(x) \in R[x, \alpha]/\langle x^n + 1 \rangle$, there are $r_1(x), r_2(x) \in F_q[x, \theta_t]$ such that

$$(1+v)k_1(x) = (1+v)r_1(x),\ (1-v)k_2(x) = (1-v)r_2(x).$$

This means that $\langle (1+v)g_1(x), (1-v)g_2(x) \rangle \subseteq C$. Hence $\langle (1+v)g_1(x), (1-v)g_2(x) \rangle = C$. Since $|C| = |C_1||C_2|$, $|C| = q^{2n-deg(g_1(x))-deg(g_2(x))}$.

Theorem 5 *Let C_{1-v} be a skew cyclic code over F_q and C_2 be a skew negacyclic codes over F_q with monic generator polynomials $g_1(x)$ and $g_2(x)$, respectively. If $C = (1+v)C_{1-v} \oplus (1-v)C_{1+v}$ is a v-skew constacyclic code of length n over R, then there is a unique polynomial $g(x) \in R[x, \alpha]$ such that $C = \langle g(x) \rangle$ and $g(x)$ is a right divisor of $x^n - v$, where $g(x) = (1+v)g_1(x) + (1-v)g_2(x)$.*

Proof By Theorem 4, we may assume that

$$C = \langle (1+v)g_1(x), (1-v)g_2(x) \rangle,$$

where $g_1(x)$ and $g_2(x)$ are the monic generator polynomials of C_{1-v} and C_{1+v}, respectively. Let $g(x) = (1+v)g_1(x) + (1-v)g_2(x)$. Clearly, $\langle g(x) \rangle \subseteq C$. Note that

$$(1+v)g_1(x) = (1+v)g(x),$$

and

$$(1-v)g_2(x) = (1-v)g(x),$$

and hence $C \subseteq \langle g(x) \rangle$, that is $C = \langle g(x) \rangle$. Since $g_1(x)$ is a monic right divisor of $x^n - 1$ and $g_2(x)$ is a monic right divisor of $x^n + 1$, there are $r_1(x) \in F_q[x, \theta_t]/\langle x^n -$

$1\rangle$ and $r_2(x) \in F_q[x, \theta_t]/\langle x^n + 1\rangle$ such that

$$x^n - 1 = r_1(x)g_1(x),\ x^n + 1 = r_2(x)g_2(x).$$

This implies that

$$x^n - v = [(1+v)r_1(x) + (1-v)r_2(x)]g(x).$$

Hence, $g(x)|x^n - v$. The uniqueness of $g(x)$ can be followed from that of $g_1(x)$ and $g_2(x)$.

In order to study the generator polynomials of the dual codes of v-skew constacyclic codes over R, we give the following definition:

Definition 2 Let $g(x) = g_0 + g_1x + \cdots + g_rx^r$ and $h(x) = h_0 + h_1x + \cdots + h_{n-r}x^{n-r}$ be polynomials in $R[x, \alpha]$ such that $x^n - v = h(x)g(x)$ and C' be a v-skew constacyclic code generated by $g(x)$ in $R[x, \alpha]/\langle x^n - v\rangle$. Then the dual code of C' is a v-skew constacyclic code generated by the polynomial $\bar{h}(x) = h_{n-r} + \alpha(h_{n-r-1})x + \cdots + \theta_t^{n-r}(h_0)x^{n-r}$.

In view of Theorems 1 and 2, we have the following corollary:

Corollary 2 *Let C_{1-v} be a skew cyclic code over F_q and C_{1+v} be a skew negacyclic codes over F_q and $g_1(x)$ and $g_2(x)$ be their generator polynomials such that*

$$x^n - 1 = h_1(x)g_1(x),\ x^n + 1 = h_2(x)g_2(x) \in F_q[x, \theta_t].$$

If $C = (1+v)C_{1-v} \oplus (1-v)C_{1+v}$, then

$$C^\perp = \langle (1+v)\bar{h_1}(x) + (1-v)\bar{h_2}(x)\rangle$$

and $|C^\perp| = q^{deg(g_1(x))+deg(g_2(x))}$.

Now, we give the following examples in support of our results:

Example 4.1 Let $R = F_9 + vF_9$ be the ring with $v^2 = 1$ and θ be the Frobenius automorphism over F_9, that is, $\theta(r) = r^3$ for any $r \in F_9$, where $F_9 = F_3[\omega],\ \omega^2 + 1 = 0$. Then

$$x^4 - 1 = (x^2 - 1)(x + \omega)(x + 2\omega) \in F_9[x, \theta],$$

$$x^4 + 1 = (2 + x + x^2)(2 + 2x + x^2) \in F_9[x, \theta].$$

If $g_1(x) = x^2 - 1,\ g_2(x) = (2 + x + x^2)$, then $C_{1-v} = \langle g_1(x)\rangle$ is a skew cyclic code over F_9 with parameters [4, 2, 2] and $C_{1+v} = \langle g_2(x)\rangle$ is a skew negacyclic code over F_9 of parameters [4, 2, 2]. Therefore, the code $C = \langle (1+v)g_1(x) + (1-v)g_2(x)\rangle$

is a v-skew constacyclic code of length 4 over R. Further, the Gray image $\phi(C)$ of C is a skew cyclic code over F_9 of length 8.

Example 4.2 Let $R = F_9 + vF_9$ be the ring with $v^2 = 1$ and θ be the Frobenius automorphism over F_9, that is, $\theta(r) = r^3$ for any $r \in F_9$, where $F_9 = F_3[\omega],\ \omega^2 + \omega + 2 = 0$. Then

$$\begin{aligned} x^6 - 1 &= (2 + (2+\omega)x + (1+2\omega)x^3 + x^4)(1 + (2+\omega)x + x^2) \\ &= (2 + x + (2+2\omega)x^2 + x^3)(1 + x + 2\omega x^2 + x^3), \\ x^6 + 1 &= (1 + x^2)^3 \in F_9[x, \theta]. \end{aligned}$$

If $g_1(x) = 2 + (2+\omega)x + (1+2\omega)x^3 + x^4$ and $g_2(x) = (1+x^2)^3$, then $C_{1-v} = \langle g_1(x)\rangle$ is a skew cyclic code of length 6 over F_9, $C_{1+v} = \langle g_2(x)\rangle$ is a skew negacyclic code of length 6 over F_9. Thus the code

$$C = \langle (1+v)g_1(x) + (1-v)g_2(x)\rangle$$

is a v-skew constacyclic code of length 6 over R. Also, the Gray image $\phi(C)$ of C is a skew cyclic code over F_9 of length 12.

Example 4.3 Let $R = F_9 + vF_9$ be the ring with $v^2 = 1$ and θ be the Frobenius automorphism over F_9, that is, $\theta(r) = r^3$ for any $r \in F_9$, where $F_9 = F_3[\omega],\ \omega^2 + 1 = 0$. Then

$$x^9 - 1 = (x+2)^9,\ x^9 + 1 = (x+1)^9 \in F_9[x, \theta].$$

Let $g_1(x) = x + 2,\ g_2(x) = x + 1$. Then $C_{1-v} = \langle g_1(x)\rangle$ is a skew cyclic code of length 9 over F_9 and $C_{1+v} = \langle g_2(x)\rangle$ is a skew negacyclic code of length 9 over F_9. Therefore, the code $C = \langle (1+v)g_1(x) + (1-v)g_2(x)\rangle$ is a v-skew constacyclic code of length 9 over R. Also, the Gray image $\phi(C)$ of C is a skew cyclic code of length 18 over F_9.

Example 4.4 Let $R = F_{25} + vF_{25}$ be the ring with $v^2 = 1$ and θ be the Frobenius automorphism over F_{25}, that is, $\theta(r) = r^5$ for any $r \in F_{25}$, where $F_{25} = F_5[\omega],\ \omega^2 + \omega + 1 = 0$. Then

$$x^4 - 1 = (x+2)(x+3)(x+\omega)(x+\omega+1),\ x^4 + 1 = (x^2-2)(x^2+2) \in F_{25}[x, \theta].$$

Let $g_1(x) = x + 2,\ g_2(x) = x^2 + 2$. Then $C_{1-v} = \langle g_1(x)\rangle$ is a skew cyclic code of parameters $[4, 3, 2]$ over F_{25} and $C_{1+v} = \langle g_2(x)\rangle$ is a skew negacyclic code of parameters $[4, 2, 2]$ over F_{25}. Therefore, the code $C = \langle (1+v)g_1(x) + (1-v)g_2(x)\rangle$ is a v-skew constacyclic code of length 4 over R. Also, the Gray image $\phi(C)$ of C is a skew cyclic code of parameters $[8, 5, 2]$ over F_{25}.

5 Applications

In this section, we give a example of skew cyclic codes and their Gray images over GF(49) using Plotkin Sum. Before giving a example, we first give the definition of Plotkin Sum. Let $C \oplus_P D$ denote the Plotkin Sum of two linear codes C and D, also called $(u|u+v)$ construction, where $u \in C,\ v \in D$. Let $C = (1+v)C_{1-v} \oplus (1-v)C_{1+v}$ be a linear code of length n over R, then its Gray image $\phi(C)$ is none other than $(C_{1-v} \oplus_P C_{1+v})$. We construct skew cyclic codes over GF(49) with some conditions. If C_{1-v} is a [20, 19, 2] code, C_{1+v} is a [20, 18, 3] code, then the Gray image of C has parameters [40, 37, 2] over GF(49).

The following table contains some v-skew constacyclic codes over the ring $R = F_{25} + vF_{25}$, where $v^2 = 1$. First column of the table denotes the length of cyclic codes over R, second and third columns denote the parameters of C_{1-v} and C_{1+v} over F_{25}, respectively, and column four denotes the parameters of the Gray images of v-skew constacyclic codes over R.

n	C_{1-v}	C_{1+v}	$\phi(C)$
10	[10, 9, 2]	[10, 9, 2]	[20, 18, 2]
12	[12, 10, 3]	[12, 10, 2]	[24, 20, 2]
15	[15, 13, 3]	[15, 13, 3]	[30, 26, 3]
22	[22, 20, 2]	[22, 20, 2]	[44, 40, 2]
24	[24, 22, 3]	[24, 20, 3]	[48, 42, 3]
25	[25, 24, 2]	[25, 24, 2]	[50, 48, 2]

6 Conclusion

In this paper, we have studied the structural properties of v-skew constacyclic codes over the ring $F_q + vF_q$ where $v^2 = 1,\ q = p^m$ and p is an odd prime by taking the automorphism $\theta_t : a + vb \mapsto a^{p^t} + vb^{p^t}$. We have proved that the Gray image of a v-skew constacyclic code of length n over $F_q + vF_q$ is a skew cyclic code of length $2n$ over F_q. Further, we have obtained some examples of v-skew constacyclic codes over $F_q + vF_q$. For the future work, it may be interesting to study the other classes of skew constacyclic codes over this ring.

References

1. Abualrub, T., Aydin, N., Seneviratne, P.: On θ-cyclic codes over $F_2 + vF_2$. Aust. J. Combin. **54**, 115–126 (2012)
2. Al-Ashker, M.M., Mahmoud Abu-Jafar, A.Q.: Skew constacyclic codes over $F_p + vF_p$. Palest. J. Math. **5**(2), 96–103 (2016)

3. Ashraf, M., Mohammad, G.: On skew cyclic codes over $F_3 + vF_3$. Int. J. Inf. Coding Theory **2**(4), 218–225 (2014)
4. Ashraf, M., Mohammad, G.: Some constacyclic codes over $F_p + vF_p$. Aligarh Bull. Math. **33**(1–2), 5–14 (2014)
5. Ashraf, M., Mohammad, G.: On skew cyclic codes over a semi-local ring. Discret. Math. Algorithm. Appl. **7**(4), 1550042 (2015)
6. Bhaintwal, M.: Skew quasi cyclic codes over Galois rings. Des. Codes Cryptogr. **62**(1), 85–101 (2012)
7. Boucher, D., Geiselmann, W., Ulmer, F.: Skew cyclic codes. Appl. Algebra Eng. Commun. Comput. **18**(4), 379–389 (2007)
8. Boucher, D., Sole, P., Ulmer, F.: Skew constacyclic codes over Galois ring. Adv. Math. Commun. **2**(3), 273–292 (2008)
9. Cengellenmis, Y.: On the cyclic codes over $F_3 + vF_3$. Int. J. Algebra **4**(6), 253–259 (2010)
10. Gao, J.: Skew cyclic codes over $F_p + vF_p$. J. Appl. Math. Inf. **31**, 337–342 (2013)
11. Gursoy, F., Siap, I., Yildiz, B.: Construction of skew cyclic codes over $F_q + vF_q$. Adv. Math. Commun. **8**, 313–322 (2014)
12. Hammons, A.R., Jr., Kumar, P.V., Calderbank, A.R., Sloane, N.J.A., Sole, P.: The $\mathbb{Z}_4$-linearty of Kerdock, Preparata, Goethals and related codes. IEEE. Trans. Inf. Theory **40**, 301–319 (1994)
13. Jitman, S., Ling, S., Udomkavanich, P.: Skew constacyclic codes over finite chain rings. Adv. Math. Commun. **6**, 29–63 (2012)
14. Siap, I., Abualrub, T., Aydin, N., Seneviratne, P.: Skew cyclic codes of arbitrary length. Int. J. Inf. Coding Theory **2**, 10–20 (2011)
15. Zhu, S., Wang, L.: A class of constacyclic codes over $F_p + vF_p$ and its Gray image. Discret. Math. **311**, 2677–2682 (2011)

Cyclic and LCD Codes over a Finite Commutative Semi-local Ring

Om Prakash, Habibul Islam, and Arindam Ghosh

Abstract For an odd prime p, we obtain algebraic structure of cyclic codes of length n over a finite commutative non-chain semi-local ring $\mathfrak{R} = \mathbb{F}_p[u, v, w]/\langle u^2 - u, v^2 - 1, w^2 - 1, uv - vu, vw - wv, wu - uw\rangle$. These codes of length n can be viewed as principal ideals of the quotient ring $\mathfrak{R}[x]/\langle x^n - 1\rangle$. Here, a Gray map is defined to obtain p-ary quasi-cyclic codes of length $8n$ with index 8 as $\mathbb{F}_p$-images of cyclic codes of length n over $\mathfrak{R}$. Also, we present necessary and sufficient conditions for a cyclic code to be an LCD (linear complementary dual) code over $\mathfrak{R}$. Moreover, it is shown that the Gray image of an LCD code of length n over $\mathfrak{R}$ is an LCD code of length $8n$ over $\mathbb{F}_p$. Finally, a few non-trivial examples are given in support of our derived results.

Keywords Cyclic code · Non-chain ring · Semi-local ring · Gray map · LCD code

1 Introduction

Cyclic codes over finite rings have been received great attention due to a seminal work of Hammons et al. [12] in 1994. As cyclic codes contain some classes of good error-correcting codes, they have been well investigated over finite rings for the last three decades. In 1997, Kanwar and Lopez-Permouth [23] introduced cyclic codes over $\mathbb{Z}_{p^m}$ and proved that a cyclic code of length n with $gcd(n, p) = 1$ is principally generated. They have also determined the structure of self-dual cyclic codes. In 1999, cyclic and self-dual codes over $\mathbb{F}_2 + u\mathbb{F}_2$ are studied [2]. Later, Dinh and Lopez-Permouth [5] obtained algebraic properties of cyclic and negacyclic codes

O. Prakash (✉) · H. Islam · A. Ghosh
Department of Mathematics, Indian Institute of Technology Patna, Patna, Bihar 801 106, India
e-mail: om@iitp.ac.in

H. Islam
e-mail: habibul.pma17@iitp.ac.in

A. Ghosh
e-mail: arindam.pma14@iitp.ac.in

M. Ashraf et al. (eds.), *Algebra and Related Topics with Applications*,
Springer Proceedings in Mathematics & Statistics 392,
https://doi.org/10.1007/978-981-19-3898-6_28

over finite chain rings. Meanwhile, Abualrub and Siap [1] explicitly determined the generator polynomials and minimum spanning sets for the cyclic codes of length n over $\mathbb{Z}_2 + u\mathbb{Z}_2$ and $\mathbb{Z}_2 + u\mathbb{Z}_2 + u^2\mathbb{Z}_2$, respectively. In 2010, Zhu et al. [36] revisited and discussed some results for cyclic codes over $\mathbb{F}_2 + u\mathbb{F}_2$. They have obtained generators, idempotent generator polynomials for cyclic codes and prove that these codes are principally generated. In [35], Yildiz and Karadeniz discussed the structural properties of cyclic codes of odd length and obtained some good binary linear codes from the Gray images of these codes over the finite commutative non-chain ring $\mathbb{F}_2 + u\mathbb{F}_2 + v\mathbb{F}_2 + uv\mathbb{F}_2$. Also, Dougherty et al. [7] studied cyclic codes over the local Frobenius ring $\mathbb{F}_2[u_1, u_2, \ldots, u_k]/\langle u_i^2 = 0, u_iu_j = u_ju_i\rangle$. They explored binary quasi-cyclic codes with index 2^k from cyclic codes under a Gray map. In fact, there are plenty of related works that we refer to [6, 14, 16, 18, 19, 31].

Linear complementary dual (shortly, LCD) codes were introduced by Massey [28] in 1992 and the existence of asymptotically good LCD codes was shown . Later, in 1994, Yang and Massey [34] derived necessary and sufficient conditions for cyclic codes to be LCD codes over $GF(q)$. Also, to address a question raised by Massey [28], Sendrier [30] proved that LCD codes over $GF(q)$ meet the Gilbert-Varshavov bound. In 2009, Esmaeili and Yari [8] studied LCD quasi-cyclic codes and derived necessary and sufficient conditions for maximal one-generator quasi-cyclic codes to be LCD codes over $\mathbb{F}_q$. Interested reader can see some recent studies on LCD codes in [3, 4, 11, 25, 26, 32, 33].

Here, for a prime $p > 2$, we study cyclic codes of length n over a finite commutative semi-local ring $\mathbb{F}_p[u, v, w]/\langle u^2 - u, v^2 - 1, w^2 - 1, uv - vu, vw - wv, wu - uw\rangle$. Also, we construct LCD codes from cyclic codes under certain conditions. Moreover, under a Gray map, $\mathbb{F}_p$-images of cyclic and LCD codes are discussed. The main motivation behind this paper is to determine some good relations between cyclic and LCD codes over the underlying ring.

The arrangement of the paper is as follows: In Sect. 2, we recall basic facts and define a Gray map, whereas in Sect. 3, we discuss the linear codes. We derive the structural properties of cyclic codes in Sect. 4 and LCD codes in Sect. 5, respectively. Section 6 concludes the paper.

2 Basic Facts and Gray Map

Let p be an odd prime and $\mathbb{F}_p$ a finite field, and $\mathfrak{R} := \mathbb{F}_p[u, v, w]/\langle u^2 - u, v^2 - 1, w^2 - 1, uv - vu, vw - wv, wu - uw\rangle$. Thus $\mathfrak{R}$ is a finite commutative semi-local non-chain ring (with unity) of p^8 elements which is the extension of both rings $\mathbb{F}_p[u]/\langle u^2 - u\rangle$ in [9, 10] and $\mathbb{F}_p[v, w]/\langle v^2 - 1, w^2 - 1, vw - wv\rangle$ in [14, 15]. Let $\mathfrak{C} \subseteq \mathfrak{R}^n$. Then $\mathfrak{C}$ is called a *linear code* of length n over $\mathfrak{R}$ if $\mathfrak{C}$ is an $\mathfrak{R}$-submodule of $\mathfrak{R}^n$. Also, the elements of $\mathfrak{C}$ are called *codewords*. The dual of $\mathfrak{C}$ is denoted by $\mathfrak{C}^\perp$ and defined as $\mathfrak{C}^\perp = \{\mathfrak{a} \in \mathfrak{R}^n \mid \mathfrak{a} \cdot \mathfrak{b} = 0, \ \forall\ \mathfrak{b} \in \mathfrak{C}\}$, where the (Euclidean) inner product of two codewords $\mathfrak{a} = (\mathfrak{a}_0, \mathfrak{a}_1, \ldots, \mathfrak{a}_{n-1})$, $\mathfrak{b} = (\mathfrak{b}_0, \mathfrak{b}_1, \ldots, \mathfrak{b}_{n-1}) \in \mathfrak{C}$ is $\mathfrak{a} \cdot \mathfrak{b} = \sum_{i=0}^{n-1} \mathfrak{a}_i\mathfrak{b}_i$. If $\mathfrak{C} \subseteq \mathfrak{C}^\perp$, then $\mathfrak{C}$ is called *self-orthogonal*, and if $\mathfrak{C} = \mathfrak{C}^\perp$, then it is *self-dual*. For any polynomial $f(x) \in \mathfrak{R}[x]$, its reciprocal polynomial is defined by $f^*(x) = x^{deg(f(x))} f(\frac{1}{x})$. Note that $deg(f(x)) = deg(f^*(x))$ if $f(0) \neq 0$. The polynomial $f(x)$ is said to be *self-reciprocal* if $f(x) = f^*(x)$.

Also, $\mathfrak{R}$ can be considered as $\mathbb{F}_p + u\mathbb{F}_p + v\mathbb{F}_p + w\mathbb{F}_p + uv\mathbb{F}_p + vw\mathbb{F}_p + wu\mathbb{F}_p + uvw\mathbb{F}_p$ where $u^2 = u, v^2 = 1, w^2 = 1, uv = vu, vw = wv, wu = uw$, and hence every element $\mathfrak{r} \in \mathfrak{R}$ can be written uniquely as $\mathfrak{r} = a_1 + ua_2 + va_3 + wa_4 + uva_5 + vwa_6 + wua_7 + uvwa_8$ where $a_i \in \mathbb{F}_p$ for $1 \leq i \leq 8$. Let $\kappa_1 = \frac{u(1-v)(1-w)}{4}$, $\kappa_2 = \frac{u(1-v)(1+w)}{4}$, $\kappa_3 = \frac{u(1+v)(1-w)}{4}$, $\kappa_4 = \frac{u(1+v)(1+w)}{4}$, $\kappa_5 = \frac{(1-u)(1-v)(1-w)}{4}$, $\kappa_6 = \frac{(1-u)(1-v)(1+w)}{4}$, $\kappa_7 = \frac{(1-u)(1+v)(1-w)}{4}$, and $\kappa_8 = \frac{(1-u)(1+v)(1+w)}{4}$. Then

1. $\sum_{i=1}^{8} \kappa_i = 1$;
2. $\kappa_i \kappa_j = \begin{cases} \kappa_i, & \text{if } i = j \\ 0, & \text{if } i \neq j \end{cases}$.

Therefore, by the Chinese Remainder Theorem, $\mathfrak{R} = \oplus_{i=1}^{8} \kappa_i \mathfrak{R} \cong \oplus_{i=1}^{8} \kappa_i \mathbb{F}_p$. Thus, every element $\mathfrak{r} \in \mathfrak{R}$ can be uniquely written as $\mathfrak{r} = \sum_{i=1}^{8} e_i \kappa_i$, for $e_i \in \mathbb{F}_p$, $1 \leq i \leq 8$. In that case, we define a map

$$\psi : \mathfrak{R} \longrightarrow \mathbb{F}_p^8$$

by

$$\psi(\mathfrak{r}) = (e_1, e_2, \ldots, e_8). \tag{1}$$

The map ψ can be extended to $\mathfrak{R}^n$ componentwise. Recall that for a codeword $\mathfrak{a} = (\mathfrak{a}_0, \mathfrak{a}_1, \ldots, \mathfrak{a}_{n-1}) \in \mathfrak{C}$ the Hamming weight

$$w_H(\mathfrak{a}) = |i \ : \mathfrak{a}_i \neq 0 \ \forall \ i|$$

and the distance between $\mathfrak{a}, \mathfrak{b} \in \mathfrak{C}$ is $d_H(\mathfrak{a}, \mathfrak{b}) = w_H(\mathfrak{a} - \mathfrak{b})$. The Hamming distance for a code $\mathfrak{C}$ is defined by $d_H(\mathfrak{C}) = min\{d_H(\mathfrak{a}, \mathfrak{b}) \mid \mathfrak{a} \neq \mathfrak{b}, \forall \ \mathfrak{a}, \mathfrak{b} \in \mathfrak{C}\}$. We define Gray weight of $\mathfrak{r} \in \mathfrak{R}$ by $w_G(\mathfrak{r}) = w_H(\psi(\mathfrak{r}))$ and Gray weight of $\bar{\mathfrak{r}} = (\mathfrak{r}_0, \mathfrak{r}_1, \ldots, \mathfrak{r}_{n-1}) \in \mathfrak{R}^n$ by $w_G(\bar{\mathfrak{r}}) = \sum_{i=0}^{n-1} w_G(\mathfrak{r}_i)$. The Gray distance between $\mathfrak{a}, \mathfrak{b} \in \mathfrak{C}$ is defined by $d_G(\mathfrak{a}, \mathfrak{b}) = w_G(\mathfrak{a} - \mathfrak{b})$ and Gray distance for the code $\mathfrak{C}$ is $d_G(\mathfrak{C}) = min\{d_G(\mathfrak{a}, \mathfrak{b}) \mid \mathfrak{a} \neq \mathfrak{b}, \mathfrak{a}, \mathfrak{b} \in \mathfrak{C}\}$.

3 Linear Codes over $\mathfrak{R}$

In this section, we review some important results of linear codes over $\mathfrak{R}$ which are analogous to the results in [10, 14, 15]. These results are useful to obtain the structure of cyclic codes in the subsequent section.

Theorem 1 *The map ψ from $(\mathfrak{R}^n, d_G)$ to $(\mathbb{F}_p^{8n}, d_H)$ defined in Eq. (1) is linear and distance preserving.*

Proof Let $\mathfrak{a}, \mathfrak{b} \in \mathfrak{R}^n$ and $\alpha \in \mathbb{F}_p$. It is checked that $\psi(\mathfrak{a} + \alpha\mathfrak{b}) = \psi(\mathfrak{a}) + \alpha\psi(\mathfrak{b})$. Hence, ψ is linear. Further, $d_G(\mathfrak{a}, \mathfrak{b}) = w_G(\mathfrak{a} - \mathfrak{b}) = w_H(\psi(\mathfrak{a} - \mathfrak{b})) = w_H(\psi(\mathfrak{a}) - \psi(\mathfrak{b})) = d_H(\psi(\mathfrak{a}), \psi(\mathfrak{b}))$ for all $\mathfrak{a}, \mathfrak{b} \in \mathfrak{R}^n$. Thus, ψ preserves the distance.

Theorem 2 *Let $\mathfrak{C}$ be an $[n, k, d_G]$ linear code over $\mathfrak{R}$. Then*

1. *$\psi(\mathfrak{C})$ is an $[8n, k, d_H]$ linear code over $\mathbb{F}_p$ where $d_H = d_G$.*
2. *$\psi(\mathfrak{C}^\perp) = (\psi(\mathfrak{C}))^\perp$.*
3. *$\mathfrak{C}$ is self-dual if and only if $\psi(\mathfrak{C})$ is self-dual.*

Proof 1. ψ being linear, $\psi(\mathfrak{C})$ is a linear code of length $8n$ over $\mathbb{F}_p$. Further, ψ is bijective and isometric, so $\psi(\mathfrak{C})$ is a linear code over $\mathbb{F}_p$ with parameters $[8n, k, d_H]$ and $d_H = d_G$.

2. Let $\mathfrak{a} = (\mathfrak{a}_0, \mathfrak{a}_1, \ldots, \mathfrak{a}_{n-1}) \in \mathfrak{C}$, $\mathfrak{b} = (\mathfrak{b}_0, \mathfrak{b}_1, \ldots, \mathfrak{b}_{n-1}) \in \mathfrak{C}^\perp$ with $\mathfrak{a}_j = \sum_{i=1}^{8} \kappa_i t_j^i$, $\mathfrak{b}_j = \sum_{i=1}^{8} \kappa_i m_j^i$, where $t_j^i, m_j^i \in \mathbb{F}_p$, for $1 \le i \le 8, 0 \le j \le n-1$. Since, $\mathfrak{a} \cdot \mathfrak{b} = 0$, we have $\sum_{j=0}^{n-1} \sum_{i=1}^{8} t_j^i m_j^i = 0$. On the other side, $\psi(\mathfrak{a}) \cdot \psi(\mathfrak{b}) = 8 \sum_{j=0}^{n-1} \sum_{i=1}^{8} t_j^i m_j^i = 0$. Therefore, $\psi(\mathfrak{C}^\perp) \subseteq (\psi(\mathfrak{C}))^\perp$. As ψ is bijective, $|\psi(\mathfrak{C}^\perp)| = |(\psi(\mathfrak{C}))^\perp|$. Hence, $\psi(\mathfrak{C}^\perp) = (\psi(\mathfrak{C}))^\perp$.
3. Let $\mathfrak{C}$ be self-dual. Then $\mathfrak{C}^\perp = \mathfrak{C}$, which implies that $\psi(\mathfrak{C}^\perp) = \psi(\mathfrak{C})$ and $(\psi(\mathfrak{C}))^\perp = \psi(\mathfrak{C})$. Therefore, $\psi(\mathfrak{C})$ is self-dual. Conversely, let $\psi(\mathfrak{C})$ be self-dual. Then $(\psi(\mathfrak{C}))^\perp = \psi(\mathfrak{C})$, and hence $\psi(\mathfrak{C}^\perp) = \psi(\mathfrak{C})$. As ψ is bijective, $\mathfrak{C}^\perp = \mathfrak{C}$. Thus, $\mathfrak{C}$ is self-dual.

Let $\mathcal{A}_i$ be a non-empty set for $i = 1, 2, \ldots, 8$. We denote $\bigoplus_{i=1}^{8} \mathcal{A}_i = \{a_1 + a_2 + \cdots + a_8 \mid a_i \in \mathcal{A}_i \ \forall\, i\}$ and $\bigotimes_{i=1}^{8} \mathcal{A}_i = \{(a_1, a_2, \ldots, a_4) \mid a_i \in \mathcal{A}_i \ \forall\, i\}$. For a linear code $\mathfrak{C}$ of length n over $\mathfrak{R}$, we define $\mathfrak{C}_i = \{e_i \in \mathbb{F}_p^n \mid$ there exist $e_1, \ldots, e_{i-1}, e_{i+1}, \ldots, e_8$ such that $\sum_{i=1}^{8} \kappa_i e_i \in \mathfrak{C}\}$. It is easy to verify that $\mathfrak{C}_i (1 \le i \le 8)$ are linear codes of length n over $\mathbb{F}_p$.

Theorem 3 *If $\mathfrak{C}$ is a linear code of length n over $\mathfrak{R}$, then $\psi(\mathfrak{C}) = \bigotimes_{i=1}^{8} \mathfrak{C}_i$ and $|\mathfrak{C}| = \prod_{i=1}^{8} |\mathfrak{C}_i|$.*

Proof Let $\mathfrak{a} = (\mathfrak{a}_0^1, \mathfrak{a}_1^1, \ldots, \mathfrak{a}_{n-1}^1, \mathfrak{a}_0^2, \mathfrak{a}_1^2, \ldots, \mathfrak{a}_{n-1}^2, \ldots, \mathfrak{a}_0^8, \mathfrak{a}_1^8, \ldots, \mathfrak{a}_{n-1}^8) \in \psi(\mathfrak{C})$ and $\mathfrak{r}_j = \sum_{i=1}^{8} \kappa_i \mathfrak{a}_j^i$ for $0 \le j \le n-1$. As ψ is bijective, $\mathfrak{r} = (\mathfrak{r}_0, \mathfrak{r}_1, \ldots, \mathfrak{r}_{n-1}) \in \mathfrak{C}$. Therefore, $(\mathfrak{a}_0^i, \mathfrak{a}_1^i, \ldots, \mathfrak{a}_{n-1}^i) \in \mathfrak{C}_i$ for $1 \le i \le 8$, and this implies that $\mathfrak{a} \in \bigotimes_{i=1}^{8} \mathfrak{C}_i$. Hence, $\psi(\mathfrak{C}) \subseteq \bigotimes_{i=1}^{8} \mathfrak{C}_i$.

Conversely, let $\mathfrak{a} = (\mathfrak{a}_0^1, \mathfrak{a}_1^1, \ldots, \mathfrak{a}_{n-1}^1, \mathfrak{a}_0^2, \mathfrak{a}_1^2, \ldots, \mathfrak{a}_{n-1}^2, \ldots, \mathfrak{a}_0^8, \mathfrak{a}_1^8, \ldots, \mathfrak{a}_{n-1}^8) \in \bigotimes_{i=1}^{8} \mathfrak{C}_i$. Then $\mathfrak{a}^i = (\mathfrak{a}_0^i, \mathfrak{a}_1^i, \ldots, \mathfrak{a}_{n-1}^i) \in \mathfrak{C}_i$ for $1 \le i \le 8$. In order to prove $\mathfrak{a} \in \psi(\mathfrak{C})$, we need to show $\psi(\mathfrak{b}) = \mathfrak{a}$ for some $\mathfrak{b} = \sum_{i=1}^{8} s_i \kappa_i \in \mathfrak{C}$. We choose $s_i = \sum_{j=1}^{8} \kappa_j t_{ij}$, where $t_{ij} \in \mathbb{F}_p^n$, for $1 \le i, j \le 8$ and $t_{ii} = \mathfrak{a}^i$. Then $\mathfrak{b} = \sum_{i=1}^{8} \mathfrak{a}^i \kappa_i$ and $\psi(\mathfrak{b}) = \mathfrak{a}$. Thus, $\bigotimes_{i=1}^{8} \mathfrak{C}_i \subseteq \psi(\mathfrak{C})$. Therefore, $\psi(\mathfrak{C}) = \bigotimes_{i=1}^{8} \mathfrak{C}_i$. Further, ψ is bijective, we have $|\mathfrak{C}| = |\psi(\mathfrak{C})|$. Thus, $\prod_{i=1}^{8} |\mathfrak{C}_i|$.

Corollary 1 *For a linear code* $\mathfrak{C}$*, the generator matrix is* $M = \begin{pmatrix} \kappa_1 M_1 \\ \kappa_2 M_2 \\ \vdots \\ \kappa_8 M_8, \end{pmatrix}$ *where* M_i *is a generator matrix of* $\mathfrak{C}_i$*,* $1 \leq i \leq 8$*.*

Corollary 2 *If* $\psi(\mathfrak{C}) = \bigotimes_{i=1}^{8} \mathfrak{C}_i$*, then* $\mathfrak{C} = \bigoplus_{i=1}^{8} \kappa_i \mathfrak{C}_i$*.*

Corollary 3 *Let* $\mathfrak{C} = \bigoplus_{i=1}^{8} \kappa_i \mathfrak{C}_i$ *be a linear code over* $\mathfrak{R}$ *and* $[n, k_i, d_H(\mathfrak{C}_i)]$ *be the parameters of* $\mathfrak{C}_i$ *for* $1 \leq i \leq 8$*. Then* $\psi(\mathfrak{C})$ *is an* $[8n, \sum_{i=1}^{8} k_i, d_H(\psi(\mathfrak{C}))]$ *linear code over* $\mathbb{F}_p$ *where* $d_H(\psi(\mathfrak{C})) = min\{d_H(\mathfrak{C}_i),\ 1 \leq i \leq 8\}$*.*

Theorem 4 *Let* $\mathfrak{C} = \bigoplus_{i=1}^{8} \kappa_i \mathfrak{C}_i$ *be a linear code of length n over* $\mathfrak{R}$*. Then* $\mathfrak{C}^{\perp} = \bigoplus_{i=1}^{8} \kappa_i \mathfrak{C}_i^{\perp}$*. Moreover,* $\mathfrak{C}$ *is self-dual if and only if* $\mathfrak{C}_i$ *is self-dual for* $1 \leq i \leq 8$*.*

Proof Consider $\mathcal{S}_i = \{e_i \in \mathbb{F}_p^n \mid$ there exist $e_1, \ldots, e_{i-1}, e_{i+1}, \ldots, e_8$ such that $\sum_{i=1}^{8} e_i \kappa_i \in \mathfrak{C}^{\perp}\}$. Then $\mathfrak{C}^{\perp} = \bigoplus_{i=1}^{8} \kappa_i \mathcal{S}_i$ is unique. It is easy to see that $\mathcal{S}_1 \subseteq \mathfrak{C}_1^{\perp}$. On the other side, let $r \in \mathfrak{C}_1^{\perp}$. Then $r \cdot a_1 = 0$ for all $a_1 \in \mathfrak{C}_1$. Let $z = \sum_{i=1}^{8} \kappa_i a_i \in \mathfrak{C}$. Now, $\kappa_1 r \cdot z = \kappa_1 a_1 \cdot r = 0$, and which gives that $\kappa_1 r \in \mathfrak{C}^{\perp}$. From the uniqueness of $\mathfrak{C}^{\perp}$, we have $r \in \mathcal{S}_1$. Therefore, $\mathfrak{C}_1^{\perp} \subseteq \mathcal{S}_1$. Hence, $\mathcal{S}_1 = \mathfrak{C}_1^{\perp}$. Similarly, we have $\mathfrak{C}_i^{\perp} = \mathcal{S}_i$ for $i = 2, \ldots, 8$. Thus, $\mathfrak{C}^{\perp} = \bigoplus_{i=1}^{8} \kappa_i \mathfrak{C}_i^{\perp}$.

Further, let $\mathfrak{C}$ be self-dual, i.e., $\mathfrak{C}^{\perp} = \mathfrak{C}$, which implies that $\bigoplus_{i=1}^{8} \kappa_i \mathfrak{C}_i = \bigoplus_{i=1}^{8} \kappa_i \mathfrak{C}_i^{\perp}$, and $\mathfrak{C}_i^{\perp} = \mathfrak{C}_i$ for $i = 1, 2, \ldots, 8$. Hence $\mathfrak{C}_i$ is self-dual for $1 \leq i \leq 8$. Conversely, let $\mathfrak{C}_i$ be self-dual for $1 \leq i \leq 8$. Then $\mathfrak{C} = \bigoplus_{i=1}^{8} \kappa_i \mathfrak{C}_i = \bigoplus_{i=1}^{8} \kappa_i \mathfrak{C}_i^{\perp} = \mathfrak{C}^{\perp}$. Therefore, $\mathfrak{C}$ is self-dual.

4 Cyclic Codes over $\mathfrak{R}$

This section characterizes cyclic codes over $\mathfrak{R}$ and proves that they are principally generated. In the next section, we impose few restrictions on these cyclic codes to be LCD codes.

Definition 1 Let $\mathfrak{C}$ be a linear code of length n over $\mathfrak{R}$. Then it is called cyclic if for any $\mathfrak{a} = (\mathfrak{a}_0, \mathfrak{a}_1, \ldots, \mathfrak{a}_{n-1}) \in \mathfrak{C}$, its cyclic shift $\sigma(\mathfrak{a}) = (\mathfrak{a}_{n-1}, \mathfrak{a}_0, \ldots, \mathfrak{a}_{n-2}) \in \mathfrak{C}$.

For a cyclic code $\mathfrak{C}$ of length n over $\mathfrak{R}$, we identify each codeword $\mathfrak{a} = (\mathfrak{a}_0, \mathfrak{a}_1, \ldots, \mathfrak{a}_{n-1}) \in \mathfrak{C}$ by a polynomial $a(x) \in \mathfrak{R}[x]/\langle x^n - 1\rangle$ under the correspondence $\mathfrak{a} = (\mathfrak{a}_0, \mathfrak{a}_1, \ldots, \mathfrak{a}_{n-1}) \longmapsto \mathfrak{a}(x) = (\mathfrak{a}_0 + \mathfrak{a}_1 x + \cdots + \mathfrak{a}_{n-1} x^{n-1}) \bmod (x^n - 1)$. Under this polynomial representation of $\mathfrak{C}$, we get the next result.

Theorem 5 *A linear code* $\mathfrak{C}$ *of length n over* $\mathfrak{R}$ *is cyclic if and only if it is an ideal of* $\mathfrak{R}[x]/\langle x^n - 1\rangle$*.*

Proof Straightforward.

Theorem 6 *A linear code $\mathfrak{C} = \bigoplus_{i=1}^{8} \kappa_i \mathfrak{C}_i$ of length n over $\mathfrak{R}$ is cyclic if and only if $\mathfrak{C}_i$ is cyclic for $1 \leq i \leq 8$.*

Proof Let $\mathfrak{C}$ be a cyclic code of length n over $\mathfrak{R}$. Let $\mathfrak{a}^i = (\mathfrak{a}_0^i, \mathfrak{a}_1^i, \ldots, \mathfrak{a}_{n-1}^i) \in \mathfrak{C}_i$, for $1 \leq i \leq 8$. Now, consider $\mathfrak{r}_j = \sum_{i=1}^{8} \kappa_i \mathfrak{a}_j^i$, for $0 \leq j \leq n-1$. Thus, by definition $\mathfrak{r} = (\mathfrak{r}_0, \mathfrak{r}_1, \ldots, \mathfrak{r}_{n-1}) \in \mathfrak{C}$, and $\sigma(\mathfrak{r}) = (\mathfrak{r}_{n-1}, \mathfrak{r}_0, \ldots, \mathfrak{r}_{n-2}) \in \mathfrak{C}$. Again, $\sigma(\mathfrak{r}) = \sum_{i=1}^{8} \kappa_i \sigma(\mathfrak{a}^i) \in \mathfrak{C} = \bigoplus_{i=1}^{8} \kappa_i \mathfrak{C}_i$. Then $\sigma(\mathfrak{a}^i) \in \mathfrak{C}_i$, for $1 \leq i \leq 8$. Hence, $\mathfrak{C}_i$ is cyclic for $1 \leq i \leq 8$.

Conversely, let $\mathfrak{C}_i$ be a cyclic code of length n over $\mathbb{F}_p$, for $1 \leq i \leq 8$. Let $\mathfrak{r} = (\mathfrak{r}_0, \mathfrak{r}_1, \ldots, \mathfrak{r}_{n-1}) \in \mathfrak{C}$, where $\mathfrak{r}_j = \sum_{i=1}^{8} \kappa_i \mathfrak{a}_j^i$ for some $\mathfrak{a}_j^i \in \mathbb{F}_p$, $1 \leq i \leq 8$ and $0 \leq j \leq n-1$. Then $\mathfrak{a}^i = (\mathfrak{a}_0^i, \mathfrak{a}_1^i, \ldots, \mathfrak{a}_{n-1}^i) \in \mathfrak{C}_i$, for $1 \leq i \leq 8$. Therefore, $\sigma(\mathfrak{a}^i) \in \mathfrak{C}_i$, for $1 \leq i \leq 8$. Also, $\sigma(\mathfrak{r}) = \sum_{i=1}^{8} \kappa_i \sigma(\mathfrak{a}^i) \in \bigoplus_{i=1}^{8} \kappa_i \mathfrak{C}_i = \mathfrak{C}$. Hence, $\mathfrak{C}$ is a cyclic code of length n over $\mathfrak{R}$.

Theorem 7 *Let $\mathfrak{C}$ be a cyclic code given by Theorem 6. Then there exists a polynomial $f(x) \in \mathfrak{R}[x]$ such that $\mathfrak{C} = \langle f(x) \rangle$ and $f(x) \mid x^n - 1$.*

Proof By Theorem 6, $\mathfrak{C}_i$ is a cyclic code of length n over $\mathbb{F}_p$, for $i = 1, 2, \ldots, 8$. Let $f_i(x) \in \mathbb{F}_p[x]$ such that $\mathfrak{C}_i = \langle f_i(x) \rangle$ and $f_i(x) \mid x^n - 1$, for $1 \leq i \leq 8$. Clearly, $\kappa_1 f_1(x), \kappa_2 f_2(x), \ldots, \kappa_8 f_8(x)$ generate the code $\mathfrak{C}$. We take $f(x) = \sum_{i=1}^{8} \kappa_i f_i(x) \in \mathfrak{R}[x]$, then $\langle f(x) \rangle \subseteq \mathfrak{C}$. On the other side, $\kappa_i f(x) = \kappa_i f_i(x) \in \langle f(x) \rangle$, for $1 \leq i \leq 8$. Therefore, $\mathfrak{C} \subseteq \langle f(x) \rangle$. Hence, $\mathfrak{C} = \langle f(x) \rangle$.

Further, $x^n - 1 = h_i(x) f_i(x)$ in $\mathbb{F}_p[x]$, for $1 \leq i \leq 8$. Since $f(x) \sum_{i=1}^{8} \kappa_i h_i(x) = \sum_{i=1}^{8} \kappa_i h_i(x) f_i(x) = x^n - 1$. Thus, $f(x) \mid x^n - 1$.

Corollary 4 *Let $\mathfrak{C}$ be a cyclic code given by Theorem 6. Then*

$$|\mathfrak{C}| = p^{8n - \sum_{i=1}^{8} deg(f_i(x))},$$

where $f_i(x)$ is the generator of $\mathfrak{C}_i$, for $i = 1, 2, \ldots, 8$.

Corollary 5 *The ring $\mathfrak{R}[x]/\langle x^n - 1 \rangle$ is principally generated.*

Corollary 6 *Let $\mathfrak{C}$ be a cyclic code given by Theorem 6 and $\mathfrak{C}_i = \langle f_i(x) \rangle$ such that $x^n - 1 = h_i(x) f_i(x)$ for $1 \leq i \leq 8$. Then*

1. $\mathfrak{C}^\perp = \bigoplus_{i=1}^{8} \kappa_i \mathfrak{C}_i^\perp$ *is a cyclic code and* $\mathfrak{C}^\perp = \langle \sum_{i=1}^{8} \kappa_i h_i^*(x) \rangle$.
2. $|\mathfrak{C}^\perp| = p^{\sum_{i=1}^{8} deg(f_i(x))}$.

Definition 2 Let $\mathfrak{C}$ be a linear code of length $n = ml$ over $\mathfrak{R}$, for some positive integers m, l. Then $\mathfrak{C}$ is called a quasi-cyclic code of length n and index l if $\pi_l(\mathfrak{C}) = \mathfrak{C}$, where $\pi_l : \mathfrak{R}^n \longrightarrow \mathfrak{R}^n$ is a map defined by

$$\pi_l(c^1 \mid c^2 \mid \cdots \mid c^l) = (\sigma(c^1) \mid \sigma(c^2) \mid \cdots \mid \sigma(c^l)), \tag{2}$$

for $c^i \in \mathfrak{R}^m$, $i = 1, 2, \ldots, l$ and σ is the cyclic shift operator.

Lemma 1 *Let the maps* ψ *be defined in Eq. (1) and* π_8 *be defined in Eq. (2). Then* $\psi\sigma = \pi_8\psi$.

Proof It is verified by the definition of the maps ψ and π_8.

Theorem 8 *A linear code* $\mathfrak{C}$ *of length* n *over* $\mathfrak{R}$ *is cyclic if and only if its Gray image* $\psi(\mathfrak{C})$ *is a quasi-cyclic code of length* $8n$ *and index* 8 *over* $\mathbb{F}_p$.

Proof Let $\mathfrak{C}$ be cyclic over $\mathfrak{R}$. Then $\sigma(\mathfrak{C}) = \mathfrak{C}$, and by Lemma 1, $\psi(\sigma(\mathfrak{C})) = \psi(\mathfrak{C}) = \pi_8(\psi(\mathfrak{C}))$. This yields that $\psi(\mathfrak{C})$ is a quasi-cyclic code of length $8n$ and index 8 over $\mathbb{F}_p$.

Conversely, let $\psi(\mathfrak{C})$ be a quasi-cyclic code of length $8n$ and index 8 over $\mathbb{F}_p$. Thus $\pi_8(\psi(\mathfrak{C})) = \psi(\mathfrak{C})$, and by Lemma 1, we have $\psi(\sigma(\mathfrak{C})) = \psi(\mathfrak{C})$. Again, ψ is injective, so $\sigma(\mathfrak{C}) = \mathfrak{C}$. Hence, $\mathfrak{C}$ is a cyclic code of length n over $\mathfrak{R}$.

In order to validate the technique here we give an example of cyclic code.

Example 1 Let $\mathfrak{C}$ be a cyclic code of length 13 over $\mathfrak{R} = \mathbb{F}_3[u, v, w]/\langle u^2 - u, v^2 - 1, w^2 - 1, uv - vu, vw - wv, wu - uw\rangle$. Now, in $\mathbb{F}_3[x]$, we have

$$x^{13} - 1 =(x + 2)(x^3 + 2x + 2)(x^3 + x^2 + 2)(x^3 + x^2 + x + 2)(x^3 + 2x^2 + 2x + 2).$$

Let $f_i(x) = x^3 + 2x + 2$ for $i = 1, 2, 3, 4$ and $f_5(x) = x^3 + x^2 + 2$ for $i = 5, 6, 7, 8$. In this way, $\mathfrak{C}_i = \langle f_i(x)\rangle$ is a cyclic code of length 13 over $\mathbb{F}_3$, for $i = 1, 2, \ldots, 8$. Thus, $\mathfrak{C} = \langle\sum_{i=1}^{8} \kappa_i f_i(x)\rangle$ is a cyclic code of length 13 over $\mathfrak{R}$. Moreover, $\psi(\mathfrak{C})$ is a linear code over $\mathbb{F}_3$ with parameters [104, 80, 3].

5 LCD Codes over $\mathfrak{R}$

In the present section, we obtain the structure of LCD codes via cyclic codes over $\mathfrak{R}$. These codes can be used in cryptography to prevent popular attacks, like side-channel attacks(SCA) and fault injection attacks(FIA), see [3].

Definition 3 A linear code $\mathfrak{C}$ is called a linear complementary dual (or LCD) code if it meets trivially with its dual, i.e., $\mathfrak{C} \cap \mathfrak{C}^{\perp} = \{0\}$.

Theorem 9 [34] *Let* $n = p^t s$ *with* $gcd(s, p) = 1$ *and* $\mathfrak{C} = \langle f(x)\rangle$ *be a cyclic code of length* n *over* $\mathbb{F}_p$. *Then* $\mathfrak{C}$ *is an LCD code if and only if* $f(x)$ *is self-reciprocal and all the monic irreducible factors of* $f(x)$ *have the same multiplicity in* $f(x)$ *and in* $x^n - 1$.

Theorem 10 *Let* $\mathfrak{C} = \bigoplus_{i=1}^{8} \kappa_i\mathfrak{C}_i$ *be a cyclic code of length* n *over* $\mathfrak{R}$. *Then* $\mathfrak{C}$ *is LCD if and only if* $\mathfrak{C}_i$ *is LCD over* $\mathbb{F}_p$ *for* $1 \leq i \leq 8$.

Proof It can be proved based on the fact that $\mathfrak{C} \cap \mathfrak{C}^{\perp} = \{0\}$ if and only if $\mathfrak{C}_i \cap \mathfrak{C}_i^{\perp} = \{0\}$ for $1 \leq i \leq 8$.

Corollary 7 *Let $n = p^t s$ and $gcd(s, p) = 1$. Let $\mathfrak{C} = \bigoplus_{i=1}^{8} \kappa_i \mathfrak{C}_i$ with $\mathfrak{C}_i = \langle f_i(x) \rangle$ such that $f_i(x) \in \mathbb{F}_p[x]$ and $f_i(x) \mid x^n - 1$ for $1 \le i \le 8$, be a cyclic code of length n over $\mathfrak{R}$. Then $\mathfrak{C}$ is LCD if and only if $f_i(x)$ is self-reciprocal and each monic irreducible factor of $f_i(x)$ has the same multiplicity in $f_i(x)$ and in $x^n - 1$, for $1 \le i \le 8$.*

Proof It is verified by using Theorems 9 and 10.

Lemma 2 [34] *Let $\mathfrak{C}$ satisfying $gcd(n, p) = 1$, be a cyclic code of length n over $\mathbb{F}_p$. Then $\mathfrak{C}$ is an LCD code if and only if $\mathfrak{C}$ is a reversible code.*

Theorem 11 *Let $\mathfrak{C} = \bigoplus_{i=1}^{8} \kappa_i \mathfrak{C}_i$ satisfying $gcd(n, p) = 1$ be a cyclic code of length n over $\mathfrak{R}$. Then $\mathfrak{C}$ is an LCD code if and only if $\mathfrak{C}_i$ is a reversible code of length n over $\mathbb{F}_p$ for $1 \le i \le 8$.*

Proof Combining Theorem 10 and Lemma 2, it is verified.

Corollary 8 *Let $\mathfrak{C} = \bigoplus_{i=1}^{8} \kappa_i \mathfrak{C}_i$ satisfying $gcd(n, p) = 1$, be a cyclic code of length n over $\mathfrak{R}$ where $\mathfrak{C}_i = \langle f_i(x) \rangle$ is a cyclic code of length n over $\mathbb{F}_p$ for $1 \le i \le 8$. Then $\mathfrak{C}$ is an LCD code if and only if $f_i(x)$ is self-reciprocal polynomial in $\mathbb{F}_p$, for $1 \le i \le 8$.*

Proof Note that a cyclic code $\mathfrak{C}_i = \langle f_i(x) \rangle$ with $f_i(x) \mid x^n - 1$ is reversible if and only if $f_i(x)$ is self-reciprocal in $\mathbb{F}_p$ for $1 \le i \le 8$. The rest parts follow from Theorem 11.

Lemma 3 *For a linear code $\mathfrak{C}$ of length n over $\mathfrak{R}$ and the map ψ defined in Eq. (1), $\psi(\mathfrak{C} \cap \mathfrak{C}^\perp) = \psi(\mathfrak{C}) \cap \psi(\mathfrak{C})^\perp$.*

Proof Let $\mathfrak{r} \in \psi(\mathfrak{C}) \cap \psi(\mathfrak{C}^\perp)$. Then there exists $\mathfrak{a} \in \mathfrak{C}$ and $\mathfrak{b} \in \mathfrak{C}^\perp$ such that $\psi(\mathfrak{a}) = \mathfrak{r}$ and $\psi(\mathfrak{b}) = \mathfrak{r}$. As ψ is injective and $\psi(\mathfrak{a}) = \psi(\mathfrak{b}) = \mathfrak{r}$, we have $\mathfrak{a} = \mathfrak{b} \in \mathfrak{C} \cap \mathfrak{C}^\perp$. Therefore, $\mathfrak{r} = \psi(\mathfrak{a}) \in \psi(\mathfrak{C} \cap \mathfrak{C}^\perp)$. Hence, $\psi(\mathfrak{C}) \cap \psi(\mathfrak{C}^\perp) \subseteq \psi(\mathfrak{C} \cap \mathfrak{C}^\perp)$.

On the other side, let $\mathfrak{r} \in \psi(\mathfrak{C} \cap \mathfrak{C}^\perp)$. Then there exist $\mathfrak{a} \in \mathfrak{C} \cap \mathfrak{C}^\perp$ such that $\psi(\mathfrak{a}) = \mathfrak{r}$. Also, $\mathfrak{a} \in \mathfrak{C} \cap \mathfrak{C}^\perp$ implies that $\mathfrak{a} \in \mathfrak{C}$ and $\mathfrak{a} \in \mathfrak{C}^\perp$. Therefore, $\psi(\mathfrak{a}) \in \psi(\mathfrak{C})$ and $\psi(\mathfrak{a}) \in \psi(\mathfrak{C}^\perp)$. Hence, $\mathfrak{r} = \psi(\mathfrak{a}) \in \psi(\mathfrak{C}) \cap \psi(\mathfrak{C}^\perp)$. Thus, $\psi(\mathfrak{C} \cap \mathfrak{C}^\perp) \subseteq \psi(\mathfrak{C}) \cap \psi(\mathfrak{C}^\perp)$. Now, from the above calculation, we can conclude that $\psi(\mathfrak{C}) \cap \psi(\mathfrak{C}^\perp) = \psi(\mathfrak{C} \cap \mathfrak{C}^\perp)$. Again, from Theorem 2, we know $\psi(\mathfrak{C}^\perp) = \psi(\mathfrak{C})^\perp$. Hence, $\psi(\mathfrak{C}) \cap \psi(\mathfrak{C})^\perp = \psi(\mathfrak{C} \cap \mathfrak{C}^\perp)$.

Theorem 12 *Let $\mathfrak{C}$ be a linear code of length n over $\mathfrak{R}$. Then $\mathfrak{C}$ is an LCD code if and only if $\psi(\mathfrak{C})$ is an LCD code of length $8n$ over $\mathbb{F}_p$.*

Proof Let $\mathfrak{C}$ be LCD. Then $\mathfrak{C} \cap \mathfrak{C}^\perp = \{0\}$. Now, by Lemma 3, $\psi(\mathfrak{C}) \cap \psi(\mathfrak{C})^\perp = \psi(\mathfrak{C} \cap \mathfrak{C}^\perp) = \psi(\{0\}) = \{0\}$. This implies that $\psi(\mathfrak{C})$ is an LCD code of length $8n$ over $\mathbb{F}_p$.

Conversely, suppose $\psi(\mathfrak{C})$ is an LCD code of length $8n$ over $\mathbb{F}_p$ i.e., $\psi(\mathfrak{C}) \cap \psi(\mathfrak{C})^\perp = \{0\}$. By Lemma 3, we have $\psi(\mathfrak{C} \cap \mathfrak{C}^\perp) = \psi(\mathfrak{C}) \cap \psi(\mathfrak{C})^\perp = \{0\}$. Again, ψ is injective, so $\mathfrak{C} \cap \mathfrak{C}^\perp = \{0\}$. Hence, $\mathfrak{C}$ is an LCD code of length n over $\mathfrak{R}$.

Now, we present two examples of LCD codes to support our obtained results.

Example 2 Let $\mathfrak{C}$ be a cyclic code of length 15 over $\mathfrak{R} = \mathbb{F}_3[u, v, w]/\langle u^2 - u, v^2 - 1, w^2 - 1, uv - vu, vw - wv, wu - uw\rangle$. Now, in $\mathbb{F}_3[x]$, we have

$$x^{15} - 1 = (x + 2)^3(x^4 + x^3 + x^2 + x + 1)^3.$$

Let $f_i(x) = (x^4 + x^3 + x^2 + x + 1)^3$, for $1 \leq i \leq 8$. Then $\mathfrak{C}_i = \langle f_i(x)\rangle$ is an LCD code of length 15 over $\mathbb{F}_3$, for $i \leq i \leq 8$. By Theorem 10, $\mathfrak{C} = \langle\sum_{i=1}^{8} \kappa_i f_i(x)\rangle = \langle(x^4 + x^3 + x^2 + x + 1)^3\rangle$ is an LCD code of length 15 over $\mathfrak{R}$. Moreover, $\psi(C)$ is an LCD code over $\mathbb{F}_3$ with parameters [120, 24, 5].

Example 3 Let $\mathfrak{C}$ be a cyclic code of length 9 over $\mathfrak{R} = \mathbb{F}_5[u, v, w]/\langle u^2 - u, v^2 - 1, w^2 - 1, uv - vu, vw - wv, wu - uw\rangle$. Now, in $\mathbb{F}_5[x]$, we have

$$x^9 - 1 = (x + 4)(x^2 + x + 1)(x^6 + x^3 + 1).$$

Let $f_i(x) = x^6 + x^3 + 1$ for $1 \leq i \leq 8$. Since $f_i(x)$ is self-reciprocal, $\mathfrak{C}_i = \langle f_i(x)\rangle$ is a reversible cyclic code of length 9 over $\mathbb{F}_5$ for $1 \leq i \leq 8$. Now, by Theorem 11, $\mathfrak{C} = \langle\sum_{i=1}^{8} \kappa_i f_i(x)\rangle = \langle x^6 + x^3 + 1\rangle$ is an LCD code of length 9 over $\mathfrak{R}$. Hence, $\psi(\mathfrak{C})$ is an LCD code over $\mathbb{F}_5$ with parameters [72, 24, 3].

6 Conclusion

In this paper, we determined the complete structure of cyclic codes of length n over $\mathfrak{R}$ and obtained LCD codes of length $p^t s$ where $gcd(s, p) = 1$, over $\mathfrak{R}$. Also, we constructed LCD codes of length $8n$ over $\mathbb{F}_p$ as the Gray image of LCD codes of length n over $\mathfrak{R}$. Similar to [13, 15, 20, 21], one can attempt to investigate quantum codes over $\mathbb{F}_p$ via cyclic codes over $\mathfrak{R}$ along with the map ψ in the future.

Acknowledgements The authors would like to thank the anonymous referee(s) for their careful reading and constructive suggestions to improve the presentation of the paper. The authors are also thankful to the University Grants Commission (UGC), Govt. of India for providing research grants under Sr. No. 2121540952, Ref. No. 20/12/2015(ii)EU-V dated 31/08/2016 and the Indian Institute of Technology Patna for providing research facilities.

References

1. Abualrub, T., Siap, I.: Cyclic codes over the rings $\mathbb{Z}_2 + u\mathbb{Z}_2$ and $\mathbb{Z}_2 + u\mathbb{Z}_2 + u^2\mathbb{Z}_2$. Des. Codes Cryptogr. **42**(3), 273–287 (2007)
2. Bonnecaze, A., Udaya, P.: Cyclic codes and self-dual codes over $\mathbb{F}_2 + u\mathbb{F}_2$. IEEE Trans. Inf. Theory **45**, 1250–1255 (1999)
3. Carlet, C., Guilley, S.: Complementary dual codes for counter-measures to side-channel attacks. Adv. Math. Commun. **10**(1), 131–150 (2016)

4. Carlet, C., Mesnager, S., Tang, C., Qi, Y., Pellikaan, R.: Linear Codes Over $\mathbb{F}_q$ Are Equivalent to LCD Codes for $q > 3$. IEEE Trans. Inform. Theory **64**(4), 3010–3017 (2018)
5. Dinh, H.Q., Lopez-Permouth, S.R.: Cyclic and negacyclic codes over finite chain rings. IEEE Trans. Inform. Theory **50**, 1728–1744 (2004)
6. Dertli, A., Cengellenmis, Y., Eren, S.: On the linear codes over the ring R_p. Discrete Math. Algorithm. Appl. **8**(2), 1650036 (2016)
7. Dougherty, S.T., Karadeniz, S., Yildiz, B.: Cyclic codes over R_k. Des. Codes Cryptogr. **63**(1), 113–126 (2012)
8. Esmaeili, M., Yari, S.: On complementary-dual quasi-cyclic codes. Finite Fields Appl. **15**, 357–386 (2009)
9. Gao, J., Ma, F., Fu, F.: Skew constacyclic codes over the ring $\mathbb{F}_q + u\mathbb{F}_q$. Appl. Comput. Math. **6**(3), 286–295 (2017)
10. Gursoy, F., Siap, I., Yildiz, B.: Construction of skew cyclic codes over $\mathbb{F}_q + u\mathbb{F}_q$. Adv. Math. Commun. **8**(3), 313–322 (2014)
11. Guneri, C., Ozkaya, B., Sole, P.: Quasi-cyclic complementary dual codes. Finite Fields Appl. **42**, 67–80 (2016)
12. Hammons, A.R., Jr., Kumar, P.V., Calderbank, A.R., Sloane, N.J.A., Sole, P.: The $\mathbb{Z}_4$-linearity of Kerdock, Preparata, Coethals, and related codes. IEEE Trans. Inform. Theory **40**(2), 301–319 (1994)
13. Islam, H., Prakash, O.: Quantum codes from the cyclic codes over $\mathbb{F}_p[u, v, w]/\langle u^2 - 1, v^2 - 1, w^2 - 1, uv - vu, vw - wv, wu - uw\rangle$. J. Appl. Math. Comput. **60**(1–2), 625–635 (2019)
14. Islam, H., Verma, R.K., Prakash, O.: A family of constacyclic codes over $\mathbb{F}_{p^m}[v, w]/\langle v^2 - 1, w^2 - 1, vw - wv\rangle$. Int. J. Inf. Coding Theory **5**(3–4), 198–210 (2020)
15. Islam, H., Prakash, O., Verma, R.K.: Quantum codes from the cyclic codes over $\mathbb{F}_p[v, w]/\langle v^2 - 1, w^2 - 1, vw - wv\rangle$. Springer Proceedings in Mathematics & Statistics, Springer, Singapore **307**, 67–74 (2020)
16. Islam, H., Bag, T., Prakash, O.: A class of constacyclic codes over $\mathbb{Z}_4[u]/\langle u^k\rangle$. J. Appl. Math. Comput. **60**(1–2), 237–251 (2019)
17. Islam, H., Prakash, O.: A class of constacyclic codes over the ring $\mathbb{Z}_4[u, v]/\langle u^2, v^2, uv - vu\rangle$ and their Gray images. Filomat. **33**(8), 2237–2248 (2019)
18. Islam, H., Prakash, O.: Construction of reversible cyclic codes over $\mathbb{Z}_{p^k}$. J. Discrete Math. Sci. Cryptogr. (2021). https://doi.org/10.1080/09720529.2020.1815341
19. Islam, H., Prakash, O.: New $\mathbb{Z}_4$ codes from constacyclic codes over a non-chain ring. Comput. Appl. Math. **40**(1), 1–17 (2019)
20. Islam, H., Prakash, O., Verma, R.K.: New quantum codes from constacyclic codes over the ring $R_{k,m}$. Adv. Math. Commun. (2020). https://doi.org/10.3934/amc.2020097
21. Islam, H., Prakash, O.: New quantum codes from constacyclic and additive constacyclic codes. Quantum Inf. Process. **19**(9), 1–17 (2020)
22. Kai, X., Zhu, S., Li, P.: $(1 + \lambda u)$-constacyclic codes over $\mathbb{F}_p[u]/\langle u^m\rangle$. J. Frank. Inst. **347**(7), 51–762 (2010)
23. Kanwar, P., Lopez-Permouth, S.R.: Cyclic codes over the integers modulo p^m. Finite Fields Appl. **3**, 334–352 (1997)
24. Karadeniz, S., Yildiz, B.: $(1 + v)$-Constacyclic codes over $\mathbb{F}_2 + u\mathbb{F}_2 + v\mathbb{F}_2 + uv\mathbb{F}_2$. J. Franklin Inst. **348**, 2625–2632 (2011)
25. Li, C., Ding, C., Li, S.: LCD cyclic codes over finite fields. IEEE Trans. Inf. Theory **63**(7), 4344–4356 (2017)
26. Liu, X.S., Liu, H.L.: LCD codes over finite chain rings. Finite Field Appl. **15**, 1–19 (2015)
27. Massey, J.L.: Reversible codes. Inform. and Control **7**, 369–380 (1964)
28. Massey, J.L.: Linear codes with complementary duals. Discrete Math. **106**(107), 337–342 (1992)
29. Qian, J.-F., Zhang, L.-N., Zhu, S.-X.: $(1 + u)$-constacyclic and cyclic codes over $\mathbb{F}_2 + u\mathbb{F}_2$. Appl. Math. Lett. **19**(8), 820–823 (2006)
30. Sendrier, N.: Linear codes with complementary duals meet the Gilbert-Varshamov bound. Discrete Math. **304**, 345–347 (2004)

31. Singh, A.K., Kewat, P.K.: On cyclic codes over the ring $\mathbb{Z}_p[u]/\langle u^k \rangle$. Des. Codes Cryptogr. **74**(1), 1–13 (2015)
32. Sok, L., Shi, M., Solé, P.: Constructions of optimal LCD codes over large finite fields. Finite Fields Appl. **50**, 138–153 (2018)
33. Yadav, Y., Islam, H., Prakash, O., Solé, P.: Self-dual and LCD double circulant and double negacirculant codes over $\mathbb{F}_q + u\mathbb{F}_q + v\mathbb{F}_q$. J. Appl. Math. Comput. (2021). https://doi.org/10.1007/s12190-021-01499-9 (2021)
34. Yang, X., Massey, J.L.: The condition for a cyclic code to have a complementary dual. Discrete Math. **126**, 391–393 (1994)
35. Yildiz, B., Karadeniz, S.: Cyclic codes over $\mathbb{F}_2 + u\mathbb{F}_2 + v\mathbb{F}_2 + uv\mathbb{F}_2$. Des. Codes Cryptogr. **58**(3), 221–234 (2011)
36. Zhu, S., Wang, Y., Shi, M.: Some result on cyclic codes over $\mathbb{F}_2 + v\mathbb{F}_2$. IEEE Trans. Inf. Theory **56**, 1680–1684 (2010)

Graph Theory

Some Recent Results on the Graphs of Finite-Dimensional Vector Spaces

Mohammad Ashraf and Mohit Kumar

Abstract Let $\mathbb{V}$ be a finite-dimensional vector space. In this survey, we present the results concerning the fundamental properties of the graphs associated with finite-dimensional vector spaces.

Keywords Graph · Diameter · Connected graph · Clique · Subspace

1 Definitions and Preliminaries

Let $G = (V(G), E(G))$ be a Graph, where $V(G)$ is the set of vertices and $E(G)$ is the set of edges of G. We say that G is connected if there exists a path between any two distinct vertices of G. For vertices a and b of G, $d(a, b)$ denotes the length of the shortest path from a to b. In particular, $d(a, a) = 0$ and $d(a, b) = \infty$ if there is no such path. The diameter of G is denoted by $diam(G) = sup\{d(a, b) \mid a, b \in V(G)\}$. A cycle in a graph G is a path that begins and ends at the same vertex. A cycle of length n is denoted by C_n. A graph is said to be Eulerian if it contains a cycle containing all the edges in G exactly once. The girth of G, denoted by $gr(G)$, is the length of a shortest cycle in G, $(gr(G) = \infty$ if G contains no cycle). Two graphs $G_1 = (V(G_1), E(G_1))$ and $G_2 = (V(G_2), E(G_2))$ are said to be isomorphic if there exists a bijection $\phi : V(G_1) \to V(G_2))$ such that $(u, v) \in E(G_1)$ if and only if $(\phi(u), \phi(v)) \in E(G_2)$. A complete graph G is a graph where all distinct vertices are adjacent. The complete graph with $|V(G)| = n$ is denoted by K_n. A graph $H = (V(H), E(H))$ is said to be a subgraph of G, if $V(H) \subseteq V(G)$ and $E(H) \subseteq E(G)$. Moreover, H is said to be an induced subgraph of G if $V(H) \subseteq V(G)$ and $E(H) = \{(u, v) \in E(G) \mid u, v \in V(H)\}$ and is denoted by

M. Ashraf (✉)
Department of Mathematics, Aligarh Muslim University, Aligarh, Uttar Pradesh 202002, India
e-mail: mashraf80@hotmail.com

M. Kumar
Department of Mathematics, Institute of Applied Sciences and Humanities,
GLA University Mathura, Uttar Pradesh 281406, India

M. Ashraf et al. (eds.), *Algebra and Related Topics with Applications*,
Springer Proceedings in Mathematics & Statistics 392,
https://doi.org/10.1007/978-981-19-3898-6_29

$G[V(H)]$. Also, G is called a null graph if $E(G) = \emptyset$. For a graph G, a complete subgraph of G is called a clique. The clique number, $\omega(G)$, is the greatest integer $n \geqslant 1$ such that $K_n \subseteq G$, and $\omega(G) = \infty$ if $K_n \subseteq G$ for all $n \geqslant 1$. The chromatic number $\chi(G)$ of a graph G is the minimum number of colours needed to colour all the vertices of G such that every two adjacent vertices get different colours. For a connected graph G, $\delta = \min\{deg(x) \mid x \in V(G)\}$ and $\Delta = \max\{deg(x) \mid x \in V(G)\}$. A graph G is *perfect* if for every induced subgraph H of G, $\chi(H) = \omega(H)$. A subset of $V(G)$ is called independent if any two distinct vertices are pair-wise non-adjacent in that subset. An independent set which is maximal with respect to inclusion is called as a maximal *independent set*. A subset $\mathfrak{D}$ of $V(G)$ is said to be a dominating set if each vertex in $V(G) \setminus \mathfrak{D}$ is adjacent to at least one vertex in $\mathfrak{D}$. The dominating number $\gamma(G)$ are the minimum size of a dominating set in G. Let S_k denote the sphere with k handles, where k is a nonnegative integer, that is, k is an oriented surface with k handles. The genus of a graph G, denoted by $g(G)$, is the minimal integer n such that the graph can be embedded in S_n. Intuitively, G is embedded in a surface if it can be drawn in the surface so that its edges intersect only at their common vertices. Note that a graph G is a planar iff $g(G) = 0$ and G is toroidal iff $g(G) = 1$. Graph-theoretic terms are presented as they appeared in West [25]. Note that if x is a real number, then $\lceil x \rceil$ is the least integer that is greater than or equal to x. Throughout this paper, $\mathbb{V}$ denotes a finite-dimensional vector space over a field $\mathbb{F}$ and $\mathbb{F}_q$ is a finite field of order q. Let $\mathcal{B} = \{v_1, v_2, \ldots, v_n\}$ be a basis of an n-dimensional vector space $\mathbb{V}$ over a field $\mathbb{F}$. Then any vector $v \in \mathbb{V}$ can be expressed as a linear combination of the form $v = \sum_{i=1}^{n} \alpha_i v_i$. We define skeleton of v with respect to $\mathcal{B}$ as $S_B(v) = \{v_i \mid \alpha_i \neq 0\}$ and $\mathbb{V}_B = \{v \in \mathbb{V} \mid 0 < |S_B(v)| < n\}$, $S^k = \{v \in \mathbb{V} \mid k \leq |S_B(v)| \leq n-1\}$, $[S_v^k] = \{u \in \mathbb{V} \mid |\mathcal{S}_B(u)| = k \text{ and } v \in \mathcal{S}_B(u)\}$, $[S^k] = \{u \in \mathbb{V} \mid |S_B(u)| = k\}$, $\mathbb{V}^* = \mathbb{V} \setminus \{0\}$.

2 Introduction

The study of the interrelation between the graphs and algebraic structures had been an interesting area of research. In recent years, the study of graphs associated with rings and vector spaces has gained remarkable attention from many researchers. In the graphs, which are associated with algebraic structures, much attention has been paid to studying the combinatorial properties of algebraic structures, like clique number, chromatic number, domination number, independence number, etc. During the past seven decades, there has been ongoing interest concerning the relationship between algebraic structures, namely rings and graphs connected with algebraic structures, namely zero divisor graph, total graph, ideal inclusion graph, etc.; for recent survey articles, see [4] and [3]. For the study of graphs associated with a commutative ring, see [10] and [9]. The present survey entitled "Some recent results on the graphs of finite-dimensional vector spaces" is a part of research work carried out by authors [11, 13, 15, 18] and the main purpose of this survey article is to collect some properties of graphs associated with finite-dimensional vector spaces and investigate fundamental properties (chromatic and clique number, domination number, independence number, genus, etc.) of these graphs. Section 3 initiates with the investigation of the fundamental properties of the non-zero component graph. Recently, Das [13, 18] introduced and analysed the non-zero component graph on finite-dimensional vector spaces. The non-zero component graph is an undirected graph with all non-trivial vectors of $\mathbb{V}^*$ as the set of vertices, and any two distinct vectors u and v in G are adjacent if and only if u and v share at least one α_i with non-zero coefficient in their basic representation, i.e., there exists at least one α_i along which both u and v have non-zero components. Later on, Nikandish et al. [22] investigated the colouring of $\Gamma(\mathbb{V}_\alpha)$. Further Chelvam et al. [11] investigate the genus and various properties of $\Gamma(\mathbb{V}_\alpha)$. Section 4 deals with the study of non-zero component union graphs associated with the vectors of a finite-dimensional vector space. Recently, Das [15] introduced and studied non-zero component union graph $\Gamma(\mathscr{B})$ on a finite-dimensional vector space, where the graph $\Gamma(\mathscr{B})$ is an undirected graph with vertices as elements of $\mathbb{V}^*$ and any two distinct vectors u and v in $\mathbb{V}^*$ are adjacent if and only if $\mathscr{S}_{\mathscr{B}}(u) \cup \mathscr{S}_{\mathscr{B}}(v) = \mathscr{B}$. Very recently, Ashraf et al. introduced and studied distinct component graphs on finite-dimensional vector spaces. The graphs $\mathfrak{I}^*_{\mathfrak{B}}(\mathbb{V})$ are simple undirected graphs with $\mathbb{V}^*$ as a set of vertices and any two distinct vertices u, v in $\mathfrak{I}^*_{\mathfrak{B}}(\mathbb{V})$ are adjacent if and only if $\mathfrak{S}_{\mathfrak{B}}(\mathfrak{u}) \cap \mathfrak{S}_{\mathfrak{B}}(\mathfrak{v}) = \emptyset$. In Sect. 5, we look into the properties of a distinct component graph on vector spaces which was introduced by the authors together with Parveen in [23]. Section 6 goes through the article Component intersection graphs on finite-dimensional vector spaces in which the authors defined the notion of Component intersection graphs on finite-dimensional vector spaces as well as studied several important properties of this graph. Section 7 opens with the literature of subspace inclusion graphs introduced by Das [16] and further studied by various authors which can be seen in [6]. Finally in Sect. 8, we introduce a notion of a new type of graph viz., Dimension graph on finite-dimensional vector space and find out the several basic properties of this new graph.

3 Non-zero Component Graphs on Finite-Dimensional Vector Spaces

In this section, we give several specific results of non-zero component graphs that have appeared in the literature. This illustrates the power of this unifying concept and explains why these non-zero component graphs all share common properties related to diameter and girth. We start with some basic properties of $\Gamma(\mathbb{V}_\beta)$ essentially obtained for $\Gamma(\mathbb{V}_\beta)$ in [13].

Example 1 Let $\mathbb{F}_2 \times \mathbb{F}_2$ and $\mathbb{F}_2 \times \mathbb{F}_2 \times \mathbb{F}_2$ be the vector spaces over $\mathbb{F}_2$ and $\mathcal{B}_1 = \{u_1, u_2\}$, $\mathcal{B}_2 = \{v_1, v_2, v_3\}$ be a basis of $\mathbb{F}_2 \times \mathbb{F}_2$ and $\mathbb{F}_2 \times \mathbb{F}_2 \times \mathbb{F}_2$, respectively. Clearly, $V(\Gamma(\mathbb{V}_{\mathcal{B}_1}) = \{u_1, u_2, u_1 + u_2, 2u_1, 2u_2, u_1 + 2u_2, 2u_1 + u_2\}$ and

$V(\Gamma(\mathbb{V}_{\mathcal{B}_2}) = \{v_1, v_2, v_3, v_1 + v_2, v_2 + v_3, v_1 + v_3, v_1 + v_2 + v_3\}$ and Non-zero Component graphs $\Gamma(\mathbb{V}_{\mathcal{B}_1})$ and $\Gamma(\mathbb{V}_{\mathcal{B}_2})$ are given by the following Figs. 3.1 and 3.2:

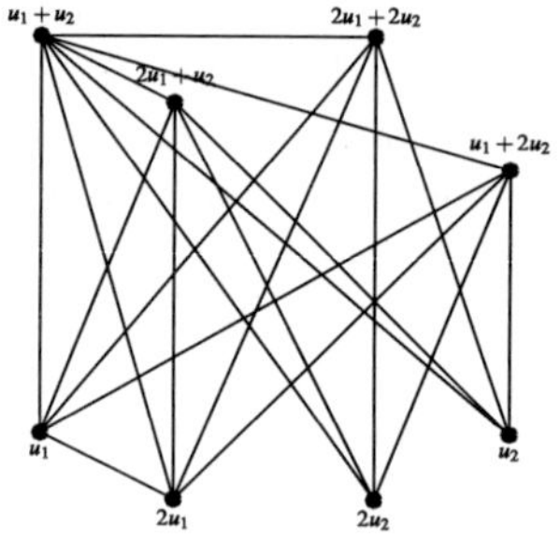

$\Gamma(\mathbb{V}_{\mathcal{B}_1})$
$\mathbb{F} = \mathbb{F}_2$
Figure 3.1

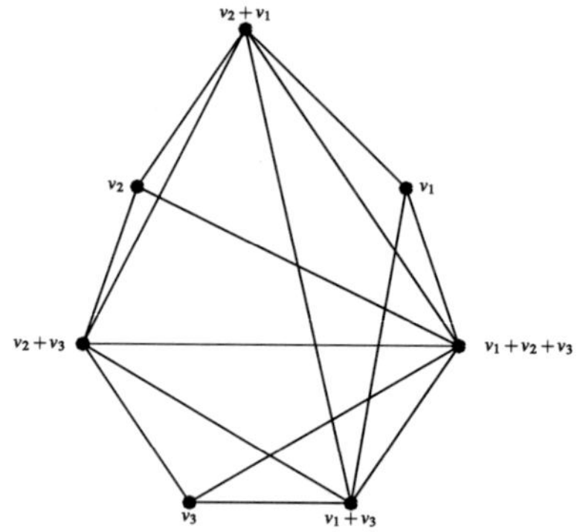

$\Gamma(\mathbb{V}_{\mathcal{B}_2})$
$\mathbb{F} = \mathbb{F}_2$
Figure 3.2

Theorem 1 ([13]) *Let* $\mathbb{V}$ *be a vector space over a field* $\mathbb{F}$. *Then the following statements hold:*
(1) *If* $\Gamma(\mathbb{V}_\alpha)$ *and* $\Gamma(\mathbb{V}_\beta)$ *are the graphs with respect to two bases* $\{\alpha_1, \alpha_2, \ldots, \alpha_n\}$ *and* $\{\beta_1, \beta_2, \ldots, \beta_n\}$, *then* $\Gamma(\mathbb{V}_\alpha)$ *and* $\Gamma(\mathbb{V}_\beta)$ *are graph isomorphic.*
(2) $\Gamma(\mathbb{V}_\alpha)$ *is connected and* $diam(\Gamma(\mathbb{V}_\alpha) = 2$.
(3) *The domination number of* $\Gamma(\mathbb{V}_\alpha)$ *is* 1.
(4) $\Gamma(\mathbb{V}_\alpha)$ *is complete if and only if* $\mathbb{V}$ *is one-dimensional.*

Note that the set $\{\alpha_1, \alpha_2, \ldots, \alpha_n\}$ is a minimal dominating set of $\Gamma(\mathbb{V}_\alpha)$. Now, the question arises: what is the maximum possible number of vertices in a minimal dominating set? The answer is given as n in view of the following results.

Theorem 2 ([13]) *If* $D = \{\beta_1, \beta_2, \ldots, \beta_\ell\}$ *is a minimal dominating set of* $\Gamma(\mathbb{V}_\alpha)$, *then* $\ell \leq n$, *i.e., the maximum cardinality of a minimal dominating set is* n. *The independence number of* $\Gamma(\mathbb{V}_\alpha)$ *is* $dim(\mathbb{V})$.

Theorem 3 ([13]) *Let* I *be an independent set in* $\Gamma(\mathbb{V}_\alpha)$, *then* I *is a linearly independent subset of* $\mathbb{V}$.

The converse of Theorem 3 is not true, in general. Consider a vector space $\mathbb{V}$, its basis $\{\alpha_1, \alpha_2, \ldots, \alpha_n\}$, and the set $L = \{\alpha_1 + \alpha_2, \alpha_2, \ldots, \alpha_n\}$. Clearly, L is linearly independent in $\mathbb{V}$, but it is not an independent set in $\Gamma(\mathbb{V}_\alpha)$ as $\alpha_1 + \alpha_2 \sim \alpha_2$.

Theorem 4 ([13]) *Let* $\mathbb{V}$ *and* $\mathbb{W}$ *be two finite-dimensional vector spaces over a field* $\mathbb{F}$. *Then the following statements hold:*
(1) *If* $\Gamma(\mathbb{V}_\alpha)$ *and* $\Gamma(\mathbb{V}_\beta)$ *are isomorphic as graphs with respect to some basis* $\{\alpha_1, \alpha_2, \ldots, \alpha_n\}$ *and* $\{\beta_1, \beta_2, \ldots, \beta_n\}$ *of* $\mathbb{V}$ *and* $\mathbb{W}$, *respectively, then* $dim(\mathbb{V}) = dim(\mathbb{W})$, *i.e.,* $n = k$.

(2) *If* $\mathbb{V}$ *and* $\mathbb{W}$ *are isomorphic as vector spaces, then for any basis* $\{\alpha_1, \alpha_2, \ldots, \alpha_n\}$ *and* $\{\beta_1, \beta_2, \ldots, \beta_n\}$ *of* $\mathbb{V}$ *and* $\mathbb{W}$*, respectively,* $\Gamma(\mathbb{V}_\alpha)$ *and* $\Gamma(\mathbb{V}_\beta)$ *are isomorphic as graphs.*
(3) *If for any basis* $\{\alpha_1, \alpha_2, \ldots, \alpha_n\}$ *and* $\{\beta_1, \beta_2, \ldots, \beta_n\}$ *of* $\mathbb{V}$ *and* $\mathbb{W}$*, respectively,* $\Gamma(\mathbb{V}_\alpha)$ *and* $\Gamma(\mathbb{V}_\beta)$ *are isomorphic as graphs, then* $\mathbb{V}$ *and* $\mathbb{W}$ *are isomorphic as vector spaces.*

Example 2 Consider a one-dimensional vector space $\mathbb{V}$ over $\mathbb{Z}_5$ generated by α (say). Then $\Gamma(\mathbb{V}_\alpha)$ is a complete graph of 4 vertices with $\mathbb{V} = \alpha, 2\alpha, 3\alpha, 4\alpha$. Consider the map $T : \Gamma(\mathbb{V}_\alpha) \to \Gamma(\mathbb{V}_\alpha)$ given by $T(\alpha) = 2\alpha, T(2\alpha) = \alpha, T(3\alpha) = 4\alpha, T(4\alpha) = 3\alpha$. Clearly, T is a graph isomorphism, but as $T(2\alpha) = \alpha 6 = 4\alpha = 2(2\alpha) = 2T(\alpha)$, T is not linear.

Theorem 5 ([13]) *Let* $\phi : \Gamma(\mathbb{V}_\alpha) \to \Gamma(\mathbb{V}_\alpha)$ *be a graph automorphism. Then,* ϕ *maps*
$\{\alpha_1, \alpha_2, \ldots, \alpha_n\}$ *to another basis* $\{\beta_1, \beta_2, \ldots, \beta_n\}$ *such that there exists* $\sigma \in S_n$*, where each* β_i *is of the form* $c_i \alpha\sigma_{(i)}$ *and each* $c_{i's}$ *is non-zero.*

It can be easily noted that ϕ maps the basis $\{\alpha_1, \alpha_2, \ldots, \alpha_n\}$ to another basis $\{\beta_1, \beta_2, \ldots, \beta_n\}$ it may not be a vector space isomorphism. It is because linearity of ϕ is not guaranteed as shown in Example 2. However, the following result is true.

Theorem 6 ([13]) *Let* ϕ *be a graph automorphism, which maps* $\alpha_i \mapsto c_{i_{j_i}} \alpha_{\sigma(i)}$ *for some* $\sigma \in S_n$*. Then, if* $c \neq 0$*,* $\phi(c\alpha_i) = d\alpha_{\sigma(i)}$ *for some non-zero* d*, more generally for all* $k \in 1, 2, \ldots, n$ *if* $c_1.c_2 \ldots c_k \neq 0$*, then*

$$\phi(c_1\alpha_{i_1} + c_2\alpha_{i_2} + \ldots + c_k\alpha_{i_k}) = d_1\alpha(i_1) + d_2\alpha(i_2) + \ldots + d_k\alpha(i_k)$$

for some $d_{i's}$ *with* $d_1.d_2 \ldots d_k \neq 0$*.*

Corollary 1 $\Gamma(\mathbb{V}_\alpha)$ *is not vertex transitive if* $dim(\mathbb{V}) > 1$*.*

The set of vertices adjacent to $\alpha_{i_1} + \alpha_{i_2} + \ldots + \alpha_{i_k}$ is the same as the set of vertices adjacent to $c_1\alpha_{i_1} + c_2\alpha_{i_2} + \ldots + c_k\alpha_{i_k}$, i.e., $N(\alpha_{i_1} + \alpha_{i_2} + \ldots + \alpha_{i_k}) = c_1\alpha_{i_1} + c_2\alpha_{i_2} + \ldots + c_k\alpha_{i_k})$ for $c_1c_2 \ldots c_k \neq 0$.

Theorem 7 ([13]) *Let* $\mathbb{V}$ *be a vector space over a finite field* $\mathbb{F}$ *with* q *elements and* Γ *be its associated graph with respect to a basis* $\alpha_1, \alpha_2, \ldots, \alpha_n$*. Then, the degree of the vertex* $\alpha_{i_1} + \alpha_{i_2} + \ldots + \alpha_{i_k}$*, where* $c_1c_2 \ldots c_k \neq 0$*, is* $(q^k - 1)q^{n-k} - 1$*.*

In view of the following theorems, the authors in [18] studied minimum degree, edge, connectivity, order and size of $\Gamma(\mathbb{V}_\alpha)$.

Theorem 8 ([18]) *Let* $\mathbb{V}$ *be a vector space over a finite field* $\mathbb{F}$ *with* q *elements.*
(1) $\Gamma(\mathbb{V}_\alpha)$ *is not Eulerian.*
(2) *Then the minimum degree* δ *of* $\Gamma(\mathbb{V}_\alpha)$ *is* $q^{n-1}(q-1) - 1$*.*
(3) *Edge connectivity of* $\Gamma(\mathbb{V}_\alpha)$ *is* $q^{n-1}(q-1) - 1$*.*
(4) *The order and size of* $\Gamma(\mathbb{V}_\alpha)$ *is* $q^n - 1$ *and* $\frac{q^{2n}-q^n+1-(2q-1)^n}{2}$*.*

Das in [18] studied the structure of maximal cliques in $\Gamma(\mathbb{V}_\alpha)$ and its clique number. In this process, he proved that $\Gamma(\mathbb{V}_\alpha)$ possess two different classes of maximal cliques. Let M be a maximal clique in $\Gamma(\mathbb{V}_\alpha)$ and $S(M) = \{\mathcal{S}_\mathcal{B}(\beta) : \beta \in M\}$ and $S[M] = \{|\mathcal{S}_\mathcal{B}(\beta)| : \mathcal{S}_\mathcal{B}(\beta) \in S(M)\}$. Since M is a clique, $\mathcal{S}_\mathcal{B}(u) \cap \mathcal{S}_\mathcal{B}(v) \neq \emptyset, \forall u, v \in M$. By maximality of M, if $u \in M$ and $\mathcal{S}_\mathcal{B}(u) \subset \mathcal{S}_\mathcal{B}(v)$ for some $v \in \Gamma(\mathbb{V}_\alpha)$, then $v \in M$. As $\emptyset \neq S[M] \subset N$, by well-ordering principle, it has a least element, say k. Then there exists some $v^* \in M$ with $|\mathcal{S}_\mathcal{B}(u^*)| = k$, where $v^* = c_1\alpha_{i_1} + c_2\alpha_{i_2} + \ldots + c_k\alpha_{i_k}$.

Theorem 9 ([18]) *Let M be a maximal clique in $\Gamma(\mathbb{V}_\alpha)$. If k is the least element of $S[M]$ and $k \leq \frac{n}{2}$, then M belongs to a family of maximal cliques $\{M_{k,i} : 1 \leq k \leq \frac{n}{2}; i \in \{1, 2, \ldots, n\}\}$ of $\Gamma(\mathbb{V}_\alpha)$ where $M_{k,i} = \{v \in \Gamma(\mathbb{V}_\alpha) : \alpha_i \mathcal{S}_\mathcal{B}(v)\}$ and $|\mathcal{S}_\mathcal{B}(v)| \geq k\}$ and*

$$|M| = (q-1) \sum_{r=k-1}^{n-1} \binom{n-1}{r} (q-1)^r.$$

It is to be noted that for same value of k and by fixing different α_i, we get different maximal cliques. Since these maximal cliques depend both on k and α_i, we get a family of maximal cliques $M_{k,i}$ where $1 \leq k \leq \frac{n}{2}$ and $i \in \{1, 2, \ldots, n\}$ and $M \in M_{k,i}$.

Theorem 10 ([18]) *Let M be a maximal clique in $\Gamma(\mathbb{V}_\alpha)$. If k is the least element of $S[M]$ and $k > \frac{n}{2}$, then $k = \lfloor \frac{n}{2} \rfloor + 1$ and $M = \{v \in \Gamma(\mathbb{V}_\alpha : |\mathcal{S}_\mathcal{B}(v)| \geq \lfloor \frac{n}{2} \rfloor + 1\}$ and*

$$|M| = \sum_{r=k}^{n} \binom{n}{r} (q-1)^r.$$

It is obvious from Theorem 9 that $|M_{k,i}|$ is maximum when $k = 1$, i.e., $M_{1,i} = \{c_1\alpha_1 + c_2\alpha_2 + \ldots + c_n\alpha_n : c_i = 0\}$ and $|M_{1,i}| = (q-1)q^{n-1}$. Thus, the clique number of $\Gamma(\mathbb{V}_\alpha)$ is

$$\omega(\Gamma(\mathbb{V}_\alpha)) = max\{(q-1)q^{n-1}, \sum_{r=\lfloor \frac{n}{2} \rfloor+1}^{n} \binom{n-1}{r} (q-1)^r\}$$

and it depends on the value of q and n. A maximal clique M contains at most one α_i, because if M contain α_i and α_j, then $\alpha_i \sim \alpha_j$ which contradicts that M is a clique. Moreover, if M is a maximal clique containing α_i, then $M = M_{1,i}$. It follows since every $v \in M$ is adjacent to α_i, i.e., every v has a non-zero component along α_i and hence $M_{1,i} \subset M$. Now, by maximality of $M_{1,i}$, it follows that $M = M_{1,i}$.

Corollary 2 ([18]) *Let $\mathbb{V}$ be a vector space over a finite field $\mathbb{F}$ with q elements.*
(1) If $q = 2$, then $\omega(\Gamma(\mathbb{V}_\alpha)) = 2^{n-1}$.
(2) If $q > 2$ and n is odd, then $\omega(\Gamma(\mathbb{V}_\alpha)) = \sum_{r=\lfloor \frac{n}{2} \rfloor+1}^{n} \binom{n-1}{r}(q-1)^r$.
(3) If $q = 2$, then $\Gamma(\mathbb{V}_\alpha)$, then $2^{n-1} \leq \chi(\Gamma(\mathbb{V}_\alpha)) \geq 2^{n-1} + 2^{n-2} - \frac{n}{2}$.
(4) If $q > 2$, then $\Gamma(\mathbb{V}_\alpha)$ is Hamiltonian.

(5) *If* $q = 2$ *and* $n \geq 3$, *then* $\Gamma(\mathbb{V}_\alpha)$ *is 2-connected.*
(6) *If* $q = 2$ *and* $n \geq 3$, *then* $\delta \geq max\{(|\Gamma(\mathbb{V}_\alpha)| + 2)/3, \alpha(\Gamma(\mathbb{V}_\alpha))\}$.
(7) *If* $q = 2$ *and* $n \geq 3$, *then* $\Gamma(\mathbb{V}_\alpha)$ *is Hamiltonian.*

While Das in [13] was mainly interested in the study of interrelationship between vector space isomorphisms and graph isomorphisms, also, the automorphism group of the non-zero component graph was studied. In the conclusion part of the above-mentioned paper, the author proposed the colouring of a non-zero component graph as a topic of further research. After that in [18], he studied the clique number and chromatic number for some particular cases and in [18] is a natural continuation of the study in this direction. The main aim of this paper is to show that $\Gamma(\mathbb{V}_\alpha)$ is a weakly perfect graph. Also, it is proved that $\chi'(\Gamma(\mathbb{V}_\alpha)) = \Delta(\Gamma(\mathbb{V}_\alpha))$. In view of the following results in [18], he proved that $\chi(\Gamma(\mathbb{V}_\alpha)) = \omega(\Gamma(\mathbb{V}_\alpha))$. Also, he gave an explicit formula for $\chi(\Gamma(\mathbb{V}_\alpha))$.

Theorem 11 ([22]) *Let* $\mathbb{V}$ *be an n-dimensional vector space over the field* $\mathbb{F}_q$. *Then* $\Gamma(\mathbb{V}_\alpha)$ *is weakly perfect.*

Let $\mathbb{V}$ be a finite-dimensional vector space over a field $\mathbb{F}$. If $\mathbb{F}$ is an infinite field. Then, it is not hard to see that the set $\{a\alpha_1\}$, where a runs over all non-zero elements of $\mathbb{F}$, is an infinite clique of $\Gamma(\mathbb{V}_\alpha)$. Thus, we may assume that $\mathbb{F}$ is a finite field of order q, where q is a power of a prime number. Let $a = a_1\alpha_{i_1} + \ldots + a_k\alpha_{i_k}$ and $b = b_1\alpha_{j_1} + \ldots + b_l\alpha_{j_l}$ be two distinct vertices of $\Gamma(\mathbb{V}_\alpha)$ ($a_1, \ldots, a_k, b_1, \ldots, b_l$ are all non-zero). Define the relation $\sim$ on $V(\Gamma(\mathbb{V}_\alpha))$ as follows: $a \sim b$ if and only if $\mathcal{S}a = \mathcal{S}b$, where for every $\beta \in V(\Gamma(\mathbb{V}_\alpha))$ the set S_β (skeleton of β) is the set of $\alpha_{i's}$ with non-zero coefficients in the basic representation of β with respect to $\{\alpha_1, \alpha_2, \ldots, \alpha_n\}$. Indeed, $a \sim b$ if and only if $\{\alpha_{i_1}, \ldots, \alpha_{i_k}\} = \{\alpha_{j_1}, \ldots, \alpha_{j_l}\}$. It is easily seen that $\sim$ is an equivalence relation on $V(\Gamma(\mathbb{V}_\alpha))$. By $[a]$, we mean the equivalence class of a.

Theorem 12 ([22]) *Let* $\mathbb{V}$ *be an n-dimensional vector space over the field* $\mathbb{F}_q$. *Then* $\Gamma(\mathbb{V}_\alpha)$ *is weakly perfect.*

The following corollaries directly follow from the above theorem.

Corollary 3 *Let* $\mathbb{V}$ *be an n-dimensional vector space over the field* $\mathbb{F}_q$.
(1) *If n is odd, then*

$$\omega(\Gamma(\mathbb{V}_\alpha)) = \chi(\Gamma(\mathbb{V}_\alpha)) = \sum_{i=0}^{\lfloor \frac{n}{2} \rfloor} \binom{n}{i} (q-1)^{n-i}.$$

(2) *If n is even, then*

$$\omega(\Gamma(\mathbb{V}_\alpha)) = \chi(\Gamma(\mathbb{V}_\alpha)) = \sum_{i=0}^{\frac{n}{2}-1} \binom{n}{i} (q-1)^{n-i} + \frac{\binom{n}{\frac{n}{2}}(q-1)^{\frac{n}{2}}}{2}.$$

In the following theorem, Nikandish et al. [22] studied the edge chromatic number of $\Gamma(\mathbb{V}_\alpha)$ and proved that, for every positive integer n, $\chi'(\Gamma(\mathbb{V}_\alpha)) = \Delta(\Gamma(\mathbb{V}_\alpha))$.

Theorem 13 ([22]) *Let $\mathbb{V}$ be an n-dimensional vector space over a field $\mathbb{F}_q$. Then the following statements hold:*
(1) $\Delta(\Gamma(\mathbb{V}_\alpha)) = q^n - 2$.
(2) $\delta(\Gamma(\mathbb{V}_\alpha)) = q^n - q^{n-1} - 1$.
(3) *The number of vertices with maximum degree in $\Gamma(\mathbb{V}_\alpha)$ is $(q-1)^n$.*
(4) *The number of vertices with minimum degree in $\Gamma(\mathbb{V}_\alpha)$ is $n(q-1)$.*
(5) $\chi'(\Gamma(\mathbb{V}_\alpha)) = \delta(\Gamma(\mathbb{V}_\alpha))$.

The vertex connectivity and the girth of $\Gamma(\mathbb{V}_\alpha)$ had been studied in [11]. If $\mathbb{V}$ is a one-dimensional vector space over a field F_q with $q \geq 2$) elements, then by Theorem 1 (5), $\Gamma(\mathbb{V}_\alpha)$ is complete and $\Gamma(\mathbb{V}_\alpha) \cong K_{q-1}$. Hence, the vertex connectivity of the non-zero component graph associated with a one-dimensional vector space over a finite field of q elements is $q-2$. In [11], Chelvam et al. were interested in the vertex connectivity of $\Gamma(\mathbb{V}_\alpha)$ corresponding to $k \geq 2$-dimensional vector spaces over a finite field.

Theorem 14 ([11]) *Let $\mathbb{V}$ be a k-dimensional vector space over a field F_q such that $k \geq 2$ and $q \geq 2$. Then the following statements hold:*
(1) *The vertex connectivity of $\Gamma(\mathbb{V}_\alpha)$ is $q^{k-1}(q-1) - (q-1)$.*
(2) *$\Gamma(\mathbb{V}_\alpha)$ is planar if and only if either $k = 1$ and $q \leq 5$ (or) $k = 2$ and $q = 2$ (or) $k = 3$ and $q = 2$.*
(3) *$\Gamma(\mathbb{V}_\alpha)$ is planar if and only if $k = 1$ and $q \leq 5$ (or) $k = 2$ and $q = 2$ (or) $k = 3$, $q = 2$.*
(4) *$\Gamma(\mathbb{V}_\alpha)$ is toroidal if and only if $k = 1$ and $q = 7, 8$.*
(5) *$g(\Gamma(\mathbb{V}_\alpha)) = 2$ if and only if either $k = 1$ and $q = 9$ or $k = 2$ and $q = 3$.*

In view of the following theorem, the authors in [2] analysed removability of $\Gamma(\mathbb{V}_\alpha)$.

Theorem 15 [2] *Let $\mathbb{V}$ be an n-dimensional vector space over a field $\mathbb{F}_q$. Then the following statements hold:*
(1) *If $q = 2$ and $n = 2$, then $dim_{\Gamma(\mathbb{V}_\alpha)} = 1$.*
(2) *If $q = 2$ and $n \geq 3$, then $dim_{\Gamma(\mathbb{V}_\alpha)} = n$.*
(3) *If $q \geq 3$, then $dim_{\Gamma(\mathbb{V}_\alpha)} = \sum_{k=1}^{n} \binom{n}{k}((q-1)^k - 1)$.*
(4) *If $q \geq 2$ and $n \geq 3$, then $pd_{\Gamma(\mathbb{V}_\alpha)} = n + (q-1)^n$.*

4 Non-zero Component Union Graph of a Finite-Dimensional Vector Space

In this section, we assemble the results of non-zero component union graph studied by the authors in [15].

Definition 1 Let $\mathcal{B}$ be a basis of finite-dimensional vector space $\mathbb{V}$ over a field $\mathbb{F}$. Then any two distinct vectors u and v of non-zero component graph $\Gamma(\mathcal{B})$ are adjacent if and only if $\mathcal{S}_{\mathcal{B}}(u) \cup \mathcal{S}_{\mathcal{B}}(v) = \mathcal{B}$.

Example 3 Let $\mathbb{F}_2 \times \mathbb{F}_2$ and $\mathbb{F}_2 \times \mathbb{F}_2 \times \mathbb{F}_2$ be the vector spaces over $\mathbb{F}_2$ and $\mathcal{B}_1 = \{u_1, u_2\}$, $\mathcal{B}_2 = \{v_1, v_2, v_3\}$ be a basis of $\mathbb{F}_2 \times \mathbb{F}_2$ and $\mathbb{F}_2 \times \mathbb{F}_2 \times \mathbb{F}_2$, respectively. Clearly, $V(\Gamma(\mathcal{B}_1) = \{u_1, u_2, u_1 + u_2, 2u_1, 2u_2, u_1 + 2u_2, 2u_1 + u_2\}$ and $V(\Gamma(\mathcal{B}_2)$ $= \{v_1, v_2, v_3, v_1 + v_2, v_2 + v_3, v_1 + v_3, v_1 + v_2 + v_3\}$ and Non-zero Component union graphs $\Gamma(\mathcal{B}_1)$ and $\Gamma(\mathcal{B}_2)$ are given by the following Figs. 4.1 and 4.2:

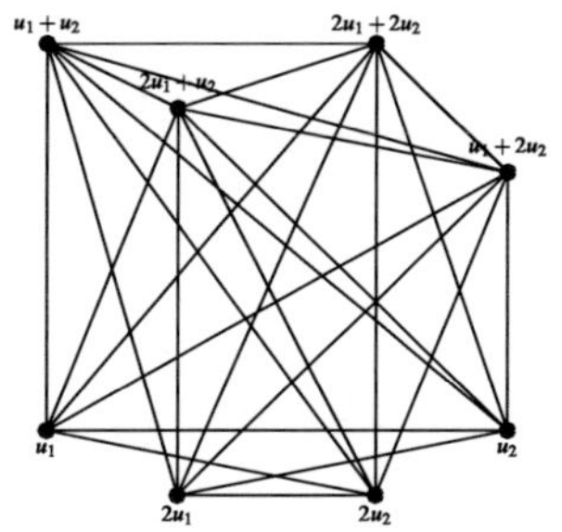

$\Gamma(\mathcal{B}_1)$
$\mathbb{F} = \mathbb{F}_2$
Figure 4.1

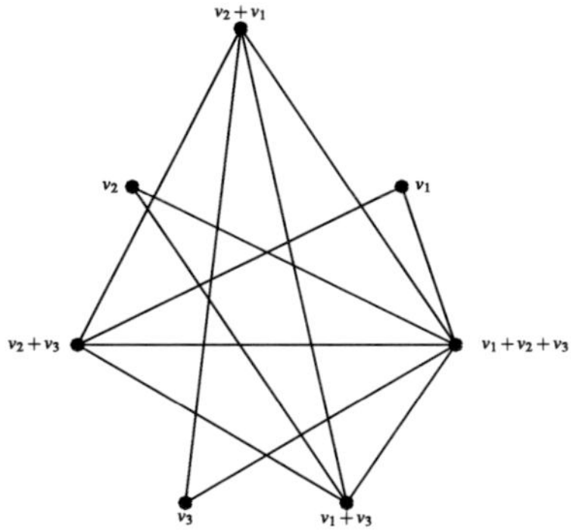

$\Gamma(\mathcal{B}_2)$
$\mathbb{F} = \mathbb{F}_2$
Figure 4.2

Theorem 16 ([15]) *Let $\mathbb{V}$ be a vector space over a field $\mathbb{F}$. Let $\Gamma(\mathcal{B}_1)$ and $\Gamma(\mathcal{B}_2)$ be the graphs associated with $\mathbb{V}$ w.r.t. two bases $\mathcal{B}_1$ and $\mathcal{B}_2$ of $\mathbb{V}$. Then $\Gamma(\mathcal{B}_1)$ and $\Gamma(\mathcal{B}_2)$ are graph isomorphic.*

The above theorem shows that the graph properties associated with G do not depend on the choice of the basis $\mathcal{B}$. However, two vertices may be adjacent with respect to one basis but non-adjacent to some other basis as shown in the following example. Let $\mathbb{V} = \mathbb{R}^2$,$\mathbb{F} = \mathbb{R}$ with two bases $\mathcal{B}_1 = \{\alpha_1 = (1, 0), \alpha_2 = (0, 1)\}$ and $\mathcal{B}_2 = \{\beta_1 = (1, 1), \beta_2 = (-1, 1)\}$. Consider $a = (1, 1)$ and $b = (2, 2)$. Clearly, $a \sim b$ in $\Gamma(\mathcal{B}_1)$, but $a \nsim b$ in $\Gamma(\mathcal{B}_2)$.

Theorem 17 ([15]) *$\Gamma(\mathcal{B})$ is connected and $diam\Gamma(\mathcal{B}) = 2$. The domination number of $\Gamma(\mathcal{B})$ is one. $\Gamma(\mathcal{B})$ is complete if and only if $\mathbb{V}$ is one-dimensional or $\mathbb{V}$ is two-dimensional and $|\mathbb{F}| = 2$. Girth of $\Gamma(\mathcal{B}) = 3$ except when $\mathbb{V}$ is one-dimensional with $|F| = 2$ or 3. Moreover, every vertex in $\Gamma(\mathcal{B})$ belongs to some triangle in $\Gamma(\mathcal{B})$.*

Let $V_n = \{\alpha \in V \ : \ |\mathcal{S}_{\mathcal{B}}(u)| = n\}$ where n is the dimension of the vector space $\mathbb{V}$. Clearly, V_n is a clique in $\Gamma(\mathcal{B})$. It is to be noted that V_n is not a maximal clique as $V_n \cup \{\alpha\}$ is also a clique for any non-zero $\alpha \in V$. Moreover, as vertices in V_n are adjacent to all vertices in $\Gamma(\mathcal{B})$, V_n is contained in any maximal clique of $\Gamma(\mathcal{B})$. In view of the following theorem, Das [15] studied maximal cliques in $\Gamma(\mathcal{B})$ and a proper colouring of $\Gamma(\mathcal{B})$. Moreover, he proved that two non-zero component union graphs are isomorphic if and only if the base vector spaces over the same field are isomorphic. Let $n > 1$ and M be a clique in $\Gamma(\mathcal{B})$ and $\alpha \in M$ with $|\mathcal{S}_{\mathcal{B}}(\alpha)| < n$. If $\beta \in \mathbb{V}$ be such that $\mathcal{S}_{\mathcal{B}}(\beta) \subseteq \mathcal{S}_{\mathcal{B}}(\alpha)$, then $\beta \nsim \alpha$. $\mathcal{S}_{\mathcal{B}}(\alpha) \subseteq \mathcal{S}_{\mathcal{B}}(\beta)$. In particular, among all vectors $\alpha \in \mathbb{V} \setminus V_n$ having the same skeleton $\mathcal{S}_{\mathcal{B}}(\alpha)$, at most one of them can be in a clique. Define $\xi_i = \Sigma_{j=1}^{n}\alpha_j - \alpha_i$, for $i = 1, 2, \ldots, n$ and let $U_{n-1} = \xi_1, \xi_2, \ldots, \xi_n$. Observe that U_{n-1} is also a clique in $\Gamma(\mathcal{B})$.

Theorem 18 ([15]) *$\mathcal{M} = V_n \cup U_{n-1}$ is a maximal clique in $\Gamma(\mathcal{B})$.*

The following theorem shows that the graph theoretical properties of $\Gamma(\mathcal{B})$ depend on the vector space.

Theorem 19 ([15]) *Let $\mathbb{V}$ and $\mathbb{W}$ be two finite-dimensional vector spaces over the same field $\mathbb{F}$ having bases $\mathcal{B}$ and $\mathcal{B}'$. Then $\mathbb{V}$ and $\mathbb{W}$ are isomorphic as vector spaces if and only if $\Gamma(\mathcal{B})$ and $\Gamma(\mathcal{B}')$ are isomorphic as graphs.*

Theorem 20 ([15]) *Let $\mathbb{V}$ be a vector space over a finite field $\mathbb{F}$ with q elements and $\Gamma(\mathcal{B})$ be its associated graph with respect to a basis $B = \{\alpha_1, \alpha_2, \ldots, \alpha_n\}$. Then, the degree of the vertex $\alpha = c_1\alpha_{i_1} + c_2\alpha_{i_2} + \cdots + c_k\alpha_{i_k}$, where $c_1c_2\ldots c_k \neq 0$, is $(q-1)^{n-k}q^k$, if $1 < k < n$ $q^n - 2$, if $k = n$.*

Corollary 4 ([15]) *The maximum and minimum degree of $\Gamma(\mathcal{B})$ where $\mathbb{F}$ is a finite field with q elements is $\Delta = q^n - 2$ and $\delta = q(q-1)n - 1$. $\Gamma(\mathcal{B})$ is Eulerian if and only if q is even.*

Note that the order of base field q is odd, and the only odd degree vertices are the vertices with maximum degree. The other vertices are of even degree. In view of the following theorems, the author in ([15]) studied order, size, clique and chromatic numbers of $\Gamma(\mathcal{B})$:

Theorem 21 ([15]) *If $\mathbb{V}$ is an n-dimensional vector space over a finite field $\mathbb{F}$ with q elements, then the order of $\Gamma(\mathcal{B})$ is $q^n - 1$ and the size m of $\Gamma(\mathcal{B})$ is $\frac{(q-1)^n[(q+1)^n-3]}{2}$.*

Theorem 22 ([15]) *Let $\mathbb{V}$ be an n-dimensional vector space over a finite field $\mathbb{F}$ with q elements. Then the clique number and chromatic number of $\Gamma(\mathcal{B})$ are both equal to $n + (q-1)^n$, i.e., $\Gamma(\mathcal{B})$ is weakly perfect.*

In view of the following theorems, the author in [15] studied the maximal independent sets in $\Gamma(\mathcal{B})$ and provide a lower bound on the independence number of $\Gamma(\mathcal{B})$.

Theorem 23 ([15]) *$I = \{\alpha \in \Gamma(\mathbb{V}_{\mathcal{B}}) : S(\alpha) \subseteq \{\alpha_1, \alpha_2, \ldots, \alpha_{n-1}\}\}$ is a maximal independent set in $\Gamma(\mathbb{V}_{\mathcal{B}})$. Moreover, if $\mathbb{F}$ is a finite field with q elements, $|I| = q^{n-1} - 1$.*

Theorem 24 ([15]) *Let I be a maximal independent set in $\Gamma(\mathbb{V}_{\mathcal{B}})$ and $S[I] = \{|S(\alpha)| : \alpha \in I\}$. If k is the maximum element of $S[I]$ with $k < n/2$, then $k = \lceil \frac{n}{2} \rceil - 1$ and $I = \{\alpha \in \Gamma(\mathbb{V}_{\mathcal{B}}) : |S(\alpha)| < \frac{n}{2}\}$. Moreover, if $\mathbb{F}$ is a finite field with q elements,*

$$|I| = \sum_{r=1}^{\lceil \frac{n}{2} \rceil - 1} \binom{n}{r} (q-1)^r.$$

Corollary 5 *Independence number of $\Gamma(\mathbb{V}_{\mathcal{B}})$ is greater than or equal to $q^{n-1} - 1$.*

5 Distinct Component Graph on Finite-Dimensional Vector Space

The authors in [23] introduced component inclusion graph $\mathcal{I}^*_{\mathcal{B}}(\mathbb{V})$ and studied various fundamental properties of $\mathcal{I}^*_{\mathcal{B}}(\mathbb{V})$. In this section, we give all fundamental results of $\mathcal{I}^*_{\mathcal{B}}(\mathbb{V})$ studied by the authors [23].

Definition 2 Let $\mathbb{V}$ be an n-dimensional vector space over a field $\mathbb{F}$. Then component inclusion graph $\mathcal{I}^*_{\mathcal{B}}(\mathbb{V})$ is a simple(undirected) graph with the set of vertices $\mathbb{V}^*$, and any two distinct vertices u and v of $\mathcal{I}_{\mathcal{B}}(\mathbb{V})$ are adjacent if and only if $\mathcal{S}_{\mathcal{B}}(u) \neq \mathcal{S}_{\mathcal{B}}(v)$.

The following example illustrates the graphical representation of $\mathbb{F}_3^2$ and $\mathbb{F}_2^3$ by distinct component graphs $\mathcal{I}^*_{\mathcal{B}_1}(\mathbb{F}_3^2)$ and $\mathcal{I}^*_{\mathcal{B}_2}(\mathbb{F}_2^3)$, respectively.

Example 4 Let $\mathbb{F}_3 \times \mathbb{F}_3$ and $\mathbb{F}_2 \times \mathbb{F}_2 \times \mathbb{F}_2$ be the vector spaces over $\mathbb{F}_3$ and $\mathbb{F}_2$, respectively, and $\mathcal{B}_1 = \{u_1, u_2\}$, $\mathcal{B}_2 = \{v_1, v_2, v_3\}$ be a basis of $\mathbb{F}_3 \times \mathbb{F}_3$ and $\mathbb{F}_2 \times \mathbb{F}_2 \times \mathbb{F}_2$, respectively. Clearly, $V(\mathcal{I}^*_{\mathcal{B}_1}(\mathbb{F}_3 \times \mathbb{F}_3)) = \{u_1, u_2, u_1 + u_2, 2u_1, 2u_2, u_1 + 2u_2, 2u_1 + u_2\}$ and $V(\mathcal{I}^*_{\mathcal{B}_2}(\mathbb{F}_2 \times \mathbb{F}_2 \times \mathbb{F}_2)) = \{v_1, v_2, v_3, v_1 + v_2, v_2 + v_3, v_1 + v_3, v_1 + v_2 + v_3\}$ and distinct component graphs $\mathcal{I}^*_{\mathcal{B}_1}(\mathbb{F}_3 \times \mathbb{F}_3)$ and $\mathcal{I}^*_{\mathcal{B}_2}(\mathbb{F}_2 \times \mathbb{F}_2 \times \mathbb{F}_2)$ are given by the following Figs. 5.1 and 5.2:

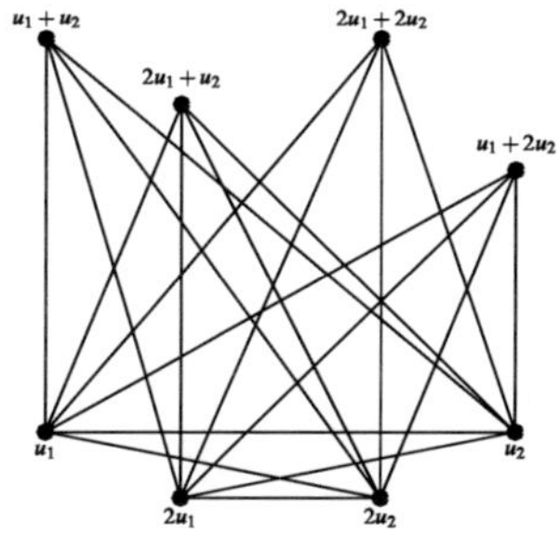

$\mathcal{I}^*_{\mathcal{B}_1}(\mathbb{F}_3 \times \mathbb{F}_3)$
$\mathbb{F} = \mathbb{F}_3$
Figure 5.1

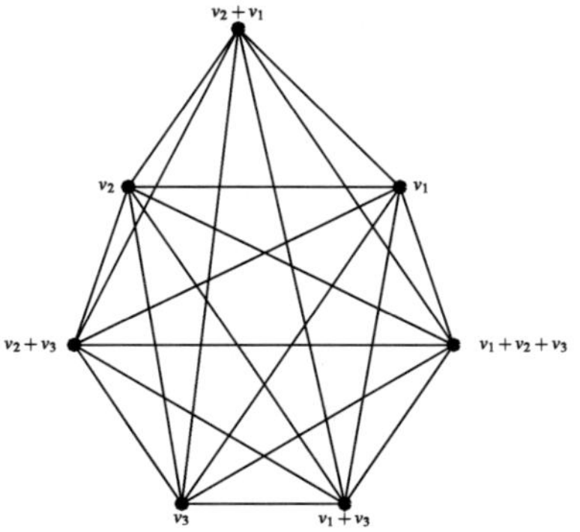

$\mathcal{I}^*_{\mathcal{B}_2}(\mathbb{F}_2 \times \mathbb{F}_2 \times \mathbb{F}_2)$
$\mathbb{F} = \mathbb{F}_2$
Figure 5.2

Theorem 25 ([23]) *Let $\mathbb{V}$ be an n-dimensional vector space over the field $\mathbb{F}$. Then the following statements hold:*
(1) *$\mathcal{I}^*_{\mathcal{B}}(\mathbb{V})$ is an edgeless graph if and only if n=1.*
(2) *$\mathcal{I}^*_{\mathcal{B}}(\mathbb{V})$ is connected and $diam(\mathcal{I}^*_{\mathcal{B}}(\mathbb{V})) \leq 2$.*
(3) *$\mathcal{I}^*_{\mathcal{B}}(\mathbb{V})$ is triangulated and hence $gr(\mathcal{I}^*_{\mathcal{B}}(\mathbb{V})) = 3$.*

Theorem 26 ([23]) *Let $\mathbb{V}$ be a finite-dimensional vector space over the field $\mathbb{F}$. Then $\mathcal{I}^*_{\mathcal{B}}(\mathbb{V}) = K_{2^n-1}$ if and only if $\mathbb{V}$ is n-dimensional with $|\mathbb{F}| = 2$.*

The following corollary gives a characterization on the diameter of $\mathcal{I}^*_{\mathcal{B}}(\mathbb{V})$.

Corollary 6 ([23]) *Let $\mathbb{V}$ be n-dimensional vector spaces over $\mathbb{F}$ such that $n \geq 2$. Then*

$$diam(\mathcal{I}^*_{\mathcal{B}}(\mathbb{V})) = \begin{cases} 1, & if \quad |\mathbb{F}| = 2 \\ 2, & if \quad |\mathbb{F}| \neq 2. \end{cases}$$

Theorem 27 ([23]) *Let $\mathbb{V}$ be n-dimensional vector spaces over $\mathbb{F}$ such that $n \geq 2$. Then*

$$\gamma(\mathcal{I}^*_{\mathcal{B}}(\mathbb{V})) = \begin{cases} 1, & if \quad |\mathbb{F}| = 2 \\ 2, & if \quad |\mathbb{F}| \neq 2. \end{cases}$$

By the above theorem, the author in [23] characterized the domination number of $\mathcal{I}^*_{\mathcal{B}}(\mathbb{V})$. Moreover, in view of the following theorem, they also studied clique and chromatic numbers of $\mathcal{I}^*_{\mathcal{B}}(\mathbb{V})$.

Theorem 28 ([23]) *Let $\mathbb{V}$ be a finite-dimensional vector space. Then the following statements hold:*
(1) $\omega(\mathcal{I}^*_{\mathcal{B}}(\mathbb{V})) = 2^n - 1$ *if and only if $\mathbb{V}$ is n-dimensional.*
(2) *If $\mathbb{V}$ is an n-dimensional, then* $\chi(\mathcal{I}^*_{\mathcal{B}}(\mathbb{V})) = 2^n - 1$.

The following theorem shows that the graph theoretical properties do not depend on the choice of the basis.

Theorem 29 ([23]) *Let $\mathbb{V}$ be a finite-dimensional vector space and $\mathcal{I}^*_{\mathcal{B}_1}(\mathbb{V})$, $\mathcal{I}^*_{\mathcal{B}_2}(\mathbb{V})$ be the graphs associated with respect to the basis $\mathcal{B}_1$ and $\mathcal{B}_2$. Then the graphs $\mathcal{I}^*_{\mathcal{B}_1}(\mathbb{V})$ and $\mathcal{I}^*_{\mathcal{B}_2}(\mathbb{V})$ are isomorphic.*

Corollary 7 ([23]) *Let $\mathbb{V}_1$ and $\mathbb{V}_2$ be two finite-dimensional vector spaces over the same field $\mathbb{F}$ with bases $\mathcal{B}_1$ and $\mathcal{B}_2$. Then $\mathbb{V}_1$ and $\mathbb{V}_2$ are isomorphic (as vector spaces) if and only if $\mathcal{I}^*_{\mathcal{B}_1}(\mathbb{V}_1)$ and $\mathcal{I}^*_{\mathcal{B}_2}(\mathbb{V}_2)$ are isomorphic (as graphs).*

Corollary 8 ([23]) *Let $\mathbb{V}$ be a vector space over a field $\mathbb{F}$ and $\mathcal{I}^*_{\mathcal{B}_1}(\mathbb{V})$, $\mathcal{I}^*_{\mathcal{B}_2}(\mathbb{V})$ be the graphs associated with $\mathbb{V}$ with respect to the bases $\mathcal{B}_1 = \{u_1, u_2, \cdots, u_n\}$ and $\mathcal{B}_2 = \{v_1, v_2, \cdots, v_n\}$ of $\mathbb{V}$, respectively. Then the graphs $\mathcal{I}^*_{\mathcal{B}_1}(\mathbb{V})$ and $\mathcal{I}^*_{\mathcal{B}_2}(\mathbb{V})$ are isomorphic.*

Let $\mathbb{V}$ be a one-dimensional vector space over $\mathbb{Z}_5$ generated by $\mathcal{B} = \{u\}$ (say). Then $\mathcal{I}^*_{\mathcal{B}}(\mathbb{V})$ is an edgeless graph of 4 vertices with $V(\mathcal{I}^*_{\mathcal{B}}(\mathbb{V})) = \{u, 2u, 3u, 4u\}$. Consider the map $\mathbb{T} : \mathcal{I}^*_{\mathcal{B}}(\mathbb{V}) \to \mathcal{I}^*_{\mathcal{B}}(\mathbb{V})$ given by $\mathbb{T}(u) = 2u$, $\mathbb{T}(2u) = u$, $\mathbb{T}(3u) = 4u$, $\mathbb{T}(4u) = 3u$. Clearly, $\mathbb{T}$ is a graph isomorphism, but as $\mathbb{T}(2u) = u \neq 4u = 2(2u) = 2\mathbb{T}(u)$ and $\mathbb{T}$ is not linear.

Theorem 30 ([23]) *Let $\mathbb{V}$ be a vector space over a finite field $\mathbb{F}$ with q elements and $\mathcal{I}^*_{\mathcal{B}}(\mathbb{V})$ be its associated graph with respect to a basis $\mathcal{B} = \{v_1, v_2, \dots, v_n\}$. Then the following statements hold:*
(1) $n(\mathcal{I}^*_{\mathcal{B}}(\mathbb{V})) = q^n - 1$.
(2) *If* $v \in V(\mathcal{I}^*_{\mathcal{B}}(\mathbb{V}))$ *such that* $|\mathcal{S}_{\mathcal{B}}(v)| = k$*, then* $deg(v) = q^n - 1 - (q-1)^k$.
(3) $\delta(\mathcal{I}^*_{\mathcal{B}}(\mathbb{V})) = q^n - 1 - (q-1)^n$ *and* $\Delta(\mathcal{I}^*_{\mathcal{B}}(\mathbb{V})) = q^n - q$.
(4) $\mathcal{I}^*_{\mathcal{B}}(\mathbb{V}))$ *is Eulerian.*
(5) $m(\mathcal{I}^*_{\mathcal{B}}(\mathbb{V})) = \frac{1}{2}(\sum\limits_{i=1}^{n} \binom{n}{i}(q^n - 1)(q-1)^i - (q-1)^{2i})$.

6 Component Intersection Graphs on Finite-Dimensional Vector Spaces

In this section, we study the component intersection graphs $\mathcal{I}^{\mathcal{B}}(\mathbb{V})$ proposed by the authors in [8]. Let $\mathcal{B}$ be a basis of an n-dimensional vector space $\mathbb{V}$. Then the component intersection graph $\mathcal{I}^{\mathcal{B}}(\mathbb{V})$ is a simple (undirected) graph with the set of vertices $\mathbb{V} \setminus \{u \in \mathbb{V} : |\mathcal{S}_{\mathcal{B}}(u)| = n\}$, and any two distinct vertices u and v of $\mathcal{I}^{\mathcal{B}}(\mathbb{V})$ are adjacent if and only if $\mathcal{S}_{\mathcal{B}}(u) \cap \mathcal{S}_{\mathcal{B}}(v) = \emptyset$. We collect some basic results concerning connectedness, diameter, completeness, clique number and chromatic number of $\mathcal{I}^{\mathcal{B}}(\mathbb{V})$.

Example 5 Let $\mathbb{F}_2 \times \mathbb{F}_2$ and $\mathbb{F}_2 \times \mathbb{F}_2 \times \mathbb{F}_2$ be the vector spaces over $\mathbb{F}_2$ and $\mathcal{B}_1 = \{\alpha_1, \alpha_2\}$, $\mathcal{B}_2 = \{\beta_1, \beta_2, \beta_3\}$ be the bases of $\mathbb{F}_2 \times \mathbb{F}_2$ and $\mathbb{F}_2 \times \mathbb{F}_2 \times \mathbb{F}_2$, respectively. Clearly, $V(\mathcal{I}^{\mathcal{B}_1}(\mathbb{F}_2 \times \mathbb{F}_2)) = \{\alpha_1, \alpha_2\}$ and $V(\mathcal{I}^{\mathcal{B}_2}(\mathbb{F}_2 \times \mathbb{F}_2 \times \mathbb{F}_2)) = \{\beta_1, \beta_2, \beta_3, \beta_1 + \beta_2, \beta_2 + \beta_3, \beta_1 + \beta_3\}$, respectively. The component intersection graphs $\mathcal{I}^{\mathcal{B}_1}(\mathbb{F}_2 \times \mathbb{F}_2)$ and $\mathcal{I}^{\mathcal{B}_2}(\mathbb{F}_2 \times \mathbb{F}_2 \times \mathbb{F}_2)$ are given by the following Figs. 6.1 and 6.2:

$\mathcal{I}^{\mathcal{B}_1}(\mathbb{F}_2 \times \mathbb{F}_2)$

Figure 6.1

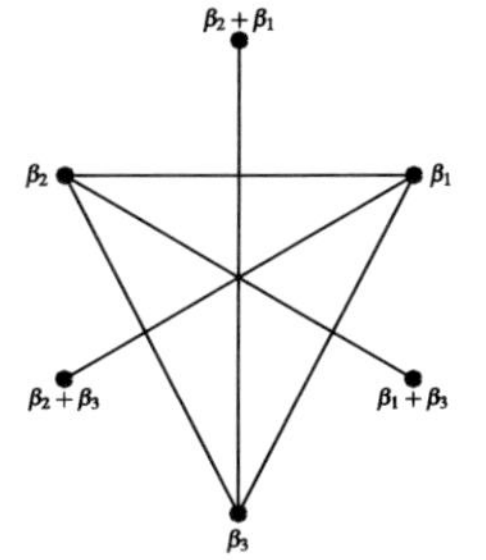

$\mathcal{I}^{\mathcal{B}_2}(\mathbb{F}_2 \times \mathbb{F}_2 \times \mathbb{F}_2)$

Figure 6.2

In view of above example, it may be noticed that the vertex set of our proposed graph $\mathcal{I}^{\mathcal{B}_1}(\mathbb{F}_2 \times \mathbb{F}_2)$ is different from the vertex set of $\Gamma((\mathbb{F}_2 \times \mathbb{F}_2)_\alpha)$ and $\Gamma(\mathbb{F}_2 \times \mathbb{F}_2)_{\mathcal{B}_1}$ introduced by Das [13, 15]. Also, the vertex set of $\mathcal{I}^{\mathcal{B}_2}(\mathbb{F}_2 \times \mathbb{F}_2 \times \mathbb{F}_2)$ is different from the vertex set of $\Gamma((\mathbb{F}_2 \times \mathbb{F}_2 \times \mathbb{F}_2)_\alpha)$ and $\Gamma(\mathbb{F}_2 \times \mathbb{F}_2 \times \mathbb{F}_2)_{\mathcal{B}_2}$. The vertices α_1 and α_2 are not adjacent in $\Gamma(\mathbb{V}_\alpha)$, but are adjacent in $\mathcal{I}^{\mathcal{B}_1}(\mathbb{F}_2 \times \mathbb{F}_2)$. Moreover,

β_1 and β_2 are not adjacent in $\Gamma(\mathbb{F}_2 \times \mathbb{F}_2 \times \mathbb{F}_2)_{\mathcal{B}_2}$ and $\Gamma((\mathbb{F}_2 \times \mathbb{F}_2 \times \mathbb{F}_2)_\beta)$, but are adjacent in $\mathcal{I}^{\mathcal{B}_2}(\mathbb{F}_2 \times \mathbb{F}_2 \times \mathbb{F}_2)$. Let $\mathcal{B} = \{\gamma_1, \gamma_2, \ldots, \gamma_n\}$ be the basis of a vector space $\mathbb{V}$ such that $n \geq 3$. The vertex set of $\mathcal{I}^{\mathcal{B}}(\mathbb{V})$ is different from the vertex set of $\Gamma(\mathbb{V})_{\mathcal{B}}$ and $\Gamma(\mathbb{V}_\gamma)$. Also, γ_1 and γ_2 are not adjacent in $\Gamma(\mathbb{V}_\gamma)$ and $\Gamma(\mathbb{V})_{\mathcal{B}}$, but are adjacent in $\mathcal{I}^{\mathcal{B}}(\mathbb{V})$.

Theorem 31 ([8]) *Let $\mathbb{V}$ be a vector space over a finite field $\mathbb{F}$ with q elements and $\mathcal{I}^{\mathcal{B}}(\mathbb{V})$ be its associated graph with respect to a basis $\mathcal{B} = \{v_1, v_2, \ldots, v_n\}$. Then the following statements hold:*
(1) $\mathcal{I}^{\mathcal{B}}(\mathbb{V})$ is a complete bipartite graph if and only if $n = 2$.
(2) $\mathcal{I}^{\mathcal{B}}(\mathbb{V})$ is complete if and only if $dim(\mathbb{V}) = 2$ and $\mathbb{F} = \mathbb{F}_2$.

The following theorems characterized diameter and girth of $\mathcal{I}^{\mathcal{B}}(\mathbb{V})$.

Theorem 32 ([8]) *Let $\mathbb{V}$ be an n-dimensional vector space over the field $\mathbb{F}$. Then*

$$diam(\mathcal{I}^{\mathcal{B}}(\mathbb{V})) = \begin{cases} 1, & \text{if } n = 2 \text{ and } \mathbb{F} = \mathbb{F}_2 \\ 2, & \text{if } n = 2 \text{ and } \mathbb{F} \neq \mathbb{F}_2 \\ 3, & \text{if } n \geq 3. \end{cases}$$

Theorem 33 ([8]) *Let $\mathbb{V}$ be an n-dimensional vector space over $\mathbb{F}$. Then*

$$gr(\mathcal{I}^{\mathcal{B}}(\mathbb{V})) = \begin{cases} \infty, & \text{if } n = 2 \text{ and } \mathbb{F} = \mathbb{F}_2 \\ 4, & \text{if } n = 2 \text{ and } \mathbb{F} \neq \mathbb{F}_2 \\ 3, & \text{if } n \geq 3. \end{cases}$$

Theorem 34 ([8]) *Let $\mathbb{V}$ be an n-dimensional vector space over the field $\mathbb{F}$ such that $n \geq 4$. Then $\mathcal{I}^{\mathcal{B}}(\mathbb{V})$ is triangulated.*

In view of the following theorem, the authors in [8] studied clique, chromatic and domination number of $\mathcal{I}^{\mathcal{B}}(\mathbb{V})$.

Theorem 35 ([8]) *Let $\mathbb{V}$ be a finite-dimensional vector space over the field $\mathbb{F}$. Then the following statements hold:*
(1)([8]) $\mathbb{V}$ is n-dimensional if and only if $\omega(\mathcal{I}^{\mathcal{B}}(\mathbb{V})) = n$.
(2)([8]) if $\mathbb{V}$ is n-dimensional, then $\chi(\mathcal{I}^{\mathcal{B}}(\mathbb{V})) = n$.
(3)([8]) $\mathbb{V}$ is n-dimensional, then $\gamma(\mathcal{I}^{\mathcal{B}}(\mathbb{V})) = n$.

Corollary 9 ([8]) *Let $\mathbb{V}_1$ and $\mathbb{V}_2$ be two finite-dimensional vector spaces over the field $\mathbb{F}$. Then $\mathbb{V}_1$ and $\mathbb{V}_2$ are isomorphic (as vector spaces) if and only if $\mathcal{I}^{\mathcal{B}_1}(\mathbb{V}_1)$ and $\mathcal{I}^{\mathcal{B}_2}(\mathbb{V}_2)$ are isomorphic (as graphs).*

Corollary 10 ([8]) *Let $\mathbb{V}$ be a vector space over a field $\mathbb{F}$ and $\mathcal{I}^{\mathcal{B}_1}(\mathbb{V})$, $\mathcal{I}^{\mathcal{B}_2}(\mathbb{V})$ be the graphs associated with $\mathbb{V}$ with respect to the bases $\mathcal{B}_1 = \{u_1, u_2, \cdots, u_n\}$ and $\mathcal{B}_2 = \{v_1, v_2, \cdots, v_n\}$ of $\mathbb{V}$, respectively. Then $\mathcal{I}^{\mathcal{B}_1}(\mathbb{V})$ and $\mathcal{I}^{\mathcal{B}_2}(\mathbb{V})$ are graph isomorphic.*

The above corollary shows that the graph theoretic properties of $\mathcal{I}^{\mathcal{B}}(\mathbb{V})$ do not depend on the choice of the basis $\mathcal{B}$.

Theorem 36 ([8]) *Let $\mathbb{V}$ be an n-dimensional vector space. Then the following statements hold:*
(1) *There exists a graph homomorphism from $\mathcal{I}^{\mathcal{B}}(\mathbb{V})$ to $\mathcal{I}^{\mathcal{B}}(\mathbb{V})[\mathcal{B}]$.*
(2) *If $n = 2k + 1$, i.e., n is odd, then $\mathcal{S}^{k+1}$ is a maximal independent set in $\mathcal{I}^{\mathcal{B}}(\mathbb{V})$.*
(3) *If $n = 2k$, i.e., n is even, then $\mathcal{S}^{k+1} \cup [\mathcal{S}^{k}_{v_i}]$ is a maximal independent set in $\mathcal{I}^{\mathcal{B}}(\mathbb{V})$.*

Corollary 11 ([8]) *Let $\mathbb{V}$ be an n-dimensional vector space over a field $\mathbb{F}_q$. Then the following statements hold:*
(1) *If v is a vertex in $\mathcal{I}^{\mathcal{B}}(\mathbb{V})$ such that $|\mathcal{S}_{\mathcal{B}}(v)| = k$, then $deg(v) = q^{n-k} - 1$.*
(2) *The order of $\mathcal{I}^{\mathcal{B}}(\mathbb{V})$ is $q^n - (q-1)^n - 1$.*
(3) *$\delta = q - 1$ and $\Delta = q^{n-1} - 1$.*
(4) *$\mathcal{I}^{\mathcal{B}}(\mathbb{V})$ is Eulerian if and only if q is even.*

7 Subspace Inclusion Graph $\mathcal{I}_n(\mathbb{V})$

In this section, we collect the fundamental results of subspace inclusion graph [16, 17]. We start our discussion with the following theorems.

Theorem 37 ([16]) *Let $\mathbb{V}$ be an n-dimensional vector space. Then the following statements hold:*
(1) *If $\mathbb{W}$ is a subspace of $\mathbb{V}$ with dimension greater than 1, then $\mathcal{I}_n(\mathbb{W})$ is a subgraph of $\mathcal{I}_n(\mathbb{V})$.*
(2) *If $n \geq 3$, then $diam(\mathcal{I}_n(\mathbb{V})) = 3$.*
(3) *If $n \geq 3$, then $\mathcal{I}_n(\mathbb{V})$ is not planar.*

Theorem 38 ([16]) *Let $\mathbb{V}$ be a finite-dimensional vector space over a field $\mathbb{F}$. $\mathbb{V}$ is an n-dimensional vector space if and only if $\omega(\mathcal{I}_n(\mathbb{V})) = \chi(\mathcal{I}_n(\mathbb{V})) = n - 1$.*

In view of the following theorem, the author in [16] studied clique and chromatic number of $\mathcal{I}_n(\mathbb{V})$.

Remark 1 Theorem 38 shows that $\mathcal{I}_n(\mathbb{V})$ is weakly perfect. In [17], the author proved that the graph $\mathcal{I}_n(\mathbb{V})$ is perfect, i.e., $\omega(H) = \chi(H)$ for every induced subgraph H of $\mathcal{I}_n(\mathbb{V})$.

Theorem 39 ([17]) *Let $\mathbb{V}$ be a finite-dimensional vector space. Then $\mathcal{I}_n(\mathbb{V})$ is perfect.*

The following theorem shows that graph theoretic properties depend on the choice of the vector spaces.

Theorem 40 ([16]) *Let $\mathbb{V}_1$ and $\mathbb{V}_2$ be two finite-dimensional vector spaces over the same field $\mathbb{F}$. Then $\mathbb{V}_1$ and $\mathbb{V}_2$ are isomorphic (as vector spaces) if and only if $\mathcal{I}_n(\mathbb{V}_1)$ and $\mathcal{I}_n(\mathbb{V}_2)$ are isomorphic (as graphs).*

Theorem 41 *Let $\mathbb{V}$ be an n-dimensional vector space over a field $\mathbb{F}_q$. Then the following holds:*
(1) *([16]) If $\mathbb{W}$ is a k-dimensional non-trivial proper subspace of $\mathbb{V}$, then*

$$deg(\mathbb{W}) = \sum_{i=1}^{k-1} \binom{k}{i}_q + \sum_{j=1}^{n-k-1} \binom{n-k}{j}_q.$$

(2) *([16]) If W_1 and W_2 be k and $(n-k)$-dimensional (respectively) non-trivial proper subspace of $\mathbb{V}$, then $deg(W_1) = deg(W_2)$.*
(3) *([17]) If q is odd, then $\mathcal{I}_n(\mathbb{V})$ is Eulerian.*
(4) *([17]) If q is even, then $\mathcal{I}_n(\mathbb{V})$ Eulerian if and only if n is even.*

In case of three-dimensional vector space $\mathbb{V}$, the following theorem gives some important properties of $\mathcal{I}_n(\mathbb{V})$.

Theorem 42 ([16, 17]) *Let $\mathbb{V}$ be an n-dimensional vector space over a field $\mathbb{F}_q$. Then the following holds:*
(1) *$\mathcal{I}_n(\mathbb{V})$ is bipartite if and only if $n = 3$.*
(2) *$\mathcal{I}_n(\mathbb{V})$ is regular if and only if $n = 3$.*
(3) *$\mathcal{I}_n(\mathbb{V})$ is edge-transitive.*
(4) *$\mathcal{I}_n(\mathbb{V})$ is vertex-transitive, vertex connectivity of $\mathcal{I}_n(\mathbb{V})$ is $q+1$ and edge connectivity of*
$\mathcal{I}_n(\mathbb{V})$ is $q+1$.
(5) *$\mathcal{I}_n(\mathbb{V})$ is a retract of a Cayley graph.*
(6) *The independence number of $\mathcal{I}_n(\mathbb{V})$ is q^2+q+1.*
(7) *If $q = 2, 3, 5, 8$ or 17, then $\mathcal{I}_n(\mathbb{V})$ is Hamiltonian.*

In case of a three-dimensional vector space, the following conjecture is posed by Das in [17].

Theorem 43 ([17, Conjecture 6.1]) *Let $\mathbb{V}$ be a three-dimensional vector space over a field $\mathbb{F}_q$. Then the following holds:*
(1) *$\mathcal{I}_n(\mathbb{V})$ is a Cayley graph.*
(2) *$\mathcal{I}_n(\mathbb{V})$ is distance regular.*
(3) *$\mathcal{I}_n(\mathbb{V})$ is distance regular.*
(4) *$\mathcal{I}_n(\mathbb{V})$ is a Hamiltonian.*

Ma et al. [19] studied the independence number of $\mathcal{I}_n(\mathbb{V})$ in view of the following theorem.

Theorem 44 ([19]) *Let $\mathbb{V}$ be an n-dimensional vector space over $\mathbb{F}_q$, where $n = 2m$ or $n = 2m+1$. Then $\alpha(\mathcal{I}_n(\mathbb{V})) = \binom{n}{m}_q$.*

The authors together with Mohammad [6] generalized $\mathcal{I}_n(\mathbb{V})$ as a subspace-based subspace inclusion graph $I_n^{\mathbb{W}}(\mathbb{V})$, where the vertex set $V(I_n^{\mathbb{W}}(\mathbb{V}))$ is the collection of all subspaces U of $\mathbb{V}$ such that $U + \mathbb{W} \neq \mathbb{V}$ and $U \nsubseteq \mathbb{W}$, i.e., $V(I_n^{\mathbb{W}}(\mathbb{V})) = \{U \subseteq \mathbb{V} \mid U + \mathbb{W} \neq V, U \nsubseteq \mathbb{W}\}$ and any two distinct vertices U_1 and U_1 of $I_n^{\mathbb{W}}(\mathbb{V})$

are adjacent if and only if either $U_1 + \mathbb{W} \subset U_2 + \mathbb{W}$ or $U_2 + \mathbb{W} \subset U_1 + \mathbb{W}$. They proved the following interesting results.

Theorem 45 ([6]) *Let $\mathbb{W}$ be a k-dimensional subspace of an n-dimensional vector space $\mathbb{V}$ over a field $\mathbb{F}$. Then the following statements hold:*
(1) *If $k = 0$, then $I_n^{\mathbb{W}}(V) = I_n(\mathbb{V})$.*
(2) *If $\mathbb{W}_1, \mathbb{W}_2$ are two vertices of $I_n^{\mathbb{W}}(\mathbb{V})$ such that $dim(\mathbb{W}_1 + \mathbb{W}) = dim(\mathbb{W}_2 + \mathbb{W})$, then $\mathbb{W}_1$ is not adjacent to $\mathbb{W}_2$, i.e., $\mathbb{W}_1 \nsim \mathbb{W}_2$ in $I_n^{\mathbb{W}}(\mathbb{V})$.*
(3) *If $n - k = 1$, then $I_n^{\mathbb{W}}(\mathbb{V})$ is an empty graph.*
(4) *If $n - k \geq 4$, then $I_n^{\mathbb{W}}(\mathbb{V})$ is triangulated.*
(5) *$I_n^{\mathbb{W}}(\mathbb{V})$ is never complete.*

In view of the following theorem, the authors in [6] studied diameter, clique and chromatic number of $I_n^{\mathbb{W}}(\mathbb{V})$.

Theorem 46 ([6]) *Let $\mathbb{W}$ be a k-dimensional subspace of an n-dimensional vector space $\mathbb{V}$ over a field $\mathbb{F}$. Then the following statements hold:*
(1) *If $n - k \geq 3$, then $diam(I_n^{\mathbb{W}}(\mathbb{V})) = 3$.*
(2) *$\chi(I_n^{\mathbb{W}}(\mathbb{V})) = n - k - 1$.*

Theorem 47 ([6]) *Let W be a subspace of a finite-dimensional vector space $\mathbb{V}$. Then $dim(\mathbb{V}) - (dim(\mathbb{W}) + 1) = m$ if and only if $\omega(I_n^{\mathbb{W}}(\mathbb{V})) = m$, where $m = dim(\mathbb{V}) - (dim(\mathbb{W}) + 1)$.*

Theorem 48 ([6]) *Let $\mathbb{W}_1$ and $\mathbb{W}_2$ be two subspaces of a finite-dimensional vector space $\mathbb{V}$. Then $I_n(\mathbb{W}_1) \simeq I_n(\mathbb{W}_2)$ if and only if $dim(\mathbb{W}_1) = dim(\mathbb{W}_2)$.*

The above theorem shows that the graph theoretic properties also depend on the choice of subspaces.

Theorem 49 ([6]) *Let $\mathbb{W}$ be a k-dimensional subspace of an n-dimensional vector space of $\mathbb{V}$ over a finite field $\mathbb{F}$ with q elements. Then the set containing those subspaces $\mathbb{U}$ of $\mathbb{V}$ such that $\mathbb{U} + \mathbb{W} = \mathbb{V}$, i.e., $\{\mathbb{U} \subseteq \mathbb{V} \mid \mathbb{U} + \mathbb{W} = \mathbb{V}\}$ has $(\sum_{r=0}^{k-1} n_r + 1)$ elements, where*

$$n_r = \frac{(q^k - 1)(q^k - q)\cdots(q^k - q^{r-1})(q^n - q^k)(q^n - q^{k+1})\cdots(q^n - q^{n-1})}{(q^{n-k+r} - 1)(q^{n-k+r} - q)\cdots(q^{n-k+r} - q^{n-k+r-1})}.$$

Theorem 50 ([6]) *Let $\mathbb{W}$ be a k-dimensional subspace of an n-dimensional vector space of $\mathbb{V}$ over a finite field $\mathbb{F}$ of order q. Then $I_n^{\mathbb{W}}(\mathbb{V})$ is a graph of order $G(n, q) - (G(k, q) + \sum_{r=0}^{k-1} n_r + 1)$, where*

$$n_r = \frac{(q^k - 1)(q^k - q)\cdots(q^k - q^{r-1})(q^n - q^k)(q^n - q^{k+1})\cdots(q^n - q^{n-1})}{(q^{n-k+r} - 1)(q^{n-k+r} - q)\cdots(q^{n-k+r} - q^{n-k+r-1})}$$

and $G(n, q)$ is the Galois number. Particularly, when $\mathbb{W} = (0)$, order of $I_n^{\mathbb{W}}(\mathbb{V})$ is $G(n, q) - 2$.

Theorem 51 ([6]) *Let $\mathbb{W}$ be a k-dimensional subspace of an n-dimensional vector space of $\mathbb{V}$ over a finite field $\mathbb{F}$ of order q and $\mathbb{U} \in V(I_n^{\mathbb{W}}(\mathbb{V}))$ such that $dim(\mathbb{U} + \mathbb{W}) = l$. Then $deg(\mathbb{U}) = \sum_{r=1}^{l-k-1} {l-k \brack r}_q (\sum_{i=0}^{k-1} n_i + 1) + \sum_{s=1}^{n-l-1} {n-l \brack s}_q (\sum_{i=0}^{k-1} p_i + 1)$, where*

$$n_i = \frac{(q^k - 1)(q^k - q)\cdots(q^k - q^{i-1})(q^{k+r} - q^k)(q^{k+r} - q^{k+1})\cdots(q^{k+r} - q^{k+r-1})}{(q^{r+i} - 1)(q^{r+i} - q)\cdots(q^{r+i} - q^{r+i-1})}$$

and

$$p_i = \frac{(q^k - 1)(q^k - q)\cdots(q^k - q^{i-1})(q^{l+s} - q^k)(q^{l+s} - q^{k+1})\cdots(q^{l+s} - q^{l+s-1})}{(q^{l+s-k+i} - 1)(q^{l+s-k+i} - q)\cdots(q^{l+s-k+i} - q^{l+s-k+i-1})}.$$

Theorem 52 ([6]) *Let $\mathbb{W}$ be a k-dimensional subspace of an n-dimensional vector space $\mathbb{V}$ over $\mathbb{F}_q$. Then the following statements hold.*
(1) If q is odd, then $I_n^{\mathbb{W}}(\mathbb{V})$ is Eulerian.
(2) If q is even, then $I_n^{\mathbb{W}}(\mathbb{V})$ is Eulerian if and only if $n - k$ is even.

Acknowledgements The authors are indebted to the referee for his/her useful suggestions and critical comments.

References

1. Abawajy, J.H., Kelarev, A.V., Chowdhury, M.: Power graphs: a survey. Electron. J. Graph Theory Appl. **1**(2), 125–147 (2013)
2. Ali, U., Bokhary, S.A., Wahid, K., Abbas, G.: On resolvability of a graph associated to a finite vector space. J. Algebra Appl. (2019). https://doi.org/10.1142/S0219498819500294
3. Anderson, D.F., Badawi, A.: On the zero-divisor graph of a commutative ring. Commun. Algebra **36**, 3073–3092 (2008)
4. Anderson, D.F., Livingston, P.S.: The zero-divisor graph of a commutative ring. J. Algebra **217**, 434–447 (1999)
5. Afkhami, M., Khashyarmanesh, K., Sakhdari, S.M.: The annihilator graph of a commutative semigroup. J. Algebra Appl. **2**(14), 1550015 (2015)
6. Ashraf, M., Kumar, M., Mohammad, G.: A subspace based subspace inclusion graph. Contrib. Discret. Math. **15**(2), 73–83 (2020)
7. Ashraf, M., Kumar, M., Mohammad, G.: Regular graph on vector spaces. Aligarh Bull. Math. **38**(1–2), 83–95 (2019)
8. Ashraf, M., Kumar, M.: Component intersection graphs on finite-dimensional vector spaces. Ars Combin. (To Appear)
9. Badawi, A.: On the annihilator graph of a commutative ring. Commun. Algebra **42**, 1–14 (2014)
10. Beck, I.: Colouring of commutative rings. J. Algebra **116**, 208–226 (1988)
11. Chelvam, T.T., Ananthi, K.P.: On the genus of graphs associated with vector spaces. J. Algebra Appl. https://doi.org/10.1142/S0219498820500863
12. Chudnovsky, M., Robertson, N., Seymour, P., Thomas, R.: The strong perfect graph theorem. Ann. Math. **164**, 51–229 (2006)
13. Das, A.: Non-zero component graph of a finite dimensional vector space. Commun. Algebra **44**, 3918–3926 (2016)

14. Das, A.: On non-zero component graph of vector spaces over finite fields. J. Algebra Appl. **16**, 1750007 (2017). https://doi.org/10.1142/S0219498817500074
15. Das, A.: Non-zero component union graph of a finite dimensional vector space. Linear Multilinear Algebra **6**, 1276–12875 (2016)
16. Das, A.: Subspace inclusion graph of a vector space. Commun. Algebra **44**, 4724–4731 (2016)
17. Das, A.: On subspace inclusion graph of a vector space. Linear Multilinear Algebra **66**(3), 554–564 (2016)
18. Das, A.: On non-zero component graph of a vector space over finite fields. Commun. Algebra **44**, 4724–4731 (2016)
19. Ma, X., Wang, D.: Independence number of subspace inclusion graph and subspace sum graph of a vector space. Linear Multilnear Algebra **66**(12), 2430–2437 (2018)
20. Plesnik, J.: Critical graphs of given diameter. Acta Fac. Rerum Natur. Univ. Comenian. Math. **30**(12), 71–93 (1975)
21. Goldman, J., Rota, G.C.: The number of Subspaces of a vector space. In: Recent Progress in Combinatorics Proceedings of Third Waterloo Conference on Combinatorics 1968, pp.75–83. Academic, New York (1969)
22. Nikandish, R., Maimani, H.R., Khaksari, A.: Coloring of a non-zero component graph associated with a finite dimensional vector space. J. Algebra Appl. **18**(8), 1750173 (2017). https://doi.org/10.1142/S0219498817501730
23. Parveen, N., Ashraf,M., Kumar, M.: Distinct component graph on a finite dimensional vector spaces. Open Math., (To Appear)
24. Syvester, J.: Proof of the Hitherto Undemonstrated Fundamental Theorem of Invariants, pp. 117–126. Chelsea, New York (NY) (1973)
25. West, D.W.: Introduction to Graph Theory, 2nd edn. Prentice Hall, Upper Saddle River (2001)
26. White, A.T.: Graphs, Groups and Surfaces. North Holland, Amsterdam (1973)

Two Value Graph Magma Algebras and Amenability

Pınar Aydoğdu, Joaquín Díaz Boils, Sergio R. López-Permouth, and Rebin A. Muhammad

Abstract We introduce *two value graph magma algebras* and examine the amenability of their bases for the commutative case. We use these algebras to construct a commutative algebra which has a unique simple basis up to mutual congeniality and no projective bases.

Keywords Amenable bases · Congeniality of bases · Proper congeniality · Simple Bases · Infinite-dimensional algebras

1 Introduction

Let A be an infinite-dimensional algebra over a field F and $\mathcal{B}$ be a basis for A. Let P be the F-vector space consisting of the direct product, indexed by $\mathscr{B}$, of copies of the field F. P may alternately be denoted by ${}_{\mathscr{B}}P$, $\prod_{b\in\mathscr{B}} Fb$ or $= F^{\mathscr{B}}$. In [2], the feasibility of a (left) A-module structure on P that is natural in the sense that it extends the module structure ${}_AA$ is discussed. To that avail, the notion of a (left) amenable basis $\mathcal{B}$ is defined by a condition that guarantees that ${}_{\mathscr{B}}P$ has such a (left)

P. Aydoğdu (✉)
Department of Mathematics, Hacettepe University, 06800 Beytepe Ankara, Turkey
e-mail: paydogdu@hacettepe.edu.tr

Center of Ring Theory and its Applications, Ohio University, Athens, OH 45701, USA

J. D. Boils
Facultad de Ciencias Exactas y Naturales, Pontificia Universidad Católica del Ecuador, Quito, Ecuador
e-mail: jdiaz987@puce.edu.ec

S. R. López-Permouth
Department of Mathematics, Ohio University, Athens, OH 45701, USA
e-mail: lopez@ohio.edu

R. A. Muhammad
Department of Mathematics and Statistics, Montgomery College, Germantown, MD, USA
e-mail: Rebin.Muhammad@montgomerycollege.edu

M. Ashraf et al. (eds.), *Algebra and Related Topics with Applications*,
Springer Proceedings in Mathematics & Statistics 392,
https://doi.org/10.1007/978-981-19-3898-6_30

A-module structure. When that is the case, one writes $P = {}_{\mathscr{B}}M$ and says that ${}_{\mathscr{B}}M$ is *the basic submodule induced by the amenable basis* $\mathscr{B}$.

The question of whether amenable bases yield isomorphic module structures arises naturally and the notion of congeniality of bases is introduced in the same work in order to investigate it. Since its inception, the study of these and other related notions has seen significant activity (e.g. [3, 9] and [8]). The notions of amenability and congeniality have even been extended to infinite-dimensional modules over arbitrary algebras in [9] and [8].

Both amenability of bases and congeniality among them boil down to a requirement that certain linear transformations have representations which are row- and column-finite. Row- and column-finite have historically been the subject of much interest in the mathematical community; the survey paper [1] is a good introduction to some of the problems pertaining to the subject.

The remarkable result that any countable family of infinite matrices can be conjugated simultaneously to make each one of its members row- and column-finite (see [12]) was instrumental to prove in [2] that every countable-dimensional algebra has at least one (left) amenable basis. That result raised the question of whether it is possible to have an algebra where all bases are amenable. In [3], it is shown that there is no algebra such that every basis is amenable; in fact, it is proved that for any left amenable basis $\mathscr{B}$, there exist infinitely many non-amenable bases which are discordant to $\mathscr{B}$ and one another.

Also in [3], certain *contracted* semigroup algebras are considered where the semigroup is induced by a graph. The construction from [6] and [14] required some adaptations because the structures considered earlier involved exclusively finite graphs while, in our context, we are only interested in graphs over infinite sets of vertices. A more detailed account of this construction opens Sect. 3.

Graph algebras have been instrumental to answer open questions regarding the amenability of bases. Non-commutative infinite-dimensional graph algebras are constructed in [3], where the notions of amenability and simplicity are not left-right symmetric. Other examples of one-sided amenable bases for algebras of infinite matrices may be found in [3]. Another question raised in [2] is whether all algebras have simple or projective bases; non-commutative graph algebras which have neither simple nor projective bases were constructed in [3].

In [2], it is shown that the algebra $F[x]$ has at least as many pairwise discordant simple bases as there are elements in F^* for a field F. It is asked if all algebras have simple bases and, if so, how many pairwise discordant simple bases can you have? Note that the same questions can be asked for projective bases as well; the situation with projective bases is a little more intriguing since we do not have any examples of an algebra with a projective basis. As mentioned above, a negative answer is given to the existence question in [3] and an example of a non-commutative algebra with a unique simple basis is given. In this work, inspired by the work in [3], we introduce a family of algebras which we call *two value graph algebras* and examine the amenability of their bases in the commutative case; we use two value graph algebras to construct a commutative algebra which has a unique simple basis up to mutual congeniality and has no projective bases (see Sect. 5). Two value magmas are also studied in [7] for different purposes.

In Sect. 2, we recall the basic notions for the convenience of the reader. In Sect. 3, we construct two value graph algebras induced by two value graph magmas. Since we are working on associative algebras, we characterize when a two value graph magma is associative. The structure of commutative two value graph algebras is investigated in Sect. 4; we determine and classify graphs that induce commutative two value magmas.

In Sect. 5, we give some necessary and sufficient conditions for bases to be amenable for those commutative two value graph magma algebras which are constructed on graphs that cannot be decomposed into two infinite graphs. Moreover, we show algebras which do not have projective bases and have unique simple bases up to mutual congeniality. Section 6 deals with the amenability of bases for the algebras that are constructed on the graphs which can be written as a direct sum of infinite graphs.

2 Preliminaries

We begin this section with some general notational conventions. Given two bases $\mathscr{B}$ and $\mathscr{C}$ of a vector space V, and a linear transformation $T : V \to V$, for an element $v \in V$, the notation $[v]_{\mathscr{B}}$ refers to the coordinates of v with respect to the basis $\mathscr{B}$, and the matrix $[T]^{\mathscr{C}}_{\mathscr{B}}$ to the representation of T with respect to the bases $\mathscr{B}$ and $\mathscr{C}$. In other words, $[T]^{\mathscr{C}}_{\mathscr{B}}$ has rows indexed by the elements of $\mathscr{C}$ and columns indexed by those of $\mathscr{B}$ in such a way that the bth column of $[T]^{\mathscr{C}}_{\mathscr{B}}$ is $[T(b)]_{\mathscr{C}}$. For all T, the matrix $[T]^{\mathscr{C}}_{\mathscr{B}}$ is column-finite. The identity $[T]^{\mathscr{C}}_{\mathscr{B}}[v]_{\mathscr{B}} = [T(v)]_{\mathscr{C}}$ holds for all $v \in V$. For any two bases $\mathscr{B}$ and $\mathscr{C}$, the matrix $Q = [I]^{\mathscr{C}}_{\mathscr{B}}$, representing the identity map I with respect to $\mathscr{B}$ and $\mathscr{C}$ serves as a change of basis matrix, in the sense that, as $Q^{-1} = [I]^{\mathscr{B}}_{\mathscr{C}}$, the identity $[T]^{\mathscr{C}}_{\mathscr{C}} = Q[T]^{\mathscr{B}}_{\mathscr{B}}Q^{-1}$ holds for any endomorphism $T : V \to V$. Note that Q is an invertible column-finite matrix having a column-finite inverse. A matrix Q is change of basis matrix if and only if it is an invertible column-finite matrix with column-finite inverse. Not all invertible column-finite matrices have a column-finite inverse. We use the simplified notation $[T]_{\mathscr{B}}$ instead of $[T]^{\mathscr{B}}_{\mathscr{B}}$.

Let $\mathcal{M}= {}_{\mathscr{B}}\mathcal{M}$ denote the left A-module structure on ${}_{\mathscr{B}}P$ which extends ${}_{A}A$. Notice that $\mathcal{M}$ may be viewed as $\{\phi : \mathscr{B} \to F\}$. Therefore, an element $\phi \in \mathcal{M}$ may be denoted by $\phi = (\phi(b))_{b\in\mathscr{B}}$ or we may also write $\phi = \sum_{b\in\mathscr{B}} \phi_b b$, with $\phi_b = \phi(b)$ whenever $b \in \mathscr{B}$. Define a product of elements of A with elements from $\mathcal{M}$. It suffices to define $b\phi$ for any $b \in \mathscr{B}$ and $\phi \in \mathcal{M}$. If all sums involved were finite, one would necessarily have that, for any $c \in \mathscr{B}$, $[b\phi(c)] = \sum_{d\in {}_b\mathscr{B}_c}(bd)_c\phi(d)$, where ${}_b\mathscr{B}_c = \{d \in \mathscr{B}|(bd)_c \neq 0\}$. This need for the requirement that the sums be finite inspires the definition of amenability given in [2].

Let S be a non-empty set. A *multiset* M with underlying set S is a set of ordered pairs

$$M = \{(s_i, n_i)|s_i \in S, n_i \in \mathbb{Z}^+, s_i \neq s_j \text{for} i \neq j\}.$$

The number n_i is referred to as the *multiplicity* of the elements s_i in M. If the underlying set is finite, then the multiset is said to be *finite*. One often writes out the elements of a multiset according to the multiplicities, as in $M = \{x_1, x_1, x_2, x_2, x_2, x_3, \ldots\}$ (see, for instance, [13]). We recall some basic notions from the work [2].

Definition 1 A basis $\mathscr{B}$ for an infinite-dimensional F-algebra A is *left amenable* if for all $r \in A$, the column-finite matrix $[l_r]_{\mathscr{B}}$, where $l_r : A \to A$ is the left multiplication by r, is also row-finite. A *right amenable* basis is defined similarly.

Definition 2 A basis $\mathscr{B}$ is said to be *congenial* to a basis $\mathscr{C}$ if for every element $c \in \mathscr{C}$ there exists only finitely many elements $b \in \mathscr{B}$ such that, when b is represented in terms of elements of $\mathscr{C}$, the representation uses a non-zero coefficient for c. Equivalently, $\mathscr{B}$ is congenial to $\mathscr{C}$ if the transition matrix $[I]_{\mathscr{B}}^{\mathscr{C}}$ is row-finite. If, in addition, the inverse $[I]_{\mathscr{C}}^{\mathscr{B}}$ is also row-finite, then we say that the bases $\mathscr{B}$ and $\mathscr{C}$ are *mutually congenial*. When $\mathscr{B}$ is congenial to $\mathscr{C}$, left multiplication by $[I]_{\mathscr{B}}^{\mathscr{C}}$ is an F-linear map from ${}_{\mathscr{B}}P$ to ${}_{\mathscr{C}}P$ which we call a *congeniality map*. When $\mathscr{B}$ and $\mathscr{B}$ are both amenable, and $\mathscr{B}$ is congenial to $\mathscr{C}$, the congeniality map is a A-module homomorphism.

While congeniality is not necessarily symmetric (see [2, Examples 3.2]), mutual congeniality is an equivalence relation. These observations give rise to the following definitions.

Definition 3 If $\mathscr{B}$ is congenial to $\mathscr{C}$ but $\mathscr{C}$ is not congenial to $\mathscr{B}$, then we say that $\mathscr{B}$ is *properly congenial* to $\mathscr{C}$. If neither $\mathscr{B}$ is congenial to $\mathscr{C}$ nor $\mathscr{C}$ is congenial to $\mathscr{B}$, then we say that $\mathscr{B}$ and $\mathscr{C}$ are *discordant*.

Relations between congeniality and amenability are established in [2, Theorem 3.5]. It is an interesting fact that congeniality maps are onto [2, Sect. 4], which yields the impression that when $\mathscr{B}$ is congenial to $\mathscr{C}$, ${}_{\mathscr{B}}P$ is larger than (or isomorphic to) ${}_{\mathscr{C}}P$. This motivates the following definitions.

Definition 4 A left amenable basis $\mathscr{B}$ is called *left simple* if it is not properly congenial to any other amenable basis, and it is called *left projective* if there does not exist any left amenable basis which is properly congenial to $\mathscr{B}$. Likewise, *right simple* and *right projective* bases are defined.

[[2, Sect. 5] provides examples of abundant mutually discordant simple bases in the algebra of polynomials with a single variable. It must be noted that the existence of projective bases is still largely hypothetical for not a single example has yet been given. We show below algebras where no simple bases can be found as well as somewhere projective bases can be proven not to exist. Recently, it has been shown that no graph magma algebra has a projective basis (see [9]). In [[2], Theorem 5.2], it is shown that the standard basis $\{x^i : i \in \mathbb{N}\}$ for the polynomial algebra $F[x]$ over a field F is simple. It is tempting to think that the standard basis $\mathscr{B} = \{x^i : i \in \mathbb{Z}\}$ for the algebra of Laurent polynomials $F[x, x^{-1}]$ over a field F is simple. However, while $\mathscr{B}$ is indeed amenable, it is not simple; that is the subject of our next proposition.

Proposition 1 *The standard basis $\mathscr{B} = \{x^i : i \in \mathbb{Z}\}$ for the algebra of Laurent polynomials $F[x, x^{-1}]$ over a field F is not simple.*

Proof Consider the basis $\mathscr{C} = \{1\} \cup \{1 + x^{-1}, 1 + x^{-1} + x, 1 + x^{-1} + x + x^{-2}, 1 + x^{-1} + x + x^{-2} + x^2, \cdots\}$ for $F[x, x^{-1}]$. It is easy to see that $\mathscr{B}$ is properly congenial to $\mathscr{C}$. We prove that $\mathscr{C}$ is amenable. Denote

$$c_m = \begin{cases} 1 + \sum_{i=1}^{p} x^i + \sum_{j=1}^{p} x^{-j} & \text{if } m = 2p, \\ 1 + \sum_{i=1}^{p-1} x^i + \sum_{j=1}^{p} x^{-j} & \text{if } m = 2p - 1 \end{cases}$$

with $c_{2p} - c_{2p-1} = x^p$ and $c_{2p+1} - c_{2p} = x^{-(p+1)}$.

In order to show that $\mathscr{C}$ is amenable, it suffices to show that $[L_x]_{\mathscr{C}}$ and $[L_{x^{-1}}]_{\mathscr{C}}$ will have finite non-zero rows and columns. That is indeed the case, as shown by the following two equations:

$$xc_m = \begin{cases} \sum_{i=m-1}^{m+1} (-1)^i c^i & \text{if } m = 2p - 1, \\ \sum_{i=m-2}^{m+2} (-1)^i c^i & \text{if } m = 2p \end{cases} \tag{1}$$

$$x^{-1}\dot{c}_m = \begin{cases} \sum_{i=m-2}^{m+2} (-1)^i c^i & \text{if } m = 2p - 1, \\ \sum_{i=m-1}^{m+1} (-1)^i c^i & \text{if } m = 2p. \end{cases} \tag{2}$$

□

Note, however, that, while we have shown that standard basis for the Laurent polynomial algebra $F[x, x^{-1}]$ is not simple, the question of whether $F[x, x^{-1}]$ has a simple basis is still open.

3 Two Value Graph Magma Algebras

Graph magma algebras are introduced in [3] and have proven to be a fertile setting for answering many questions about amenability-related questions. For the sake of comparison, we begin by reviewing here some of the basic facts about those algebras.

It is important to point out that in [6] and [14] (and all other literature that we are aware of), the authors use the expression *graph algebras* for the magmas under consideration, even in the general case when the operations are not even expected to be associative. We, on the other hand, call a *graph magma* what [6] and [14] call a graph algebra, and reserve the expression *graph algebra* for the contracted semigroup algebra over an associative graph magma. This terminology is introduced explicitly in [3] to avoid confusion and is influenced by the common trend in the literature to use the expression *magma* for binary operations considered without any further assumptions. References for contracted semigroup algebras include [5, 11] and [10].

Definition 5 1. A graph Magma Algebra is a modified version of a semigroup algebra; something that in the literature is known as a contracted semigroup algebra ([5, 10, 11], etc.). For a field F and a monoid S with an annihilator element $0 \in S$ (an element such that $0x = x0 = 0$, for all $x \in S$). The contracted semigroup algebra one gets from F and S is the quotient of the semigroup algebra with $0 \in S$ identified with $0 \in F$.

2. The semigroup that we use for Graph Magma Algebras is easily described in graph theoretic terms; consider a simple directed graph $G = (V, E)$, with infinite vertex set V and set of edges E. For our purposes, it suffices to consider the case when $V = \{v_i\}_{i=1}^{\infty}$ is countable.
Let $\mathscr{S} = \{0\} \cup \{v_i\}_{i \in \mathbb{N}}$ where $v_0 = 1$ (in other words, $\mathscr{S} = V \cup \{0, 1\}$). Then, (for all i) define $0 * v_i = 0 = v_i * 0$, $1 * v_i = v_i = v_i * 1$ and, for $i, j > 0$,

$$v_i * v_j = \begin{cases} v_i & (v_i, v_j) \in E \\ 0 & otherwise. \end{cases}$$

An additional condition is required to guarantee that this operation makes $\mathscr{S}$ into a semigroup; as shown in [3], for example, that condition is as follows:

Theorem 1 *The operation* $* : \mathscr{S} \times \mathscr{S} \to \mathscr{S}$ *defined above is associative exactly when, for all* $u, v, w \in V$, *if* $(u, v) \in E$ *then* $(u, w) \in E$ *if and only if* $(v.w) \in E$.

Two value graph magma algebras are monoid algebras; this is in direct contrast with (one value) magma algebras; there is no need for contraction here since we are not dealing with a monoid and not with a semigroup with an annihilator element 0. There are formal similarities, nonetheless; outside of that technical observation, the constructions are indeed very similar.

Definition 6 Consider a directed graph $G = (V, E)$, with a countably infinite vertex set $V = \{x_i\}_{i=1}^{\infty}$ and a set of edges E. We identify the set of edges with a subset of $V \times V$. The *two value graph magma* of G is defined to be a set with universe $V \cup \{1\}$, where the binary operation on V is defined as follows:

$$x_i \cdot x_j = \begin{cases} x_i & \text{if } (x_i, x_j) \in E, \\ x_j & \text{otherwise.} \end{cases}$$

We let $V \cup \{1\}$ be a spanning set. We call an algebra *two value graph magma algebra* (2V *graph magma algebra*, for short) if it is generated by an associative two value graph magma. Such an algebra will be denoted by $A(G)$. For convenience, the unit 1 of 2V graph magma algebra will be denoted by x_0. Note that $V \cup \{x_0\}$ is a basis for the 2V graph magma algebra $A(G)$ which we call the *standard basis*.

We will give the characterization of the associativity of two value graph magmas below. Before we mention this characterization, let us recall some notations used in [6] for readers' sake.

By a graph, we mean a directed graph without multiple edges but possibly with loops. The sum $G_1 + G_2$ of graphs $G_1 = (V_1, E_1)$ and $G_2 = (V_2, E_2)$ is the graph with vertex set $V_1 \cup V_2$ and edge set $E_1 \cup E_2 \cup E_{1,2}$, where

$$E_{1,2} = \{(x, y) | x \in V_1, y \in V_2\}.$$

If G_1 and G_2 have no common vertices, then we say that the sum is direct, and we denote it by $G_1 \oplus G_2$.

Let $G = (V, E)$ be a graph. The source of the graph G, $source(G)$, consists of those $x \in V$ such that $(x, y) \in E$ for some $y \in V$. The target of G, $target(G)$, is the set of elements $y \in V$ satisfying $(x, y) \in E$ for some $x \in V$. For an element $x \in V$, $outset(x) = \{y \in V | (x, y) \in E\}$ and $inset(x) = \{z \in V | (z, x) \in E\}$. For further information related to graph theory, we refer the reader to [4].

Now we investigate when a two value graph magma is associative. Let $G = (V, E)$ be a graph. The *inverse graph* G^{-1} is a graph with vertex set V and the edge set $E^{-1} = \{(x, y) | (y, x) \in E\}$. The *complement graph* G^c is a graph with vertex set V and the edge set $E^c = \{(x, y) | (x, y) \notin E\}$. We use the notation $x \to y$ if $(x, y) \in E$. The associativity of two value magmas is also discussed in [7].

Proposition 2 *Let $G = (V, E)$ be a (directed) graph. Then the following statements are equivalent:*

(1) The two value graph magma of G is associative.
(2) G is transitive and $a \to b$ implies that $a \to c$ or $c \to b$ for all $a, b, c \in V$.
(3) The two value graph magma of G^c is associative.
(4) The two value graph magma of G^{-1} is associative.

Proof (1) $\Rightarrow$ (2) Suppose that the two value graph magma of G is associative. Pick $a, b, c \in V$ such that $a \to b$, but $a \not\to c$ and $c \not\to b$. Then $(ac)b = cb = c$ and $a(cb) = ab = a$. But this is a contradiction because of the associativity of the two value graph magma of G. Now suppose that $a \to b \to c$. We will show $a \to c$. Assume on the contrary that $a \not\to c$. It follows that $a(bc) = ab = a$ and $(ab)c = ac = c$, a contradiction.

(2) $\Rightarrow$ (1) We will show that $(ab)c = a(bc)$ for all $a, b, c \in V$. Assume that $a \to b$. If $a \to c$, then $(ab)c = ac = a$. Since $a \to b$ and $a \to c$, we also have $a(bc) = a$. If $a \not\to c$, then $c \to b$ by assumption. If $b \to c$, then we would have $a \to c$ by transitivity. Hence, $b \not\to c$. Then we obtain $a(bc) = ac = c$ and $(ab)c = ac = c$.

Now assume that $a \nrightarrow b$. If $b \to c$, then $a(bc) = (ab)c$. Let $b \nrightarrow c$. Then $a \nrightarrow c$. We conclude that $a \nrightarrow c$, because otherwise either $a \to b$ or $b \to c$. It follows that $a(bc) = ac = c$ and $(ab)c = bc = c$.

(2) $\Rightarrow$ (3) Assume that we have $a \to b$ and $a \nrightarrow c$ in G^c for some $a, b, c \in V$. Then $a \nrightarrow b$ and $a \to c$ in G. We will prove that $c \to b$ in G^c. Assume contrarily that $c \nrightarrow b$ in G^c so that $c \to b$ in G. Transitivity of G implies that $a \to b$ which is a contradiction. Hence, $c \to b$ in G. To show the transitivity of G^c, assume that $a \to b \to c$ in G^c. Then we obtain $a \nrightarrow b \nrightarrow c$ in G. If $a \to c$ then either $a \to b$ or $b \to c$ because G is associative. But this is impossible. Thus, $a \nrightarrow c$ in G which means that $a \to c$ in G^c.

(2) $\Rightarrow$ (4) Assume that $a \to b$ in G^{-1}. Then $b \to a$ in G. It follows that either $b \to c$ or $c \to a$ in G which means that either $c \to b$ or $a \to c$ in G^{-1}. If $a \to b \to c$ in G^{-1}, then we will have $c \to b \to a$ in G. By transitivity of G, we obtain that $c \to a$ and so $a \to c$ in G^{-1}.

(3) $\Rightarrow$ (2) and (4) $\Rightarrow$ (2) are obvious. □

4 Commutative Two Value Graph Magmas

This section is devoted to understanding the structure of commutative two value graph magmas. But we will begin with a brief discussion of commutative graph magmas, for the sake of comparison, before we proceed with our consideration of commutative two value graph magmas.

For a graph magma of $G = (V, E)$, the binary operation on V is defined as

$$x_i \cdot x_j = \begin{cases} x_i & \text{if } (x_i, x_j) \in E, \\ x_j & \text{otherwise.} \end{cases}$$

A null graph is a graph without edges. The null graph of order m is denoted by N_m. We assume that complete graphs contain all edges including loops. The complete graph of order n is denoted by K_n.

The following result characterizes the commutativity of a graph magma algebra.

Proposition 3 *Let $G = (V, E)$ be a directed graph. Then the graph magma of G is commutative if and only if the connected components of G are either a null graph or K_1.*

We know from [6, Proposition 4] that the graph magma of G is associative if and only if connected components of G are either null, complete or direct sum of a null and a complete. Therefore, we obtain the following result:

Proposition 4 *Let $G = (V, E)$ be a directed graph. If the graph magma of G is commutative, then it is associative.*

By [6, Corollary 5], if a graph magma of a directed graph G is associative, then the directed graph $G = (V, E)$ is transitive. Hence, we obtain the following diagram for the graph magma $M(G)$ of a directed graph G:

$$M(G) \text{ is commutative} \Longrightarrow M(G) \text{ is associative} \Longrightarrow G \text{ is transitive}$$

Unlike commutative graph magmas, a commutative $2V$ graph magma need not be associative. We give an example of a commutative non-associative $2V$ graph magma constructed on a tournament below. A tournament is a directed graph in which every pair of distinct vertices is connected by a directed edge with any one of two possible orientations. Notice that a tournament is not transitive, but it gives a commutative $2V$ graph magma.

Example 1 Let $G = (V, E)$ be a directed graph with the edge set

$$E = \{(x_1, x_2), (x_2, x_3), (x_3, x_4), (x_4, x_1), (x_4, x_2), (x_3, x_1)\}.$$

Then the $2V$ graph magma $M(G)$ is commutative but G is not transitive, and hence $M(G)$ is not associative.

Example 2 A null graph of infinite order is transitive but the two value graph magma of such a null graph is neither associative nor commutative. A complete graph of infinite order is also transitive. The graph magma of infinite order complete graph is associative, but it is not commutative.

Now we turn our attention to commutative $2V$ graph magmas which is our main concern in this paper. Since we are dealing with associative graph magma algebras, we will consider a transitive directed graph. We start with a useful lemma which states that the set of vertices V of a commutative two value graph magma can have at most one element m_1 such that $outset(m_1) = V$ and at most one element m_2 such that $inset(m_2) = V$.

Lemma 1 *Let $G = (V, E)$ be a graph. Suppose that the two value graph magma of G is commutative. If $S_1 = \{m \in V : m \to n \text{ for each } n \in V\}$ and $S_2 = \{m \in V : m \leftarrow n \text{ for each } n \in V\}$, then $|S_1| \leq 1$ and $|S_2| \leq 1$.*

Proof Suppose that S_1 has at least two elements, say m_1 and m_2. Then $m_1 \to m_2$ and $m_2 \to m_1$. But since $A(G)$ is commutative, this is impossible. Similarly, $|S_2| \leq 1$.

Remark 1 Notice that the two value graph magma of a graph $G = (V, E)$ is commutative if and only if for any two distinct elements a and b in V we have either $(a, b) \in E$ or $(b, a) \in E$.

The next theorem describes the structure of a commutative two value graph magma. We note that $x_\infty, x_{-\infty} \in V$ indicate elements with $inset(x_\infty) = V$ and $outset(x_{-\infty}) = V$, respectively.

Theorem 2 *Let $G = (V, E)$ be a transitive directed graph. If the two value graph magma of G is commutative, then G must be in one of the following form or a direct sum of these forms:*

(1) V does not have any element x with $inset(x) = V$ or $outset(x) = V$, and the graph is denoted by (G).
(2) V has an element x such that $outset(x) = V$, but does not have any element y such that $inset(y) = V$, and the graph is denoted by $[G)$.
(3) V has an element x such that $inset(x) = V$, but does not have any element y such that $outset(y) = V$, and the graph is denoted by $(G]$.
(4) $G = (G_1) \oplus \{x_\infty\}$ for some graph G_1 with $V_{G_1} = V - \{x_\infty\}$.
(5) $G = \{x_{-\infty}\} \oplus (G_1)$ for some graph G_1 with $V_{G_1} = V - \{x_{-\infty}\}$.
(6) $G = [G_1) \oplus \{x_\infty\}$ for some graph G_1 with $V_{G_1} = V - \{x_\infty\}$.
(7) $G = \{x_{-\infty}\} \oplus (G_1]$ for some graph G_1 with $V_{G_1} = V - \{x_{-\infty}\}$.
(8) $G = \{x_{-\infty}\} \oplus (G_1) \oplus \{x_\infty\}$ for some graph G_1 with $V_{G_1} = V - \{x_\infty, x_{-\infty}\}$.

Proof Let S_1 and S_2 be as in Lemma 1.

Case 1 : Suppose that $|S_1| = |S_2| = 0$. Then $x_1 \notin S_1 \cup S_2$ for $x_1 \in V$. Since $A(G)$ is commutative, $outset(x_1) \neq \emptyset$ and $inset(x_1) \neq \emptyset$. Then there exist $x_2, x_3 \in V$ such that $x_3 \to x_1 \to x_2$. By the same reason, $outset(x_2) \neq \emptyset$ and $inset(x_3) \neq \emptyset$. Therefore, we must have $x_5 \to x_3 \to x_1 \to x_2 \to x_4$ for some $x_4, x_5 \in V$. Continuing in this way, we obtain the graph (G):

Case 2 : Let $|S_1| = 1$ and $|S_2| = 0$. Without loss of generality, assume that $x_1 \in S_1$. Then $x_1 \to x_2$. Since $x_2 \notin S_2$, there exist elements $x_3 \in outset(x_2)$. Therefore, we have $x_1 \to x_2 \to x_3$. Now consider the graph

for n elements of V. Pick $x_{n+1} \in V$. Let $\mathscr{A} = \{j | x_{n+1} \to x_j;\ j = 1, 2, \cdots n\}$. If $\mathscr{A} = \emptyset$, then we obtain

Now suppose that $\mathscr{A} \neq \emptyset$ and let $m = min(\mathscr{A})$. Let x_m be the first element satisfying $x_{n+1} \to x_m$ then we get

On the other hand, $outset(x_j) \neq \emptyset$ for any $x_j \in V$, because $S_2 = \emptyset$. Continuing in this way, if $|inset(x_2)| < \infty$ we obtain the graph $[G)$:

In case $|inset(x_2)| = \infty$, we will get $\{x_{-\infty}\} \oplus (G)$.

Case 3 : If $|S_1| = 0$ and $S_2 = 1$, then using a similar argument in Case 2, we get either $(G]$ when $|outset(x_2)| < \infty$ or $(G) \oplus \{x_\infty\}$ when $|outset(x_2)| = \infty$.

Case 4 : Let $|S_1| = |S_2| = 1$. Let's say $x_1 \in S_1$ and $x_2 \in S_2$. Since $x_1 \notin S_2$, $outset(x_1) \neq \emptyset$. If $x_3 \in outset(x_1)$, then we will have $x_1 \to x_3 \to x_2$. Suppose that $|inset(x_3)| < \infty$. Then $|outset(x_3)|$ must be infinite. Arguing like in Case 2, we obtain the graph $[G) \oplus \{x_\infty\}$. If $|outset(x_3)| < \infty$, then $|inset(x_3)|$ must be infinite. By a similar argument in Case 3, we get the graph $\{x_{-\infty}\} \oplus (G]$. If both $|outset(x_3)|$ and $|inset(x_3)|$ are infinite, then we get $\{x_{-\infty}\} \oplus (G) \oplus \{x_\infty\}$. □

5 Amenable Bases for Commutative Two Value Graph Magma Algebras

A graph $G = (V, E)$ is said to be *basic* if its direct sum decomposition does not contain more than one infinite subgraph; that is to say, if $G = G_1 \oplus G_2$ for some graphs $G_1 = (V_1, E_1)$ and $G_2 = (V_2, E_2)$, then either $|V_1|$ is finite or $|V_2|$ is finite. The graphs $[G)$, $(G]$, $[G) \oplus \{x_\infty\}$ and $\{x_{-\infty}\} \oplus (G]$ are basic, whereas the remaining graphs in Theorem 2 are not basic since we can decompose $(G) = (G_1] \oplus [G_2)$ for some infinite subgraphs G_1 and G_2. In this section, we will investigate the conditions under which the bases for those commutative $2V$ graph magma algebras that are constructed on basic graphs are amenable.

Let $G = (V, E)$ be a directed graph with the countably infinite vertex set $V = \{x_i\}_{i=1}^{\infty}$ and a set of edges E. Throughout this section, suppose that G is transitive and the two value graph magma of G is commutative. If $\mathscr{C}$ is a basis for a $2V$ graph magma algebra, then we define the set $\mathscr{C}_{x_i} = \{c \in \mathscr{C} | (c)_{x_i} \neq 0\}$, where $x_i \in V \cup \{x_0\}$.

Theorem 3 *A basis $\mathscr{C}$ for the commutative $2V$ graph magma algebra $A([G))$ is amenable if and only if $|\mathscr{C}_{x_j}| < \infty$ for all $x_j \in V$ and $\sum_{i=0}^{n_j} \alpha_{i_j} \neq 0$ for finitely many j, where $c_j = \sum_{i=0}^{n_j} \alpha_{i_j} x_{i_j} \in \mathscr{C}$.*

Proof Let $S = \{j | \sum \alpha_{i_j} \neq 0,\ c_j = \sum \alpha_{i_j} x_{i_j}\}$. Assume that $|S| = \infty$ and that $\mathscr{C}$ is amenable. Consider the set $\mathscr{D} = \{c_1, c_2, \ldots\}$, where $c_j = \sum \alpha_{i_j} x_{i_j}$ and $\sum \alpha_{i_j} \neq 0$. For $c_k \in \mathscr{D}$, we have that $x_1 c_k = \sum \alpha_{i_k} x_1 = (\sum \alpha_{i_k}) x_1$.

Since $\sum \alpha_{i_k} \neq 0$ for each $c_k \in \mathscr{D}$, $[l_{x_1}]_{\mathscr{C}}$ is not row-finite. It follows from this contradiction that S must be finite.

Now assume that there exists $x_j \in V$ such that $|\mathscr{C}_{x_j}| = \infty$. Assume further that x_j is the first element of the ordered standard basis satisfying this condition. Let $\mathscr{C}_{x_j} = \{c_1, c_2, \ldots\}$. Consider the set $\mathscr{C}^*_{x_j} = \mathscr{C}_{x_j} \setminus \cup_{i=1}^{j-1} \mathscr{C}_{x_i}$. The set $\cup_{i=1}^{j-1} \mathscr{C}_{x_i}$ is finite. For any $c_k \in \mathscr{C}^*_{x_j}$, we may write $c_k = \alpha_k x_0 + \sum_{s=0}^{n_k} \beta_s x_{s+j}$ for some scalars α_k, β_s and $n_k \in \mathbb{N}$. It follows that

$$x_{j+1} c_k = \alpha_k x_{j+1} + \beta_0 x_j + \sum_{s=1}^{n_k} \beta_s x_{j+1} = \beta_0 x_j + \left(\alpha_k + \sum_{s=1}^{n_k} \beta_s \right) x_{j+1}.$$

Since $c_k \in \mathscr{C}^*_{x_j}$, $\beta_0 \neq 0$ which implies that $x_{j+1}c_k \neq 0$. Therefore, the multiset $x_{j+1}\mathscr{C}^*_{x_j}$ has infinitely many non-zero elements. If at least one of the elements of $x_{j+1}\mathscr{C}^*_{x_j}$ has infinite multiplicity up to scalar multiplication, then we get that the matrix $[l_{x_{j+1}}]_{\mathscr{C}}$ is not row-finite, which contradicts the amenability of the basis $\mathscr{C}$. It follows that every element of $x_{j+1}\mathscr{C}^*_{x_j}$ has finite multiplicity up to scalar multiplication. Hence, the multiset $x_{j+1}\mathscr{C}^*_{x_j}$ has infinitely many different non-zero elements.

Now let $\mathscr{C}^* = \{c_1^*, c_2^*, \cdots\}$ be the set which consists of different non-zero elements of $x_{j+1}\mathscr{C}^*_{x_j}$. Since each c_i^* can be expressed as a linear combination of x_j and x_{j+1}, the set $\mathscr{C}^*$ spans a two-dimensional vector space. Then $\mathscr{C}^*$ must be linearly dependent.

On the other hand, x_j and x_{j+1} can be written as a linear combination of the elements of the basis $\mathscr{C}$. Suppose that $\{d_1, d_2, \ldots, d_m\}$ be the elements of $\mathscr{C}$ used in the expressions of x_j and x_{j+1}. Then we obtain $\mathscr{C}^* = \{\beta_{11}d_1 + \cdots + \beta_{m1}d_m, \beta_{12}d_1 + \cdots + \beta_{m2}d_m, \beta_{13}d_1 + \cdots + \beta_{m3}d_m, \ldots\}$ for some scalars β_{ij}, where $1 \leq i \leq m$, $j \geq 1$. If there were finitely many linearly dependent elements in $\mathscr{C}^*$, then we would get an infinite-dimensional subspace of a finite-dimensional vector space, which is impossible. Hence, $\mathscr{C}^*$ has infinitely many linearly dependent elements which means that at least one of the elements in $\{d_1, \ldots, d_m\}$ must appear infinitely many times in the expression of the elements in $\mathscr{C}^*$. Consequently, the matrix $[l_{x_{j+1}}]_{\mathscr{C}}$ is not row-finite. This contradiction implies that $|\mathscr{C}_{x_j}|$ must be finite.

To prove the necessity, pick $x_k \in V$. Consider the set $S_k = \mathscr{C} \setminus \cup_{j=1}^{k-1} \mathscr{C}_{x_j}$. By assumption, $\cup_{j=1}^{k-1}\mathscr{C}_{x_j}$ is finite. Then S_k is infinite. Since $x_k c_l = \sum_{i=0}^{n_l} \alpha_i x_k$ for each $c_l \in S_k$ and $\sum_{i=0}^{n_l} \alpha_i \neq 0$, for finitely many l, the matrix $[l_{x_k}]_{\mathscr{C}}$ is row-finite. □

Lemma 2 *Let A be an infinite-dimensional algebra over a field K, $\mathscr{B}$ be the standard basis and $\mathscr{C}$ be an amenable basis for A. Then $|{}_l\mathscr{C}_a| < \infty$, where ${}_l\mathscr{C}_a = \{c \in \mathscr{C} | x_l c = c + ka$ for some $0 \neq k \in K\}$ for $a \in A$ and $x_l \in \mathscr{B}$.*

Proof Write $a = \alpha_1 c_1 + \cdots + \alpha_k c_k$ for some $k \in \mathbb{N}$. Consider the set $S = \{c_1, \ldots, c_k\}$. If $|{}_l\mathscr{C}_a|$ is infinite, then $|{}_l\mathscr{C}_a \setminus S|$ will be infinite, too. But then the matrix $[l_{x_l}]_{\mathscr{C}}$ cannot be row-finite. Hence, $|{}_l\mathscr{C}_a|$ must be finite. □

Theorem 4 *A basis $\mathscr{C}$ for the commutative $2V$ graph magma algebra $A((G])$ is amenable if and only if $|\mathscr{C}_{x_i}| < \infty$ for all $x_i \in V \cup \{x_0\}$.*

Proof Assume that $\mathscr{C}$ is an amenable basis for $A((G])$ and that $|\mathscr{C}_{x_0}| = \infty$. One can observe that there exists a non-zero scalar α_j such that $x_1 c_j = c_j + \alpha_j(x_1 - x_0)$ for each $c_j \in \mathscr{C}_{x_0}$. It follows from Lemma 2 that $|\mathscr{C}_{x_0}|$ must be finite.

Now assume that $|\mathscr{C}_{x_k}| = \infty$ for some $x_k \in V$ and that $|\mathscr{C}_{x_j}| < \infty$ for all $j = 1, \ldots, k-1$. Let $S = \cup_{j=0}^{k-1}\mathscr{C}_{x_j}$. Then S is a finite set. Therefore, $\mathscr{C}_{x_k} \setminus S$ is an infinite set. If $c_i \in \mathscr{C}_{x_k} \setminus S$, then we obtain $c_i x_{k+1} = c_i + \alpha(x_{k+1} - x_k)$ for some non-zero scalar α. Again by Lemma 2, $|\mathscr{C}_{x_k}| < \infty$, too.

For the necessity part, suppose that $|\mathscr{C}_{x_j}| < \infty$ for all $x_j \in V \cup \{x_0\}$. Pick $x_k \in V \cup \{x_0\}$. Then $x_j x_k = x_k$ for all $j \leq k$ and $x_j x_k = x_j$ for all $j > k$ $(x_i x_j = x_{max(i,j)})$. Consider the set $S_k = \cup_{j=0}^{k}\mathscr{C}_{x_j}$. By assumption, S_k is a finite set. For any $c_j \in \mathscr{C} \setminus S_k$, we have that $x_k c_j = c_j$. Consequently, we get the following row-finite matrix:

$$
[l_{x_k}]_{\mathscr{C}} = \begin{array}{c} \\ S_k \\ (\mathscr{C}\backslash S_k) \end{array} \begin{array}{c} \begin{array}{cc} x_k S_k & x_k(\mathscr{C}\backslash S_k) \end{array} \\ \left(\begin{array}{c|c} * & 0 \\ 0 & Id_{\mathbb{N}\times\mathbb{N}} \end{array}\right) \end{array}
$$

□

In [2, Corollary 5.3], it is shown that the algebra $F[x]$ has at least as many pairwise discordant simple bases as there are elements in F^*, where F is a field. Also, it is known from [3] that there are examples of graph magma algebras which do not have any simple or projective bases. Moreover, in [9], the authors are able to show that no graph magma algebras have a projective basis. Inspired by the characterization of amenable bases for the $2V$ graph magma algebra $A((G])$ in Theorem 4, we are able to observe the following remarkable result which indicates that there are algebras whose simple bases do exist and they are mutually congenial but they do not have any projective bases.

Theorem 5 *Let A be an infinite-dimensional algebra over a field and let $\mathscr{B} = \{b_1, b_2, \ldots\}$ be a basis of A such that for any basis $\mathscr{C}$ of A, $\mathscr{C}$ is amenable if and only if $|\mathscr{C}_b| < \infty$ for each $b \in \mathscr{B}$. Then A has a unique simple basis (up to mutual congeniality) and has no projective basis.*

Proof We claim that B is a simple basis of A. First note that by assumption $\mathscr{B}$ is amenable. Suppose $\mathscr{B}$ is congenial to $\mathscr{C}$, where $\mathscr{C}$ is an amenable basis of A. It follows that $|\mathscr{C}_b| < \infty$ for each $b \in \mathscr{B}$. But this implies that $\mathscr{C}$ is congenial to $\mathscr{B}$. Hence, $\mathscr{B}$ and $\mathscr{C}$ are mutually congenial. It also follows that $\mathscr{B}$ is unique up to mutual congeniality. Since proper congeniality matrices always exist, it is easy to observe that A does not have a projective basis. □

Theorem 5 together with Theorem 4 gives the following immediate consequence:

Corollary 1 *The commutative $2V$ graph magma algebra $A((G])$ has a unique simple basis (up to mutual congeniality) and does not have any projective bases.*

Theorem 6 *A basis $\mathscr{C}$ for the commutative $2V$ graph magma algebra $A(\{x_{-\infty}\} \oplus (G])$ is amenable if and only if $|\mathscr{C}_{x_j}| < \infty$ for all $x_j \in (V \cup \{x_0\})\backslash\{x_{-\infty}\}$ and $\sum \alpha_{i_j} \neq 0$ for finitely many j, where $c_j = \sum \alpha_{i_j} x_{i_j} \in \mathscr{C}$.*

Proof $\Longrightarrow$: Let $\mathscr{C} = \{c_0, c_1, \ldots\}$ be an amenable basis for the given graph algebra. Then the set $x_{-\infty}\mathscr{C}$ consists of those elements $\sum \alpha_{i_j} x_{-\infty}$ such that $c_j = \sum \alpha_{i_j} x_{i_j}$ for $j = 0, 1, \ldots$. Since $\mathscr{C}$ is amenable, $\sum \alpha_{i_j} \neq 0$ for finitely many j. Now assume that $|\mathscr{C}_{x_0}| = \infty$. By a proof similar to that of Theorem 4, one can see that $\mathscr{C}_{x_0}$ must be a finite set. Let $S = \mathscr{C}_{x_1} \backslash \mathscr{C}_{x_0}$. Then for any $d_i \in S$, we obtain that $x_2 d_i = d_i + \alpha_{1i}(x_2 - x_1) + \alpha_i(x_2 - x_{-\infty})$ for some scalars α_{1i} and α_i. We ensure that α_{1i} is non-zero for each i, because it is the coefficient of x_1 in the expression of $d_i \in S$ for each i. Hence, by Lemma 2, we must have that $|\mathscr{C}_{x_1}|$ is finite. By induction, one can obtain that $\mathscr{C}_{x_i}$ is a finite set for each $x_i \neq x_{-\infty}$.

$\Longleftarrow$: Write $x = x_{-\infty}$ and $\mathscr{C} = \mathscr{C}_x \cup \mathscr{C}_x^c$. Pick $x_k \in V$. Then $x_k \mathscr{C}_x$ contains elements of the form $\sum \alpha_{i_j} x$, where $c_j = \sum \alpha_{i_j} x_{i_j} \in \mathscr{C}_x$. But since $\sum \alpha_{i_j} \neq 0$ for finitely many j, the multiset $x_k \mathscr{C}_x$ has finitely many non-zero elements. Consider the set $\mathscr{C}_x^c = (\cup_{i=1}^{k-1} \mathscr{C}_{x_i}) \cup S$, where $S = \mathscr{C}_{x_\infty}^c \setminus \cup_{i=1}^{k-1} \mathscr{C}_{x_i}$. By the assumption, $\cup_{i=1}^{k-1} \mathscr{C}_{x_i}$ is finite. Moreover, $x_k S = S$. This gives that the matrix $[l_{x_k}]$ is row-finite. □

The next example shows that finiteness of $|\mathscr{C}_{x_{-\infty}}|$ is not a necessary condition for a basis $\mathscr{C}$ for $A(\{x_{-\infty}\} \oplus (G])$ to be amenable.

Example 3 Let $x = x_{-\infty}$. Consider the basis $\mathscr{C} = \{x, x_0, x - x_1, x - x_2, x - x_3, \cdots\}$ for the commutative $2V$ graph magma algebra $A(\{x_{-\infty}\} \oplus (G])$. Then $|\mathscr{C}_x| = \infty$. But $\mathscr{C}$ is amenable since $x\mathscr{C} = \{x, x, 0, 0, \ldots\}$ and $x_i \mathscr{C} = \{x, x_i, x - x_i, \ldots, x - x_i, x - x_{i+1}, x - x_{i+2}, x - x_{i+3}, \ldots\}$ for all i.

Theorem 7 *A basis $\mathscr{C}$ for the commutative $2V$ graph magma algebra $A([G) \oplus \{x_\infty\})$ is amenable if and only if $|\mathscr{C}_{x_j}| < \infty$ for all $x_j \in (V \cup \{x_0\}) \backslash \{x_\infty\}$ and $\sum \alpha_{i_j} \neq 0$ for finitely many j, where $c_j = \sum \alpha_{i_j} x_{i_j} \in \mathscr{C}$.*

Proof $\Longleftarrow$: First consider the element $x = x_\infty \in V$. Reorder $\mathscr{C} = \mathscr{C}_{x_0} \cup \mathscr{C}_{x_0}^c$. Then $x\mathscr{C} = x\mathscr{C}_{x_0} \cup x\mathscr{C}_{x_0}^c$. By assumption, $x\mathscr{C}_{x_0}$ is finite. On the other hand, $x\mathscr{C}_{x_0}^c = \mathscr{C}_{x_0}^c$. Hence, the matrix $[l_x]_{\mathscr{C}}$ is row-finite.

Now pick any $x_i \in V$, where $x_i \neq x_\infty$ and $x_i \neq x_0$. Again by assumption, $x_i \mathscr{C}_{x_0}$ is finite. Let $S = (\cup_{j=1}^{i} \mathscr{C}_{x_j}) \cap \mathscr{C}_{x_0}^c$. S is a finite set and we may reorder $\mathscr{C}_{x_0}^c = S \cup (\mathscr{C}_{x_0}^c \backslash S)$. It follows that $x_i \mathscr{C}_{x_0}^c = x_i S \cup x_i(\mathscr{C}_{x_0}^c \backslash S)$. By the hypothesis, $x_i S$ is finite. On the other hand, since $\sum \alpha_{i_j} \neq 0$ for finitely many j, where $c_j = \sum \alpha_{i_j} x_{i_j} \in \mathscr{C}$, we obtain that $x_i(\mathscr{C}_{x_0}^c \backslash S)$ is infinite with all but finitely many non-zero elements. This implies that the matrix $[l_{x_i}]_{\mathscr{C}}$ is row-finite.

$\Longrightarrow$: Consider the set $S = \{c \in \mathscr{C} | c = \sum \alpha_{i_j} x_{i_j} \text{and} \sum a_{i_j} \neq 0\}$. Suppose that $|S| = \infty$. Since $x_1 c_j = x_1(\sum \alpha_{i_j})$ for any $c_j \in S$, the multiset $x_1 S$ contains infinitely many x_1 with non-zero coefficients. This gives that $[l_{x_1}]_{\mathscr{C}}$ is not row-finite. But this leads to a contradiction since $\mathscr{C}$ is amenable.

Now assume that $|\mathscr{C}_{x_1}| = \infty$. Let $\mathscr{C}_{x_1} = \{c_1, c_2, \ldots\}$. Write $c_i = \alpha_{i_0} x_0 + \alpha_{i_1} x_1 + \sum_{j=2}^{n_i} \alpha_{i_j} x_j$. Then we have that

$$x_2 c_i = \alpha_{i_0} x_2 + \alpha_{i_1} x_1 + \sum_{j=2}^{n_i} \alpha_{i_j} x_2 = \alpha_{i_1} x_1 + \left(\alpha_{i_0} + \sum_{j=2}^{n_i} \alpha_{i_j} \right) x_2,$$

where $\alpha_{i_1} \neq 0$ for each i. It follows that the matrix $[l_{x_2}]_{\mathscr{C}}$ is not row-finite, a contradiction. Hence, $|\mathscr{C}_{x_1}| < \infty$. In a similar way, one can show that $|\mathscr{C}_{x_i}| < \infty$ for any $x_i \in V$ such that $x_i \neq x_\infty$.

Further assume that $|\mathscr{C}_{x_0}| = \infty$ and let $\mathscr{C}_{x_0} = \{d_1, d_2, \ldots\}$. If we write $d_i = \sum_{j=0}^{n_i} \alpha_{i_j} x_j$ for each i, then we will have

$$x_\infty d_i = \alpha_{i_0} x_\infty + \sum_{j=1}^{n_i} \alpha_{i_j} x_j = d_i - \alpha_{i_0} x_0 + \alpha_{i_0} x_\infty = d_i + \alpha_{i_0}(x_\infty - x_0)$$

for each i. Since $\alpha_{i_0} \neq 0$ for each i, the matrix $[l_{x_\infty}]_{\mathscr{C}}$ is not row-finite. This contradiction gives that $|\mathscr{C}_{x_0}|$ must be finite. □

6 Further Results on Commutative Two Value Graph Magma Algebras

In this section, we will deal with some commutative $2V$ graph magma algebras which are constructed on the direct sum of basic graphs.

Let $\mathscr{C}$ be a set. Denote by $\mathscr{C}_{<x_j>}$ the set of elements $c_i \in \mathscr{C}$ such that $c_i = \alpha_l x_l + \alpha_k x_k + S$ for some $l < j < k$, where $\alpha_l \neq 0$, $\alpha_k \neq 0$ and $S = \sum_{i \neq j,k,l} \alpha_i x_i$. The following result holds for any infinite-dimensional algebra.

Lemma 3 *Let $\mathscr{C}$ be a basis for an infinite-dimensional algebra A. If there exists a j such that $|\mathscr{C}_{<x_j>}| < \infty$, and $|\mathscr{C}_{x_i}| < \infty$ for all $x_i \in V$, then $|\mathscr{C}_{<x_i>}| < \infty$ for all $x_i \in V$.*

Proof By hypothesis, we can assume that l is the smallest integer satisfying $|\mathscr{C}_{<x_l>}| < \infty$. Then consider the set $T = \cup_{i=1}^{l} \mathscr{C}_{x_i}$ which is finite by assumption.

Pick an integer $j < l$. Let $T^* = T \cap \mathscr{C}_{<x_j>}$. We can write $\mathscr{C}_{<x_j>} = T^* \cup (\mathscr{C}_{<x_j>} \backslash T^*)$. It follows that $|\mathscr{C}_{<x_j>}| = \infty$. Then we must have that $|\mathscr{C}_{<x_j>} \backslash T^*| = \infty$. But we also have $\mathscr{C}_{<x_j>} \backslash T^* \subseteq \mathscr{C}_{<x_l>}$ which gives a contradiction since $|\mathscr{C}_{<x_j>}| < \infty$. Hence, $|\mathscr{C}_{<x_j>}| < \infty$. A similar contradiction will be obtained if we pick $j > l$. Thus, $|\mathscr{C}_{<x_j>}| < \infty$ for all $x_j \in V$. □

Proposition 5 *Suppose that the two value graph magma of a transitive directed graph G is commutative. Let $\mathscr{C}$ be a basis for the commutative $2V$ graph magma algebra $A((G))$. Assume that $|\mathscr{C}_x| < \infty$ for all $x \in V \cup \{x_0\}$ and there exists $x_j \in V$ such that $|\mathscr{C}_{<x_j>}| < \infty$ and $\sum \alpha_{i_j} \neq 0$ for finitely many j, where $c_j = \sum \alpha_{i_j} x_{i_j} \in \mathscr{C}$. Then C is amenable.*

Proof Reorder $\mathscr{C} = \mathscr{C}_{x_j} \cup \mathscr{C}_{x_j}^c$. Let $S = \mathscr{C}_{x_0} \cap \mathscr{C}_{x_j}^c$. Then we write $\mathscr{C}_{x_j}^c = S \cup (\mathscr{C}_{x_j}^c \backslash S)$. S is a finite set by assumption. We also have $\mathscr{C}_{x_j}^c \backslash S = \mathscr{C}_{x<x_j} \cup \mathscr{C}_{x>x_j} \cup \mathscr{C}_{<x_j>}$, where $\mathscr{C}_{x<x_j} = \{c = \sum \alpha_k x_k \in \mathscr{C} \mid x_k \in inset(x_j)\}$ and $\mathscr{C}_{x>x_j} = \{c = \sum \alpha_k x_k \in \mathscr{C} \mid x_k \in outset(x_j)\}$. Then $x_j \mathscr{C}_{x<x_j} = \mathscr{C}_{x<x_j}$ and $x_j \mathscr{C}_{x>x_j} = \{\sum \alpha_{i_j} x_j \in \mathscr{C} \mid \alpha_{i_j} \in \mathbb{F}\}$. Hence, $[l_{x_j}]_{\mathscr{C}}$ is row-finite. By Lemma 3, we conclude that $[l_{x_i}]_{\mathscr{C}}$ is row-finite for all $x_i \in V$. Thus, $\mathscr{C}$ is an amenable basis for $A((G))$. □

Proposition 6 *Suppose that the two value graph magmas of transitive directed graphs G_i are commutative for each $i = 1, \ldots, n$. Let $G = \oplus_{i=1}^{n}[G_i)$, $V = \bigsqcup_i^n V_i$ and $\mathscr{C} = (\bigsqcup_{i=1}^{n} \mathscr{C}_i) \bigsqcup \mathscr{C}_\pi$ where V_i is the set of vertices of G_i, $\mathscr{C}_i$ is an amenable basis for the commutative 2V graph magma algebra $A([G_i))$ and $\mathscr{C}_\pi$ is a set of multinomial in which the terms come from different vertex sets V_i. If $|\mathscr{C}_\pi| < \infty$, $|\mathscr{C}_{x_0}| < \infty$ and $\sum \alpha_{i_j} \neq 0$ for finitely many j, where $c_j = \sum \alpha_{i_j} x_{i_j} \in \mathscr{C}$ and $x_{i_j} \in V \cup \{x_0\}$, then $\mathscr{C}$ is an amenable basis for $A(G)$.*

Proof Let $x_j \in V$. Write $\mathscr{C} = \mathscr{C}_{x_0} \cup \mathscr{C}_{x_0}^c$. Since $|\mathscr{C}_{x_0}| < \infty$, $x_j \mathscr{C}_{x_0}$ is finite, too. Also, we may write $\mathscr{C}_{x_0}^c = T \bigcup (\mathscr{C}_{x_0}^c \backslash T)$, where $T = \mathscr{C}_{x_0}^c \bigcap \mathscr{C}_\pi$. By assumption, $x_j T$ is finite. On the other hand, we have that $\mathscr{C}_{x_0}^c \backslash T = \bigsqcup_{i=1}^{n} S_i$ such that $S_i \subseteq V_i$ for each $i = 1, \cdots, n$. It follows that $x_j S_i = S_i$ for each $i < j$. Since $\sum \alpha_{i_j} \neq 0$ for finitely many $c_j = \sum \alpha_{i_j} x_{i_j} \in \mathscr{C}$, $x_j S_i$ has finitely many non-zero elements for each $i \geq j$. Hence, the matrix $[l_{x_j}]_{\mathscr{C}}$ is row-finite. □

The next example shows that Proposition 6 is not true if $G = \oplus_{i \in \mathbb{N}}[G_i)$.

Example 4 Suppose that the 2V graph magmas of transitive directed graphs G_i are commutative for each $i \in \mathbb{N}$. Let $G = \oplus_{i \in \mathbb{N}}[G_i)$ and $V_1 = V - \bigsqcup_{p \in \mathbb{P}} V_p$, where $\mathbb{P}$ is the set of prime numbers and $V_p = \{x_p, x_{p^2}, x_{p^3}, \ldots\}$. Consider the set $\mathscr{C}_1 = \{x_0, x_0 - x_1, x_1 - x_6, x_6 - x_{10}, \ldots\}$, where i and j cannot be written as a power of a prime in the expressions $x_i - x_j$ and they are consecutive numbers of this type. Define the sets $\mathscr{C}_p = \{x_p, x_p - x_{p^2}, x_{p^2} - x_{p^3}, \ldots\}$ for $p \in \mathbb{P}$. $\mathscr{C}_p \cup \{x_0\}$ is an amenable basis for $A([G_p))$. Also, $\mathscr{C} = \mathscr{C}_1 \bigsqcup_{p \in \mathbb{P}} \mathscr{C}_p$ is a basis for $A(G)$. Observe that $|\mathscr{C}_{x_0}|$ is finite and $|\mathscr{C}_\pi| = 0$. But since the multiset $x_1 \mathscr{C}$ contains infinitely many x_1, $\mathscr{C}$ is not amenable for $A(G)$.

Theorem 8 *Suppose that the two value graph magmas of transitive directed graphs G_i are commutative for each $i \in \mathbb{N}$. Let $G_\infty = \oplus_{i \in \mathbb{N}}[G_i)$ and consider the algebra $A(G_\infty)$. Suppose that $\mathscr{B} = \bigcup_{i \in \mathbb{N}} \mathscr{B}_i$ is a basis for $A(G_\infty)$ and that $\mathscr{B}_i$ is a basis for the commutative 2V graph magma algebra $A([G_i)$ for each i. Then $\mathscr{B}$ is amenable if and only if $|\mathscr{B}_{x_i}|$ is finite for each i and $\sum \alpha_{i_j} \neq 0$ for finitely many j, where $b_j = \sum \alpha_{i_j} x_{i_j} \in \mathscr{B}$ and $x_{i_j} \in V \cup \{x_0\}$.*

Proof The necessity follows from Theorem 3. For the sufficiency suppose that the set $S = \{b \in \mathscr{B} | \sum \alpha_i \neq 0$, where $b = \sum \alpha_i x_i\}$ is infinite and contains elements from $\mathscr{B}_i$ for different indices i. For such an i, there exists $b_i = \sum \alpha_i x_{j_i} \in \mathscr{B}_i$ such that $\sum \alpha_i \neq 0$. It follows from Theorem 3 that $\mathscr{B}_i$ is an amenable basis for the commutative $2V$ graph magma algebra $A([G_i))$. On the other hand, $x_1 \mathscr{B}$ contains infinitely many non-zero elements which contradicts the amenability of $\mathscr{B}$. □

Theorem 9 *Let $G_{-\infty} = \oplus_{i \in \mathbb{N}}(G_i]$, where each G_i is a transitive directed graph. Assume that the graph magmas of the graphs G_i are commutative. Consider the algebra $A(G_{-\infty})$. Suppose that $\mathscr{B} = \bigcup_{i \in \mathbb{N}} \mathscr{B}_i$ is a basis for $A(G_{-\infty})$ and that $\mathscr{B}_i$ is a basis for the commutative 2V graph magma algebra $A([G_i)$ for each i. Then $\mathscr{B}$ is amenable if and only if $\mathscr{B}_{x_i}$ and S_i are finite sets for each $i \in \mathbb{N}$, where $S_i = \{\sum \alpha_i x_i \in \mathscr{B}_i | \sum \alpha_i \neq 0\}$.*

Proof The necessity follows from Theorem 4. To prove the sufficiency, assume that $\mathscr{B}$ is amenable and there exists an $i \in \mathbb{N}$ such that S_i is infinite. Take $x \in V_{i+1}$. Note that $xy = x$ for every $x \in V_{i+1}$ and $y \in V_i$. Then the multiset xS_i contains infinitely many non-zero elements which are all multiples of x. This contradicts the fact that $\mathscr{B}$ is amenable. The rest of the proof is similar to that of Theorem 4. □

Corollary 2 *If a basis $\mathscr{B} = \bigcup_{i\in\mathbb{N}} \mathscr{B}_i$ is amenable in $A(G_\infty)$, then it is also amenable in $A(G_{-\infty})$.*

We end the section with Example 5 which indicates that the converse of Proposition 2 need not be true.

Example 5 Consider the commutative $2V$ graph magma algebras $A([G_i]))$, where graph magmas of transitive directed graphs G_i are commutative, and the corresponding basis $\mathscr{B}_i = \{x_0, x_{1_i}, x_{1_i} - x_{2_i}, x_{2_i} - x_{3_i}, \ldots\}$ for each $i \in \mathbb{N}$. Then S_i is a finite set for each $i \in \mathbb{N}$, where $S_i = \{\sum \alpha_i x_i \in \mathscr{B}_i | \sum \alpha_i \neq 0\}$. Also, $|\mathscr{B}_{x_i}| \leq 2$ for each $i \in \mathbb{N}$. But $\mathscr{B} = \bigcup_{i\in\mathbb{N}} \mathscr{B}_i$ is not an amenable basis for the commutative $2V$ magma algebra $A(G_\infty)$, where $G_\infty = \oplus_{i\in\mathbb{N}}[G_i)$, because each $\mathscr{B}_i$ contains x_{1_i}.

Acknowledgements This work was done during the visit of the first author to the Ohio University, Center of Ring Theory and its Applications. She would like to thank all the members of the Center for their warm hospitality. She also acknowledges the financial support she received from Hacettepe University, Scientific Research Projects Coordination Unit, Project No: FBI-2017-14952.

References

1. Abrams, G., Simón, J.J: Isomorphisms between infinite matrix rings: a survey. Algebra and its Applications (Athens, OH, 1999). Contemp. Math. Am. Math. Soc. 259, 1–12 (2000)
2. Al-Essa, L.M., López-Permouth, S.R., Muthana, N.M.: Modules over infinite-dimensional algebras. Linear and Multilinear Algebra **66**(3), 488–496 (2018)
3. Aydoğdu, P., López-Permouth, S.R., Muhammad, Rebin A.: Infinite-dimensional algebras without simple bases. Linear and Multilinear Algebra **68**(12), 2390–2407 (2020). https://doi.org/10.1080/03081087.2019.1585414
4. Diestel, R.: Graph theory. Graduate Texts in Mathematics. Springer, Berlin, Heidelberg
5. Dixit, V.N., Singhal, P.K., Weinert, H.J.: On quasi-ideals in contracted semigroup rings. Semigroup Forum **33**(1), 431–439 (1986)
6. Kelarev, A.V., Sokratova, O.V.: Syntactic semigroups and graph algebras. Bull. Austral. Math. Soc. **62**(3), 471–477 (2000)
7. López-Permouth, S.R., Owusu-Mensah, I., Rafieipour, A.: Binary Operations induced by graphs. preprint (2020)
8. López-Permouth, S.R., Stanley, B.: Topological characterizations of amenability and congeniality of bases. Appl. General Topol. **21**(1), 1–15 (2020)
9. López-Permouth, S.R., Stanley, B.: Graph Magma Algebras have no projective bases. To appear in Linear and Multilinear Algebra. https://doi.org/10.1080/03081087.2019.1652552
10. Munn, W.D.: Moore-Penrose inversion in complex contracted inverse semigroup algebras. Aust. Math. Soc. J. Ser. A. Pure Math. Stat. **66**(3), 297 (1999)
11. Munn, W.D.: A class of contracted inverse semigroup rings. Proc. Roy. Soc. Edinb.: Sect. A Math. **107**, 175–196 (1987)

12. Nielsen, P.P.: Row and column finite matrices. In: Proceedings of the American Mathematical Society, 9 (2007)
13. Roman, S.: Advanced Linear Algebra. [electronic resource]. Springer, New York (2008)
14. Oates-Williams, S.: On the variety generated by Murskii's algebra. Algebra Univ. **18**(2), 175–177 (1984)

Graph of Linear Transformations Over $\mathbb{R}$

Ayman Badawi and Yasmine El-Ashi

Abstract In this paper, we study a connection between graph theory and linear transformations of finite dimensional vector spaces over $\mathbb{R}$ (the set of all real numbers). Let R^m, R^n be finite vector spaces over R, and let L be the set of all non-trivial linear transformations from R^m into R^n. An equivalence relation $\sim$ is defined on L such that two elements $f, k \in L$ are equivalent, $f \sim k$, if and only if $\ker(f) = \ker(k)$. Let $m, n \geq 1$ be positive integers and $V_{m,n}$ be the set of all equivalence classes of $\sim$. We define a new graph, $G_{m,n}$, to be the undirected graph with vertex set equal to $V_{m,n}$, such that two vertices, $[x], [y] \in V_{m,n}$ are adjacent if and only if $\ker(x) \cap \ker(y) \neq 0$. The relationship between the connectivity of the graph $G_{m,n}$ and the values of m and n has been investigated. We determine the values of m and n so that $G_{m,n}$ is a complete graph. Also, we determine the diameter and the girth of $G_{m,n}$.

Keywords Zero-divisor graph · Total graph · Unitary graph · Dot product graph · Annihilator graph · Linear transformations graph

1 Introduction

Let R be a commutative ring with $1 \neq 0$. Recently, there has been considerable attention in the literature to associating graphs with commutative rings (and other algebraic structures), as well as studying the interplay between ring-theoretic and graph-theoretic properties; see the survey articles [10, 11, 19, 39, 46]. In particular, as in [17], the *zero-divisor graph* of R is the (simple) graph $\Gamma(R)$ with vertices $Z(R) \setminus \{0\}$, and distinct vertices x and y are adjacent if and only if $xy = 0$. This concept is due to Beck [29], who let all the elements of R be vertices and was mainly interested in coloring. The zero-divisor graph of a ring R has been studied extensively by many authors, for example, see [2–9, 12, 22, 23, 38–44, 47–54, 58]. David. F.

A. Badawi (✉) · Y. El-Ashi
Department of Mathematics & Statistics, The American University of Sharjah, P.O. Box 26666, Sharjah, United Arab Emirates
e-mail: abadawi@aus.edu

M. Ashraf et al. (eds.), *Algebra and Related Topics with Applications*,
Springer Proceedings in Mathematics & Statistics 392,
https://doi.org/10.1007/978-981-19-3898-6_31

Anderson and the first-named author [13] introduced the *total graph* of R, denoted by $T(\Gamma(R))$. We recall from [13] that the total graph of a commutative ring R is the (simple) graph $\Gamma(R)$ with vertices R, and distinct vertices x and y are adjacent if and only if $x + y \in Z(R)$. The total graph (as in [13]) has been investigated in [5–8, 35, 46, 48, 52, 56], and several variants of the total graph have been studied in [4, 14–16, 21, 28, 31–34, 36, 37, 45].

Let $a \in Z(R)$ and let $ann_R(a) = \{r \in R \mid ra = 0\}$. In 2014, A. Badawi [27] introduced the annihilator graph of R. We recall from [27] that the annihilator graph of R is the (undirected) graph $AG(R)$ with vertices $Z(R)^* = Z(R) \setminus \{0\}$, and two distinct vertices x and y are adjacent if and only if $ann_R(xy) \neq ann_R(x) \cup ann_R(y)$. See the survey article [24]. It follows that each edge (path) of the classical zero-divisor of R is an edge (path) of $AG(R)$. For further investigations of $AG(R)$, see [20, 51, 57]. In 2015, A. Badawi investigated the *total dot product graph* of R [26]. In this case, $R = A \times A \times \cdots \times A$ (n times), where A is a commutative ring with non-zero identity, and $1 \leq n < \infty$ is an integer. The *total dot product graph* of R is the (undirected) graph denoted by $TD(R)$, with vertices $R^* = R \setminus \{(0, 0, \ldots 0)\}$. Two distinct vertices are adjacent if and only if $x \cdot y = 0 \in A$, where $x \cdot y$ denote the normal dot product of x and y. The *zero-divisor dot product graph* of R is the induced subgraph $ZD(R)$ of $TD(R)$ with vertices $Z(R)^* = Z(R) \setminus \{(0, 0, \ldots, 0)\}$. It follows that each edge (path) of the classical zero-divisor graph $\Gamma(R)$ is an edge (path) of $ZD(R)$. In [26], both graphs $TD(R)$ and $ZD(R)$ are studied. The total dot product graph was recently further investigated in [1]. Other types of graphs attached to groups and rings were studied (for example) in [6, 8, 28, 38, 40, 44, 45].

Let G be a graph. Two vertices v_1, v_2 of G are said to be *adjacent* in G if v_1, v_2 are connected by an edge of G and we write $v_1 - v_2$. For vertices x and y of G, we define $\mathrm{d}(x, y)$ to be the length of a shortest path from x to y ($\mathrm{d}(x, x) = 0$ and $\mathrm{d}(x, y) = \infty$ if there is no path). Then the *diameter* of G is $\mathrm{diam}(G) = \sup\{ \mathrm{d}(x, y) \mid x$ and y are vertices of $G\}$. The *girth* of G, denoted by $gr(G)$, is the length of a shortest cycle in G ($gr(G) = \infty$ if G contains no cycles).

We say G is *connected* if there is a path in G from u to v for every $u, v \in V$. Therefore, a graph is said to be *disconnected*, if there exist at least two vertices $u, v \in V$ that are not joined by a path. We say that G is *totally disconnected* if no two vertices of G are adjacent. We denote the complete graph on n vertices by K_n, and recall that a graph G is called complete if every two vertices of G are adjacent.

In this paper, we introduce a connection between graph theory and linear transformations of finite dimensional vector spaces over $\mathbb{R}$ (the ring of all real numbers). Let U and W be finite dimensional vector spaces over $\mathbb{R}$, such that $m = dim(U)$ and $n = dim(W)$. Since every finite dimensional vector space over $\mathbb{R}$ with dimension k is isomorphic to $\mathbb{R}^k$, we conclude that U is isomorphic to $\mathbb{R}^m$ and W is isomorphic to $\mathbb{R}^n$. Let $m, n \geq 1$ be positive integers and $L = \{t : \mathbb{R}^m \to \mathbb{R}^n \mid t$ is a non-trivial linear transformation from $\mathbb{R}^m$ into $\mathbb{R}^n\}$. If $s, t \in L$, then we say that s is equivalent to t, and we write $s \sim t$ if and only if $Ker(s) = Ker(t)$. Clearly, $\sim$ is an equivalence relation on L. For each $t \in L$, the set $[t] = \{s \in L | s \sim t\}$ is called the *equivalence class* of t. Let $V_{m,n}$ be the set of all equivalence classes of $\sim$. For positive integers $m, n \geq 1$, let

$G_{m,n}$ be a simple undirected graph with vertex set $V_{m,n}$ such that two distinct vertices $[f], [k] \in V_{m,n}$ are adjacent if and only if $Ker(f) \cap Ker(k) \neq \{(0, \ldots, 0)\} \subset \mathbb{R}^m$.

2 Results

Remark 1 If a graph G has one vertex, then we say that G is totally disconnected. Note that some authors state that such a graph is connected.

We have the following result.

Theorem 1 *The undirected graph $G_{m,1}$ is totally disconnected if and only if $m = 1$ or $m = 2$. Furthermore, if $m = 1$, then $V_{1,1} = \{[t]\}$ for some $t \in L$.*

Proof Assume $m = 1$. Let $[t] \in V_{1,1}$. Since $t \in L$ (i.e., t is a non-trivial linear transformation from $\mathbb{R}$ into $\mathbb{R}$), we conclude that $dim(Range(t)) = 1$. Since $dim(Ker(t)) + dim(Range(t)) = m = 1$ and $dim(Range(t)) = 1$, we conclude that $Ker(t) = \{0\}$. Thus, $f \in [t]$ for every $f \in L$. Hence, $V_{1,1} = \{[t]\}$ for some $t \in L$. Thus, $G_{1,1}$ is totally disconnected by Remark 1.

Assume $m = 2$. Let $[t], [f] \in V_{2,1}$ be two distinct vertices. Since $t, f \in L$ (i.e., t, f are non-trivial linear transformations from $\mathbb{R}^2$ into $\mathbb{R}$), we conclude that $dim(Range(t)) = dim(Range(t)) = 1$. Since $dim(Ker(t)) + dim(Range(t)) = m = 2$ and $dim(Range(t)) = 1$, we conclude that $dim(Ker(t)) = 1$. Similarly, $dim(Ker(f)) = 1$. Since $t, f \in L$, and $dim(Ker(t)) = dim(Ker(f)) = 1$, we conclude that $Ker(t)$ and $Ker(f)$ are distinct lines passing through the origin (0, 0). Thus $Ker(t) \cap Ker(f) = \{(0, 0)\}$. Hence $[t], [f]$ are nonadjacent. Thus $G_{2,1}$ is totally disconnected.

Now assume $m > 2$. We show that $G_{m,1}$ is connected. Let, $[t], [w] \in V_{m,1}$ be two distinct vertices. We show that $\ker(f) \cap \ker(k) \neq \{(0, \ldots, 0)\}$ for some $f \in [t]$ and $k \in [w]$. Let $\mathbf{M}_f$ be the standard $1 \times m$ matrix representation of f for some $f \in [t] \in V_{m,1}$ and $\mathbf{M}_k$ be the standard $1 \times m$ matrix representation of k for some $k \in [w] \in V_{m,1}$. By hypothesis, $\mathbf{M}_f$ is not *row-equivalent* to $\mathbf{M}_k$. Say, $\mathbf{M}_f = \begin{bmatrix} f_{11} \; f_{12} \cdots f_{1m} \end{bmatrix}$ and $\mathbf{M}_k = \begin{bmatrix} k_{11} \; k_{12} \cdots k_{1m} \end{bmatrix}$.

Let $\mathbf{M}_{fk} = \begin{bmatrix} \mathbf{M}_f \\ \mathbf{M}_k \end{bmatrix}$ and consider the system, $\mathbf{M}_{fk}\mathbf{x} = \mathbf{0}$, that is,

$$\begin{bmatrix} f_{11} \; f_{12} \cdots f_{1m} \\ k_{11} \; k_{12} \cdots k_{1m} \end{bmatrix} \begin{bmatrix} x_1 \\ x_2 \\ \vdots \\ x_m \end{bmatrix} = \begin{bmatrix} 0 \\ 0 \\ \vdots \\ 0 \end{bmatrix}.$$

Since $m > 2$, the number of equations < the number of unknown variables. Hence, the system $\mathbf{M}_{fk}\mathbf{x} = \mathbf{0}$ has infinitely many solutions. Therefore, $\ker(f) \cap \ker(k) \neq \mathbf{0}$, that is, the vertices $[t]$ and $[w]$ are adjacent. Further, since $[t], [w]$ were chosen randomly, we conclude that the graph $G_{m,1}$ is complete for $m > 2$.

Theorem 2 *For $m = 1$ or $m = 2$, the undirected graph $G_{2,n}$ is totally disconnected for every positive integer $n \geq 1$.*

Proof Assume $m = 1$ and $n \geq 1$ be a positive integer. Then by the proof of Theorem 1, we conclude that $V_{1,n} = \{[t]\}$ for some $t \in L$. Hence, $V_{1,n}$ is totally disconnected by Remark 1.

Assume $m = 2$, and let $[t], [w] \in V$ be two distinct vertices. We want to show $\ker(f) \cap \ker(k) = 0$ for some $f \in [t]$ and $k \in [w]$. We may assume that neither $Ker(f) = 0$ nor $Ker(k) = 0$. Hence, $dim(Ker(f)) = dim(Ker(k)) = 1$. Thus, $Ker(f) \cap Ker(k) = \{(0, 0)\}$. Since $[f]$, $[k]$ were chosen randomly, we conclude that the graph $G_{2,n}$ is totally disconnected for $m = 2$.

Theorem 3 *The graph $G_{m,n}$ is complete if and only if $m \geq 2n + 1$.*

Proof Let $[t], [w] \in V$ such that $Ker(f) \neq 0$ and $Ker(k) \neq 0$ for some $f \in [t]$ and $k \in [w]$. Let $\mathbf{M}_f$ be the standard $n \times m$ matrix representation of $[f]$, $\mathbf{M}_k$ be the standard $n \times m$ matrix representation of $[k]$, and let $\mathbf{M}_{fk} = \begin{bmatrix} \mathbf{M}_f \\ \mathbf{M}_k \end{bmatrix}$.

Assume, $(x_1, x_2, \ldots, x_m) \in \mathbf{R}^m$ is a solution to $\mathbf{M}_{fk}\mathbf{x} = 0$, that is,

$$\begin{bmatrix} \mathbf{M}_f \\ \mathbf{M}_k \end{bmatrix}_{2n \times m} \begin{bmatrix} x_1 \\ x_2 \\ \vdots \\ x_m \end{bmatrix}_{m \times 1} = \begin{bmatrix} 0 \\ 0 \\ \vdots \\ 0 \end{bmatrix}_{2n \times 1}.$$

Let $r = \operatorname{rank}\left(\mathbf{M}_{fk}\right)$.

Assume $m \geq 2n + 1$. We show $\ker(f) \cap \ker(k) \neq 0$. Since $r \leq 2n$ and $m \geq 2n + 1$, we have number of equations < number of unknown variables. Hence, the system $\mathbf{M}_{fk}\mathbf{x} = 0$ has infinitely many solutions, or null $\left(\mathbf{M}_{fk}\right) \neq 0$. Therefore, $\ker(f) \cap \ker(k) \neq 0$, that is, the vertices $[t]$ and $[w]$ are adjacent. Since $[t]$ and $[w]$ are chosen randomly, we conclude that the graph $G_{m,n}$ is complete for $m \geq 2n + 1$.

Conversaly, assume that $G_{m,n}$ is complete. We show that $m \geq 2n + 1$. Suppose that $m < 2n + 1$. We show that $G_{m,n}$ is not complete. Let $[t], [w] \in V$ such that $Ker(f) \neq 0$ and $Ker(k) \neq 0$ for some $f \in [t]$ and $k \in [w]$.

Case I: Suppose $r = m$.
We conclude that $\mathbf{M}_{fk}$ has m independent rows, say $R_1, R_2, \ldots, R_m$.
Consider the system,

$$\begin{bmatrix} R_1 \\ R_2 \\ \vdots \\ R_m \end{bmatrix} \begin{bmatrix} x_1 \\ x_2 \\ \vdots \\ x_m \end{bmatrix} = \begin{bmatrix} 0 \\ 0 \\ \vdots \\ 0 \end{bmatrix}.$$

Since $\left[\, R_1\ R_2 \cdots R_m \,\right]^T$ is an invertible $m \times m$ matrix, we have $\text{null}\left(\left[\, R_1\ R_2 \cdots R_m \,\right]\right)^T = (0, 0, \ldots, 0)$. Thus, $\ker(t) \cap \ker(w) = 0$. Hence, the vertices $[t]$ and $[w]$ are nonadjacent

Case II: Suppose $r < m$. Thus, we have the following system:

$$\begin{bmatrix} R_1 \\ R_2 \\ \vdots \\ R_r \end{bmatrix} \begin{bmatrix} x_1 \\ x_2 \\ \vdots \\ x_m \end{bmatrix} = \begin{bmatrix} 0 \\ 0 \\ \vdots \\ 0 \end{bmatrix}.$$

Since number of equations < number of unknown variables, we conclude that $\text{null}\left(\left[\, R_1\ R_2 \cdots R_r \,\right]^T\right) \neq (0, 0, \ldots, 0)$. This implies $\ker(f) \cap \ker(k) \neq 0$. Hence, the vertices $[t]$ and $[w]$ are adjacent.

Since the vertices $[t]$ and $[w]$ can either be adjacent or nonadjacent, we conclude that the graph $G_{m,n}$ is not complete for every $1 \leq m < 2n + 1$.

Theorem 4 *Consider the undirected graph $G_{m,n}$. Assume $m \leq n$ and $m \neq 1$ or $m \neq 2$. Then $G_{m,n}$ is connected and $diam(G_{m,n}) = 2$.*

Proof Let $[t], [w] \in V$ such that $[t]$ and $[w]$ are nonadjacent. Choose $f \in [t]$ and $k \in [w]$. Then $\text{rank}\left(M_f\right) \neq m$ and $\text{rank}\,(M_k) \neq m$, where M_f and M_k are the standard matrix representations of f and k, with size $n \times m$.

Assume $\text{rank}\left(M_f\right) = m - i$, where $i \in \mathbf{N}, i \neq 1$, and $\text{rank}\,(M_k) = m - j$, where $j \in \mathbf{N},\ j \neq 1$. Then choose any non-zero row from M_f or M_k, say Y, to form the $n \times m$ matrix M_d, where

$$M_d = \begin{bmatrix} Y \\ 0 \\ \vdots \\ 0 \end{bmatrix}$$

is the standard matrix representation of some $d \in [h] \in V_{m,n}$, such that $[t] - [h] - [w]$.

Assume that $\text{rank}\left(M_f\right) = m - 1$ and $\text{rank}\,(M_k) = m - 1$. Then M_f has $m - 1$ independent rows, $R_1, R_2, \ldots, R_{m-1}$. Since $[t]$ and $[w]$ are nonadjacent, M_k has one row say R such that $\{R_1, R_2, \ldots, R_{m-1}, R\}$ is an independent set which forms a basis for $\mathbf{R}^m$. Let $K \neq R$ be a non-zero row in M_k. Hence, $K \in \text{rowspace}\,(M_k)$. Since $K \in \mathbf{R}^m$, we have

$$K = c_1 R_1 + c_2 R_2 + \cdots + c_{m-1} R_{m-1} + c_m R.$$

Let $Y = K - c_m R$. Thus, $Y \in \text{rowspace}(M_k)$ (since both K and $c_m R$ are $\in$ rowspace (M_k)) and $Y \in \text{rowspace}(M_f)$. Let $M_d = \begin{bmatrix} Y \\ 0 \\ \vdots \\ 0 \end{bmatrix}_{n \times m}$ be the standard matrix representation of some $d \in [h] \in V_{m,n}$. Since $Y \in \text{rowspace}(M_f)$, Y becomes a zero row through row operations using the rows in M_f. Thus, $\text{null}(M_{fd}) \neq 0$, since $\text{rank}(M_{fd}) = m - 1$. Hence, $\ker(f) \cap \ker(d) \neq 0$. Hence, $[t]$, $[h]$ are connected by an edge. Similarly, since $Y \in \text{rowspace}(M_k)$, Y becomes a zero row through row operations using the rows in M_k. Thus, $\text{null}(M_{kd}) \neq 0$, since $\text{rank}(M_{kd}) = m - 1$. Hence, $\ker(d) \cap \ker(k) \neq 0$. Thus, $[h]$ and $[w]$ are adjacent. Therefore, we have $[t] - [h] - [w]$.

Example 1 Suppose $m = 3$ and $n = 4$. So we are considering the graph $G([t] : \mathbf{R}^3 \to \mathbf{R}^4)$, where $m \leq n$, and $m \neq 1$ or $m \neq 2$, as given in Theorem 4. Let $[T], [L] \in V$, such that $[T]$ and $[L]$ are not adjacent ($\ker(T) \cap \ker(L) = 0_{m=3}$), and $[T] \neq 0$, $[L] \neq 0$. Let $f \in [T]$ and $k \in [L]$. Since $[T]$ and $[L]$ are non-trivial vertices, then $\text{rank}(M_f) \neq m$ and $\text{rank}(M_k) \neq m$, where M_f and M_k are the standard matrix representations of f and k.
Suppose

$$M_f = \begin{bmatrix} 1\,0\,0 \\ 0\,1\,1 \\ 0\,0\,0 \\ 0\,0\,0 \end{bmatrix}_{4\times3}, M_k = \begin{bmatrix} 0\,0\,0 \\ 0\,0\,1 \\ 1\,1\,0 \\ 0\,0\,0 \end{bmatrix}_{4\times3}.$$

Let $M_{fk} = \begin{bmatrix} M_f \\ M_k \end{bmatrix}_{8\times3}$.
It can be easily seen that $\text{rank}(M_{fk}) = 3$, which implies that $\text{null}(M_{fk}) = 0$. Therefore, $\ker(f) \cap \ker(k) = 0$, that is, the vertices $[T]$ and $[L]$ are not adjacent. We have $\text{rank}(M_f) = 2 = 3 - 1 = m - 1$, and $\text{rank}(M_k) = 2 = 3 - 1 = m - 1$.

Then M_f has two independent rows R_1 and R_2, such that $R_1 = [1\;0\;0]$ and $R_2 = [0\;1\;1]$. The vertices $[T]$ and $[L]$ are not adjacent, thus M_k has one row R, such that $\{R_1, R_2, R\}$ are independent and form a basis for $\mathbf{R}^m$, where $m = 3$. In this example, $R = [0\;0\;1]$. Let $K \neq R$ be a non-zero row in M_k, $K = [1\;1\;0]$. $K \in \text{rowspace}(M_k)$ and since $K \in \mathbf{R}^3$, it can be written as a linear combination of $\{R_1, R_2, R\}$ as follows:

$$K = 1.R_1 + 1.R_2 - R = [1\;0\;0] + [0\;1\;1] - [0\;0\;1] = [1\;1\;0].$$

Let $Y = K - (-1).R = K + R = [1\;1\;0] + [0\;0\;1] = [1\;1\;1]$.

This implies $Y \in$ rowspace (M_k) and $Y \in$ rowspace $\left(M_f\right)$. Let $M_d = \begin{bmatrix} Y \\ 0 \\ 0 \\ 0 \end{bmatrix}_{4\times 3} =$

$\begin{bmatrix} 1\,1\,1 \\ 0\,0\,0 \\ 0\,0\,0 \\ 0\,0\,0 \end{bmatrix}_{4\times 3}$ be the standard matrix representation of some $d \in [W]$.

Since $Y \in$ rowspace $\left(M_f\right)$, Y becomes a zero row through row operations using the rows in M_f. Thus, null $\left(M_{fd}\right) \neq 0$ since rank $\left(M_{fd}\right) = 2$. Hence, ker $(T) \cap$ ker $(W) \neq 0$. Hence, $[T], [W]$ are adjacent. Similarly, since $Y \in$ rowspace (M_k), Y becomes a zero row through row operations using the rows in M_k. Hence, null $(M_{kd}) \neq 0$ since rank $(M_{kd}) = 2$. Thus, ker $(L) \cap$ ker $(W) \neq 0$. Thus, $[W], [L]$ are adjacent. Therefore, we have $[T] - [W] - [L]$.

Theorem 5 *Consider the undirected graph $G_{m,n}$. Assume that $n < m \leq 2n$ and $m \neq 1$ or $m \neq 2$. Then $G_{m,n}$ is connected and $diam(G_{m,n}) = 2$.*

Proof Let $[T], [L] \in V$, such that $[T]$ and $[L]$ are not adjacent (ker $(T) \cap$ ker $(L) = 0_m$), and $[T] \neq 0$, $[L] \neq 0$. Let $f \in [T]$ and $k \in [L]$, then rank $\left(M_f\right) < m$ and rank $(M_k) < m$, where M_f and M_k are the standard matrix representations of f and k, with size $n \times m$.

Assume that $n + 1 < m \leq 2n$. Then rank $\left(M_f\right) = n - i$, where $i = 0, 1, 2, \ldots$, and rank $(M_k) = n - j$, where $j = 0, 1, 2, \ldots$. Thus, we can choose any non-zero row from M_f or M_k, say Y, to form the $n \times m$ matrix M_d, where

$$M_d = \begin{bmatrix} Y \\ 0 \\ \vdots \\ 0 \end{bmatrix}$$

is the standard matrix representation of some $d \in [W]$, such that $[T] - [W] - [L]$.

Assume that $m = n + 1$. Then we have three cases. **Case I**. Assume that rank $\left(M_f\right) = n = m - 1$, and rank $(M_k) = n - j$, where $j = 1, 2, \ldots$. Then we can choose any non-zero row, say Y from M_f (Note that M_f is the matrix with the higher rank) to form the $n \times m$ matrix M_d, where

$$M_d = \begin{bmatrix} Y \\ 0 \\ \vdots \\ 0 \end{bmatrix}$$

is the standard matrix representation of some $d \in [W]$, such that $[T] - [W] - [L]$. **Case II**. Assume that rank $\left(M_f\right) = n - i$, where $i = 1, 2, \ldots$ and rank $(M_k) = n -$

j, where $j = 1, 2, \ldots$. In this case, any non-zero row Y can be chosen either from M_f or M_k, to form M_d, where

$$M_d = \begin{bmatrix} Y \\ 0 \\ \vdots \\ 0 \end{bmatrix}$$

is the standard matrix representation of some $d \in [W]$, such that $[T] - [W] - [L]$.
Case III. Assume that rank $\left(M_f\right) = n$ and rank $(M_k) = n$. Then M_f has n independent rows $R_1, R_2, \ldots, R_n$. Since $[T]$ and $[L]$ are not adjacent, M_k has one row say R such that $\{R_1, R_2, \ldots, R_{m-1}, R\}$ is an independent set which forms a basis for $\mathbf{R}^m = \mathbf{R}^{n+1}$. Let $K \neq R$ be a non-zero row in M_k. Hence, $K \in$ rowspace (M_k). Since $K \in \mathbf{R}^{n+1}$, we have

$$K = c_1 R_1 + c_2 R_2 + \cdots + c_n R_n + c_{n+1} R.$$

Let $Y = K - c_{n+1} R$. Hence, $Y \in$ rowspace (M_k) (since both K, $c_{n+1} R \in$ rowspace (M_k)) and $Y \in$ rowspace $\left(M_f\right)$. Let $M_d = \begin{bmatrix} Y \\ 0 \\ \vdots \\ 0 \end{bmatrix}_{n \times m}$ be the standard matrix representation of some $d \in [W]$.

Since $Y \in$ rowspace $\left(M_f\right)$, Y becomes a zero row through row operations using the rows in M_f, null $\left(M_{fd}\right) \neq 0$ since rank $\left(M_{fd}\right) = n$. Hence, ker $(T) \cap$ ker $(W) \neq 0$. Thus, $[T], [W]$ are adjacent. Similarly, since $Y \in$ rowspace (M_k), Y becomes a zero row through row operations using the rows in M_k. Hence, null $(M_{kd}) \neq 0$ since rank $(M_{kd}) = n$. Thus, ker $(L) \cap$ ker $(W) \neq 0$. Thus, $[W], [L]$ are adjacent. Therefore, we have $[T] - [W] - [L]$.

Example 2 Suppose $m = 4$ and $n = 3$ and consider the graph $G_{4,3}$. Note that $n < m \leq 2n$, $m \neq 1, 2$ and and $m = n + 1$. Thus m, n satisfy the given hypothesis in Theorem 5. Let $[T], [L] \in V$, such that $[T]$ and $[L]$ are not adjacent. Let $f \in [T]$, and $k \in [L]$. Then rank $\left(M_f\right) < m$ and rank $(M_k) < m$, where M_f and M_k are the standard matrix representations of f and k, with size $n \times m = 3 \times 4$. Suppose,

$$M_f = \begin{bmatrix} 1\,0\,0\,0 \\ 0\,1\,0\,1 \\ 0\,0\,1\,0 \end{bmatrix}_{3\times4}, \; M_k = \begin{bmatrix} 1\,0\,0\,0 \\ 0\,1\,0\,0 \\ 0\,0\,0\,1 \end{bmatrix}_{3\times4}.$$

Let $M_{fk} = \begin{bmatrix} M_f \\ M_k \end{bmatrix}_{6\times4}$. It can be easily seen that rank $\left(M_{fk}\right) = 4$, which implies that null $\left(M_{fk}\right) = 0$. Therefore, ker $(f) \cap$ ker $(k) = 0$, that is, the vertices $[T]$ and $[L]$ are not adjacent. Hence, rank $\left(M_f\right) = 3 = n$, and rank $(M_k) = 3 = n$. Then M_f has three independent rows R_1, R_2, and R_3, such that $R_1 = \left[1\;0\;0\;0\right]$, $R_2 = \left[0\;1\;0\;1\right]$, and $R_3 = \begin{bmatrix} 0 & 0 & 1 & 0 \end{bmatrix}$. The vertices $[T]$ and $[L]$ are not adjacent, thus M_k has one row, $R = \begin{bmatrix} 0 & 0 & 0 & 1 \end{bmatrix}$, such that $\{R_1, R_2, R_3, R\}$ is an independent set which forms a basis for $\mathbf{R}^4$. Let $K \neq R$ be a non-zero row in M_k, $K = \begin{bmatrix} 0 & 1 & 0 & 0 \end{bmatrix}$. Since $K \in \text{rowspace}(M_k)$ and $K \in \mathbf{R}^4$, it can be written as a linear combination of $\{R_1, R_2, R_3, R\}$ as follows:

$$K = 0.R_1 + 1.R_2 + 0.R_3 + (-1).R = \begin{bmatrix} 0 & 1 & 0 & 1 \end{bmatrix} - \begin{bmatrix} 0 & 0 & 0 & 1 \end{bmatrix} = \begin{bmatrix} 0 & 1 & 0 & 0 \end{bmatrix}.$$

Let $Y = K - (-1).R = K + R = \begin{bmatrix} 0 & 1 & 0 & 0 \end{bmatrix} + \begin{bmatrix} 0 & 0 & 0 & 1 \end{bmatrix} = \begin{bmatrix} 0 & 1 & 0 & 1 \end{bmatrix}$.

This implies $Y \in \text{rowspace}(M_k)$ and $Y \in \text{rowspace}(M_f)$. Let $M_d = \begin{bmatrix} Y \\ 0 \\ 0 \end{bmatrix}_{3\times 4} = \begin{bmatrix} 0\,1\,0\,1 \\ 0\,0\,0\,0 \\ 0\,0\,0\,0 \end{bmatrix}_{3\times 4}$ be the standard matrix representation of some $d \in [W]$.

Since $Y \in \text{rowspace}(M_f)$, Y becomes a zero row through row operations using the rows in M_f. Thus, $\text{null}(M_{fd}) \neq 0$, since $\text{rank}(M_{fd}) = 3$. Hence, $\ker(T) \cap \ker(W) \neq 0$. Thus, $[T], [W]$ are adjacent. Similarly, since $Y \in \text{rowspace}(M_k)$, Y becomes a zero row through row operations using the rows in M_k. Thus, $\text{null}(M_{kd}) \neq 0$ since $\text{rank}(M_{kd}) = 3$. Hence, $\ker(L) \cap \ker(W) \neq 0$. Thus, $[W], [L]$ are adjacent. Therefore, we have $[T] - [W] - [L]$.

Theorem 6 *Assume that $G_{m,n}$ is connected. Then $gr(G_{m,n})) = 3$.*

Proof $[T], [L] \in V$, such that $[T]$ and $[L]$ are adjacent, $\ker(T) \cap \ker(L) \neq 0$ and $[T] \neq 0$, $[L] \neq 0$. Let $f \in [T]$ and $k \in [L]$, then M_f and M_k are the standard matrix representations of f and k with size $n \times m$. Suppose that each matrix M_f and M_k is composed of only one row, R_f and R_k that are independent of each other since f and k are in different equivalence classes $[T]$ and $[L]$. M_f and M_k can be written as follows:

$$M_f = \begin{bmatrix} R_f \\ 0 \\ \vdots \\ 0 \end{bmatrix}_{n\times m}, M_k = \begin{bmatrix} R_k \\ 0 \\ \vdots \\ 0 \end{bmatrix}_{n\times m}.$$

Let $Y = R_f + R_k$. Since Y is a linear combination of two linearly independent rows, then the set $\{Y, R_f, R_k\}$ is also linearly independent.

Let $M_d = \begin{bmatrix} Y \\ 0 \\ \vdots \\ 0 \end{bmatrix}_{n\times m}$ be the standard matrix representation of some non-trivial linear transformation d. Since Y is independent of both R_f and R_k, M_d is not row-equivalent to either M_f or M_k, hence d is in a different equivalence class from both f and k, say $d \in [W]$. Since ker $(T) \cap$ ker $(L) \neq 0$, we have null $\left(M_{fk}\right) \neq 0$, which implies null $\left(M_{fd}\right) \neq 0$ and null $(M_{kd}) \neq 0$. Therefore, we have $[T] - [L] - [W] - [T]$. This forms the shortest possible cycle. Hence, $gr(G_{m,n})) = 3$.

Acknowledgements The second-named author would like to thank the Graduate Office at the American University of Sharjah for the continuous support.

References

1. Abdulla, M., Badawi, A.: On the dot product graph of a commutative ring II. 25 Int. Electron. J. Algebra **28**, 61–74 (2020)
2. Akbari, S., Maimani, H.R., Yassemi, S.: When a zero-divisor graph is planar or a complete r-partite graph. J. Algebra **270**, 169–180 (2003)
3. Akbari, S., Mohammadian, A.: On the zero-divisor graph of a commutative ring. J. Algebra **274**, 847–855 (2004)
4. Abbasi, A., Habib, S.: The total graph of a commutative ring with respect to proper ideals. J. Korean Math. Soc. **49**, 85–98 (2012)
5. Akbari, S., Heydari, F.: The regular graph of a non-commutative ring. Bull. Aust. Math. Soc. (2013). https://doi.org/10.1017/S0004972712001177
6. Akbari, S., Aryapoor, M., Jamaali, M.: Chromatic number and clique number of subgraphs of regular graph of matrix algebras. Linear Algebra Appl. **436**, 2419–2424 (2012)
7. Akbari, S., Jamaali, M., Seyed Fakhari, S.A.: The clique numbers of regular graphs of matrix algebras are finite. Linear Algebra Appl. **43**, 1715–1718 (2009)
8. Akbari, S., Kiani, D., Mohammadi, F., Moradi, S.: The total graph and regular graph of a commutative ring. J. Pure Appl. Algebra **213**, 2224–2228 (2009)
9. Anderson, D.D., Naseer, M.: Beck's coloring of a commutative ring. J. Algebra **159**, 500–514 (1993)
10. Anderson, D.F., Axtell, M., Stickles, J.: Zero-divisor graphs in commutative rings. In: Fontana, M., Kabbaj, S.E., Olberding, B., Swanson, I. (eds.) Commutative Algebra Noetherian and Non-Noetherian Perspectives, pp. 23–45. Springer, New York (2010)
11. Anderson, D.F., Badawi, A.: The zero-divisor graph of a commutative semigroup: a survey. In: groups, modules, and model theory surveys and recent developments. In: Manfred, D., László, F., Goldsmith, B., Lutz Strüngmann, 23–39. Germany/NewYork: Springer (2017). https://doi.org/10.1007/978-3-319-51718-6_2
12. Anderson, D.F., Badawi, A.: On the zero-divisor graph of a ring. Comm. Algebra **36**, 3073–3092 (2008)
13. Anderson, D.F., Badawi, A.: The total graph of a commutative ring. J. Algebra **320**, 2706–2719 (2008)
14. Anderson, D.F., Badawi, A.: The total graph of a commutative ring without the zero element. J. Algebra Appl. (2012). https://doi.org/10.1142/S0219498812500740
15. Anderson, D.F., Badawi, A.: The generalized total graph of a commutative ring. J. Algebra Appl. (2013). https://doi.org/10.1142/S021949881250212X
16. Anderson, D.F., Fasteen, J., LaGrange, J.D.: The subgroup graph of a group. Arab. J. Math. **1**, 17–27 (2012)
17. Anderson, D.F., Livingston, P.S.: The zero-divisor graph of a commutative ring. J. Algebra **217**, 434–447 (1999)
18. Anderson, D.F., Mulay, S.B.: On the diameter and girth of a zero-divisor graph. J. Pure Appl. Algebra **210**, 543–550 (2007)
19. Anderson, D.F., Asir, T., Badawi, A., Tamizh Chelvam, T.: Graphs from rings. Springer (2021)

20. Afkhami, M., Khashyarmanesh, K., Sakhdari, S. M.: The annihilator graph of a commutative semigroup. J. Algebra Appl. **14**, 14 (2015). https://doi.org/10.1142/S0219498815500152
21. Atani, S.E., Habibi, S.: The total torsion element graph of a module over a commutative ring. An. Stiint. Univ.'Ovidius Constanta Ser. Mat. **19**, 23–34 (2011)
22. Axtel, M, Coykendall, J., Stickles, J.: Zero-divisor graphs of polynomials and power series over commutative rings. Comm. Algebra **33**, 2043–2050 (2005)
23. Axtel, M., Stickles, J.: Zero-divisor graphs of idealizations. J. Pure Appl. Algebra **204**, 235–243 (2006)
24. Badawi, A.: Recent results on annihilator graph of a commutative ring: a survey. In: Prasad, K., et al. (eds.) Nearrings, Nearfields, and Related Topics (11 pages), World Scientific, New Jersey (2017)
25. Badawi, A.: On the total graph of a ring and its related graphs: a survey. In: Fontana, et al. (eds.), Commutative algebra: recent advances in commutative rings, integer-valued polynomials, and polynomial functions, edited by M. Fontana et al. (eds.), pp. 39–54. Springer Science (2014),. https://doi.org/10.1007/978-1-4939-0925-43
26. Badawi, A.: On the dot product graph of a commutative ring. Comm. Algebra **43**, 43–50 (2015)
27. Badawi, A.: On the annihilator graph of a commutative ring. Comm. Algebra **42**(1), 108–121 (2014). https://doi.org/10.1080/00927872.2012.707262
28. Barati, Z., Khashyarmanesh, K., Mohammadi, F., Nafar, K.: On the associated graphs to a commutative ring. J. Algebra Appl. (2012). https://doi.org/10.1142/S021949881105610
29. Beck, I.: Coloring of commutative rings. J. Algebra **116**, 208–226 (1988)
30. Bollaboás, B.: Graph Theory. An Introductory Course, Springer, New York (1979)
31. Tamizh Chelvam, T., Asir, T.: Domination in total graph on $\mathbb{Z}_n$. Discrete Math. Algorithms Appl. **3**, 413–421 (2011)
32. Tamizh Chelvam, T., Asir, T.: Domination in the total graph of a commutative ring. J. Combin. Math. Combin. Comput. 87, 147–158 (2013)
33. Tamizh Chelvam, T., Asir, T.: Intersection graph of gamma sets in the total graph. Discuss. Math. Graph Theory 32, 339–354 (2012)
34. Tamizh Chelvam, T., Asir, T.: On the Genus of the Total Graph of a Commutative Ring. Comm. Algebra **41**, 142–153 (2013)
35. Tamizh Chelvam, T., Asir, T.: On the total graph and its complement of a commutative ring. Comm. Algebra (2013). doi: https://doi.org/10.1080/00927872.2012.678956
36. Tamizh Chelvam, T., Asir, T.: The intersection graph of gamma sets in the total graph. I. J. Algebra Appl. (2013). doi: https://doi.org/10.1142/S0219498812501988
37. Tamizh Chelvam, T., Asir, T.: The intersection graph of gamma sets in the total graph II: J. Algebra Appl. (2013). doi: https://doi.org/10.1142/S021949881250199X
38. Chiang-Hsieh, H.-J., Smith, N.O., Wang, H.-J.: Commutative rings with toroidal zerodivisor graphs. Houston J. Math. **36**, 1–31 (2010)
39. Coykendall, J., Sather-Wagstaff, S.: Sheppardson, L., Spiroff, S.: On zero divisor graphs. In: Francisco, C. et al. (eds.)Progress in Commutative Algebra 2: Closures, finiteness and factorization. Walter Gruyter, Berlin, pp. 241–299 (2012)
40. DeMeyer, F., DeMeyer, L.: Zero divisor graphs of semigroups. J. Algebra **283**, 190–198 (2005)
41. DeMeyer, F., McKenzie, T., Schneider, K.: The zero-divisor graph of a commutative semigroup. Semigroup Forum. **65**, 206–214 (2002)
42. DeMeyer, F., Schneider, K.: Automorphisms and zero divisor graphs of commutative rings. In: Commutative Rings. Nova Science Publications, Hauppauge, NY, pp. 25–37 (2002)
43. DeMeyer, L., D'Sa, M., Epstein, I., Geiser, A., Smith, K.: Semigroups and the zero divisor graph. Bull. Inst. Combin. Appl. **57**, 60–70 (2009)
44. DeMeyer, L., Greve, L., Sabbaghi, A., Wang, J.: The zero-divisor graph associated to a semigroup. Comm. Algebra **38**, 3370–3391 (2010)
45. Khashyarmanesh, K., Khorsandi, M.R.: A generalization of the unit and unitary Cayley graphs of a commutative ring. Acta Math. Hungar. **137**, 242–253 (2012)
46. Maimani, H.R., Pouranki, M.R., Tehranian, A., Yassemi, S.: Graphs attached to rings revisited. Arab. J. Sci. Eng. **36**, 997–1011 (2011)

47. Maimani, H.R., Pournaki, M.R., Yassemi, S.: Zero-divisor graph with respect to an ideal. Comm. Algebra **34**, 923–929 (2006)
48. Maimani, H.R., Wickham, C., Yassemi, S.: Rings whose total graphs have genus at most one. Rocky Mountain J. Math. **42**, 1551–1560 (2012)
49. Mojdeh1, D.A., Rahimi, A.M: Domination sets of some graphs associated to commutative ring. Comm. Algebra **40**, 3389–3396 (2012)
50. Mulay, S.B.: Cycles and symmetries of zero-divisors. Comm. Algebra **30**, 3533–3558 (2002)
51. Nikandish, R., Nikmehr, M.J., Bakhtyiari, M.: Coloring of the annihilator graph of a commutative ring. J. Algebra Appl. **15**(7) (2016). https://doi.org/10.1142/S0219498816501243
52. Pucanović, Z., Petrović, Z.: On the radius and the relation between the total graph of a commutative ring and its extensions. Publ. Inst. Math.(Beograd)(N.S.) **89**, 1–9 (2011)
53. Redmond, S.P.: An ideal-based zero-divisor graph of a commutative ring. Comm. Algebra **31**, 4425–4443 (2003)
54. Smith, N.O.: Planar zero-divisor graphs. Comm. Algebra **35**, 171–180 (2007)
55. Sharma, P.K., Bhatwadekar, S.M.: A note on graphical representations of rings. J. Algebra **176**, 124–127 (1995)
56. Shekarriz, M.H., Shiradareh Haghighi, M.H., Sharif, H.: On the total graph of a finite commutative ring. Comm. Algebra **40**, 2798–2807 (2012)
57. Visweswaran, S., Patel, H.D.: A graph associated with the set of all nonzero annihilating ideals of a commutative ring. Discrete Math. Algorithm. Appl. **6**, 22 (2014). https://doi.org/10.1142/S1793830914500475
58. Wickham, C.: Classification of rings with genus one zero-divisor graphs. Comm. Algebra **36**, 325–345 (2008)

On Distance Laplacian (Signless) Eigenvalues of Commuting Graphs of Dihedral and Dicyclic Groups

S. Pirzada, Bilal A. Rather, Rezwan ul Shaban, and Imran Bhat

Abstract For a finite group $\mathcal{G}$ with identity e, let X be a nonempty subset of $\mathcal{G}$. The commuting graph $G = \mathcal{C}(\mathcal{G}, X)$ is a simple connected graph with vertex set X, where two vertices $x, \ y \in X$ are adjacent if and only if x and y commute in X. In this article, we find the distance Laplacian and distance signless Laplacian eigenvalues of the commuting graph associated to dihedral group, semi-dihedral group and dicyclic group. We show that the commuting graphs of the dihedral group, semi-dihedral group and dicyclic group are distance Laplacian integral.

Keywords Distance Laplacian matrix · Distance signless Laplacian matrix · Commuting graph · Dihedral group · Semi-dihedral group · Dicyclic group

1 Introduction

All graphs considered in this article are connected, undirected, simple and finite. A graph is denoted by $G(V(G), E(G))$ (or simply by G), where $V(G)$ is the vertex set and $E(G)$ is the edge set of G. The *order* and the *size* of G are the cardinalities of $V(G)$ and $E(G)$, respectively. The *degree* of a vertex v in G is the number of edges incident with v and is denoted by $d_G(v)$ (or simply by d_v). The *neighborhood* of a vertex v, denoted by $N(v)$, is the set of vertices of G adjacent to v, so that $d_v = |N(V)|$. A graph G is called *r-regular* if degree of each vertex is r.

The adjacency matrix $A(G) = (a_{ij})$ of G is a (0, 1)-square matrix of order n whose (i, j)-entry is equal to 1 if v_i is adjacent to v_j and equal to 0, otherwise. Since $A(G)$ is real symmetric, so we take adjacency spectrum as $\lambda_n \leq \lambda_{n-1} \leq \cdots \leq \lambda_1$. Let $Deg(G) = diag(d_1, d_2, \ldots, d_n)$ be the diagonal matrix of vertex degrees $d_i = d_G(v_i), i = 1, 2, \ldots, n$ associated to G. The matrices $L(G) = Deg(G) - A(G)$ and $Q(G) = Deg(G) + A(G)$ are, respectively, the Laplacian and the signless Laplacian

S. Pirzada (✉) · B. A. Rather · R. Shaban · I. Bhat
Department of Mathematics, University of Kashmir, Srinagar, J & K, India
e-mail: pirzadasd@kashmiruniversity.ac.in

M. Ashraf et al. (eds.), *Algebra and Related Topics with Applications*,
Springer Proceedings in Mathematics & Statistics 392,
https://doi.org/10.1007/978-981-19-3898-6_32

matrices, and their spectrum are, respectively, the Laplacian spectrum and signless Laplacian spectrum of the graph G. These matrices are real symmetric and positive semi-definite. We take $0 = \mu_n \leq \mu_{n-1} \leq \cdots \leq \mu_1$ to be the Laplacian eigenvalues of the graph G. More literature about these matrices can be found in the book [7].

In G, the *distance* between two vertices $u, v \in V(G)$, denoted by $d(u, v)$, is defined as the length of the shortest path between u and v. The *distance matrix* of G is denoted by $\mathcal{D}(G)$ and is defined as $\mathcal{D}(G) = (d_{uv})$, where $d_{uv} = d(u, v)$, if u is adjacent to v and zero otherwise. For more about $\mathcal{D}(G)$, we refer the reader to [5]. The *transmission* $Tr_G(v)$ of a vertex v is defined to be the sum of the distances from v to all other vertices in G, that is, $Tr_G(v) = \sum_{u \in V(G)} d(u, v)$. For any vertex $v_i \in V(G)$, the transmission $Tr_G(v_i)$ is called the *transmission degree*, shortly denoted by Tr_i and the sequence $\{Tr_1, Tr_2, \ldots, Tr_n\}$ is called the *transmission degree sequence* of G.

Let $Tr(G) = diag(Tr_1, Tr_2, \ldots, Tr_n)$ be the diagonal matrix of vertex transmissions of G. Aouchiche and Hansen [6] introduced the distance Laplacian $\mathcal{D}^L(G) = Tr(G) - \mathcal{D}(G)$ and the distance signless Laplacian $\mathcal{D}^Q(G) = Tr(G) + \mathcal{D}(G)$ for the distance matrix of a connected graph G. Since the matrices $\mathcal{D}^L(G)$ and $\mathcal{D}^Q(G)$ are real symmetric positive semi-definite (definite in case of $\mathcal{D}^Q(G)$), we denote their eigenvalues by $0 = \rho_n^L \leq \rho_{n-1}^L \leq \cdots \leq \rho_1^L$ and $\rho_n^Q \leq \rho_{n-1}^Q \leq \cdots \leq \rho_1^Q$, respectively. The eigenvalues ρ_1^L and ρ_{n-1}^L are called the distance Laplacian spectral radius and second smallest distance Laplacian eigenvalue of graph G. Similarly, ρ_1^Q is known as distance signless Laplacian spectral radius of G. More about distance Laplacian and distance signless Laplacian matrices can be found in [6, 11, 17] and the references therein.

Let $\mathcal{G}$ be a finite group and X be a non empty subset of $\mathcal{G}$. The *commuting graph*, denoted by $\mathcal{C}(\mathcal{G}, X)$, is defined with X as vertex set and two vertices x and y are adjacent if and only if x and y commute in X. Commuting graphs of matrix rings and semirings over finite fields were studied in [1, 8]. Metric dimension, resolving polynomial, clique number and chromatic number of commuting graphs on dihedral groups were discussed in [3, 23]. Recent results on the commuting graph of generalized dihedral groups can be found in [14] and the references therein. The connectivity and spectral radius of adjacency matrix of commuting graphs were studied in [4], and Laplacian and signless Laplacian spectra of commuting graphs on dihedral group were investigated in [2]. For other spectral and energies of commuting graphs, we refer to [9, 12] and the references therein. Also, the investigation of spectra in zero divisor graphs can be seen in [18–22].

2 Distance Laplacian Spectra of the Commuting Graphs of D_{2n} and Q_{4n}

Consider an $n \times n$ matrix

$$M = \begin{pmatrix} A_{1,1} & A_{1,2} & \cdots & A_{1,s} \\ A_{2,1} & A_{2,2} & \cdots & A_{2,s} \\ \vdots & \vdots & \ddots & \vdots \\ A_{s,1} & A_{s,2} & \cdots & A_{s,s} \end{pmatrix},$$

whose rows and columns are partitioned according to a partition $P = \{P_1, P_2, \ldots, P_m\}$ of the set $X = \{1, 2, \ldots, n\}$. The quotient matrix $\mathcal{Q}$ (see [7]) is an $s \times s$ matrix whose entries are the average row sums of the blocks $A_{i,j}$ of M. The partition P is said to be *equitable* if each block $A_{i,j}$ of M has constant row (and column) sum, and in this case, the matrix $\mathcal{Q}$ is called as *equitable quotient matrix*. In general, the eigenvalues of M interlace the eigenvalues of $\mathcal{Q}$. In case the partition is equitable, we have the following lemma.

Lemma 1 *[7] If the partition P of X of matrix M is equitable, then each eigenvalue of $\mathcal{Q}$ is an eigenvalue of M.*

Assume that a graph G has a kind of symmetry so that its associated matrix is written in the form

$$M = \begin{pmatrix} X & \beta & \cdots & \beta & \beta \\ \beta^T & B & \cdots & C & C \\ \vdots & \vdots & \ddots & \vdots & \vdots \\ \beta^T & C & \cdots & B & C \\ \beta^T & C & \cdots & C & B \end{pmatrix}, \tag{1}$$

where $X \in R^{t \times t}$, $\beta \in R^{t \times s}$ and $B, C \in R^{s \times s}$, such that $n = t + cs$, where c is the number of copies of B. Then the spectrum of this matrix can be obtained as the union of the spectrum of smaller matrices using the following technique given in [10].

Lemma 2 *Let M be a matrix of the form given in* (1), *with $c \geq 1$ copies of the block B. Then*

(i) $\sigma(B - C) \subseteq \sigma(M)$ *with multiplicity* $c - 1$;
(ii) $\sigma(M) \setminus \sigma^{(c-1)}(B - C) = \sigma(M')$ *is the set of the remaining* $t + s$ *eigenvalues of* M, *where* $M' = \begin{pmatrix} X & \sqrt{c}.\beta \\ \sqrt{c}.\beta^T & B + (c-1)C \end{pmatrix}$, *where* $\sigma^{(k)}(Y)$ *indicates the multi-set formed by* k *copies of the spectrum of* Y, *denoted by* $\sigma(Y)$.

The following lemma gives the equivalent method of finding the determinant (det) of a matrix.

Lemma 3 *[13] Let M_1, M_2, M_3 and M_4 be, respectively, $p \times p$, $p \times q$, $q \times p$ and $q \times q$ matrices with M_1 and M_4 invertible. Then*

$$det\begin{pmatrix} M_1 & M_2 \\ M_3 & M_4 \end{pmatrix} = det(M_1)det(M_4 - M_3M_1^{-1}M_2)$$
$$= det(M_4)det(M_1 - M_2M_4^{-1}M_3),$$

where $M_4 - M_3M_1^{-1}M_2$ and $M_1 - M_2M_4^{-1}M_3$ are called the Schur complement of M_1 and M_4, respectively.

Let $G(V, E)$ be a graph of order n and $G_i(V_i, E_i)$ be graphs of order n_i, where $i = 1, \ldots, n$. The *joined union* $G[G_1, \ldots, G_n]$ is the graph $H(W, F)$ with

$$W = \bigcup_{i=1}^{n} V_i \quad \text{and} \quad F = \bigcup_{i=1}^{n} E_i \cup \bigcup_{\{v_i, v_j\} \in E} V_i \times V_j.$$

In other words, the joined union is the union of graphs $G_1, \ldots, G_n$ together with the edges $v_{ik}v_{jl}$, $v_{ik} \in G_i$ and $v_{jl} \in G_j$, whenever v_iv_j is an edge in G.

The following theorem gives the distance Laplacian spectrum of the joined union of graphs $G_1, G_2, \ldots, G_n$, in terms of Laplacian spectrum of the graphs $G_1, G_2, \ldots, G_n$.

Theorem 1 *Let[11] G be a graph of order n having vertex set $V(G) = \{v_1, \ldots, v_n\}$. Let G_i be a graph of order m_i and Laplacian eigenvalues $\mu_{i1} \geq \mu_{i2} \geq \ldots \geq \mu_{im_i}$, where $i = 1, 2, \ldots, n$. The distance Laplacian spectrum of the joined union $G[G_1, \ldots, G_n]$ consists of the eigenvalues $2m_i - \mu_{ik} + \alpha_i$ for $i = 1, \ldots, n$ and $k = 1, 2, 3, \ldots, m_i - 1$, where $\alpha_i = \sum_{k=1, k \neq i}^{n} m_k d_G(v_i, v_k)$. The remaining n eigenvalues are given by the matrix*

$$M = \begin{pmatrix} \alpha_1 & -m_2d_G(v_1, v_2) & \ldots & -m_nd_G(v_1, v_n) \\ -m_1d_G(v_2, v_1) & \alpha_2 & \ldots & -m_nd_G(v_2, v_n) \\ \vdots & \vdots & \ddots & \vdots \\ -m_1d_G(v_n, v_1) & -m_2d_G(v_n, v_2) & \ldots & \alpha_n \end{pmatrix}.$$

We assume all our groups are finite with identity element denoted by e. For notations and definitions, we follow [15]. The presentation of dihedral group D_{2n}, $n > 2$, is given by

$$D_{2n} = \langle a, b : a^n = e = b^2, aba = b \rangle.$$

Clearly, the last condition is equivalent to $ab = ba^{-1} = ba^{n-1}$. Similarly, the presentation of semi-dihedral SD_{8n} of order $8n$ and dicyclic group Q_{4n} of order $4n$ are given by

$$SD_{8n} = \langle a, b : a^{4n} = e = b^2, ab = ba^{2n-2} \rangle,$$

and

$$Q_{4n} = \langle a, b : a^{2n} = e = b^4, b^2 = a^n, ab = ba^{-1} \rangle,$$

respectively. The center of $\mathcal{G}$, denoted by $Z(\mathcal{G})$, is defined by

$$Z(\mathcal{G}) = \{z \in \mathcal{G} : za = az, \text{ for each } a \in \mathcal{G}.\}$$

We suppose that any finite cyclic group $\mathcal{G}$ of order n is isomorphic to integers modulo group $\mathbb{Z}_n$. Clearly, the commuting graph $G = \mathcal{C}(\mathbb{Z}_n, \mathbb{Z}_n)$ is the complete graph K_n, as every element of $\mathbb{Z}_n$ commutes with every other element. The distance Laplacian spectrum of $\mathcal{C}(\mathbb{Z}_n, \mathbb{Z}_n)$ is $\{0, n^{[n-1]}\}$, where $[n-1]$ represents the algebraic multiplicity (or multiplicity) of the eigenvalue n. It easily follows that $Z(D_{2n}) = \{e\}$, for odd n and $Z(D_{2n}) = \left\{e, a^{\frac{n}{2}}\right\}$, for even n. Also, $Z(Q_{4n}) = \{e, a^n\}$ is the center of dicyclic group. For the commuting graph $G = \mathcal{C}(D_{2n}, Z(D_{2n}))$ (see [3]), G is K_1, for odd n and G is K_2, for even n. So, the commuting graphs $\mathcal{C}(\mathcal{G}, Z(\mathcal{G}))$ have simple structures and their spectral properties follow easily, so are omitted here.

The next result can be found in [3], which gives the structure of D_{2n}, where X is D_{2n} itself.

Lemma 4 *For the commuting graph $G = \mathcal{C}(D_{2n}, D_{2n})$ of dihedral group D_{2n}, we have*

$$G = \begin{cases} K_1 \nabla \left(K_{n-1} \cup \overline{K}_n\right), & \text{if } n \text{ is odd,} \\ K_2 \nabla \left(K_{n-2} \cup \frac{n}{2} K_2\right), & \text{if } n \text{ is even.} \end{cases}$$

The following result can be found in [24].

Lemma 5 *For the commuting graph $G = \mathcal{C}(SD_{8n}, D_{8n})$ of dihedral group D_{2n}, we have*

$$G = \begin{cases} K_4 \nabla \left(K_{4n-4} \cup n K_4\right), & \text{if } n \text{ is odd,} \\ K_2 \nabla \left(K_{4n-2} \cup 2n K_2\right), & \text{if } n \text{ is even.} \end{cases}$$

In the following result, we find the distance Laplacian eigenvalues of dihedral group.

Theorem 2 *For the commuting graph $\mathcal{C}(D_{2n}, D_{2n})$ of dihedral group D_{2n}, the following hold.*

(i) If n is odd, then the distance Laplacian spectrum of $\mathcal{C}(D_{2n}, D_{2n})$ is

$$\left\{0, 2n, (3n)^{[n-2]}, (4n-1)^{[n]}\right\}.$$

(ii) If n is even, then the distance Laplacian spectrum of $\mathcal{C}(D_{2n}, D_{2n})$ is

$$\left\{0, (2n)^{[2]}, (4n-2)^{[n/2]}, (4n-4)^{[n/2]}, (3n)^{[n-3]}\right\}.$$

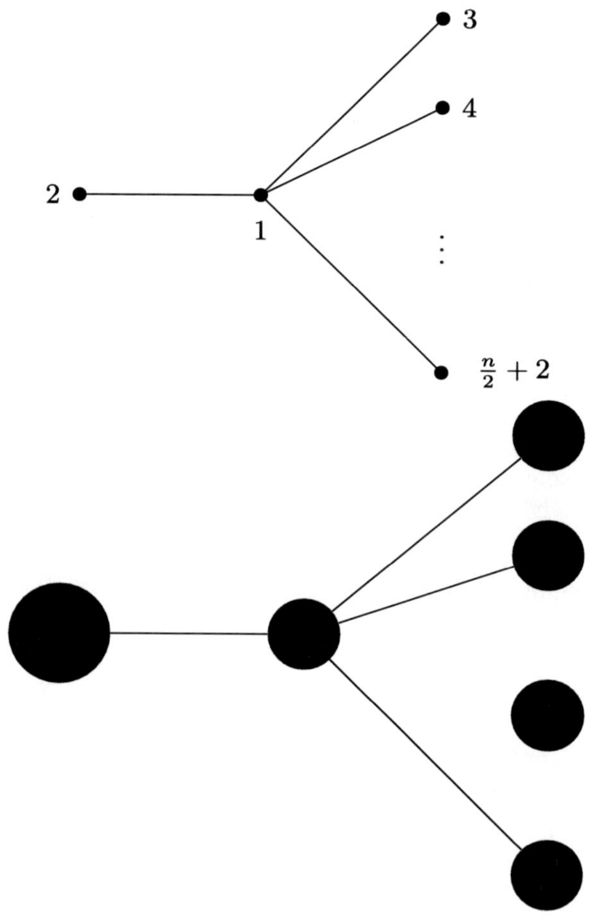

Fig. 1 Graph S and the commuting graph $\mathcal{C}(D_{2n}, D_{2n})$ with even n

Proof **(i).** By Lemma 4, the commuting graph of D_{2n} is

$$\mathcal{C}(D_{2n}, D_{2n}) = K_1 \nabla \big(K_{n-1} \cup \overline{K}_n\big) = P_3[\overline{K}_n, K_1, K_{n-1}],$$

that is, $\mathcal{C}(D_{2n}, D_{2n})$ is the *pineapple graph* (a graph obtained by appending pendent edges to a vertex of complete graph). By using Theorem 1, the α_i's are

$$\alpha_1 = 1 + 2(n-1) = 2n-1, \alpha_2 = n + n - 1 = 2n - 1 \text{ and } \alpha_3 = 2n + 1.$$

Again, by Theorem 1, the distance Laplacian eigenvalues of $\mathcal{C}(D_{2n}, D_{2n})$ are the eigenvalue $2m_1 - \lambda_{1k} + \alpha_1 = 2n + 2n - 1 = 4n - 1$ with multiplicity $n-1$, the eigenvalue $3n$ with multiplicity $n-2$, and the remaining three eigenvalues are the eigenvalues of following quotient matrix

$$M = \begin{pmatrix} 2n-1 & -1 & -2(n-1) \\ -n & 2n-1 & -(n-1) \\ -2n & -1 & 2n+1 \end{pmatrix}.$$

Since each row sum of M is zero, so 0 is the simple eigenvalue and the remaining two eigenvalues are $2n$ and $4n-1$.

(ii). Let n be even. Then by Lemma 4, the commuting graph $\mathcal{C}(D_{2n}, D_{2n})$ of D_{2n} is

$$\mathcal{C}(D_{2n}, D_{2n}) = K_2 \nabla \Big(K_{n-2} \cup \frac{n}{2} K_2\Big) = S[K_2, K_{n-2}, \underbrace{K_2, K_2, \ldots, K_2}_{\frac{n}{2}}],$$

where S is shown in Fig. 1. The value of α_i's are

$$\alpha_1 = n - 2 + 2\frac{n}{2} = 2n - 2, \alpha_2 = 2n + 2, \alpha_3 = \cdots = \alpha_{\frac{n}{2}+2} = 4n - 6.$$

By Theorem 1, the distance Laplacian spectrum of $\mathcal{C}(D_{2n}, D_{2n})$ consists of the simple eigenvalue $2m_1 - \lambda_{1k} + \alpha_1 = 4 + 2n - 2 - 2 = 2n$, the eigenvalue $3n$ with multiplicity $n-3$, the eigenvalue $4n-4$ with multiplicity $\frac{n}{2}$ and the eigenvalues of the matrix

$$M = \left(\begin{array}{cc|cccc} 2n-2 & -n+2 & -2 & -2 & \ldots & -2 \\ -2 & 2n+2 & -4 & -4 & \ldots & -4 \\ \hline -2 & -2n+4 & 4n-6 & -4 & \ldots & -4 \\ -2 & -2n+4 & -4 & 4n-6 & \ldots & -4 \\ \vdots & \vdots & \vdots & \vdots & \ddots & \vdots \\ -2 & -2n+4 & -4 & -4 & \ldots & 4n-6 \end{array}\right) = \left(\begin{array}{c|c} A_{2\times 2} & B_{2\times\frac{n}{2}} \\ \hline C_{\frac{n}{2}\times 2} & D_{\frac{n}{2}\times\frac{n}{2}} \end{array}\right).$$

Clearly, D is invertible; therefore, by Lemma 3, we have

$$det(M - xI) = det(D - xI)det((A - xI) - C(D - xI)^{-1}B),$$

where

$$D - xI = \begin{pmatrix} 4n-6-x & -4 & \ldots & -4 \\ -4 & 4n-6-x & \ldots & -4 \\ \vdots & \vdots & \ddots & \vdots \\ -4 & -4 & \ldots & 4n-6-x \end{pmatrix}.$$

Now, applying Lemma 2 to the matrix $D - xI$, with $X = [\mathbf{0}]$, $\beta = [\mathbf{0}]$, $B = [4n - 6 - x]$ and $C = [-4]$, we see that $4n-2$ is the distance Laplacian eigenvalue of $\mathcal{C}(D_{2n}, D_{2n})$ with multiplicity $\frac{n}{2}$. Since 0 is the simple eigenvalue of the matrix M, thus

$$\mu_1 + \frac{n}{2}2(2n-1) = 2n - 2 + 2n + 2 + \frac{n}{2}2(2n-3)$$

implies that $\mu_1 = 2n$ is the distance Laplacian eigenvalue of M. This proves the result in this case. □

In the following result, we find the distance Laplacian eigenvalues of semi-dihedral group.

Theorem 3 *For the commuting graph $C(SD_{8n}, SD_{8n})$ of dihedral group SD_{8n}, the following hold.*

(i) If n is odd, then the distance Laplacian spectrum of $C(SD_{8n}, SD_{8n})$ is

$$\left\{0, (8n)^{[4]}, (12n)^{[4n-5]}, (16n-8)^{[3n]}, (16n-4)^{[n]}\right\}.$$

(ii) If n is even, then the distance Laplacian spectrum of $C(SD_{8n}, SD_{8n})$ is

$$\left\{0, (8n)^{[2]}, (16n-4)^{[2n]}(12n)^{[4n-3]}, (16n-2)^{[2n]}\right\}.$$

Proof By using Lemma 5, the proof of this result is similar to Theorem 2. □

In the next result, we will find the distance Laplacian eigenvalues of dicyclic group Q_{4n}.

Theorem 4 *The Laplacian spectrum of the commuting graph $C(Q_{4n}, Q_{4n})$ of dicyclic group Q_{4n} of order $4n$ is*

$$\left\{0, (4n)^{[2]}, (6n)^{[2n-3]}, (8n-2)^{[n]}, (8n-2)\right\}.$$

Proof The commuting graph $C(Q_{4n}, Q_{4n})$ of Q_{4n} [4] is given below

$$C(Q_{4n}, Q_{4n}) = K_2 \nabla \big(K_{2n-2} \cup nK_2\big) = H[K_2, K_{2n-2}, \underbrace{K_2, K_2, \ldots, K_2}_{n}],$$

where underlying graph H is the star graph on $n+2$ vertices. Using Theorem 1, the value of α_i's are

$$\alpha_1 = 2n - 2 + 2n = 4n - 2, \alpha_2 = 4n + 2, \alpha_3 = \cdots = \alpha_{n+2} = 8n - 4.$$

Again, using Theorem 1, the distance Laplacian spectrum of $C(Q_{4n}, Q_{4n})$ consists of the simple eigenvalue $2m_1 - \lambda_{1k} + \alpha_1 = 4 + 4n - 2 - 2 = 4n$, the eigenvalue $6n$ with multiplicity $2n-3$, the eigenvalue $8n-2$ with multiplicity n and the eigenvalues of following matrix

$$M = \left(\begin{array}{cc|cccc} 4n-2 & -2n+2 & -2 & -2 & \dots & -2 \\ -2 & 4n+2 & -4 & -4 & \dots & -4 \\ \hline -2 & -4n+4 & 8n-6 & -4 & \dots & -4 \\ -2 & -4n+4 & -4 & 8n-6 & \dots & -4 \\ \vdots & \vdots & \vdots & \vdots & \ddots & \vdots \\ -2 & -4n+4 & -4 & -4 & \dots & 8n-6 \end{array}\right) = \left(\begin{array}{c|c} A_{2\times 2} & B_{2\times n} \\ \hline C_{n\times 2} & D_{n\times n} \end{array}\right).$$

Since D is invertible, so by Lemma 3, we have

$$det(M - xI) = det(D - xI)det((A - xI) - C(D - xI)^{-1}B),$$

where

$$D - xI = \begin{pmatrix} 8n-6-x & -4 & \dots & -4 \\ -4 & 8n-6-x & \dots & -4 \\ \vdots & \vdots & \ddots & \vdots \\ -4 & -4 & \dots & 8n-6-x \end{pmatrix}.$$

Now, applying Lemma 2 to the matrix $D - xI$, with $X = [\mathbf{0}]$, $\beta = [\mathbf{0}]$, $B = [8n - 6 - x]$ and $C = [-4]$, we see that $8n - 2$ is the distance Laplacian eigenvalue of $\mathcal{C}(Q_{4n}, Q_{4n})$ with multiplicity n. It is well known that 0 is the simple eigenvalue of the distance Laplacian matrix, so

$$\mu_1 + 8n^2 = 8n + 8n^2 - 4n.$$

Thus, $\mu_1 = 4$ is the remaining distance Laplacian eigenvalue of M. □

A matrix $M \in M_n(\mathbb{F})$ over field $\mathbb{F}$ is called integral if its spectrum consists of only integers. Similarly, the distance Laplacian matrix $\mathcal{D}^L(G)$ of G is integral if all eigenvalues of $\mathcal{D}^L(G)$ are integers.

From Theorems 2, 3 and 4, we have the following consequence.

Proposition 1 *The commuting graphs of dihedral group D_{2n}, semi-dihedral group SD_{8n} and dicyclic graphs Q_{4n} are distance Laplacian integral.*

3 Distance Signless Laplacian Spectra of the Commuting Graphs of D_{2n} and Q_{4n}

In this section, we obtain the distance signless Laplacian spectra of the commuting graphs of dihedral, semi-dihedral and the dicyclic groups.

The following result gives the distance signless Laplacian spectra of the joined union of graphs $G_1, G_2, \dots, G_n$ in terms of the adjacency spectrum of the graphs $G_1, G_2, \dots, G_n$ and the eigenvalues of quotient matrix.

Theorem 5 *Let*[17] *G be a graph of order n having vertex set $V(G) = \{v_1, \ldots, v_n\}$. Let G_i be r_i-regular graphs of order m_i having adjacency eigenvalues $\lambda_{i1} = r_i \geq \lambda_{i2} \geq \ldots \geq \lambda_{in_i}$, where $i = 1, 2, \ldots, n$. The distance signless Laplacian spectrum of the joined union graph $G[G_1, \ldots, G_n]$ of order $\sum_{i=1}^{n} n_i$ consists of the eigenvalues $2m_i + \alpha_i - r_i - \lambda_{ik} - 4$ for $i = 1, \ldots, n$ and $k = 2, 3, \ldots, n_i$, where $\alpha_i = \sum_{k=1,k\neq i}^{n} n_k d_G(v_i, v_k)$. The remaining n eigenvalues are given by the equitable quotient matrix*

$$Q = \begin{pmatrix} 4m_1 + \alpha_1 - 2r_1 - 4 & m_2 d_G(v_1, v_2) & \ldots & m_n d_G(v_1, v_n) \\ m_1 d_G(v_2, v_1) & 4m_2 + \alpha_2 - 2r_2 - 4 & \ldots & m_n d_G(v_2, v_n) \\ \vdots & \vdots & \ddots & \vdots \\ m_1 d_G(v_n, v_1) & m_2 d_G(v_n, v_2) & \ldots & 4m_n + \alpha_n - 2r_n - 4 \end{pmatrix}.$$

In the following result, we find the distance signless Laplacian eigenvalues of commuting graph of dihedral group.

Theorem 6 *For the commuting graph $C(D_{2n}, D_{2n})$ of dihedral group D_{2n}, the following hold.*

(i) If n is odd, then the distance signless Laplacian spectrum of $C(D_{2n}, D_{2n})$ consists of the eigenvalue $3n - 2$ with multiplicity $n - 2$, the eigenvalue $4n - 5$ with multiplicity $n - 1$ and the eigenvalues of following cubic polynomial

$$x^3 - (12n - 9)x^2 + \left(40n^2 - 62n + 24\right)x - 40n^3 + 94n^2 - 74n + 20.$$

(ii) If n is even, then the distance signless Laplacian spectrum of $C(D_{2n}, D_{2n})$ is

$$\left\{2n - 2, (3n - 2)^{[n-3]}, (4n - 6)^{\left[\frac{n}{2}\right]}, (4n - 8)^{\left[\frac{n}{2}\right]}\right\},$$

together with the eigenvalues of (2)

Proof **(i)**. By Theorem 5, the α_i's are

$$\alpha_1 = 1 + 2(n - 1) = 2n - 1, \alpha_2 = n + n - 1 = 2n - 1 \text{ and } \alpha_3 = 2n + 1.$$

So, the distance signless Laplacian eigenvalues of $C(D_{2n}, D_{2n})$ are the eigenvalues $2m_1 + \alpha_1 - r_1 - \lambda_{1k} - 4 = 2n + 2n - 1 - 0 - 0 - 4 = 4n - 5$ with multiplicity $n - 1$, the eigenvalue $3n - 2$ with multiplicity $n - 2$ and the remaining three eigenvalues are the eigenvalues of the following quotient matrix

$$M = \begin{pmatrix} 6n - 5 & 1 & 2(n - 1) \\ n & 2n - 1 & n - 1 \\ 2n & 1 & 4n - 3 \end{pmatrix}.$$

(ii). By using Theorem 5, the distance signless Laplacian spectrum of $\mathcal{C}(D_{2n}, D_{2n})$ consists of the simple eigenvalue $2m_1 + \alpha_1 - r_1 - \lambda_{1k} - 4 = 4 + 2n - 2 - 1 + 2 - 4 = 2n - 2$, the eigenvalue $3n - 2$ with multiplicity $n - 3$, the eigenvalue $4n - 6$ with multiplicity $\frac{n}{2}$ and the eigenvalues of matrix

$$M = \left(\begin{array}{cc|cccc} 2n & n-2 & 2 & 2 & \dots & 2 \\ 2 & 4n-4 & 4 & 4 & \dots & 4 \\ \hline 2 & 2n-4 & 4n-4 & 4 & \dots & 4 \\ 2 & 2n-4 & 4 & 4n-4 & \dots & 4 \\ \vdots & \vdots & \vdots & \vdots & \ddots & \vdots \\ 2 & 2n-4 & 4 & 4 & \dots & 4n-4 \end{array}\right) = \left(\begin{array}{c|c} A_{2\times 2} & B_{2\times \frac{n}{2}} \\ \hline C_{\frac{n}{2}\times 2} & D_{\frac{n}{2}\times\frac{n}{2}} \end{array}\right).$$

Clearly, D is invertible, thus by Lemma 3, we have

$$det(M - xI) = det(D - xI)det((A - xI) - C(D - xI)^{-1}B),$$

where

$$D - xI = \begin{pmatrix} 4n-4-x & 4 & \dots & 4 \\ 4 & 4n-4-x & \dots & 4 \\ \vdots & \vdots & \ddots & \vdots \\ 4 & 4 & \dots & 4n-4-x \end{pmatrix}.$$

Now, applying Lemma 2 to the matrix $D - xI$, with $X = [\mathbf{0}]$, $\beta = [\mathbf{0}]$, $B = [4n - 4 - x]$ and $C = [4]$, we see that $4n - 8$ is the distance Laplacian eigenvalue of $\mathcal{C}(D_{2n}, D_{2n})$ with multiplicity $\dfrac{n}{2}$. The remaining two eigenvalues of $\mathcal{C}(D_{2n}, D_{2n})$ are the zeros of the polynomial

$$det((A - xI) - C(D - xI)^{-1}B). \tag{2}$$

□

In the following results, we find the distance Laplacian spectra of the commuting graphs of semi-dihedral SD_{8n} and dicyclic group Q_{4n}. The proof is similar to that of Theorem 6.

Theorem 7 *For the commuting graph $\mathcal{C}(SD_{8n}, SD_{8n})$ of semidihedral group SD_{8n}, the following hold.*

(i) If n is odd, then the distance signless Laplacian spectrum of $\mathcal{C}(SD_{8n}, SD_{8n})$ consists of the eigenvalue $8n - 2$ with multiplicity 3, the eigenvalue $12n - 2$ with multiplicity $4n - 5$, the eigenvalue $16n - 10$ with multiplicity $3n$, the eigenvalue $16n - 14$ with multiplicity $n - 1$ and the remaining three eigenvalues are the zeros of following polynomial $det((A - xI) - C(D - xI)^{-1}B)$, where

$$A = \begin{pmatrix} 8n-6 & 4n-4 \\ 4 & 16n-10 \end{pmatrix}, B = \begin{pmatrix} 4 & 4 & \dots & 4 \\ 8 & 8 & \dots & 8 \end{pmatrix}, C = \begin{pmatrix} 4 & 4 & \dots & 4 \\ 8n-8 & 8n-8 & \dots & 8n-8 \end{pmatrix}^T$$

$$\text{and} \quad D = \begin{pmatrix} 16n-6 & 8 & \dots & 8 \\ 8 & 16n-6 & \dots & 8 \\ \vdots & \vdots & \ddots & \vdots \\ 8 & 8 & \dots & 16n-6 \end{pmatrix}.$$

(ii) If n is even, then the distance signless Laplacian spectrum of $C(SD_{8n}, SD_{8n})$ consists of the simple eigenvalue $8n-2$, the eigenvalue $12n-2$ with multiplicity $4n-3$, the eigenvalue $16n-6$ with multiplicity $2n$, the eigenvalue $16n-8$ with multiplicity $2n-1$ and the remaining three eigenvalues are the zeros of following polynomial
$det((A - xI) - C(D - xI)^{-1}B)$, where

$$A = \begin{pmatrix} 8n & 4n-2 \\ 2 & 16n-4 \end{pmatrix}, B = \begin{pmatrix} 2 & 2 & \dots & 2 \\ 4 & 4 & \dots & 4 \end{pmatrix}, C = \begin{pmatrix} 2 & 2 & \dots & 2 \\ 8n-4 & 8n-4 & \dots & 8n-4 \end{pmatrix}^T$$

$$\text{and} \quad D = \begin{pmatrix} 16n-4 & 4 & \dots & 4 \\ 4 & 16n-4 & \dots & 4 \\ \vdots & \vdots & \ddots & \vdots \\ 4 & 4 & \dots & 16n-4 \end{pmatrix}.$$

Theorem 8 *The signless Laplacian spectrum of the commuting graph $C(Q_{4n}, Q_{4n})$ of dicyclic group Q_{4n} of order $4n$ consists of the simple eigenvalue $4n-2$, the eigenvalue $6n-2$ with multiplicity $2n-3$, the eigenvalue $8n$ with multiplicity $n-1$, the eigenvalue $8n-6$ with multiplicity n and the remaining three eigenvalues are the zeros of following polynomial*

$$det((A - xI) - C(D - xI)^{-1}B),$$

where

$$A = \begin{pmatrix} 4n & 2n-2 \\ 2 & 8n-6 \end{pmatrix}, B = \begin{pmatrix} 2 & 2 & \dots & 2 \\ 4 & 4 & \dots & 4 \end{pmatrix}, C = \begin{pmatrix} 2 & 2 & \dots & 2 \\ 4n-4 & 4n-4 & \dots & 4n-4 \end{pmatrix}^T$$

$$\text{and} \quad D = \begin{pmatrix} 8n-2 & 4 & \dots & 4 \\ 4 & 8n-2 & \dots & 4 \\ \vdots & \vdots & \ddots & \vdots \\ 4 & 4 & \dots & 8n-2 \end{pmatrix}.$$

Acknowledgements The research of S. Pirzada is supported by the SERB-DST research project number MTR/2017/000084.

References

1. Abdollahi, A.: Commuting graph of full matrix rings over finite fields. Linear Algebra Appl. **428**, 2947–2954 (2008)
2. Abdussakir, A., Rivatul, R.E., Muflihatun, N.: On the spectra of commuting and non commuting graph on dihedral groups. Cauchy **4**, 176–182 (2017)
3. Ali, F., Salman, M., Huang, S.: On the commuting graph of dihedral group communications. Algebra **44**(6), 2389–2401 (2016)
4. Ali, F., Li, Y.: The connectivity and the spectral radius of commuting graphs on certain finite groups. Linear Multilinear Algebra (2019). https://doi.org/10.1080/03081087.2019.1700893
5. Aouchiche, M., Hansen, P.: Distance spectra of graphs: a survey. Linear Algebra Appl. **458**, 301–386 (2014)
6. Aouchiche, M., Hansen, P.: Two Laplacians for the distance matrix of a graph. Linear Algebra Appl. **439**, 21–33 (2013)
7. Cvetković, D., Doob, M., Sachs, H.: Spectra of graphs. In: Theory and Applications, Pure and Applied Mathematics, vol. 87. Academic Press, Inc. New York (1980)
8. Dolžan, D., Oblak, P.: Commuting graph of matrices over semirings. Linear Algebra Appl. **435**, 1657–1665 (2011)
9. Fasfous, W.N.T., Nath, R.K., Sharafdini, R.: Various spectra and energy of commuting graphs of finite rings. Hacettepe J. Math. Stat. **49**(6), 1915–1925 (2020)
10. Fritscher, E., Trevisan, V.: Exploring symmetries to decompose matrices and graphs preserving the spectrum. SIAM J. Matrix Anal. Appl. **37**(1), 260–289 (2016)
11. Ganie, H.A.: On distance Laplacian spectrum (energy) of graphs. Discrete Math. Algorithms Appl. **2050061**, 16 (2020)
12. Ghorbani, M., Alkhansari, Z.G., Bashi, A.Z.: On the eigenvalue of non commuting graphs of groups. J. Algebr. Struct. Appl. **4**(2), 27–38 (2017)
13. Horn R., Johnson, C.: Matrix Analysis, 2nd edn. Cambridge University Press (2013)
14. Kakkar, V., Rawat, G.S.: On commuting graph of generalized dihedral groups. Discrete Math. Algorithms Appl. **11**(2), 1950024 (2019). https://doi.org/10.1142/S1793830919500241
15. Nicholson, W.K.: Introduction to Abstract Algebra, 4th edn. Wiley, New Jersey (2012)
16. Pirzada, S.: An Introduction to Graph Theory. Universities Press, Orient BlackSwan, Hyderabad (2012)
17. Pirzada, S., Rather, Bilal, A., Aijaz, M., Chishti, T.A.: On distance signless Laplacian spectrum of graphs and spectrum of zero divisor graphs of $\mathbb{Z}_n$. Linear Multilinear Algebra (2020). https://doi.org/10.1080/03081087.2020.1838425
18. Pirzada, S., Wani, Bilal, A., Somasundaram, A.: On the eigenvalues of zero divisor graph associated to finite commutative ring $Z_{p^M q^N}$. Int. J. Graphs Comb. https://doi.org/10.1080/09728600.2021.1873060
19. Pirzada, S., Rather, Bilal, A., Chishti, T.A.: On distance Laplacian spectrum of zero divisor graphs of $\mathbb{Z}_n$. Carpathian Math. Publ. **13**(1), 48–57 (2021)
20. Pirzada, S., Rather, Bilal, A., Chishti, T.A.: On normalized Laplacian spectrum of zero divisor graphs of commutative ring Z_n. Electron. J. Graph Theory Appl. **9**(2) (2021) 331–345
21. Pirzada, S., Rather, Bilal, A., Ul Shaban, R.: Merajuddin: on signless Laplacian spectrum of the zero divisor graph of the ring Z_n. Korean J. Math. **29**(1), 13–24 (2021)
22. Rather, Bilal, A. , Pirzada, S., Naikoo, T.A., Shang, Y.: On Laplacian eigenvalues of the zero-divisor graph associated to the ring of integers modulo n. Mathematics **9**(5, 482) (2021)
23. Tamizh Chelvam, T., Selvakumar, K., Raja, S.: Commuting graph on dihedral group. J. Math. Comp. Sci. **2**(2), 402–406 (2011)
24. Torktaz, M., Ashrafi, A.R.: Spectral properties of the commuting graphs of certain finite groups. AKCE Int. J. Graphs Comb. **16**, 300–309 (2019)

Spectrum of Graphs over Rings: A Survey

V. Rabikka, T. Asir, and M. Evangeline Prathibha

Abstract This article is a survey of results concerning the eigenvalues of graphs constructed from commutative rings. Specifically, we consider the zero-divisor graphs, unitary Cayley graphs, unit graphs, and total graphs.

Keywords Eigenvalue · Spectrum · Unitary cayley graph · Unit graph · Total graph

1 Introduction

Spectral graph theory is the study and exploration of graphs through the eigenvalues and eigenvectors of matrices naturally associated with those graphs. The study on adjacency matrix of a graph and its eigenvalues has a long history. Historically, the first relation between the spectrum and the structure of a graph was discovered in 1876 by Kirchhoff when he proved his famous matrix-tree theorem. The eigenvalues of a graph have been used in several areas of mathematical research and are playing a significant role in the development of various physical and chemical theories. There are several existing books concerning graph eigenvalues, for example [15, 19, 20].

In recent decades, the graphs constructed from algebraic structures have been extensively studied by many authors and have become a major field of research. The idea of constructing a graph from an algebraic structure was originated by Arthur Cayley in 1878. In recent years, assigning graphs to rings is playing an important role in the study of the structure of rings. The benefit of studying these graphs is that one may find some results about the algebraic structures and vice versa. In the

V. Rabikka · T. Asir (✉)
Department of Mathematics-DDE, Madurai Kamaraj University, Madurai 625 021, Tamil Nadu, India
e-mail: asirjacob75@gmail.com

M. E. Prathibha
Department of Mathematics, Lady Doak College, Madurai 625 002, Tamil Nadu, India
e-mail: evangelineprathibham@ldc.edu.in

M. Ashraf et al. (eds.), *Algebra and Related Topics with Applications*, Springer Proceedings in Mathematics & Statistics 392,
https://doi.org/10.1007/978-981-19-3898-6_33

literature, there are many papers on assigning a graph to a commutative ring. Some of the graphs to be mentioned are zero-divisor graphs, unitary Cayley graphs, unit graphs, and total graphs.

This article will consist of results on eigenvalues and energy of graphs over commutative rings. In particular, eigenvalues of graphs constructed from commutative rings have gained attention due to their prominent roles in algebraic graph theory as well as in some other areas like quantum computing. The second section deals with eigenvalues of graphs from rings whereas the third section deals with the Laplacian eigenvalues of graphs from rings.

Throughout the article, R will be a commutative ring with identity $1 \neq 0$. We denote $Z(R)$, $U(R)$, and $Nil(R)$ by the set of zero-divisors of R, the set of units of R, and the nilradical of a ring R, respectively. Also, we denote the ring of integers modulo n by Z_n. Further K_n, $K_{m,n}$, P_n, and C_n denote respectively the complete graph on n vertices, complete bipartite graph with a bipartition into vertex sets of cardinality m and n, path on n vertices and cycle with n vertices. Moreover, the neighborhood set of a vertex v in G is $N_G(u) = \{v \in V(G) : v \text{ is adjacent to } u \text{ in } G\}$. All the definitions related to algebra are from Dummit and Foote [23], and the definitions related to graph theory are from Chartrand and Zhang [17].

2 Spectrum of Graphs Associated to Rings

For a undirected graph G of order n, the *adjacency matrix* $A(G)$ is an $n \times n$ matrix with both rows and columns as indexed as vertices such that $(i, j)-$entry is the number of edges joining i and j with each loop being counted as two edges. Also, the eigenvalues of $A(G)$ are called the *eigenvalues* of G. Note that the adjacency matrix $A(G)$ of a undirected graph G is a real symmetric matrix and its eigenvalues are only real numbers. The *multiplicity* (in fact, algebraic multiplicity) is the number of times an eigenvalue appears in a characteristic polynomial of $A(G)$. The eigenvalues of G with its multiplicity are called as *spectrum* of a graph G. If $\lambda_1, \ldots, \lambda_k$ are the eigenvalues of G with corresponding multiplicities $m_1, \ldots, m_k$, then we denote spectrum of G either by

$$Spec(G) = \begin{pmatrix} \lambda_1 & \ldots & \lambda_k \\ m_1 & \ldots & m_k \end{pmatrix},$$

or simply by $\lambda_1^{[m_1]}, \ldots, \lambda_k^{[m_k]}$ with m_i omitted if it is equal to 1.

The graph energy is a graph-spectrum-based quantity, introduced in the 1970s. The first paper in which graph the *energy* was defined as the sum of absolute values of the eigenvalues of the adjacency matrix, namely as $\sum_{i=1}^{n} |\lambda_i|$, where $\lambda_1, \ldots, \lambda_n$ are the eigenvalues of G, appeared in 1978 [60].

Example 1 Let us consider the following graph, cycle on 4 vertices (Fig. 1):

Then the adjacency matrix

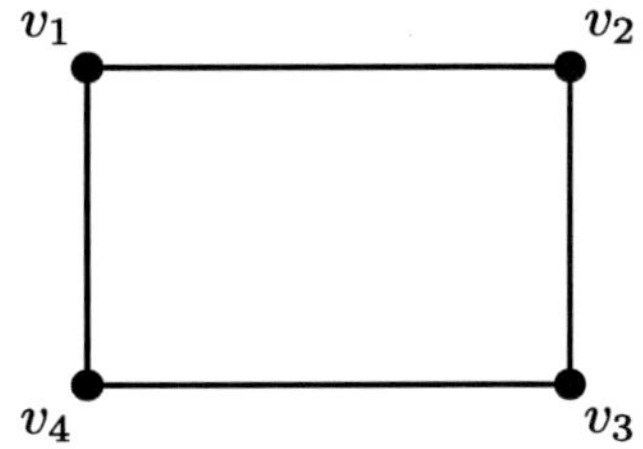

Fig. 1 The graph C_4

$$A(C_4) = \begin{bmatrix} 0 & 1 & 0 & 1 \\ 1 & 0 & 1 & 0 \\ 0 & 1 & 0 & 1 \\ 1 & 0 & 1 & 0 \end{bmatrix}.$$

The characteristic equation of $A(C_4)$ is $\lambda^4 - 4\lambda^2 = 0$. So that the eigenvalues of graph C_4 is $0^{[2]}, \pm 2$.

Some of the results of the eigenvalues for standard graphs are listed below.

Remark 1 [16] Let $n, m \in \mathbb{Z}^+$. Then

(i). The spectrum of P_n is $2cos\left(\frac{\pi j}{n+1}\right)$ for $j = 1, 2, \ldots, n$.
(ii). The spectrum of C_n is $2cos\left(\frac{2\pi j}{n}\right)$ for $j = 0, 1, \ldots, n-1$.
(iii). The spectrum of K_n is $(n-1)^{[1]}, (-1)^{[n-1]}$.
(vi). The spectrum of $K_{m,n}$ is $0^{[m+n-2]}, \pm\sqrt{mn}$.

2.1 *Zero-Divisor Graphs*

The idea of a graph associated with the zero-divisors of a commutative ring R was introduced by Beck [12] in 1988. The present definition, along with the name for the zero-divisor graph, was introduced by Anderson and Livingston [7] in 1999, after modifying Beck's definition. Note that Beck [12] took all elements of the commutative ring R as vertices of the graph $\Gamma_0(R)$. The modified definition of the zero-divisor graph is as follows.

Definition 1 ([7]) Let R be a commutative ring with identity and $Z(R)$ be its set of zero-divisors. The *zero-divisor graph* of R, denoted by $\Gamma(R)$, is the simple graph with vertex set $Z(R)^* = Z(R) \setminus \{0\}$, and two distinct vertices x and y are adjacent if $xy = 0$.

Example 2 In the Figure below, zero-divisor graphs for several rings are given.

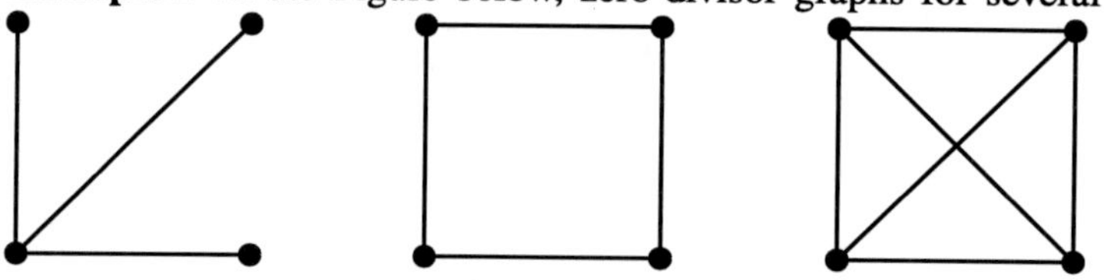

$\Gamma(\mathbb{Z}_2 \times \mathbb{F}_4)$ $\Gamma(\mathbb{Z}_3 \times \mathbb{Z}_3)$ $\Gamma(\mathbb{Z}_{25})$ or $\Gamma(\frac{\mathbb{Z}_5(x)}{<x^2>})$

(a) The adjacency matrix for $\Gamma(\mathbb{Z}_2 \times \mathbb{F}_4)$ is

$$A(G) = \begin{bmatrix} 0\,1\,1\,1 \\ 1\,0\,0\,0 \\ 1\,0\,0\,0 \\ 1\,0\,0\,0 \end{bmatrix}.$$

The characteristic polynomial of $\Gamma(\mathbb{Z}_2 \times \mathbb{F}_4)$ is $\lambda^2(\lambda^2 - 3) = 0$. Hence the eigenvalues are $0^{[2]}$, $\sqrt{3}^{[2]}$.

(b) By Example 1, we get the eigenvalues of $\Gamma(\mathbb{Z}_3 \times \mathbb{Z}_3)$ is $0^{(2)}$, ± 2.

(c) The adjacency matrix for $\Gamma(\mathbb{Z}_{25})$ or $\Gamma(\mathbb{Z}_5(x)/\langle x^2 \rangle)$ is

$$\begin{bmatrix} 0\,1\,1\,1 \\ 1\,0\,1\,1 \\ 1\,1\,0\,1 \\ 1\,1\,1\,0 \end{bmatrix}.$$

Here the characteristic polynomial is $\lambda^4 - 6\lambda^2 - 8\lambda - 3 = 0$ and so the eigenvalues are $-3, 1, 1, 1$.

The concept of the energy of zero-divisor graph was initiated by Ahmadi et al. in [3]. In particular, they considered the zero-divisor graph of $\mathbb{Z}_n$ for $n = p^2$ or pq where p and q are primes.

Theorem 1 ([3] Theorem 1) *If p is a prime number, then the energy of $\Gamma(\mathbb{Z}_{p^2})$ is* $2p - 4$.

Theorem 2 ([3] Theorem 2) *If p and q are two prime numbers, then the non-zero eigenvalues of $\Gamma(\mathbb{Z}_{pq})$ are $\pm\sqrt{(p-1)(q-1)}$. In particular, $E(\Gamma(\mathbb{Z}_{pq})) =$* $2\sqrt{(p-1)(q-1)}$.

Motivated by the above work, Surendranath Reddy et al. [50] extended the study of energy of the zero-divisor graph of $\mathbb{Z}_n$.

Theorem 3 ([50] Theorem 4.1) *If p is a prime number, then the non-zero eigenvalues of the zero-divisor graph $\Gamma(\mathbb{Z}_{p^3})$ are* $\frac{(p-1)(1 \pm \sqrt{1+4p})}{2}$.

Theorem 4 ([50] Theorem 4.3) *Let $n = p^2q$ with p and q primes. If $\lambda \neq 0$ is a non-zero eigenvalues of the zero-divisor graph $\Gamma(\mathbb{Z}_n)$, then $\lambda^4 - (p-1)\lambda^3 - 2p(p-1)(q-1)^2\lambda^2 + p(p-1)^2(q-1)\lambda + p(p-1)^3(q-1)^2 = 0$.*

Further results were found by Young [59] in 2015.

Theorem 5 ([59] Corollary 2.9) *Let $n = \prod_{i=1}^{s} p_i^{t_i}$ where p_i's are primes. Then the multiplicity of the eigenvalue 0 of $A(\Gamma(\mathbb{Z}_n))$ is*

$$n - \phi(n) - \prod_{i=1}^{s}(t_i + 1) + 2.$$

Theorem 6 ([59] Proposition 3.3) *Let λ_1 be the largest eigenvalue of $A(\Gamma(\mathbb{Z}_n))$:*

(i). If n is a product containing two or more distinct primes, then $\lambda_1 \geq \sqrt{\phi(n)}$.
(ii). If $n = p^t$, then $\lambda_1 \geq p^{\lceil t/2 \rceil} - 1$.

Theorem 7 ([59] Theorem 3.4) *For any positive integer k, there exists only a finite number of integers n such that all the eigenvalues of $A(\Gamma(\mathbb{Z}_n))$ are less or equal than k.*

Theorem 8 ([59] Theorem 3.7) *Suppose the rank of $A(\Gamma(\mathbb{Z}_n))$ is r. Then $A(\Gamma(\mathbb{Z}_n))$ has $\lceil r/2 \rceil$ positive eigenvalues and $\lfloor r/2 \rfloor$ negative eigenvalues.*

Further Sharma et al. [61] made some observations on the adjacency matrices and eigenvalues of the graphs $\Gamma(\mathbb{Z}_p \times \mathbb{Z}_p)$ and $\Gamma(\mathbb{Z}_p[i] \times \mathbb{Z}_p[i])$.

In 2019, Katja Mönius [45] precisely determined the spectra of $\Gamma(\mathbb{Z}_p \times \mathbb{Z}_p \times \mathbb{Z}_p)$ and $\Gamma(\mathbb{Z}_p \times \mathbb{Z}_p \times \mathbb{Z}_p \times \mathbb{Z}_p)$ in terms of the prime number p. They also provided the characteristic polynomials of $\Gamma(\mathbb{Z}_{p^2} \times \mathbb{Z}_p)$ and $\Gamma(\mathbb{Z}_p \times \mathbb{Z}_p \times \mathbb{Z}_q)$ for primes $q \neq p$. In what follows, by the nullity $\eta(G)$ of a graph G we mean the multiplicity of the eigenvalue 0 of G. Obviously, we have that $\eta(G) = \dim A(G) - \text{rank } A(G)$.

Theorem 9 ([45] Theorem 4.2) *Let $R \cong \mathbb{Z}_{p_1^{t_1}} \times \ldots \times \mathbb{Z}_{p_r^{t_r}}$ for prime numbers p_j and $r, t_j \in \mathbb{N}$. Then the zero-divisor graph $\Gamma(R)$ has $\prod_{i=1}^{r}(t_i + 1) - 2$ non-zero eigenvalues, and the nulity of $\Gamma(R)$ equals*

$$\eta(\Gamma(R)) = \prod_{i=1}^{r} p_i^{t_i - 1}\left(\prod_{i=1}^{r} p_i - \prod_{i=1}^{r}(p_i - 1)\right) - \prod_{i=1}^{r}(t_i + 1) + 1$$

Let us see some examples.

Example 3 ([45] *Example 4.3*) Let p be a prime number and $R \cong \mathbb{Z}_p \times \mathbb{Z}_p \times \mathbb{Z}_p$. Then the multiplicity of the eigenvalue 0 is $3(p+1)(p-2)$ and eigenvalues are $\lambda_{1,2} = \frac{1}{2}(1 - p \pm (p-1)\sqrt{4p-3})$ and $\lambda_{3,4} = p - 1 \pm \sqrt{p - 2p^2 + p^3}$. Therefore the $\text{Spec}(\Gamma(R)) = \{\lambda_1^{[2]}, \lambda_2^{[2]}, \lambda_3^{[1]}, \lambda_4^{[1]}, 0^{[3(p+1)(p-2)]}\}$.

Example 4 (*Example 4.4* [45]) Let p be a prime number and $R \cong \mathbb{Z}_p \times \mathbb{Z}_p \times \mathbb{Z}_p \times \mathbb{Z}_p$. Then the multiplicity of the eigenvalue 0 is $p^4 - (p-1)^4 - 2^4 - 1$ and eigenvalues are $\lambda_1 = (p-1)^2$, $\lambda_2 = -p^2 + p - 1$, $\lambda_{3,4} = \frac{1}{2}(-2p^2 + 3p - 1 \pm (p-1)\sqrt{4p-3})$ and $\lambda_{5,6} = \frac{1}{2}(2p^2 - p - 1 \pm \sqrt{3}\sqrt{4p^3 - 9p^2 + 6p - 1})$. Thus the $\text{Spec}(\Gamma(R)) = \{\lambda_1^{[5]}, \lambda_2^{[1]}, \lambda_3^{[3]}, \lambda_4^{[3]}, \lambda_5^{[1]}, \lambda_6^{[1]}, 0^{[p^4-(p-1)^4-2^4-1]}\}$.

3 Laplacian Spectrum of Graphs Associated to Rings

In this section, we concentrate on the Laplacian eigenvalues of zero-divisor graph of a ring. The Laplacian matrix of a simple graph is the difference of the diagonal matrix of vertex degree and the adjacency matrix. In the past decades, the Laplacian spectrum has received much more attention, since it has been applied to several fields, such as randomized algorithms, combinatorial optimization problems, and machine learning. The Laplacian matrix has a long history. The first celebrated result is attributable to Kirchhoff [32] in 1847 paper concerned with electrical networks. However, it didn't receive much attention until the work of Fiedler, which appeared in 1973 [24] and 1975 [25]. Mohar in his survey [43] argued that, because of its importance in various physical and chemical theories, the spectrum of the Laplacian matrix is more natural and important than the more widely studied adjacency spectrum. In [5], Alon used the smallest positive eigenvalue of the Laplacian matrix to estimate the expander and magnifying coefficients of graphs. For more background and motivation on research of the Laplacian matrix, the reader may refer to the above book and [28, 41–44].

The eigenvalues of Laplacian and signless Laplacian matrix are called Laplacian eigenvalue and signless Laplacian eigenvalue. Let $D(G)$ be the $n \times n$ diagonal matrix with the diagonal entires as the degrees of the vertex. Then, the matrices $L(G) = D(G) - A(G)$ and $Q(G) = D(G) + A(G)$ are *Laplacian* and *signless Laplacian* matrices of a graph G respectively. The elements of $L(G)$ are given by

$$L_{i,j} = \begin{cases} deg(v_i) & \text{if } i = j, \\ -1 & \text{if } i \neq j \text{ and } v_i \text{ is adjacent to } v_j, \\ 0 & \text{otherwise} \end{cases}$$

where $deg(v_i)$ is degree of the vertex v_i.

The *normalized Laplacian matrix* is denoted by $\mathcal{L}(G)$ and defined as

$$\mathcal{L}_{i,j} = \begin{cases} 1 & \text{if } i = j \text{ and } deg(v_i) \neq 0, \\ -\frac{1}{\sqrt{deg(v_i)deg(v_j)}} & \text{if } i \neq j \text{ and } v_i \text{ is adjacent to } v_j, \\ 0 & \text{otherwise.} \end{cases}$$

If $\mu_1, \ldots, \mu_t$ are the distinct Laplacian eigenvalues of $L(G)$ with respective multiplicities $m_1, \ldots, m_t$, then we shall denote the *Laplacian spectrum* of $L(G)$ by

$$\sigma_L(G) = \begin{Bmatrix} \mu_1 & \mu_2 & \dots & \mu_t \\ m_1 & m_2 & \dots & m_t \end{Bmatrix}$$

or simply by $\mu_1^{[m_1]}, \dots, \mu_t^{[m_t]}$. Similarly, the spectrum of $Q(G)$, is called the *signless Laplacian spectrum* and it is denoted by $\sigma_Q(G)$ and the spectrum of $\mathcal{L}(G)$ is called the *normalized Laplacian spectrum* of G and it is denoted by $\sigma_{\mathcal{L}}(G)$.

Since $L(G)$ is a real, symmetric, and positive semidefinite matrix, all its eigenvalues are real and nonnegative. Since the sum of the entries in each row of $L(G)$ is zero, the smallest eigenvalue of $L(G)$ is 0. The second smallest eigenvalue of $L(G)$, denoted by $\mu(G)$, is called the *algebraic connectivity* of G. The largest eigenvalue of $L(G)$, denoted by $\lambda(G)$, is called the *Laplacian spectral radius* of G.

Some of the results of Laplacian spectrum for standard graphs are listed below.

Remark 2 [16] Let $n, m \in \mathbb{Z}^+$. Then

(i). The Laplacian spectrum of P_n is $2 - 2cos\left(\frac{\pi j}{n}\right)$ for $j = 0, 1, \dots, n-1$.
(ii). The Laplacian spectrum of C_n is $2 - 2cos\left(\frac{2\pi j}{n}\right)$ for $j = 0, 1, \dots, n-1$.
(iii). The Laplacian spectrum of K_n is $0^{[1]}, n^{[n-1]}$.
(iv). The Laplacian spectrum of $K_{m,n}$ is $0^{[1]}, m^{[n-1]}, n^{[m-1]}, (m+n)^{[1]}$.

Remark 3 [1] The signless Laplacian spectrum and normalized Laplacian spectrum of the complete graph K_n are known. Indeed,

$$\sigma_Q(K_n) = \begin{Bmatrix} 2n-2 & n-2 \\ 1 & n-1 \end{Bmatrix} \text{ and } \sigma_{\mathcal{L}}(K_n) = \begin{Bmatrix} 0 & \frac{n}{n-1} \\ 1 & n-1 \end{Bmatrix}.$$

Recently, Chattopadhyay et al. [18] studied the Laplacian eigenvalues of the zero-divisor graph of $\mathbb{Z}_n$. As a first step, the authors of [18] explored the structure of $\Gamma(\mathbb{Z}_n)$ by using the proper divisors of n. In particular, they defined a graph Υ_n, which made the work to determine the Laplacian eigenvalues of $\Gamma(\mathbb{Z}_n)$ easier. The following remark describes the graph Υ_n.

Remark 4 [18]

(i). The simple graph denoted by Υ_n is a graph with the proper divisors $d_1, d_2, \dots, d_k$ of n as vertices in which two distinct vertices d_i and d_j are adjacent if and only if n divides $d_i d_j$.
(ii). Let $d_1, d_2, \dots, d_k$ be the proper divisors of n. For $1 \le i \le k$, we assign the weight $\phi(\frac{n}{d_i}) = |A_{d_i}|$ to the vertex d_i of the graph Υ_n where $A_{d_i} = \{x \in \mathbb{Z}_n : (x, n) = d_i\}$. Define $M_{d_j} = \sum\limits_{d_i \in N_{\Upsilon_n}(d_j)} \phi(\frac{n}{d_i})$ for $1 \le j \le k$. The $k \times k$ vertex weighted Laplacian matrix $\mathbf{L}(\Upsilon_n)$ of Υ_n is given by

$$\mathbf{L}(\Upsilon_n) = \begin{bmatrix} M_{d_1} & -t_{1,2} & \dots & -t_{1,k} \\ -t_{2,1} & M_{d_2} & \dots & -t_{2,k} \\ \dots & \dots & \dots & \dots \\ -t_{k,1} & -t_{k,2} & \dots & M_{d_k} \end{bmatrix}$$

where, for $1 \leq i \neq j \leq k$,

$$t_{i,j} = \begin{cases} \phi(\frac{n}{d_j}) & \text{if } d_i \text{ is adjacent to } d_j \text{ in } \Upsilon_n \\ 0 & \text{otherwise.} \end{cases}$$

The following Lemma is useful for further discussion.

Lemma 1 ([18] Lemma 2.7) *Let $\Gamma(A_{d_i})$ be the induced subgraph of $\Gamma(\mathbb{Z}_n)$ on the vertex set A_{d_i} for $1 \leq i \leq k$. Then $\Gamma(\mathbb{Z}_n) = \Upsilon_n[\Gamma(A_{d_1}), \Gamma(A_{d_2}), \ldots, \Gamma(A_{d_k})]$.*

The notation in the above lemma is explained in the following example.

Example 5 ([1] *Example 2.4*) Consider the ring $\mathbb{Z}_{12}$. We have $d_1 = 2, d_2 = 3, d_3 = 4$ and $d_4 = 6$. Then G_{12} is the graph $2 - 6 - 4 - 3$, which is isomorphic to P_4. Now by lemma 1,

$$\Gamma(\mathbb{Z}_{12}) = \Upsilon_{12}[\Gamma(A_2), \Gamma(A_3), \Gamma(A_4), \Gamma(A_6)],$$

where $\Gamma(A_2) = \overline{K_2}$, $\Gamma(A_3) = \overline{K_2}$, $\Gamma(A_4) = \overline{K_2}$ and $\Gamma(A_6) = K_1$

The following theorem describes the Laplacian spectrum of the zero-divisor graph of $\mathbb{Z}_n$.

Theorem 10 ([18] Theorem 3.3) *If $d_1, d_2, \ldots, d_k$ are the proper divisors of n, then the Laplacian spectrum of $\Gamma(\mathbb{Z}_n)$ is given by*

$$\sigma_L(\Gamma(\mathbb{Z}_n)) = \bigcup_{j=1}^{k}(M_{d_j} + (\sigma_L(\Gamma(A_{d_j})) \setminus \{0\})) \bigcup \sigma(\mathbf{L}(\Upsilon_n))$$

where $\sigma(\mathbf{L}(\Upsilon_n))$ denotes the spectrum of $\mathbf{L}(\Upsilon_n)$ and $M_{d_j} + (\sigma_L(\Gamma(A_{d_j})) \setminus \{0\})$ means that M_{d_j} is added to each element of the multiset $\sigma_L(\Gamma(A_{d_j})) \setminus \{0\}$.

For instance, consider the following example.

Example 6 ([18] *Example 3.4*) The Laplacian spectrum of $\Gamma(\mathbb{Z}_n)$ for $n = pq$ or p^2q where p and q are distinct primes are as follows,

$$\sigma_L(\Gamma(\mathbb{Z}_{pq})) = \begin{Bmatrix} p-1 & q-1 & 0 & p+q-2 \\ q-2 & p-2 & 1 & 1 \end{Bmatrix}$$

and

$$\sigma_L(\Gamma(\mathbb{Z}_{p^2q})) = \begin{Bmatrix} p-1 & pq-1 & p^2-1 & q-1 \\ \phi(pq)-1 & \phi(p)-1 & \phi(q)-1 & \phi(p^2)-1 \end{Bmatrix} \bigcup \sigma(\mathbf{L}(\Upsilon_{p^2q})).$$

Observe that, by Theorem 10, out of the $n - \phi(n) - 1$ number of Laplacian eigenvalues of $\Gamma(\mathbb{Z}_n)$, $n - \phi(n) - 1 - k$ of them are known to be nonzero integer values.

The remaining k Laplacian eigenvalues of $\Gamma(\mathbb{Z}_n)$ will come from the spectrum of $\mathbf{L}(\Upsilon_n)$. So the following result is an immediate consequence of Theorem 10. Before proceeding to the theorem recall that a graph G is called *Laplacian integral* if all the Laplacian eigenvalues of G are integers.

Theorem 11 ([18] Theorem 4.1) *The zero-divisor graph $\Gamma(\mathbb{Z}_n)$ is Laplacian integral if and only if all the eigenvalues of $\mathbf{L}(\Upsilon_n)$ are integers.*

Theorem 12 ([18] Theorem 4.3) *Let $n = p^t$ where p is a prime and $t \geq 2$ is a positive integer. Then the following hold:*

(i). If $t = 2$, then the Laplacian spectrum of $\Gamma(\mathbb{Z}_n)$ is given by

$$\begin{Bmatrix} 0 \\ 1 \end{Bmatrix} \text{ or } \begin{Bmatrix} p-1 & 0 \\ p-2 & 1 \end{Bmatrix}$$

according to $p = 2$ or $p \geq 3$.

(ii). If $t = 2m$ for some integer $m \geq 2$, then the Laplacian spectrum of $\Gamma(\mathbb{Z}_n)$ is given by

$$\begin{Bmatrix} p^{2m-1}-1 & p^{2m-2}-1 & \dots & p^{m+1}-1 & p^m-1 & p^{m-1}-1 & \dots & p-1 & 0 \\ \phi(p) & \phi(p^2) & \dots & \phi(p^{m-1}) & \phi(p^m)-1 & \phi(p^{m+1}) & \dots & \phi(p^{2m-1}) & 1 \end{Bmatrix}$$

(iii). If $t = 2m+1$ for some integer $m \geq 1$, then the Laplacian spectrum of $\Gamma(\mathbb{Z}_n)$ is given by

$$\begin{Bmatrix} p^{2m}-1 & p^{2m-1}-1 & \dots & p^{m+1}-1 & p^m-1 & p^{m-1}-1 & \dots & p-1 & 0 \\ \phi(p) & \phi(p^2) & \dots & \phi(p^m) & \phi(p^{m+1})-1 & \phi(p^{m+2}) & \dots & \phi(p^{2m}) & 1 \end{Bmatrix}$$

As a consequence of Theorems 11 and 12, one can have the following results.

Corollary 1 ([18] Corollary 4.4) *If p is a prime and $t \geq 2$, then $\Gamma(\mathbb{Z}_{p^t})$ is Laplacian integral and so all the eigenvalues of $\mathbf{L}(\Upsilon_{p^t})$ are integers.*

Corollary 2 ([18] Corollary 4.5) *Let $n = p^t$ for some prime p and positive integer t with $n \neq 4$. Then $\lambda(\Gamma(\mathbb{Z}_{p^t})) =| \Gamma(\mathbb{Z}_{p^t}) |$.*

Note that for any simple graph, $\lambda(G) \leq |V(G)|$. Now, the following theorem characterizes the values of n for which equality holds when $G = \Gamma(\mathbb{Z}_n)$.

Proposition 1 ([18] Proposition 5.4) $\lambda(\Gamma(\mathbb{Z}_n)) = |\Gamma(\mathbb{Z}_n)|$ *if and only if n is a product of two distinct primes or n is a prime power with $n \neq 4$.*

Theorem 13 ([18] Theorem 5.8) *The following hold:*

(i). If n is not a prime power nor a product of two distinct primes, then $\mu(\Gamma(\mathbb{Z}_n))$ *is the second smallest eigenvalue of* $\mathbf{L}(\Upsilon_n)$.
(ii). If n is not a prime power, then $\lambda(\Gamma(\mathbb{Z}_n))$ *is the largest eigenvalue of* $\mathbf{L}(\Upsilon_n)$.

Further, the authors of [1] determined the bounds for the smallest Laplacian eigenvalue $\mu(\Gamma(\mathbb{Z}_n))$ and the largest Laplacian eigenvalue $\lambda(\Gamma(\mathbb{Z}_n))$ in terms of $\delta(\Gamma(\mathbb{Z}_n))$ and $\Delta(\Gamma(\mathbb{Z}_n))$, respectively.

Theorem 14 ([1] Theorem 4.5) *Let* $n = p_1^{t_1} p_2^{t_2} \cdots p_k^{t_k}$ *where* $p_1 < p_2 < \ldots < p_k$ *are prime numbers. Then the following statements hold:*

(i). If $k \geq 2$ *or* $k = 1$ *with* $t_1 > 2$, *then* $\mu(\Gamma(\mathbb{Z}_n)) < p - 1$. *Otherwise,* $\mu(\Gamma(\mathbb{Z}_n)) < p - 2$
(ii). $\mu(\Gamma(\mathbb{Z}_n)) = 0$ *if and only if* $n \in \{4, 8, 9, 4q, pq\}$, *where p and q are distinct prime numbers.*

Theorem 15 ([1] Theorem 4.8)

(i). Let $n = p_1^{k_1} p_2^{k_2} \cdots p_s^{k_s}$, *where* $s > 1$, $p_1 < p_2 < \ldots < p_s$ *are distinct prime numbers and* $k_i \geq 1$. *Then the largest eigenvalue* $\lambda(\Gamma(\mathbb{Z}_n)) \leq \frac{2n}{p_1} - 2$ *if* $k_1 = 1$, *and* $\lambda(\Gamma(\mathbb{Z}_n)) \leq \frac{2n}{p_1} - 4$ *if* $k_1 > 1$.
(ii). If $n = p^k$, *where p is a prime number and* $k \geq 2$, *then* $\lambda(\Gamma(\mathbb{Z}_n)) \leq \frac{2n}{p} - 4$.

In 2020, Afkhami et al. [1] described the signless Laplacian and normalized Laplacian spectrum of the zero-divisor graph $\Gamma(\mathbb{Z}_n)$.

Notice that, by Lemma 1, $\Gamma(\mathbb{Z}_n) = \Upsilon_n[\Gamma(A_{d_1}), \Gamma(A_{d_2}), \ldots, \Gamma(A_{d_k})]$ where $\Gamma(A_{d_i})$ is either the complete graph $K_{\phi(\frac{n}{d_i})}$ or its complement graph $\overline{K}_{\phi(\frac{n}{d_i})}$. So $\Gamma(A_{d_i})$ is either $(\phi(\frac{n}{d_i}) - 1)$−regular or 0−regular. Now one can easily see that $\mathcal{Q}(\Upsilon_n)$ reads as follows:

$$\mathcal{Q}(\Upsilon_n) = \begin{bmatrix} 2r_1 + M_{d_1} & t_{1,2} & \ldots & t_{1,k} \\ t_{2,1} & 2r_2 + M_{d_2} & \ldots & t_{2,k} \\ \ldots & \ldots & \ldots & \ldots \\ t_{k,1} & t_{k,2} & \ldots & 2r_k + M_{d_k} \end{bmatrix}$$

where r_j is equal to $\phi(\frac{n}{d_j}) - 1$ or 0, and $M_{d_j} = \sum_{d_i \in N_{\Upsilon_n}(d_j)} \phi(\frac{n}{d_i})$, for $1 \leq j \leq k$, and also

$$t_{i,j} = \begin{cases} \sqrt{\phi(\frac{n}{d_i})\phi(\frac{n}{d_j})} & \text{if } d_i \text{ is adjacent to } d_j \text{ in } \Upsilon_n \\ 0 & \text{otherwise} \end{cases}$$

for $1 \leq i \neq j \leq k$.

The next result, determined the signless Laplacian spectrum of $\Gamma(\mathbb{Z}_n)$ in terms of the spectrum of $\mathcal{Q}(\Upsilon_n)$.

Theorem 16 ([1] Theorem 3.2) *Let* $d_1, d_2, \ldots, d_k$ *be the proper divisors of n. Then the signless Laplacian spectrum of* $\Gamma(\mathbb{Z}_n)$ *is given by*

$$\sigma_Q(\Gamma(\mathbb{Z}_n)) = \left(\bigcup_{j=1}^{k}(M_{d_j} + (\sigma_Q(\Gamma(A_{d_j})) \setminus \{2r_j\}))\right) \bigcup \sigma(\mathcal{Q}(\Upsilon_n)),$$

where r_j is equal to $\phi(\frac{n}{d_j}) - 1$ or 0, and $M_{d_j} + (\sigma_Q(\Gamma(A_{d_j})) \setminus \{2r_j\})$ means that M_{d_j} is added to each element of the multiset $\sigma_Q(\Gamma(A_{d_j})) \setminus \{2r_j\}$.

The illustration for the signless Laplacian spectrum of $\Gamma(\mathbb{Z}_n)$ for some special cases of n is given in the following example.

Example 7 (*Example 3.3* [1]) Let p and q be distinct prime numbers. Then the signless Laplacian spectrum of $\Gamma(\mathbb{Z}_n)$ for $n = pq$ and $n = p^2q$ are

$$\sigma_Q(\Gamma(\mathbb{Z}_{pq})) = \begin{pmatrix} 0 & p-1 & q-1 & p+q-2 \\ 1 & q-2 & p-2 & 1 \end{pmatrix}$$

and

$$\sigma_Q(\Gamma(\mathbb{Z}_{p^2q})) = \begin{pmatrix} p-1 & pq-3 & p^2-1 & q-1 \\ pq-p-q & p-2 & q-2 & p^2-p-1 \end{pmatrix} \bigcup \sigma(\mathcal{Q}(\Upsilon_{p^2q}))$$

where the characteristic polynomial of $\mathcal{Q}(\Upsilon_{p^2q}))$ is

$$\Big((x-p^2+1)(x-q+1) - \phi(q)\phi(p^2)\Big)\left((x-p-pq+4)(x-p+1) - \phi(pq)\phi(p)\right)$$
$$-\left((x-\phi(q))(x-\phi(p))\phi(p)\phi(q)\right).$$

Next, we move on to the normalized Laplacian spectrum of the zero-divisor graph of $\mathbb{Z}_n$.

From Lemma 1, $\Gamma(\mathbb{Z}_n) = \Upsilon_n[\Gamma(A_{d_1}), \Gamma(A_{d_2}), \ldots, \Gamma(A_{d_k})]$ where $\Gamma(A_{d_i})$ is either $(\phi(\frac{n}{d_i}) - 1)$–regular or 0–regular. Therefore it is clear that

$$\mathcal{L}(\Upsilon_n) = \begin{bmatrix} \frac{M_{d_1}}{r_1+M_{d_1}} & t_{1,2} & \ldots & t_{1,k} \\ t_{2,1} & \frac{M_{d_2}}{r_2+M_{d_2}} & \ldots & t_{2,k} \\ \ldots & \ldots & \ldots & \ldots \\ t_{k,1} & t_{k,2} & \ldots & \frac{M_{d_k}}{r_k+M_{d_k}} \end{bmatrix}$$

where r_j is equal to $\phi(\frac{n}{d_j}) - 1$ or 0, and $M_{d_j} = \sum_{d_i \in N_{\Upsilon_n}(d_j)} \phi(\frac{n}{d_i})$, for $1 \le j \le k$, and also

$$t_{i,j} = \begin{cases} -\sqrt{\frac{\phi(\frac{n}{d_i})\phi(\frac{n}{d_j})}{(r_i+M_{d_i})(r_j+M_{d_j})}} & \text{if } d_i \text{ is adjacent to } d_j \text{ in } \Upsilon_n \\ 0 & \text{otherwise} \end{cases}$$

for $1 \le i \ne j \le k$.

The next result determined the normalized Laplacian spectrum of $\Gamma(\mathbb{Z}_n)$ in terms of the spectrum of $\mathcal{L}(\Upsilon_n)$.

Theorem 17 ([1] Theorem 5.2) *Let $d_1, d_2, \ldots, d_k$ be the proper divisors of n. Then the normalized Laplacian spectrum of $\Gamma(\mathbb{Z}_n)$ is given by*

$$\sigma_{\mathcal{L}}(\Gamma(\mathbb{Z}_n)) = \left(\bigcup_{j=1}^{k}\left(\frac{M_{d_j}}{r_j + M_{d_j}} + \frac{r_j}{r_j + M_{d_j}} + (\sigma_{\mathcal{L}}(\Gamma(A_{d_j})) \setminus \{0\})\right)\right) \bigcup \sigma(\mathcal{L}(\Upsilon_n)),$$

where r_j is equal to $\phi(\frac{n}{d_j}) - 1$ or 0.

The exact value of $\sigma_{\mathcal{L}}(\Gamma(\mathbb{Z}_{pq}))$ for primes $p \neq q$ is calculated in the following example.

Example 8 ([1] *Example 5.3*) Let p and q be distinct prime numbers. Then the normalized Laplacian spectrum of $\Gamma(\mathbb{Z}_n)$ for $n = pq$ and $n = p^2q$ are

$$\sigma_{\mathcal{L}}(\Gamma(\mathbb{Z}_{pq})) = \begin{pmatrix} 0 & 1 & 2 \\ 1 & p+q-4 & 1 \end{pmatrix}$$

and

$$\sigma_{\mathcal{L}}(\Gamma(\mathbb{Z}_{p^2q})) = \begin{pmatrix} 1 & \frac{p^2q-3p+2}{pq-2} \\ pq-4 & p-2 \end{pmatrix} \bigcup \sigma(\mathcal{L}(\Upsilon_{p^2q}))$$

where the characteristic polynomial of $\mathcal{L}(\Upsilon_{p^2q}))$ is

$$\frac{\left((p+1)(x-1)^2 - p\right)((x-1)(x(pq-2) - pq + p) - \phi(pq)) - (q-1)(x-1)^2}{(pq-2)(p+1)}.$$

The following result is about the second largest eigenvalue of $\mathcal{L}(\Gamma(\mathbb{Z}_n))$.

Theorem 18 ([1] Theorem 5.5) *Assume that λ_{n-1} is the second largest eigenvalue of $\mathcal{L}(\Gamma(\mathbb{Z}_n))$. Then $\lambda_{n-1} \geq 1$. Moreover $\lambda_{n-1} = 1$ if and only if $n \in \{8, 9, pq\}$, where p and q are distinct prime numbers.*

We close the section by mentioning some results on the energy of line graph of zero-divisor graph of $\mathbb{Z}_n$. One can associate a given graph G with its line graph, denoted by $L(G)$, such that each vertex of $L(G)$ represents an edge of G, and any two vertices of $L(G)$ are adjacent if and only if their corresponding edges in G share a common vertex. One important theorem, in [56], due to Whitney about line graphs is that with one exceptional case, $L(G) = K_3$, the structure of any connected graph can be recovered from its line graph, i.e., there is a one-to-one correspondence between the class of connected graphs and the class of connected line graphs. With the class of zero-divisor graphs at hand, it is natural to keep an eye on the properties of their line graphs and seek any relation between them. The importance of line graphs stems from the fact that the line graph transforms the adjacency relation on edges to adjacency relation on vertices (see [29]).

The line graph $L(\Gamma(\mathbb{Z}_n))$ of $\Gamma(\mathbb{Z}_n)$ is defined to be the graph whose vertex set constitutes of the edges of $\Gamma(\mathbb{Z}_n)$, where two vertices are adjacent if the corresponding edges have a common vertex in $\Gamma(\mathbb{Z}_n)$.

Theorem 19 ([52] Theorem 3.1) *If $p = 2$ and q is an odd prime number, then energy of $L(\Gamma(\mathbb{Z}_{2q}))$ is $2q - 4$.*

Theorem 20 ([52] Theorem 3.2) *If $p = 3$ and $q > 3$ is any prime number, then energy of $L(\Gamma(\mathbb{Z}_{3q}))$ is $4q - p - 5$.*

Theorem 21 ([52] Theorem 3.3) *If p and $2 < q$ are distinct primes, then energy of $L(\Gamma(\mathbb{Z}_{pq}))$ is $4pq - 8p - 8q + 16$.*

Theorem 22 ([52] Theorem 3.4) *Let $n = p^2$, where $p \geq 5$ be any prime number. Then $E(L(\Gamma(\mathbb{Z}_{p^2}))) = 2p^2 - 10p + 8$.*

3.1 Unitary Cayley Graphs

The definition of Cayley graph was introduced by Arthur Cayley in 1878 to explain the concept of abstract groups which are described by a set of generators. In the last 50 years, the theory of Cayley graphs has grown into a substantial branch of algebraic graph theory. The unitary Cayley graph of a ring was initially investigated for $\mathbb{Z}_n$ by Dejter and Giudici in [21] where some properties of $G_{\mathbb{Z}_n}$ are presented. In 2009, Akhtar et al. [4] generalized the unitary Cayley graph $G_{\mathbb{Z}_n}$ to G_R for a finite ring R and obtained various characterization results regarding connectedness, chromatic index, diameter, girth, and planarity of G_R.

Definition 2 [4] Let R be a commutative ring and $U(R)$ be the group of unit of R. The *unitary Cayley graph* of R, denoted by G_R, is a simple graph whose vertex set is R and two distinct vertices x and y are adjacent if $x - y \in U(R)$.

Note that G_R is the special case of Cayley graph, in fact $G_R = Cay(R, U(R))$. Moreover, if R is a local ring with a unique maximal ideal $\mathfrak{m}$, then $U(R) = R \setminus \mathfrak{m}$.

Example 9 (a). If n is a prime, then $G_{\mathbb{Z}_n}$ is the complete graph on n vertices. Therefore, by Remark 1, the spectrum of $G_{\mathbb{Z}_n}$ is $(n-1)^{[1]}, (-1)^{[n-1]}$.

(b). If n is a power of 2, then $G_{\mathbb{Z}_n}$ is the complete bipartite $K_{\frac{n}{2},\frac{n}{2}}$ and so by Remark 1, the spectrum of $G_{\mathbb{Z}_n}$ is $0^{[n-2]}, \pm\sqrt{\frac{n^2}{4}}$. In particular, the spectrum of $G_{\mathbb{Z}_4}$ is $0^{[2]}, \pm 2$.

First, let us consider the unitary Cayley graph of $\mathbb{Z}_n$. Note that $U(\mathbb{Z}_n) = \{x \in \mathbb{Z}_n : gcd(x, n) = 1\}$. Thus $G_{\mathbb{Z}_n}$ has the vertex set $V(G_{\mathbb{Z}_n}) = \mathbb{Z}_n$ and edge set $E(G_{\mathbb{Z}_n}) =$

$\{(x, y) : x, y \in \mathbb{Z}_n$ and $gcd(a - b, n) = 1\}$. For some results on $G_{\mathbb{Z}_n}$, one can refer the reader to [13, 14, 26, 33].

Some basic properties of the eigenvalues of the unitary Cayley graphs of $\mathbb{Z}_n$ have been illustrated by Klotz and Sander, see Theorems 13 and 15 in [33]. In 2009, Aleksandar Ilić [30] and Ramaswamy et al. [49] independently calculated the energy of unitary Cayley graphs and have establish the necessary and sufficient conditions for $G_{\mathbb{Z}_n}$ to be hyperenergetic. Note that the graphs $G_{\mathbb{Z}_n}$ have integral spectrum and play an important role in modeling quantum spin networks supporting the perfect state transfer.

Theorem 23 ([30] Theorem 2.3, [49] Theorem 3.7) *The energy of unitary Cayley graph $G_{\mathbb{Z}_n}$ equals $2^k\phi(n)$, where k is the number of distinct prime factors dividing n.*

In what follows, we mean a graph G with order n is hyperenergetic if its energy exceeds the energy of the complete graph K_n; that is $E(G) > 2n - 2$.

Theorem 24 ([30] Theorem 2.4, [49] Theorem 3.10) *Let k be the number of distinct prime factors dividing n. Then the unitary Cayley graph $G_{\mathbb{Z}_n}$ is hyperenergetic if and only if $k > 2$ or $k = 2$ and n is odd.*

The next two results deal with the complement of unitary Cayley graph $\overline{G_{\mathbb{Z}_n}}$.

Theorem 25 ([30] Theorem 3.1) *Let $m = p_1p_2\cdots p_k$ be the largest square-free number that divides n. The energy of the complement of unitary Cayley graph $\overline{G_{\mathbb{Z}_n}}$ equals*

$$E(\overline{G_{\mathbb{Z}_n}}) = 2n - 2 + (2^k - 2)\phi(n) - \prod_{i=1}^{k} p_i + \prod_{i=1}^{k}(2 - p_i).$$

Theorem 26 ([30] Theorem 3.2) *The complement of unitary Cayley graph $\overline{G_{\mathbb{Z}_n}}$ is hyperenergetic if and only if n has at least two distinct prime factors and $n \neq 2p$, where p is a prime number.*

A number of other papers also have considered the eigenvalue properties of unitary Cayley graphs of $\mathbb{Z}_n$, see [22, 31, 51, 57].

In [34], Lanski and Maróti considered the unitary Cayley graph of an Artinian ring R and showed that G_R contains 2^{k-1} connected components, each of which is a bipartite graph, where k is the number of summands isomorphic to $\mathbb{Z}_2$ in $R/J(R)$. In [31], Kiani and Aghaei investigated the isomorphism problem for unitary Cayley graphs associated with finite (commutative) rings. They proved that if $G_R \cong G_S$ where R, S are finite rings, then $G_{R/J(R)} \cong G_{R(S/J(S)}$, and if, in addition, R and S are commutative, then $R/J(R) \cong S/J(S)$. They also proved that if $G_{R(M_n(F))} \cong G_R$ where F is a finite filed, then $R \cong M_n(F)$.

Akhtar et al. [4] studied and obtained eigenvalues of the unitary Cayley graph G_R when R is a local ring.

Proposition 2 ([31] Proposition 2.1, [4] Proposition 10.2) *Let R be a finite local ring with maximal ideal $\mathfrak{m}$ of size m. Then*

$$Spec(G_R) = \begin{pmatrix} |U(R)| & -m & 0 \\ 1 & \frac{|U(R)|}{m} & \frac{|R|}{m}(m-1) \end{pmatrix}$$

In particular, if F is the field with q elements, then

$$Spec(G_F) = \begin{pmatrix} q-1 & -1 \\ 1 & q-1 \end{pmatrix}$$

Since the eigenvalues of G_F are all nonzero when F is a field, the formula for the spectrum of G_R becomes quite complicated when many of the local factors of R are fields. However, if none of the local factors of R are fields, the formula takes on a rather appealing form:

Theorem 27 ([4] Corollary 10.3) *Let R be a finite ring and suppose R has t local factors, none of which are fields. Then*

$$Spec(G_R) = \begin{pmatrix} (-1)^t|Nil(R)| & 0 \\ |R_{red}| & |R| - |R_{red}| \end{pmatrix}$$

where $R_{red} = R_1/\mathfrak{m}_1 \times \cdots \times R_s/\mathfrak{m}_s$.

It is well known that, see [10], every finite commutative ring can be expressed as a direct product of finite local rings, and this decomposition is unique to permutations of such local rings. Using this fact, the following result, provides all the eigenvalues of the unitary Cayley graph of finite commutative ring.

Theorem 28 ([31] Lemma 2.3) *Let R be a finite commutative ring, where $R = R_1 \times \cdots \times R_s$ and R_i is a local ring with maximal ideal $\mathfrak{m}_i$ of size m_i for all $i \in \{1, 2, \ldots, s\}$. Then the eigenvalues of G_R are*

(i). $(-1)^{|C|}\frac{|U(R)|}{\prod_{j\in C}|U(R_j)|/m_j}$ *with multiplicity* $\prod_{j\in C}|U(R_j)|/m_j$ *for all subsets of C of $\{1, 2, \ldots, s\}$ and*

(ii). 0 *with multiplicity* $|R| - \prod_{i=1}^{s}\left(1 + \frac{|U(R_i)|}{m_i}\right)$.

We next proceed to the result of the energy of the unitary Cayley graph of a finite commutative ring R.

Theorem 29 ([31] Theorem 2.4) *Let R be a finite commutative ring, where $R = R_1 \times \cdots \times R_s$ and R_i is a local ring with maximal ideal $\mathfrak{m}_i$ of size m_i for all $i \in \{1, 2, \ldots, s\}$. Then the energy of G_R is $2^s|U(R)|$.*

Recall that a graph G with n vertices is called *hyperenergetic* if $E(G) > 2(n-1)$ where $E(G)$ is the energy of G.

Theorem 30 ([31] Theorem 2.5) *Let R be a finite commutative ring, where $R = R_1 \times \cdots \times R_s$ and R_i is a local ring with maximal ideal $\mathfrak{m}_i$ of size m_i for all $i \in \{1, 2, \ldots, s\}$. Assume that $|R_1|/m_1 \le \ldots \le |R_s|/m_s$. Then*

(i). *For $s = 1$, G_R is not hyperenergetic.*
(ii). *For $s = 2$, G_R is hyperenergetic if and only if $|R_1|/m_1 \ge 3$ and $|R_2|/m_2 \ge 4$.*
(iii). *For $s \ge 3$, G_R is hyperenergetic if and only if $(|R_{s-2}|/m_{s-2} \ge 3)$ or $(|R_{s-1}|/m_{s-1} \ge 3$ and $|R_s|/m_s \ge 4)$.*

The authors [31] have covered the energy of the complement of unitary Cayley graphs $\overline{G}_R$. They have given the spectrum of $\overline{G}_R$ which consists of eigenvalues $|R| - |U(R)| - 1, -1 - \lambda_2, \ldots, -1 - \lambda_{|R|}$, where λ_i is an eigenvalue of G_R not associated with **1** for all $i \in \{2, 3, \ldots, |R|\}$

Theorem 31 ([31] Theorem 4.1) *Let R be a finite ring, where $R = R_1 \times \cdots \times R_s$, and R_i is a local ring with maximal ideal $\mathfrak{m}_i$ of size m_i for all $i \in \{1, 2, \ldots, s\}$. Then $E(\overline{G}_R) = 2|R| - 2 + (2^s - 2)|U(R)| - \prod_{i=1}^{s} |R_i|/m_i + \prod_{i=1}^{s}(2 - |R_i|/m_i)$.*

The notion of the quadratic unitary Cayley graph of $\mathbb{Z}_n$ was introduced by Beaudrap in [11]. He defined the *quadratic unitary Cayley graph* of $\mathbb{Z}_n$, whose vertex set is the ring $\mathbb{Z}_n$, and where residues x, y modulo n are adjacent if and only if their difference is a quadratic residue.

In 2015, Liu and Zhou [58] generalized the concept of quadratic unitary Cayley graph from $\mathbb{Z}_n$ to a finite commutative ring. Moreover, they focused on the spectral properties of the following family of Cayley graphs on finite commutative rings. The quadratic unitary Cayley graph was defined as follows.

Definition 3 ([58]) Given a finite commutative ring R, the *quadratic unitary Cayley graph* of R, denoted by $\mathcal{G}_R$, is defined as the Cayley graph $Cay(R, T_R)$ on the additive group of R with respect to $T_R = Q_R \cup (-Q_R)$, where $Q_R = \{u^2 : u \in U(R)\}$. That is, $\mathcal{G}_R$ has vertex set R such that $x, y \in R$ are adjacent if and only if $x - y \in T_R$.

Notice that the quadratic unitary Cayley graphs are also generalizations of the well-known Paley graphs. In fact, in the special case where $R = \mathbb{F}_q$ is a finite field, where $q \equiv 1 (\mathrm{mod}\ 4)$ is a prime power, $\mathcal{G}_{\mathbb{F}_q}$ is exactly the Paley graph $P(q)$, which by definition is the graph with vertex set $\mathbb{F}_q$ such that $x, y \in \mathbb{F}_q$ are adjacent if and only if $x - y$ is a non-zero square of $\mathbb{F}_q$.

Recall that every finite commutative ring can be written as the direct product of local ring. By using this, the authors made an assumption as follows:

Assumption 32 Whenever we consider a finite commutative ring $R = R_1 \times \cdots \times R_s$ with unit element $1 \ne 0$, we assume that each R_i, $1 \le i \le s$, is a local ring with maximal ideal $\mathfrak{m}_i$ of order m_i such that

$$|R_1|/m_1 \le |R_2|/m_2 \le \ldots \le |R_s|/m_s.$$

Further,

$$|U(R)| = \prod_{i=1}^{s}(|R_i| - m_i) = \prod_{i=1}^{s} m_i((|R_i|/m_i) - 1) = |R|\prod_{i=1}^{s}\left(1 - \frac{1}{|R_i|/m_i}\right).$$

The following result deals with the spectral properties of $\mathcal{G}_R$ when R is a local ring.

Theorem 33 ([58] Theorem 2.4) *Let R be a local ring with maximal ideal $\mathfrak{m}$ of order m.*

(i). *If* $|R|/m \equiv 1 (\text{mod } 4)$, *then*

$$Spec(\mathcal{G}_R) = \begin{pmatrix} \frac{|R|-m}{2} & \frac{m(-1+\sqrt{|R|/m})}{2} & \frac{m(-1-\sqrt{|R|/m})}{2} & 0 \\ 1 & (|R|/m-1)/2 & (|R|/m-1)/2 & |R|-|R|/m \end{pmatrix}.$$

(ii). *If* $|R|/m \equiv 3 (\text{mod } 4)$, *then*

$$Spec(\mathcal{G}_R) = \begin{pmatrix} |R|-m & -m & 0 \\ 1 & |R|/m-1 & |R|-|R|/m \end{pmatrix}.$$

Authors of [58] have also defined a new notation as follows:
If A and B are disjoint subsets of $\{1, 2, \ldots, s\}$, then

$$\lambda_{A,B} = (-1)^{|B|}\frac{|U(R)|}{2^s \prod_{i\in A}\left(\sqrt{|R_i|/m_i}+1\right)\prod_{j\in B}\left(\sqrt{|R_j|/m_j}-1\right)}.$$

In particular, $\lambda_{\emptyset,\emptyset} = |U(R)|/2^s$.

Theorem 34 ([58] Theorem 2.6) *Let R be as in Assumption 32 such that $|R_i|/m_i \equiv 1 (\text{mod } 4)$ for $1 \le i \le s$. Then the eigenvalues of $\mathcal{G}_R$ are,*

(i). $\lambda_{A,B}$, *repeated* $\frac{1}{2^{|A|+|B|}}\prod_{k\in A\cup B}(|R_k|/m_k - 1)$ *times, for all pairs* (A, B) *of subsets of* $\{1, 2, \ldots, s\}$ *such that* $A \cap B = \emptyset$; *and*

(ii). 0 *with multiplicity* $|R| - \sum_{\substack{A,B\subseteq\{1,\ldots,s\}\\ A\cap B=\emptyset}}\left(\frac{1}{2^{|A|+|B|}}\prod_{k\in A\cup B}(|R_k|/m_k - 1)\right)$.

Theorem 35 ([58] Theorem 2.7) *Let R be as in Assumption 32 such that $|R_i|/m_i \equiv 1 (\text{mod } 4)$ for $1 \le i \le s$, and let R_0 be a local ring with maximal ideal $\mathfrak{m}_0$ of order m_0 such that $|R_0|/m_0 \equiv 3 (\text{mod } 4)$. Then the eigenvalues of $\mathcal{G}_{R_0\times R}$ are*

(i). $|U(R_0)| \cdot \lambda_{A,B}$, *repeated* $\frac{1}{2^{|A|+|B|}}\prod_{k\in A\cup B}(|R_k|/m_k - 1)$ *times, for all pairs* (A, B) *of subsets of* $\{1, 2, \ldots, s\}$ *such that* $A \cap B = \emptyset$;

(ii). $-\frac{|U(R_0)|}{|R_0|/m_0-1} \cdot \lambda_{A,B}$, *repeated* $\frac{|R_0|/m_0-1}{2^{|A|+|B|}}\prod_{k\in A\cup B}(|R_k|/m_k - 1)$ *times, for all pairs* (A, B) *of subsets of* $\{1, 2, \ldots, s\}$ *such that* $A \cap B = \emptyset$; *and* (iii). 0 *with multiplicity* $|R| - \sum_{\substack{A,B\subseteq\{1,\ldots,s\}\\ A\cap B=\emptyset}} \left(\frac{|R_0|/m_0}{2^{|A|+|B|}} \prod_{k\in A\cup B} (|R_k|/m_k - 1)\right).$

The following three results obtain the eigenvalues of $\mathcal{G}_{\mathbb{Z}_n}$. Let $n = p_1^{\alpha_1} p_2^{\alpha_2} \cdots p_s^{\alpha_s}$ be an integer in canonical factorization, where $p_1 < p_2 < \ldots < p_s$ are primes. It is well known that $\mathbb{Z}_n \cong \mathbb{Z}_{p_1^{\alpha_1}} \times \mathbb{Z}_{p_2^{\alpha_2}} \times \cdots \times \mathbb{Z}_{p_s^{\alpha_s}}$, where each $\mathbb{Z}_{p_i^{\alpha_i}}$ is a local ring with unique maximal ideal $(p_i^{\alpha_i})$ of order $p_i^{\alpha_i - 1}$.

Corollary 3 ([58] Corollary 2.8)

(i). *If* $p \equiv 1 \pmod 4$ *is a prime and* $\alpha \geq 1$ *an integer, then*

$$Spec(\mathcal{G}_{\mathbb{Z}_{p^\alpha}}) = \begin{pmatrix} p^{\alpha-1}(p-1)/2 & p^{\alpha-1}(-1+\sqrt{p})/2 & 0 & p^{\alpha-1}(-1-\sqrt{p})/2 \\ 1 & (p-1)/2 & p^\alpha - p & (p-1)/2 \end{pmatrix}$$

(ii). *If* $p \equiv 3 \pmod 4$ *is a prime and* $\alpha \geq 1$ *an integer, then*

$$Spec(\mathcal{G}_{\mathbb{Z}_{p^\alpha}}) = \begin{pmatrix} p^{\alpha-1}(p-1) & -p^{\alpha-1} & 0 \\ 1 & (p-1) & p^\alpha - p \end{pmatrix}$$

In what follows, ϕ denotes Euler's totient function.

Corollary 4 ([58] Corollary 2.9) *Let* $n = p_1^{\alpha_1} \cdots p_s^{\alpha_s}$ *be an integer in canonical factorization such that each* $p_i \equiv 1 \pmod 4$. *Then the eigenvalues of* $\mathcal{G}_{\mathbb{Z}_n}$ *are*

(i). $(-1)^{|B|} \cdot \frac{\phi(n)}{2^s \prod_{i\in A}(\sqrt{p_i}+1) \prod_{j\in B}(\sqrt{p_j}-1)}$, *repeated* $\frac{1}{2^{|A|+|B|}} \prod_{k\in A\cup B} (p_k - 1)$ *times, for all pairs* (A, B) *of subsets of* $\{1, 2, \ldots, s\}$ *such that* $A \cap B = \emptyset$; *and*

(ii). 0 *with multiplicity* $n - \sum_{\substack{A,B\subseteq\{1,\ldots,s\}\\ A\cap B=\emptyset}} \left(\frac{1}{2^{|A|+|B|}} \prod_{k\in A\cup B} (p_k - 1)\right).$

Corollary 5 (Corollary 2.10 [58]) *Let* $n = p^\alpha p_1^{\alpha_1} \cdots p_s^{\alpha_s}$ *be an integer in canonical factorization such that each* $p \equiv 3 \pmod 4$ *and each* $p_i \equiv 1 \pmod 4$. *Then the eigenvalues of* $\mathcal{G}_{\mathbb{Z}_n}$ *are*

(i). $(-1)^{|B|} \cdot \frac{\phi(n)}{2^s \prod_{i\in A}(\sqrt{p_i}+1) \prod_{j\in B}(\sqrt{p_j}-1)}$, *repeated* $\frac{1}{2^{|A|+|B|}} \prod_{k\in A\cup B} (p_k - 1)$ *times, for all pairs* (A, B) *of subsets of* $\{1, 2, \ldots, s\}$ *such that* $A \cap B = \emptyset$;

(ii). $(-1)^{|B|+1} \cdot \frac{\phi(n)}{2^s (p-1) \prod_{i\in A}(\sqrt{p_i}+1) \prod_{j\in B}(\sqrt{p_j}-1)}$, *repeated* $\frac{p-1}{2^{|A|+|B|}} \prod_{k\in A\cup B} (p_k - 1)$ *times, for all pairs* (A, B) *of subsets of* $\{1, 2, \ldots, s\}$ *such that* $A \cap B = \emptyset$; *and*

(iii). 0 *with multiplicity* $n - \sum_{\substack{A,B\subseteq\{1,\ldots,s\}\\ A\cap B=\emptyset}} \left(\frac{p}{2^{|A|+|B|}} \prod_{k\in A\cup B} (p_k - 1)\right).$

By using the results on spectra of quadratic unitary Cayley graphs, Liu and Zhou determined the energies of $\mathcal{G}_R$. Moreover they found out when such a graph is hyperenergetic. The following is an immediate consequence of Theorem 33.

Theorem 36 ([58] Theorem 3.1) *Let R be a local ring with maximal ideal $\mathfrak{m}$ of order m.*

(i). *If $|R|/m \equiv 1 \pmod 4$, then $E(\mathcal{G}_R) = \left(\sqrt{|R|/m} + 1\right)|U(R)|/2$;*
(ii). *If $|R|/m \equiv 3 \pmod 4$, then $E(\mathcal{G}_R) = 2|U(R)|$.*

Theorem 37 ([58] Theorem 3.2) *Let R be as in Assumption 32 such that $|R_i|/m_i \equiv 1 \pmod 4$ for $1 \leq i \leq s$, and R_0 be a local ring with maximal ideal $\mathfrak{m}_0$ of order m_0 such that $|R_0|/m_0 \equiv 3 \pmod 4$. Then*

(i). $E(\mathcal{G}_R) = \frac{|U(R)|}{2^s} \prod_{i=1}^{s} \left(\sqrt{|R_i|/m_i} + 1\right)$;
(ii). $E(\mathcal{G}_{R_0 \times R}) = \frac{|U(R_0)||U(R)|}{2^{s-1}} \prod_{i=1}^{s} \left(\sqrt{|R_i|/m_i} + 1\right)$.

3.2 Unit Graphs and Total Graphs

The unit graph was first investigated by Grimaldi for $\mathbb{Z}_n$ in [27]. In 2010, Ashrafi et al. [8] generalized the unit graph $G(\mathbb{Z}_n)$ to $G(R)$ for an arbitrary ring R. Many other papers are also devoted to this topic (see [9, 36–39]). A survey of the study of unit graphs can be found in [35].

Definition 4 Let R be a commutative ring with nonzero identity and $U(R)$ be the set of all units in R. The *unit graph* of R, denoted by $G(R)$, has vertex set as the set of all elements of R, for distinct vertices x and y are adjacent if and only if $x + y \in U(R)$.

In 2014, Pranjali and Mukti Acharya [48] gave a MATLAB program to determine the energy of $G(\mathbb{Z}_n)$. Using that program, they have calculated the energy of $G(\mathbb{Z}_n)$ for some specific values of n.

Theorem 38 ([48] Theorem 3.2) *For the unit graph $G(\mathbb{Z}_n)$, the energy $E(G(\mathbb{Z}_n)) = n$, when $n = 2^k, k > 1$.*

Theorem 39 ([48] Theorem 3.4) *For an odd prime p, the energy of unit graph $G(\mathbb{Z}_{2p})$ is $E(G(\mathbb{Z}_{2p})) = 2^2\phi(p)$.*

Theorem 40 ([48] Theorem 3.5) *The energy of unit graph $G(\mathbb{Z}_n)$ is never an integer, where $n = p^k, k > 1$ and p is an odd prime.*

In variation to the concept of zero-divisor graph, Anderson and Badawi [6] introduced the total graph of a commutative ring. In recent years, many research articles have been published on the total graph of commutative rings (see [2, 6, 40, 54, 55]).

Definition 5 Let R be a commutative ring and $Z(R)$ be the set of all zero-divisors of R. The *total graph* of R is a simple graph with all the elements of a ring as the vertices in which distinct $x, y \in R$ are adjacent if and only if $x + y \in Z(R)$.

In 2017, Sheela Suthar and Om Prakash [53] studied the energy of total graph of $\mathbb{Z}_n$. Specifically, the following main results are proved.

Theorem 41 ([53] Theorem 2.1) *If $p = 2$ and q is an odd prime number, then energy of $T_\Gamma(\mathbb{Z}_{2q})$ is $4(q-1)$.*

Theorem 42 ([53] Theorem 2.2) *If $p = 2$ and q is an odd prime number, then energy of $T_\Gamma(\mathbb{Z}_{p^2q})$ is $10q - 6$.*

Theorem 43 ([53] Theorem 2.3) *If p is an odd prime number, then energy of $T_\Gamma(\mathbb{Z}_{p^2})$ is $(p-1)(p+2)$.*

Theorem 44 ([53] Theorem 2.4) *If p is an odd prime number, then energy of $T_\Gamma(\mathbb{Z}_{p^3})$ is $p^3 + p^2 - 2$.*

The authors of [53] offered few results for Laplacian energy of total graph on $\mathbb{Z}_n$. The Laplacian energy, denoted by $LE(G)$, is defined as

$$LE(G) = \sum_{i=1}^{n} |\lambda_i - \frac{2m}{n}|,$$

where m is the number of edges and n is the number of vertices.

Theorem 45 ([53] Theorem 3.1) *If $p = 2$ and q is an odd prime number, then the Laplacian energy of $T_\Gamma(\mathbb{Z}_{2q})$ is $2q^2$.*

Theorem 46 ([53] Theorem 3.2) *If $p = 2$ and q is an odd prime, then the Laplacian energy of $T_\Gamma(\mathbb{Z}_{p^2q})$ is $4q(2q+1)$.*

Another energy of a graph is the distance energy. The *distance energy* of a graph is the sum of absolute values of the eigenvalues of the distance matrix. The distance matrix of a connected graph, denoted by $\Delta(G)$, is defined in a similar way as the adjacency matrix: the entry in the ith row, jth column is the distance between the ith and jth vertex.

Theorem 47 ([53] Theorem 3.3) *If $p = 2$ and q is an odd prime, then the distance energy of $T_\Gamma(\mathbb{Z}_{2q})$ is $2(2q + q - 2)$.*

Theorem 48 ([53] Theorem 3.4) *If $p = 2$ and $q > 2$ is any prime number, then the distance energy of $T_\Gamma(\mathbb{Z}_{p^2q})$ is $14q - 8$.*

Theorem 49 ([53] Theorem 3.5) *If p is any prime number and $n \in \mathbb{N}$, then the distance energy of $T_\Gamma(\mathbb{Z}_{p^n})$ is ∞.*

References

1. Afkhami, M., Barati, Z., Khashyarmanesh, K.: On the signless Laplacian and normalized Laplacian spectrum of the zero-divisor graphs. Ric. Mat. (2020). https://doi.org/10.1007/s11587-020-00519-3

2. Afkhami, M., Barati, Z., Khashyarmanesh, K.: When the unit, unitary and total graphs are ring graphs and outer planar. Rocky Mountain J. Math. **44**(3), 705–716 (2014)
3. Ahmadi, M.R., Nezhad, R.J.: Energy and Wiener index of zero-divisor graphs. Iranian. J. Math. Chem. **2**(1), 45–51 (2011)
4. Akhtar, R., Boggess, M., Henderson, T.J., Jiménez, I., Karpman, R., Kinzel, A., Pritikin, D.: On the unitary Cayley graph of a finite ring. Electron. J. Comb. **16**(R117), 13 (2009)
5. Alon, N.: Eigenvalues of expanders. Combinatorica **6**(2), 83–96 (1986)
6. Anderson, D.F., Badawi, A.: The total graph of commutative ring. J. Algebra **320**(7), 2706–2719 (2008)
7. Anderson, D.F., Livingston, P.S.: The zero-divisor graph of a commutative ring. J. Algebra **217**(2), 434–447 (1999)
8. Ashrafi, N., Maimani, H.R., Pournaki, M.R., Yassemi, S.: Unit graphs associated with rings. Comm. Algebra **38**(8), 2851–2871 (2010)
9. Asir, T., Tamizh Chelvam T.: On the genus of generalized unit and unitary Cayley graphs of a commutative ring. Acta Math. Hungar. **142**(2), 444–458 (2014)
10. Atiyah, M.F., Macdonald, I.G.: Introduction to Commutative Algebra. Addison-Wesley Publishing Co., Reading, MA-London-Don Mills, ON (1969)
11. Beaudrap, N.: de: on restricted unitary Cayley graphs and symplectic transformations modulo n. Electron. J. Combin. **17**(1), R69, 26 (2010)
12. Beck, I.: Coloring of commutative rings. J. Algebra **116**(1), 208–226 (1988)
13. Berrizbeitia, P., Giudici, R.E.: On cycles in the sequence of unitary Cayley graphs. Discrete Math. **282**(1–3), 239–243 (2004)
14. Berrizbeitia, P., Giudici, R.E.: Counting pure k-cycles in sequences of Cayley graphs. Discrete Math. **149**(1–3), 11–18 (1996)
15. Biggs, N.L.: Algebraic Graph Theory, 2nd edn. Cambridge University Press, Cambridge (1993)
16. Brouwer, A.E., Haemers, W.H.: Spectra of Graphs. Springer, New York Dordrecht Heidelberg London (2012)
17. Chartrand, G.: Lesniak Graphs and Digraphs. Chapman & Hall/CRC (2000)
18. Chattopadhyay, S., Patra, K.L., Sahoo, B.K.: Laplacian eigenvalues of the zero-divisor graph of the ring $\mathbb{Z}_n$. Linear Algebra Appl. **584**, 267–286 (2020)
19. Chung, F.R.K.: Spectral graph theory. In: Conference Board of the Mathematical Sciences. American Mathematical Society (1994)
20. Cvetković, D., Rawlinson, P., Simić, S.: An Introduction to the Theory of Graph Spectra. Cambridge University Press (2010)
21. Dejterand, I.J., Giudici, R.E.: On unitary Cayley graphs. J. Combin. Math. Combin. Comput. **18**, 121–124 (1995)
22. Droll, A.: A classification of Ramanujan unitary Cayley graphs. Electron. J. Combin. **17**(1), 6 (2010)
23. Dummit, D.S., Foote, R.M.: Abstract Algebra. Wiley-New York (2003)
24. Fiedler, M.: Algebra connectivity of graphs. Czechoslovake Math. J. **23**(98), 298–305 (1973)
25. Fiedler, M.: A property of eigenvectors of nonnegative symmetric matrices and its application to graph theory. Czechoslovake Math. J. **25**(100), 619–633 (1975)
26. Fuchs, E.D.: Longest induced cycles in circulant graphs. Electron. J. Combin. **12**, 12 (2005)
27. Grimaldi, R.P.: Graphs from rings. In: Proceedings of the 20th Southeastern Conference on Combinatorics. Graph Theory and Computing (Boca Raton, FL), Congress Number 71, pp. 95–103 (1990)
28. Grone, R.: On the geometry and Laplacian of a graph. Linear Algebra Appl. **150**, 167–178 (1991)
29. Hsieh, H.C.J., Lee, P.F., Wang, H.J.: The embedding of line graphs associated to the zero-divisor graphs of commutative rings. Israel J. Math. **180**, 193–222 (2010)
30. Ilić, A.: The energy of unitary Cayley graphs. Linear Algebra Appl. **431**(10), 1881–1889 (2009)
31. Kiani, D., Haji Aghaei, M.M., Meemark, Y., Suntornpoch, B.: Energy of unitary Cayley graphs and GCD-graphs. Linear Algebra Appl. **435**(6), 1336–1343 (2011)

32. Kirchhoff, G.: Uber die Auflosung der Gleichungen, auf welche man bei der Untersuchung der linearen Verteilung galvanischer Strome gefuhrt wird. Ann. Phys. Chem. **72**, 497–508 (1847)
33. Klotz, W., Sander, T.: Some properties of unitary Cayley graphs. Electron. J. Combin. **14**, 12 (2007)
34. Lanski, C., Maróti, A.: Ring elements as sums of units. Cent. Eur. J. Math. **7**(3), 395–399 (2009)
35. Maimani, H.R., Pournaki, M.R., Tehranian, A., Yassemi, S.: Graphs attached to rings revisited. Arab. J. Sci. Eng. **36**(6), 997–1011 (2011)
36. Maimani, H.R., Pournaki, M.R., Yassemi, S.: A class of weakly perfect graphs. Czechoslovak Math. J. **60**(135, 4), 1037–1041 (2010)
37. Maimani, H.R., Pournaki, M.R., Yassemi, S.: Necessary and sufficient conditions for unit graphs to be Hamiltonian. Pacific J. Math. **249**(2), 419–429 (2011)
38. Maimani, H.R., Pournaki, M.R., Yassemi, S.: Rings which are generated by their units: a graph theoretical approach. Elem. Math. **65**(1), 17–25 (2010)
39. Maimani, H.R., Pournaki, M.R., Yassemi, S.: Weakly perfect graphs arising from rings. Glasg. Math. J. **52**(3), 417–425 (2010)
40. Maimani, H.R., Wickham, C., Yassemi, S.: Rings whose total graphs have genus at most one. Rocky Mountain J. Math. **42**(5), 1551–1560 (2012)
41. Merris, R.: A survey of graph Lapalacians. Linear Multilinear Algebra **39**(1–2), 19–31 (1995)
42. Merris, R.: Laplacian matrices of graphs: a survey. Linear Algebra Appl. **197**(198), 143–176 (1994)
43. Mohar, B.: The Laplacian spectrum of graphs. In: Alavi, Y., et al. (eds.) Proceedings of the Sixth Quadrennial International Conference on the Theory and Applications of Graphs, Western Michigan University, Kalamazoo, 1988, Wiley, New York, pp. 871–898 (1991)
44. Mohar, B.: Laplace eigenvalues of graphs-a survey. Discret. Math. **109**(1–3), 171–183 (1992)
45. Monius, K.: Eigenvalues of zero-divisor graphs of finite commutative rings (2019). arXiv:191012567v1 [math.CO]
46. Nikmehr, M.J., Heidarzadeh, L., Soleimani, N.: Calculating different topological indices of total graph of $\mathbb{Z}_n$. Studia Sci. Math. Hungar. **51**(1), 1339–1343 (2014)
47. Pirzada, S., Aijaz, M., Imran Bhat, M.: On zero-divisor graphs of the rings $\mathbb{Z}_n$. Afr. Mat. **31**(3–4), 727–737 (2020)
48. Acharya, M.: Energy and Wiener index of unit graph. Appl. Math. Inf. Sci. **9**(3), 1339–1343 (2015)
49. Ramaswamy, H.N.: Veena: on the energy of unitary Cayley graphs. Electron. J. Combin. **16**(1), N24, 8 (2009)
50. Surendranath Reddy, B., Jain, R.S., Laxmikanth, N.: Eigenvalues and Wiener index of the zero-divisor graph $\Gamma(\mathbb{Z}_n)$. arXiv:1707.05083 [math.RA]
51. Sander, T.: Eigenspaces of Hamming graphs and unitary Cayley graphs. Ars Math. Contemp. **3**(1), 13–19 (2010)
52. Sheela, S., Prakash, O.: Adjacency matrix and energy of the line graph of $\Gamma(\mathbb{Z}_n)$ (2018). arXiv:180608944v1 [math.CO]
53. Sheela, S., Prakash, O.: Energy and Wiener index of total graph over ring $\mathbb{Z}_n$. Electron. Notes Discrete Math. **63**, 485–495 (2017)
54. Tamizh Chelvam, T., Asir, T.: On the genus of the total graph of a commutative ring. Comm. Algebra **41**(1), 142–153 (2013)
55. Tamizh Chelvam, T., Asir, T.: On the total graph and its complement of a commutative ring. Comm. Algebra **41**(10), 3820–3835 (2013)
56. Whitney, H.: Congruent graphs and connectivity of graphs. Amer. J. Math. **54**(1), 150–168 (1932)
57. Liu, X., Zhou, S.: Spectral properties of unitary Cayley graphs of finite commutative rings. Electron. J. Combin. **19**(4), 19 (2012)
58. Liu, X., Zhou, S.: Quadratic unitary Cayley graphs of finite commutative rings. Linear Algebra Appl. **479**, 73–90 (2015)

59. Young, M.: Adjacency matrices of zero-divisor graphs of integers modulo n. Involve **8**(5), 753–761 (2015)
60. Gutman, I.: The energy of a graph. Ber. Math. Statist. Sekt. Forsch. Graz. **103**, 1–22 (1978)
61. Sharma, P., Sharma, A., Vats., R.K.: Analysis of adjacency matrix and neighborhood associated with zero-divisor graph of finite commutative rings. Int. J. Comput. Appl. **14**(3), 38–42 (2011)

On a New Extension of Annihilating-Ideal Graph of Commutative Rings

Nadeem ur Rehman, Mohd Nazim, and Junaid Nisar

Abstract Let R be a commutative ring with unity and $A(R)$ be the set of annihilating-ideals R. In this paper, we introduced and studied the *extended sum annihilating-ideal graph* of R, denoted by $AG_{\Omega}(R)$, with vertex set $A(R)^*$ and two vertices I_1 and I_2 are adjacent if and only if either $I_1 I_2 = 0$ or $I_1 + I_2$ is an annihilating-ideal R. We prove that $AG_{\Omega}(R)$ is a connected graph with diameter at most two and girth exactly three. We classify all the Artinian commutative rings R for which $AG_{\Omega}(R)$ is isomorphic to some well-known graphs. Finally, we characterized Artinian commutative rings for which $AG_{\Omega}(R)$ has genus one.

Keywords Annihilating-ideal graph · Sum annihilating-ideal graph · Zero-divisor graph · Complete graph · Planar graph

1 Introduction

Throughout this paper, all rings are commutative with unity. For a commutative ring R, we use $\mathbb{I}(R)$ to denote the set of ideals of R and $\mathbb{I}(R)^* = \mathbb{I}(R) \setminus \{0\}$. An ideal I_1 of R is said to *annihilating-ideal* of R if there exists a nonzero ideal I_2 of R such that $I_1 I_2 = 0$. The set of annihilating-ideals of R is denoted by $A(R)$ and $A(R)^* = A(R) \setminus \{0\}$. For any undefined notation or terminology in ring theory, we refer the reader to [11].

Let G be a graph with vertex set $V(G)$. The $distance$ between two vertices u and v of G denoted by $d(u, v)$, is the smallest path from u to v. If there is no such path, then $d(u, v) = \infty$. The $diameter$ of G is defined as $diam(G) = \sup\{d(u, v) : u, v \in V(G)\}$. A $cycle$ is a closed path in G. The $girth$ of G denoted by $gr(G)$ is the length of a shortest cycle in G ($gr(G) = \infty$ if G contains no cycle). A graph is said to be *complete graph* if all its vertices are adjacent. A complete graph with n vertices is denoted by K_n. If G is a graph such that the vertices of G can be partitioned into

N. Rehman (✉) · M. Nazim · J. Nisar
Department of Mathematics, Aligarh Muslim University, Aligarh 202002, India
e-mail: nu.rehman.mm@amu.ac.in

M. Ashraf et al. (eds.), *Algebra and Related Topics with Applications*,
Springer Proceedings in Mathematics & Statistics 392,
https://doi.org/10.1007/978-981-19-3898-6_34

two nonempty disjoint sets U_1 and U_2 such that vertices u and v are adjacent if and only if $u \in U_1$ and $v \in U_2$, then G is called a *complete bipartite graph.* A complete bipartite graph with disjoint vertex sets of size m and n, respectively, is denoted by $K_{m,n}$. We write $K_{n,\infty}$ (respectively, $K_{\infty,\infty}$) if one (respectively, both) of the disjoint vertex sets is infinite. A complete bipartite graph of the form $K_{1,n}$ is called a *star graph.*

A connected graph G is said to be a *tree* if it does not contain any cycle. A graph G is said to be *unicycle* if it contains unique cycle. A graph G is *a split graph* if the vertex set can be partitioned into a clique and an independent set. A graph G is said to be *planar* if it can be drawn in the plane so that its edges intersect only at their ends. A subdivision of a graph is a graph obtained from it by replacing edges with pairwise internally-disjoint paths. A remarkably simple characterization of planar graphs was given by Kuratowski in 1930. Kuratowski's Theorem says that a graph G is planar if and only if it contains no subdivision of K_5 or $K_{3,3}$. An undirected graph G is said to be *outerplanar* if it can be embedded in the plane in such a way that all the vertices lie on the unbounded face of the drawing. The *genus* of a graph G, denoted by $\gamma(G)$, is the minimum integer k such that the graph can be drawn without crossing itself on a sphere with k handles (i.e., an oriented surface of genus k). Thus, a planar graph has genus 0, because it can be drawn on a sphere without self-crossing. For more details on graph theory, we refer the reader to see [18, 19].

The concept of *zero-divisor graph* of a commutative ring R, denoted by $\Gamma(R)$, was introduced by Beck [12]. The vertex set of $\Gamma(R)$ is $Z(R)^*$ and two vertices x_1 and x_2 are adjacent if and only if $x_1x_2 = 0$, for details see [5, 8, 10].

In [13], Behboodi generalized the zero-divisor graph to ideals by defining *the annihilating-ideal graph* $AG(R)$, with vertex set $A(R)^*$ and two vertices I_1 and I_2 are adjacent if and only if $I_1I_2 = 0$. For more details on annihilating-ideal graph, we refer the reader to see [1–4, 7, 14].

Recently, Visweswaran and Patel [17] have introduced and investigated the *sum annihilating-ideal graph* of a commutative ring R, denoted by $\Omega(R)$, whose vertex set is $A(R)^*$ and two distinct vertices I_1 and I_2 are adjacent if and only if $I_1 + I_2 \in A(R)$. For more details, see [15].

In this paper, we introduced the *extended sum annihilating-ideal graph* of a commutative ring R. It is an undirected graph denoted by $AG_\Omega(R)$, whose vertex set is $A(R)^*$ and two distinct vertices I_1 and I_2 are adjacent if and only if either $I_1I_2 = 0$ or $I_1 + I_2 \in A(R)^*$. We prove that $AG_\Omega(R)$ is a connected graph with diameter at most two and girth exactly three. Further, all the Artinian commutative rings are characterized for which $AG_\Omega(R)$ is a complete graph. We investigated some situations under which $AG_\Omega(R)$ is identical to both $AG(R)$ and $\Omega(R)$. Finally, we classify all the Artinian commutative rings under which $AG_\Omega(R)$ is a tree, a unicycle, a split graph, an outerplanar graph, a planar graph, and a toroidal graph.

The following result is the direct consequence of the definition of the extended sum annihilating-ideal graph.

Theorem 1 *Let R be a commutative ring. Then $AG(R)$ and $\Omega(R)$ both are spanning subgraph of $AG_\Omega(R)$.*

The following examples shows that $AG(R)$, $\Omega(R)$ and $AG_{\Omega}(R)$ are all different.

Example 1 Consider $R = \mathbb{Z}_2 \times \mathbb{Z}_4$, then $AG(R)$, $\Omega(R)$ and $AG_{\Omega}(R)$ are given as

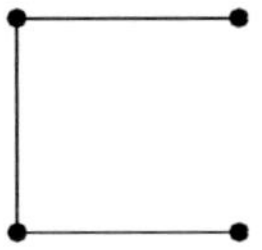

$AG(\mathbb{Z}_2 \times \mathbb{Z}_4)$

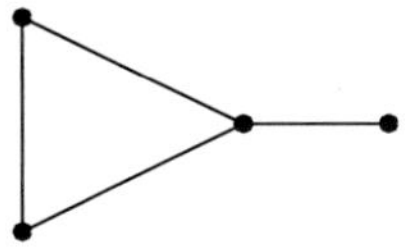

$\Omega(\mathbb{Z}_2 \times \mathbb{Z}_4)$

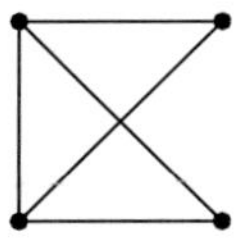

$AG_{\Omega}(\mathbb{Z}_2 \times \mathbb{Z}_4)$

Example 1 Consider $R = \mathbb{Z}_2 \times \mathbb{Z}_2 \times \mathbb{Z}_2$, then $AG(R)$, $\Omega(R)$ and $AG_{\Omega}(R)$ are given as

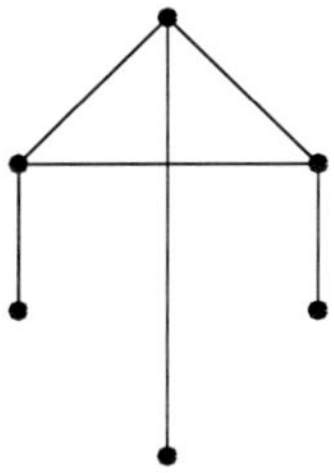

$AG(\mathbb{Z}_2 \times \mathbb{Z}_2 \times \mathbb{Z}_2)$

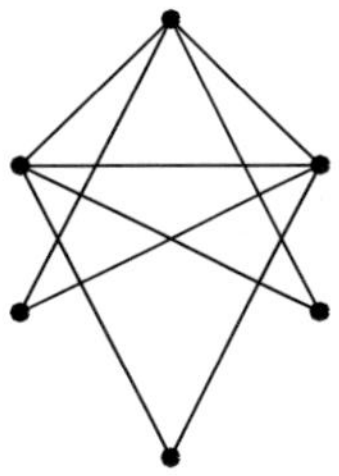

$\Omega(\mathbb{Z}_2 \times \mathbb{Z}_2 \times \mathbb{Z}_2)$

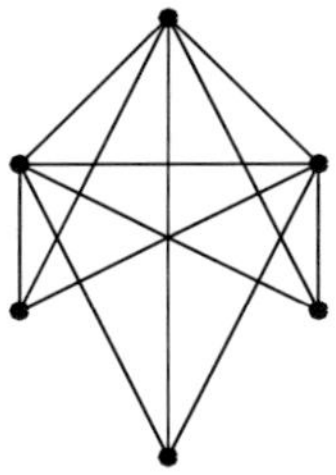

$AG_{\Omega}(\mathbb{Z}_2 \times \mathbb{Z}_2 \times \mathbb{Z}_2)$

2 Connectedness of Extended Sum Annihilating-Ideal Graph

In this section, first we discuss the connectedness of $AG_{\Omega}(R)$. Then we characterized all the Artinian rings for which $AG_{\Omega}(R)$ is complete. Finally, we discuss some situations under which $AG_{\Omega}(R)$ is identical to $AG(R)$ and $\Omega(R)$.

Theorem 2 *Let R be a commutative ring. Then* $AG_{\Omega}(R)$ *is connected with* $diam(AG_{\Omega}(R)) \leq 2$.

Proof Let I_1 and I_2 be distinct vertices of $AG_{\Omega}(R)$. We have the following cases:
Case(i) : $I_1I_2 = 0$. Then $I_1 \sim I_2$ is a path in $AG_{\Omega}(R)$.
Case(ii) : $I_1I_2 \neq 0$, $I_1^2 = 0$ and $I_2^2 = 0$. Then $I_1 \sim I_1I_2 \sim I_2$ is a path in $AG_{\Omega}(R)$.
Case(iii) : $I_1I_2 \neq 0$, $I_1^2 = 0$ and $I_2^2 \neq 0$. Then there exists $K_2 \in A(R)^*$ such that $I_2K_2 = 0$. If $I_1K_2 = 0$, then $I_1 \sim K_2 \sim I_2$ is a path in $AG_{\Omega}(R)$. If $I_1K_2 \neq 0$, then $I_1 \sim I_1K_2 \sim I_2$ is a path in $AG_{\Omega}(R)$.
Case(iv) : $I_1I_2 \neq 0$, $I_1^2 \neq 0$ and $I_2^2 = 0$. Using the similar argument as used in Case(iii).
Case(v) : $I_1I_2 \neq 0$, $I_1^2 \neq 0$ and $I_2^2 \neq 0$. Then there exist $K_1, K_2 \in A(R)^*$ such that $I_1K_1 = 0$ and $I_2K_2 = 0$. If $K_1 = K_2$, then $I_1 \sim K_1 \sim I_2$ is a path in $AG_{\Omega}(R)$. If $K_1 \neq K_2$ and $K_1K_2 = 0$, then $K_1(K_2 + I_1) = 0$. Thus, $K_2 + I_1 \in A(R)^*$, which implies that $I_1 \sim K_2 \sim I_2$ is a path in $AG_{\Omega}(R)$. If $K_1 \neq K_2$ and $K_1K_2 \neq 0$, then $K_1K_2(I_1 + I_2) = 0$. Thus, $I_1 \sim I_2$ in $AG_{\Omega}(R)$.
In all the above cases, $d(I_1, I_2) \leq 2$. Since I_1 and I_2 are arbitrary vertices of $AG_{\Omega}(R)$. Hence, $AG_{\Omega}(R)$ is connected and $diam(AG_{\Omega}(R)) \leq 2$.

Theorem 3 *Let R be a commutative ring with at least three nonzero annihilating ideals. Then* $AG_{\Omega}(R)$ *contains a cycle and* $gr(AG_{\Omega}(R)) = 3$.

Proof If $AG_{\Omega}(R)$ is a complete graph, then $gr(AG_{\Omega}(R)) = 3$. Suppose that there exist two vertices I_1 and I_2 such that $I_1 \nsim I_2$. Since $I_1, I_2 \in A(R)^*$, there exist $K_1, K_2 \in A(R)^*$ such that $I_1K_1 = 0$ and $I_2K_2 = 0$. If $K_1 = K_2$, then $K_1(I_1 + I_2) = 0$, implies that I_1 and I_2 are adjacent, a contradiction. Since $K_1K_2(I_1 + I_2) = 0$ and $I_1 + I_2 \notin A(R)^*$, therefore, $K_1K_2 = 0$. If $I_1 = K_1$, then $K_2(I_1 + I_2) = 0$. This implies that $I_1 \sim I_2$, a contradiction. Thus, $I_1 \neq K_1$. Also, if $I_1 = K_2$, then $I_1I_2 = K_2I_2 = 0$, implies that I_1 and I_2 are adjacent. again a contradiction. Thus, $I_1 \neq K_2$. On the other hand $K_1(K_2 + I_1) = 0$. Thus, $I_1 \sim K_1 \sim K_2 \sim I_1$ is a cycle of length three in $AG_{\Omega}(R)$. Hence, $gr(AG_{\Omega}(R)) = 3$.

Theorem 4 ([15] Theorem 2.1) *Let R be an Artinian commutative ring. Then* $\Omega(R)$ *is complete if and only if R is local.*

Theorem 5 *Let R be an Artinian commutative ring. Then* $AG_{\Omega}(R)$ *is complete if and only if either R is Artinian local or* $R \cong F_1 \times F_2$, *where* F_1 *and* F_2 *are fields.*

Proof Suppose R is an Artinian local ring, then by Theorem 4, $\Omega(R)$ is complete and hence $AG_{\Omega}(R)$ is complete. If $R \cong F_1 \times F_2$, then $AG_{\Omega}(R) \cong K_2$.

Conversely, suppose $AG_\Omega(R)$ is complete. If R is an Artinian local ring, then we are done. Thus, we assume that R is non-local. Since R is Artinian ring, $R \cong R_1 \times R_2 \times \cdots \times R_n$, where each R_i is Artinian local and $n \geq 2$. We claim that R_i is field for each $1 \leq i \leq n$. Suppose on contrary that R_i is not a field with nonzero maximal ideal $\mathfrak{m}_i$ for some $1 \leq i \leq n$. Consider $I_1 = R_1 \times R_2 \times \cdots \times \mathfrak{m}_i \times \cdots \times R_n$ and $I_2 = (0) \times (0) \times \cdots \times R_i \times \cdots \times (0)$. Then I_1 and I_2 are vertices of $AG_\Omega(R)$ such that $I_1 I_2 \neq 0$ and $I_1 + I_2 \notin A(R)^*$. Thus, I_1 and I_2 are not adjacent in $AG_\Omega(R)$, which is a contradiction to the completeness of $AG_\Omega(R)$. Hence, R_i is a field for each $1 \leq i \leq n$. If $n \geq 3$, then $I_3 = F_1 \times F_2 \times (0) \times F_4 \times \cdots \times F_n$ and $I_4 = F_1 \times (0) \times F_3 \times \cdots \times F_n$ are vertices of $AG_\Omega(R)$ such that $I_3 I_4 \neq 0$ and $I_3 + I_4 \notin A(R)^*$. Thus, I_3 and I_4 are not adjacent in $AG_\Omega(R)$, a contradiction. Hence, $n = 2$. This completes the proof.

Corollary 1 *Let R be a reduced Artinian commutative ring which is not a field. Then $AG_\Omega(R)$ is complete if and only if $R \cong F_1 \times F_2$, where F_1 and F_2 are fields.*

Corollary 2 *Let n be any positive integer. Then $AG_\Omega(\mathbb{Z}_n)$ is complete if and only if either $n = p^m$, where p is prime number and m is positive integer or $n = pq$, where p and q are distinct prime numbers.*

Theorem 6 *Let R be an Artinian commutative ring. Then $AG_\Omega(R) = AG(R)$ if and only if $AG(R)$ is complete.*

Proof Suppose that $AG(R)$ is complete. Since $AG(R)$ is spanning subgraph of $AG_\Omega(R)$, therefore, $AG_\Omega(R) = AG(R)$.

Conversely, let $AG_\Omega(R) = AG(R)$. Since R is Artinian, $R \cong R_1 \times R_2 \times \cdots \times R_n$, where R_i is Artinian local for each i. If $n \geq 2$, then $I_1 = (0) \times R_2 \times (0) \times \cdots \times (0)$ and $I_2 = (0) \times R_2 \times R_3 \times (0) \times \cdots \times (0)$ are adjacent in $AG_\Omega(R)$ but not adjacent in $AG(R)$, a contradiction. Thus, $n \leq 2$. Hence,the following two cases occur:

Case(i): $n = 1$. Then R is Artinian local. Thus, by Theorem 5, $AG_\Omega(R)$ is complete and hence $AG(R)$ is complete.

Case(ii): $n = 2$. We claim that R_1 and R_2 both are field. Suppose on contrary that R_1 is not a field with nonzero maximal ideal $\mathfrak{m}_1$. Then $I_3 = (0) \times R_2$ and $I_4 = \mathfrak{m}_1 \times R_2$ are annihilating-ideals of R such that I_3 and I_4 are adjacent in $AG_\Omega(R)$ but not adjacent in $AG(R)$, a contradiction. Hence, $R \cong F_1 \times F_2$, which implies that $AG(R)$ is complete.

Corollary 3 *Let R be an Artinian ring. Then $AG_\Omega(R) = AG(R)$ if and only if either $R \cong F_1 \times F_2$, where F_1 and F_2 are fields or R is local with exactly two nonzero proper ideals.*

Corollary 4 *Let n be a positive integer. Then $AG_\Omega(\mathbb{Z}_n) = AG(\mathbb{Z}_n)$ if and only if either $n = p^m$, where p is prime number, m is positive integer and $m = 1, 2, 3$ or $n = pq$, where p and q are distinct prime numbers.*

Theorem 7 *Let R be an Artinian commutative ring. Then $AG_\Omega(R) = \Omega(R)$ if and only if R is local.*

Proof Suppose R is local, then by Theorem 4, $\Omega(R)$ is complete. Since $\Omega(R)$ is spanning subgraph of $AG_{\Omega}(R)$, therefore $AG_{\Omega}(R) = \Omega(R)$.

Conversely, suppose that $AG_{\Omega}(R) = \Omega(R)$. Since R is Artinian, $R \cong R_1 \times R_2 \times \cdots \times R_n$, where each R_i is Artinian local. If $n \geq 2$, then $I_1 = R_1 \times (0) \times R_3 \times \cdots \times R_n$ and $I_2 = (0) \times R_2 \times (0) \times \cdots \times (0)$ are vertices of $AG_{\Omega}(R)$ such that I_1 and I_2 are adjacent in $AG_{\Omega}(R)$ but not adjacent in $\Omega(R)$, a contradiction. Hence, $n = 1$, which implies that R is local.

Corollary 5 *Let n be a positive integer. Then $AG_{\Omega}(\mathbb{Z}_n) = \Omega(\mathbb{Z}_n)$ if and only if $n = p^m$, where p is prime number, m is positive integer.*

3 Extended Sum Annihilating-Ideal Graph as Some Special Type of Graphs

In this section, we characterized all the Artinian rings R for which $AG_{\Omega}(R)$ is a tree, a unicycle graph, a spit graph, an outerplanar graph, and a planar graph.

Theorem 8 *Let R be an Artinian commutative ring. Then $AG_{\Omega}(R)$ is a tree if and only if one of the following holds:*

1. *$R \cong F_1 \times F_2$, where F_1 and F_2 are fields.*
2. *R is local with at most two nonzero proper ideals.*

Proof Suppose $AG_{\Omega}(R)$ is tree. Since R is Artinian, $R \cong R_1 \times R_2 \times \cdots \times R_n$, where R_i is Artinian local for each $1 \leq i \leq n$. Let $n \geq 3$. Consider $I_1 = R_1 \times (0) \times (0) \times \cdots \times (0)$, $I_2 = (0) \times R_2 \times (0) \times \cdots \times (0)$, and $I_3 = (0) \times (0) \times R_3 \times (0) \times \cdots \times (0)$. It is easy to see that $I_1 \sim I_2 \sim I_3 \sim I_1$ is a cycle in $AG_{\Omega}(R)$, a contradiction. Hence, $n \leq 2$. The following two cases occur:
Case(i) : $n = 2$. We claim that R_1 and R_2 both are fields. Suppose on contrary that R_1 is not a field with nonzero maximal ideal $\mathfrak{m}_1$. Consider $J_1 = \mathfrak{m}_1 \times (0)$, $J_2 = \mathfrak{m}_1 \times R_2$ and $J_3 = (0) \times R_2$. Then $J_1 \sim J_2 \sim J_3 \sim J_1$ is a cycle in $AG_{\Omega}(R)$, which is a contradiction. Hence, R_1 and R_2 both are fields.
Case(ii) : $n = 1$. Then R is an Artinian local ring. Thus, by Theorem 5, $AG_{\Omega}(R)$ is complete. Since $AG_{\Omega}(R)$ is tree, R has at most three nonzero proper ideals.

Converse is clear.

Theorem 9 *Let R be an Artinian commutative ring. Then $AG_{\Omega}(R)$ is unicycle if and only if R is local with exactly three nonzero proper ideals.*

Proof Suppose $AG_{\Omega}(R)$ is unicycle. Since R is Artinian, $R \cong R_1 \times R_2 \times \cdots \times R_n$, where R_i is Artinian local for each $1 \leq i \leq n$. Let $n \geq 3$. Consider $I_1 = R_1 \times (0) \times (0) \times \cdots \times (0)$, $I_2 = (0) \times R_2 \times (0) \times \cdots \times (0)$, $I_3 = (0) \times (0) \times R_3 \times (0) \times \cdots \times (0)$ and $I_4 = R_1 \times (0) \times R_3 \times (0) \times \cdots \times (0)$. Then $I_1 \sim I_2 \sim I_3 \sim I_1$ as well as $I_1 \sim I_4 \sim I_2 \sim I_1$ are two distinct cycles in $AG_{\Omega}(R)$, a contradiction. Hence, $n \leq 2$. The following two cases occur:

Case(i) : $n = 2$. We claim that R_1 and R_2 both are fields. Suppose on contrary that R_2 is not a field with nonzero maximal ideal $\mathfrak{m}_2$. Consider $J_1 = R_1 \times (0)$, $J_2 = (0) \times R_2$, $J_3 = (0) \times \mathfrak{m}_2$ and $J_4 = R_1 \times \mathfrak{m}_2$. It is easy to see that $J_1 \sim J_2 \sim J_3 \sim J_1$ and $J_1 \sim J_4 \sim J_3 \sim J_1$ are two different cycles in $AG_\Omega(R)$, which is a contradiction. Thus, R_1 and R_2 both are fields. This implies that $AG_\Omega(R) \cong K_2$, again a contradiction.
Case(ii) : $n = 1$. Then R is an Artinian local ring. Thus, by Theorem 5, $AG_\Omega(R)$ is complete. Since $AG_\Omega(R)$ is unicycle, R has exactly three nonzero proper ideals.

Converse is clear.

Lemma 1 *[18] Let G be a connected graph. Then G is a split graph if and only if G contains no induced subgraph isomorphic to $2K_2$, C_4, C_5.*

Theorem 10 *Let R be an Artinian commutative ring. Then $AG_\Omega(R)$ is a split graph if and only if one of the following holds:*

1. *$R \cong F_1 \times F_2$, where F_1 and F_2 are fields.*
2. *R is local with at most three nonzero proper ideals.*

Proof Suppose $AG_\Omega(R)$ is split graph. Since R is Artinian, $R \cong R_1 \times R_2 \times \cdots \times R_n$, where R_i is Artinian local for each $1 \leq i \leq n$. Let $n \geq 3$. Consider $I_1 = R_1 \times (0) \times (0) \times \cdots \times (0)$, $I_2 = (0) \times R_2 \times (0) \times \cdots \times (0)$, $I_3 = (0) \times (0) \times R_3 \times (0) \times \cdots \times (0)$ and $I_4 = R_1 \times (0) \times R_3 \times (0) \times \cdots \times (0)$. It is easy to see that $I_1 \sim I_2 \sim I_3 \sim I_4 \sim I_1$ is a cycle of length four in $AG_\Omega(R)$, which is a contradiction by Lemma 1. Hence, $n \leq 2$. The following two cases occur:
Case(i) : $n = 2$. We claim that R_1 and R_2 both are fields. Suppose on contrary that R_1 is not a field with nonzero maximal ideal $\mathfrak{m}_1$. Consider $J_1 = R_1 \times (0)$, $J_2 = \mathfrak{m}_1 \times (0)$, $J_3 = \mathfrak{m}_1 \times R_2$ and $J_4 = (0) \times R_2$. Then $J_1 \sim J_2 \sim J_3 \sim J_4 \sim J_1$ is a cycle of length four in $AG_\Omega(R)$, a contradiction. Hence, R_1 and R_2 both are fields.
Case(ii) : $n = 1$. Then R is an Artinian local ring. Thus, by Theorem 5, $AG_\Omega(R)$ is complete. Since $AG_\Omega(R)$ is a split graph, R has at most three nonzero proper ideals.

Converse is clear.

Lemma 2 *[19] A graph G is outerplanar if and only if it does not contain a subdivision of K_4 or $K_{2,3}$.*

Theorem 11 *Let R be an Artinian commutative ring. Then $AG_\Omega(R)$ is outerplanar if and only if one of the following holds:*

1. *$R \cong F_1 \times F_2$, where F_1 and F_2 are fields.*
2. *$R \cong F_1 \times R_2$, where F_1 is field and $(R_2, \mathfrak{m}_2)$ is an Artinian local ring with unique nonzero proper ideal $\mathfrak{m}_2$.*
3. *R is local with at most three nonzero proper ideals.*

Proof Suppose $AG_\Omega(R)$ is outerplanar. Since R is Artinian, $R \cong R_1 \times R_2 \times \cdots \times R_n$, where R_i is Artinian local for each $1 \leq i \leq n$. Let $n \geq 3$. Consider $I_1 = R_1 \times (0) \times (0) \times \cdots \times (0)$, $I_2 = (0) \times R_2 \times (0) \times \cdots \times (0)$, $I_3 = (0) \times (0) \times R_3 \times$

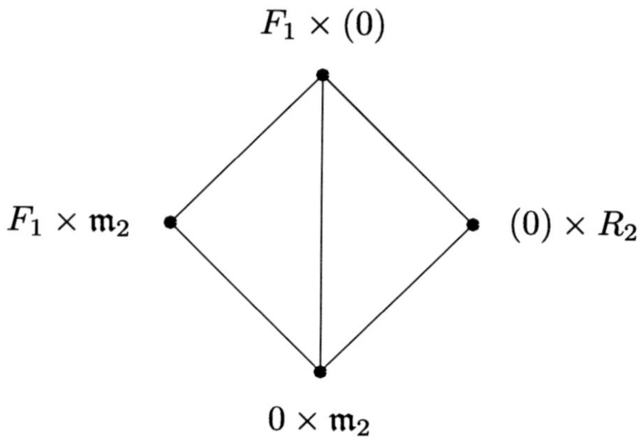

Fig. 1 The graph $AG_{\Omega}(F_1 \times R_2)$, where $\mathfrak{m}_2$ is the only nonzero proper ideal of R_2

$(0) \times \cdots \times (0)$, $J_1 = R_1 \times R_2 \times (0) \times \cdots \times (0)$ and $J_2 = R_1 \times (0) \times R_3 \times (0) \times \cdots \times (0)$. One can see that the subgraph induced by the set $\{I_1, I_2, I_3, J_1, J_2\}$ contains a copy of $K_{2,3}$ in $AG_{\Omega}(R)$, which is a contradiction by Lemma 2. Hence, $n \leq 2$. The following two cases occur:

Case(i) : $n = 2$. We claim that R_1 or R_2 is a field. Suppose on contrary that R_1 and R_2 both are not field with nonzero maximal ideals $\mathfrak{m}_1$ and $\mathfrak{m}_2$, respectively. Consider $I_1 = \mathfrak{m}_1 \times (0)$, $I_2 = (0) \times \mathfrak{m}_2$, $I_3 = \mathfrak{m}_1 \times \mathfrak{m}_2$, $J_1 = R_1 \times (0)$ and $J_2 = (0) \times R_2$. Then the subgraph induced by the set $\{I_1, I_2, I_3, J_1, J_2\}$ contains a subdivision of $K_{2,3}$ in $AG_{\Omega}(R)$, a contradiction by Lemma 2.

If R_1 and R_2 both are fields, then we are done. Suppose R_2 is not a field with nonzero maximal ideal $\mathfrak{m}_2$. Let I be a nonzero ideal of R_2 such that $I \neq \mathfrak{m}_2$. Consider $K_1 = R_1 \times (0)$, $K_2 = (0) \times \mathfrak{m}_2$, $L_1 = R_1 \times \mathfrak{m}_2$, $L_2 = R_1 \times I$ and $L_3 = (0) \times I$.

Then the subgraph induced by the set $\{K_1, K_2, L_1, L_2, L_3\}$ contains a copy of $K_{2,3}$ in $AG_{\Omega}(R)$, again a contradiction. Hence, $\mathfrak{m}_2$ is the only nonzero proper ideal of R_2.

Case(ii) : $n = 1$. Then R is an Artinian local ring. Thus, by Theorem 5, $AG_{\Omega}(R)$ is complete. Since $AG_{\Omega}(R)$ is outerplanar, R has at most three nonzero proper ideals.

Converse is clear by Theorem 5, Lemma 2, and Fig. 1.

Lemma 3 (Kuratowski's Theorem) *A graph G is planar if and only if it contains no subdivision of K_5 or $K_{3,3}$.*

Theorem 12 *Let R be an Artinian commutative ring. Then $AG_{\Omega}(R)$ is planar if and only if one of the following holds:*

1. *$R \cong F_1 \times F_2$, where F_1 and F_2 are fields.*
2. *$R \cong F_1 \times R_2$, where F_1 is field and $(R_2, \mathfrak{m}_2)$ is an Artinian local ring with unique nonzero proper ideal $\mathfrak{m}_2$.*
3. *R is local with at most four nonzero proper ideals.*

Proof Suppose $AG_{\Omega}(R)$ is planar. Since R is Artinian, $R \cong R_1 \times R_2 \times \cdots \times R_n$, where R_i is Artinian local for each $1 \leq i \leq n$. Let $n \geq 3$. Consider $I_1 = R_1 \times (0) \times (0) \times \cdots \times (0)$, $I_2 = (0) \times R_2 \times (0) \times \cdots \times (0)$, $I_3 = (0) \times (0) \times R_3 \times (0) \times \cdots \times (0)$, $J_1 = R_1 \times R_2 \times (0) \times \cdots \times (0)$, $J_2 = R_1 \times (0) \times R_3 \times (0) \times \cdots \times (0)$ and $J_3 = (0) \times R_2 \times R_3 \times (0) \times \cdots \times (0)$. It is easy to see that the subgraph induced by the set $\{I_1, I_2, I_3, J_1, J_2, J_3\}$ contains a subdivision of $K_{3,3}$ in $AG_{\Omega}(R)$, a contradiction by Lemma 3. Hence, $n \leq 2$. The following two cases occur:
Case(i) : $n = 2$. We claim that R_1 or R_2 is a field. Suppose on contrary that R_1 and R_2 both are not field with nonzero maximal ideals $\mathfrak{m}_1$ and $\mathfrak{m}_2$, respectively. Consider $I_1 = \mathfrak{m}_1 \times R_2$, $I_2 = R_1 \times \mathfrak{m}_2$, $I_3 = (0) \times R_2$, $J_1 = \mathfrak{m}_1 \times \mathfrak{m}_2$, $J_2 = (0) \times \mathfrak{m}_2$ and $J_3 = \mathfrak{m}_1 \times (0)$. Then the subgraph induced by the set $\{I_1, I_2, I_3, J_1, J_2, J_3\}$ contains a subdivision of $K_{3,3}$ in $AG_{\Omega}(R)$, which is a contradiction by Lemma 3.

If R_1 and R_2 both are fields, we are done. Suppose R_2 is not a field with nonzero maximal ideal $\mathfrak{m}_2$. Let I be a nonzero ideal of R_2 such that $I \neq \mathfrak{m}_2$. Consider $K_1 = R_1 \times (0)$, $K_2 = (0) \times \mathfrak{m}_2$, $K_3 = R_1 \times \mathfrak{m}_2$, $K_4 = R_1 \times I$ and $K_5 = (0) \times I$. Then the subgraph induced by the set $\{K_1, K_2, K_3, K_4, K_5\}$ contains a subdivision of K_5 in $AG_{\Omega}(R)$, again a contradiction by Lemma 3. Hence, $\mathfrak{m}_2$ is the only nonzero proper ideal of R_2.
Case(ii) : $n = 1$. Then R is an Artinian local ring. Thus, by Theorem 5, $AG_{\Omega}(R)$ is complete. Since $AG_{\Omega}(R)$ is planar, R has at most four nonzero proper ideals.

Converse is clear by Theorem 5, Lemma 3, and Fig. 1.

4 Genus of Extended Sum Annihilating-Ideal Graph

In this section, we characterized all the Artinian commutative rings R for which $AG_{\Omega}(R)$ is toroidal.

Lemma 4 [19] $\gamma(K_n) = \lceil \frac{1}{12}(n-3)(n-4) \rceil$, *where* $\lceil x \rceil$ *is the least integer that is greater than or equal to* x. *In particular,* $\gamma(K_n) = 1$ *if* $n = 5, 6, 7$.

Lemma 5 *[19]* $\gamma(K_{m,n}) = \lceil \frac{1}{4}(m-2)(n-2) \rceil$, *where* $\lceil x \rceil$ *is the least integer that is greater than or equal to* x. *In particular,* $\gamma(K_{4,4}) = \gamma(K_{3,n}) = 1$ *if* $n = 3, 4, 5, 6$. *Also,* $\gamma(K_{5,4}) = \gamma(K_{6,4}) = \gamma(K_{3,m}) = 2$, *if* $m = 7, 8, 9, 10$.

Lemma 6 (Proposition 4.4.4 [16]) *Let* G *be a connected graph with* q *edges and* $m \geq 3$ *vertices. Then*

$$\gamma(G) \geq \left\lceil \frac{q}{6} - \frac{m}{2} + 1 \right\rceil.$$

Theorem 13 *Let* R *be an Artinian local commutative ring which is not a field. Then* $\gamma(AG_{\Omega}(R)) = 1$ *if and only if* R *has at least five and at most seven nonzero proper ideals.*

Proof The proof follows from Theorem 5 and Lemma 4.

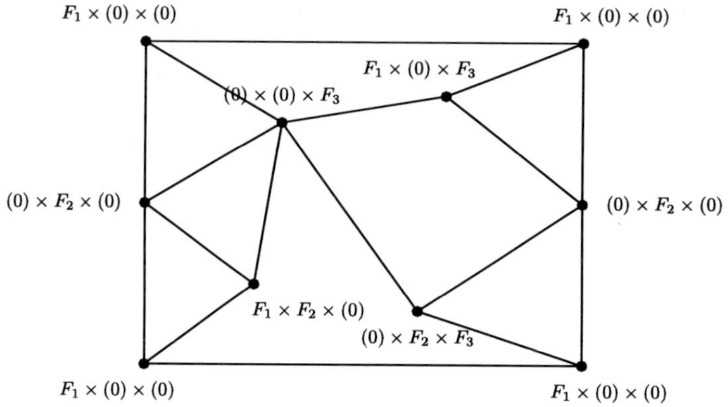

Fig. 2 Toroidal embedding of $AG_{\Omega}(F_1 \times F_2 \times F_3)$

Theorem 14 *Let $R \cong F_1 \times F_2 \times \cdots \times F_n$ be an Artinian commutative ring, where each F_i is a field and $n \geq 3$. Then $\gamma(AG_{\Omega}(R)) = 1$ if and only if $n = 3$.*

Proof Suppose $\gamma(AG_{\Omega}(R)) = 1$. If $n \geq 4$, then the subgraph induced by the set $\{F_1 \times (0) \times \cdots \times (0), F_1 \times F_2 \times (0) \times \cdots \times (0), F_1 \times (0) \times F_3 \times (0) \times \cdots \times (0), F_1 \times (0) \times (0) \times F_4 \times (0) \times \cdots \times (0), (0) \times F_2 \times F_3 \times (0) \times \cdots \times (0), (0) \times F_2 \times (0) \times F_4 \times (0) \times \cdots \times (0), (0) \times (0) \times F_3 \times F_4 \times (0) \times \cdots \times (0)\} \cup \{(0) \times F_2 \times (0) \times \cdots \times (0), (0) \times (0) \times F_3 \times (0) \times \cdots \times (0), (0) \times (0) \times (0) \times F_4 \times (0) \times \cdots \times (0)\}$ contains a copy of $K_{3,7}$ is $AG_{\Omega}(R)$. Thus, by Lemma 5, $\gamma(AG_{\Omega}(R)) > 1$, a contradiction. Hence $n = 3$.

Converse is clear by Fig. 2.

Theorem 15 ([15] Theorem 4.7) *Let $R \cong R_1 \times R_2 \times \cdots \times R_n$, where each $(R_i, \mathfrak{m}_i)$ is Artinian local ring with $m_i \neq 0$ and $n \geq 2$. Then $\gamma(\Omega(R)) = 1$ if and only if $n = 2$ and $\mathfrak{m}_i$ is the only nonzero proper ideal of R_i.*

Theorem 16 *Let $R \cong R_1 \times R_2 \times \cdots \times R_n$, where each $(R_i, \mathfrak{m}_i)$ is Artinian local ring with $m_i \neq 0$ and $n \geq 2$. Then $\gamma(AG_{\Omega}(R)) = 1$ if and only if $n = 2$ and $\mathfrak{m}_i$ is the only nonzero proper ideal of R_i.*

Proof Since $\Omega(R)$ is a spanning subgraph of $AG_{\Omega}(R)$, therefore, the proof follows from Theorem 15 and Fig. 3.

Theorem 17 ([15] Theorem 4.9) *Let $R \cong R_1 \times R_2 \times \cdots \times R_n \times F_1 \times F_2 \times \cdots \times F_m$ be an Artinian commutative ring, where $(R_i, \mathfrak{m}_i)$ is Artinian local ring with $\mathfrak{m}_i \neq 0$, F_j is field and $n, m \geq 1$. Then $\gamma(\Omega(R)) = 1$ if and only if $n = m = 1$ and R_1 has exactly two or three nonzero proper ideals.*

Theorem 18 *Let $R \cong F_1 \times F_2 \times \cdots \times F_m \times R_1 \times R_2 \times \cdots \times R_n$ be an Artinian commutative ring, where $(R_i, \mathfrak{m}_i)$ is Artinian local ring with $\mathfrak{m}_i \neq 0$, F_j is field and $n, m \geq 1$. Then $\gamma(AG_{\Omega}(R)) = 1$ if and only if $n = m = 1$ and R_1 has exactly two nonzero proper ideals.*

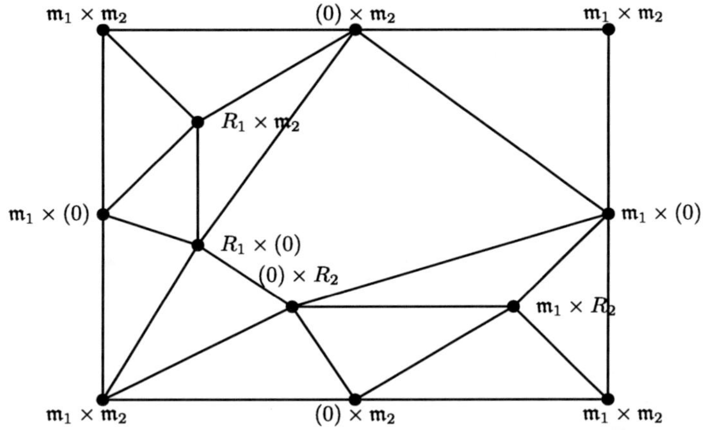

Fig. 3 Toroidal embedding of $AG_{\Omega}(R_1 \times R_2)$, at (8, −3.3) **where** R_i **has exactly one nonzero proper ideal for** $i = 1, 2$

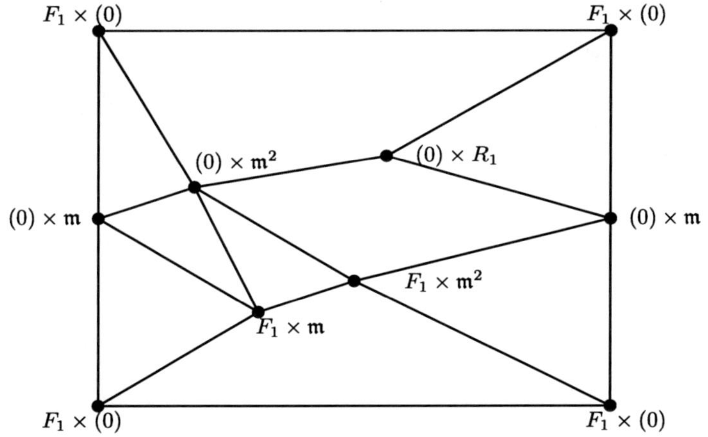

Fig. 4 Toroidal embedding of $AG_{\Omega}(F_1 \times R_1)$, at (8, −3.4) **where** R_1 **has exactly two nonzero proper ideals**

Proof Since $\Omega(R)$ is a spanning subgraph of $AG_{\Omega}(R)$, the proof follows from Theorem 17, Lemma 6, and Fig. 4.

References

1. Aalipour, G., Akbari, S., Behboodi, M., Nikandish, R., Nikmehr, M.J., Shaveisi, F.: The classification of the annhilating-ideal graph of commutative rings. Algebra Colloq. **21**(2), 249–256 (2014)
2. Aalipour, G., Akbari, S., Nikandish, R., Nikmehr, M.J., Shaveisi, F.: On the coloring of the annhihilating-ideal graph of a commutative ring. Discret. Math. **312**, 2620–2626 (2012)
3. Aalipour, G., Akbari, S., Nikandish, R., Nikmehr, M.J., Shaveisi, F.: Minimal prime ideals and cycles in annhilating-ideal graphs. Rocky Mountain J. Math. **5**, 1415–1425 (2013)
4. Ahrari, M., Safari Sabet, S.A., Amini, B.: On the girth of the annihilating-ideal graph of a commutative ring. J. Linear and Top. Algebra 209–216 (2015)
5. Akbari, S., Mohammadian, A.: On the zero-divisor graph of a commutative ring. J. Algebra **274**, 847–855 (2004)
6. Akbari, S., Mohammadian, A.: Zero-divisor graph of a non-commutative rings. J. Algebra **296**, 462–479 (2006)
7. Aliniaeifard, F., Behboodi, M., Mehdi-Nezhad, E., Rahimi, A.M.: On the Diameter and Girth of an Annihilating-Ideal Graph (2014). arXiv:14114163v1 [math.RA]
8. Anderson, D.F., Livingston, P.S.: The zero-divisor graph of a commutative ring. J. Algebra **217**, 434–447 (1999)
9. Anderson, D.F., Grace McClurkin, D.F.: Generalizations of the zero-divisor graph. Int. Electron. J. Algebra **27**, 237–262 (2020)
10. Anderson, D.D., Naseer, M.: Beck's coloring of a commutative ring. J. Algebra **159**, 500–514 (1993)
11. Atiyah, M.F., Macdonald, I.G. : Introduction to Commutative Algebra. Addison-Wesley Publishing Company (1969)
12. Beck, I.: Coloring of commutative rings. J. Algebra **116**, 208–226 (1988)
13. Behboodi, M., Rakeei, Z.: The annihilating-ideal graph of commutative ring I. J. Algebra App. **10**, 727–739 (2011)
14. Behboodi, M., Rakeei, Z.: The annihilating-ideal graph of commutative ring II. J. Algebra App. **10**, 741–753 (2011)
15. Kala, R., Kavitha, S.: Sum annihilating ideal graph of a commutative ring. Palestine J. Math **5**(2), 282–290 (2016)
16. Mohar, B., Thomassen, C.: Graphs on Surfaces. The Johns Hopkins University Press, Baltimore and London (1956)
17. Visweswaran, S., Patel, H.D.: A graph associated with the set of all nonzero annihilating ideals. Discret. Math., Algorithms Appl. **6**(4), 1450047, 22 (2014)
18. West, D.B.: Introduction to Graph Theory, 2nd edn. Prentice-Hall of India, New Delhi (2002)
19. White, A.T.: Graphs. Groups and Surfaces, North-Holland, Amsterdam (1973)

Spectrum of the 3-zero-divisor Hypergraph of Some Classes of Local Rings

K. Selvakumar, J. Beautlin Jemi, and Nadeem ur Rehman

Abstract In this paper, we initiate the study of the spectrum of the 3-zero-divisor hypergraph of commutative rings. We first compute the adjacency matrix of this hypergraph for some classes of local rings. Also, we discuss the spectrum of the 3-zero-divisor hypergraph $\mathscr{H}_3(R)$ of some local rings R. Furthermore, we discuss the Laplacian matrix and their spectrum of $\mathscr{H}_3(R)$.

Keywords k-zero-divisor hypergraph · Adjacency matrix · Laplacian matrix · Spectrum · Energy

1 Introduction

The study linking commutative ring theory with graph theory has been started with the concept of the zero-divisor graph of a commutative ring. This definition was introduced by Beck, Anderson, and Livingston in [2, 3]. In view of this, Eslahchi and Rahimi [6] have introduced and investigated a graph called the k-zero-divisor hypergraph of a commutative ring and later was studied extensively in [8–10]. In this paper, we initiate the study of the spectrum of the 3-zero-divisor hypergraph of commutative rings. We first compute the adjacency matrix of this hypergraph for some classes of local rings. Also, we discuss the spectrum of the 3-zero-divisor hypergraph $\mathscr{H}_3(R)$ of some local rings R. Furthermore, we discuss the Laplacian matrix and their spectrum of $\mathscr{H}_3(R)$.

K. Selvakumar (✉) · J. B. Jemi
Department of Mathematics, Manonmaniam Sundaranar University, Abishekapatti, Tirunelveli 627 012, Tamil Nadu, India
e-mail: selva_158@yahoo.co.in

J. B. Jemi
e-mail: beautlinjemi09@gmail.com

N. Rehman
Department of Mathematics, Aligarh Muslim University, Aligarh 202002, India
e-mail: nu.rehman.mm@amu.ac.in

M. Ashraf et al. (eds.), *Algebra and Related Topics with Applications*,
Springer Proceedings in Mathematics & Statistics 392,
https://doi.org/10.1007/978-981-19-3898-6_35

A hypergraph $\mathscr{H}$ is a pair $(V(\mathscr{H}), E(\mathscr{H}))$ of disjoint sets, where $V(\mathscr{H})$ is a non empty finite set whose elements are called vertices and the elements of $E(\mathscr{H})$ are nonempty subsets of $V(\mathscr{H})$ called edges. The hypergraph $\mathscr{H}$ is called k-uniform whenever every edge e of $\mathscr{H}$ is of size k. The number of edges containing a vertex $v \in V(\mathscr{H})$ is its degree $d_{\mathscr{H}}(v)$. The *adjacency matrix* $A(\mathscr{H})$ of $\mathscr{H}$ is a square matrix whose rows and columns are indexed by the vertices of $\mathscr{H}$ and for all $x, y \in V(\mathscr{H})$, whose entry $a_{x,y}$ is defined as follows:

$$a_{x,y} = \begin{cases} |\{e \in E(\mathscr{H}) \ : \ x, y \in e\}| & x \neq y \\ 0 & \text{otherwise} \end{cases}$$

The spectrum of $\mathscr{H}$ is defined as

$$\sigma(\mathscr{H}) = (\ \lambda_1(\mathscr{H}),\ \lambda_2(\mathscr{H}),\ \ldots\ ,\ \lambda_t(\mathscr{H})\ )$$

where $\lambda_1(\mathscr{H}) \leq \lambda_2(\mathscr{H}) \leq \cdots \leq \lambda_t(\mathscr{H})$ are the eigenvalues of $A(\mathscr{H})$ and $t = |V(\mathscr{H})|$. If $\lambda_1, \lambda_2, \ldots, \lambda_s$ are distinct eigenvalues of $A(\mathscr{H})$ and their multiplicities are $m(\lambda_1), \cdots, m(\lambda_s)$, then we shall write

$$Spec(\mathscr{H}) = \begin{pmatrix} \lambda_1 & \lambda_2 & \cdots & \lambda_s \\ m(\lambda_1) & m(\lambda_2) & \cdots & m(\lambda_s) \end{pmatrix}$$

The *Laplacian matrix* of $\mathscr{H}$ is defined as $L(\mathscr{H}) = D(\mathscr{H}) - A(\mathscr{H})$, where

$$D(\mathscr{H}) = diag\ (\ D(x_1),\ D(x_2),\ \ldots\ ,\ D(x_n)\ ) \text{ and } D(x) = \sum_{y \in V} a_{x,y}$$

is the degree of x in $\mathscr{H}$. A hypergraph is called *integral hypergraph* if all the eigenvalues of its adjacency matrix are integers. The *energy* of a hypergraph $\mathscr{H}$ is defined as the sum of the absolute values of all the eigenvalues. Also, A^t denotes the transpose matrix of A. For basic definitions on hypergraphs, one may refer [1, 4].

2 Adjacency Matrix of $\mathscr{H}_3(\mathbb{Z}_{p^n})$

First, we obtain the adjacency matrix of the 3-zero-divisor hypergraph $\mathscr{H}_3(\mathbb{Z}_{p^n})$ when $3 \leq n \leq 7$, $n \in \mathbb{N}$ and p is a prime number.

From the definition of k-zero-divisor, for any $k \geq 3$, $Z(\mathbb{Z}_8, k) = \emptyset$ and so $\mathscr{H}_k(\mathbb{Z}_8)$ is the empty graph. In view of this, throughout this paper, we assume that $\mathbb{Z}_{p^n}$ is a ring, where p is a prime number, $p^n \neq 8$ and $n \geq 3$.

Proposition 1 ([8]) *Let* $R = \mathbb{Z}_{p^n}$, *where* p *is a prime number,* $p^n \neq 8$ *and* $n \geq 3$. *Then* $|Z(R, 3)| = p\left(p^{n-2} - 1\right)$.

Theorem 1 *Let $R = \mathbb{Z}_{p^3}$, where $p \geq 3$. Then the adjacency matrix of $\mathscr{H}_3(R)$ is*

$$A(\mathscr{H}_3(R)) = \begin{bmatrix} 0 & p(p-1)-2 & \cdots & p(p-1)-2 \\ p(p-1)-2 & 0 & \cdots & p(p-1)-2 \\ \vdots & \vdots & \ddots & \vdots \\ p(p-1)-2 & p(p-1)-2 & \cdots & 0 \end{bmatrix}$$

Proof By Proposition 1, $Z(R,3) = \{kp : 1 \leq k \leq p^2 - 1 \text{ and } p \nmid k\}$ with cardinality $m = p(p-1)$. For any distinct $x, y, z \in Z(R,3)$, $xyz = 0$ and by definition of 3-zero-divisor, $\{x, y, z\}$ is an edge of $\mathscr{H}_3(R)$ and so $\mathscr{H}_3(R)$ is complete. Hence, all the entries in the $m \times m$ adjacency matrix of $\mathscr{H}_3(R)$ other than the diagonal entries are $p(p-1)-2$.

Theorem 2 *Let $R = \mathbb{Z}_{p^4}$, where $p \geq 2$. Then the adjacency matrix of $\mathscr{H}_3(R)$ is*

$$A(\mathscr{H}_3(R)) = \left[\begin{array}{cccc|c} 0 & p(p-1) & \cdots & p(p-1) & \\ p(p-1) & 0 & \cdots & p(p-1) & K_{p^2(p-1)\times p(p-1)} \\ \vdots & \vdots & \ddots & \vdots & \\ p(p-1) & p(p-1) & \cdots & 0 & \\ \hline & & K^t_{p(p-1)\times p^2(p-1)} & & O_{p(p-1)\times p(p-1)} \end{array}\right]$$

where O is the $p(p-1) \times p(p-1)$ zero matrix, $K = [b_{ij}]$ is the $p^2(p-1) \times p(p-1)$ matrix with $b_{ij} = p^2(p-1) - 1$ for all i, j and K^t denotes the transpose matrix of K.

Proof By Proposition 1, $Z(R,3) = \{kp : 1 \leq k \leq p^3 - 1 \text{ and } p^2 \nmid k\}$ with cardinality $p(p^2-1)$. Consider $Z(R,3) = P \cup Q$, where $P = \{kp : 1 \leq k \leq p^3 - 1 \text{ and } p \nmid k\}$, and $Q = \{\ell p^2 : 1 \leq \ell \leq p^2 - 1 \text{ and } p^2 \nmid \ell\}$. Then $|P| = p^2(p-1)$ and $|Q| = p(p-1)$. Also, $P \cap Q = \emptyset$.

Note that $E(\mathscr{H}_3(R)) = \{\, \{x, y, z\} : x, y \in P, z \in Q \,\}$. Let $x, y \in Z(R,3)$. If $x, y \in P$, then $a_{x,y} = |\{\{x, y, z\} : z \in Q\}| = p(p-1)$. If $x \in P$, $y \in Q$ or $x \in Q$, $y \in P$, then $a_{x,y} = |\{\{x, y, z\} : z \in P\}| = p^2(p-1) - 1$. If $x, y \in Q$, then $xy = 0$. Hence, $x, y \notin e$ for all $e \in E(\mathscr{H}_3(R))$ and so $a_{x,y} = 0$. Hence, we rearrange the 3-zero-divisors such that all the elements of P appear first and then the elements of Q, we get the required adjacency matrix.

Theorem 3 *Let $R = \mathbb{Z}_{p^5}$, where $p \geq 2$. Then the adjacency matrix of $\mathscr{H}_3(R)$ is*

$$A(\mathscr{H}_3(R)) = \begin{bmatrix} A & B & C \\ B^t & D & O \\ C^t & O & O \end{bmatrix}$$

where $A = [a_{ij}]$ is the $p^3(p-1) \times p^3(p-1)$ with $a_{ii} = 0$ and $a_{lt} = p(p-1)$ for $l \neq t$,
$B = [b_{ij}]$ is the $p^3(p-1) \times p^2(p-1)$ matrix with $b_{ij} = p^2(p-1) - 1$,
$C = [c_{ij}]$ is the $p^3(p-1) \times p(p-1)$ matrix with $c_{ij} = p^3(p-1) - 1$ and

$D = [d_{ij}]$ is the $p^2(p-1) \times p^2(p-1)$ matrix with $d_{lt} = p^3(p-1) + p^2(p-1) - 2$ for all $t \neq l$, $d_{ii} = 0$, and O denotes the zero matrix.

Proof By Proposition 1, $Z(R,3) = \{kp : 1 \leq k \leq p^4 - 1 \text{ and } p^3 \nmid k\}$ with cardinality $p(p^3-1)$. Consider $Z(R,3) = P_1 \cup P_2 \cup P_3$, where $P_1 = \{kp : 1 \leq k \leq p^4 - 1 \text{ and } p \nmid k\}$, $P_2 = \{\ell p^2 : 1 \leq \ell \leq p^3 - 1 \text{ and } p^2 \nmid \ell\}$, and $P_3 = \{mp^3 : 1 \leq m \leq p^2 - 1 \text{ and } p^3 \nmid m\}$. Then $P_i \cap P_j = \emptyset$ for all $i \neq j$. Also $|P_1| = p^3(p-1)$, $|P_2| = p^2(p-1)$ and $|P_3| = p(p-1)$.

Let $x, y \in Z(R,3)$. If $x, y \in P_1$, then $a_{x,y} = |\{\{x,y,z\} : z \in P_3\}| = p(p-1)$. If $x \in P_1, y \in P_2$ or $x \in P_2, y \in P_1$, then $a_{x,y} = |\{\{x,y,z\} : z \in P_2\}| = p^2(p-1) - 1$. If $x \in P_1, y \in P_3$ or $x \in P_3, y \in P_1$, then $a_{x,y} = |\{\{x,y,z\} : z \in P_1\}| = p^3(p-1) - 1$. If $x, y \in P_2$, then $a_{x,y} = |\{\{x,y,z\} : z \in P_1\}| + |\{\{x,y,z\} : z \in P_2\}| = p^3(p-1) + p^2(p-1) - 2$. If $x \in P_2, y \in P_3$ or $x \in P_3, y \in P_2$ or $x, y \in P_3$, then $xy = 0$. Hence, $x, y \notin e$ for all $e \in E(\mathcal{H}_3(R))$ and so $a_{x,y} = 0$. Now we rearrange the 3-zero-divisors such that all the elements of P_1 appear first, P_2 appear second, and then the elements of P_3, we get the required adjacency matrix.

By modifying the proof of Theorems 2 and 3, we can prove Theorems 4 and 5.

Theorem 4 *Let $R = \mathbb{Z}_{p^6}$, where $p \geq 2$. Then the adjacency matrix of $\mathcal{H}_3(R)$ is*

$$A(\mathcal{H}_3(R)) = \begin{bmatrix} A & B & C & D \\ B^t & E & F & O \\ C^t & F^t & O & O \\ D^t & O & O & O \end{bmatrix}$$

where $A = [a_{ij}]$ is the $p^4(p-1) \times p^4(p-1)$ with $a_{ii} = 0$ and $a_{lt} = p(p-1)$ for $l \neq t$,
$B = [b_{ij}]$ is the $p^4(p-1) \times p^3(p-1)$ matrix with $b_{ij} = p^2(p-1)$,
$C = [c_{ij}]$ is the $p^4(p-1) \times p^2(p-1)$ matrix with $c_{ij} = p^3(p-1)$,
$D = [d_{ij}]$ is the $p^4(p-1) \times p(p-1)$ matrix with $d_{ij} = p^4(p-1) - 1$,
$E = [e_{ij}]$ is the $p^3(p-1) \times p^3(p-1)$ matrix with $e_{lt} = p^2(p-1) + p^3(p-1) - 2$ for all $t \neq l$ and $e_{ii} = 0$,
$F = [f_{ij}]$ is the $p^3(p-1) \times p^2(p-1)$ matrix with $f_{ij} = p^4(p-1) + p^3(p-1) - 1$,
$O = [o_{ij}]$ is the zero matrix.

Theorem 5 *Let $R = \mathbb{Z}_{p^7}$ where $p \geq 2$. Then the adjacency matrix of the hypergraph $\mathcal{H}_3(R)$ is*

$$A(\mathcal{H}_3(R)) = \begin{bmatrix} A & B & C & D & E \\ B^t & F & G & H & O \\ C^t & G^t & K & O & O \\ D^t & H^t & O & O & O \\ E^t & O & O & O & O \end{bmatrix}$$

where $A = [a_{ij}]$ is the $p^5(p-1) \times p^5(p-1)$ with $a_{ii} = 0$ and $a_{lt} = p(p-1)$ for $l \neq t$,

$B = [b_{ij}]$ is the $p^5(p-1) \times p^4(p-1)$ matrix with $b_{ij} = p^2(p-1)$,
$C = [c_{ij}]$ is the $p^5(p-1) \times p^3(p-1)$ matrix with $c_{ij} = p^3(p-1) - 1$,
$D = [d_{ij}]$ is the $p^5(p-1) \times p^2(p-1)$ matrix with $d_{ij} = p^4(p-1)$,
$E = [e_{ij}]$ is the $p^5(p-1) \times p(p-1)$ matrix with $e_{ij} = p^5(p-1) - 1$,
$F = [f_{ij}]$ is the $p^4(p-1) \times p^4(p-1)$ matrix with $f_{lt} = p^2(p-1) + p^3(p-1)$ for all $t \neq l$ and $f_{ii} = 0$,
$G = [g_{ij}]$ is the $p^4(p-1) \times p^3(p-1)$ matrix with $f_{ij} = (p^3(p-1) - 1) + (p^4(p-1) - 1)$,
$H = [h_{ij}]$ is the $p^4(p-1) \times p^2(p-1)$ matrix with $h_{ij} = p^5(p-1) + p^4(p-1) - 1$,
$K = [k_{ij}]$ is the $p^3(p-1) \times p^3(p-1)$ matrix with $k_{lt} = p^5(p-1) + p^4(p-1) + p^3(p-1) - 2$ for all $t \neq l$ and $k_{ii} = 0$, O denotes the zero matrix.

3 Spectrum of $\mathscr{H}_3(R)$

In this section, we discuss the spectrum of the 3-zero-divisor hypergraph $\mathscr{H}_3(\mathbb{Z}_{p^n})$ for $n = 3$ and $n = 4$. Furthermore, we determine the energy of $\mathscr{H}_3(\mathbb{Z}_{p^n})$.

Theorem 6 ([7]) *Let M_1, M_2, M_3 and M_4 be, respectively, $p \times p$, $p \times q$, $q \times p$, and $q \times q$ matrices with M_1 and M_4 invertible. Then*

$$\det \begin{bmatrix} M_1 & M_2 \\ M_3 & M_4 \end{bmatrix} = \det(M_4)\det(M_1 - M_2M_4^{-1}M_3) = \det(M_1)\det(M_4 - M_3M_1^{-1}M_2)$$

Theorem 7 *Let $R = \mathbb{Z}_{p^3}$, where $p \geq 3$. Then*

$$Spec(\mathscr{H}_3(R)) = \begin{pmatrix} (p^2-p-2)(p^2-p-1) & -(p^2-p-2) \\ 1 & p(p-1)-1 \end{pmatrix}$$

and $\mathscr{H}_3(R)$ is integral.

Proof By Theorem 1, the adjacency matrix of $\mathscr{H}_3(R)$ is given by

$$A(\mathscr{H}_3(R)) = \begin{bmatrix} 0 & p(p-1)-2 & \cdots & p(p-1)-2 \\ p(p-1)-2 & 0 & \cdots & p(p-1)-2 \\ \vdots & \vdots & \ddots & \vdots \\ p(p-1)-2 & p(p-1)-2 & \cdots & 0 \end{bmatrix}$$

Let λ be a non-zero eigenvalue of $A(\mathscr{H}_3(R))$. Then

$$\det(A(\mathscr{H}_3(R)) - \lambda I) = \det\begin{bmatrix} -\lambda & p(p-1)-2 & \cdots & p(p-1)-2 \\ p(p-1)-2 & -\lambda & \cdots & p(p-1)-2 \\ \vdots & \vdots & \ddots & \vdots \\ p(p-1)-2 & p(p-1)-2 & \cdots & -\lambda \end{bmatrix} = 0$$

Using the properties of the determinant, we get

$$n \quad \det\begin{bmatrix} -1 & 0 & \cdots & 0 & 1 & 0 & \cdots & 0 \\ 0 & -1 & \cdots & 0 & 0 & 1 & \cdots & 0 \\ \vdots & \vdots & \ddots & \vdots & \vdots & \vdots & \ddots & \vdots \\ 0 & 0 & \cdots & -1 & 0 & 0 & \cdots & 1 \\ 1 & 1 & \cdots & 1 & \frac{-\lambda}{p^2-p-2} & 1 & \cdots & 1 \\ 1 & 1 & \cdots & 1 & 1 & \frac{-\lambda}{p^2-p-2} & \cdots & 1 \\ \vdots & \vdots & \ddots & \vdots & \vdots & \vdots & \ddots & \vdots \\ 1 & 1 & \cdots & 1 & 1 & 1 & \cdots & \frac{-\lambda}{p^2-p-2} \end{bmatrix} = 0 \quad \text{where} \quad n =$$

$(p^2 - p - 2)^{p(p-1)}\left(\frac{\lambda}{p^2-p-2} + 1\right)^{\frac{p(p-1)}{2}}$. Then by Theorem 6,

$\det(A - \lambda I) = \left(p^2 - p - 2\right)^{p(p-1)} \left(\frac{\lambda}{p^2-p-2} + 1\right)^{\frac{p(p-1)}{2}} \det\begin{bmatrix} P & Q \\ R & S \end{bmatrix}$. Then

$$\det(A - \lambda I) = \left(p^2 - p - 2\right)^{p(p-1)} \left(\frac{\lambda}{p^2-p-2} + 1\right)^{\frac{p(p-1)}{2}}$$
$$\det(P)\det(S - RP^{-1}Q) = 0.$$

Since P is a scalar matrix of order $\frac{p(p-1)}{2}$, we get $\det(P) = (-1)^{\frac{p(p-1)}{2}}$ and

$$RP^{-1}Q = \begin{bmatrix} -1 & -1 & \cdots & -1 \\ -1 & -1 & \cdots & -1 \\ \vdots & \vdots & \ddots & \vdots \\ -1 & -1 & \cdots & -1 \end{bmatrix}.$$

$$\text{Thus } \det(S - RP^{-1}Q) = \det\begin{bmatrix} \frac{-\lambda}{p^2-p-2} + 1 & 2 & \cdots & 2 \\ 2 & \frac{-\lambda}{p^2-p-2} + 1 & \cdots & 2 \\ \vdots & \vdots & \ddots & \vdots \\ 2 & 2 & \cdots & \frac{-\lambda}{p^2-p-2} + 1 \end{bmatrix}$$

Using the properties of the determinant, we get
$\det(S - RP^{-1}Q)$

$$= (-1)^{\frac{p(p-1)}{2}-3}\left(\frac{\lambda}{p^2-p-2} + 1\right)^{\frac{p(p-1)}{2}-1}\left[\frac{-\lambda}{p^2-p-2} + p\,(p-1) - 1\right].$$

Therefore,

$$\det(A - \lambda I) = \left(\lambda + p^2 - p - 2\right)^{p(p-1)-1}\left(\lambda - p\,(p-1)\left(p^2 - p - 2\right)\right.$$
$$\left. + p^2 - p - 2\right) = 0.$$

and so $\left(\lambda + p^2 - p - 2\right)^{p(p-1)-1} = 0$ or $\lambda - p\,(p-1)\left(p^2 - p - 2\right) + p^2 - p -$

$2 = 0$.

Hence, $Spec(\mathscr{H}_3(R)) = \begin{pmatrix} \left(p^2-p-2\right)\left(p^2-p-1\right) & -\left(p^2-p-2\right) \\ 1 & p(p-1)-1 \end{pmatrix}$.

Corollary 1 *Let $R = \mathbb{Z}_{p^3}$, where $p \geq 3$. Then the energy of $\mathscr{H}_3(\mathbb{Z}_{p^3})$ is*

$$2\left(p^2-p-1\right)\left(p^2-p-2\right).$$

Proof By Theorem 7, the Energy of $\mathscr{H}_3(\mathbb{Z}_{p^3})$

$$\begin{aligned} &= |\lambda_1| + |\lambda_2| + \cdots + \left|\lambda_{p(p-1)}\right| \\ &= \underbrace{\left|-\left(p^2-p-2\right)\right| + \cdots + \left|-\left(p^2-p-2\right)\right|}_{p(p-1)-1 \ times} + \left|\left(p^2-p-2\right)\left(p^2-p-1\right)\right| \\ &= (p(p-1)-1)\left(p^2-p-2\right) + \left(p^2-p-2\right)\left(p^2-p-1\right) \\ &= 2\left(p^2-p-1\right)\left(p^2-p-2\right). \end{aligned}$$

Theorem 8 *Let $R = \mathbb{Z}_{p^4}$, where $p \geq 2$. Then*

$$Spec(\mathscr{H}_3(R)) = \begin{pmatrix} 0 & a & b & -p(p-1) \\ p(p-1)-1 & 1 & 1 & p^2(p-1)-1 \end{pmatrix}$$

where $a = \frac{p(p-1)[p^2(p-1)-1]}{2}\left(1+\sqrt{1+4p}\right)$ *and* $b = \frac{p(p-1)[p^2(p-1)-1]}{2}\left(1-\sqrt{1+4p}\right)$.

Proof By Theorem 2, the adjacency matrix of $\mathscr{H}_3(R)$ is given by,

$$A(\mathscr{H}_3(R)) = \left[\begin{array}{cccc|c} 0 & p(p-1) & \cdots & p(p-1) & \\ p(p-1) & 0 & \cdots & p(p-1) & K_{p^2(p-1)\times p(p-1)} \\ \vdots & \vdots & \ddots & \vdots & \\ p(p-1) & p(p-1) & \cdots & 0 & \\ \hline & K^t_{p(p-1)\times p^2(p-1)} & & & O_{p(p-1)\times p(p-1)} \end{array}\right]$$

where O is the $p(p-1) \times p(p-1)$ zero matrix, $K = [b_{ij}]$ is the $p^2(p-1) \times p(p-1)$ matrix with $b_{ij} = p^2(p-1) - 1$ for all i, j.

Let λ be a non-zero eigenvalue of $A(\mathscr{H}_3(\mathbb{Z}_{p^4}))$. Then by Theorem 6,

$$\det(A - \lambda I) = \det\begin{bmatrix} P & Q \\ R & S \end{bmatrix} = \det(S)\det(P - QS^{-1}R) = 0$$

where $P = \begin{bmatrix} -\lambda & p(p-1) & \cdots & p(p-1) \\ p(p-1) & -\lambda & \cdots & p(p-1) \\ \vdots & \vdots & \ddots & \vdots \\ p(p-1) & p(p-1) & \cdots & -\lambda \end{bmatrix}$; $S = \begin{bmatrix} -\lambda & 0 & \cdots & 0 \\ 0 & -\lambda & \cdots & 0 \\ \vdots & \vdots & \ddots & \vdots \\ 0 & 0 & \cdots & -\lambda \end{bmatrix}$

$$Q = R^t = \begin{bmatrix} p^2(p-1)-1 & p^2(p-1)-1 & \cdots & p^2(p-1)-1 \\ p^2(p-1)-1 & p^2(p-1)-1 & \cdots & p^2(p-1)-1 \\ \vdots & \vdots & \ddots & \vdots \\ p^2(p-1)-1 & p^2(p-1)-1 & \cdots & p^2(p-1)-1 \end{bmatrix}$$

Clearly, $\det(S) = (-\lambda)^{p(p-1)}$ and $QS^{-1}R =$

$$\begin{bmatrix} \frac{-(p(p-1))}{\lambda}\left[(p^2(p-1)-1)^2\right] & \frac{-(p(p-1))}{\lambda}\left[(p^2(p-1)-1)^2\right] & \cdots & \frac{-(p(p-1))}{\lambda}\left[(p^2(p-1)-1)^2\right] \\ \frac{-(p(p-1))}{\lambda}\left[(p^2(p-1)-1)^2\right] & \frac{-(p(p-1))}{\lambda}\left[(p^2(p-1)-1)^2\right] & \cdots & \frac{-(p(p-1))}{\lambda}\left[(p^2(p-1)-1)^2\right] \\ \vdots & \vdots & \ddots & \vdots \\ \frac{-(p(p-1))}{\lambda}\left[(p^2(p-1)-1)^2\right] & \frac{-(p(p-1))}{\lambda}\left[(p^2(p-1)-1)^2\right] & \cdots & \frac{-(p(p-1))}{\lambda}\left[(p^2(p-1)-1)^2\right] \end{bmatrix}$$

and hence

$$\det(P - QS^{-1}R) = \det \begin{bmatrix} x & y & \cdots & y \\ y & x & \cdots & y \\ \vdots & \vdots & \ddots & \vdots \\ y & y & \cdots & x \end{bmatrix}$$

where $x = -\lambda + \frac{p(p-1)}{\lambda}[(p^2(p-1)-1)^2]$ and $y = p(p-1) + \frac{p(p-1)}{\lambda}[(p^2(p-1)$ $-1)^2]$.

Using the properties of the determinant, we get
$\det(P - QS^{-1}R) = (-1)^{p^2(p-1)-2}[\lambda + p(p-1)]^{p^2(p-1)-1} \quad \frac{1}{\lambda}[\lambda^2 - p(p-1)$ $[p^2(p-1)-1]\lambda - p(p-1)[p^2(p-1)]\,(p^2(p-1)-1)^2]$.

Therefore,
$\det(A - \lambda I) = (-\lambda)^{p(p-1)-1}\,[\lambda + p(p-1)]^{p^2(p-1)-1}$
$[\lambda^2 - p(p-1)[p^2(p-1)-1]\lambda - p(p-1)[p^2(p-1)]\,(p^2(p-1)$ $-1)^2] = 0$.

Since $\lambda \neq 0$, we have

$$[\lambda + p(p-1)]^{p^2(p-1)-1} = 0$$

or

$$\lambda^2 - p(p-1)(p^2(p-1)-1)\lambda - p(p-1)(p^2(p-1))\,(p^2(p-1)-1)^2 = 0.$$

Hence

$$Spec(\mathscr{H}_3(R)) = \begin{pmatrix} 0 & a & b & -p\,(p-1) \\ p(p-1)-1 & 1 & 1 & p^2(p-1)-1 \end{pmatrix}$$

where $a = \frac{p(p-1)[p^2(p-1)-1]}{2}\left(1+\sqrt{1+4p}\right)$ and $b = \frac{p(p-1)[p^2(p-1)-1]}{2}\left(1-\sqrt{1+4p}\right)$.

Corollary 2 *Let $R = \mathbb{Z}_{p^4}$, where $p \geq 2$. Then the energy of $\mathscr{H}_3(\mathbb{Z}_{p^4})$ is*

$$2p\,(p-1)\left[p^2\,(p-1)-1\right]$$

Proof By Theorem 8, the Energy of $\mathscr{H}_3(\mathbb{Z}_{p^4})$ is

$$
\begin{aligned}
&= |\lambda_1| + |\lambda_2| + \cdots + \left|\lambda_{p^2(p-1)+p(p-1)}\right| \\
&= \underbrace{|-p\,(p-1)| + \cdots + |-p\,(p-1)|}_{p^2(p-1)-1 \ times} + \left|\frac{p(p-1)[p^2(p-1)-1]}{2}\left(1+\sqrt{1+4p}\right)\right| + \\
&\quad \left|\frac{p(p-1)[p^2(p-1)-1]}{2}\left(1-\sqrt{1+4p}\right)\right| \\
&= 2p\,(p-1)\left[p^2\,(p-1)-1\right].
\end{aligned}
$$

4 Laplacian Spectrum of $\mathscr{H}_3(R)$

In this section, we discuss the Laplacian spectrum of $\mathscr{H}_3(\mathbb{Z}_{p^n})$ for $n = 3$ and $n = 4$.

Theorem 9 *Let $R = \mathbb{Z}_{p^3}$, where $p \geq 3$. Then the Laplacian spectrum of $\mathscr{H}_3(\mathbb{Z}_{p^3})$ is*

$$
LSpec(\mathscr{H}_3(R)) = \begin{pmatrix} 0 & (p(p-1)-2)\,(p(p-1)) \\ 1 & p(p-1)-1 \end{pmatrix}
$$

Proof By Theorem 1, $D(x) = (p(p-1)-1)\,(p(p-1)-2)$ for all $x \in Z(R, 3)$ and by definition of $D(\mathscr{H}_3(R))$,

$$
D = \begin{bmatrix}
(p(p-1)-1)\,(p(p-1)-2) & 0 & \cdots & 0 \\
0 & (p(p-1)-1)\,(p(p-1)-2) & \cdots & 0 \\
\vdots & \vdots & \ddots & \vdots \\
0 & 0 & \cdots & (p(p-1)-1)\,(p(p-1)-2)
\end{bmatrix}
$$

Since $L(\mathscr{H}_3(\mathbb{Z}_{p^3})) = D - A(\mathscr{H}_3(\mathbb{Z}_{p^3}))$ and by Theorem 1, the Laplacian matrix of $\mathscr{H}_3(R)$ is given by

$$
L(\mathscr{H}_3(\mathbb{Z}_{p^3})) = \begin{bmatrix}
y & -(p(p-1)-2) & \cdots & -(p(p-1)-2) \\
-(p(p-1)-2) & y & \cdots & -(p(p-1)-2) \\
\vdots & \vdots & \ddots & \vdots \\
-(p(p-1)-2) & -(p(p-1)-2) & \cdots & y
\end{bmatrix}
$$

where $y = (p(p-1)-1)\,(p(p-1)-2)$.
Let λ be a non-zero eigenvalue of $L(\mathscr{H}_3(\mathbb{Z}_{p^3}))$. Then $\det(L - \lambda I) = 0$ and so

$$
\det \begin{bmatrix}
y & -(p(p-1)-2) & \cdots & -(p(p-1)-2) \\
-(p(p-1)-2) & y & \cdots & -(p(p-1)-2) \\
\vdots & \vdots & \ddots & \vdots \\
-(p(p-1)-2) & -(p(p-1)-2) & \cdots & y
\end{bmatrix} = 0
$$

where $y = (p(p-1)-1)\,(p(p-1)-2)] - \lambda$.
Using the properties of the determinant, we get

$$m\ \det\left[\begin{array}{cccc|ccc} -1 & 0 & \cdots & 0 & 1 & \cdots & 0 \\ 0 & -1 & \cdots & 0 & 0 & \cdots & 0 \\ \vdots & \vdots & \ddots & \vdots & \vdots & \ddots & \vdots \\ 0 & 0 & \cdots & -1 & 0 & \cdots & 1 \\ \hline 1 & 1 & \cdots & 1 & \frac{\lambda-(p(p-1)-1)(p(p-1)-2)}{p(p-1)-2} & \cdots & 1 \\ 1 & 1 & \cdots & 1 & 1 & \cdots & 1 \\ \vdots & \vdots & \ddots & \vdots & \vdots & \ddots & \vdots \\ 1 & 1 & \cdots & 1 & 1 & \cdots & \frac{\lambda-(p(p-1)-1)(p(p-1)-2)}{p(p-1)-2} \end{array}\right] = 0.$$

where $m = [p(p-1)]^{p(p-1)}\left(1 - \frac{[\lambda-(p(p-1)-1)(p(p-1)-2)]}{p(p-1)-2}\right)^{\frac{p(p-1)}{2}}$

Let $\det(L-\lambda I) = [p(p-1)]^{p(p-1)}\left(1 - \frac{[\lambda-(p(p-1)-1)(p(p-1)-2)]}{p(p-1)-2}\right)^{\frac{p(p-1)}{2}} \det\begin{bmatrix} P & Q \\ R & S \end{bmatrix}$.

Then $\det(L-\lambda I) =$

$$[p(p-1)]^{p(p-1)}\left(1 - \frac{[\lambda-(p(p-1)-1)(p(p-1)-2)]}{p(p-1)-2}\right)^{\frac{p(p-1)}{2}} \det(P)\det(S - RP^{-1}Q) = 0.$$

Since P is a scalar matrix of order $\frac{p(p-1)}{2}$, we get $\det(P) = (-1)^{\frac{p(p-1)}{2}}$. Also

$$RP^{-1}Q = \begin{bmatrix} -1 & -1 & \cdots & -1 \\ -1 & -1 & \cdots & -1 \\ \vdots & \vdots & \ddots & \vdots \\ -1 & -1 & \cdots & -1 \end{bmatrix} \text{ and so } \det(S - RP^{-1}Q) =$$

$$\det\begin{bmatrix} \frac{\lambda-(p(p-1)-1)(p(p-1)-2)}{p(p-1)-2}+1 & 2 & \cdots & 2 \\ 2 & \frac{\lambda-(p(p-1)-1)(p(p-1)-2)}{p(p-1)-2}+1 & \cdots & 2 \\ \vdots & \vdots & \ddots & \vdots \\ 2 & 2 & \cdots & \frac{\lambda-(p(p-1)-1)(p(p-1)-2)}{p(p-1)-2}+1 \end{bmatrix}$$

Using the properties of the determinant, we get

$\det(S - RP^{-1}Q) = 2\,(-1)^{\frac{p(p-1)}{2}-3}\left(\frac{\lambda-(p(p-1)-1)(p(p-1)-2)}{p(p-1)-2} - 1\right)^{\frac{p(p-1)}{2}-1}\left[\frac{\lambda}{2(p(p-1)-2)}\right]$.

Therefore, $\det(L-\lambda I) = \lambda\,[\lambda - (p(p-1)-2)\,(p(p-1))]^{p(p-1)-1} = 0$.

Hence, $LSpec\,(\mathscr{H}_3(R)) = \begin{pmatrix} 0 & (p(p-1)-2)\,(p(p-1)) \\ 1 & p(p-1)-1 \end{pmatrix}$.

Theorem 10 *Let $R = \mathbb{Z}_{p^4}$, where $p \geq 2$. Then the Laplacian spectrum of $\mathscr{H}_3(\mathbb{Z}_{p^4})$ is*

$$LSpec\,(\mathscr{H}_3(R)) = \begin{pmatrix} a & b & c & d \\ 1 & 1 & p^2(p-1)-1 & p(p-1)-1 \end{pmatrix}$$

where $a = \frac{p(p-1)[p^2(p-1)-1]}{2}(3+\sqrt{1+4p})$, $b = \frac{p(p-1)[p^2(p-1)-1]}{2}(3-\sqrt{1+4p})$, $c = p(p-1) + 2p(p-1)\left(p^2(p-1)-1\right)$ *and* $d = 2p(p-1)\left(p^2(p-1)-1\right)$.

Proof By Theorem 2, $D(x) = 2p(p-1)\left(p^2(p-1)-1\right)$ for all $x \in Z(R,3)$ and by definition of $D(\mathscr{H}_3(R))$,

$$D = \begin{bmatrix} 2p(p-1)\left(p^2(p-1)-1\right) & 0 & \cdots & 0 \\ 0 & 2p(p-1)\left(p^2(p-1)-1\right) & \cdots & 0 \\ \vdots & \vdots & \ddots & \vdots \\ 0 & 0 & \cdots & 0 \\ 0 & 0 & \cdots & 2p(p-1)\left(p^2(p-1)-1\right) \end{bmatrix}$$

Since $L(\mathscr{H}_3(\mathbb{Z}_{p^3})) = D - A(\mathscr{H}_3(\mathbb{Z}_{p^3}))$ and by Theorem 2, the Laplacian matrix of $\mathscr{H}_3(R)$ is given by

$$L(\mathscr{H}_3(\mathbb{Z}_{p^4})) = \left[\begin{array}{cccc|c} x & -p(p-1) & \cdots & -p(p-1) & \\ -p(p-1) & x & \cdots & -p(p-1) & K_{p^2(p-1)\times p(p-1)} \\ \vdots & \vdots & \ddots & \vdots & \\ -p(p-1) & -p(p-1) & \cdots & x & \\ \hline & K^t_{p(p-1)\times p^2(p-1)} & & & C_{p(p-1)\times p(p-1)} \end{array}\right]$$

where $x = 2p(p-1)\left(p^2(p-1)-1\right)$, $C=[d_{ij}]$ is the $p(p-1)\times p(p-1)$ matrix with $d_{lk} = 2p(p-1)(p^2(p-1)-1)$ for all $l \neq k$ and $d_{ii} = 0$, $K = [b_{ij}]$ is the $p^2(p-1)\times p(p-1)$ matrix with $b_{ij} = -(p^2(p-1)-1)$ for all i, j and K^t denotes the transpose matrix of K.

Let λ be a non-zero eigenvalue of $L(\mathscr{H}_3(\mathbb{Z}_{p^4}))$.

Then $\det(L-\lambda I) = \det\begin{bmatrix} P & Q \\ R & S\end{bmatrix} = \det(S)\det(P - QS^{-1}R) = 0$, where

$$P = \begin{bmatrix} x & -p(p-1) & \cdots & -p(p-1) \\ -p(p-1) & x & \cdots & -p(p-1) \\ \vdots & \vdots & \ddots & \vdots \\ -p(p-1) & -p(p-1) & \cdots & x \end{bmatrix}, \quad x = 2p(p-1)\left(p^2(p-1)\right.$$

$\left.-1\right) - \lambda.$

$$S = \begin{bmatrix} x & 0 & \cdots & 0 \\ 0 & x & \cdots & 0 \\ \vdots & \vdots & \ddots & \vdots \\ 0 & 0 & \cdots & x \end{bmatrix}, x = 2p(p-1)\left(p^2(p-1)-1\right) - \lambda.$$

$$Q = R^t = \begin{bmatrix} -\left(p^2(p-1)-1\right) & -\left(p^2(p-1)-1\right) & \cdots & -\left(p^2(p-1)-1\right) \\ -\left(p^2(p-1)-1\right) & -\left(p^2(p-1)-1\right) & \cdots & -\left(p^2(p-1)-1\right) \\ \vdots & \vdots & \ddots & \vdots \\ -\left(p^2(p-1)-1\right) & -\left(p^2(p-1)-1\right) & \cdots & -\left(p^2(p-1)-1\right) \end{bmatrix}$$

Clearly, $\det(S) = \left(2p(p-1)\left(p^2(p-1)-1\right)-\lambda\right)^{p(p-1)}$; $QS^{-1}R =$

$$\begin{bmatrix} \frac{p(p-1)\left[\left(p^2(p-1)-1\right)^2\right]}{2p(p-1)\left(p^2(p-1)-1\right)-\lambda} & \frac{p(p-1)\left[\left(p^2(p-1)-1\right)^2\right]}{2p(p-1)\left(p^2(p-1)-1\right)-\lambda} & \cdots & \frac{p(p-1)\left[\left(p^2(p-1)-1\right)^2\right]}{2p(p-1)\left(p^2(p-1)-1\right)-\lambda} \\ \frac{p(p-1)\left[\left(p^2(p-1)-1\right)^2\right]}{2p(p-1)\left(p^2(p-1)-1\right)-\lambda} & \frac{p(p-1)\left[\left(p^2(p-1)-1\right)^2\right]}{2p(p-1)\left(p^2(p-1)-1\right)-\lambda} & \cdots & \frac{p(p-1)\left[\left(p^2(p-1)-1\right)^2\right]}{2p(p-1)\left(p^2(p-1)-1\right)-\lambda} \\ \vdots & \vdots & \ddots & \vdots \\ \frac{p(p-1)\left[\left(p^2(p-1)-1\right)^2\right]}{2p(p-1)\left(p^2(p-1)-1\right)-\lambda} & \frac{p(p-1)\left[\left(p^2(p-1)-1\right)^2\right]}{2p(p-1)\left(p^2(p-1)-1\right)-\lambda} & \cdots & \frac{p(p-1)\left[\left(p^2(p-1)-1\right)^2\right]}{2p(p-1)\left(p^2(p-1)-1\right)-\lambda} \end{bmatrix}$$

and hence $\det(P - QS^{-1}R) = \det \begin{bmatrix} x & y & \cdots & y \\ y & x & \cdots & y \\ \vdots & \vdots & \ddots & \vdots \\ y & y & \cdots & x \end{bmatrix}$

where $x = 2p(p-1)\left(p^2(p-1)-1\right) - \lambda - \frac{p(p-1)\left[\left(p^2(p-1)-1\right)^2\right]}{2p(p-1)\left(p^2(p-1)-1\right)-\lambda}$ and

$y = -p(p-1) - \frac{p(p-1)\left[\left(p^2(p-1)-1\right)^2\right]}{2p(p-1)\left(p^2(p-1)-1\right)-\lambda}$.

Using the properties of the determinant, we get

$\det(P - QS^{-1}R) = (-1)^{p^2(p-1)-1}[p(p-1) + 2p(p-1)\left(p^2(p-1)-1\right) - \lambda]^{p^2(p-1)-1} \cdot x$

where $x = \frac{[\lambda^2 - 3p(p-1)[p^2(p-1)-1]\lambda + p(p-1)\left(p^2(p-1)-1\right)^2[2p(p-1) - p^2(p-1)]]}{2p(p-1)\left(p^2(p-1)-1\right)-\lambda}$.

Therefore, $\det(L - \lambda I) = 0$ which implies that

$$\left(2p(p-1)\left(p^2(p-1)-1\right)-\lambda\right)^{p(p-1)-1}[p(p-1)+2p(p-1)\left(p^2(p-1)-1\right)-\lambda]^{p^2(p-1)-1}$$
$$[\lambda^2 - 3p(p-1)[p^2(p-1)-1]\lambda + p(p-1)\left(p^2(p-1)-1\right)^2[2p(p-1) - p^2(p-1)]] = 0.$$

Since $\lambda \neq 0$, we have $\left(2p(p-1)\left(p^2(p-1)-1\right)-\lambda\right)^{p(p-1)-1} = 0$

$$[\lambda^2 - 3p(p-1)[p^2(p-1)-1]\lambda + p(p-1)\left(p^2(p-1)-1\right)^2[2p(p-1) - p^2(p-1)]] = 0,$$
$$[p(p-1)+2p(p-1)\left(p^2(p-1)-1\right)-\lambda]^{p^2(p-1)-1} = 0.$$

Hence

$$LSpec\ (\mathscr{H}_3(R)) = \begin{pmatrix} a & b & c & d \\ 1 & 1 & p^2(p-1)-1 & p(p-1)-1 \end{pmatrix}$$

where $a = \frac{p(p-1)[p^2(p-1)-1]}{2}\left(3+\sqrt{1+4p}\right)$, $b = \frac{p(p-1)[p^2(p-1)-1]}{2}\left(3-\sqrt{1+4p}\right)$, $c = p(p-1) + 2p(p-1)\left(p^2(p-1)-1\right)$ and $d = 2p(p-1)\left(p^2(p-1)-1\right)$.

Problem 1 *1. Find the Spectrum of $\mathscr{H}_3(\mathbb{Z}_{p^n})$ for $n = 5, 6,$ and 7.*
2. Find the Laplacian Spectrum of $\mathscr{H}_3(\mathbb{Z}_{p^n})$ for $n = 5, 6,$ and 7.

Acknowledgements The authors are deeply grateful to the referee for careful reading of the manuscript and helpful suggestions. This research work is supported by the DST-SERB Project No. EEQ/2016/000367 of Science and Engineering Research Board, Government of India, for the first author.

References

1. Bretto, A.: Hypergraph Theory An Introduction. Springer, Londan (2013)
2. Anderson, D.F., Livingston, P.S.: The zero-divisor graph of a commutative ring. J. Algebra **217**, 434–447 (1999)
3. Beck, I.: Coloring of commutative rings. J. Algebra **116**, 208–226 (1988)
4. Berge, C.: Graphs and Hypergraphs. North-Holland Publishing Company, Londan (2003)
5. Cvetković, D., Rowlinson, P., Simić, S.: An Introduction to the Theory of Graph Spectra. Cambridge University Press, New York (2010)
6. Eslahchi, C., Rahimi, A.M.: The *k*-zero divisor hypergraph of a commutative ring. Int. J. Math. Math. Sci. **50875**, 15 (2007)
7. Silvester, J.R.: Determinants of block matrices. Math. Gaz. **84**(501), 460–467 (2000)
8. Selvakumar, K., Beautlin Jemi, J.: The Wiener Index of the 3-zero-divisor Hypergraph. In: AIP Conference Proceedings 2261 (2020)
9. Selvakumar, K., Ramanathan, V.: Classification of nonlocal rings with genus one 3-zero-divisor hypergraphs. Comm. Algebra **45**(1), 275–284 (2016)
10. Tamizh Chelvam, T., Selvakumar, K., Ramanathan, V.: On the planarity of the *k*-zero-divisor hypergraphs. AKCE Int. J. Graphs Combin. **12**(2), 169–179 (2015)

Complement of the Generalized Total Graph of Commutative Rings – A Survey

T. Tamizh Chelvam

Abstract There are so many graph constructions from algebraic structures. In particular, graphs from commutative rings are extensively studied. Zero-divisor graphs from commutative rings are the first graph construction in this regard. In the zero divisor graph of a commutative ring, edges are constructed through multiplication of the underlying ring. In variation to this, several graphs are constructed using addition of a commutative ring. The first graph construction using addition is the total graph and later generalized total graphs from commutative rings are introduced and studied. In this paper, we make a survey of results obtained on the complement of the generalized total graph of commutative rings as well as fields.

Keywords Total graph · Cayley graph · Unitary cayley graph · Generalized total graph · Domination number · Dominating sets · Intersection graphs

1 Introduction

Construction of graphs from commutative rings was started with the study initiated by Beck [14] through zero-divisor graphs of commutative rings. Since then, many graphs are constructed from commutative rings and through these graphs various properties of commutative rings are studied by several authors [5, 14]. There are so many ways to construct graphs from ring structures. Through these constructions, the interplay between algebraic structures and graphs are studied. Indeed, it is worthwhile to relate algebraic properties of commutative rings to combinatorial properties of derived graphs. In the case of zero-divisor graphs, multiplication of the ring is used for adjacency(edges) of vertices in the graph. In variation to this, the addition of the ring is used to construct edges in the total graph. Several authors studied the total graph of commutative rings [3, 4, 7–12, 21, 23–30].

T. Tamizh Chelvam (✉)
Department of Mathematics, Manonmaniam Sundaranar University,
Tirunelveli 627012, Tamil Nadu, India
e-mail: tamche59@gmail.com

M. Ashraf et al. (eds.), *Algebra and Related Topics with Applications*,
Springer Proceedings in Mathematics & Statistics 392,
https://doi.org/10.1007/978-981-19-3898-6_36

2 Additive Graphs from Commutative Rings

Throughout this paper, R denotes a commutative with identity 1, $Z(R)$ be its set of all zero-divisors, $U(R)$ be the multiplicative group of unit elements of R. Cayley graphs is the first graph construction from finite groups. In similar to the Cayley graph of finite groups, several graphs are defined through addition of R. Cayley graph of commutative rings were introduced and studied by Akhtar et al. [2]. Actually the Cayley graph of R, denoted by $Cay(R, U(R))$, is the simple undirected graph with vertex set R and two distinct vertices x and y are adjacent if $x - y \in U(R)$. Subsequently several authors studied about Cayley graph from commutative rings [1, 2, 19, 20]. Further properties about unitary Cayley graph are studied in [22]. Later on, Ashrafi et al. [6] introduced the *unit graph* of R, denoted by $G(R)$. Actually $G(R)$ is the graph obtained by setting all the elements of R to be the vertices and defining distinct vertices x and y to be adjacent if $x + y \in U(R)$. Anderson and Badawi [3] introduced and studied the total graph $T_\Gamma(R)$ of a commutative ring R. Actually the *total graph* of R is the undirected graph whose vertices are the elements in R and two distinct vertices x and y are adjacent if $x + y \in Z(R)$. The complement of the total graph is denoted by $\overline{T_\Gamma(R)}$. Note that two distinct vertices x and y in $\overline{T_\Gamma(R)}$ are adjacent if $x + y \in Reg(R)$. Several properties of $\overline{T_\Gamma(R)}$ were studied by the authors in [31–33, 38].

Khashyarmanesh et al. [18] introduced the graph $\Gamma(R, G, S)$ where G is a multiplicative subgroup of $U(R)$ and S is a non-empty subset of G such that $S^{-1} = \{s^{-1} : s \in S\} \subseteq S$. The graph $\Gamma(R, G, S)$ is a simple graph with vertex set R and two distinct vertices x and y are adjacent if there exists $s \in S$ such that $x + sy \in G$. Note that all these graphs are not isomorphic. But there are relations between these graphs. For instance, if $S = \{1\}$, then $\Gamma(R, S, U)$ is same as the unit graph $G(R)$ and if $S = \{-1\}$, then $\Gamma(R, S, U)$ is the unitary Cayley graph $Cay(R, U(R))$ of R. Further if R is finite, then the complement $\overline{T_\Gamma(R)}$ of the total graph $T_\Gamma(R)$ is nothing but the unit graph $G(R)$. In case of an infinite ring, $G(R)$ is a subgraph of $\overline{T_\Gamma(R)}$. The classes like $\Gamma(R, U, U)$ were extensively studied by Tamizh Chelvam et al. [34–37].

3 Generalized Total Graphs

A subset H of R to be a *multiplicative-prime* subset of R if the following two conditions hold: (i) $ab \in H$ for every $a \in H$ and $b \in R$; (ii) if $ab \in H$ for $a, b \in R$, then either $a \in H$ or $b \in H$. For example, H is multiplicative-prime subset of R if H is a prime ideal of R, H is a union of prime ideals of R, $H = Z(R)$, or $H = R \setminus U(R)$. In fact, it is easily seen that H is a multiplicative-prime subset of R if and only if $R \setminus H$ is a saturated multiplicatively closed subset of R. Thus H is a multiplicative-prime subset of R if and only if H is a union of prime ideals of R. Note that if H is a multiplicative-prime subset of R, then $Nil(R) \subseteq H \subseteq R \setminus U(R)$; and if H

is also an ideal of R, then H is necessarily a prime ideal of R. In particular, if $R = Z(R) \cup U(R)$ (e.g., R is finite), then $Nil(R) \subseteq H \subseteq Z(R)$.

The notion of the generalized total graph of a commutative ring was introduced and studied by Anderson and Badawi [5]. For a multiplicative-prime subset H of R, the *generalized total graph* of R, denoted by $GT_H(R)$, as the (simple) graph with all elements of R as vertices, and for distinct $x, y \in R$, the vertices x and y are adjacent if and only if $x + y \in H$. For $A \subseteq R$, let $GT_H(A)$ be the induced subgraph of $GT_H(R)$ with all elements of A as the vertices. For example, $GT_H(R \setminus H)$ is the induced subgraph of $GT_H(R)$ with vertices $R \setminus H$. When $H = Z(R)$, we have that $GT_H(R)$ is the so-called total graph. The concept of generalized total graph, unlike the earlier concept of total graph, allows us to study graphs of integral domains. For a prime ideal P of a commutative ring R which is not an integral domain with $|P| = \alpha$ and $|R/P| = \beta$, we have $|P| = |a_i + P| \geq 2$ for $a_i + P \in R/P$ for $1 \leq i \leq \beta$. Since R contains identity 1, we denote $1 + 1$ by 2. Let us give examples for the generalized total graph now.

Example 1 Let $R = \mathbb{Z}_6$. Then, $< 2 >= \{0, 2, 4\}$ and $< 3 >= \{0, 3\}$ are the two different multiplicative prime subsets of $\mathbb{Z}_6$. The generalized total graph of $\mathbb{Z}_6$ with respect to $< 2 >$, $< 3 >$ are given below (Figs. 1 and 2):

Fig. 1 $GT_{<2>}(\mathbb{Z}_6)$

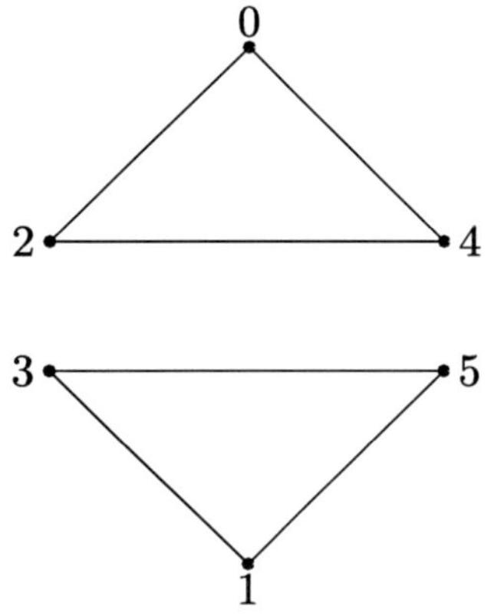

Fig. 2 $GT_{<3>}(\mathbb{Z}_6)$

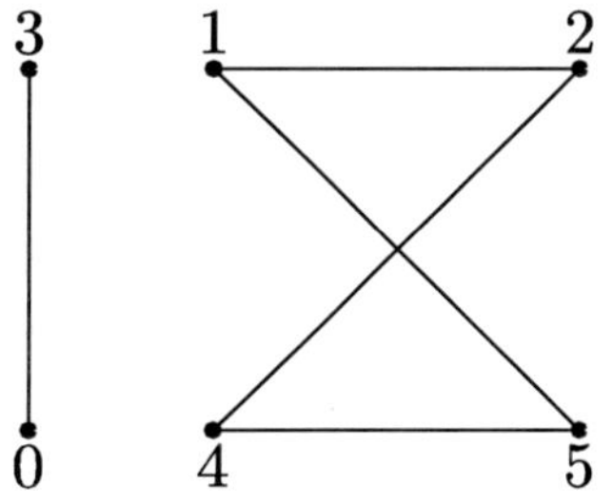

Certain results obtained by Anderson and Badawi [5] on a generalized total graph are given below.

Theorem 1 (Theorem 2.2, [5]) *Let P be a prime ideal of a commutative ring R and let $|P| = \alpha$ and $|R/P| = \beta$.*

(i) *If $2 \in P$, then $GT_P(R)$ is the union of $\beta - 1$ disjoint K_α's;*
(ii) *If $2 \notin P$, then $GT_P(R)$ is the union of $\frac{\beta-1}{2}$ disjoint $K_{\alpha,\alpha}$'s*

Theorem 2 (Theorem 2.3, [5]) *Let P be a prime ideal of a commutative ring R. Then*

(i) *$GT_P(R)$ is complete if and only if either $R/H \cong \mathbb{Z}_2$ or $R \cong \mathbb{Z}_3$;*
(ii) *$GT_P(R)$ is connected if and only if either $R/H \cong \mathbb{Z}_2$ or $R/H \cong \mathbb{Z}_3$;*
(iii) *$GT_P(R)$ (and hence $GT_P(P)$ and $GT_P(R)$) is totally disconnected if and only if $P = \{0\}$ (thus R is an integral domain) and $char(R) = 2$.*

Theorem 3 (Theorem 2.5, [5]) *Let P be a prime ideal of a commutative ring R. Then*

(i) (a) *$diam(GT_P(R)) = 0$ if and only if $R \cong \mathbb{Z}_2$;*
(b) *$diam(GT_P(R)) = 1$ if and only if either $R/P \cong \mathbb{Z}_2$ and $R \ncong \mathbb{Z}_2$; (i.e., $R/P \cong \mathbb{Z}_2$ and $|P| \geq 2$), or $R \cong \mathbb{Z}_3$;*
(c) *$diam(GT_P(R)) = 2$ if and only if either $R/P \cong \mathbb{Z}_3$ and $R \ncong \mathbb{Z}_3$; (i.e., $R/P \cong \mathbb{Z}_3$ and $|P| \geq 2$);*
(d) *Otherwise, $diam(GT_P(R)) = \infty$.*
(ii) (a) *$gr(GT_P(R)) = 3$ if and only if $2 \in GT_P(R)$ and $|P| \geq 3$;*
(b) *$gr(GT_P(R)) = 4$ if and only if $2 \notin P$ and $|P| \geq 2$;*
(c) *Otherwise, $gr(GT_P(R)) = \infty$.*
(iii) (a) *$gr(GT_H) = 3$ if and only if $2 \in GT_P(R)$ and $|P| \geq 3$;*
(b) *$gr(GT_P(R)) = 4$ if and only if $2 \notin P$ and $|P| \geq 2$;*
(c) *Otherwise, $gr(GT_P(R)) = \infty$.*

Theorem 4 (Theorem 2.8, [5]) *Let P be a prime ideal of a commutative ring R. Then the following statements are equivalent.*

(i) *$GT_P(R)$ is connected;*
(ii) *Either $x + y \in P$ or $x - y \in P$ for all $x, y \in R \setminus P$;*
(iii) *Either $x + y \in P$ or $x + 2y \in P$ for all $x, y \in R \setminus P$. In particular, either $2x \in P$ or $3x \in P$ (but not both) for all $x \in R \setminus P$;*
(iv) *Either $R/P \cong \mathbb{Z}_2$ or $R/P \cong \mathbb{Z}_3$.*

Theorem 5 (Theorem 3.1, [5]) *Let R be a commutative ring and H a multiplicative prime subset of a commutative ring R that is not an ideal of R. If $GT_H(R)$ is connected, then $GT_H(R)$ connected.*

Theorem 6 (Theorems 3.2 & 3.4, [5]) *Let R be a commutative ring and H a multiplicative prime subset of a commutative ring R that is not an ideal of R. Then $GT_H(R)$ is connected if and only if $1 = z_1 + \cdots + z_n$ for some $z_1, \ldots, z_n \in H$. $(Z(R)) = R$*

(i.e., $R = (z_1, \ldots, z_n)$) for some $z_1, \ldots, z_n \in Z(R)$). In particular, if H is not a ideal of R and either $dim(R) = 0$ (i.e., R is finite) or R is an integral domain with $diam(R) = 1$, then $GT_H(R)$ is connected. Let $n \geq 2$ be the least integer such that $1 = z_1 + \cdots + z_n$ for some $z_1, \ldots, z_n \in H$. Then $diam(GT_H(R)) = n$. In particular, H is not an ideal of R and either $dim(R) = 0$ (i.e., R is finite) or R is an integral domain with $dim(R) = 1$, then $GT_H(R)$ is connected.

Corollary 1 (Corollary 3.5, [5]) *Let R be a commutative ring and H a multiplicative prime subset of a commutative ring R that is not an ideal of R such that $GT_H(R)$ is connected.*

(i) *$diam(GT_H(R)) = d(0, 1)$;*
(ii) *If $diam(GT_H(R)) = n$, then $diam(GT_H(R)) \geq n - 2$.*

Theorem 7 (Theorem 3.14, [5]) *Let R be a commutative ring and H a multiplicative prime subset of a commutative ring R that is not an ideal of R.*

(i) *Either $gr(GT_H(H)) = 3$ or $gr(GT_H(H)) = \infty$. Moreover, if $gr(GT_H(H)) = \infty$, then $R \cong \mathbb{Z}_2\mathbb{Z}_2$ and $H = Z(R)$; so $GT_H(H)$ is a $K_{1,2}$ is a star graph with center 0;*
(ii) *$gr(GT_H(R)) = 3$ if and only if $gr(GT_H(R)) = 3$;*
(iii) *$gr(GT_H(R)) = 4$ if and only if $gr(GT_H(H)) = \infty$ (if and only if $R \ncong \mathbb{Z}_2\mathbb{Z}_2$);*
(iv) *If $char R = 2$, then $gr(GT_P(R)) = 3$ or ∞. In particular, $gr(GT_P(R)) = 3$ if $char R = 2$ and $GT_P(R)$ contains a cycle;*
(v) *$gr(GT_P(R)) = 3, 4,$ or ∞. In particular, $gr(GT_P(R)) \leq 4$ if $GT_P(R)$ contains a cycle.*

4 Complement of the Generalized Total Graph

Tamizh Chelvam and Balamurugan [31–33, 38] studied about the complement of the generalized total graph $GT_P(R)$ extensively. In this section, we present results on some graph theoretical properties of $\overline{GT_P(R)}$. More specifically, we present results on girth, Eulerian nature of $\overline{GT_P(R)}$, independence number, clique number and chromatic number of $\overline{GT_P(R)}$. Let us start with some basic properties of $\overline{GT_P(R)}$.

For a prime ideal P of a commutative ring R which is not an integral domain with $|P| = \alpha$ and $|R/P| = \beta$, we have $|P| = |a_i + P| \geq 2$ for $a_i + P \in R/P$ for $1 \leq i \leq \beta$. Assume that R is finite and $2 \in P$. Since $|R|$ is finite, $R \cong R_1 \times \cdots \times R_q$ where each R_i is a local ring. Note that $P = P_1 \times \cdots \times P_q$ where $P_i \subseteq R_i$ for $1 \leq i \leq q$. Since P is a prime ideal in R, $P_i = Z(R_i)$ for exactly one i, $1 \leq i \leq q$ and $P_j = R_j$ for $1 \leq j \neq i \leq q$. Since $2 \in P \subseteq Z(R)$, $2 \in P_i \subseteq Z(R_i)$ for some i, $1 \leq i \leq q$. By [18, Corollary 2.3], $|R_i| = 2^{\alpha_i}$ where α_i is a positive integer. This implies that β is even. Hence we get that β is even if $2 \in P$ and β is odd if $2 \notin P$. The following partition of R into distinct cosets $a_0 + P, a_1 + P, \ldots, a_{\beta-1} + P$ of P with $a_0 \in P$ is considered through out this paper.

(i) If $2 \in P$, then $R = P \bigcup (\bigcup_{i=1}^{\beta-1} a_i + P)$;

(ii) If $2 \notin P$, then $R = P \bigcup (\bigcup_{i=1}^{\frac{\beta-1}{2}} a_i + P)(\bigcup_{i=1}^{\frac{\beta-1}{2}} -a_i + P)$.

Now, using Theorem 1, we see the degrees of vertices in $\overline{GT_P(R)}$.

Lemma 1 (Lemma 1, [32]) *Let R be a finite commutative ring which is not an integral domain and P be a prime ideal in R. Assume that $|P| = \alpha$ and $|R/P| = \beta$. Then the following are true in $\overline{GT_P(R)}$.*

(i) *If $2 \in P$, then $deg(v) = (\beta - 1)\alpha$ for every $v \in R$;*

(ii) *If $2 \notin P$, then* $deg(v) = \begin{cases} (\beta-1)\alpha & for\ v \in P; \\ (\beta-1)\alpha - 1 & for\ v \in R \setminus P. \end{cases}$

The following is an immediate consequence of Lemma 2.

Lemma 2 (Lemma 2, [32]) *Let R be a finite commutative ring which is not an integral domain and P be a prime ideal in R. Assume that $|P| = \alpha$ and $|R/P| = \beta$. Then the following hold:*

(i) *$\overline{GT_P(R)}$ contains no isolated vertex;*
(ii) *$\overline{GT_P(R)}$ contains no vertex of degree $|R| - 1$;*
(iii) *$\overline{GT_P(R)}$ is complete β-partite if and only if $2 \in P$;*
(iv) *$\overline{GT_P(R)}$ is connected bi-regular if and only if $2 \notin P$. Moreover, $\Delta(\overline{GT_P(R)}) = \delta(\overline{GT_P(R)}) + 1$;*
(v) *$\overline{GT_P(R)}$ is connected.*

The following lemma follows from Theorem 1.

Lemma 3 (Lemma 3, [32]) *Let R be a finite commutative ring which is not an integral domain and P be a prime ideal in R. Assume that $|P| = \alpha$ and $|R/P| = \beta$. Then the following are true in $\overline{GT_P(R)}$.*

(i) *Let $2 \in P$. Two distinct vertices x and y are adjacent in $\overline{GT_P(R)}$ if and only if x and y are not in the same coset of P;*
(ii) *Let $2 \notin P$. Two distinct vertices x and y are adjacent in $\overline{GT_P(R)}$ if and only if $x \in a_i + P$ and $y \in R \setminus (-a_i + P)$ for some i, $0 \le i \le \frac{\beta-1}{2}$.*

The following lemma gives the girth of $\overline{GT_P(R)}$.

Lemma 4 (Lemma 4, [32]) *Let R be a finite commutative ring which is not an integral domain and P be a prime ideal in R. Assume that $|P| = \alpha$ and $|R/P| = \beta$. Then*

$$gr(\overline{GT_P(R)}) = \begin{cases} 4 & if\ 2 \in P\ and\ \beta = 2; \\ 3 & if\ 2 \in P\ and\ \beta \ge 3; \\ 3 & if\ 2 \notin P. \end{cases}$$

Note that, a *clique* in a graph G is a complete subgraph of G. The order of the largest clique in a graph G is its *clique number*, which is denoted by $\omega(G)$. The following lemma gives the clique number of $\overline{GT_P(R)}$.

Lemma 5 (Lemma 5, [32]) *Let R be a finite commutative ring which is not an integral domain and P be a prime ideal in R. Assume that $|P| = \alpha$ and $|R/P| = \beta$. Then*

$$\omega(\overline{GT_P(R)}) = \begin{cases} \beta & if\ 2 \in P; \\ (\frac{\beta-1}{2})\alpha + 1 & if\ 2 \notin P. \end{cases}$$

An assignment of colors to the vertices of a graph G so that adjacent vertices are assigned different colors is called a *proper coloring* of G. The smallest number of colors in any proper coloring of a graph G is called the *chromatic number* of G and is denoted by $\chi(G)$. A set of vertices in a graph G is an *independent* if no two vertices in the set are adjacent. In the following lemma, we see the chromatic number of $\overline{GT_P(R)}$.

Lemma 6 (Lemma 6, [32]) *Let R be a finite commutative ring which is not an integral domain and P be a prime ideal in R. Assume that $|P| = \alpha$ and $|R/P| = \beta$. Then*

$$\chi(\overline{GT_P(R)}) = \begin{cases} \beta & if\ 2 \in P; \\ (\frac{\beta-1}{2})\alpha + 1 & if\ 2 \notin P. \end{cases}$$

Corollary 2 ([15]) *A nontrivial connected graph G is Eulerian if and only if every vertex of G has even degree.*

Using Corollary 2, a characterization for $\overline{GT_P(R)}$ to be Eulerian is obtained and the same is given below.

Lemma 7 (Lemma 7, [32]) *Let R be a finite commutative ring which is not an integral domain and P be a prime ideal in R. Assume that $|P| = \alpha$ and $|R/P| = \beta$. Then the following are true:*

(i) *If $2 \in P$, then $\overline{GT_P(R)}$ is Eulerian if and only if α is even;*
(ii) *If $2 \notin P$, then $\overline{GT_P(R)}$ is not Eulerian.*

The *vertex independence number* (or the *independence number*) $\beta(G)$ of a graph G is the maximum cardinality of an independent set of vertices in G. In the following lemma, we obtain the independence number of $\overline{GT_P(R)}$.

Lemma 8 (Lemma 8, [32]) *Let R be a finite commutative ring which is not an integral domain and P be a prime ideal in R. Assume that $|P| = \alpha$ and $|R/P| = \beta$. Then the independence number $\beta(\overline{GT_P(R)}) = \alpha$.*

The *edge independence number* $\beta_1(G)$ of a graph G is the maximum cardinality of an independent set of edges. In the following lemma, we obtain the edge independence number of $\overline{GT_P(R)}$.

Lemma 9 (Lemma 9, [32]) *Let R be a finite commutative ring which is not an integral domain and P be a prime ideal in R. Assume that* $|P| = \alpha$ *and* $|R/P| = \beta$. *Then the edge independence number*

$$\beta_1(\overline{GT_P(R)}) = \begin{cases} \frac{|R|}{2} & if\ 2 \in P; \\ (\frac{\beta-1}{2})\alpha + \frac{\lfloor \alpha \rfloor}{2} & if\ 2 \notin P. \end{cases}$$

Tamizh Chelvam and Balamurugan [32] obtained certain characterizations like when $\overline{GT_P(R)}$ is claw-free, unicylic, pancyclic or perfect. A graph G is said to be *unicyclic* if G contains exactly one cycle. A graph G is a *claw-free* if G does not have $K_{1,3}$(a claw) as the induced subgraph.

Theorem 8 (Theorem 2, [32]) *Let R be a finite commutative ring which is not an integral domain and P be a prime ideal in R with* $|P| = \alpha$ *and* $|R/P| = \beta$. *Then the following hold:*

(i) $\overline{GT_P(R)}$ *is claw-free if and only if* $|P| = 2$;
(ii) $\overline{GT_P(R)}$ *is unicyclic if and only if R is isomorphic to either of the rings* $\mathbb{Z}_2 \times \mathbb{Z}_2$, $\mathbb{Z}_4$, $\frac{\mathbb{Z}_2[x]}{<x^2>}$.

Note that a graph G is perfect if and only if both G and $\overline{G}$ have no induced subgraph that is an odd cycle of length at least 5 [39, 8.1.2]. Using this, a characterization for $\overline{GT_P(R)}$ to be perfect is obtained and the same is given below.

Theorem 9 (Theorem 3, [32]) *Let R be a finite commutative ring which is not an integral domain and P be a prime ideal in R with* $|P| = \alpha$ *and* $|R/P| = \beta$. *Then* $\overline{GT_P(R)}$ *is a perfect graph.*

A graph G of order $m \geq 3$ is *pancyclic* [13, Definition 6.3.1] if G contains cycles of all lengths from 3 to m.

Theorem 10 (Theorem 4, [32]) *Let R be a finite commutative ring which is not an integral domain and P be a prime ideal in R, where* $|P| = \alpha$ *and* $|R/P| = \beta$. *Then* $\overline{GT_P(R)}$ *is pancyclic if and only if either* $2 \notin P$ *or* $2 \in P$ *with* $\beta > 2$.

Now, let us see some results concerning planarity and outerplanarity of $\overline{GT_P(R)}$. Using the known famous characterizations for planarity, Tamizh Chelvam and Balamurugan obtained the following.

Theorem 11 (Theorem 7, [32]) *Let R be a finite commutative ring which is not an integral domain and P be a prime ideal in R, where* $|P| = \alpha$ *and* $|R/P| = \beta$. *Then the following hold:*

(i) *If* $2 \in P$, *then* $\overline{GT_P(R)}$ *is planar if and only if R is isomorphic to any one of* $\mathbb{Z}_4$, $\mathbb{Z}_2\mathbb{Z}_2$, $\frac{\mathbb{Z}_2[x]}{<x^2>}$;
(ii) *If* $2 \notin P$, *then* $\overline{GT_P(R)}$ *is planar if and only if* $R = \mathbb{Z}_6$;
(iii) $\overline{GT_P(R)}$ *is outerplanar if and only if R is either* $\mathbb{Z}_4$, $\mathbb{Z}_2 \times \mathbb{Z}_2$ *or* $\frac{\mathbb{Z}_2[x]}{<x^2>}$.

Now let us list out some of the results concerning various domination parameters of $\overline{GT_P(R)}$. More specifically, we present results on γ_t, γ_c, γ_{cl}, γ_p, γ_s, γ_w and γ_i of $\overline{GT_P(R)}$.

A nonempty subset S of V is called a *dominating set* if every vertex in $V \setminus S$ is adjacent to at least one vertex in S. A subset S of V is called a *total dominating set* if every vertex in V is adjacent to some vertex in S. A dominating set S is called a *connected (or clique) dominating* set if the subgraph induced by S is connected (or complete). A dominating set S is called an *independent dominating set* if no two vertices of S are adjacent. A dominating set S is called a *perfect dominating set* if every vertex in $V \setminus S$ is adjacent to exactly one vertex in S. A dominating set S is called a *strong (or weak) dominating set*, if for every vertex $u \in V \setminus S$ there is a vertex $v \in S$ with $deg(v) \geq deg(u)(ordeg(v) \leq deg(u))$ and u is adjacent to v. The *domination number* γ of G is defined to be the minimum cardinality of a dominating set in G and the corresponding dominating set is called as a γ-set of G. Similar definition is applicable for the *total domination number* γ_t, *connected domination number* γ_c,*clique domination number* γ_{cl}, *independent domination number* γ_i, *perfect domination number* γ_p, *strong domination number* γ_s *and the weak domination number* γ_w. For all these definitions, one can refer Haynes et al., [17]. In the following Lemma, we present the domination number of $\overline{GT_P(R)}$.

Lemma 10 (Lemma 10, [32]) *Let R be a finite commutative ring which is not an integral domain and P be a prime ideal in R. Assume that $|P| = \alpha$ and $|R/P| = \beta$. Then $\gamma(\overline{GT_P(R)}) = 2$.*

In view of Lemma 10, the following characterization provides all γ-sets in $\overline{GT_P(R)}$.

Lemma 11 (Lemma 11, [32]) *Let R be a finite commutative ring which is not an integral domain and P be a prime ideal in R. Then $S = \{x, y\} \subseteq V(\overline{GT_P(R)})$ is a γ-set in $\overline{GT_P(R)}$ if and only if x, y are in two distinct cosets of P in R.*

Corollary 3 (Corollary 2, [32]) *Let R be a finite commutative ring which is not an integral domain and P be a prime ideal in R. Assume that $|P| = \alpha$ and $|R/P| = \beta$. Then $\gamma_t(\overline{GT_P(R)}) = \gamma_c(\overline{GT_P(R)}) = \gamma_{cl}(\overline{GT_P(R)}) = 2$.*

A graph G is called *excellent* if, for every vertex $v \in V(G)$, there is a γ-set S containing v. Using Lemma 11, the following is obtained.

Corollary 4 (Corollary 3, [32]) *Let R be a finite commutative ring which is not an integral domain and P be a prime ideal in R. Assume that $|P| = \alpha$ and $|R/P| = \beta$. Then $\overline{GT_P(R)}$ is excellent.*

Lemma 12 (Lemma 12, [32]) *Let R be a finite commutative ring which is not an integral domain and P be a prime ideal in R. Assume that $|P| = \alpha$ and $|R/P| = \beta$. Then $\overline{GT_P(R)}$ has a perfect dominating set if and only if $2 \in P$ and $\beta = 2$.*

Lemma 13 (Lemma 13, [32]) *Let R be a finite commutative ring which is not an integral domain and P be a prime ideal in R. Assume that $|P| = \alpha$ and $|R/P| = \beta$. Then the following are true:*

(i) *If $2 \in P$, then $\gamma_s(\overline{GT_P(R)}) = \gamma_w(\overline{GT_P(R)}) = 2$;*
(ii) *If $2 \notin P$, then $\gamma_s(\overline{GT_P(R)}) = \alpha$ and $\gamma_w(\overline{GT_P(R)}) = 2$.*

In the following lemma, the independent domination number of $\overline{GT_P(R)}$ is obtained.

Lemma 14 (Lemma 14, [32]) *Let R be a finite commutative ring which is not an integral domain and P be a prime ideal in R. Assume that $|P| = \alpha$ and $|R/P| = \beta$. Then*

$$\gamma_i(\overline{GT_P(R)}) = \begin{cases} \alpha & if\ 2 \in P; \\ 2 & if\ 2 \notin P. \end{cases}$$

Note that, a graph G is said to be *well-covered* if $\beta(G) = \gamma_i(G)$. The following lemma provides a characterization for $\overline{GT_P(R)}$ to be well covered.

Lemma 15 (Lemma 15, [32]) *Let R be a finite commutative ring which is not an integral domain and P be a prime ideal in R. Assume that $|P| = \alpha$ and $|R/P| = \beta$. Then*

(i) *If $2 \in P$, then $\overline{GT_P(R)}$ is well-covered;*
(ii) *If $2 \notin P$, then $\overline{GT_P(R)}$ is well covered if and only if $|P| = 2$.*

A *domatic partition* of G is a partition of $V(G)$ into dominating sets of G. The maximum number of sets in a domatic partition of G is called the *domatic number* of G and the same is denoted by $d(G)$.

Lemma 16 (Lemma 16, [32]) *Let R be a finite commutative ring which is not an integral domain and P be a prime ideal in R. Assume that $|P| = \alpha$ and $|R/P| = \beta$. Then*

$$d(\overline{GT_P(R)}) = \begin{cases} \frac{|R|}{2} & if\ 2 \in P; \\ \frac{|R|-\alpha}{2} + 1 & if\ 2 \notin P. \end{cases}$$

5 Complement of the Generalized Total Graph of $\mathbb{Z}_n$

Tamizh Chelvam and Balamurugan[33] further studied the complement of the generalized total graph of $\mathbb{Z}_n$. Especially they have obtained characterizations for Eulerian and Hamiltonian, and further obtained the independence number and covering numbers. In the main result, it is proved that the conjecture on coloring of graphs from commutative rings proposed by Beck [14] is true for the complement of the generalized total graph for $\mathbb{Z}_n$. Also, they have obtained some domination parameters for the complement of the generalized total graph for $\mathbb{Z}_n$. Further, they studied some distance properties for the complement of the generalized total graph for $\mathbb{Z}_n$.

Let $n = p_1^{\alpha_1} p_2^{\alpha_2} \ldots p_k^{\alpha_k}$ where $p_1 < p_2 < \cdots < p_k$, $p_j' s$ are prime, $\alpha_j' s$ are positive integers and $P =< p_j >$ for some j. Then P is a prime ideal of $\mathbb{Z}_n$ and hence one can have $\overline{GT_P(\mathbb{Z}_n)}$

Lemma 17 (Lemma 2.2, [33]) *Let $n = p_1^{\alpha_1} p_2^{\alpha_2} \ldots p_k^{\alpha_k}$ where $p_1 < p_2 < \cdots < p_k$, $p_j' s$ are prime, $\alpha_j' s$ are positive integers and $P =< p_j >$ for some j. Then the following are true in $\overline{GT_P(\mathbb{Z}_n)}$*

(i) *If $n = 2$, then $deg(v) = 1$, for every $v \in \mathbb{Z}_n$;*

(ii) *If n is an odd prime p, then $deg(v) = \begin{cases} n-1 & \text{if } v = 0 \\ n-2 & \text{if } v \neq 0; \end{cases}$*

(iii) *If $2 \in P$, then $deg(v) = \frac{n}{2}$, for every $v \in \mathbb{Z}_n$;*

(iv) *If $2 \notin P$, then $deg(v) = \begin{cases} n - \frac{n}{p_j} & \text{for } v \in P \\ n - \frac{n}{p_j} - 1 & \text{for } v \in \mathbb{Z}_n \setminus P. \end{cases}$*

In view of Lemma 17, the following was proved by Tamizh Chelvam and Balamurugan [33].

Lemma 18 (Lemma 2.3, [33]) *Let $n = p_1^{\alpha_1} p_2^{\alpha_2} \ldots p_k^{\alpha_k}$ where $p_1 < p_2 < \cdots < p_k$, $p_j' s$ are prime, $\alpha_j' s$ are positive integers and $P =< p_j >$ for some j. Then*

(i) *$\overline{GT_P(\mathbb{Z}_n)}$ contains no isolated vertex;*
(ii) *$\overline{GT_P(\mathbb{Z}_n)}$ contains a vertex of degree $n-1$ if and only if n is a prime integer.*
(iii) *$\overline{GT_P(\mathbb{Z}_n)}$ is regular if and only if $n = 2$ or $2 \in P$;*
(iv) *$\overline{GT_P(\mathbb{Z}_n)}$ is biregular if and only if n is odd. Moreover in this case, $\Delta(\overline{GT_P(\mathbb{Z}_n)}) = \delta(\overline{GT_P(\mathbb{Z}_n)}) + 1$;*
(v) *$\overline{GT_P(\mathbb{Z}_n)}$ is a nontrivial connected graph.*

Example 2 The graph $\overline{GT_{<2>}(\mathbb{Z}_6)}$ is given in Fig. 3.

In view of Lemma 18 and Example 2, the structure of the complement of the generalized total graph of $\mathbb{Z}_n$ when $P =< 2 >$ is given in the following lemma.

Lemma 19 (Lemma 2.4, [33]) *Let $n = p_1^{\alpha_1} p_2^{\alpha_2} \ldots p_k^{\alpha_k}$ where $p_1 < p_2 < \cdots < p_k$, $p_j' s$ are prime, $\alpha_j' s$ are positive integers and $P =< p_j >$ for some j. If $p_j = 2$, then $\overline{GT_P(\mathbb{Z}_n)} = K_{\frac{n}{2}, \frac{n}{2}}$.*

Fig. 3 $\overline{GT_{<2>}(\mathbb{Z}_6)}$

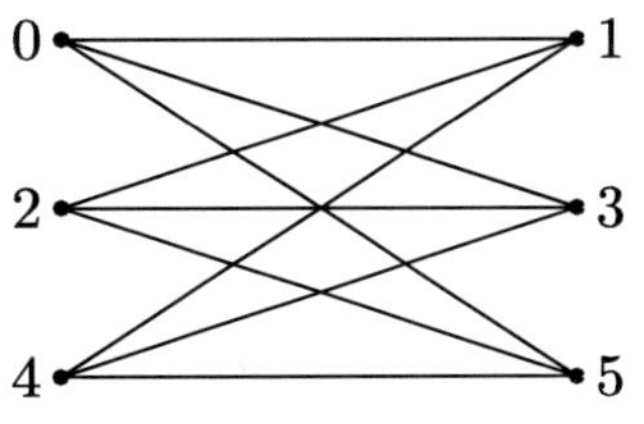

Remark 1 (*Remark 2.5,* [33]) In Lemma 5, if p_j is an odd prime, then two distinct elements x and y are adjacent in $\overline{GT_P(\mathbb{Z}_n)}$ if and only if $x \in i + P$ and $y \in \mathbb{Z}_n \setminus (p_j - i + P)$ for some i and $1 \leq i < p_j$.

The following is proved regarding distances between vertices in $\overline{GT_P(\mathbb{Z}_n)}$.

Lemma 20 *Let $n = p_1^{\alpha_1} p_2^{\alpha_2} \ldots p_k^{\alpha_k}$ where $p_1 < p_2 < \cdots < p_k$, p_j's are prime, α_j's are positive integers and $P =< p_j >$ for some j. Let x, y be two distinct vertices in $\mathbb{Z}_n$. Then the following are true in $\overline{GT_P(\mathbb{Z}_n)}$.*

(i) *If $n = 2$, then $d(x, y) = 1$;*
(ii) *If n is an odd prime, then*
$$d(x, y) = \begin{cases} 1 & \text{if } x + y \notin P; \\ 2 & \text{otherwise.} \end{cases}$$
(iii) *If n is composite and $p_j = 2$, then*
$$d(x, y) = \begin{cases} 1 & \text{if } x \in P \text{ and } y \in 1 + P; \\ 2 & \text{if either } x, y \in P \text{ or } x, y \in 1 + P. \end{cases}$$
(iv) *If n is composite and $p_j \neq 2$, then*
$$d(x, y) = \begin{cases} 1 & \text{if } x \in P \text{ and } y \in \mathbb{Z}_n \setminus P; \\ 2 & \text{if } x, y \in P; \\ 2 & \text{if } x \in i + P \text{ and } y \in (p_j - i) + P; \\ 1 & \text{otherwise.} \end{cases}$$

In view of Lemma 20, authors observed the following regarding eccentricity, radius, self-centered, and periphery. In the following lemma, we see the eccentricity of all the vertices in the complement of generalized total graph for $\mathbb{Z}_n$.

Lemma 21 *Let $n = p_1^{\alpha_1} p_2^{\alpha_2} \ldots p_k^{\alpha_k}$ where $p_1 < p_2 < \cdots < p_k$, p_j's are prime, α_j's are positive integers and $P =< p_j >$ for some j. Let $x \in \mathbb{Z}_n$. Then the following are true in $\overline{GT_P(\mathbb{Z}_n)}$:*

(i) *If $n = 2$, then $e(x) = 1$;*
(ii) *If n is an odd prime, then* $e(x) = \begin{cases} 1 & \text{if } x = 0; \\ 2 & \text{if } x \in \mathbb{Z}_n \setminus \{0\}. \end{cases}$
(iii) *If n is composite and $p_j = 2$, then $e(x) = 2$;*
(iv) *If n is composite and $p_j \neq 2$, then $e(x) = 2$.*

The following lemma gives the radius and diameter of the complement of generalized total graph for $\mathbb{Z}_n$.

Lemma 22 *Let $n = p_1^{\alpha_1} p_2^{\alpha_2} \ldots p_k^{\alpha_k}$ where $p_1 < p_2 < \cdots < p_k$, p_j's are prime, α_j's are positive integers and $P =< p_j >$ for some j. Let $x \in \mathbb{Z}_n$. Then the following are true:*

(i) *If $n = 2$, then $rad(\overline{GT_P(\mathbb{Z}_n)}) = diam(\overline{GT_P(\mathbb{Z}_n)}) = 1$;*
(ii) *If n is an odd prime, then $rad(\overline{GT_P(\mathbb{Z}_n)}) = 1$;*

(iii) *If* n *is an odd prime, then* $diam(\overline{GT_P(\mathbb{Z}_n)}) = 2$;
(iv) *If* n *is composite, then* $rad(\overline{GT_P(\mathbb{Z}_n)}) = 2$;
(v) *If* n *is composite, then* $diam(\overline{GT_P(\mathbb{Z}_n)}) = 2$.

In the following Corollary, a characterization for the complement of generalized total graph for $\mathbb{Z}_n$ to be self-centered.

Corollary 5 *Let* $n = p_1^{\alpha_1} p_2^{\alpha_2} \dots p_k^{\alpha_k}$ *where* $p_1 < p_2 < \cdots < p_k$, p_j's *are prime,* α_j's *are positive integers and* $P =< p_j >$ *for some* j. *Then the following are true:*

(i) *Let* n *be prime. Then* $\overline{GT_P(\mathbb{Z}_n)}$ *is self-centered if and only if* $n = 2$;
(ii) *If* n *is composite, then* $\overline{GT_P(\mathbb{Z}_n)}$ *is self-centered.*

The following lemma gives the relation between the complement of generalized total graph for $\mathbb{Z}_n$ and its periphery graph.

Lemma 23 *Let* $n = p_1^{\alpha_1} p_2^{\alpha_2} \dots p_k^{\alpha_k}$ *where* $p_1 < p_2 < \cdots < p_k$, p_j's *are prime,* α_j's *are positive integers and* $P =< p_j >$ *for some* j. *Then the following are true:*

(i) *If* n *is prime, then* $Per(\overline{GT_P(\mathbb{Z}_n)}) = \overline{GT_P(\mathbb{Z}_n)}$ *if and only if* $n = 2$;
(ii) *If* n *is composite, then* $Per(\overline{GT_P(\mathbb{Z}_n)}) = \overline{GT_P(\mathbb{Z}_n)}$;
(iii) *If* n *is an odd prime, then*

$$Per(\overline{GT_P(\mathbb{Z}_n)}) = \begin{cases} K_1 \cup K_1 & \text{if } n = 3; \\ (n-3) \text{ regular connected graph} & \text{if } n > 3. \end{cases}$$

In the following lemma, a characterization for $\overline{GT_P(\mathbb{Z}_n)}$ to be Eulerian is obtained.

Lemma 24 (Lemma 2.7, [33]) *Let* $n = p_1^{\alpha_1} p_2^{\alpha_2} \dots p_k^{\alpha_k}$ *where* $p_1 < p_2 < \cdots < p_k$, p_j's *are prime,* α_j's *are positive integers and* $P =< p_j >$ *for some* j. *Then the following are true:*

(i) *If* $P =< p_1 >$ *and* $p_1 = 2$, *then* $\overline{GT_P(\mathbb{Z}_n)}$ *is Eulerian if and only if* $n = 4k$ *for some positive integer* k;
(ii) *If* $P =< p_j >$ *and* $p_j \neq 2$ *or* n *is prime, then* $\overline{GT_P(\mathbb{Z}_n)}$ *is not Eulerian.*

The following lemma provides a situation where the complement of the generalized total graph for $\mathbb{Z}_n$ is Hamiltonian.

Lemma 25 (Lemma 2.9, [33]) *Let* $n = p_1^{\alpha_1} p_2^{\alpha_2} \dots p_k^{\alpha_k} > 3$ *where* $p_1 < p_2 < \cdots < p_k$, p_j's *are prime,* α_j's *are positive integers and* $P =< p_j >$ *for some* j. *Then* $\overline{GT_P(\mathbb{Z}_n)}$ *is Hamiltonian.*

The vertex independence number of $\overline{GT_P(\mathbb{Z}_n)}$ was obtained through the following lemma.

Lemma 26 (Lemma 2.10, [33]) *Let* $n = p_1^{\alpha_1} p_2^{\alpha_2} \dots p_k^{\alpha_k}$ *where* $p_1 < p_2 < \cdots < p_k$, $p_j's$ *are prime,* $\alpha_j's$ *are positive integers and* $P =< p_j >$ *for some* j. *Then*

$$\beta(\overline{GT_P(\mathbb{Z}_n)}) = \begin{cases} 1 & \text{if } n = 2; \\ 2 & \text{if } n \text{ is an odd prime}; \\ \frac{n}{2} & \text{if } n \text{ is a composite integer and } p_j = 2; \\ \frac{n}{p_j} & \text{if } n \text{ is a composite integer and } p_j \neq 2. \end{cases}$$

The vertex covering number $\alpha(\overline{GT_P(\mathbb{Z}_n)})$ is given by the following corollary.

Corollary 6 (Corollary 2.12, [33]) *Let* $n = p_1^{\alpha_1} p_2^{\alpha_2} \dots p_k^{\alpha_k}$ *where* $p_1 < p_2 < \cdots < p_k$, $p_j's$ *are prime,* $\alpha_j's$ *are positive integers and* $P =< p_j >$ *for some* j. *Then the vertex covering number*

$$\alpha(\overline{GT_P(\mathbb{Z}_n)}) = \begin{cases} 1 & \text{if } n = 2; \\ n-2 & \text{if } n \text{ is an odd prime}; \\ \frac{n}{2} & \text{if } n \text{ is a composite integer and } p_j = 2; \\ n - \frac{n}{p_j} & \text{if } n \text{ is a composite integer and } p_j \neq 2. \end{cases}$$

Note that a graph G is planar if and only if G does not contain either K_5 or $K_{3,3}$. Using this, a characterization for $\overline{GT_P(\mathbb{Z}_n)}$ to be planar was obtained and the same is stated below.

Theorem 12 (Theorem 3.3, [33]) *Let* $n = p_1^{\alpha_1} p_2^{\alpha_2} \dots p_k^{\alpha_k}$ *where* $p_1 < p_2 < \dots < p_k$, $p_j's$ *are prime and* $\alpha_j's$ *are positive integers. Then the following are true:*

(i) *Let* n *be composite,* $p_1 = 2$ *and* $P =< p_1 >$. *Then* $\overline{GT_P(\mathbb{Z}_n)}$ *is planar if and only if* $n = 4$;
(ii) *Let* n *be composite,* $p_j \neq 2$ *and* $P =< p_j >$. *Then* $\overline{GT_P(\mathbb{Z}_n)}$ *is planar if and only if* $n = 6$;
(iii) *Let* n *be prime,* $n = p$ *and* $P =< p >$. *Then* $\overline{GT_P(\mathbb{Z}_p)}$ *is planar if and only if* $p \in \{2, 3, 5\}$.

Now, we state characterization for $\overline{GT_P(\mathbb{Z}_n)}$ to be toroidal.

Theorem 13 (Theorem 3.4, [33]) *Let* $n = p_1^{\alpha_1} p_2^{\alpha_2} \dots p_k^{\alpha_k}$ *where* $p_1 < p_2 < \dots < p_k$, $p_j's$ *are prime and* $\alpha_j's$ *are positive integers. Then the following are true:*

(i) *Let* n *be composite,* $p_1 = 2$ *and* $P =< p_1 >$. *Then* $\overline{GT_P(\mathbb{Z}_n)}$ *is toroidal if and only if* $n \in \{6, 8\}$;
(ii) *Let* $n \geq 9$ *be composite,* $p_j \neq 2$ *and* $P =< p_j >$. *Then* $\overline{GT_P(\mathbb{Z}_n)}$ *is not toroidal.*
(iii) *Let* n *be prime and* $P =< n >$. *Then* $\overline{GT_P(\mathbb{Z}_p)}$ *is toroidal if and only if* $n = 7$.

6 Domination Properties of $\overline{GT_P(\mathbb{Z}_n)}$

The concepts of dominating sets and domination numbers are very important concepts in graph theory. In this section, we present results concerning the domination number of the complement of the total graph of a commutative ring through ring theoretic properties. In the following lemma, we present the value of the domination number of $\overline{GT_P(\mathbb{Z}_n)}$.

Lemma 27 (Lemma 4.1, [33]) *Let* $n = p_1^{\alpha_1} p_2^{\alpha_2} \dots p_k^{\alpha_k}$ *where* $p_1 < p_2 < \dots < p_k$, p_j'*s are prime,* α_j'*s are positive integers and* $P =< p_j >$ *for some* j*. Then*

$$\gamma(\overline{GT_P(\mathbb{Z}_n)}) = \begin{cases} 1 & \text{if } n \text{ is a prime integer;} \\ 2 & \text{if } n \text{ is a composite integer.} \end{cases}$$

The following characterization of γ-sets in $\overline{GT_P(\mathbb{Z}_n)}$ was proved in [33].

Theorem 14 (Theorem 4.2, [33]) *Let* $n = p_1^{\alpha_1} p_2^{\alpha_2} \dots p_k^{\alpha_k}$ *where* $p_1 < p_2 < \dots < p_k$, p_j'*s are prime,* α_j'*s are positive integers and* $P =< p_j >$ *for some* j*.*

(i) *Let* n *be composite and* $S = \{a, b\} \subseteq \mathbb{Z}_n$*. Then* S *is a* γ*-set if and only if* a, b *are in two distinct cosets of* P *in* $\mathbb{Z}_n$*;*
(ii) *The set* $S = \{0\}$ *is a* γ*-set in* $\overline{GT_P(\mathbb{Z}_n)}$ *if and only if* n *is a prime number.*

Corollary 7 (Corollary 4.3, [33]) *Let* n *be composite. Then* $\gamma_t(\overline{GT_P(\mathbb{Z}_n)}) = \gamma_c(\overline{GT_P(\mathbb{Z}_n)}) = \gamma_{cl}(\overline{GT_P(\mathbb{Z}_n)}) = 2$.

Recall that when $p_j = 2$, $\overline{GT_P(\mathbb{Z}_n)}$ is a complete bi-partite graph. Using this along with Theorem 14, the following was proved.

Lemma 28 (Lemma 4.4, [33]) *Let* $n = p_1^{\alpha_1} p_2^{\alpha_2} \dots p_k^{\alpha_k}$ *where* $p_1 < p_2 < \dots < p_k$, p_j'*s are prime,* α_j'*s are positive integers and* $P =< p_j >$ *for some* j*. Then the following are true:*

(i) *If* n *is a composite integer, then* $\overline{GT_P(\mathbb{Z}_n)}$ *is excellent;*
(ii) *Let* n *be prime. Then* $\overline{GT_P(\mathbb{Z}_n)}$ *is excellent if and only if* $n = 2$.

In the following Lemma, a characterization for the complement graph of $\mathbb{Z}_n$ to have a perfect domination set was obtained.

Lemma 29 (Lemma 4.5, [33]) *Let* $n = p_1^{\alpha_1} p_2^{\alpha_2} \dots p_k^{\alpha_k}$ *where* $p_1 < p_2 < \dots < p_k$, p_j'*s are prime,* α_j'*s are positive integers and* $P =< p_1 >$ *. Then the following are true:*

(i) *If* n *is a prime integer, then* $\{0\}$ *is a perfect dominating set in* $\overline{GT_P(\mathbb{Z}_n)}$*;*
(ii) *If* n *is a composite integer, then perfect dominating set exists in* $\overline{GT_P(\mathbb{Z}_n)}$ *if and only if* $p_1 = 2$.

Corollary 8 (Corollary 4.6, [33]) *Let $n = p_1^{\alpha_1} p_2^{\alpha_2} \dots p_k^{\alpha_k}$ be an integer where $p_1 < p_2 < \dots < p_k$, p_j's are prime, α_j's are positive integers. Then*

$$\gamma_p(\overline{GT_P(\mathbb{Z}_n)}) = \begin{cases} 1 & \text{if } n \text{ is a prime}; \\ 2 & \text{if } n \text{ is a composite and } p_1 = 2; \\ 0 & \text{if } n \text{ is a composite and } p_1 \neq 2. \end{cases}$$

Lemma 30 (Lemma 4.7, [33]) *Let $n = p_1^{\alpha_1} p_2^{\alpha_2} \dots p_k^{\alpha_k}$ be a composite integer where $p_1 < p_2 < \dots < p_k$, p_j's are prime and α_j's are positive integer. Then the following are true:*

(i) *If n is composite, $p_1 = 2$ and $P =< p_1 >$, then $\gamma_s(\overline{GT_P(\mathbb{Z}_n)}) = \gamma_w(\overline{GT_P(\mathbb{Z}_n)}) = 2$;*
(ii) *If n is composite, $p_j \neq 2$ and $P =< p_j >$ for some j, then $\gamma_s(\overline{GT_P(\mathbb{Z}_n)}) = \frac{n}{p_j}$ and $\gamma_w(\overline{GT_P(\mathbb{Z}_n)}) = 2$.*

Lemma 31 (Lemma 4.8, [33]) *Let $n > 1$ be a prime integer. Then the following are true:*

(i) *If $n = 2$, then $\gamma_s(\overline{GT_P(\mathbb{Z}_n)}) = \gamma_w(\overline{GT_P(\mathbb{Z}_n)}) = 1$.*
(ii) *If $n \neq 2$, then $\gamma_s(\overline{GT_P(\mathbb{Z}_n)}) = 1$ and $\gamma_t(\overline{GT_P(\mathbb{Z}_n)}) = \gamma_c(\overline{GT_P(\mathbb{Z}_n)}) = \gamma_{cl}(\overline{GT_P(\mathbb{Z}_n)}) = \gamma_w(\overline{GT_P(\mathbb{Z}_n)}) = 2$.*

In the following lemma, authors obtained the independent dominating number of $\overline{GT_P(\mathbb{Z}_n)}$.

Lemma 32 (Lemma 4.9, [33]) *Let $n = p_1^{\alpha_1} p_2^{\alpha_2} \dots p_k^{\alpha_k}$ where $p_1 < p_2 < \dots < p_k$, p_j's are prime, α_j's are positive integers and $P =< p_j >$ for some j. Then*

$$\gamma_i(\overline{GT_P(\mathbb{Z}_n)}) = \begin{cases} 1 & \text{if } n \text{ is a prime}; \\ \frac{n}{2} & \text{if } n \text{ is composite and } p_j = 2; \\ 2 & \text{if } n \text{ is composite and } p_j \neq 2. \end{cases}$$

Corollary 9 (Corollary 4.10, [33]) *If $n > 1$ is prime, then $\gamma_{eff}(\overline{GT_P(\mathbb{Z}_n)}) = 1$.*

A graph G is *well-covered* if $\beta(G) = \gamma_i(G)$. In the following lemma, authors discussed when $\overline{GT_P(\mathbb{Z}_n)}$ is well-covered.

Lemma 33 (Lemma 4.11, [33]) *Let $n = p_1^{\alpha_1} p_2^{\alpha_2} \dots p_k^{\alpha_k}$ where $p_1 < p_2 < \dots < p_k$, p_j's are prime and α_j's are positive integers. Then the following are true:*

(i) *If $n = 2$, $p_1 = 2$ and $P =< p_1 >$, then $\overline{GT_P(\mathbb{Z}_n)}$ is well-covered;*
(ii) *If n is composite, $p_1 = 2$ and $P =< p_1 >$, then $\overline{GT_P(\mathbb{Z}_n)}$ is well-covered;*
(iii) *Let n be composite, $p_j \neq 2$ and $P =< p_j >$. Then $\overline{GT_P(\mathbb{Z}_n)}$ is well covered if and only if $n = 2p_j$.*

Lemma 34 (Lemma 4.12, [33]) *Let* $n = p_1^{\alpha_1} p_2^{\alpha_2} \dots p_k^{\alpha_k}$ *where* $p_1 < p_2 < \cdots < p_k$, $p_j's$ *are prime,* $\alpha_j's$ *are positive integers and* $P = < p_j >$ *for some* j. *Then*

$$d(\overline{GT_P(\mathbb{Z}_n)}) = \begin{cases} 2 & \text{if } n = 2; \\ \frac{n+1}{2} & \text{if } n \text{ is an odd prime}; \\ \frac{n}{2} & \text{if } n \text{ is composite and } p_j = 2; \\ \dfrac{n - \frac{n}{p_j}}{2} + 1 & \text{if } n \text{ is composite and } p_j \neq 2. \end{cases}$$

In the following lemma, authors obtained when the complement of the generalized total graph of $\mathbb{Z}_n$ is domatically full.

Lemma 35 (Lemma 4.13, [33]) *Let* $n = p_1^{\alpha_1} p_2^{\alpha_2} \dots p_k^{\alpha_k}$ *where* $p_1 < p_2 < \cdots < p_k$, $p_j's$ *are prime,* $\alpha_j's$ *are positive integers and* $P = < p_j >$ *for some* j. *Then* $\overline{GT_P(\mathbb{Z}_n)}$ *is domatically full if and only if* $n = 2, 3$.

7 Complement of the Generalized Total Graph of Fields

Throughout this section F denotes a finite field. In a field F, $\{0\}$ is the only prime ideal. When R is the field F and $H = \{0\}$, Tamizh Chelvam and Balamurugan [31] studied about generalized total graph as $GT(F)$. In this section, we present several graph theoretical properties of the generalized total graph $GT(F)$ and its complement $\overline{GT(F)}$. In particular, we provide the structure of $GT(F)$ and $\overline{GT(F)}$. Further we present results concerning the domination number of $GT(F)$ and $\overline{GT(F)}$ and gamma sets in $GT(F)$ and $\overline{GT(F)}$.

Note that $GT(F)$ is the generalized total graph of the field F with the unique multiplicative prime subset $\{0\}$. If F is a field with of characteristic 2, then $x + x = 0$ for every $x \in F$. When the characteristic of the field F is greater than 2, for any $0 \neq x \in F, x \neq -x$ and $x + (-x) = 0$. In view of these, one can have the following structure for $GT(F)$.

Lemma 36 (Lemma 2.2, [31]) *Let* F *be a finite field. Then*

$$GT(F) = \begin{cases} \underbrace{K_1 \cup \cdots \cup K_1}_{|F| copies} & \text{if } char(F) = 2; \\ K_1 \cup \underbrace{K_{1,1} \cup \cdots \cup K_{1,1}}_{\frac{|F|-1}{2} copies} & \text{if } char(F) > 2. \end{cases}$$

The following lemma follows from Lemma 36.

Lemma 37 (Lemma 2.3, [31]) *Let* F *be a finite field. Then the following are true:*

(i) $\omega(GT(F)) = \begin{cases} 1 & \text{if } char(F) = 2; \\ 2 & \text{if } char(F) > 2. \end{cases}$

(ii) $\chi(GT(F)) = \begin{cases} 1 & \textit{if } char(F) = 2; \\ 2 & \textit{if } char(F) > 2. \end{cases}$

Lemma 38 (Lemma 2.4, [31]) *Let F be a finite field. Then*

(i) *The vertex independence number*

$$\beta(GT(F)) = \begin{cases} |F| & \textit{if } char(F) = 2; \\ \frac{|F|+1}{2} & \textit{if } char(F) > 2. \end{cases}$$

(ii) *If $char(F) > 2$, then the edge independence number, $\beta_1(GT(F)) = \frac{|F|-1}{2}$.*

Lemma 39 (Lemma 2.7, [31]) *Let F be a finite field. Then the following are true:*

(i) *The vertex covering number* $\alpha(GT(F)) = \begin{cases} 0 & \textit{if } char(F) = 2; \\ \frac{|F|-1}{2} & \textit{if } char(F) > 2. \end{cases}$

(ii) *The edge covering number, $\alpha_1(GT(F)) = 0$.*

In the following Lemma, authors obtained the domination number of the generalized total graph $GT(F)$.

Lemma 40 (Lemma 2.8, [31]) *Let F be a finite field. Then the following are true:*

(i) $\gamma(GT(F)) = \begin{cases} |F| & \textit{if } char(F) = 2; \\ \frac{|F|+1}{2} & \textit{if } char(F) > 2. \end{cases}$

(ii) *$GT(F)$ is an excellent graph;*

(iii) $\gamma_i(GT(F)) = \gamma_p(GT(F)) = \gamma_{eff}(GT(F)) = \begin{cases} |F| & \textit{if } char(F) = 2; \\ \frac{|F|+1}{2} & \textit{if } char(F) > 2. \end{cases}$

(iv) *$GT(F)$ is well-covered;*

(v) $\gamma_s(GT(F)) = \gamma_w(GT(F)) = \begin{cases} |F| & \textit{if } char(F) = 2; \\ \frac{|F|+1}{2} & \textit{if } char(F) > 2. \end{cases}$

Now we present results concerning some graph theoretical properties like diameter, girth, radius, Eulerian, and Hamiltonian of $\overline{GT(F)}$. The following lemma provides the structure of $\overline{GT(F)}$.

Lemma 41 (Lemma 3.1, [31]) *Let F be a finite field. Then the following are true:*

(i) *If $char(F) = 2$, then $\overline{GT(F)} = K_{|F|}$;*

(ii) *If $char(F) > 2$, then $\overline{GT(F)}$ is a connected bi-regular graph with $\Delta = |F| - 1$ and $\delta = |F| - 2$.*

Lemma 42 (Lemma 3.2, [31]) *Let F be a finite field. Then* $gr(\overline{GT(F)}) = \begin{cases} \infty & \textit{if } |F| = 2, 3; \\ 3 & \textit{if } |F| \geq 5. \end{cases}$

Lemma 43 (Lemma 3.3, [31]) *Let F be a finite field. Then*

$$\omega(\overline{GT(F)}) = \begin{cases} |F| & \textit{if } char(F) = 2; \\ \frac{|F|+1}{2} & \textit{if } char(F) > 2. \end{cases}$$

The chromatic number of $\overline{GT(F)}$ is proved in the following lemma.

Lemma 44 (Lemma 3.4, [31]) *Let F be a finite field.*
Then $\chi(\overline{GT(F)}) = \begin{cases} |F| & \text{if } char(F) = 2; \\ \frac{|F|+1}{2} & \text{if } char(F) > 2. \end{cases}$

Corollary 10 (Corollary 3.5, [31]) *Let F be a finite field. Then $\overline{GT(F)}$ is weakly perfect.*

Lemma 45 (Lemma 3.6, [31]) *Let F be a finite field. Then $\overline{GT(F)}$ is not Eulerian.*

Lemma 46 (Lemma 3.8, [31]) *Let F be a finite field and $|F| > 3$. Then $\overline{GT(F)}$ is Hamiltonian.*

Lemma 47 (Lemma 3.9, [31]) *Let F be a finite field. Then* $\beta(\overline{GT(F)}) = \begin{cases} 1 & \text{if } char(F) = 2; \\ 2 & \text{if } char(F) > 2. \end{cases}$

Corollary 11 (Corollary 3.10, [31]) *Let F be a finite field. Then* $\alpha(\overline{GT(F)}) = \begin{cases} |F| - 1 & \text{if } char(F) = 2; \\ |F| - 2 & \text{if } char(F) > 2. \end{cases}$

Lemma 48 (Lemma 3.11, [31]) *Let F be a finite field. Then the edge independence number* $\beta_1(\overline{GT(F)}) = \left\lfloor \frac{|F|}{2} \right\rfloor$.

Corollary 12 (Corollary 3.12, [31]) *Let F be a finite field. Then the edge covering number*
$\alpha_1(\overline{GT(F)}) = |F| - \left\lfloor \frac{|F|}{2} \right\rfloor$.

In the following results, we present about various domination parameters of $\overline{GT(F)}$. More specifically, we discuss about $\gamma_t, \gamma_c, \gamma_{cl}, \gamma_p, \gamma_{eff}, \gamma_s, \gamma_w$ and independence domination number of $\overline{GT(F)}$. In the following Lemma, we obtain the domination number of $\overline{GT(F)}$.

Lemma 49 (Lemma 4.1, [31]) *Let F be a finite field. Then $\gamma(\overline{GT(F)}) = 1$.*

Lemma 50 (Lemma 4.2, [31]) *Let F be a finite field. Then the following hold:*

(i) *The set $S = \{v\}, \ v \in V(\overline{GT(F)})$ is a γ-set in $\overline{GT(F)}$ if and only if $char(F) = 2$.*
(ii) *The set $S = \{0\}$, is the γ-set in $\overline{GT(F)}$ if and only if $char(F) > 2$.*

Lemma 51 (Lemma 4.3, [31]) *Let F be a finite field. Then $\overline{GT(F)}$ is excellent if and only if $char(F) = 2$.*

Lemma 52 (Lemma 4.4, [31]) *Let F be a finite field. Then the following are true:*

(i) $\gamma_p(\overline{GT(F)}) = \gamma_i(\overline{GT(F)}) = 1$.

(ii) *If* $char(F) = 2$, *then* $\gamma_s(\overline{GT(F)}) = \gamma_w(\overline{GT(F)}) = 1$;
(iii) *If* $char(F) > 2$, *then* $\gamma_s(\overline{GT(F)}) = 1$ *and* $\gamma_t(\overline{GT(F)}) = \gamma_c(\overline{GT(F)}) = \gamma_{cl}(\overline{GT(F)}) = \gamma_w(\overline{GT(F)}) = 2$.

Corollary 13 (Corollary 4.5, [31]) *Let F be a finite field. Then* $\gamma_{eff}(\overline{GT(F)}) = 1$.

Lemma 53 (Lemma 4.6, [31]) *Let F be a finite field. Then* $\overline{GT(F)}$ *is well-covered if and only if* $char(F) = 2$.

Lemma 54 (Lemma 4.7, [31]) *Let F be a finite field. Then*

$$d(\overline{GT(F)}) = \begin{cases} F & \text{if } char(F) = 2; \\ \frac{|F|+1}{2} & \text{if } char(F) > 2. \end{cases}$$

Lemma 55 (Lemma 4.8, [31]) *Let F be a finite field. Then the following are true:*

(i) *If* $char(F) = 2$, *then* $\overline{GT(F)}$ *is domatically full.*
(ii) *If* $char(F) > 2$, *then* $\overline{GT(F)}$ *is domatically full if and only if* $|F| = 3$.

Lemma 56 (Lemma 2.3, [38]) *Let F be a finite field. Then the following are true:*

(i) *If* $char(F) = 2$, *then* $\overline{GT(F)} = K_{|F|}$;
(ii) *If* $char(F) > 2$, *then* $\overline{GT(F)}$ *is a connected bi-regular graph with* $\Delta = |F| - 1$ *and* $\delta = |F| - 2$.

Note that, a graph G is said to be *unicyclic* if G contains exactly one cycle.

Theorem 15 (Theorem 2.4, [38]) *Let F be a finite field. Then the following hold:*

(i) $\overline{GT(F)}$ *is bipartite if and only if either* $F \cong \mathbb{Z}_2$ *or* $F \cong \mathbb{Z}_3$;
(ii) $\overline{GT(F)}$ *is neither a cycle nor an unicyclic graph.*

Recall that, a *chordal graph* is a simple graph G in which every cycle in G of length four and greater has a *cycle chord.* Also, a *split graph* is a graph in which the vertices can be partitioned into a clique and an independent set.

Theorem 16 (Theorem 2.6, [38]) *Let F be a finite field. Then the following are equivalent:*

(i) *Either* $char(F) = 2$ *or* $F \cong \mathbb{Z}_3$;
(ii) $\overline{GT(F)}$ *is a split graph;*
(iii) $\overline{GT_P(R)}$ *is a chordal graph.*

A graph G is a *claw-free* if G does not have the claw $K_{1,3}$ as the induced subgraph of G.

Theorem 17 (Theorem 2.7, [38]) *Let F be a finite field. Then* $\overline{GT(F)}$ *is a claw-free graph.*

A graph G is *perfect* if and only if no induced subgraph of G is an odd cycle of length at least five or the complement of one.

Theorem 18 (Theorem 2.8, [38]) *Let F be a finite field. Then $\overline{GT(F)}$ is a perfect graph.*

Theorem 19 (Theorem 2.9, [38]) *Let F be a finite field. Then $\overline{GT_P(R)}$ is a pancyclic if and only if $|F| > 3$.*

Corollary 14 (Corollary 2.10, [38]) *Let F be a finite field. Then $\overline{GT_P(R)}$ is a vertex-pancyclic if and only if $|F| > 3$.*

Note that, an *edge clique cover* of a graph G is a collection of cliques $L_1, L_2, \ldots, L_k$ such that $E(G) = \bigcup_{i=1}^{k} E(L_i)$. The minimum cardinality of an edge clique cover of G is called *the edge-clique covering number* of G and is denoted by $\theta_1(G)$.

The following lemma provides the clique number of $\overline{GT(F)}$.

Lemma 57 (Lemma 2.11, [38]) *Let F be a finite field. Then*

$$\omega(\overline{GT(F)}) = \begin{cases} |F| & \text{if } char(F) = 2; \\ \frac{|F|+1}{2} & \text{if } char(F) > 2. \end{cases}$$

Theorem 20 (Theorem 2.12, [38]) *Let F be a field. Then*

$$\theta_1(\overline{GT_P(R)}) = \begin{cases} 1 & \text{if } char(F) = 2; \\ 2 & \text{if } F \cong \mathbb{Z}_3; \\ 2 + \frac{|F|-1}{2} & \text{otherwise.} \end{cases}$$

Now we present a result about planarity and outerplanarity of $\overline{GT(F)}$.

Theorem 21 (Theorem 3.3, [38]) *Let F be a finite field. Then the following hold:*

(i) *$\overline{GT(F)}$ is planar if and only if $|F| \leq 5$;*
(ii) *$\overline{GT(F)}$ is outer planar if and only if $|F| \leq 3$.*

Acknowledgements This research work is supported by CSIR Emeritus Scientist Scheme (No. 21 (1123)/20/EMR-II) of Council of Scientific and Industrial Research, Government of India.

References

1. Aalipour, G., Akbari, S.: Some properties of a Cayley graph of a commutative ring. Comm. Algebra **42**(4), 1582–1593 (2014). https://doi.org/10.1080/00927872.2012.745866
2. Akhtar, R., Boggess, M., Jackson-Henderson, T., Jimenez, I., Karpman, R., Kinzel, A., Pritikin, D.: On the unitary Cayley graph of a finite ring. Electron. J. Combin. **16**(1), 117–130 (2009)
3. Anderson, V., Badawi, A.: The total graph of a commutative ring. J. Algebra **320**, 2706–2719 (2008)
4. Anderson, D.F., Badawi, A.: On the total graph of a commutative ring without the zero element. J. Algebra Appl. **11**(4), 1250074, 18 (2012). https://doi.org/10.1142/S0219498812500740

5. Anderson, D.F., Badawi, A.: The generalized total graph of a commutative ring. J. Algebra Appl. **12**(5), 1250212, 18 (2013). https://doi.org/10.1142/S021949881250212X
6. Ashrafi, N., Maimani, H.R., Pournaki, M.R., Yassemi, S.: Unit graphs associated with rings. Comm. Algebra **38**, 2851–2871 (2010)
7. Asir, T., Tamizh Chelvam, T.: On the total graph and its complement of a commutative ring. Comm. Algebra **41**, 3820–3835 (2013). https://doi.org/10.1080/00927872.2012.678956
8. Asir, V., Tamizh Chelvam, T.: On the intersection graph of gamma sets in the total graph II. J. Algebra Appl. **12**(4), 1250199, 14 (2013). https://doi.org/10.1142/S021949881250199X
9. Asir, T., Tamizh Chelvam, T.: On the genus of generalized unit and unitary Cayley graphs of a commutative ring. Acta Math. Hungar. **142**, 444–458 (2014)
10. Asir, T., Tamizh Chelvam, T.: On the genus two characterizations unit, unitary Cayley and comaximal graphs. ARS Combinatoria **138**, 77–91 (2018)
11. Asir, T., Maimani, H.R., Pournaki, M.R., Tamizh Chelvam, T.: Some bounds for the genus of a class of graphs arising from rings. Houston J. Math. **45**(2), 371–384 (2019)
12. Badawi, A.: On the total graph of a ring and its related graphs: a survey. In: Fontana, M., et al. (eds.) Commutative Algebra: Recent Advances in Commutative Rings, Integer-Valued Polynomials, and Polynomial Functions, pp. 39–54. Springer, New York (2014)
13. Balakrishnan, R., Ranganathan, K.: A Text Book of Graph Theory. Springer (2000)
14. Beck, I.: Coloring of commutative rings. J. Algebra **116**, 208–226 (1988)
15. Chartrand, G., Zhang, P.: Introduction to Graph Theory. Tata McGraw-Hill Edition (2006)
16. Golumbic, M.C.: Algorithmic Graph Theory and Perfect Graphs, 2nd edn. Elsevier B.V, Amsterdam (2004)
17. Haynes, T.W., Hedetniemi,S.T., Slater,P.J.: Fundamentals of Domination in Graphs, Marcel Dekker (1998)
18. Khashyarmanesh, K., Khorsandi, M.R.: A generalization of the unit and unitary Cayley graphs of a commutative ring. Acta Math. Hungar. **137**(4), 242–253 (2012)
19. Klotz, W., Sander, T.: Some properties of unitary Cayley graphs. Electron. J. Combin. **14**(1), 45, 12 (2007)
20. Liu, X., Zhou, S.: Spectral properties of unitary Cayley graphs of finite commutative rings. Electron. J. Combin. **19**(4), 13, 19 (2012)
21. Nazzal, K.: Total graphs associated to a commutative ring. Palestine. J. Math. **5**(1), 108–126 (2016)
22. Su, H., Zhou, V.: On the girth of the unit graph of a ring. J. Algebra Appl. **13**(2), 1350082, 12 (2014)
23. Tamizh Chelvam, T., Asir, T.: A note on total graph of $\mathbb{Z}_n$,. J. Discrete Math. Sci. Cryptography **14**(1), 1–7 (2011). https://doi.org/10.1142/S1793830911001309
24. Tamizh Chelvam, T., Asir, T.: Domination in total graph on Z_n,. Discrete Math. Algorithms Appl. **3**(4), 413–421 (2011). https://doi.org/10.1142/S1793830911001309
25. Tamizh Chelvam, T., Asir, T.: Intersection graph of gamma sets in the total graph. Discuss. Math. Graph Theory **32**, 339–354 (2012). https://doi.org/10.7151/dmgt.1611
26. Tamizh Chelvam, T., Asir, T.: On the intersection graph of gamma sets in the total graph I. J. Algebra Appl. **12**(4), 1250198, 18 (2013). https://doi.org/10.1142/S0219498812501988
27. Tamizh Chelvam, T., Asir, T.: Domination in the total graph of a commutative ring. J. Combin Math. Combin. Comput. **87**, 147–158 (2013)
28. Tamizh Chelvam, T., Asir, T.: On the genus of the total graph of a commutative ring. Comm. Algebra **41**, 142–153 (2013)
29. Tamizh Chelvam,T., Asir, T.: Distances in zero-divisor and total graphs from commutative rings: a survey. AKCE Int. J. Graphs Comb. **13**, 290–298 (2016). https://doi.org/10.1016/j.akcej.2016.11.009
30. Tamizh Chelvam, T., Asir, T., Selvakumar, K.: On the domination in graphs from commutative rings. In: Rizvi, T., et al. (eds.), Algebra and its Applications. Springer Proceedings in Mathematics and Statistics, vol. 174, S. Springer Science (2016). https://doi.org/10.1007/978-981-10-1651-6

31. Tamizh Chelvam, T., Balamurugan, M.: On the generalized total graph of fields and its complement. Palestine J. Math. **7**(2), 450–457 (2018)
32. Tamizh Chelvam, T., Balamurugan, M.: On the complement of the generalized total graph of commutative rings. J. Anal. **27**, 539–553 (2019). https://doi.org/10.1007/s41478-018-0093-6
33. Tamizh Chelvam. T., Balamurugan, M.: Complement of the generalized total graph of $\mathbb{Z}_n$ FILOMAT **33**(18), 6103–6113 (2019). https://doi.org/10.2298/FIL1918103T
34. Tamizh Chelvam, V., Anukumar Kathirvel, S.: Note on generalized Cayley graphs of finite rings and its complement. J. Anal. **27**, 555–566 (2019). https://doi.org/10.1007/s41478-018-0094-5
35. Tamizh Chelvam, T., Anukumar Kathirvel, S.: Generalized unit and unitary Cayley graphs of finite rings. J. Algebra Appl. 1950006, 21 (2019). https://doi.org/10.1142/S0219498819500063
36. Tamizh Chelvam, T., Anukumar Kathirvel, S., Balamurugan, M.: Domination in generalized Cayley graphs of finite rings. Indian J. Pure Appl. Math. **51**(2), 533–556 (2020). https://doi.org/10.1007/s13226-020-0415-7
37. Tamizh Chelvam, T., Anukumar Kathirvel, S., Balamurugan, M.: Intersection graph of gamma sets in generalized Cayley graphs of finite rings. Houston J. Math. **46**(3), 561–582 (2020)
38. Tamizh Chelvam, T., Balamurugan, M.: Complement of the generalized total graph of fields. AKCE J. Graph Theory Comb. **17**(3), 730–733 (2020). https://doi.org/10.1016/j.akcej.2019.12.005
39. West, D.B.: Introduction to Graph Theory. 2nd edn, (2007)